Treasure Your Exceptions

Alan G. Cock • Donald R. Forsdyke

Treasure Your Exceptions

The Science and Life of William Bateson

2nd ed. 2022

Springer

Alan G. Cock
School of Biological Sciences
University of Southampton
Southampton, UK

Donald R. Forsdyke
Dept. Biomedical & Molecular Sciences
Queen's University
Kingston, ON, Canada

ISBN 978-3-030-92101-9 ISBN 978-3-030-92099-9 (eBook)
https://doi.org/10.1007/978-3-030-92099-9

This Springer imprint is published by the registered company Springer Nature Switzerland AG
The registered company address is: Gewerbestrasse 11, 6330 Cham, Switzerland

To past, present, and future Christiana Herringhams and Eliza Savages, who treasure those "the system" will not.

Prologue to First Edition

> *Nevertheless, if I may throw out a word of council to beginners, it is: Treasure your exceptions! When there are none, the work gets so dull that no one cares to carry it further. Keep them always uncovered and in sight. Exceptions are like the rough brickwork of a growing building which tells that there is more to come and shows where the next construction is to be.*
>
> *William Bateson (1908)*

Part I by Donald Forsdyke

To understand evolution we must first understand the historical development of ideas on evolution. But to understand its history, we must first understand evolution. This paradox implies that the study of evolution and of its history must go hand in hand. There can hardly be a better example of this than the life of Gregor Mendel, the founder of Genetics. Mendel died in 1884. Any biography written between 1884 and 1899 would have described his life as a monk in an Augustinian monastery in the city of Brünn in Moravia. There would have been plenty to write about. As Prelate he was much involved in community affairs. Under his predecessor the monastery had become a major intellectual and cultural centre, and Mendel encouraged this. He became chairman of the Moravian Mortgage Bank and fought bitterly with the German Liberal Party, which had imposed severe taxes on monasteries. Oh, and yes, he did dabble in research for a few years before becoming Prelate in 1868. He crossed different lines of peas and scored the inheritance of various characters among the offspring – even published a paper or two.

As has been often told, in the 1890s and unaware of Mendel's work, various botanists in continental Europe (de Vries, Correns and Tschermak) began to think along the same lines. Since the publication of Charles Darwin's great book in 1859 [1], Darwinism had dominated the biological sciences. Now, as if scales had dropped from their eyes, researchers began to push beyond Darwin and understand, for the first time, a paper that Mendel had published in 1865 in an obscure journal, the *Verhandlungen des naturforschenden Vereines in Brunn*, copies of which has been distributed to many academic centres [2]. In 1900 Mendel's work was found and confirmed (i.e. "rediscovered"). Studies of evolution between 1865 and that date were defective because studies of the history of evolution in that period had been defective. And studies of the history of evolution were defective, because studies of evolution in that period had been defective. Evolutionists had not alerted historians to the significance of the new approach Mendel had pioneered, and the few historians who came across the work had not effectively communicated it to evolutionists. It fell to William Bateson, who had also begun researching along these lines, to communicate Mendel's ideas to the English-speaking world, and to fight for their acceptance.

By 1909 the battle was over. Mendelism was accepted and Mendel's paper had been translated and distributed world-wide. Holder of the first Chair in Biology at Cambridge University, Bateson was at the height of his power and influence. In 1906 one of his students had spread the word in a popular textbook – *Variation, Heredity and Evolution* – which ran to several editions. In 1907 Bateson had given the Silliman Lecture at Yale, and in 1909 his advanced treatise, *Mendel's Principles of Heredity*, was published by Cambridge University Press. To commemorate the centenary of the birth of Charles Darwin, a call to contribute essays to a volume entitled *Darwin and Modern Science* had gone to leading figures in the field. Bateson took this as an opportunity, not only to pay tribute to Darwin and Mendel, but also to push the agenda of science ahead and beyond them. His essay – "Heredity and Variation in Modern Lights" – encapsulated in its clearest form, a new view of evolution he had developed from foundations laid by Francis Galton decades earlier.

It might just as well have been 1865. Unlike Mendel on that date, Bateson in 1909 was not an obscure monk in a distant monastery. He, more than any other in the biological sciences, was centre stage. This was his hour. Yet, as with Mendel, his words fell on fallow ground. In a way, he was partly responsible for his own demise, as he later came to recognize. He, more than any other, had travelled the land, indeed, the world, trumpeting Mendelism – the idea that the various characters of living organisms were represented by distinct units (later known as genes), half of which came to a child from its mother and half of which came to a child from its father. In essence this was quite a simple message. Then as now, in marketing, simple messages worked. Then as now, the same applied to the marketing of scientific ideas. Accordingly, subtle scientific ideas tended to lose out to simple scientific ideas, and subtle scientists tended to lose out to the unsubtle. The appeal of Darwinism had been so seductive and had suggested so many interesting lines of research, that Mendel's abstractions (after all, no one had ever seen a Mendelian unit) would very likely not have been understood in 1865. Likewise, the appeal of Mendelism was now so seductive that Bateson's abstractions, to be discussed in the chapters ahead, suffered a similar fate in 1909.

Genes explained so much that the temptation to believe that they could explain everything was overwhelming. And the genic paradigm spawned an eminently marketable research agenda. What were genes? How were they passed from generation to generation? How was the information contained in a gene expressed in an organism carrying that gene? Could "good" genes be transferred to organisms with "bad" genes, so curing genetic defects? With so much to be done there was little incentive to look beyond genes. The genic juggernaut moved off, gaining a momentum that would take it through the twentieth century and into the twenty-first.

Nevertheless, Bateson's voice was not a lone one. As set out in my speciation text, *The Origin of Species Revisited*, Darwin's research associate, George Romanes, had anticipated Bateson in 1886 [3]. Subsequently, in *Evolutionary Bioinformatics*, I showed how biologist Richard Goldschmidt had taken a similar position in the 1930s [4]. Paleontologist and biohistorian Stephen Jay Gould struggled for decades to make sense of Goldschmidt, but in 2002 in his final work *The Structure of Evolutionary Theory* he admitted defeat [5]. Gould confessed his "relative ignorance" of the flood of new information on genomes emerging from various genome sequencing projects. It turned out that, just as Bateson and the Continental botanists had to push beyond Darwinism in the 1890s to "rediscover" the work of Mendel and found a new science – Genetics, so those analyzing the vast quantities of DNA sequence information made available in the 1990s had to push beyond genocentrism to "rediscover" the work of Romanes, Bateson and Goldschmidt, and found a new science – Evolutionary Bioinformatics (EB).

Yet, there are remarkable differences between the "rediscoveries" of the 1890s and the 1990s, which make the task of writing a biography of Bateson particularly fascinating. Mendel quietly wrote his few papers that, although of major import, were narrowly focussed on his experimental work. He then meekly (regarding science) retired to his cloister, was "rediscovered" within two decades of his death and was celebrated henceforth. Bateson, on the other hand, took the whole of biology (and much more) as his domain, and was far from meek in arguing his case. Both before and long after his death, he was attacked by the genocentrists (labelled by Gould as "ultra-Darwinian fundamentalists"), and he was not "rediscovered" until many decades after his death [3]. Whereas Mendel's eclipse was for 35 years, Bateson's was for a century.

Up to the present time, while his early contributions to genetics are recognized, his fundamental contributions to evolutionary biology have never been celebrated – indeed, the very opposite – he has been condemned as a conservative who slowed scientific advance. Biohistorians have acknowledged their difficulties in determining his role. John Lesch observed [6] that "The development of evolutionary theory in the two decades from Darwin's death to the turn of the century remains very largely *terra incognita* for the historian," while William Provine lamented [7] that "Evolutionary biology in the period 1859–1925 is extraordinarily complex." To understand these years we have to understand Bateson.

Which brings us to this book, and its dual authorship by Alan Cock and myself. The reader will find below part two of this three-part Prologue, which was written by Alan around 1980. The typescript had a pencilled date indicating that he reread and approved it in 1984. A quarter of a century separates it from my parts. Briefly, the story is as follows. When Bateson died in 1926 he was director of the John Innes Institute at Merton in Surrey. His wife, Caroline Beatrice Durham (hereafter

"Beatrice"), removed the bulk of his personal scientific papers and correspondence. To these she added various personal and family papers, and transcripts of letters sent to her by some of his correspondents. She used them when writing and editing *William Bateson, F. R. S. Naturalist. His Essays and Addresses Together with a Short Account of his Life* [8], and when editing *Letters from the Steppe Written in the Years 1886-1887 by William Bateson* [9]. Both were published in 1928. Her goals were limited. She "attempted only to sketch a rare personality" and hoped that a "more competent hand" would later complete her work.

At her death in 1941 the papers were held in storage until 1954 when they were transferred to her son Gregory Bateson, who, with his wife Margaret Mead, had gained a reputation as an anthropologist and was based in the USA. The papers eventually found their way to the loft of an out-house at the summer-home of their daughter, Mary Catherine Kassarjian, in the small New Hampshire town of Hancock, near Boston. Here they were inspected by historian William Coleman in 1964 and he borrowed a selection (perhaps 20% of the whole), which he microfilmed at Johns Hopkins University. The microfilm became part of the collection of the American Philosophical Society in Philadelphia [10], and assisted William Provine when composing *The Origins of Theoretical Population Genetics*, 1971 [11]. Despite its daunting title, this was also a mini-biography of Bateson. Coleman wrote a long article entitled "Bateson and Chromosomes: Conservative Thought in Science," which was published in 1970 [12]. In 1975 some of the papers were examined by David Lipset, who had travelled with Gregory in Asia in the early 1970s. This led to a comprehensive biography *Gregory Bateson. The Legacy of a Scientist*, in 1980 [13]. It was also, in many respects, a splendid biography of William, but Lipset mentioned in the Preface that a more complete biography was being written by Alan Cock.

After graduating in Zoology at Cambridge University in 1947, Alan worked at the Poultry Genetics Unit in the School of Agriculture at Cambridge. He was assistant to Michael Pease, who had himself been an assistant to Reginald Punnett, who had been Bateson's assistant. So Alan could rightly claim to be a scientific great-grandson of William Bateson. Remarkably, Pease was still using the same system of short-hand notation and record-keeping that Bateson and Punnett had developed in earlier decades. In 1957 Alan went to the Poultry Research Centre at the University of Edinburgh where he obtained a doctorate in genetics in 1962. In 1964 he was appointed lecturer in zoology at the University of Southampton. In the 1970s his interests turned to biohistory: "My personal interest in Bateson (evoked by beginning my research career in a Unit directly descended from his Cambridge operations) developed in 1972 into a serious intention to write a book about him." That year Alan obtained a copy of Coleman's microfilm and in 1973 he examined the remaining materials that Beatrice had not removed from the John Innes Institute – now relocated to Norwich. He also explored the Bateson-Punnett research notebooks then held in the Department of Genetics in Cambridge. His first biographical paper was entitled "William Bateson, Mendelism and Biometry" [14]. In December 1974 Alan visited America on a short-term travel grant from the Wellcome Trust to examine the remaining papers. He reported back to the Trust in February:

> My original plan was to remove the papers temporarily to Harvard University and there to obtain as full and complete a permanent record of them as was possible within the six weeks available by a combination of note-taking and selective Xerox-copying. The practical side of this programme was radically altered (in an almost wholly welcome sense) during the first few days of my visit by the decision of Bateson's only surviving son (the anthropologist Gregory Bateson) and granddaughter (Mrs. Mary Kassarjian) to allow the papers to be sent on indefinite loan to the University Library, Cambridge, England. The papers are in fact now on their way across the Atlantic. In view of the rarity with which historical papers travel eastwards across the Atlantic, I feel rather proud of having influenced their fate. Dr. Stephen Gould, of the Agassiz Museum of Comparative Zoology, Harvard, very kindly offered me working space in the museum. Only six days were spent away from Harvard: two in collecting the papers from Mrs. Kasserjian's house at Hancock, and four in visiting Dr. Gregory Bateson at Santa Cruz, California.

Elsewhere he noted that "Gregory Bateson very generously agreed to my suggestion that the entire body of papers be returned to Britain. Cambridge University Library was agreed upon as their final home, once I had finished with them." Thus, in 1975, with help from Mary Catherine Kasserjian, he and David Lipset sorted through the Bateson papers in Hancock. Old family papers and ones relating to Gregory Bateson were to stay in the USA (at the University of California, Santa Cruz). The remaining papers were repatriated, and the long task of cataloguing and indexing began at the University of Southampton [15].

At the outset Alan began submitting proposals for the book to various publishing houses. For unclear reasons, clues to which may emerge here, there was no great enthusiasm. Fortunately, Lipset did not have this problem, and his book on Gregory Bateson was shortly to appear. But the shadows were lengthening. Alan had long been prone to periods of depression and in 1984 he was operated on for a tumor. Although he was able to complete the cataloguing, and produced several valuable papers, the prospects for the ultimate book were not good.

Meanwhile in the 1970s there had been great advances in sequencing technology. The genetic information that passes between generations in the form of a sequence of bases in DNA could be directly read. In the 1980s the technology became automated and sequencing machines began to gush forth so much data that only computers could handle them. In the 1990s bioinformatic analyses in my laboratory and elsewhere began to suggest a re-evaluation of William Bateson's work. Like Alan two decades earlier, I was increasingly drawn away from science *per se*, and towards its history. Yet, in my readings of the historical literature I repeatedly encountered papers, such as that of Coleman [12], that disparaged Bateson. Unfortunately, it was not until *The Origin of Species Revisited* was in press that I encountered the thoughtful, meticulous, and sympathetic analyses that Alan has tucked away in predominantly low-profile journals.

Having greatly enjoyed Abraham Pais's *The Science and Life of Albert Einstein* [16], I toyed with the idea of something similar about Bateson. A major difference was that, whereas Pais had been obliged to restrict Einstein's science to segments that only those with advanced knowledge would comprehend, I felt certain that I could explain evolution from the new perspective in a way that generally educated readers would follow. Through the biography of a major originator of the new perspective, evolution, with all its subtleties, might be made intelligible to the educated layperson. More than this, the field of evolutionary biology had split into factions, personalized in the forms of Richard Dawkins (advocate of Darwin's natural selection as the supreme agency in evolution), and of Stephen Jay Gould (advocate of hierarchical agencies). Resolving this rift was one of the goals of a text nearing completion [4]. A new, less technical, work might further help.

However, there was a major stumbling block. I was still much engaged in research. The prospect of foraging through dusty archives for materials with which to reconstruct Bateson's life, while appealing, was not something for which I was particularly well trained. So I continued the development of some webpages on Bateson and let the matter rest. Then I learned from David Lipset's biography of Gregory that Alan Cock was writing a full biography of William. Here was the "more competent hand" that Beatrice had hoped for. Since Lipset's book was published in 1980, Alan's should already be published? But a search revealed no trace.

In 2001 I emailed Lipset asking what he knew of Alan's project. "I doubt that Cock ever finished the project on WB. He was an up and down sort of man." Alan was no longer working at the University of Southampton, but I traced him to London where he was living in an apartment close to his eldest daughter. In the spring of 2004, a year before his death, I met him and two of his daughters, and had an opportunity to examine his files. With his and their approval, his personal collection of Bateson-related materials and many of his own personal papers, were shipped to me in the fall of 2004. It did not take long to decide, as I checked his copies of the Coleman microfilms and rummaged through copies of the Hancock papers, that, although still a major undertaking, a dual-authored work was feasible without unduly trespassing on my other research interests.

In the pages that follow I present my gleanings from the incomplete chapters and other jottings I found among Alan's papers. Sometimes Alan's words are directly attributed to him, particularly when large segments are transposed from his files. Two chapters are his alone (apart from minor editing). Five chapters are mine alone. However, to spell out at every point the relative contributions of Alan and myself would be too distracting. All I can say is that the entire collection of Alan's papers (together with his copy of the Bateson papers) is now deposited in the Archives of Queen's University where they are available for those who might wish to pursue the matter.

A question of much interest was whether Alan's sympathy for William Bateson reflected a deeper understanding of his work than Alan's scientific contemporaries had been capable of. The answer seems to be – as indicated in Alan's own words above and below – that he was interested in Bateson, because of his connection with the same Cambridge laboratory, because of their common interest in fowl genetics, and last and most importantly, because Bateson was a fascinating figure in his own right. The revolution in EB of the 1990s was too late to inform Alan's account, as it did mine. However, we both had long careers as productive "card carrying" scientists, rather than as historians of science. As will become apparent here, this difference in perspective makes our account different from previous biographies.

For a scientist looking at history there are three overriding double questions: What did he (she) know and when did he know it? What did he think and when did he think it? What did he do and why did he do it? Answers to these questions give us information on *process* – the process of scientific discovery. If the elements of that process have remained essentially unchanged since Bateson's day – an assertion assumed and not argued here – then a better understanding of ways this outstanding contributor to scientific progress operated may help us improve the process of discovery as it now operates [17].

For many reasons Bateson's life is ideal for this purpose. Before the days of email, correspondence was not entered into lightly, and was usually composed with care. Often there was at least one preliminary draft, and neither this nor the final copy could be destroyed by merely pressing "delete." Bateson's letters were treasured and kept. Beatrice was able to recover many from his correspondents after his death. So it is now possible to reconstruct, in blow-by-blow fashion, the genesis of his ideas and discoveries. Finally, whereas the ultimate value of much contemporary work still remains to be determined, a century later relatively clear endpoints have emerged so it should now be easier to reach a consensus on the merit of his contributions. Historians tend to label this as "Whig history," but we are unapologetic [18].

Part II by Alan Cock

Why is Bateson interest-worthy? A distinguished Victorian-Edwardian scientist – but there are many such. He was one of the founders of the modern science of genetics, but the layman is likely to ask: what really important discoveries did he make? The only answer possible is a rather lame one: two near-misses. He came very close to rediscovering the epoch-making work of Mendel, done thirty-five years earlier but buried in neglect. Bateson *was* the first to show that Mendel's laws apply to animals as well as to plants. He discovered the important phenomenon of genetic linkage, though the explanation of it that he offered proved to be widely off the mark. Near misses, it will be said, are rightly soon forgotten, together with their perpetrators. He invented much of the basic terminology of modern genetics (including the term "genetics" itself), but terminology is a matter for the specialist, and a rather pedestrian affair at that. He demonstrated that acorn worms are evolutionary relatives of the vertebrates. But what are acorn-worms, and who cares?

So far Bateson probably sounds eminently worthy but eminently dull: the last kind of person whose biography one would want to read – or write. What I have stated baldly above is true but put thus it gives a very misleading picture. To his contemporaries he was quite the opposite of this: not dull, but fiery and always the centre of controversy, and to many not even worthy, since he had cast doubt on doctrines such as Darwin's theory of evolution by natural selection, and the chromosome theory of heredity, which in many minds had crept gradually from the status of scientific theories to that of sacred dogmas.

The figures of the past tend to acquire a patina which, instead of giving them added lustre, makes them appear more solemn, even pompous, than they really were. Late in life, Bateson was offered, and declined, a knighthood. This tells us several things about him, one of which is that he set no inordinate store by pomp and circumstance. Certainly, he took life seriously – in science, in art, in family affairs and, on the rare occasions when he involved himself with such matters, in politics. But he did not ordinarily lapse from seriousness into solemnity. Examples of his dry humor and often deadly wit are scattered through these pages. For sheer light spirits, intermingled with the humdrum routine of scientific work, the story of his bets is as good an illustration as any. When sorting through his young experimental chickens, he used to lay small bets with his colleague (Reginald Punnett) on the sex of dubious individuals. These bets – and their settlement – are recorded in their experimental notebooks, alongside scientific notes about things like plumage color and comb type.

The things about Bateson which most interest and attract me fall into two categories which, although conceptually separable, in practice interlock a good deal. The first category concerns his influence on the development of genetics – and of biology generally. The second devolves around his personality and character. Over a period, which can be centred very crudely but conveniently on the year 1900, biology underwent a transition from being mainly an observational and descriptive science to being mainly an experimental and analytic one. Bateson's career nicely spans this divide, and not just in a chronological sense. He began in an entirely observational way, in the field of comparative anatomy and embryology. Though his work there brought him the beginnings of a "sound" reputation – even some fame – he soon became disillusioned with the limited scope and logical weakness of its interpretative aspects. Switching to a new – and far less fashionable – field, the study of variations, he still remained for long at an essentially observational level. Not until 1897-8 did he take the vital step of using experimental breeding to investigate the heredity of his variants. He still lacked any general analytical scheme for interpreting his results, but all this was changed dramatically in 1900 by the rediscovery of Mendel's work. Almost overnight, genetics became an analytical science, and broad and exciting new horizons were opened.

In the years that followed, Bateson played a prominent part in developing and extending Mendel's work. However, no list of his discoveries could convey the strength and breadth of Bateson's influence. On his colleagues and students he exerted an almost magnetic effect. They felt an intense loyalty and affection. There was a distinctive Batesonian style or approach to genetics, of which one characteristic was his strong concern for the interrelationships between genetics and the rest of biology. So great was the rate of progress in genetics that many other geneticists of the period tended to regard genetics as a self-contained and autonomous empire. His concern for biology-as-a-whole did sometimes help to lead Bateson into trouble, notably his opposition (which he eventually admitted was erroneous) to the chromosome theory of heredity.

As to his personality and character, those well-worn adjectives, complex and contradictory, are nevertheless singularly apposite. Self-critical, iconoclastic, no glad sufferer of fools, aesthetic, reticent on emotional matters, … but such lists tend to be unconvincing. A single story can perhaps do more. This atheist son of a Doctor of Divinity used to read passages from the Bible to his three sons at breakfast – lest they grow up to be empty-headed atheists.

Anyone intending to write at length about Bateson faces at once an obstacle and an encouragement in the 160-page biographical *Memoir* (1928) by his widow Beatrice [8]. This is so well done as to be, within its self-imposed limitations, unsurpassable. The most obvious limitation is that she never delves at all deeply into his scientific work, though naturally it is always popping in and out of the narrative. There are other omissions too (perhaps imposed by limitations on length), and one or two places where Beatrice conceals more than she reveals. The laconic single sentence with which she introduces herself ("In June 1896 we were married, and I began to learn what life may be.") hides a story worthy of a place in any "Selected Readings in Romantic Engagements."

They met in the winter of 1888–1889 in Dresden on an opera-going expedition. There was a whirlwind courtship and an engagement party at St. John's College, Cambridge, where William got somewhat inebriated. Mrs. Durham thought William might have alcoholic tendencies and decided that one alcoholic in the family (her husband) was quite enough. William was told, with no shred of explanation, that he was not to communicate with Beatrice again. Beatrice herself was carefully chaperoned by her mother and an elder sister. They were thus kept apart for six years. In 1896, Beatrice published, as a "come-on" signal to William, a short story in the *English Illustrated Magazine* which was her own story thinly disguised. William made contact *via* an older married woman (who played fairy godmother to him again 10 years later). They were promptly married and lived happily ever after (they really did, so far as I can tell).

The encouragement to be gleaned from Beatrice's memoir is in her preface, where she writes; "Later a more competent hand may, I hope, undertake a full biography and account of William Bateson's work." This was not just a pious expression. She did something practical about it, by preserving his papers, together with a fairly substantial collection of his letters, borrowed and copied from friends. Both of these are now with me.

Part III by Donald Forsdyke

Beatrice placed advertisements for letters in *The Times* and the international journal *Nature*. Soon they came from far afield, some of them unfortunately too late to be included in her *Memoir*. But, despite her efforts, many letters could not be located. Writing in 1927 to Charles Hurst, whom we shall meet in Chap. 10, she noted [19]:

> I am very glad you have your letters, the loss of the others was a dreadful blow to me. I am of course collecting for a qualified biographer and historian later on. … I am not intending to make any immediate use of the material I am now collecting. I am simply putting together

> all that I can collect with a view to a future biography and full history. The Press has in hand, with [a selection of] my husband's [scientific] papers, a short personal memoir.

Commenting on her *Memoir* she noted: "It gives a fair guide to any future biographer should we all be underground, and it is quite uncontentious. At least that is what it is aimed at." Her decisions of what to add or omit were not made alone. For example, her journalist sister-in-law, Margaret, wrote (June 22, 1926):

> There are many little allusions to persons which they were never meant to see and which would or might give pain to them or their friends. Also there are passages relating to my own affairs which I should not wish to be published; and similarly allusions to the affairs of other members of the family. But, when all omissions are allowed for, there remains much which is entertaining, humorous and interesting.

When Alan Cock and David Lipset examined the papers in January 1976, they were essentially as Beatrice had left them, done up in bundles, sometimes with an added note, but otherwise uncatalogued. Alan ruefully noted that the price of the "coup" of bringing the papers West-to-East across the Atlantic, would be some fairly basic, and time-consuming, archival work, carried out with shoe-string support from UK funding organizations. Furthermore, many people who had known Bateson were, at that time, still alive. To really fulfill Beatrice's mandate, people such as Cyril Darlington, who had been a cytologist at the John Innes Institute, and Darwin's granddaughter, Nora Barlow, a volunteer worker at the Institute, would have to be interviewed. Again, this was all expensive and time-consuming.

However, now we are in the twenty-first century and William Bateson and those who knew him are, indeed, "underground". There are few people to interview, and the underlying science is better understood. The time would seem propitious for the completion of Beatrice's (and Alan's) work. In Chap. 1, some key characters are introduced, and early family connections between the Darwins, Butlers and Batesons are noted. There is description of William's birth to a world of great privilege – one of six diversely gifted siblings. After aimless "self-satisfied and desultory" years at Rugby school, he flowered at Cambridge in animal morphology, rather than in physiology. Among earlier graduates of the school of physiology were George Romanes (animals) and Francis Darwin (plants), who were both research associates of Charles Darwin in the years before his death. In various ways they were to greatly influence Bateson. However, it was the enthusiasm of his fellow St. John's College student, Raphael Weldon, that first led him to morphology.

Chapter 2 describes Bateson's successful studies on the embryology of acorn-worms – which won him a college fellowship – but this was followed by disillusionment with the embryological approach. Searching for a new theme there followed solitary expeditions to the Aral Sea and the Nile delta. Noting a frequent lack of correlation between an organism's appearance and the pressures of its environment, Bateson began increasingly to question the potency of Darwinian natural selection. Having lost faith in Darwin, he turned to Darwin's cousin, Francis Galton, whose theoretical insights had moved far beyond Darwin, and who was now pressing for statistical studies of biological variation. Because of the importance of

Galton's influence, he is accorded an entire chapter at an early stage (Chap. 3). Others accorded chapters at appropriate points in our narrative are Romanes (Chap. 5), and Samuel Butler (Chap. 19).

Returning empty handed from foreign travels, and with ideas not in accord with the conventional wisdom, Bateson's position was precarious. While Weldon went from strength to strength – to a Chair at University College in London, and to a Fellowship of the Royal Society (FRS) – Bateson became Steward of the St. John's College kitchens. His wooing of Beatrice having failed, he buried himself in a heroic attempt to collate reports of biological variants in the medico-scientific literature. A massive book, *Materials for the Study of Variation* resulted (Chap. 4). Here he placed particular emphasis on the fact that, whereas under Darwin's natural selection progressive changes over many generations (i.e. many steps) were required to generate new and perfect types, such perfect types (e.g. a type with one perfect extra finger), could actually appear in only one generation (one step). This appearance in one generation he described as the "discontinuity of variation," to contrast it with the "continuity of variation" needed for successive steps.

However, he went further, holding that "discontinuity of variation" was of high importance for understanding the "discontinuity of species" – namely, that organisms divide into discrete groups (species). The members of a species share common characters and vary *within limits* about a mean type. But how do they vary *beyond* the limits to create a new mean type (i.e. a new species)? Weldon and others correctly scorned Bateson's claim that his circus of oddities was somehow of relevance to the question of the origin of species. Yet his catalogue of deformities greatly interested physicians and horticulturalists, many of whom were to aid him in the years ahead. At this time Bateson helped Mrs. Herringham, the wealthy wife of a St. Bartholomew's Hospital physician, through a personal crisis. She was to become the "fairy godmother" mentioned above by Alan. His growing estrangement from Weldon was later to broaden into a major dispute between "Mendelians" and "Biometricians."

Despite its generally poor reception, *Materials* won high praise from Galton, and Bateson joined Weldon as a FRS. However, it was to Weldon that the Royal Society (RS) gave financial support in the form of grants for his "biometric" studies carried out with mathematical advice from Karl Pearson. Chapter 6 outlines Bateson's indirect attack on Weldon through the RS Committee responsible for Weldon's funding, which was chaired by Galton. There was also a direct attack. Week after week throughout the 1890s the Victorians scurried to read the latest scientific controversy in the pages of *Nature*, to which Bateson was often a prominent contributor with Weldon his predominant target. This led to a major coup in 1897, the reorientation of the RS Committee to include Bateson and some like-minded biologists, its renaming as the Evolution Committee, and the extension of funding to Bateson and his research colleague Rebecca Saunders. There was even a glimmer of professional advancement – he was appointed Deputy to the Professor of Animal Morphology in 1899 – the year he and Beatrice moved to Grantchester with their family of two (Chap. 7).

The early breeding studies of Bateson and Saunders, and the joyous "rediscovery" of Mendel's work are the subject of Chap. 8. Bateson lamented the overzealous focus on Darwin that had left Mendel in the shadows. As "Mendel's Bulldog" (Chap. 9), and with the staunch support of Hurst (Chap. 10), Bateson met the Biometricians (Weldon and Pearson) head on. Soon there collected around him an eager, albeit poorly financed, band – the Mendelians – many of whom were brilliant women for whom prevailing attitudes made academic advancement difficult. By 1906 the battle was over. Bateson was "On Course" (Chap. 11) and a Cambridge professorship in Biology soon followed. 1909 was his triumphal year, the year of the Darwin Centenary celebrations, the year when he articulated more clearly than ever before bold concepts that, as with Mendel decades earlier, were to create not the slightest ripple on the intellectual waters of his time (Chap. 12).

Around the time of the Mendelian "rediscovery" it had been observed that the movement of chromosomes during cell division was precisely what would be predicted if they were the carriers of Mendel's units. Although acknowledging the probable correctness of this (indeed he was one of the first to note it), to the end of his life Bateson was highly critical of the supporting evidence (Chap. 13). While many commentators have construed this as pig-headedness, later chapters suggest it was Bateson's deep reading of Galton, together with his apparently subconscious adoption of Romanes, that had led him to demand more of chromosomes than the mere transfer of character units (genes).

But the term of his Professorship was for only five years and facilities for extensive breeding studies still escaped him. His appointment as the first Director of the John Innes Institute in 1910 changed all this (Chap. 14). He was no longer fighting the establishment, he *was* the establishment. Moving on from Mendelism, he now turned to the study of the many "exceptions" that he had long "treasured" – rogue peas, variegation and chimaeras – in the hope that they might somehow lead to his goal, an understanding of the origin of species. Despite many distractions – the escalating demands of the Eugenics movement (Chap. 15) and growing militarism among German scientists (Chap. 16) – he was able to complete another major book, *Problems of Genetics* (1913). The outbreak of the First World War found him presiding at the British Association (BA) in Australia. His two eldest sons were soon to die, one on the battlefield, one a suicide.

Chapter 17 describes Bateson's struggles in a bizarre post-war world that, despite his Victorian-Edwardian past, he largely understood. Unfortunately, it did not understand him. In many quarters his salutation of the work of the school of Thomas Morgan in America and his recognition that the chromosomes were carriers of Mendel's units, were seen as capitulations. However, his speeches were skillfully framed to reveal his abiding dissatisfaction. Still unresolved was the problem of how the limits of a species were maintained yet could be transgressed for the production of a new species. In these final years, his life-long interest in art – William Blake, Old Master prints, Japanese drawings – flourished, and his expertise was recognised by election to the British Museum's Board of Trustees (Chap. 18).

The remaining chapters deal with issues that do not fit neatly into the earlier chronology. Lipset's biography of Bateson's son, Gregory, began with an entire chapter on Samuel Butler [13]: "Butler's ghost will haunt this book – not only in [Gregory] Bateson's thought, but in the family culture from which it grew." The same applies to the present biography, except that, in this case, Butler comes late (Chap. 19) because, while he was always there, William Bateson unfortunately failed to note his writings until late in life. Chapter 20 describes the various visits that Bateson made to the Brünn monastery where Mendel had worked, while Chap. 21 describes Bateson's long-standing opposition to the view that characters acquired in a single generation could be transmitted to offspring (Lamarckism). Finally, there were his forays into politics, both of post-war reconstruction (Chap. 22) and of his university – the long battle for women's rights (Chap. 23). After his death in 1926 there was a continuing failure to understand, and hence to correctly represent, Bateson and his work (Chap. 24). In the Epilogue the conclusion is drawn that to Darwin's tall shadow had been added that of Mendel. Just as Darwin had once enshrouded Mendel, they now collectively enshrouded the work of many others, including that of Bateson himself. Indeed, it is argued that the history of the biological sciences in the twentieth century would have been transformed had the works of several pre- and post-Mendelian "Mendels" been better understood.

There are many quotations. We have not hesitated to let Bateson speak for himself. His rich prose conveys the story far better than we can. Unless otherwise stated, all italicized emphasis is Bateson's. However, the meaning of language has and is changing. What Bateson meant by a word may not coincide with what we now understand it to mean. Wherever we foresee ambiguity, explanatory remarks have been inserted within square parentheses. A few spelling differences – he wrote "shew" rather than "show" – have been altered without comment.

Many inviting avenues emerge in the course of a work of this nature. But the exigencies of time have dictated that many stones be left unturned. We have been led, partly by intuition and partly by serendipity, to explore some avenues rather than others, and to turn some relatively inconspicuous stones while ignoring some prominent ones that others might consider more worthy. Where we have erred it has been in favor of nineteenth century authors. Where we have omitted to mention, or have unwittingly duplicated, relevant ideas of modern authors, we offer our apologies. But there is a certain logic in this. The writings of nineteenth century authors provide the raw materials for later historians; we saw an error in the former as more likely to impair our story than an error in the latter. For example, John van Wyhe has recently exposed as "entirely absent from the primary evidence" the long held notion that Darwin deliberately delayed publication to keep his ideas on natural selection a secret [20]:

> The myth of Darwin's delay has remained unquestioned for far too long. It generates a cascade of subtle errors that ultimately accumulate to a distorted picture of the man and his science and indeed to early Victorian scientific communities. A varied and overwhelming array of evidence demonstrates that Darwin did not avoid publishing his theory for 20 years. ... And this is the ultimate lesson to be drawn from the myth of Darwin's delay: the

> danger of confirmation bias lurks over historians just as much as scientists, if not more so. Once we, as historians, come to believe a story, it is easy to find apparent confirmations and, when the evidence contradicts it, difficult to let it go.

It appears that much concerning Bateson has been "difficult to let go." As far as we can determine, so different from those of others are the interpretations of Bateson that we offer, so central was Bateson in the post-Darwinian period, and so central was this period to biohistory in general, that our book has come to represent not only Bateson's life and science, but also a revised history of the biosciences.

Dept. Biomedical & Molecular Sciences
Queen's University
Kingston, ON, Canada

Donald Forsdyke

Prologue to Second Edition

> *The history of evolutionary theory over the last century has historiographic implications. For one thing, it suggests that we need a new history of population genetics and raises questions about its origins.*
>
> *Mark Adams [21]*

The first edition Prologue (2008) declared that the book was calling for a revised history of the biosciences. In the bicentenary year of Mendel's birth (2022), the case for this appears more compelling. Indeed, independent support from historians of science is now clear. Too late to mention in the first edition, I came across Laura Otis's work on "organic memory" [22] that anticipated and greatly extends Chap. 19 where nineteenth-century information concepts relate genetic memory to brain memory. Her preface states:

> In this study, I begin by analyzing self-consciously scientific formulations of the organic memory idea, especially those of Ewald Hering, Samuel Butler, and Théodule Ribot. I consider the motivation of scientists who conflated two unknowns: how people could store and pass on genetic information and how people could acquire and store new memories.

Furthermore, attention was drawn in Chap. 24 to the paper "Little Evolution, Big Evolution" published in French (1990) by Mark Adams [23]. Here he set out the case for macroevolution as had been argued by William Bateson and Richard Goldschmidt. In a privately circulated 34-page English version, a copy of which I received in 2003, Adams wrote:

> Alas, my own recent work has convinced me that much of this traditional view of the history of population genetics and its role in the evolutionary synthesis is fundamentally mistaken. Many of the details that I will deal with, of course, are well known to historians of the subject who work with original sources, and I think that the factual details in my account will not be especially controversial. But it seems to me that the standard interpretation has all too often ignored these facts and their implications, or else interpreted them in a Whiggish, wishful, or willful way. Properly understood, it seems to me, those facts demand a radically new interpretation of the history of Darwinism, population genetics, and the evolutionary synthesis.

Although Adams received much encouragement from readers of his English "samizdat," formal publication was not achieved until 2021 [21]! Regarding the underlying science, it remains a 1990 paper in that it has not benefitted from subsequent research. Nevertheless, Adams's profound insights, influenced by his deep knowledge of the works of early Russian evolutionists, accord well with present considerations. For example, regarding what he describes as "Darwin's central problem," Adams notes [21]:

> In order to get from genetics (and population genetics) to evolution, one must see a link between the origin and character of intraspecific varieties and the origin and character of species and higher systematic categories (genus, family, order, etc.); and to do this, it is necessary to consider the variation within species (intraspecific) and the variation between species (interspecific) as qualitatively the same. If intraspecific variation and interspecific variation differ not in kind, but only in degree, then it is possible, by extension, to envision selection as the creator of new species. But if varieties are fundamentally different from species – if the fundamental character of intraspecific and interspecific variation is essentially different – then the effect of selection on a population cannot explain evolution.

Similar views were expressed by Provine, who's influential 1971 text (11) was republished in 2001 with a retrospective "afterword" [24]: "In 1970 I could see the origins of theoretical population genetics as an unalloyed good for evolutionary biology, and thus obviously a great subject for a historian. Now I see these same theoretical models of the early 1930s, still widely used today, as an impediment to understanding evolutionary biology, and their amazing persistence in textbooks and classrooms as a great topic for other historians." I have summarized the mounting evidence supporting Bateson's science in a textbook [4] and in some post-2008 papers, some of which [25, 26] are included in Part VI, which now replaces the first edition's brief Appendix. Here, there is also an outline of Mendel "basics" [27], which might serve as a first port of call for those less familiar with Mendel's work and associated controversies, e.g., the Biometricians versus the Mendelians – a dispute sometimes characterized as natural versus group selectionists, or microevolutionists versus macroevolutionists. For those less interested in historical details, Part VI sufficiently repeats Parts I–V to permit an independent introductory reading. As before, Chaps. 21 and 23 are my coauthor's work. The rest are dual-authored, except Chaps. 1, 5 and 19, and those in Parts V and VI.

While deeply engaged in science, Bateson was also politically active both locally and internationally. In view of worldwide post-2008 socio-political turbulences, his caveats concerning implications of Mendel's discoveries for human cultural and racial issues today appear even more relevant. Indeed, we should note that 2022 was, as with Mendel, also a bicentenary year for the birth of the coiner of "eugenics" – Darwin's cousin, Francis Galton. According to our publisher's tabulation of first edition chapter downloads, those for the eugenics chapter have far exceeded others. Galton has been among the many targets of social reformers who have seldom shrunk from moral indictments as seen from the comfortable perspective of their own times, but with little regard for the times of their targets – a phenomenon sometimes known as "presentism."

Hopefully, our biography here provides insight into Bateson's "present." Although Bateson was a creature of his times, in the words of a close colleague, his "fearless pronouncements" as "one of the most clear-sighted and critical minds" of those times are clearly validated. While going through our book it will be easy – for those with a mind to do so – to cherry-pick statements that indicate otherwise, but the broad overview strongly suggests the opinion of Bateson's colleague is sound [28]. Similarly, Galton's eugenic statement that "demagogues who 'played to the gallery' would play to a more sensible gallery than at present" has a modern ring. Shortly after the First World War, Bateson held that "Genetic science … gives no clear sanction to these [eugenic] proposals" and predicted that "our immediate posterity will learn something of the consequences of un-applied biology."

As far as I am aware, there are no Bateson statues, and no buildings or professorships named after him. Rather than pulling down statues, and de-naming streets and buildings [29, 30], perhaps we should, more positively, be engaging our unsung heroes in the reverse. Indeed, positive reviews of our first edition [31–35] feed a distant hope that the text may one day assist an enterprising playwright intent on an accurate TV historical drama series.

Dept. Biomedical & Molecular Sciences
Queen's University
Kingston, ON, Canada

Donald Forsdyke

William Bateson, 1905

Acknowledgments

The "repatriation" of the Bateson papers was approved by Gregory Bateson and by Mary Catherine Bateson. With Stephen Jay Gould and David Lipset, Mary assisted Alan Cock in the collecting and sorting for shipment to the UK. The papers first home was the University of Southampton where cataloguing, supported by the Wellcome Trust, was carried out by Alan with secretarial help from Carol Newhouse and Patricia Hughes. Sources of other Bateson-relevant materials included the American Philosophical Society (Baltimore; the William Coleman microfilm), St. John's College (Cambridge) the Galton Collection at University College (London) the University of Cambridge Library and the Cambridge Department of Genetics. Arrangements were made for the transfer of materials to the John Innes Centre (Norwich), where photocopying and further cataloguing and indexing were carried out in collaboration with Brian Harrison, Peter Young, and Rosemary Harvey. Alan was granted interviews by Nora Barlow, Patrick Bateson and his uncle F. W. Bateson (a nephew of William Henry Bateson), Cyril Darlington, Edmund B. Ford, Geoffrey Keynes, Lily Newton (wife of W. C. Frank Newton), and Helen Pease. His studies of the Pearson-Weldon-Galton correspondence in the Watson Library at University College were facilitated by Maxine Merrington.

Alan's wife, Marta Cock, helped me trace him in 2004 to his address in London, where he and his four children – Sybil, Hannah, Christina, and Oliver – approved the transfer of his unfinished manuscript and his copy of the Bateson papers to Canada. Sybil did the packing and arranged the shipment. Christina authorized documentation relating to Alan's estate. In 2005, the Librarian at the John Innes Institute, Kenneth Dick, facilitated my examination of the Bateson Library, which the Trustees of the Institute had purchased from relatives in the USA. In the course of this flying visit, I came across an unpublished five-volume treatise – *William Bateson and the Emergence of Genetics* (2000) – composed by Rosemary Harvey (retired). The index indicated a different approach to the present work. My plan to photocopy the Harvey treatise did not come about, and I reluctantly decided that this would be one promising stone I would leave unturned. Hopefully, it will one day be scanned and made available online. However, the second edition greatly benefited from Rosemary's input that followed her careful reading of our first.

Among those who helped me with the relevant literature, particular thanks are due to Mark Adams and Marsha Richmond for making available advanced copies of various texts, Christiane Nivet for comments on Chap. 20, and Patrick Bateson for advanced copies of papers. Joan Horton kindly searched Bateson family UK census records. My getting up to speed in the history of science was assisted by Gerard Wyatt's donation of books from his collection. Permission to quote from Rona Hurst's unpublished text was given by Patrick Zutshi on behalf of the Syndics of Cambridge University Library.

The figures were either drawn especially for this volume or were taken from my previous works with the approval of the publishers (McGill-Queen's University Press, Montreal, and Springer, New York). For the second edition (Chaps. 27, 28 and 29), papers were reproduced with permission from Wiley (Encyclopedia of Life Sciences), the Royal Society, and Springer. Various photographs were either taken from sources where copyright had expired or were part of Alan Cock's set of the Bateson papers which I have now deposited in the Archives of Queen's University at Kingston. The photograph of Samuel Butler is reproduced with the permission of the Master and Fellows of St. John's College. Queen's University provided invaluable library resources and hosts my webpages where further information on Bateson and his fellow "Mendels" may be found.

For the first edition, Saladi Gunabala with her team at SPi Technologies in Pondicherry, India, and Andrea Macaluso, Jeffrey Ciprioni, James Russo, and Melanie Wilichinsky of Springer (New York) smoothed the passage from typescript to final copy. Likewise, the second edition was fostered by Shina Harshavardhan with her team (Gomathi Mohanarangan) at Straive Publishing, Puducherry, India, and by Kenneth Teng of Springer (New York). Charlotte Forsdyke assisted me with the copy editing. Polly Forsdyke advised on French and German usages. Sara Forsdyke and Caroline Falkner advised on Latin and Greek usages. Ruth Forsdyke provided helpful texts that I would not have otherwise found. My wife Patricia was a constant source of inspiration and advice. Alan was close to his family. Had he written an acknowledgment, I am sure it would have revealed an indebtedness to them that is no less than I to my own.

Dept. Biomedical & Molecular Sciences
Queen's University
Kingston, ON, Canada

Donald Forsdyke

Abbreviations

BA	The British Association
CBSS	Conjoint Board of Scientific Societies
Defence	Bateson's *Mendel's Principles of Heredity: A Defence* (1902)
EB	Evolutionary Bioinformatics
FRS	Fellow of the Royal Society
IRC	International Research Council
IUBS	International Union of Biological Societies.
Letters	*Letters from the Steppe*, edited by Beatrice Bateson (1928)
Materials	Bateson's *Materials for the Study of Variation* (1894)
MBA	Marine Biological Association
Memoir	*Essays and Addresses* with a Bateson biography (1928)
Principles	Bateson's *Mendel's Principles of Heredity* (1909)
Problems	Bateson's *Problems in Genetics* (1913)
RHS	The Royal Horticultural Society
RS	The Royal Society
RSM	The Royal Society of Medicine
RS-Report	*Report to Evolution Committee of the Royal Society*
SEB	Society for Experimental Biology

Contents

Part II Mendelism

Part I
Genesis of a Geneticist

Chapter 1
A Cambridge Childhood (1861–1882)

In the beginning was the Word, ... And the Word was made flesh ....

Gospel according to St. John.

There are stories within stories. To the extent that the story of William Bateson can be demarcated, it begins in 1809 in the city of Shrewsbury in England. It ends, perhaps, in 2009, in no specific location, but in the minds of people throughout the world when celebrating the two hundredth anniversary of Darwin's birth. By that time, computer analyses of DNA sequences (bioinformatics) had revolutionized evolutionary biology. The informational basis of heredity ("the word") was explicit. As will be related here, this seems to have been an essential precondition for the full recognition of Bateson and his work. Our story begins with a general overview and an introduction of the main characters.

1809–2009

The two hundred years 1809–2009 divides conveniently into four equal parts with three major chronological milestones—1859, 1909, and 1959. In 1809, Jean-Baptiste Lamarck published *Philosophie Zoologique* in France, and Charles Darwin was born in Shrewsbury, England. For several years, Darwin was a boarder at Shrewsbury School, where Dr. Samuel Butler was the headmaster. The father of William Bateson, William Henry Bateson (1812–1881), also attended Shrewsbury School and was accepted by St. John's College, Cambridge, in 1829, where he was a contemporary of Butler's son Thomas. Both were contemporaries of Darwin who was at Christ's College from 1828 to 1831. Thomas Butler and Darwin were acquainted [1]. All three were intended for the church so it is possible that they met

A. G. Cock, D. R. Forsdyke, *Treasure Your Exceptions*,
https://doi.org/10.1007/978-3-030-92099-9_1

around 1831, and may even have discussed, but almost certainly not understood as we do now, the above line from the gospel according to St. John.

Thomas Butler became Canon at Langar Rectory in Nottinghamshire, where his son Samuel was born in 1835. Neglecting his religious studies, Darwin turned to extracurricular biology before graduating and setting off on his famous voyage on The Beagle. William Henry Bateson continued his religious studies and graduated in 1836. After serving in various parishes, he returned to St. John's with a fellowship that required that he not marry (a rule not revoked until 1882). Among his pupils (1838–1841) was Charles Kingsley who in 1860 became Professor of Modern History at Cambridge. In 1846, William Henry was elected Senior Bursar, and in 1857, Master. He was acquainted with the younger Samuel Butler who, in response to paternal pressures, began studies at St. John's in 1854 with the intention of entering the church.

1859 saw the publication of Darwin's *The Origin of Species by Natural Selection, or The Preservation of Favoured Races in the Struggle for Life*. Here, Darwin proposed natural selection as a major force in evolution, but with the caveat: "I am convinced that natural selection has been the main but not the exclusive means of modification." Since man was deemed to have derived from lesser forms, the book was widely perceived as challenging religious orthodoxy. Thomas Huxley, who taught natural history in London at the School of Mines (now the Imperial College of Science and Medicine), soon became known as "Darwin's bulldog." He traveled the length of the land spreading the Darwinian message. In 1859, the younger Samuel Butler emigrated to New Zealand, omitting for the first time in his life to say his prayers at bedtime. He soon acquired Darwin's book and was to become, as later described by William Bateson, "the most brilliant, and by far the most interesting of Darwin's opponents."

Jumping ahead to the next milestone, 1909, Darwin and the younger Butler have passed away. Darwinism is triumphant and the Centenary of Darwin's birth is celebrated with the volume *Darwin and Modern Science* to which many evolutionists contribute. We can, with hindsight, now assign to Bateson, Professor of Biology at Cambridge, the role of the most brilliant and by far the most interesting of Darwin's opponents, and his contribution to the volume reflects this. Not surprisingly, many of his contemporaries consider him by far the most annoying of Darwin's opponents. He has converted the world to Mendelism and Wilhelm Johannsen now suggests that the factors responsible for the Mendelian character units be referred to as "genes." Across the Atlantic, Edmund Wilson and Thomas Morgan are enthralling freshman classes at Columbia University, which include Calvin Bridges and Alfred Sturtevant. A popular textbook is *Variation, Heredity and Evolution*, authored by one of Bateson's students. Over the next decade, Bridges and Sturtevant, together with Herman Muller, will flesh out Mendel's genic abstractions under Morgan's patient tutelage.

Bateson is at the height of his power and influence. Yet, his essay in the commemorative volume is neither self-congratulatory, nor filled with heady optimism. Darwin failed to show, as the title of his great book suggests, that natural selection can originate a species. For this, we must look beyond genes. This is not to say that genes are not involved in the critical events that differentiate one species from another. But for the spark that originates, we must often look elsewhere. Bateson's message is lost on his contemporaries and most who come after. The more he enunciates it, the more his estrangement. Acknowledging some bewilderment, the German evolutionist August Weismann diffidently declares that the task of estimating "the influence of Darwin's theories on his time and on the future" might be "better accomplished on the 200th than on the 100th anniversary of his birth" [2].

Now it is 1959, and a new cast of actors are on the evolutionary stage. We are in the television era. More importantly, we are in the era of extensive public support for biomedical research. The evolutionists meet at Chicago to celebrate the 150th anniversary of Darwin's birth and a three-volume text emerges. The politician Adlai Stevenson joins a television debate. Who better to start the proceedings than Julian Huxley, no less laudatory of Darwin than his famous grandfather [3]:

> The emergence of Darwinism … covered the fourteen-year period from 1858 to 1872; and it was in full flower until the 1890s, when Bateson initiated the anti-Darwinian reaction. This in turn lasted for about a quarter of a century, to be succeeded by the present phase of Neo-Darwinism, in which the central Darwinian concept of natural selection has been successfully related to the facts and principles of modern genetics, evolution and palaeontology.

Among the evolutionists present, there is much complacency. A "modern synthesis" of Darwin and Mendel has been achieved. Had Bateson been there, he would have referred to Voltaire's hero Dr. Pangloss for whom "everything is for the best in the best of all possible worlds." Now, with the help of the molecular biologists, who have recently announced the structure of DNA, it seems just a matter of working out the details. Most of the distinguished contributors—including Ledyard Stebbins, Alfred Emerson, Theodosius Dobzhansky, and Sewall Wright—do not mention Bateson. If there is a hero, it is Bernard Kettlewell of Oxford, whose studies with the peppered moth Edmund Ford declares to have shown "evolution in action." The Agassiz Professor of Zoology at Harvard, Ernst Mayr, refers to what he calls the "mutationism of De Vries and Bateson," which was "wrong," and "meaningless in the light of our new genetic insight" [4].

Early Years

A narrower demarcation of Bateson's story begins with his birth in 1861 and ends with his death in 1926. We cover here his life up to 1882 when he graduated. In 1857, it became clear to William Henry Bateson that he would be elected Master of St. John's and so would no longer by bound by the restrictions of his fellowship. He wrote to his sister Margaret (Feb. 26):

> I have an important announcement to make to you, no less than that I am going to be married, and, as I hope, without many weeks delay. You will remember that I wrote to you the other day saying that I might shortly visit Liverpool. The fact is that, so soon as I had it clear before me that I should be made Master of the College, I inquired of Honora [their sister]

> whether Annie Aiken, her great friend, upon whom I had long secretly set my heart, was free and at all likely to accept any advances from me. I had never made it known to anyone before, much less to the Lady herself, that such was the state of my affection. Honora could make out nothing except that she was free – so as faint heart, etc. down I came last Tuesday week and by incessant dedication of myself to the object of my mission I am delighted to tell you that I am now an accepted suitor. I have thought the matter over with all the prudence and coolness that I could command and I can say *deliberately* that such another wife could not be found. There is literally nothing wanting to complete the group of female virtues and fascination and I am as happy as possible. It was only yesterday afternoon that the final Yes was said – so you have the earliest intelligence. I think you hardly know her, but I am persuaded that you will see all her excellence at first sight and in no long time that you will regard each other with all the love and affection of sisters.

So on June 11, 1857, William Henry Bateson was married to Anna Aiken (1829–1918) in St. Bride's Church, Liverpool, by the Reverend Thompson the Regius Professor of Greek at Cambridge. Their first child died shortly after birth in January 1859. Margaret was born in 1860. This was followed a year later by William, then three daughters (Anna, 1863; Mary, 1865; Edith, 1867), and finally in 1868 another son, Edward. Unlike some others who enter this story—Grant Allen, Arthur Balfour, William Brooks, Henry Martin, George Romanes, and Hugo de Vries—who were all born in 1848, William Bateson's date of birth allowed time for those who might influence his education to digest their Darwin and loosen the ties of religious orthodoxy.

If one had to prescribe the ancestors, the time of birth, and the location, of someone who might best carry the Darwinian torch forward and into the twentieth century, Bateson's would be hard to match. The parents of William Henry Bateson and of Anna Aiken were vigorous and wealthy. Richard Bateson (1770–1863) was a Liverpool cotton merchant, and in 1806, he married a soldier's daughter, Lucy Wheler Gordon (1781–1866), who gave birth to twelve children. James Aiken (1792–1878) was a Liverpool shipping merchant and a Justice of the Peace, described by Beatrice as a "staunch liberal." In the course of his voyages, he met Anna Elizabeth Harrison of Charleston, South Carolina, and they were married in Liverpool in 1815.

William Henry and Anna were intent on bringing about reform in their domain, the Cambridge colleges. In 1850, William Henry had been secretary to a Royal Commission on the education system. The mastership gave him the power to implement its recommendations, leading to the transformation of the universities from relative monasticism to major centers of scholarship. The Master of St. John's was required to be in holy orders. That changed during William Henry's term. His religious beliefs and practices were of the broad-church variety, as is evident from his surviving sermons. His liberal attitude to religious and moral matters was reflected in his choice of the notorious Bishop Colenso of Natal to conduct the consecration service for a new college chapel. Colenso had challenged doctrine and had not insisted that polygamous Zulu chieftains give up all but one of their wives upon conversion to Christianity. Even worse, he had encouraged notions of self-government among the Zulus!

Mrs. Anna Bateson was quietly formidable and indefatigable. Apart from successfully rearing six children, she fulfilled the not inconsiderable duties of Master's wife. This still left her with surplus energies, expended in the cause of the women's section of the Liberal Party and, above all, of the movement to secure university education for women. She was in this movement from its first beginnings in the 1860s when it aimed at no more than establishing courses of lectures on university premises, given mainly by university personnel. Lest the University be prematurely alarmed, there was no talk at that time of women entering it or being given degrees. "Annie" acted as treasurer (later secretary) of the organizing committee, distributed lecture tickets, solicited funds, and organized tea parties in the Master's Lodge for supporters. She continued her active support through several campaigns up to the time of her death, but she died with victory incomplete (it did not come until 1947), although much had been achieved (Chap. 23). Her prominent obituary in the suffragette paper *The Common Cause* (Aug. 2, 1918) was headlined "One of Our Pioneers." However, the obituary noted:

> Mrs. Bateson was never an out-and-out feminist in the modern sense. She rated men's capacities very highly, having both in youth and during her early married life been influenced by the many men of ability she had known. At times she questioned whether women, even when freed from all the heavy handicaps which foolish and unjust laws and custom have imposed on them, could 'make good' equally with men.

Her enthusiasm for women's higher education infected her family. William Henry was a member of the original governing board of Newnham College (then an unofficial appendage of the university), and most of the children were involved in one way or another.

The young William grew up in the intellectually stimulating environment of the Master's Lodge, surrounded by scholars and students. It can hardly be an exaggeration to say that he imbibed art, culture, and Darwinism with his mother's milk. And there was the further stimulus of his brothers and sisters (Fig. 1.1).

These were a distinguished group of diverse abilities. Margaret (1860–1938) became one of the earliest woman journalists. She lived in Bloomsbury, London, and after a brief period on the staff of a new magazine, *The Hour Glass* (which failed), she became Employment Correspondent of *The Queen*. In 1895, she published *Professional Women and their Professions* [5]. In it, the word "profession" was interpreted liberally, from occupations of low pay and esteem to ones of high pay and esteem, from ones staffed largely or even wholly by women to ones where a few isolated women were now, greatly daring, striving for a foot in the door. For each profession, there was an interview with a particular woman. Margaret was the only one of the four sisters to marry, and that was late in her life, to William Heitland (1847—1935), a classical historian at St. John's [6].

William Bateson's birthplace (Aug. 8, 1861) is recorded as 20 St. Hilda's Terrace, Ruswarp, Whitby, Yorkshire. But his siblings, save Mary, were all born in the Master's Lodge. Sixteen months after William came Anna, with whom he was very close, and two years later, Mary, the most academic of the girls. She read history at Newnham and took first-class honors. She launched immediately upon a career of

Fig. 1.1 William Bateson

historical research under the tutelage of Frederic Maitland, a Professor of Law. It is not surprising, therefore, that much of her research, dealing mainly with the medieval period, had a legal slant. By the time of her death in 1906, she had produced several books (some in multivolumes) and numerous papers. Mary is commemorated at Newnham College by a bust in the courtyard and a scholarship in her name. Of significance for our story is that at some point she befriended Christiana Herringham, a wealthy patron of the arts, the adventures of whom may have suggested the character of Mrs. Moore to E. M. Forster for his novel *Passage to India* (Chap. 14). Edith, the youngest girl, seems to have played the role typically allotted to the youngest daughter in Victorian families: she stayed at home to look after her mother. The baby of the family, Edward, broke a little with family tradition by going to King's College, Cambridge. This was perhaps because he was to read Law, not then a strong subject at St. John's.

An early influence was the family governess, Miss Lakin, who fostered William's interests in animals and plants. People with such interests were referred to as "naturalists," and the seven-year-old boy is said to have once so designated, with respectful awe, a tramp rummaging in a ditch. Concerning his return home from a preparatory school at East Sheen for the Easter vacation his mother wrote to a friend (March 23, 1875):

> They were amazed … by seeing the four children all decked out in ribbons and garlands of ivy for the purpose of giving a welcome to Willie who is expected in an hour. Maggie [Margaret] does not come [back from school] until tomorrow and on Thursday morning we are all off to Liverpool. … Well, I am getting Willie a *stunning* little book by Sir J. Lubbock [later Lord Avebury] on the influence of insects in the development of Flowers. I will see if

> he likes it. I expect him to burrow down in it like a bee in a flower itself, and if he *does*, I shall send a copy forthwith to your Robert. Pabulum, stuff to feed the mind, comes to be what one wants, and I shall always regret that Natural History has been such a sealed book to me. Anna is picking up a nice knowledge of Botany.

At the end of the Easter vacation, she wrote again: "Poor Willie went back to school on Tuesday and he quite broke down again as he does every time. It is very trying! We are soon to part with Maggie who is on the eve of going to Metcalfe's at Hendon near London, but *she* looks cheerily forward to the change." As for a future career, at an early stage, Bateson declared his intention to be a naturalist adding: "If I am good enough … if not, I suppose I shall have to be a doctor." Although it is documented that he became much attached to Gray's *Anatomy*, there is no record as to when he first came across Darwin's great book.

After preparatory school, at age 14 he went on a scholarship to Rugby, one of the great private schools of England. His much-admired scholarship essay took the form of a Socratic dialog concerning the misplaced emphasis schools placed on sport. However, he later described this period "as a time of scarcely relieved weariness, mental starvation and despair" (address to the Salt Schools, 1915). His school reports attest to this. Mysteriously, year after year, in almost every subject, he was at or near the bottom of his class. After the first year, the scholarship was not renewed. His letters home to his father consisted of excuses for monetary expenditures and accounts of his ranking (the boys were ranked fortnightly based on their marks): "I know I have no pluck. Fellows have kicked that out of me long ago." Letters to his mother were more forthcoming (March 10, 1878):

> I have read a great deal this term and been happier than usual. And I don't want you to tell this [to others]. But somehow I never feel really happy, I suppose I never shall; at least I don't think I ever did except when I got my scholarship. And when Sidgwick says I am doing better, it never seems any better. Is anyone happy? I don't think I shall be. You will say this is all morbid nonsense, but it is not to me and it is true. I never get on with anybody for long: at home even, I am always in some scrape except when I am alone. And don't please write back that I am foolish and that, and then not tell me how to cure it.

In July 1879, the Headmaster declared his father "right in removing him early, for he is very self-satisfied and desultory – even indolent in most things – here." Thus, at age 18, Bateson entered St. John's College in September 1879 as a Cambridge undergraduate. However, in his first term, he failed an examination in mathematics (the "Littlego") at his first attempt. He later noted: "Being destined for Cambridge I was specially coached in mathematics at school. Arriving here [Cambridge], I was again coached, but failed. Coached once more I passed, having wasted, not one, but several hundred hours in that study. Needless to say, my knowledge of mathematics is *nil*." This exaggeration is reminiscent of the Scarecrow in *The Wizard of Oz*; while in 1905 (*Nature*) Bateson was declaring his lack of mathematical prowess, mathematical concepts were underlying much of the science he practiced. An early preference for qualitative rather than quantitative approaches can be considered as a possible factor delaying the commencement of his experimental breeding studies (Chap. 6). We do not know whether this was due to a real lack of ability, or to a lack of confidence in that ability implanted by his examiners, or to a snobbish early disregard for a subject that could be used "in trades and professions for the making of money" and not in "another world, where 'utility' does not count."

Physiology

Bateson grew up as a prospective "naturalist" in a scientifically more welcoming environment than, say, the Canadian-born Romanes, or Hugo de Vries in Holland. They were both aged eleven when *The Origin of Species* appeared. Romanes grew up to wealth and personal freedom, living both in continental Europe and in London (Chap. 5). But to the end of his days, religious questions were of major concern. Bateson, on the other hand, took religion pragmatically as part of the society he lived in. He was atheistic rather than agnostic and it seems that Beatrice thought likewise.

However, Romanes, who in 1867 became a student at Gonville and Caius College, had a singular advantage that was denied Bateson. Among his contemporaries were Francis ("Frank") Balfour and, one of Charles Darwin's sons, Francis (another "Frank"). When Romanes' interests turned to evolutionary biology, who better to discuss them with than Charles Darwin himself [7]. For eight years, until Darwin died in 1882, Romanes was in practice, if not in name, Darwin's research associate, an honor he shared with Francis Darwin. Romanes worked mainly with animals and Francis worked mainly with plants.

Romanes came under the influence of Michael Foster who had been appointed Praelector at Trinity College in 1870 and gave lectures in physiology and embryology (developmental biology). Romanes became an experimental physiologist (i.e., if you wanted to know what a nerve did, you cut it and watched to see what functions were impaired). Balfour, on the other hand, was more interested in embryology and comparative anatomy (i.e., he would describe when a nerve first appeared in an embryo, what path it took through the tissues, and how it varied in different types of organism).

A key difference was that physiologists tended to work on live organisms (hence they were vulnerable to attack by the anti-vivisectionists), whereas "morphologists" tended to work on the dead. But there was more to it than this. Physiology was an experimental science, whereas morphology was an observational science. In general, morphologists were content to examine "Nature's experiments" and then construct hypotheses to explain what they had seen. Physiologists constructed hypotheses and then, to test the hypotheses, experimented on the materials Nature had made available.

Romanes moved to London in 1874, as had Francis Darwin who qualified in medicine at St. George's Hospital in 1875 and studied the physiology of vascular dilatation at the Brown Institute, which was associated with University College. Romanes became associated at University College with William Sharpey under whom Foster had studied prior to his Cambridge appointment, and with John Burdon Sanderson who had taken Foster's position at the College, and would later move to Oxford where Romanes would join him. Also based in University College was a protégé of Huxley, Edwin Ray Lankester (1847–1929), holder of the Jodrell Chair of Zoology from 1872 until 1891. Balfour stayed on at Cambridge, working closely with Foster around whom the famous School of Physiology was to emerge [8, 9]. Foster and Balfour had strong ties to Huxley. Foster assumed prime responsibility for teaching physiology and histology (the microscopic structure of normal tissues) and Balfour taught morphology and embryology.

Morphology

When William Bateson began his Cambridge studies in 1879, the School of Physiology was well established. However, strongly influenced by a student who had entered St. John's College a year earlier, Walter Frank Raphael Weldon (1860–1906), he was drawn to Balfour's Morphological Laboratory, in the domain of the Professor of Zoology and Comparative Anatomy, Alfred Newton, an authority on birds. Among fellow students were Sidney Harmer (1862–1950), Walter Heape (1855–1929), Arthur Shipley (1861–1927), and his new friend Weldon. Balfour was killed in a climbing accident in 1882, the year of the death of Charles Darwin. As we shall see, Balfour's uncle (the Marquis of Salisbury) and siblings (Arthur, Eleanor, Evelyn, Gerald) came to exert a considerable influence in the biological sciences.

Bateson learned his lessons well, and it would seem, unquestioningly. He gained first-class degrees in both the first (1882) and second (Zoology; 1883) parts of the Natural Sciences Tripos. His father, who had despaired at his lack of success at school, did not witness this. He died in 1881. Bateson later (Toronto, 1922) describes this as a time when:

> Morphology was studied because it was the material believed to be most favourable for the elucidation of the problems of evolution, and we all thought that in embryology, the quintessence of morphological truth was most palpably presented. Therefore, every aspiring zoologist was an embryologist, and the one topic of professional conversation was evolution. … I wonder if there is a single place where the academic problems of morphology which we discussed with such avidity can now arouse a moment's concern.

The reasons for the excitement are not hard to discern. In Berlin, Robert Remak had proposed that all cells were derived by the division of pre-existing cells ("*Omnis cellula e cellula*," was Virchow's handy phrase). So all cells in an organism stemmed from the single fertilized egg cell [10]. This cell, and its division following fertilization, was described in mammals by Karl Ernst von Baer and others in the 1820s. The temptation to draw a parallel between the development of a multicellular embryo from a single cell over days, weeks or months, and the branching evolution of multicellular organisms from primitive, amoeba-like organisms over millions of years, became irresistible.

Indeed, many believed with Fritz Müller that "ontogeny recapitulates phylogeny"—namely, that a study of developing embryos (ontogeny) might provide insight into the relationships between groups of animals (phylogeny). In the words of Ernst Haeckel [11]: "Ontogeny (embryology or the development of the individual) is a concise and compressed recapitulation of phylogeny (the paleontological or genealogical series) conditioned by laws of heredity and adaptation." So it seemed that embryology would aid the classification of organisms based on lines of descent. As late as 1888, Lankester wrote in the *Encyclopaedia Britannica* [12]: "It was the application of Fritz Müller's law of recapitulation which gave the chief stimulus to recent embryological investigations; and though it is now recognized that 'recapitulation' is vastly and bewilderingly modified by special adaptations in every case, yet the principle has served, and still serves, as a guide of great value."

Cambridge Life

With the removal of college restrictions on marriage, there emerged a rich, albeit somewhat inward-looking, world of academic family life, as is related by Gwen Raverat, one of Charles Darwin's granddaughters, in *A Period Piece. A Cambridge Childhood* [13]. She held her uncles, the five sons of Darwin, to be "the most unself-conscious people that ever lived." The name of Darwin turned heads, and many of those heads were female. Yet, William, the eldest son, did not marry until age 34 (to Sara Sedgwick). George Darwin (1845–1912), the eldest academic son, did not marry until age 38 (to the niece of the Professor of Greek). Among their children were Gwen, the authoress of *A Period Piece*, and Margaret. The latter came to marry the surgeon Geoffrey Keynes, who had an early passion for insects, attended Bateson's "bible classes" on genetics, and shared his admiration for the works of William Blake.

Francis Darwin (1848–1925), held to be "the most charming of the brothers," was married at age 26 to the first of three wives, the last of whom, was the widow of Frederic Maitland. Leonard (1850–1943), later Major Darwin, was 32 at the time of his first marriage. The youngest Darwin son, Horace (1851–1928), married at age 29. His daughter, Nora, also attended the "bible classes" and became one of the many who assisted Bateson's researches.

Of Bateson's college days, his brother-in-law reminisced [14]:

> It must have been the Michaelmas term of 1879, when I became conscious of the presence in College of a large and rather untidy undergraduate, who was pointed out to me as the son of our honoured Master, Dr. Bateson. I did not at once get to know this youth, but I was from the first attracted by his appearance, even in his bodily movements unconventional. He seemed a sort of living protest against the 'average' quality of his contemporaries. Acquaintance soon confirmed the suggestions of his outer bearing, and I found myself in touch with a man of frank independence in thought word and deed.

With the death of his father, Bateson came to play a greater role in family affairs. There was a move from the Master's lodge, first to Queen Anne Terrace and then to 8 Harvey Road. Nearby at 6 Harvey Road were the newly-wed John Neville Keynes and Florence Ada Brown, who were about to produce three children: Maynard

(1883) the future economist, Margaret (1885) the future wife of physiologist A. V. Hill, and Geoffrey (1887), who is mentioned above. The year 1882 must have been particularly busy since, apart from studies, Bateson worked for the Liberals in the Cambridge parliamentary elections, a one-time foray into "purely ephemeral" party politics. At this time, he visited Dresden with one of his sisters. He later recalled the visit in glowing terms in a letter (Jan. 1887) to his sister Anna who was there in December 1886:

> In my mind, Dresden and Antwerp are the two best places outside Italy (excepting Gt. Britain and Ireland)! It was in Dresden that I first saw pictures; first heard Wagner; and first made the acquaintance of Browning; so that I have some cause to remember it .... I went to Dresden never having seen a good Raphael, and never having cared much about the engravings of the 'San Siste,' and fancying that Raphael must be rot, and the 'San Siste' gave me a dreadful turn. I never shall forget turning the corner into the little room at the end, not knowing anything particular was there, and then coming face to face with that picture.

Anna and a companion had not shared William's delight:

> Well! Lord spare you to see it again and may it please Him to soften your hard hearts! You say that your companion is as bad as you are – I hope that this is a libel, but if it isn't, I think you ought both to have been put out. I wonder if you discussed the price of beer, "*per litre*", in the 'San Siste' room, as I once heard two Yanks do!

Bateson returned to the theme in his next letter (Jan. 1887):

> If you want to find out what a picture is good for, I mean, what it is worth to you personally, and it does not strike you at first glance, you should go by yourself and look at it for a good while, and fancy that it is looking at you – and then that the figure in the picture is there with you, and that you are one of the company – and then, perhaps, you suddenly become aware that the people in the picture are a new thing to your conception of people and things, and have come down from Heaven. It is monstrous that such a process should be necessary with the 'San Siste', but sometime try it. Fancy you are where St. Barbara is, and that the woman has just come into the room carrying the child. Would you think it an ordinary experience to be in the presence of such a person? What do you think you would say to her? You would then feel that you are in the presence of most surpassing beauty, such as you had never conceived.

Other than Weldon, did Bateson have any particular friends during this period? We have it from her son David (Chap. 2) that one was Constance Black (later Garnett). A few months younger than Bateson, she entered college (Newnham) in September 1879. Here, she met Florence Ada Brown (who was to become a neighbor of the Bateson's; see above), Ellen Crofts (a lecturer in English literature who in 1883 became Francis Darwin's second wife), Alice Lloyd (who became a journalist and married Horace Darwin's colleague, Albert Dew-Smith), and Edith Sharpley (a Classics student). Constance had originally intended to study Natural Sciences, but she opted for Classics and Mathematics and soon dropped the latter to concentrate on Classics. In 1883, she emerged, having endured written examinations for six hours daily on six consecutive days, to share top place with Edith Sharpley in part II of the Classics Tripos [15].

Summary

Some key characters are introduced. Attention is drawn to family connections between the Darwins, Butlers, and Batesons. William was born to great privilege, son of the Master of St. John's College, Cambridge, and one of six diversely gifted siblings. After painful years at Rugby, he flowered at Cambridge. The religious views of his parents were pragmatic, and he became atheistic rather than agnostic. Under the influence of a college friend, Raphael Weldon, he was drawn to morphology (Balfour) rather than physiology (Foster) and took first-class honors in 1882. Another friend was a Newnham College student Constance Black. In a visit to Dresden, he "first saw pictures; first heard Wagner" and was greatly enthused. Following the death of his father in 1881, the family became neighbors of the Keynes in Harvey Road. Among early graduates of Foster's school of physiology were George Romanes (animals) and Francis Darwin (plants). They were both research associates of Charles Darwin in the eight years before his death in 1882 and were to have a deep impact on Bateson's life.

Chapter 2
From Virginia to the Aral Sea (1883–1889)

> *If it is true, in each egg, all the functions and faculties of a definite mature animal lie hidden, without any corresponding organs, must we not regard heredity as a mystery too great for solution?*
>
> *William Brooks, 1883 [1].*

When Foster went from London to Cambridge in 1870, he took with him Henry Martin whose prowess had become evident soon after he entered University College at the age of 16. In the Natural Sciences Tripos of 1873, Martin took first place ahead of Francis Balfour. Martin then stayed on at Cambridge, while Balfour spent several months at Anton Dohrn's new Statione Zoologica in Naples, where marine life forms were readily available. In 1876, Martin was appointed by Daniel Gilman (see Chap. 9) to the Chair of Biology at the newly established Johns Hopkins University in Baltimore. Paralleling the division of labor between Foster and Balfour at Cambridge, at Johns Hopkins, Martin was primarily engaged in physiology research, with his assistant, William K. Brooks, acting as morphologist.

Brooks

Although perhaps tempted to follow the well-trodden path from Cambridge to Naples, acting on the advice of Raphael Weldon and Balfour's assistant Adam Sedgwick (1854–1913), who now ran the Morphological Laboratory, it was to Brooks and the embryology of the acorn worm, *Balanoglossus*, that Bateson turned after his Cambridge graduation. Among many young American biologists who came under Brooks' influence were Thomas Morgan and Edmund Wilson. Brooks had organized an annual mobile marine biological station—Chesapeake Zoological Laboratory—created anew each summer by renting space by the sea. Bateson had seen in a Johns Hopkins circular that the worm had been observed at the marine

A. G. Cock, D. R. Forsdyke, *Treasure Your Exceptions*,
https://doi.org/10.1007/978-3-030-92099-9_2

station and joined in the summers of 1883 (Hampton, Virginia) and 1884 (Beaufort, North Carolina).

The collaboration was by no means one-sided. Bateson brought with him some state-of-the-art technology, namely the Jung microtome, which facilitated the cutting of thin sections of tissues for microscopic examination. Continuing in the orthodox tradition of tracing evolutionary relationships through comparative morphology and embryology, Bateson concluded from dissections and microscopic analyses that acorn worms were lowly relatives of the vertebrates. He saw that the group to which it belonged (*Enteropneusta*) could provide an evolutionary link between lower marine invertebrates that were without spinal columns (*Echinodermata*) and higher animals that had spinal columns (*Chordata*; named by Balfour in 1880). Lankester had divided the *Chordata* into three classes, the urochordates (tunicates), the cephalochordates (amphioxus), and the vertebrates. In several papers published from 1884 to 1887, Bateson identified the acorn worm and its allies as a fourth class (*Hemichordata* [2]). The perceived importance of this was described by a later collaborator [3]:

> Among the questions most keenly disputed was that connected with the origin of the great vertebrate group. From what group of invertebrates could they be supposed to descend? … Between the lowliest vertebrate and everything else there seemed, even in speculation, an unbridgeable gap. … Bateson … showed that, judged by the canons of research then in vogue, Balanoglossus must be regarded as the humblest member of the group to which the vertebrates belong, thus opening up a fresh view of the relation of this great group to the rest of the animal kingdom.

Today, Bateson's Hemichordates are considered as a separate prechordate group. Despite his success, he became disillusioned with the embryological approach. Brooks was largely responsible for this. Happy hours were spent chatting with Brooks lying in shirtsleeves and "full of novelty, suggestion, and humorously inventive thought" and "the earnest solemnity of philosophical speculation." There was a meeting with his wife, and Bateson wrote to his mother (July 20, 1883): "Mrs. Brooks is just to hand, rather a nice little person with a good deal of refinement, though a Southerner."

Brooks had published his *Handbook of Invertebrate Zoology* in 1882 and had just completed *The Law of Heredity—A Study of the Cause of Variation, and the Origin of Living Organisms* [1]. Here, he argued that the ovum in females is responsible for the conserving element (heredity) in reproduction and the sperm in males for the progressive element (variation). In an amazing *tour de force,* a large body of evidence was marshaled in support of a false theory. Nevertheless, Bateson was greatly stimulated [4]:

> It was through Brooks that I first came to realize the problem which for years had been my chief interest and concern. At Cambridge in the eighties morphology held us in its spell. That part of biology was concrete. The discovery of definite, incontrovertible fact is the best kind of scientific work, and morphological research was still bringing up new facts in quantity. It scarcely occurred to us that the supply of that particular class of fact was exhaustible, still less that facts of other classes might have a wider significance. In 1883 Brooks was just finishing his book '*Heredity*,' and naturally his talk used to turn largely on this subject. He used especially to recur in his ideas on the nature and causes of variation, and to the conception … that the functions of the male and female germ cells are distinct. The leading thought

> was … that 'the obscurity and complexity of the phenomena of heredity afford no ground for the belief that the subject is outside the legitimate province of scientific enquiry.' He deplored the fact that he had no opportunity for the requisite experiments in breeding, but he saw plainly that such experiments were the first necessity for progress in biology. To me the whole province was new. Variation and heredity with us had stood as axioms. For Brookes they were problems.

It is not clear to what extent, if at all, Bateson digested his mentor's book, which is in Bateson's library, now preserved in Norwich at the John Innes Centre, but we can take it as a guide to what they discussed. The Preface sketched out a plan that Bateson was later to adopt [1]:

> Many experiments have suggested themselves to me, but as most of them involve the cultivation and hybridization, for many generations, of such animals and plants as will thrive and multiply in confinement, they can only be carried out by some one who has the means for experimental researches, and who has also a permanent home in the country, where organisms of many kinds may be kept under observation for years, and where many specimens of hybrids between various wild and domesticated species can be reared to maturity.

Heredity dealt with many topics that could have led Bateson to a better understanding of Butler's work (Chap. 19). These included the following: Ernst Haeckel's idea that "the biogenetic process" (or "perigenesis,") is "a periodic motion, which we can best picture to ourselves as a wave motion"; Ewald Hering's idea that inheritance is a form of memory; St. George Jackson Mivart's ideas derived from observing the lines of distribution of sand on vibrating plates; and Charles Naudin's ideas on segregation. Brooks wrote of "the existence, in a simple, unorganized egg, of a power to produce a definite adult animal." The term "*a power to produce*," rather than the modern "information for," recurs in Bateson's own writings.

Brooks noted that Comte Georges Louis Leclerc de Buffon (1707–1788) had proposed that "the embryo is built up by the union of organic particles which are given off from every part of the body of the parent, and which, assembling in the sexual secretions, assume in the body of the offspring positions like those which they occupied in the parent." A minimal interpretation of this idea, similar to a suggestion of Herbert Spencer (Chap. 19), is that gametes contain millions of submicroscopic, sub-cellular, distinctive "bricks," each able to track to distinct sites in the developing embryo and there adhere together in specific ways (like pieces of a jig-saw puzzle) to reconstruct components of the parental "houses" from which they were derived. At some stage, they would be able to multiply to generate a multicellular organism from the single-celled zygote that resulted from union of male and female gametes. The bricks were "organic particles" with properties for which, even now, no analogous molecular forms are known. The most versatile of today's macromolecules, the proteins, do not have all these properties. *Something is missing*. We should note that, in itself, Brook's phrase "power to produce" implies "information for," but only in a mechanical sense as when a key, by virtue of its shape, has information for opening a lock. We will search for more subtle meanings among the writings of Bateson and others in the pages ahead.

Bateson's work on *Balanoglossus* was well received and he was elected a Fellow of St. John's College in 1885 with a small grant for research. A letter a year later (Nov. 22, 1886) revealed his growing disillusionment with the science he had been taught, and his optimistic conviction that it would only take 5 years to shatter the prevailing orthodoxy:

> *Entre nous*, the *Balanoglossus* business was a very easy victory, and wasn't much work at all. The thing did itself. Of course, the *Kudos* turned up most substantial trumps, but the thing isn't valuable really. Five years hence no one will think anything of that kind of work, which will be very properly despised. It hasn't any bearing whatever on the things we want to know. It came to me at a lucky moment and was sold at the top of the market – presently steam will be introduced into Biology and wooden ships of this class won't sell well.

The conclusion of Bateson's studies on the acorn worm marked a methodological shift away from microscopic work and toward fieldwork using the naked eye or hand lens. He would never again feel comfortable with complex instrumentation. It also marked a shift toward elaborating, from first principles, a theoretical basis for evolution. While determined no longer to be enslaved by the conventional wisdom, it is likely he sensed that this aspect of his work should remain covert. For, as the above speculations of Brooks attest, the times were rich in speculation and sparse in fact. The duty of researchers, particularly young researchers, was seen to be one of gathering facts and to exercise great caution in extrapolating beyond them.

The Steppes

Bateson's first large-scale fieldwork was an 18-month expedition to the Aral Sea region of Siberia (now the state of Kazakhstan). This was partly intended as a period of "meditation in the wilderness," to discover for himself new scientific goals [5] (Fig. 2.1).

It was a solitary expedition to the extent that he set off alone, having taken some riding lessons and read Galton's *The Art of Travel.* However, he was accustomed to being assisted by servants, and these, camels, horses, and even a tarantass (a cruel springless carriage), were acquired en route. Furthermore, travel in those times being more bureaucratic than today, and foreign travelers in Russia being rare, there was a need to charm local dignitaries. This meant that, in addition to scientific apparatus (e.g., microscope, hydrometers), his luggage included a wardrobe fit for formal occasions (court suit, cocked hat, and sword). There were also photographic equipment, various medicaments (zinc sulfate, quinine) with which he was able to treat his Kirghiz staff and various peasantry, and a revolver which he had cause to flourish from time to time.

Early news that the trip was financially viable came in a letter (May 20, 1886) to Bateson's mother from Michael Foster: "You will be pleased to hear that the Gov. Grant Ctee. (R. S.) have granted £200 to Willie for his expedition." By that time, Bateson was already in St. Petersburg! The official notification, signed by Foster as Secretary of the RS, was dated May 29. Thus, with the blessing of the RS and a £100 grant from a Cambridge travel fund for recent graduates, with letters of introduction from Foster and Huxley, and with a recommendation from the Curator of the Cambridge Botanic Gardens, R. Erwin Lynch, as to scientists he should consult

Fig. 2.1 William Bateson circa 1889

in Moscow, the 24-year-old Bateson set off in the spring of 1886 and returned in the autumn of 1887.

Even from the depths of Siberia, communications by letter or "wire" were good, and his home support "staff"—namely his mother and sisters—were quick to respond to his needs (e.g., copies of Punch and "light literature," and even a recipe for plum-pudding). Also, there was Francis Darwin. It was to him Bateson turned to make inquiries when he became financially "stranded waiting for the tide to lift me off"—the tide normally being provided by the London banking house of Matlock, where he joked that "the wealth of India" was in his personal bank account (Letter to Mary; Sept. 6, 1887).

Bateson's purposes were to obtain evidence from lake and marine organisms (i) of the existence of an "Asiatic Mediterranean Sea" perhaps extending so far as to take in Lake Balkhash and the Aral, Caspian, and Black Seas, and (ii) of environment effects, especially the degree of salinity, on animal forms. This was explained in a formal letter, written in the early part of his travels, to His Excellency, The

Governor General of the Province of Turkestan, which he had translated into the local languages. The letter was written when he had gone off his intended route having "been misled by an old map as to the frontier." He offered "apologies for having trespassed upon forbidden territory" and requested permission to spend the winter in Kazalinsk. Presuming that the letter might be passed to someone with scientific knowledge, he began by outlining the scientific nature of the expedition and his anticipation of an ancient marine communication with the Arctic Ocean (rather than with the Mediterranean Sea):

> Sir, I have the honour to inform you that I am sent jointly by The Royal Society of England and by The University of Cambridge for the purpose of studying during two years certain biological questions arising out of the gradual desiccation of the Aralo-Caspian Steppes. It is no doubt well known to you that it is generally held by geologists that the whole plain in which these seas exist, ... in past times formed one sea, which probably communicated with the Arctic Ocean. Upon the drying up of the sea for some unexplained reason, it is to be expected that the marine animals inhabiting it would be isolated in the various basins which would then be separated. The question, then, that I am engaged in investigating is firstly, whether such traces of marine life do exist in these waters, and secondly what variations they have undergone. This is to be determined by a comparison of the existing forms in various waters with each other, and with such sub-fossil remains as can be found. Also, I am at the same time observing what effect the high percentage of saline matter in the wells and lakes of these Steppes has on the freshwater fauna inhabiting them. The geology of the region has been lately treated by M. Mousliketeff, but the biological side of this work is practically new in kind in all its aspects, and a study of these variations will prove of extreme importance in determining upon what animal variation in general depends.

These purposes were in keeping with the prevailing Darwinian paradigm that, due to the operation of natural selection, organisms and their environments were closely related. Regarding Bateson's first purpose, a sea is an area of geographical continuity smaller than an ocean. At different points in a sea, one might expect to find the same range of organisms. However, if in the distant past one part of the sea had separated from another part, from then onwards the number of similar organisms should have declined, unless remedied by migrations—an unlikely event. Even so, there should still be traces of similarity between the organisms, due to their common ancestry. On the other hand, if the seas had arisen separately then, from the start, ancestral forms could have been different and fewer similarities between modern forms would be expected. This would be particularly so if the environments had always been and had remained, different, so the ancestral forms would have remained different. But what if the environments had become similar? Among many confounding factors that Bateson would have been aware of was the possibility of convergent evolution if the environments had become similar.

This brings us to Bateson's second purpose. To what extent could environments affect the forms of organisms? Different seas, lakes, and pools differed in factors such as degree of salinity or acidity. Such environments should provide a powerful demonstration of Darwinian natural selection. If, perchance, the forms were not changed in different environments, then this would tend to downplay the role of natural selection. Bateson was aware of prior studies on the brine shrimp *Artemia salina* by a Russian biologist, Schmankewitsch, as described by Brooks in *Heredity* [1]:

> The change of Artemia into Branchippus, by rearing it in fresh water, is one of the most remarkable instances of definite modification due to a change of external conditions. *Artemia salina* is a small crustacean, found in the salt lakes of America, Europe, and Africa. When this species is kept in water in which the quantity of salt is gradually diminished, it becomes transformed, in a few generations, into what has been described as a distinct species – *Artemia Milhausenii* – and if the process of dilution with fresh water is continued until it finally becomes perfectly fresh, the Artemia becomes changed into the well-known fresh-water form Branchippus, which has always been considered a distinct genus.

A covert purpose might have been to check on these studies. Could changes in salinity really transform one species into another? Bateson later summarized the matter in his first book (*Materials*): "While being in no sense desirous of disparaging the value of Schmankewitsch's very interesting observation, I think it misleading to describe the change effected as a transformation of one species into another. Schmankewitsch himself expressly said that he did not so consider it, and it is unfortunate that such a description has been applied in this case."

Throughout his travels, letters rich in descriptions of the land, its peoples and their culture, were sent home to his mother and sisters, with hand-drawn illustrations. When his mother died in 1918, they were found in a trunk. After Bateson's death Beatrice used them, together with his original field notebooks containing fragments of a personal diary, to produce *Letters from the Steppe* (hereafter "*Letters*"). The field notebooks reveal that he was often unwell with fevers and gastrointestinal problems, but this was not always mentioned in the letters. He did report in one letter (Oct. 9, 1886, from Kamishli-bar Lake, north of Kazalinsk) that illness had forced him to "lay on my back two days." In this time, he had "read Shakespeare which is unexplored ground for me." He read *Hamlet* "again and again," but he was "disappointed" with *Love's Labour Lost* and *Measure for Measure*. To the family's subsequent chiding letters, he replied: "Don't you all be so 'dim' proud that you have read 'Hamlet' before me. I don't believe you know it any better."

At an early point, he reported having "taken Galton's advice and given up washing." Contemplating the winter, he wrote (Oct. 9) to "Mamma" and all who were at, or visited, 8 Harvey Road, including Anna, who shared his biological interests (Fig. 2.2):

> I have an idea that in winter I will make an attempt to analyse [the] possibility of treating the evolution, development, progress, or whatever one calls it (meaning thereby, the passage of races across the earth, the succession of forms and so on), as if it was motion or no. One is accustomed to metaphors which assume such possibility (as, for example, that protoplasm goes in lines of least resistance, etc.) and I think it would be worth while to make a rigid examination if there is any truth to this feeling. If there is, then, can biological forces be represented by lines and treated geometrically – and if not, why not? I have often wished to think about this seriously and I think a winter at Kazalinsk should give complete and suitable, if not exactly Academic leisure for such a purpose.

As the departed senior male member of the Harvey Road household, he did not hesitate to lecture concerning his younger brother's intention to apply for entry either to the Imperial College of Science in London, or to a "crammer" named "Wrens" (Oct. 9, 1886):

Fig. 2.2 Edward Bateson

> Heitland tells me that Ned talks of I. C. S.. I know next to nothing about I. C. S. except that it is an excellent opening, if one gets it, but seeing that one would have practically to give a University education for the very doubtful chance of getting in, it is a serious matter. I don't know to what extent he would have to chuck Cambridge, but I take it that having entered his name he would have to. Next, Wrens is very costly; is not only not an education, but at Ned's age and having in view the fact that he has learned very little hitherto and has no decided turn for anything special, such a year's cram with a blackguard lot [of] fellows such as hang around a London crammer's would about prevent him ever getting an education at all. On the top of it, his chance of getting in must be very small. I take it that the fellows that get into I. C. S. are nearly all boys who have developed early and done well at school, not fellows with general undeveloped capacity such as Ned. Of course you will get proper advice about it from people who know, but this is how it strikes me in my ignorance.

"Mamma" herself was not spared his wagging finger:

> And now, my dear lady, why don't you chuck Harvey Road, No.8, and go to town? [London] You are very clearly at daggers drawn with all conventional Cambridge and with half unconventional Cambridge too by now … and if you will let me say so, this can only become worse if you go on doing the things you like doing and wish to do, and which therefore it is only right that you should do. You are doing a certain set of things, politically, socially, etc., which your present surroundings don't sympathize with, and I can't see why out of sheer *inertia* you should submit to the additional grind of having them to fight, as well as the various Causes to push on. It seems to be clear enough that now is the time to go. M. is gone. E. is gone. Ned is out of the house, and M' too. (Anna can very well take a room at Newnham Croft with some of her acquaintance.) … In town you can gather of every kind and needn't feel any old obligation to see the people you hate. If you stop you have to take a boarder, which is neither good for you, nor for the house.

Mary was later admonished about lack of economy in not using both sides of the writing paper, and Margaret likewise about misprints in *The Hour Glass*. He may have been taken to task on his remarks about Ned, since in a later letter (Nov. 22) he begged their pardon and praised Ned's decision to go to I. C. S., rather than

"Wrens ... a most blackguardly establishment," as "a good move and a very spirited one." Later Ned returned to Cambridge to study law.

By October 10, snow had fallen, but Bateson was "determined not to settle into winter quarters so long as I can possible keep at work." Having initially been subject to bureaucratic delays in St. Petersburg – and at one point there were newspaper headlines proclaiming that an "English spy" had been captured in Turkestan—he lamented having made little progress toward his goals: "You see I was 2 months late out of [the] five available at most; I had two languages to learn [Russian and Kirghiz], and all methods of collecting and traveling, so I really had a good hill to work up at starting." Eventually, he learned to write "most gorgeous and ornamental Kirghis."

The correspondence indicates that Bateson had not planned to read the scientific literature during the expedition. He hoped to catch up with new developments on his return. However, in a letter from the Hotel Morosov, Kazalinsk, in December, he reported that among the many letters awaiting him on arrival had been four from Anna, with three successive copies of the scientific journal *Nature*. These contained a three-part article by Romanes based on a lecture given at the Linnean Society in May (see Chap. 5). Presumably, the decision to send the article had been made by Anna, perhaps with encouragement from Francis Darwin. Bateson replied:

> My thanks ... for ... numerous letters, which all teem with suggestions that I hope will bud and flower and fruit when summer comes around again. By the way, who ever originated that ridiculous piece of bad logic about variations due to environmental change seeming not to be 'permanent'? How, the deuce, should they be, on any hypothesis which supposes that they result from change, which, when reversed or withdrawn, leads naturally to a return to [the] former state. If iron in soil makes Hydrangeas blue, why is this to be regarded as a false variation? Because the same hydrangea without iron is *not* blue?

Anna was quite up to checking some of his flights of fancy while continuing to keep him informed on her work and the Cambridge gossip. He wrote to her (Dec 4, 1886):

> I am beginning to get my *Brachipus* in order, and I think that they make for Schmankewitsch's view, but they don't vary so much as his did. When you have time, please send me quite rough sketches of the series of tails, with the densities they occurred in, given in *Baume*, ... No hurry about this. *Do – Do try to spell better! Weldon* has only one "l" in his name! etc., etc.. Good Luck on the strawberries. I don't feel sure that such a problem is best worked at first in a complicated case – such as ♂, ♂ + ♀ [hermaphrodite] and ♀. I should have thought that several parallel and simultaneous experiments with ♂, ♀ things would have been easier – but, "*God knows best*" as the Kirghiz say all day long. ... Do not jeer at my 'geometric biology'! Writing such a thing for one's own satisfaction won't hurt anybody, and as for there being no possibility of accurately working over such a thing, I reply that that means no more than that the sources of error must be looked for and classified: and I am inclined to think that if everybody would take the trouble to do this, we shouldn't have the slipshod logic which one commonly sees in biological speculation.

From the Tara Ghul Silver Mine, Balkash Shore, the following June, he declared that one of his primary goals had not been achieved:

> The Balkhash episode is over. We leave it behind us tomorrow, and with it, I leave all hope of getting any marine things at all. It was a poor chance, although I clung to it until it was clearly hopeless. ... But I did not much expect to get sea things here, seeing that the Aral

> had failed to show any, so that though very vexed, I don't feel exactly disappointed. So the dream of finding a 'Tertiary *fauna*' holding on here must be awakened from. But while this side of the venture is all loss, I think I can stand to win on the other half of it.

He was then 8 days travel from Karkaralinsk and was optimistic that thereafter his quest would be more successful:

> I have already got lots of freshwater things living in salt water and have no doubt whatever that when I work them up I shall find something worth having. Moreover, it is only on leaving Karkaralinsk that I begin to get into the real salt water country. Here these salt water lagoons are rare, but there they will be in quantity. Last year I hardly saw any of them. Such pools are densely crowded with life, generally only 3 or 4 species in each, but these in millions, … every cubic inch of water teems with life. The tow-net dragged some 20 yards brings up a tablespoonful or two of solid Copepods, or other Crustaceans. Generally Copepods – sometimes *Cladocera*, sometimes Rotifers – but whatever there is, is always by the million. Now this must mean something, and I think there is a reasonable chance of getting at part of the meaning of it… . The whole thing is a question of measurements, counting bristles, etc., and whether it turns out that these are constant variations or not, will be well worth knowing.

And where could this lead? Would his studies shed new light on the power of Darwin's natural selection? As usual, he was optimistic:

> When I count up the number of points that this opens up, I begin to feel quite hopeful. And by the way, if a few million beasts of one kind (*A*) live in a small place (*a*) where also live some millions of another beast (*B*), while in locality (*b*) also live some millions of *As* but no *Bs*, it seems obvious that the conditions of the Struggle must be very different in *a* and *b*. *A* and *B* presumably eat the same food (this at least may be determined) for both eat diatoms (I am thinking of Calanidae living with Daphnias). Now nothing will be easier than to find out whether there is any difference between '*A*' from (*a*) and '*A*' from (*b*). I believe this is worth knowing. So far as I know, large numbers of the same beasts living in very small *habitats* have never been touched and it is nearly certain that this is workable.

He told Anna (July 5, 1887) of a meeting south of Pavlodar with a "Steppe Kirghis" who spoke of "Professor *Aflatum*" (Plato) and "Professor *Aristolis*" (Aristotle), and "knew that America was discovered by 'Professor Colum,' (Columbus) whom he regarded as a specially great professor." The letter gave another indication of Bateson's growing distaste for the Darwinian orthodoxy:

> Yesterday a gentleman who lives at an *Aul* two days away, rode in to inspect me, having heard of my fame. … He said he knew that the English were a learned people, but he thought that they would soon know too much. 'You have had a very great Professor … I know. His name was Darbin – he wrote a false book that your people believe,' he added. I told him that I was in hopes that my present work would tend to disprove this 'false book,' which I reckoned would raise me in his estimation.

The following, written near the end of his travels, is still optimistic despite his having fallen short of his target (Sept 7, 1887):

> I said I should have the fauna of 600 waters, but I don't think it likely now that the total will exceed 450. I have already 300. Of these 230 were obtained since 8 July O. S. [a reference to the calendar he was using], which is pretty good. I had two or three days of bad luck, one big group of lakes being only saltpans, … I had hoped that they would be salt-water lakes. Anyhow if there is anything, in my view 450 will show it nearly as well as 600. There are

three main types of saline lakes: salt lakes, bitter lakes and soapy lakes. Between these the gradations are many; that is to say there are salty-soapy and salt-bitter, etc., and there are of each examples varying greatly in intensity. The stiffest 'soapy' lake in which I have seen life was 1.03 [density measurement] which was densely crowded with beasts. Among salt and salt-bitter lakes the densities to which Copepods and Cladocera can live in comfort and plenty, range up to 1.07, after that there are never anything but Phyllopods, which occur even in saturated solutions.

But all this is shop. I shall be more intelligible if I say that I shall undoubtedly have enough facts to show me whether these beasts do or do not show variations proportional to the salinity of their habitats. If they do show such variations I shall be supremely happy and it will serve as the basis of any number of life works, and if they don't, that will be worth knowing too, and I shan't feel that I have thrown this time quite away, though that will mean chucking overboard cherished convictions and looking out for a new basis for all those life-works.

A more despondent letter (Sept 8, 1887) was sent to Sidney Harmer in Cambridge:

I came … to look for two things: firstly, beasts which had lived in the *Asiatic Mediterranean* and which might be lingering on here; and, secondly, to get the fauna of a great variety of isolated waters in order to ascertain whether these differences of environment produce constant change of form in their fauna. In my first quest I failed. Of the thirteen months during which I shall actually have traveled in the Steppe (as opposed to actual wintering, touring, etc.), nine months were devoted almost exclusively to this object, and were therefore completely lost. The failure is due to the simple fact that, with the exception of the Aral cockle and the Aral *Dreissenia, Acacna,* and *Neritina,* there are no such marine survivals. … I played for a big stake and – well, I lost it.

On the second count, I think, I won. I have collected already from a good number of such isolated waters. When winter comes, I shall have about 500 such. The lakes are small and manageable and charmingly homogenous. In less than half an hour, it is possible to get at the chief conditions of life in any one lake—as, for example, size, depth, vegetation, density, bottom—and to collect many thousands of beasts that live in it—almost entirely free-swimming Crustacea.

Bateson's doubts about the future were expressed in a letter from Omsk to Margaret (Sept. 4, 1887):

I don't think I can bring myself to work under Sedgwick again. The work was never very tasteful to me, and I learn that he has recently introduced still more detail into the Elementary men's work. This detail is in any case, in the state of our knowledge, meaningless, even to the Advanced man, and to the Elementary man it simply results in his losing the few main principles which are valuable; and to the teacher, it is so much cram. It is a small matter and if I find that obvious expediency makes it necessary for me to teach there a little, I shall pocket my pride and do it; but otherwise I shall look out for some other opening.

From this, he was moved to contemplate "a beggar's life in Cambridge" and concluded that in that event it might be better to move "to the metropolis," but a little further west than Margaret's exotic Bloomsbury location. Writing again to her from Petropavlovsk, there was a suggestion that he might be able to reproduce Schmankewitsch's work, though not to the extent of producing a new species (Sept 16, 1887):

With my work all goes smoothly. Every day I get a few more 'sample faunas,' and trust in my good genius that there will be something in them when I begin to work over them. It seems to me to stand to reason that they will show something in the way of a variation. Of course if they do I shall at once start trying to produce the same variety in an aquarium, and

> if this comes off I believe it will be the first instance of a 'natural' variety produced artificially and ought to mean a perfect revolution in Biology. It will be like the Synthesis of Indigo in chemistry, only more so. And so on. It is very pleasant castle-building anyhow. If this resource were cut off one would be pretty dull in the wilderness.

Shortly thereafter, he returned to Cambridge. At least one other Bateson was traveling in 1887. His mother went to Ireland as a member of the English Home Rule Committee and spoke at public meetings. And—perhaps following his advice—in 1890, she moved from Harvey Road to London where she became an active member of the Marylebone Women's Liberal Association. Around 1897, she returned to Cambridge.

Egypt

In the spring of 1888, Bateson set off by way of Algiers and Malta for the lakes around the Nile Delta region—including a sight-seeing trip to Cairo and the pyramids. A warm letter from Weldon (March 7) drew his attention to some works in German by Theodor Eimer. Noting Bateson's impending visit to "the Dark Continent," Weldon, then based in Plymouth, offered to come aboard the P & O line steamer at Plymouth to see him off. Bateson's goals were the same—to examine the adaptation of forms to different lake environments. A letter to Margaret from the Hotel Abbat in Alexandria (April 5, 1888) demonstrated again his keen eye for the land and its peoples:

> Egypt has simply turned my head. ... I got here just in time to see the funeral procession of the Khedive's brother .... It is hopeless to attempt to make anybody realize the scene. Every nation under the sun, Copts, Sudanese, Negroes, Arabs, Bedouins in blankets, Greeks, Turks, Jews, and every race in Europe, all mixed together, dressed up to their eyes. At one point was a great heap of ruins, left since the bombardment. It was covered with women of every hue and costume, – chiefly Fellaheen women in their blue dresses, and dark blue burmous, with the extraordinary veil hanging from the mystical reed which is borne on the forehead, and amongst them negresses and *Parisiennes* mixed indiscriminately. The streets were full of flags and a perfect blaze of colour. Then came the procession: first about five hundred Arabs from various convents dressed in gorgeous robes with the green scarf bound round their fezzes and carrying banners bearing scraps from the Koran. They all kept up a doggerel chant, 'El Illah la Illahia!' After them the Egyptian troops. Then the Royal Irish in black uniforms and helmets and after them, the Turkish in white and red, and so on. I never enjoyed such a pomp. I can almost forgive you now for having seen the Princess of Wales enter Cambridge in 1864. Later in the day I called on Cookson, the Consul, who is an old bachelor, exactly, line for line, like Ann Jane in male attire. It would strike you in a moment. He has been awfully good and feeds me continually on the best. At his house I met Sir Evelyn Baring who had come for the funeral. I have been in a perfect whirl of gaiety ever since, that is to say, compared to my usual. The people go on picnics to see me collect Cockles and Copepods in the surrounding lakes. ... [and written on the envelope] I got some Cockles in *fresh water* today!

Returning from an eight-day trip to Cairo, he wrote to Anna from Alexandria (April 18, 1888):

> I never felt the vanity of human wishes as here in Egypt. The leading motive of these people was to cheat death and in some shape or other to hold on for ever, and they did so as nearly as flesh and blood may. And what has come of it all? Rameses the second is there still, sitting with his hands on his knees waiting and waiting, century by century, and the Americans go and spit on him and knock their pipes out on his toes and write their names on his shins, and that is all he has got by his waiting. I think I had sooner clear out altogether when one's time comes.

Bateson arrived back in England in early May. In June, Weldon sent a friendly note from the Savile Club conveying his amusement at seeing reported in *The St. James' Gazette* that in Australia "the rabbit, sometimes, though rarely, interbreeds with several of the colonial marsupials." In September, Bateson wrote to Anna from St. John's College relating his plans to sail to Bordeaux later in the month "to look at lagoons" of the Landes with a companion nicknamed "Skins." The latter was considered "a sharp little thing" who "thinks that it takes 2 generations to make a 'gentleman,' this being the number employed in his own manufacture." There followed a discourse on the topic of variation:

> My brain boils with evolution. It is becoming a perfect nightmare to me. I believe now that it is an axiomatic truth that no variation, however small, can occur in any part without other variations occurring in correlation to it in all other parts; or rather that no system in which a variation of one part had occurred without such correlated variation in all other parts, could continue to be a system. This follows from what one knows of the nature of an 'individual,' whatever that may be.
>
> If then it is true that no variation could occur, if it were not arranged that other variations should occur in correlation with it in all parts, all these correlated variations are dictated by the initial variation acting as an environmental change. Therefore the occurrence of any variation in a system is a proof that all parts have the power of changing with environmental change and must of necessity do so. Further, any variation must always consist chiefly of the secondary correlated variations and to an infinitely small degree of an original primary variation.
>
> You will observe that if any variation occurring in one system is acting through the mechanism of correlation as a cause of further variation, it would then happen that on the occurrence of one variation, general variation must be expected, for if all the parts are to work in with the new variation, a long time must elapse before the whole organism is again a system. … I am sure that something would be gained if it were thus possible to separate any variation into its primary and secondary parts or in any case keeping in mind the fact that any variation in a hitherto 'fixed' form, must of necessity be made up of these 2 parts. The accommodatory mechanism is the thing to go for. I don't believe it is generally recognized as existing, though when stated it seems obvious.

The Balfour Business

On the crest of a wave from his acorn worm studies, he applied in 1886 for the Studentship that had been endowed by the Balfour family following Francis Balfour's death. However, the award was made to the existing holder, William Caldwell [6]. Bateson's comments in a letter to his mother and sisters, written after allowing a considerable period for cooling off (June 11, 1887), give some insight into Cambridge politics, and his attitude to them:

> I do not complain that they [the members of the Balfour Studentship Committee] did not elect me, though of course I felt a bit disappointed, but I do complain that they elected Caldwell, which I hold was improper and a breach of trust. Had I not been a candidate, I should, as a subscriber to the Fund, have made some kind of public protest. In so doing they chose a notoriously wealthy man. They chose a man who had been tried for a long period and during that period had not succeeded in producing any tangible result; though he had already abandoned the work which he originally began, being too lazy to finish it. All this they knew. Caldwell, during his former tenure, had also defied the condition under which he was elected, *viz.* that he should reside at least one year in Cambridge. To this wealthy person, whose working expenses were already provided for, they gave trust-money to provide again for the same expenses. Worst of all it was not even pretended that he should devote himself to his work during his year of tenure. His sole business, [Professor] Newton himself says, in returning to Australia was to get married. This also, therefore, they knew. They even knew him to be such a slippery customer that they thought it necessary to get from him a written agreement that he would reside a part of his year in Cambridge. Lastly, they knew that he had spent an integral part of the time of his previous tenure in speculating in gold – an occupation which is not one of the objects which the subscribers meant to promote.
>
> All this seems to me very bad, and to justify one in stating openly that it is clear that the Electors were guided by considerations other than those which the subscribers intended should guide them. My own belief is that the choice fell upon Caldwell simply because they were too cowardly to elect me. Of course had my work been certain to succeed, they would have elected me, but they declined to take the risk. Now after this I decline absolutely to give them any share in the profits [from my expedition]. If the thing comes off, the profits will be far greater than anything which they have to offer; and if it fails, I shall not have deceived anybody. For the £200 from the R. S. and the £100 from the University are not more than even a general account of the country is worth, which I can in any case give. On the other hand, they were too proud to elect someone who would turn them out a given number of quires of print treating of ordinary topics, as Shipley, etc., would gladly do. It therefore remained to them to elect Caldwell. Under these circumstances I feel I am probably happier without it. To submit anything so sacred as one's work to the assessment of such persons as they have shown themselves to be, would merely degrade oneself, insult one's work and lower it for ever. To enter their Ring would be to reduce oneself from artist to journeyman.
>
> You may say the grapes are sour! This is perfectly true; but so, also, is what I have just said. Had they been attainable it is likely enough that I should not have discovered that the Committee was contemptible, but that discovery is none the less true, and I am glad that I have made it, and one day I shall be glad that the grapes were sour, I expect.

Nearer to the time of announcement of the results of the competition (November 22, 1886), his remarks were somewhat more moderate, perhaps because he had less inside information than at the later date:

> About the 'Balfour' business, I feel this way. Of course I should have been glad to get it; but looked at from an outsider's standpoint, I expect it is a good thing I haven't. I have been much too successful pecuniarily and so on lately, and I have really been getting more than I had worked for, which is a bad thing … . Pecuniarily I am well enough off and next year will be very cheap, really all told not dearer than living in Cambridge if as dear. In the absence of any information as to Caldwell's motives and justifications, *I* don't feel hurt. At the same time without such special knowledge (as may perhaps explain his conduct) I think he has done a most despicable thing. I don't impugn the Managing Committee, who prob-

ably acted properly enough on their information, but I can't see how it can be that Caldwell has done other than embezzle the funds to pay his matrimonial expenses, and I care very little who knows that I think so. I wrote to Newton yesterday and said as much, (of course, very tactfully).

A little later he wrote to Anna from his winter retreat in Kazalinsk (Dec. 22, 1886):

I wish I understood the Balfour Studentship Election. I fear from Weldon's account it must have been a thorough job. Personally I don't much care ... but I think Caldwell must be a proper sneak. Unless the thing gets cleaned up, I shan't think of standing next year. It's no honour to take anything at the hands of people who are guided by such hopeless corruption as seems to have been at work this time. Besides this, I don't think I should get it if I did apply. I expect it was Sedgwick and Foster that jobbed it – Clark siding with the strongest party of course. From a remark of Newton's in a late letter to me, I fancy he was at first opposed to Caldwell's claim.

The matter smoldered on in letter after letter. Caldwell seemed to have had all the luck: "Goldfields, French Cookery, heiresses and shilling cigars have not crossed my path *à la mode* Caldwell, I regret to say" (Nov 22, 1886). Anna chided him for demanding "implicit confidence" in his own honesty, while at the same time being unwilling to have confidence in the honesty of the Balfour electors. He replied from Omsk (Sept. 6, 1887) by again acknowledging that his own work "was all in embryo," but still that did not justify Caldwell. Nevertheless, we should note that the first automatic microtome suitable for cutting a block of tissue into a continuous series of sections that were displayed, on order, on a microscope slide was made in 1883 in the university workshops of Cambridge, from a design by R. Threlfall and Caldwell, who had studied under Francis Balfour. Only a single machine was made, but in 1884 twelve motor-driven machines were made by the Cambridge Instrument Company that had been co-founded by Horace Darwin and Albert Dew-Smith (1848–1903), who had worked with Romanes in the Foster laboratory. Each machine could deliver per minute over a hundred consecutive tissue sections one four thousandth of an inch in thickness. This was a major technological breakthrough. Ironically, Bateson himself utilized the Caldwell microtome to great effect in his acorn worm studies [2]. When stressing the need for work in the field as well as in the laboratory, Lankester [7] gave Caldwell [8] high praise:

Now, however, the enterprising zoologist goes to the native land of an interesting animal, there to study it as fully as possible. The most important of these voyages has been that of W. H. Caldwell of Cambridge to Australia (1885–1886) for the purpose of studying the embryology of the *Monotrema* and of *Ceratodus*, the fish-like *Dipnoon*, which has resulted in the discovery that the *Monotrema* are oviparous. Similarly, Adam Sedgwick proceeded to the Cape in order to study *Peripatus*, Bateson to the coast of Maryland to study *Balanoglossus*, and the brothers Sarassin to Ceylon to investigate the embryology of the *Coecilia*.

Bateson was further praised:

A classification which expresses the probabilities of genealogical relationships ... is that at which every teacher of Zoology now aims. That which at the present moment commends itself to me is represented by the genealogical tree .... The chief points in this classification are the inclusion of *Balanoglossus* and the *Tunicata* in the phylum *Vertebrata*, the association of the *Rotifera* and the *Chaetopoda* with the *Anthropoda*, and the total abandoning of the indefinite and indefensible group of '*Vermes*'.

Expressing a growing alienation, in his letter of November 22nd (1886), Bateson still toyed with reapplying for the Balfour Studentship:

> For next year I have no particular expectation of success, indeed I doubt as at present minded, if I shall compete. My work is of a kind with which the Cambridge people have little sympathy, and by next October, while the Kudos got from the Balanoglossus will be spent, any to be derived from my present occupation will be still in the future; and this is the kind of thing that the Committee, being humans, will be guided by. My chance was far better this year than it will be for many years to come. But I don't repine. I went into this business with open eyes, knowing that chance of promotion from any quarter was thereby indefinitely deferred, and I am prepared to go on, on that understanding. How shockingly egotistical one does get in these monthly budgets! But when one hasn't said a word of one's affairs for Lord knows how long perhaps it is excusable.

The passage of a year on the Steppes with little to show for it did not induce him to change his mind about reapplying for the Balfour (June 11, 1887):

> I have decided not to go in for the Balfour Studentship. Please don't discuss this with Weldon or anybody else. I wrote him my feelings about it – he doesn't agree with me and I did not expect he would. Our views of life differ more and more widely as time goes on. My reasons for not competing are as follows: (1) I shouldn't get it if I did. (2) I should however stand; send a preliminary account of my work, and fight out the thing and take my defeat, were it not that, after last year's election I heartily despise the Committee, and will seek neither favour nor rebuff at their hands.

A letter to his sister Margaret again reveals his deep, almost childish, bitterness (June 23, 1887):

> Have you seen the new Mrs. Caldwell? Anna says she is a beauty but very dull, but I should like more particulars. As I mentioned in a family screed, I have decided not to compete for the 'Balfour'. If I did I shouldn't get it and in doing so I should lose my liberty and degrade myself more than a self-respecting person can do. If this business comes off it will be firstly its own reward to any extent and secondly will ensure my getting my Fellowship renewed. If on the other hand it fails I am just as well without this Studentship to look back on, and [without] money got on false pretenses as it would almost be. I hope Harmer will see his way to going in for it. He is a most deserving little soul and would turn out a large quantity of solid and uninteresting work which is what our people rather like, I fancy.

Beatrice in her *Memoir* observes: "He was elected to the Balfour Studentship in November 1887." The importance of this may have been that it provided a financial shield permitting him to move away from purely embryological interests. His confidence received a further boost when in 1888 he shared Oxford University's Rolleston Prize with W. Gardiner. Concerning his travels, he reported to Newton (March 8, 1889):

> It appears therefore that on the whole the results are negative and that no variations can be found which can be shown to be correlated with the constitution of the waters. Seeing that animals taken from some of these lakes die when placed in the waters of another, it is clear that some change of organization must occur, yet these changes do not appear to lead to palpable variation of structure .... As this is so, I feel justified in embarking on a new course of work.

In 1890, his Balfour studentship was not renewed, going instead to Walter Heape, who would become a firm ally in the years ahead. Bateson wrote to Anna (July 19, 1890):

> Sedgwick tells me that he would not wish me to have Weldon's lectureship if W. goes to University College. He says, as I expected, that I have gone too far afield, and that my things are a 'fancy subject.' On the other hand, there is some chance of getting a renewal of [the college] Fellowship, though not a very good one. Newton told me he would try to get this for me. He was very pleasant; Sedgwick was very un – . So you see I have been 'considering my position,' like Sir Julian Troombley. N[ewton] thinks that I ought to be able to turn a considerable penny by regular literature – magazine articles on C[entral] Asia, the S. States, etc.. It was his suggestion and he urged me to try. Of course, I should do so only to maintain connection between body and soul, but it must be thought of.

There was a further exchange with Sedgwick that neatly displays Bateson's prickly nature and Sedgwick's warm tolerance. Bateson wrote to Sedgwick (Oct. 8, 1890) offering to vacate his room at the Morphological Laboratory. Sedgwick replied the next day:

> It is alright about the room. … With regard to your work, it is entirely your own idea that I 'attach little importance to it.' Such an idea is very far from the fact. I think you are a trifle touchy on this point, perhaps a little morbidly so. I also think that your views on Zoology – on the morphological side – are stupid and narrow, but that is a very different thing from thinking that your work is stupid and unprofitable.

Bateson replied: "You must let me thank you for your kind note. I am sincerely glad I was mistaken, but I thought that you meant yesterday to give me a hint to clear out. I am afraid what you say is partly true! always barring the 'stupidity'."

The Garnetts

From the point of view of an academic career, the Russia trip was a disaster. Most of Bateson's observations were highly significant, but negative. Environment and adaptation were not closely related. The only paper from this period likely to gain some recognition was that on *Cardium edule* (cockles) where he had detected an environmental effect on shells. Bateson not being a FRS, it was communicated on his behalf for publication in the *Philosophical Transactions* of the RS by Adam Sedgwick. The paper made the familiar Darwinian case:

> The fact that no palpable difference can be found between the conditions in several localities [in which different forms are found] is not proof that they do not exist. While these differences in condition are usually evasive and hard to detect, it is best to begin to investigate their relation to variations in animals by collecting cases in which the change in conditions is unequivocal, and proceed from this starting point to seek for correlated variation in the forms of life subjected to them.

In this instance, the difference was degree of salinity of lakes. Bateson had observed a variation that correlated with salinity. He noted that in one location, where the shells had collected in terraces, there was "an almost unique opportunity for beholding the gradual succession of these changes." However, he had not investigated the permanency of variation (i.e., whether it could be reversed by changing the degree of salinity), as, for example, skin tanning is reversed when white people move from equatorial to polar climates.

Bateson had not gone to the Steppes to find a bride, but in a tongue-in-cheek August 1887 letter from Pavlodar to his family on his 26th birthday (it "being the day of my nativity"), he suggested that the possibility had crossed his mind:

> I am particularly fond of good voices and had half-thought of inviting one young lady, the daughter of a judge, to share my future fortunes. I should probably have done so but in a lucid interval the line 'Would you advise me to sell my thorough-bred broomstick and give up my little trip to the moon?' came to my head. I reflected that the lady would probably never get to the point of seeing the full humour of this remark, which consideration was a fatal objection to her, so I went no further in the matter. Excuse this personal digression.

So why did Bateson go to the Steppes? At face value, the reasons are obvious. In their formative years, his heroes—Darwin, Huxley, Hooker, Galton—had traveled far in the cause of science. But what drew Bateson to Russia in particular? He knew that before he could make much progress, he would first have to learn two languages. One possible answer is that his interest in Russia had been fostered by another person. Perhaps Brooks had kindled the idea of repeating Schmankewitsch's work in Schmankewitsch's own territory. But surely similar environmental challenges could be found elsewhere? Brine shrimps were ubiquitous.

Perhaps the other person was his friend Constance Black (Chap. 1). Her grandfather, Peter Black (1783–1831) had operated ships out of St. Petersburg and for a brief period prior to his death was in the Russian merchant navy. At some point, it is likely she met the classicist and Russian enthusiast Miss Jane Harrison (1850–1928), who had met Ivan Turgenev when he visited Newnham College in 1878. After a brief period as Lecturer at Newnham, for which she was not suited, Black went to London in 1884 where three of her sisters were settled and earned a living tutoring. Through her sisters, she met the playwright George Bernard Shaw (their neighbor) who took her to Fabian Society meetings, Richard Garnett (1835–1906) the Superintendent of the British Museum Library reading room, and Eleanor Marx a daughter of the recently deceased Karl Marx. At that time, Eleanor was one of those attending meetings of the "Men and Women's Club" organized by Karl Pearson and others of a socialist persuasion, which in 1885 began "free and unreserved discussion of all matters in any way connected with the mutual position and relation of men and women."

In the winter of 1885–86, while Bateson was preparing for his Russian trip, Constance met Richard Garnett's son, Edward, 6 years younger than herself, and they married in 1889. Edward was an author and publisher's reader. Prior to her marriage, she ran a library at the People's Palace and wrote an article entitled "New Career for Women: Librarians" for *The Queen*, which suggests acquaintance with Margaret Bateson. Richard Garnett provided an entrée into the literary world of London, and a link with Samuel Butler. Constance expanded her interest in Russian literature, becoming the major English translator of the works of Dostoevsky,

Chekhov, Tolstoy, and Turgenev. In 1894, 2 years after the birth of her son David (1892–1981), she visited Russia and met Tolstoy.

David developed an interest in biology, but later noted that "my parents could not afford to send me to Cambridge," where "I should have studied Genetics under my mother's old friend, William Bateson." In 1910, he went to the Imperial College of Science in South Kensington, to which Adam Sedgwick had just moved. Here, there were lectures by the anti-Darwinian protozoologist Clifford Dobell (1886–1949), the botanist John Farmer (1865–1944), and the zoologist Ernest MacBride (1866–1940). A close friend was a zoology demonstrator, H. G. Newth, who "had a devotion to Samuel Butler and took Butler's views on evolution seriously."

At this time, David often visited Bateson at weekends at the John Innes Institute: "He was, perhaps, particularly kind to me, because of his friendship with my mother when he was an undergraduate at Cambridge." Once when they were passing a pen of poultry and "saw a cock treading a hen," Bateson "remarked in a reflective manner: 'Every hen has a cock *underneath*,' and he explained that femaleness was due to the presence of one extra chromosome, absent in the male and that, sometimes, the female would develop male characteristics and begin to crow." It later got back to David that Bateson had said to a friend: "I don't think David Garnett will stick to Science. He has an artist's temperament." This turned out to be accurate.

David did well in his examinations, won a prize and a scholarship, and opted to begin graduate work with Dobell whose "brilliant" lectures "revealed a clear, firm iconoclastic mind." However, David soon found the study of sexual conjugation in unicellular organisms "extremely boring." Ever present were his lively literary friends of the famous "Bloomsbury set." On occasions, he bumped into Bateson, perhaps having been lured to London from his John Innes sanctuary by his sister Margaret: "We went to the Omega Workshops to see a performance .... In the intervals, and after the performance, I met and talked to Professor Bateson, the great geneticist, Lady Ottoline Morrell and Clive and Vanessa [Bell]." The success of David's first novel led to a permanent departure from science [9].

Beatrice

Bateson married relatively late, at age thirty-five, to a lady who definitely would not "begin to crow." Indeed, as mentioned in the Prologue, behind a laconic statement of Beatrice in her *Memoir* lies a story of frustrated love worthy of the pages of *Woman's Own*. Bateson first met Beatrice at Dresden in the winter of 1888, where she was meant to have been under the watchful eye of Mr. Field (an Anglican clergyman) and his wife. The tall, blue-eyed, self-assured Bateson, and the tall, shy, Beatrice, did not waste much time. They became unofficially engaged on January 11th. It seems to have been love at first sight. A letter dated January 13, 1889, that Bateson began writing on board ship on the first day of the voyage back to England, stated his matrimonial intentions unequivocally. The letter sums up so much of Bateson that it is here quoted extensively. None of his writings display so well his attitude toward religion—an attitude, as far as we have evidence, held unwaveringly throughout his life. There is also a hint of regret that he and his father had never come to an understanding. Some remarks are elliptical, presumably reflective of an experience in Dresden to which we are not party. All italics are Bateson's, and he inserted an apology about his writing due to the "vessel rolling a bit" (Fig. 2.3):

> Madonna! …This being my Op. 1 in love-letter writing you will have to be very lenient – Besides, grapes don't grow on thorns and you mustn't look for literary nicety in the clumsy *Naturforschen* [researcher of Nature]. Now how has fate been to you? I suppose I ought to say that my heart bled for you in your solitude, or rather in your excess of social opportunities last night; but if I said so, it would be most untrue; for I fear that I am so puffed up with my own happiness that I have very little room left for grieving for any lady, or for anything else but self-exultation. I simply walk on air as the old phrase goes.
>
> At dawn this morning I took up a French novel of Henry Gréville …. My friends all around me are preparing [for bad weather] …. I feel wildly romantic. I am devouring underdone beef and drinking Bass out of a glass which has been cleaned with a camp cloth, an accident which I will at once suggest shall not occur in our *ménage*. I hear an excited Yank exclaim at intervals "I tell yer, they dawn't know heow tu cook it. That's what gat it!" My *umgebung* [surroundings] being so strikingly poetical is an excuse for the ointment being spread so thickly in this letter. If that Yank would only retire to his bunk I might perhaps do better: till then I shall most certainly play the fool, and besides a little fooling is an excellent thing, and to laugh even if one is laughing alone, far from one's heart's sum is a sunshiny pleasant thing. … I feel very much like putting my head on my plate and sobbing – why, I know not. Oh I do yearn for you! I don't know much Greek, but I remember a word in the new testament that struck me very much as a boy – apokaradokia [earnest expectation] – which is translated "the yearning of the creature" – but which means literally "the stretching of the head towards." Isn't it a beautiful old word? And so I stretch my head towards you, Beatrice.
>
> Have those rough Yanks been unkind to you my sweet, gentle, creature? I have hated them for years but I loathe the thought of them now. If those women were to chaff you I feel as if I could hunt them down and stamp on them. I hope so much that you slept last night. You looked so tired when I went away, it made me feel rather anxious but I suppose Mr. Field will apply "the proper remedies". Have you found out what a shocking fist I write? I hope you do too, as then it will excuse a little extra time spent on your letters. Signals of distress are audible in various directions now, though the sea is lovely. I do love the sea.

Fig. 2.3 Beatrice Bateson

When I was miserable as a boy, I used to tell all my troubles to the sea – it was my greatest friend and I shall have to go on deck and tell it to make room for another now.

At a later point in the voyage, the letter was continued:

As we were lying at Flushing it was freezing hard; here on the sea it is thawing – I wonder if the thaw will reach you tomorrow – anyhow, the frost has lasted our time. When I get home I shall hang up my old skates and use them no more except on great occasions, as for instance, when we make our *début* at the Welsh Harp together. I shall see my sister Margaret in town tomorrow and shall tell her. Thus I shall get a lecture I know well. M. does not approve of the married state – why I don't exactly know. I believe she thinks it particularly inexcusable in a woman. Though in a man not amounting to positive 'error' it is still a 'defect'. From your shockingly irreligious bringing up you probably are not aware that this is a technical term used by the High Church people to denote any practice which, though not forbidden by canon is yet contrary to the custom and universal use of Holy Church. You will find that my acquaintance with Theology is 'extensive and peculiar.'

Now we are on this topic, though the question is, I fear, premature, just tell me frankly if you are bent on being married in a church – I am bent on not being, but will yield to pressure. Aesthetically and dramatically I love the Church – Churches, Mosques and religious pictures I adore and reverence, but in daily life I detest it and ministers of religion, as such, one and all. For me and for most other people in this year of grace I believe the practice of religion to be an outward and visible sign of inward and spiritual duplicity. Of course there are a very few men who feel things heavenly as vividly as things earthly, but they are very rare. For me to be married in a church would be acting a lie and though I love the old services as I do, the old buildings, as some of the fairest things left to us in an age of pollution and shoddy[ness], yet my feeling is that it is ours as a trust somehow from our forefathers, in which we have no part lot, and I should feel just as false if I went to Church and took credit for sanctity as I should if I passed a false cheque.

Life is all compromises and if it should be found for some reason very important for our happiness that we should go through this form, then it must be done, but the fewer of these shams we have the better. Now tell me in a few words just what you think. If I followed my *feelings* only, I should like to be married and buried especially in the Church. The burial service is to me as beautiful as a symphony, but men like me have no longer any sight of these things.

It seems so strange that Field, who shares with the frost, the post I honour, in having brought us together, should be a parson. Of course, my father was a parson – D. D. and all the rest of it, but in his day it was different – questioning of things was rarer and to a man with a practical nature as he was, probably with the leisure for, and interest in these things never came. I wish you had known my father. He died while I was an undergraduate, before ever I knew how rare such men as he are in this world. He was the most unworldly of men, so thoughtful and gentle and yet a strong, clever man too. I know now that his ways must have been quite of the *old régime*, courtly and delicate – and yet he was essentially a man of action. Most of the changes made in Cambridge, bringing in the things of the new knowledge, have been more or less helped by him – a very few weeks before he died he sat day after day at the Arts School where the changes of the last Commission were being debated, and stood up and tackled the enemy on point after point till at last no one could be got up against him. I have heard many men say that it was a regular rout; of course I knew nothing about these things then.

Now I perceive it is 1:40 a.m. and as I may have to stick up for myself tomorrow [against your father] I think I will get a little sleep. I shall send a note by hand to your father tomorrow morning and ask him to fix a time to see me – I will say that it has been agreed between us that he shall have both our letters before I see him. If therefore he has got yours on Sunday morning he will send for me on Sunday, otherwise I shall not see him until Monday.

> But of course you will get the great telegram before this letter, so that these details will be out of date when this letter comes to you.
>
> Farewell then, dearest lady! How nice it is that your name is Beatrice. It is a name of good omen and most hallowed too. Though I can't read a line of Italian, that other Beatrice has always been a heroine of mine. Keep up a strong heart and we shall soon get things straight. Give yourself a good time and obey the Fields to whom my thankful heart expands. Yours for ever, William Bateson.

The letter was addressed to Beatrice at her Dresden address. Later Bateson organized a dinner party in his fiancée's honor at St. John's College. Beatrice's father, Arthur Durham, senior surgeon at Guy's Hospital, was an alcoholic and Mrs. Mary Durham thought she detected this in Bateson. He was promptly informed that all communication between him and Beatrice must cease forthwith. With the active cooperation of her eldest daughter Edith—letters were intercepted—she contrived that the ban remained effective.

Decades later Gregory Bateson, their youngest son, learned of the ban from Mrs. Whitehead (wife of the philosopher Alfred North Whitehead) who had been a neighbor at Granchester. Nevertheless, it is unlikely to have been for lack of opportunity that Bateson remained faithful to Beatrice. Cambridge was full of social interactions. In March 1890, after dinner at the Sidgwicks where Bateson had expressed skepticism regarding psychical research, there was an exchange of letters with the lady of the house, Eleanor Sidgwick (Arthur Balfour's sister and soon to become Principle of Newnham). She and her husband supported the Society for Psychical Research [10]. And there were interactions beyond Cambridge. It was probably his friendship with zoologist Joseph Jackson Lister of St. John's College that led to an invitation to visit Upton House, the home of the surgeon Lord Lister, of antiseptic fame. In a letter to Anna (Sept 20, 1891), Bateson wrote:

> I stopped last night with Lister's family at Leytonstone the other day. They are rather nice people, Quakers, you know. At dinner we sat down without having any standing grace, so I supposed it was all clear and took out my bread. I found however that everyone was bowing over their napkins, so I did the same. Nothing however was said, which somehow took me aback as it was so unexpected and I fear I suffered slightly out of 'nervousness.' I do wish people would conform.

Plymouth

Bateson's letters indicate a scorn for "trade." But there was a need for funding which, at that time, would most likely come from trade. The establishment of the Marine Biological Association (MBA) in 1884 with the support of the fishing industry, and with Huxley as President and the Prince of Wales as Patron, led to creation of a Biological Station at Plymouth in 1888. The founding staff included a "Mr. Cunningham, M. A. Oxon" as "naturalist" [11]. Weldon and Bateson were soon to avail themselves of the excellent facilities there provided.

It seems that Lankester, or some other (perhaps Walter Heape), had persuaded the industry that a better understanding of the senses of smell and taste in fish might lead to a better form of bait and that Mr. Bayly of Plymouth had provided funds for an appropriate investigation. Thus, his scientific aims uncertain after the Aral Sea expedition, Bateson responded (Feb. 17, 1889) to what may have been a verbal communication from Lankester:

> I have now discussed the proposal that you made to me the other day with Sedgwick, and also with Newton and Foster, and upon further consideration I am very much disposed to offer to take the work in hand. I find that no objection is likely to be raised by the Manager of the Balfour Studentship. In the first instance, however, I wish that there should be a clear understanding with the giver of the endowment as to the manner in which I should propose to conduct the investigation and especially that, though I should constantly keep in view the desirability of attaining a practical result, yet I should endeavour to reach it by scientific methods; as, for example, the study of the structure and physiology of the sense organs, the influence of drugs upon them, etc.

Lankester replied the same day with a warning:

> We [the MBA] have accepted money on the understanding that we will as far as possible procure that information which will be useful for fisheries, and our leading article of faith is that what is called scientific knowledge is the necessary basis for improvement of practical operations … . Weldon came to see me two days ago and told me he was afraid you would take very strongly the line that all aiming at practical results was beneath the dignity of the scientific man and that you would openly and freely condemn such an attitude and thus put the Association in a false position. I hope there may be some misunderstanding in this. … It would be very injurious to the Association if you should say to people in Plymouth, whether laymen or students in the laboratory, that you don't care a straw whether your researches lead to a bait or not, or that you think such objects either despicable or ridiculous.

Bateson would seem to have acquiesced, and he was informed on March 1st that the MBA Council had resolved:

> That in pursuance of the purpose of the gift of £500 from Mr. Bayly to the Association on June 30th 1888, Mr. William Bateson of St. John's College, Cambridge, be appointed for six months from April 1st 1889 at the rate of £200/year to investigate the physiology of the sense organs of fishes and especially the physiological action of various odiferous and saporous substances upon fishes.

From a "lodging house" in 21 Atheneum Street, Plymouth, Bateson wrote to Anna (May 1, 1889): "How is Edith? I am going to the R. S. soiree on Wed. 8 May and hope to go down to Canterbury to see the Fields. Will Edith be in London then?" Presumably, he here referred to his sister Edith and the Fields who had malchaperoned Beatrice at Dresden. The letter indicated unhappiness with his accommodation, and by 31st July, Bateson had moved to 6 Windsor Terrace. This also did not seem to work out, and by the 5th of November, he was reporting to Anna a move to 5 Lansdown Place "but with the same landlady." Here, he noted that "Weldon is to give a popular lecture on variation at the Institute on Thursday and is preparing gorgeous diagrams."

Bateson's work was productive. The following year two papers appeared in the *Journal of the Marine Biological Association*. The first, "Notes on the Senses and Habits of some Crustacea," indicated that he had permitted his attention to wander a little from the mark. He had turned physiologist to the extent that, in a study to investigate sensitivity to light, he had ablated the corresponding organs, the eyes, and observed how the unfortunate creatures imprisoned in fish tanks managed their diurnal rhythms. The second paper, "The Sense Organs and Perceptions of Fishes: with Remarks on the Supply of Bait," indicated that he had taken his mandate seriously. In October 1889 he was offered the possibility of continuing for a further 6 months and given the sum Mr. Bayly had made available, he might have anticipated further renewals. However, he replied: "I am anxious and am to some extent bound to go on with the work of my own which was previously undertaken." So, in November his visit to Plymouth terminated and he returned to St. John's College.

However, he was still afflicted with wanderlust. Asking that his communication be kept confidential, in December 1889, he applied to serve as naturalist on an expedition to the Arctic by Nansen sponsored by the Norwegian Government, with the object of reaching the North Pole. Nothing came of it. So, at last, Bateson was home in Cambridge, relaying (Jan. 16, 1890) various items of family news to Anna who, although in Wales, he still regarded as a member of his staff (asking, parenthetically, that she collect water snails for him, examine the symmetry of ivy leaves, and provide an address of a firm that might supply hybrid *Narcissi*):

> I am working *a l'Africaine* and have rarely slaved so hard. My lectures are to begin on Tuesday or Wednesday next. Our Lady [mother] leaves on Saturday and Mary goes into lodging (2 Panton Street) tomorrow. … Margaret gives me a luncheon on Sunday 26 January to meet the Earl … of whom we have heard so much. … A week ago I made a trip to Brighton and got a good many notes. Staying at the Brighton Grand Hotel I dined at *table d'hôte*. The people were as horrible a collection of broken crockery as I have seen. Damaged military men and almost everyone decrepit in mind or body. Amongst this human rubbish were J. J. Thomson and his bride! We did not greet each other. … I am now a member of the Press Syndicate and am about to supervise the sale of the Scriptures and Liturgy with a view to getting hold of some of the profits for the purposes of Biological Research – otherwise they go to the building of Laboratories for Medical Students which is clearly inappropriate.

Beyond these activities, there was no option but to work on the specimens he had gathered during his travels, catch up with the expanding biological literature, and confront the issue of "Variation" (which in his writings he often capitalized). A massive book, *Materials for the Study of Variation*, would emerge in 1894.

Summary

Working within the current morphological paradigm, Bateson's summertime studies of the acorn worm with William Brooks in Virginia established his reputation and led to a college fellowship. However, under Brooks' guidance he began to look afresh at evolutionary problems, especially the nature and causes of variation. His first application for the prestigious Balfour Studentship was thwarted when it was awarded to Caldwell, a designer of the first automatic microtome. Bateson vowed that he would not reapply. However, his second application was successful. Expeditions to the Russian Steppes and Egypt were relatively fruitless but drew his attention to the frequent lack of correlation between an organism's appearance and its environment. This anti-Darwinian observation did not augur well for academic advancement. Since the difficulties of working in Russia had been entirely predictable, it is speculated that he had been encouraged by Brooks to repeat Russian studies indicating that changes in salinity could transform one aquatic species into another. He may also have been encouraged by his friend Constance, who as Constance Garnett was later to become a leading translator of classical Russian authors. Bateson's disdain for "trade" research distressed Lankester, who had secured a position for him at the Marine Biological Laboratory in Plymouth. He met his future wife at Dresden in the winter of 1888, but the maneuverings of her family were to prevent their union for several years.

Chapter 3
Galton

Your many terms ... 'fertile' and 'sterile' germs, ... 'stirp,' ... 'residue,' etc., etc., quite confounded me. ... Unless you make several parts clearer, I believe ... that very few will endeavour or succeed in fathoming your meaning.

Charles Darwin to Galton (Nov. 7, 1875).

It was time for Bateson to catch up. Several works of the eminent, written in their dotage, appeared in 1889. There was *Darwinism* by Alfred Russel Wallace (1823–1913) and *Natural Inheritance* by Francis Galton (1822–1911). These expanded on ideas the authors had developed in earlier years [1, 2]. While Bateson was traveling, August Weismann (1834 -1914) had spoken to the British Association (BA) in Manchester, and now, an English translation of his *Essays upon Heredity* was available [3]. Bateson could hardly have failed to notice these. A reference to *Darwinism* was in notes Bateson made at that time, but, reflecting disenchantment with natural selection, Wallace was not cited in *Materials*, the book Bateson authored in 1894. *Natural Inheritance* was cited, and Weismann's new coinage "panmixia" was held with scorn as a "flower of speech" the ideas associated with which were "as false to the laws of life as the word to the laws of language" [4]. There will be more on panmixia in Chap. 5. Here, our concern is Galton's view of evolution. Since Darwin himself could not fathom Galton's meaning, we must anticipate some difficulties. If abstract thinking with mathematical equations is not your forte, you will not miss much by skipping to the Summary. Those who read on can take heart that once over this hurdle the rest is downhill!

A. G. Cock, D. R. Forsdyke, *Treasure Your Exceptions*,
https://doi.org/10.1007/978-3-030-92099-9_3

Latent and Patent Elements

Natural Inheritance had a strong influence on Bateson and an even stronger one on Weldon. It built upon ideas Galton had expounded in a little-noticed article published in 1872 [5]. Here, "by fair reasoning from acknowledged facts," he had set out to analyze "the complicated connection that binds an individual, hereditarily, to his parents and to his brothers and sisters, and, therefore, by an extension of similar links, to his more distant kinsfolk." Because his arguments were derived from first principles, they have an underlying robustness that can be recognized today, even though they were lost on his Victorian contemporaries. Indeed, it was perhaps because Galton attempted to be as general as possible, refusing to spell out detail, that his abstract but basically commonsensical arguments were lost. A penchant for abstruse metaphors probably impeded more than it helped. And new ideas came with new terminology—either new words had to be coined, or old words had to be used in new ways. Galton had given a preliminary outline in an article in *Macmillan's Magazine* in 1865 [6]:

> We shall … take an approximately correct view of the origin of our life if we consider our own embryos to have sprung immediately from those embryos whence our parents were developed, and these from embryos of *their* parents, and so on for ever. We should in this way look on the nature of mankind, and perhaps on that of the whole animate creation, as one continuous system, …. The father transmits, on an average, one-half of his nature, the grandfather one-fourth, the great-grandfather one-eighth; the share decreasing step by step, in a geometrical ratio, with great rapidity.

Under this scheme, transmission was "on an average," so it might not be possible to predict in detail a child's characters from a knowledge of its ancestors. The first "acknowledged fact" Galton addressed in his 1872 article was that a character seen in one generation could skip several generations before being seen again in offspring. He considered that whatever was responsible for the character had not been created afresh; rather, it had been masked in intermediate generations [5]:

> From the well-known circumstance that an individual may transmit to his descendents ancestral qualities which he does not himself possess, we are assured that they could not have been altogether destroyed in him, but must have maintained their existence in latent form. Therefore, each individual may properly be conceived as consisting of two parts, one of which is *latent* and only known to us by its effects on posterity, while the other is *patent*, and constitutes the person manifest in our senses.

Here, Galton edged toward what later became known as "phenotype" and "genotype" (Chap. 10). Phenotype refers to characters that *correspond to* (i.e., exist on a one-to-one basis with) "patent elements." These characters may be evident (patent) in the embryo, and/or in the adult, and/or in the gamete produced by the adult (i.e., a gamete's microscopically observed shape and mode of movement are gametic characters). A finger might be visible in an embryo and in an adult, but not in a gamete. Fingers are embryonic and adult characters, not gametic characters. Genotype refers to the "patent and latent elements" *themselves*, not something that corresponds to them. These “elements” are present in embryo, adult, and gamete—all three—but only the patent are actually manifest as characters at one or more of these stages.

Latent “qualities” (characters) were held by Galton to “have maintained their existence.” But in what form were these qualities “maintained?” As “elements.” But what were “elements,” and by what processes were they maintained? Direct microscopy of gametes did not help. Galton addressed the fact that, at magnifications then feasible, microscopical studies revealed virtually no internal structures within ovum or spermatozoon. So, a *structureless* ovum would be fertilized by a *structureless* spermatozoon, to create a *structureless* “newly impregnated ovum.” The latter (the single-celled “zygote”) would contain what he called the “primary elements.” These would include the “embryonic elements” that “contain materials out of which structure is evolved.” Galton proposed that these embryonic elements would, in some way and at some time, be subtracted from the totality of the primary elements, so becoming “segregated” and leaving behind a “residue” of the latent elements. Thus:

$$\text{Primary elements} = \text{Patent elements}(\text{embryonic}) + \text{Latent elements} \quad (3.1)$$

An adult bearing the patent elements and latent elements would, in turn, transfer them to structureless gametes from which the next generation would arise:

> The adjacent and, in a broad sense, separate lines of growth in which the patent and latent elements are situated, *diverge* from a common group and *converge* to a common contribution, because they were both evolved out of elements contained in a structureless ovum [gamete], and they, jointly, contributed the elements which form the structureless ova [gametes] of their offspring.

Other facts Galton addressed were that the tendency for a character to reappear wanes as the number of generations increases and that gametes are of finite size. If all elements corresponding to parental characters were retained each generation, then, since two parents contribute to make one child, as the generations passed, the gametes would get progressively larger. Since gametes remain of uniform size from generation to generation, Galton deduced that many elements corresponding to characters are lost (on average, half of each parent's elements would be lost per generation). Within the space limitation, he supposed that, on average, an individual would receive half of the elements from each parent, who in turn would each have received half of the elements from each of their parents. Galton saw this in terms of a competition among elements for a limited number of places (seats, slots, loci) in the gametes that would form the next generation:

> The embryonic elements are segregated from among [the primary elements] .... On what principle are they segregated? Since for each place there have been many unsuccessful but qualified competitors, it must have been on some principle whose effects may be described as those of ‘*Class Representation*,’ using that phrase in a perfectly general sense as indicating a mere fact, and avoiding any hypothesis or affirmation on points of detail, about most, if not all, of which we are profoundly ignorant.

To explain "class representation," he considered the citizens of a town (constituency) who would choose, from among the candidates of different political parties, one who would represent the town in parliament:

> I give as broad a meaning to the expression as a politician would give to the kindred one, a 'representative assembly.' By this he means to say that the assembly consists of representatives from various constituencies, which is a distinct piece of information as far as it goes, although it deals with no matter of detail; it says nothing about the number of electors, their qualifications, or the motives by which they are influenced; it gives no information as to the number of seats; it does not tell us how many candidates there are usually for each seat, nor whether the same person is eligible for, or may represent at the same time, more than one place, nor whether the result of the elections at one place may or may not influence those at another (on the principle of correlation).

Through this political metaphor, Galton acknowledged elements (corresponding to classes of observable characters), which would compete for sets of places in gametes traveling through the generations. The modern reader may picture a group of genes within a constituency, which are of the same type to the extent that they can represent only that one constituency. These genes compete for access to a particular "seat" on a chromosome, and if there are two chromosomes of the same kind, then they have potential access to either of the two "seats" (i.e., they can be alternative or "allelic" genes). In an individual with two chromosomes of the same kind (i.e., a diploid individual with one chromosome set from its father and one chromosome set from its mother), there would be only two seats available. So, although there might be many allelic genes *in the population of individuals* (potential alternative candidates as representatives from a particular genic constituency), it would be possible to "elect" a maximum of only two to occupy the chromosomal seats *in a particular diploid individual.*

Today, we refer to this as polymorphism (polyallelism); a maximum of two candidates from a polymorphic family of genes (many candidates in a population) can occupy two allelic chromosomal "seats" in any one diploid individual. So, regarding that genic constituency, the population of individuals would be declared to be "polymorphic." If there were no more than two candidates available to act as potential representatives for a particular genic constituency (character), then the population would not be considered polymorphic (polyallelic) for that genic constituency.

In this context, Galton introduced an idea we will be considering in later chapters—that an element *represents* a character, and representation implies *information.* Galton did not actually say "information for" a character, nor that an element "determines," or confers "the power to produce," a character, but did go on to imply an active role for his elements in "*development*" of the embryo into an adult:

> There can, I trust, be no difficulty in accepting my definition of the general character of the relation between the embryonic and the structureless [primary] elements, that the former are the result of election from the latter in some method of Class Representation. The embryonic elements are *developed* into the adult person. 'Development' is a word whose meaning is quite as distinct in respect to form, and as vague in respect to detail, as the phrase we have just been considering [i. e. Class Representation]; it embraces the combined effects of growth and multiplication, as well as those modifications in quality and proportion under both internal and external influences. If we were able to obtain an approximate

knowledge of the original elements, statistical experiences would no doubt enable us to predict the average value of the form [phenotype] into which they would become developed.

The latent elements that were not manifest were considered "greatly more varied" than the patent elements that became manifest. Galton noted that after eight generations, an individual could be considered to have 256 (i.e., 2^8) ancestors; yet, even after that time a character present in an early one of those 256 might reappear, implying that it had escaped several diluting generational steps.

Out of the structureless ovum the embryonic elements are taken by *Class Representation*, and these are *developed* (a) into the visible adult individual; on the other hand, returning to our starting point at the structureless ovum, we find, after the embryonic elements have been segregated, the large *Residue* is *developed* (b) into the latent elements contained in the adult individual.

The latent elements would be subject to competition for space like the patent elements. To make clear that their development might differ from that of the patent elements, Galton distinguished two paths, (a) and (b), between which there might be some interaction (Fig. 3.1).

Thus, since some characters (e.g., secondary sexual characters) develop late in life, they could be considered to have arisen because certain latent elements had become patent elements. However, the reverse process might imply that patent elements associated with *changed* phenotypic characters (e.g., increased muscular strength of a blacksmith) could transform into latent elements. The new latent elements, resulting from this flow from patent to latent, would accumulate during an individual's lifetime, and so would be more likely to be transferred to the offspring of elderly parents. Galton would not countenance this:

The two processes are not wholly distinct; on the contrary, the embryo, and even the adult to some degree, must receive supplementary contributions derived from their contemporary latent elements, because ancestral qualities indicated in early life frequently disappear and yield place to others. The reverse process is doubtful; it may exist in the embryonic stage, but it certainly does not exist in a sensible degree in the adult stage, else the later children of a union would resemble their parents more nearly than the earlier ones.

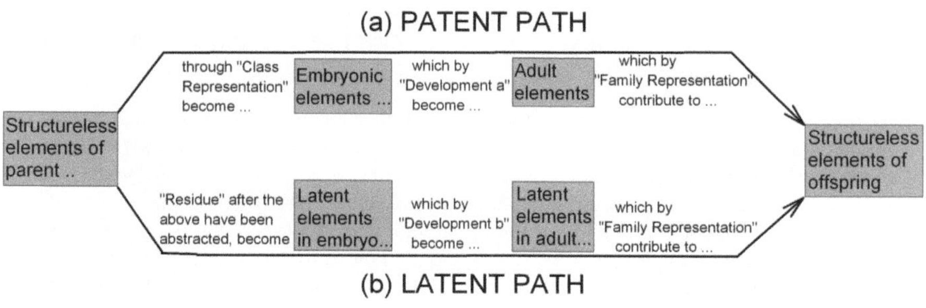

Fig. 3.1 Galton's diagram (1872). "The general chain of heredity extends from one structureless stage to another, and not from person to person." The patent path (**a**) was considered a "fainter stream of heredity" than the latent path (**b**), which reflected "a collateral kinship and very distant," but was "one of large … relative importance"

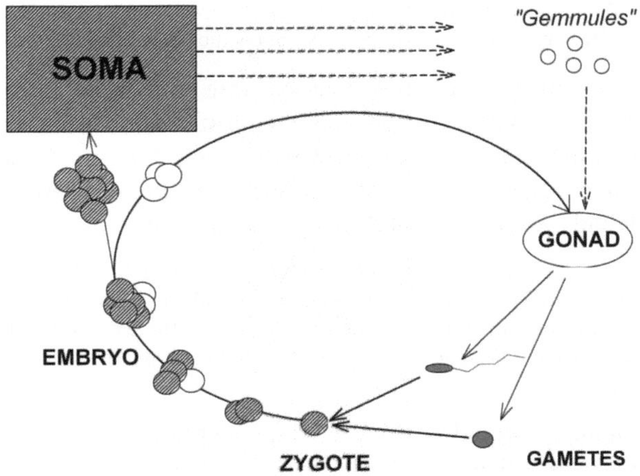

Fig. 3.2 "Curiously circuitous" relationship between parent and child. The parent consists of a disposable soma which houses the gonad containing the germline from which gametes are developed. Parental gametes unite as the unicellular zygote, which divides to form an embryo and hence a child consisting of a disposable soma which houses the gonad containing the germline from which, when the child becomes an adult, gametes are again developed. Bateson found it convenient to refer to the three stages—zygote, embryo, and adult—as "zygote." So, as far as he was concerned, zygote and gamete alternate ad infinitum. Dashed arrows indicate an aspect of Darwin's hypothesis of pangenesis by which "gemmules" transfer acquired information from soma to germline

In this way, Galton advanced the doctrine that August Weismann was to more precisely articulate—the doctrine of the separateness of the potentially immortal *germ-line* (stored in the ovary or testis of animals) and the *soma* (the mortal remainder of the body; Fig. 3.2).

Galton divided elements into classes corresponding to groups of characters. To the subclass of elements that became part of an individual (i.e., were *personal* to that individual), he gave the name "family." Thus, it was these "Family Representations" of adult patent and latent elements that would be forwarded to each offspring (Fig. 3.1). Galton next considered the processes by which some patent and latent elements gain access to the gametes that will form the next generation, and some do not:

> As regards the large variety of adult latent elements, they cannot all be transmitted, for the following obvious reason – the corresponding qualities of no two parents can be considered exactly alike; therefore the accumulation of subvarieties, if they were all preserved as the generations rolled onwards, would exceed in multitude the wildest flights of rational theory. The heritage of peculiarities through the contributions of 1000 consecutive generations, even supposing a great deal of ancestral intermarriage, must far exceed what could be packed in a single ovum. The contributions from the adult latent elements are therefore no more than *Representative*; but they have to furnish all the various members of each Class whence its representatives have afterwards to be drawn. Therefore, … we are driven to suppose, as in the previous case [i.e. the case of patent elements], a "*Family Representation,*" the similar elements contributed by the two parents ranking, of course, as the same family.

So, a personal "family" of elements is transferred to the offspring. Would this transfer have a random component, or would it be ordered in some way? Galton thought that even if random, it would still have some order. Switching now to a military metaphor, he considered that "the most probable *modus operandi*" would involve a largely random element ("indiscriminate conscription" rather than discriminate or "general" conscription):

> The phrase [Family Representation] ... does not mean that each and every family has just one representative, for it is absolutely reticent on all such matters of detail as [in the case of] those I enumerated when speaking of Class Representation .... It is in reality a large selection made out of larger and not out of smaller constituencies than those I have called "classes," similar to that which would be obtained by an indiscriminate conscription: thus, if a large army be drawn from the provinces of a country by a general conscription, its constitution, according to the laws of chance, will reflect with surprising precision the qualities of the population whence it was taken; each village will be found to furnish a contingent, and the composition of the army will be sensibly the same as if it had been due to a system of immediate representation from the several villages.

Galton's views gained wider attention from an address to the Anthropological Institute in 1875 [7, 8]. Despite his circumlocutions, the "patent" versus "latent" contrast proved useful. Romanes in a description of Weismann's views [9] noted that since the patent soma eventually dies, the patent elements determining that soma would die with it. They would have been "consumed." So, some patent elements could be considered, like Galton's latent elements, to be contributed to a "residue" that would pass to the germ-line:

> Suppose that we take a certain *quantum* of germ-plasm as this occurs in any individual organism of today. A minute portion of this germ-plasm, when mixed with a similarly minute portion from another individual, goes to form a new individual. But, in doing so, only a portion of this minute portion is consumed [made manifest]; the residue is stored up in the germinal cells of the new individual, in order to secure that continuity of the germ-plasm which Weismann assumes.

In his 1896 textbook, *The Cell in Development and Inheritance* [10], Edmund Wilson held that our "real problem" was: "How do the adult characteristics lie latent in the germ cell; and how do they become patent as development proceeds?"

Pangenesis

Galton was not alone in speculating from first principles. As we shall see in Chap. 19, in the 1870s his ideas were to some extent anticipated and extended beyond recognition, by Ewald Hering and, more articulately, by Samuel Butler. Common to Galton, Hering and Butler was the recognition that each character of an organism could have some form of "element" *representing* it in the germline. It was far from clear what "representing" meant, but it implied a one-to-one relationship between a character and the element. In his "provisional theory of pangenesis," Darwin in 1868 [11] had named such elements "gemmules" (derived from the Latin root "gemma" meaning a bud). These were held to exist *within* cells. But Darwin went further. He postulated that they had the potential to be changed during the life of an individual. Furthermore, they could leave their cell of origin (be budded off by the cell) and "circulate freely throughout the system." As such, they would be "thoroughly diffused; nor does this seem improbable considering their minuteness, and the steady circulation of fluids throughout the body." He believed some would enter the gonads and thence the gametes (spermatozoa and ova) so that they could be passed to offspring (Fig. 3.2). In short, a character *acquired* during an individual's lifetime could, by virtue of the transfer of the corresponding appropriately *educated* gemmules, be passed on to its children. Such characters would include mental characters:

> With respect to mental habits or instincts, we are so profoundly ignorant of the relation between brain and the power of thought that we do not know positively whether a fixed habit induces any change in the nervous system, though this seems highly probable; but when such habit or other mental attribute, or insanity, is inherited, we must believe that some actual modification is transmitted; and this implies … that gemmules derived from modified nerve-cells are transmitted to the offspring.

When pangenesis was discussed in 1870 at the Anthropological Society, a surgeon, Alfred Sanders, agreed that gemmules might be responsible for memory, but thought that brain gemmules would usually remain within the confines of the brain so that a child would not inherit the memories of its parents [12].

Eager to examine pangenesis, Galton transferred body fluids containing the postulated gemmules from one variety of rabbit to another, in the hope that the offspring of the recipient would display some of the distinctive characters of the donor. When he reported to the RS in 1871 that transfusion of blood between varieties had not achieved this result [13], Darwin argued in the journal *Nature* that there were other body fluids besides blood, so this aspect of pangenesis (mobility of gemmules) was not disproven [14]. Shortly thereafter, Darwin was able to recruit Romanes to further investigate pangenesis. Their approach was to transplant an ovary from a rabbit of variety A, where it would have been surrounded by A-type gemmules, to a rabbit of variety B, which would act as a surrogate mother. Within her body, the A ovary would be surrounded by B-type gemmules. The prediction was that the surrogate mother rabbit B would produce A-type offspring with some B-type characters, due to the entry of B-type gemmules into the A germline.

But ovarian transplantation had not been achieved at that time. Skills acquired in Foster's Physiology Department had prepared Romanes for the challenge. For several years, he struggled to overcome the technical difficulties, but without success [15]. In the 1890s, another Foster student, Walter Heape, who had also been with Bateson and Sedgwick in the Morphological Laboratory in the early 1880s, was able to surmount the problems. A-type children were born of a B mother, but *none* of these had any B-type characters [16]. Darwin, having died in 1882, was no longer around to argue that mobility of gemmules had not been disproven.

With their common interest in proving/disproving pangenesis, and their common location in London, it is not surprising that Romanes and Galton were acquainted. Galton discussed his "theory of heredity" with Romanes in 1875 (Chap. 25), and they were both members of the "Psychological Club." In 1885, Galton stayed with Romanes at his summer residence in Scotland, where he had his own marine biological laboratory. In correspondence with Edward Poulton (Chap. 6), Romanes suggested Galton for an honorary Oxford degree. Galton's 1875 theory was mentioned in correspondence with Thiselton-Dyer (Sept. 26, 1893) and in *An Examination of Weismannism* Romanes noted that much of Weismann's work had been anticipated by Galton. To the extent that gemmules existed *within* cells, Galton accepted pangenesis. In *Natural Inheritance* [2], he wrote:

> We appear, then, to be severally built up out of a host of minute particles of whose nature we know nothing, any one of which may be derived from any one progenitor, but which are usually transmitted in aggregates, considerable groups being derived from the same progenitor. It would seem that while the embryo is developing itself, the particles more or less qualified for each new post wait, as it were in competition, to obtain it. … The particle that succeeds, must owe its success partly to accident of position and partly to being better qualified than any equally well placed competitor to gain a lodgment.

Galton here spoke of an embryo "developing itself," thus hinting at an autonomy which will be discussed later (Chap. 19). Furthermore, he considered that the competition among particles (gemmules) for transfer to the next generation had both random ("accident of position") and non-random ("better qualified") components. Finally, he saw the particles competing while *waiting* within the embryo, rather than competing within the parental gonad prior to emergence in gametes that would form the embryo. In *Natural Inheritance,* he had also observed the relationship between parent and child to be "curiously circuitous":

> There is no direct hereditary relationship between the personal parents and the personal child, … but … the main line of hereditary connection unites the sets of *elements* out of which the personal parents had been evolved with the set out of which the personal child was evolved. The main line may be rudely likened to the chain of a necklace, and the personalities to pendants attached to its links.

There is here a hint at two levels of genetic information—that provided by the "chain" and that provided by the "pendants," and the modern reader may even interpret the waiting "in competition" as a hint of the "selfish gene" ideas of George Williams, William Hamilton, and Richard Dawkins. As will be discussed (Chap. 13), Galton later incorporated into his scheme the phenomenon that had been observed in gonads as the reduction division of the chromosomes—a process that came to be called "meiosis" [17]:

> The chief, if not the sole, line of descent runs from germ to germ and not from person to person. The person may be accepted on the whole as a fair representative of the germ, and, being so, the statistical laws which apply to the person would apply to the germs also, though with less precision in individual cases. Now this law is strictly consonant with the observed binary subdivisions of the germ cells, and the concomitant extrusion and loss of one-half of the several contributions from each of the two parents to the germ-cell of the offspring.

Jenkin's Sphere

In Scotland, in 1867, a professor of engineering, Fleeming Jenkin, saw members of a biological species as enclosed within a sphere (Fig. 3.3). For a particular character, say height, the average member of the species would be at the center of the sphere. Since most members were of average height, most members were at, or near, the center. Jenkin noted that, under the then prevailing view that parental characters were blended in their children, there would always be a tendency for individuals to vary toward the center, rather than away from it [18]:

> A given animal ... appears to be contained ... within a sphere of variation; one individual lies near one portion of the surface; another individual, of the same species, near another part of the surface; the average animal at the centre. Any individual may produce descendents varying in any direction, but is more likely to produce descendents varying towards the centre of the sphere, and the variations in that direction will be greater in amount than the variations towards the surface. Thus a set of racers of equal merit indiscriminately breeding will produce more colts and foals of inferior than of superior speed, and the falling off of the degenerate will be greater than the improvement of the select.

Galton referred to this as the "regression to the mean" (i.e., regression to the average for the population) that would invariably be observed when individuals of

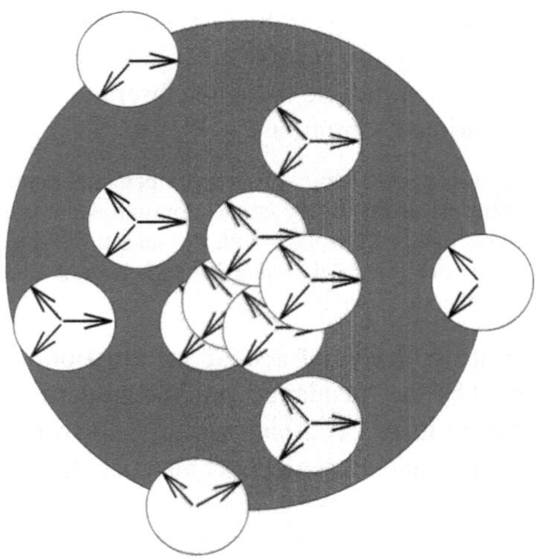

Fig. 3.3 Jenkin's sphere of species variation (gray). Members of a species and the ranges of variation of their children are represented as white circles with arrows representing the extents and directions of variations. Most individuals are clustered at the center (average phenotypes) where their children vary in any direction. Those at the precise center will tend to vary away from the center. The few individuals at the periphery (rare phenotypes with exceptional qualities) tend to vary toward the center. Thus, their children cannot escape the confines of the species (i.e., their exceptional qualities cannot develop further). The variations Jenkin refers to reflect differences in *nature*, not nurture. He is not referring to variations acquired through interaction with the environment within an individual lifetime (e.g., increased muscular strength in an individual that exercises, or increased weight of a bean seed advantageously placed on a bean plant)

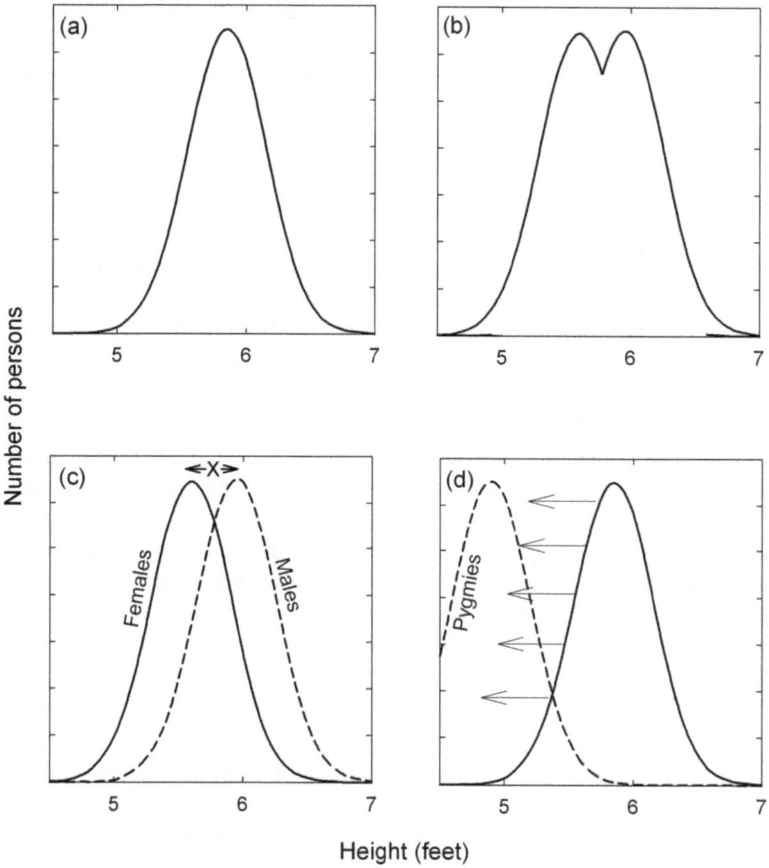

Fig. 3.4 Bell curves of the distribution of heights in a population. In (**a**), the number of individuals in small height ranges is plotted against height. In (**b**), it is shown that by carefully plotting values from a sufficient number of individuals an apparently unimodal plot can sometimes be shown to be bimodal. This may result from a difference between the sexes. In (**c**), it is shown that each point in the left curve (for females) differs by an amount "X inches" from the corresponding point on the right curve (for males). In (**d**), the leftward-pointing arrows indicate natural selection acting on height so that the average height would decrease as a population of pygmies emerged

mixed pedigree were crossed. Instead of a sphere, he preferred to draw a bell-shaped curve representing the distribution of a quantitative character such as height among members of a population (Fig. 3.4a). There would be a few individuals of much below average height, then progressively more individuals slightly below average height. Most would be of average height (the peak of the bell curve). There would then be slightly fewer individuals of slightly above average height, and very few individuals of extreme height. Through crossing, say between a tall individual and a short individual to produce children of intermediate heights, there would be an ongoing regression to the mean. Unless compensated for, either by selective mating (tall with tall, short with short), or by fresh variations that created more tall and short individuals that differed from the average, the peak of the bell curve would progressively increase in size and the flanks would shrink. In other words, *under blending inheritance* there would be a constant need for new variation to keep up the range of variability. However, in *Natural Inheritance* Galton noted the paradox that: "The large do not always beget the large, nor the small the small, and yet the observed proportions between the large and small in each degree of size and in every quality, hardly varies from one generation to another."

As shown in Fig. 3.4a, the curve can be quite smooth, or unimodal. But if many individual measurements are taken, sometimes the curves reveal bimodality (Fig. 3.4b). This might be obtained if one were to measure the heights of members of a population without discriminating between the sexes; then a bimodal ("dimorphic") curve might be obtained. In a species where females are, on average, smaller than males, they collectively have a different height distribution (Fig. 3.4c). Looked at in another way, if one did not know that there were two distinct types present in a population, a careful analysis of an appropriate parameter (in this case height) might reveal the existence of the two types. Thus, one can use measurements of this kind as an analytical tool to show that there are sub-groups within a population.

Since the two curves shown in Fig. 3.4c are the same shape, then it would be possible to "correct" female heights by addition of a constant amount to each female height (X inches in the figure). This would move the curve for females to the right, so making female heights part of the same curve as male heights. In this way, Galton was able to treat two parents as one for the purposes of comparing their characteristics with those of their children. Rather than directly using the fact, say, that a father of 68 inches and a mother of 64 inches produced children of various heights, the mother's height would be adjusted by the above factor X to, say 66 inches. Then, the father's height and the mother's adjusted height would be averaged to 67 inches—the "mid-parent" height. This value—the mid-parent value—could then be compared with the heights of the children.

Natural Selection

Nature herself, in the form of natural selection, can bring about movements of bell curves. Thus, in an environment where height greater than the average became advantageous, the entire curve (Fig. 3.4a) would move to the right. In an environment where height less than the average became advantageous, the entire curve would move to the left. The evolution of the neck of the giraffe (i.e., measurements of the heights of the members of the giraffe population at successive evolutionary periods) could be represented by a progressive movement of the curve to the right. Similarly, human races having become geographically isolated from each other in different environments in the past (thus preventing their interbreeding), some races have ended up being higher than the average, some smaller. African pygmies are an example of the latter. In this case, it is possible that selection acted, as Darwin originally proposed, on small differences in height, so that the curves progressively moved to the left (Fig. 3.4d). A more likely explanation is that small people, being less able to defend themselves against average-sized people, migrated deeper, and deeper into the jungle where they encountered other small people who had migrated similarly. Thus, there came about the selective breeding of small with small. This *reproductive selection* came about, not because average and tall people were eliminated by natural selection (negative selection), but because the geographical isolation of the small allowed them to positively select each other for reproductive purposes.

If selection were acting simply on the height character, and that character in no way impacted reproductive ability, then members of a geographically isolated pygmy race, if reunited with the main group, would still be able to reproduce with members of that group. When this occurred, the difference in height would tend to disappear among their offspring. Galton was inclined to believe that, because of such prohibitive "blending" effects, many moves of bell curves would not be gradual, but jerky, due to the sudden production of extreme variants, or "sports," that would appear independently of natural selection [2]:

> Some variations are so large and otherwise remarkable, that they seem to belong to a different class. They are known among breeders as 'sports.' ... Sometimes a sport may occur of such marked peculiarity and stability as to rank as a new type, capable of becoming the origin of a new race with very little assistance on the part of natural selection.

He did not entirely rule out the possibility of small steps, noting: "A new type may be reached without any large single stride, but through a fortunate and rapid succession of many small ones." However, whatever the size of the steps, because of blending effects he held that natural selection played only a minor role [2]:

> The theory of Natural Selection might dispense with a restriction, for which it is difficult to see either the need or the justification, namely, that the course of evolution always proceeds by steps that are severally minute, and that become effective only through accumulation. That the steps *may* be small and that they *must* be small are very different views; it is only to the latter that I object, and only when the indefinite word 'small' is used in the sense of 'barely discernible,' or as small compared with such large sports as are known to have been the origins of new races. An apparent ground [of argument] ... is ... the fact that whenever a search is made for intermediate forms ... a long and orderly series can usually be made out .... But it does not at all follow that because these intermediate forms have been found to exist, that they are the very stages that were passed through evolution.

In a book on the inheritance of fingerprints [19], Galton noted that they could be classified in terms of "loops" and "arches" and that, although individually distinct, patterns tended to be inherited without reference to natural selection. It was unlikely that fingerprint patterns would assist survival or were a factor in choice of sexual partner:

> It would be absurd ... to assert that in the struggle for existence, a person with, say, a loop on his right middle finger has a better chance of survival, or a better chance of early marriage, then one with an arch. Consequently, genera and species are hence seen to be formed without the slightest aid from either Natural or Sexual Selection. ... The results of panmixia in finger markings corroborates the argument I have used in *Natural Inheritance* and

> elsewhere to show that 'organic stability' is the primary factor by which the distinctions between genera are maintained, consequently the progress of evolution is not a smooth and uniform progression, but one that proceeds by jerks, through successive 'sports' … each in turn being favoured by Natural Selection.

Galton's doctrine of "organic stability" will be returned to in Chap. 4. In *Natural Inheritance* [2], he suggested that an organism might have a certain stability so that internal changes might build up over several generations without any obvious effect until there was a sudden change to a new position of stability, and visible differences could then appear.

Ancestral Heredity

Like Darwin, Galton was constantly reworking his ideas. His attempts more precisely to formulate inheritance in quantitative terms underwent modifications over the years, with later input from Karl Pearson, who in 1884 at age 27 had been appointed Professor of Applied Mathematics and Mechanics at University College, London. In 1900, the second edition of Pearson's monumental *The Grammar of Science* extended Galton's ideas to the "Principle of Homotyposis" (Chap. 6). Galton's thinking was neatly summarized by one of Pearson's associates, G. Udny Yule, whom we shall meet in Chap. 9 [20]. A key idea was expressed in the form of an equation that showed the relationship between the dimensions of a character X in a parent, and the average dimensions of the same character Y in its children.

If a father of height X (which equals 72 inches) had three sons of heights 66, 68, and 70 inches, then their average height (Y) is 68 inches. They are not the same height as their father, and they vary among themselves. If there is any consistency here, which holds for other father-son combinations, then can this consistency be captured in some way? If so, can we make use of what we have captured to predict future son heights if we know a father's height? If we can, then we may have discovered something fundamental about the inheritance of the height character—indeed, perhaps about the inheritance of any character. As a simple starting point, one can assume the relationship:

$$\text{Average son height} = 68 = \text{A} + (\text{Bx}72) \tag{3.2}$$

In words, average son height can be written as equal to an unknown constant (A) plus another unknown constant (B) that is multiplied by the father's height. Mathematicians call this a model. Whether it is a good or bad model is discovered when real-world data (heights of parents and their children) are fed into the equation. In more general form, it can be written:

$$\text{Y} = \text{A} + (\text{BxX}) \tag{3.3}$$

The value of the children's character Y is equal to a constant A plus another constant B which is multiplied by the value of the parent's character X. If a large sample of fathers and sons are examined (i.e., known values of X and Y are entered into the equation for each father-sons case), and if there is some consistency in the relationship, then values for A and B can be calculated.

Galton showed that, if you know the height of a father (X) and the values of A and B, then you can predict with some accuracy the likely height of a son (Y). Therefore, height is at least partially an inherited character. If the value of B in a study is significantly different from zero, then there is said to be a "regression" of Y on X, and the character under investigation is at least partially inherited. This is Galton's Law of Regression. Note here that heights of mothers are disregarded. Likewise, a separate study could disregard heights of fathers and determine the regression of sons' heights (Y) on mothers' heights (X). Alternatively, the heights of each mother and father could be combined as the "mid-parent" value (see above).

Galton went beyond the relationship between parent and immediate offspring, to consider inheritance from a more remote ancestry. Instead of X referring to a father, let it refer to a grandfather. In this case, the constant B gets less, but if it is still significantly greater than zero, then partial inheritance from the grandfather has been demonstrated. This can be continued to great-grandparents and then further down the line as long as the necessary data can be found. At each step, the value of B gets less until eventually, after several generations, it is not significantly different from zero, and no hereditary relationship is discernible for the character under consideration. It turns out that the best predictor of the measurement of a son's character is not just the measurement of the same character of the father, but the combined measurements from a collectivity of his recent ancestors. This constitutes Galton's Law of Ancestral Heredity: "The mean character of the offspring can be calculated with the more exactness, the more extensive our knowledge of the corresponding characters of the ancestry."

The law was not confined just to precisely quantifiable characters such as height, but also to quantal, all-or-none, attributes, such as seeds with either green cotyledons (say G) or yellow cotyledons (say Y). In this case: "The percentage of G's and Y's amongst the offspring can be calculated with the more exactness the more extensive our knowledge of the corresponding characters of the ancestry." In other words: "The degree to which a parental character *serves as a basis for estimating* the character of an offspring, depends not only upon its development in the individual parent, but on its degree of development in the ancestors of that parent" (my italics).

The key here is "serves as a basis for estimating." The "law" is a numerical exercise that deals with averages so as to make a prediction that is better than just tossing a coin. It is an index of a relationship, but not of the path of that relationship, which may be either direct or indirect. It tells of a relationship's existence, but not of the underlying physiological process. On average, red skies at night predict good weather the next day. There is a relationship, but not a direct one. The redness of the evening sky does not directly *cause* the good weather the following day.

Unrealistic expectations of the power of the method were raised by the term "law" and by authoritative statements of the form that: "The two parents contribute between them *on the average* one half or $(0.5)^1$ of the total heritage of the offspring;

the four grandparents, one quarter, or $(0.5)^2$; the eight great-grandparents, one eighth, or $(0.5)^3$, and so on. Thus, the sum of the ancestral contributions is expressed by the series $(0.5)^1 + (0.5)^2 + (0.5)^3$, etc., which, being equal to 1, accounts for the whole heritage" [17]. Each ancestor had *some* effect, although vanishing little as the heritage got more ancient.

The approach is of general applicability, depending on what are chosen for X and Y. It is not limited to determining correlations *between* individuals. It can also be used to determine correlations *within* individuals, such as between foot size (X) and pelvis size (Y). Since people with large feet tend to have large pelvises, in this case the constant B is usually significantly different from zero.

As we shall see, in *Defence* (Chap. 9) Bateson came to disagree with Galton declaring that: "The Law of Ancestral Heredity, and all modifications of it yet proposed, falls short in … that *it does not directly attempt to give any account of the distribution of the heritage among the gametes* of any one individual." So, "a good prediction may be made as to any given group of zygotes, but the possible constitutions of the gametes are not explicitly treated." Bateson's use of the word "zygote" has to be understood in context. Sometimes it referred to offspring as the freshly fertilized ovum (the normal usage of the word) and sometimes to offspring as the adult that developed from that ovum. He regarded this as a "natural extension," implying that it should not be beyond the sophistication of his audience. Thus, in *Principles* (Chap. 12) he stated: "The term zygote is usually restricted to the single cell which results from the process of fertilization, but by a natural extension the word may be used for the individual which develops by somatic divisions from that cell." In *Natural Inheritance* [2], Galton admitted that a character that did not blend in the offspring, be it considered as zygote or adult, would emerge "undiluted" in later generations, "giving it repeated chances of holding its own in the struggle for existence." In *Defence*, Bateson was to point out that this was at odds with the law of ancestral inheritance.

De Vries and Darwin's "Great Truth"

Given the volumes of English texts before him, it is small wonder that Bateson in 1889 overlooked what was probably the most important book of the decade, written in German by his Dutch contemporary Hugo de Vries [21]. This work, *Intracellular Pangenesis*, updated various aspects of Darwin's theory of pangenesis to bring it into line with new work, especially with microscopic observations of dividing cells. As the title suggested, de Vries was not concerned with how "formative elements" corresponding to characters acquired during an individual's lifetime might be transferred from cells of origin to gametes and hence to children. He held that the "pangens," as he called Darwin's "gemmules," were strictly intracellular, multiplying as cells multiplied (by mitosis) so that their quantity per cell would remain constant. In essence, they had all the properties that we now attribute to genes.

It is not clear when Bateson first read the book. Although he did not have "complete mastery" of German, Beatrice notes in her *Memoir* that he knew enough German "to read without difficulty." The book did not escape Romanes who had spent much time in Germany as a child. He cited it in *An Examination of Weismannism*. In correspondence with de Vries in 1900, Bateson asked for a copy which de Vries may have brought to his attention. De Vries replied from Amsterdam (Oct 10, 1900):

> I very much regret to have no more copies of my little book on Intracellular Pangenesis. I hope that you will see in my [new book] *Mutationstheorie* proofs for the Pangenesis, and that I will succeed in persuading you, what I tried to do at Cambridge, that your Discontinuity and Darwin's Pangenesis are founded on exactly the same principle. At the time I have been very much disappointed that my endeavours to defend Darwin's *God Pan*, as he called it, have found so little sympathy in England. I feel quite sure that Darwin, if he could have read my little book, would have approved my conception of his Pangenesis.

To emphasize this, to his next letter (Oct 25, 1900) de Vries added a footnote: "Theory of Mutations = Pangenesis, you know!"

An English translation of *Intracellular Pangenesis* did not become available until 1910. Referring to Darwin's pangenesis, the translator's preface noted that: "Many men discerned the weak features of the hypothesis, but to Hugo de Vries belongs the credit of having detected the 'great truth' it contained." Darwin's truth was that all the anatomical and physiological complexities of an organism—subsumed under the name "characters"—ultimately derived, on a one-on-one basis, from a multiplicity of very minute, discrete, "formative elements." Thus, an organism could be viewed as a set of characters and that set of characters could derive from a set of formative elements. Darwin held that the latter, as "gemmules," could *move beyond* cells. De Vries held that the latter, as "pangens," were *confined to* cells. Indeed, nuclei acted as "reservoirs" for pangens, which "assimilate and take nourishment and thereby grow, and then multiply by division; two new pangens, like the original one, usually originate at each cleavage. Deviations from this rule form a starting point for the origin of variations and species."

In de Vries' view, the "two main factors of variability" were deemed to be "an altered numerical relation of the pangens already present, and the formation of new kinds of pangens." Furthermore, at the level of the individual, pangens could be differentially expressed. Unlike Weismann, who had considered that "every somatic cell receives, at the time of its origination, only those hereditary elements which will be needed by itself and its descendents" (i.e., sets of specialized pangens were *parceled out* to different cell lineages), de Vries held that the phenomena of development, from zygote to embryo to adult, were determined by the *differential expression* of different pangens—*all* pangens usually being present in every cell [21]: "The differentiation of the organs must be due to the fact that the individual pangens or groups of them develop more vigorously than others. The more a certain group predominates, the more pronounced becomes the character of the respective cell." The correctness of de Vries' interpretation was evident to botanists who had seen for themselves that entire plants could be generated from cuttings taken from leaves or other parts of a plant. Thus, leaf cells, while expressing the pangens necessary for determining the characteristics of a leaf, retained the potential to express pangens necessary for determining other tissues. Differentiation was not accompanied by the differential elimination or differential consumption of non-expressed (i.e., latent) pangens. By 1904, Bateson was citing *Intracellular Pangenesis* (Chap. 9).

Bateson's gradual estrangement from de Vries is apparent from his declining in 1906 a request from the fern expert, Charles Druery, who had made a translation of Mendel for the RHS in 1901. Druery asked for Bateson's support in getting a publisher to agree to an English translation of *Mutationstheorie*. Bateson replied (Oct 27, 1906): "If the matter proceeds you must kindly take it that my part is limited to a friendly desire to see the book in English and that I cannot accept any responsibility direct or indirect."

Johannsen

To distinguish the relative roles of "nature" (variability transferred from generation to generation) and "nurture" (fluctuating, non-inherited, variability within each generation), ideally clones (to which identical twins approximate) are required. Failing this, in the early 1900s, Wilhelm Johannsen (1857–1927), the professor of plant physiology at the University of Copenhagen, generated "pure lines" using the "pedigree method" of the French breeder Louis de Vilmorin [22]. A population of human pygmies can be looked upon as a "pure line" to the extent that, with respect to height, they breed true always producing offspring of pygmy height. If outcrossed with a member of a non-pygmy human population, then the height character would appear to blend among the offspring, and the consistency of heights within the pygmy range would be lost. For more on pure lines, see Chap. 27.

Of course, it is not possible to carry out such experiments on humans. Certain plant populations where individual plants predominantly self-fertilize are more suitable. Johannsen chose beans. In the pods from a single plant (itself derived from a single bean), one can find large and small bean seeds. This partly depends on where in a pod a bean is located, and where on the plant the pod is located. Nevertheless, if one has established a "pure line" from a single bean seed of known weight, the collection of beans from a *single* plant should show a distribution of weights, ranging from small to large, with an *average* weight that would remain *constant* from generation to generation. Most beans would be around the average weight, some would be less, some more. The "pure line" criterion means that this average, and the distribution about the average, *persists* from generation to generation.

Over several seasons Johanssen selected various pure lines that were grown under essentially identical conditions. He obtained successive weight measurements for each line. Within each line, weight tended to remain typical for that line, so he could predict with some degree of accuracy (i.e., the average and distribution about the average) the weight of future beans produced in that line. A line of large beans always produced offspring with the same high *average* weight (say X grams). A line of small beans always produced offspring with the same small *average* weight (say Y grams). Even if among the offspring of an initially small bean line, there was a bean larger than the rest, when that large bean was sown, then among its progeny in the next generation the *average* weight would be unchanged (Y grams). Even though the bean larger than the rest had looked different and had a different weight from the average (i.e., it was phenotypically different), something *essential* about the nature of the bean—its ability to continue the line and produce an offspring population of small average weight—was found to be unchanged when it was sown (i.e., we would now say it was *genotypically* similar to other members of the line). Thus, the variation that had caused the parental bean in the small bean line to be large was due to *environmental* influences (e.g., place in the pod), which had brought about a "fluctuation" in size.

Johanssen went further. He tried to simulate the case of a mixed population as considered by Galton. He prevented self-fertilization and crossed several pure lines with each other. Now the averages changed and a regression to the mean could be demonstrated, presumably due to the blending of various positive and negative factors contributing to bean weight in the populations so created. The key result was that the variation *within* a pure line was observed to be due to environmental influences and was not inheritable. Thus, variation among individual organisms was due to two components, an inherited component ("nature") and an environmental component ("nurture"). One could have two beans of equal weight, one from the high end of the small bean line weight distribution, and one from the low end of the large bean line weight distribution. The two beans would be phenotypically identical, yet when sown their offspring would have the average weight characteristic of the corresponding lines.

Johannsen's famous account in 1903 [23] was dedicated to "the creator of the exact science of heredity, Francis Galton F. R. S. in respect and gratitude." There is an English translation of an accompanying review [24]. Johannsen's work was attacked by Weldon and Pearson [25] but was praised by Yule [26]. George Shull, at that time in the Biometric camp (Chap. 27), also praised Johannsen's work at a meeting in Philadelphia in December 1904 [27]. Shull did not see a conflict with Galton and began himself to articulate "the hereditary unit" (later known as the "genotype") in terms of a "general condition of the character":

> The weight or size of an individual seed [the character itself] is not the hereditary unit, but the character of all the seeds of each plant [in a pure line] considered as a whole [is the hereditary unit]. A plant which produces small seeds in general, may produce some seeds which are larger than the smallest seeds of another plant which produces large seeds in general, so that when the student of heredity wishes to use seed-characters or presumably any other repeated character, he must seek the *general condition of the character* in question in each plant and not depend upon the character of single seeds. [my italics]

In other words, there is *something* fundamental concerned with the inheritance of a character. The nearest we can get to detecting this *something* is to take the average value of the character in the population under study. By doing this, we tend to eliminate environmental influences. The notion of pure lines turned out to be of more than academic significance. In the 1920s, Bateson was called as an expert witness in a case concerning the pedigree of a line of commercial peas (the Gradus Case;

Hamm versus Berry Barclay and Company). Here, Bateson declared (Letter to Walton; October 31, 1921): "1. Small seeds *from a plant bearing on the whole large seeds* will give results as good as the large seeds of the same plant. But. 2. Small seeds from a plant *bearing more of less uniformly small seeds* cannot be expected to give offspring with large seeds."

Decoys

The works of the elder evolutionists were to some extent decoys that distracted Bateson from other works, such as those of de Vries (see above) and Butler (Chap. 19). Among Bateson's writings, there was no mention of Romanes' three volume *Darwin, and After Darwin*, the first volume of which was published in 1891 (Chap. 5). These amplified and extended Romanes' address to the Linnean Society in 1886 that Bateson had read in the copies of *Nature* Anna had sent during his Steppe travels (Chap. 2). But for Bateson everything was a decoy—the world, Cambridge life, literature, and art. In his letters from the Steppes, he had pleaded for the latest novels and the latest issues of *Punch*. Issues such as degrees for women students, and compulsory Greek for scientists (Chap. 23), drew him deeply into university affairs. Furthermore, he had to raise independent funds for his research and secure his position. He became a member of the Cambridge Press Syndicate with a view to diverting funds for research (Chap. 2), and this was one motive of his study of, and speculation in, artworks. But his next project, *Materials*, was to impose more focus. A publisher's deadline concentrates the mind!

Summary

While aware of the works of various evolutionists in their dotage (Galton, Wallace, Weismann), initially Bateson is likely to have overlooked Hugo de Vries' *Intracellular Pangenesis*. De Vries modified Darwin's hypothesis of "pangenesis," according to the "elements," or "pangens" that corresponded to individual characters, attributes that we now know to be those of genes. Darwin thought a character acquired during an individual's lifetime, could, by virtue of the transfer of the corresponding *educated* pangens (gemmules), be passed on to its children. However, experiments by Galton and Heape disproved his belief that the pangens could move from normal tissue cells to gonadal cells. Like Weismann, Galton saw the potentially immortal *germline* (stored in the ovary or testis) as distinct from the *soma* (the mortal remainder of the body). Since gametes remained the same size from generation to generation, then each parent could on average only transmit half its elements to a child, the other half being lost. This meant that there might be competition between elements for representation in future generations. "Ancestral" characters that disappeared and later reappeared were due to "latent" (hidden) elements. These were distinguished from the "patent" (overt) elements that determined characters regularly seen in the offspring. Both elements were in the gametes as "primary elements." The latent elements constituted a "residue" that remained after separation of "patent elements" from the primary elements. Galton downplayed the role of Darwin's natural selection. Major evolutionary processes, with transitions from one position of "organic stability" to another, were primarily driven by "sports" (mutations). Following his proposed "laws" of "ancestral hereditary" and "regression," Galton calculated, in terms of decreasing contributions of ancestors as they became more remote, the average contributions of each ancestral generation to contemporary organisms. In Denmark, Johannsen gave the distinction between "nature" and "nurture" experimental support.

Chapter 4
Variation (1890–1894)

Variation, whatever may be its cause, and however it may be limited, is the essential phenomenon of Evolution. Variation, in fact, is Evolution. The readiest way ... of solving the problem of Evolution is to study the facts of Variation.

William Bateson, Materials.

By 1890, Bateson had settled down to the project which was to occupy much of his life—variation, its nature, scope, causes, and significance. In his view, the main post-*Origin of Species* pursuits, "the Embryological Method" and "the Study of Adaptation," had run their respective courses and failed. Yet his position was precarious. His years "in the wilderness" had produced only one paper that could gain general approval. He no longer held the Balfour Scholarship. His sole source of professional income was his St. John's College fellowship (£200/annum). This was to be supplemented, from 1892 onwards, by stewardship of the college kitchens, gardens, and farm (£200/annum). Thomas Huxley, the *éminence grise* of British biosciences, was now in his dotage. The influences of Foster at Cambridge and of Lankester and Thiselton-Dyer in London were becoming increasingly felt. Directly or indirectly, the futures of members of the new generation of biologists were in their hands. But Bateson's views challenged their authority. He needed allies. Here, his strength appears to have been his connection with the Darwins and, especially, Francis Darwin.

Francis and Romanes

Long exposed to evolutionary ideas, it is not surprising that Francis Darwin's name appears among those with first-class degrees in the Cambridge Natural Science Tripos of 1870. It is also not surprising that Romanes' name appears on the same list

A. G. Cock, D. R. Forsdyke, *Treasure Your Exceptions*,
https://doi.org/10.1007/978-3-030-92099-9_4

with a second-class degree. Romanes entered Cambridge in 1867 and took mathematics with the intent of entering the Church [1]. Up to that point, he had read none of Darwin's works. At the end of his second year, he won a scholarship to read Natural Sciences. This he completed in eighteen months before joining Foster's physiology laboratory. Since there had been little time to prepare, the second-class degree should not have shaken his confidence. But in the spring of 1872, he was stricken with "faintness and incessant latitude" and could not continue with his post-graduate studies. After some months, "he broke down and was declared to be suffering from typhoid fever." Whether this was the correct diagnosis or not, his "long and weary convalescence" was "beguiled in part by writing an essay on 'Christian Prayer and General Laws,' the subject assigned for the Burney Prize Essay in 1973." He won. With confidence bolstered, he returned to Cambridge "resolved to devote himself to scientific research" [2]. However, being branded a second-class man could fatally impede an academic career. Mitigating circumstances were seldom recognized and would do little to still wagging tongues. Romanes left Cambridge in 1874.

Shortly after his Burney success, the twenty-five-year-old Romanes submitted a carefully reasoned letter on color variation in fish to *Nature* [3]. Fish that lie flat on the seabed are presumed to have evolved from free-swimming fish. Some species of flat fish lie on their left side and others on their right side. Depending on this, one side—the under-side—becomes white, and the other side—the top-side—assumes a coloration like that of the seabed. A left-lying flat fish that shows partial right-lying coloring (i.e., some white on top) might be regarded as a mutation ("sport"). Romanes, however, noted that, within a species, entirely reversed individuals sometime occurred (i.e., completely white on top). Reflecting his training in experimental physiology, he suggested that "a cross between a normal and a reversed individual of the same species" might sometimes result in partial color characteristics. More importantly, he suggested breeding experiments that should "not rest satisfied with mere simple crosses, however numerous, but also try various complex and reciprocal crosses." Three more letters appeared a year later, on the decrease in size of organs that appear no longer functional. These anticipated what Weismann was later to call "panmixia" (Chap. 5).

According to his biographer [2], it was Romanes' *Nature* correspondence, rather than his friendship with Francis Darwin, that in 1874 "attracted Mr. Darwin's attention, and caused him to send a friendly little note to the youthful writer." This led to a series of visits to Darwin's residence—Down House, at Downe in Kent—and their collaboration in research. It is possible that Romanes was able to engage Darwin at a higher intellectual level than Francis. Furthermore, Romanes' father having recently died, there may have arisen an emotional attachment that was more than merely that of a junior scientist for a senior. To what extent, if any, Francis Darwin sensed an intellectual and emotional rivalry is not clear. Francis was appointed Lecturer in Botany at Cambridge in 1884 and was Reader from 1888 until 1904. Over the decades, there was much correspondence between Romanes and Francis. Prior to Romanes' death in 1894, he sent Francis some of his poems but asked him to keep the fact of his poetry-writing private [4].

Francis and Anna

Bateson, having been labeled a failure at school, should have been pleased with a first-class degree in the Natural Science Tripos of 1882. He later reminisced in his Spencer Lecture (1912): "As a boy in Cambridge I learnt that if a man got a first-class degree he might be happy; if he got a second class he would be unhappy; if he got a third class, nothing but misery and a colonial life awaited him." The high regard with which he held a first-class degree is evident from his correspondence with his sister Anna and may have deeply influenced her life, while perhaps cementing his own relationship with Francis Darwin (Fig. 4.1).

Writing from St. Petersburg (May 1, 1886), Bateson showed an appropriate fraternal interest regarding Anna's final Tripos examinations: "Good luck to you next month! Don't bother yourself about it and all will go smooth." That said, one would have thought he would have dropped the subject and then given congratulations or condolences after the event. However, moving deeper toward the Steppes, he wrote to Anna from Samara en route to Orenburg (May 22):

> I have been with you in spirit, and seen the papers given out though I am not certain that my spirit has not taken a premature flight – but June 1 is the day isn't it? I hope that the spirit of Christian fortitude combined with the pagan quality of Stoicism may be with you, and that you will, in the last resort, remember that you have promised not to take it very seriously whatever evil fate may be in store. ... But all this is assuming that your way is to be cloudy, while there is no reason to suppose that they won't send the angels down after all. Mind you give me full particulars about the questions and all such matters. But they are poor things these exams. Whether it be joy or lamentation that follows it will abide only for a very little season. And so fight on! 'We have wished you good luck in the name of the Lord'.

Fig. 4.1 Anna Bateson

A few days later, to "Mamma" he wrote: "I hourly await news of Anna's exam. I do hope she won't work herself up too much if she doesn't do well. I am afraid that it will be a very anxious time for a bit, but she ought to do well enough in her own work when she takes to it. Of course you know the list in all probability by now, or at least have begun 'to wait for it'."

If all the distractions of foreign travel could not keep her brother off the topic, one can only speculate at the frenzy around Anna back at home. She ended up, as did many of her classmates, with a second-class degree—a cause in many quarters for hearty celebration. Bateson wrote to Anna from Orenburg (June 5): "Your sad telegram reached me at 1.30 last night. Despairing of getting any idea of a reply into the head of the employee [who operated the telegraph service] I retired to rest again and send an answer this morning." There followed many lines of condolence, ending with: "I am awfully sorry but I don't think it matters a little d---. Just remember that everybody respects you for what you are worth, which is, if you will allow me, a good deal, and that this thing won't interfere with your work in the least. So cheer up and let me have a cheerful letter as soon as you please." There then followed a long, light-hearted description of his travels which concluded: "So comfort you, comfort you my people! Mind you remind F. Darwin of his offer; he won't mind about your degree at all." Anna told him that she would be traveling with Ned to Norway. He replied from Irghiz (June 31, 1886):

> Good luck to you in your wanderings! I think you are quite right not to give up on account of this disaster, and I was awfully pleased that you could write so cheerfully. Of course you have had a rough fortnight of it, but I hope that is past now. You ask me what I think you ought to do. Well, you know what my advice always is in such a case. Do as you feel you would like best and beyond this it is not for an outsider to say. However, as you ask me what I think, I will try to put myself in your place and consider the situation. It seems to me that you have a turn for science most decidedly, but for the abstract side of it.

That said, one would expect encouraging advice on the exciting prospect of working with Darwin. But there followed a finger-wagging exercise on her not having "much facility for the practical," and "no eye for detail," which finally led on to:

> But I see no reason why you should not do well in any particular work that you take up, *if it interests you*. This is everything, I am sure. Whatever interests one, one is sure to do well in, which is a truism. I do not exactly know what F. Darwin proposes for you to do, but I should suppose that he is a stimulating person to work under and whatever he sets you to do is likely to lead somewhere. ... I don't think F. D. is a man who will mind at all about your 2nd class, and I should suppose that you will only find this a drawback in event of it being desirable for you to get teaching work over the head of some insignificant person who has got a first.

The fact of her second-class degree was not just a family matter. In a letter to "Mamma" in August from Irghiz, Bateson quoted a letter from Adam Sedgwick: "I am very sorry your sister missed her first, but she failed completely, or nearly so, in the vertebrate questions; some of her other answers were decidedly good." Bateson added: "So that though the punishment is heavy, the sin was not very grievous after all." In a separate letter to Anna he wrote (Aug 14, 1886):

> Now about your work. I am very glad you are going on [with biology]. Of course you must go on. Also, it should be a good lift for you to be allowed to work in [the] Botanical Gardens. … I should suppose that it would be worth while to stick to F. Darwin's work, if only to get a status in the Museum buildings, which is worth the trouble of acquiring. I lay stress on this. It is awfully stupid [of me] to know so little Botany here [in Russia], where all is 'new.' … Glad you and Ned liked Norway; I bet, however, that I have learnt more horsemanship than you, with all respect. I fire from saddle with comparative *sang-froid* now, and accept the somersaults which ensue with stoicism.

Anna considered a complete break, even emigration. In a letter to "Mamma" (Sept. 1886), Bateson wrote:

> I haven't got the letter telling me about Margaret's prospects [in journalism], but I am awfully glad she is looking up, and the town [London] should be a good move. Also, I am right glad Anna didn't go to America, which would have been sheer folly; also that Mary is to be the Historian of the future. … I am afraid the Good Lady [his mother] has been getting into a dreadful pickle about Mr. G. [Gladstone], though I quite agree that on this occasion the thing was worth fighting for. Though for me the shame of the rout is deadened by the fact that the G. O. M. [grand old man] has got his '*quietus*' which I am sure is an excellent thing for the radical cause. If I had been at home of course I should have worked on your side.

So, in 1886, Anna began botanical studies with Francis Darwin whose primary interests were the mysterious way plants could curve toward light and away from gravity, and how they controlled water loss. She soon did well. She was awarded a Bathurst Studentship and an Assistant Demonstratorship at Newnham College. In an October letter, Bateson wrote: "I am awfully glad Anna is getting some results – I can thoroughly sympathize with her delight in the feeling of blessed leisure which one has when the accursed Tripos is over." In December, Anna and a companion toured in Germany (Leipzig and Dresden), and in January 1887, she presented at the Linnean Society the results of a laboratory study on "The Effect of Stimulation on Turgescent Vegetable Tissues" [5]. Bateson wrote congratulating her "on the Bathurst, which is really extremely pleasant." To Margaret, he wrote from Kazalinsk (Feb. 2): "Anna's success at the Linnean is most delightful. Upon my word, she does seem to have blossomed forth. In every letter to me she alludes to some new scheme of work. She really seems on a fair way to do something very considerable." In June, having received a reprint from Anna, he wrote: "Anna's paper has just come to hand. Though, of course, it is partially unintelligible to me, I 'perused it with the greatest interest.' If that kind of work is as rare on the Vegetable side, as it is in the Animal Department, this alone should make it uncommonly good." To Anna, he expressed his general disdain for laboratory-based studies:

> Many thanks for your '*Separat Abdruck.*'…. It certainly must be a bit of very high class work, but as you know, I too am a little doubtful as to the value of oracular inscription on recording drums. I don't know how it may be in Botany, but in Physiology they seem to have led to very little. Of course your subject, record of growth, is of rather a different class, in-as-much as it is a really natural phenomenon with which you are dealing, and not an almost entirely artificial one as that of contraction of a dying muscle …. You are going ahead at such a rate that I feel dreadfully stranded.

In September 1887, he wrote: "I congratulate you on being on to teach at Newnham again, and also on the enormous quantity of work you seem to be getting through." So, Anna was plowing ahead with research and teaching at a Cambridge women's college. And, although Margaret's magazine was not thriving, and Ned had not done well in his examinations, his sister Mary was beginning to make a mark in medieval history. Her examination results (first class) reached Bateson in Omsk (Sept. 5, 1887):

> All hail. O historienne! This really is an achievement! Of course I wired at once. How any body on the day of such a glorious success can talk solemnly of settling to Medieval Latin, beats me. I drank the lady's health in Sherry that might have served the turn of Andrew Liang's Duchess, but it was my duty and I did. To Ned all sympathy if he feels the want of it. All the same I don't think he ought to. He has escaped a beastly uncomfortable and in most cases morally degrading career, and for my part I read his fate without a pang. *God*, if one may be permitted to *pry* into that inscrutable *Future* of which He is the *Sole Manager and Lessee*, has probably cast him for a very good *Role* in Life's Comedy if he will [be] so good to stay on the British stage .... Of course it is rather unpleasant not to get what one tries for, yet when a man tries for something more than commonly beastly, I don't think he ought to feel much hurt if his friends don't weep much.

Poor Ned! But Bateson was facing the prospect that the talents of two of his younger sisters might prove greater than his own. The paper Anna had presented in 1887 was published in 1888 in the botany section of the *Linnean Journal* with herself as first author [5]. In a second paper "On a Method of Studying Geotropism" [6], which was clearly in Francis's area of study, she was also first author. But she was the sole author of a third paper—"On the Cross-Fertilization of Inconspicuous Flowers" [7]. Here, she acknowledged Francis only for "very kind assistance," and since she did not acknowledge William, it may be assumed that the conception, execution, and writing were largely her own. The paper investigated a point raised by Charles Darwin, that even where flowers appear small and inconspicuous and are not seen to be frequently visited by insects, there would still be a benefit from intercrossing (rather than fertilization with their own pollen). Anna showed this to be so. The paper did not go unnoticed. Carl Correns commented on it when first writing to Bateson in 1900 (Chap. 6), and Francis cited it in *More Letters of Charles Darwin*, which he edited in 1903: "Miss Bateson showed that *Senecio vulgaris* clearly profits by cross-fertilization; *Stellaria media* and *Capsella bursa-pastoris* less certainly."

There followed another single-author paper in 1889 from the Botanic Garden Laboratory, Cambridge, "On the Change of Shape Exhibited by Turgescent Pith in Water," which Anna first presented at a meeting of the Cambridge Philosophical Society [8]. This followed up on earlier work of de Vries that "Charles Darwin assisted by Francis Darwin" had discussed in *The Power of Movement in Plants* (1880). Thus, this was in Francis's area, and she referred to the experiments as having been "undertaken at the suggestion of Mr. F. Darwin."

At some point, she collaborated with Bateson on a field study, and she was a coauthor in a paper that appeared in 1891 [9]. Here, brother and sister studied several common wild and garden plants: *Linaria*, *Veronica*, *Streptocarpus,* and *Gladiolus*. These all had bilaterally symmetrical flowers, as opposed to the simple radial symmetry of flowers such as the daisy. They combed the stubble fields and gardens of Cambridgeshire looking for abnormal flowers with more or fewer petals than normal. In *Veronica* alone, they found 118 such abnormal flowers. The change in number of petals disrupted the system of symmetry in the flower and in some cases led to radial symmetry. These seemingly abrupt major changes in symmetry were of major interest to Bateson, forming a large part of the book he was writing. The paper constituted a first cohesive attack on Darwin's natural selection, invoking two classes of variation:

> By the elaborate researches of Galton, it has been shown that the frequency of occurrence of certain Variations obeys the Law of Error; that is to say, that, speaking generally, the greater the departure from the normal form, the rarer will be the variation. Galton has shown that this is true of several Variations in size, etc., of Man; and Weldon has further established the same for Variations in proportional measurements of the shrimp (*Crangon vulgaris*), etc. Though these are the only Variations which have been properly investigated by a statistical method, it may be seen by inspection that the resulting proposition cannot be true of the Variations which we have been considering. For in these cases of symmetrical Variations, as we have shown, the Variation is frequently complete and seldom incomplete, and the perfection of the Variation is out of all proportion to the frequency of its occurrence. …
>
> We wish, then, to insist upon the fact that there are at least two classes of Variation; it is suggested that this is a fact of great importance. It may be remarked that, if, as may be alleged, there is little evidence that species may arise by what may be called discontinuous Variation – a Variation in kind – there is still less evidence that new forms can arise by those Variations in degree which at any given moment are capable of being arranged in a curve of Error; and no one as yet has ever indicated the way by which such variations could lead to the constitution of new forms, at all events under the sole guidance of Natural Selection. Whatever may be hereafter determined as the scope of either of these classes of Variations in the constitution of Species, it is of the first consequence to recognise that these two classes of Variation exist; and the problem of the history of any given form or structure will never be solved until it shall have been first determined whether it is the result of the one class of Variation or of the other, and whether the changes which produced it were continuous or discontinuous.

The considerable surviving correspondence between Francis Darwin and Bateson reveals a closeness, both professionally and socially, that endured. Thus, it is likely that he was a frequent visitor at Darwin's house, Wychfield, on Huntingdon Road. Here, he would have met young Bernard, the son of Francis Darwin's first wife, and young Frances, a daughter born in 1886 three years after Darwin's marriage to Ellen Crofts. Another visitor was George Darwin's daughter, Gwen, who years later described Francis Darwin ("Uncle Frank") as of "melancholy temperament," and Aunt Ellen as "not a happy person." Ellen would sit smoking on the veranda at Wychfield with her friends Jane Harrison and Alice Dew-Smith, who all seemed "wonderfully up-to-date and literary." Ellen died in 1903, and many years later, Harrison reminisced: "I could not be sorry when Ellen died, because she was so unhappy, though I don't know why. She loved her husband and the child, and had everything she wanted in the world" [10].

It was not easy for Anna sandwiched as she was in the family sequence between two academic stars, William and Mary. Her doubts are clear from a letter Bateson wrote to her from Plymouth (July 31, 1889):

> On the whole I think you have done well, but I don't understand what is to be the next step. I rather gather … you think of going into some sort of business occupation with a garden. It seems to me rather a pity to do this and it could hardly be other than unprofitable. … Also, I think it always a 'regrettable incident' when persons whose parents have got clear of trade, relapse into it. However, I know that you could not easily get the control of a large garden establishment, if at all, that was not on some sort of commercial footing – so I think that must be faced. The only practical suggestion I can make is that perhaps Mrs. Peile may put you on the track of what you want. … You had better write to her and ask.

Thus, after four years with Darwin, in 1890 Anna became an apprentice market gardener in Wales, and in 1892, she opened her own business at Bashley in Hampshire. It was not just a move from academia to "trade," but a move away from a life centered in Cambridge. Margaret may very well have been a prompting influence. Although her book *Professional Women and their Professions* did not appear until 1895, the articles upon which it was based had been appearing in *The Queen* for several years past, and the examples they gave of a spirit of adventure and determination may well have given Anna the necessary stimulus to do likewise. Anna's place with Darwin was taken by Dora Pertz from Newnham. Pertz also worked with Bateson on hybrid *Cineraria* (Chap. 6).

Along with Darwin, Anna remained a ready sounding-board for Bateson's ideas. He wrote (Sept. 14, 1891) in great excitement about "my new VIBRATORY THEORY of REPETITION of Parts in Animals and Plants." This drew an analogy between various biological observations, such as stripes on a zebra, and the observation that sand sprinkled on a horizontal metal plate that is made to vibrate will settle in lines of minimal movement. His bubbling enthusiasm over this "Undulatory Hypothesis" was communicated to Francis Darwin, "and he thinks it 'really is very neat, upon my word'," and to "Gaskell and Sedgwick and both of them take it kindly." He wrote (Sept. 26): "I am tremendously pleased with the IDEA. F. D. came round today and advised me to make a feature for it for the book. I have also had some talk to Lankester about it and he gave it a very decent *accueil.* You'll see – it will be a commonplace for education, like the multiplication-table or Shakespeare, before long!" A draft paper entitled "Vibratory Theory of Linear and Radial Segmentation as Found in Living Bodies" was not published, but the vibratory theme resurfaced in later years. In a post-script, he mused: "Of course, Heredity becomes quite a simple phenomenon in the light of this." The notion that heredity information might be communicated from molecule to molecule by means of vibrations was already evident in the writings of St. George Mivart, Ewald Hering, and Samual Butler (Chap. 19), but it seems that, despite Brooks' *Heredity*, Bateson did not come across their work until much later.

Lankester

By 1889, early signs of ill-health were afflicting Romanes, and "he began to weary of London and the distractions of London life. By degrees, his thoughts and inclinations turned strongly in the direction of Oxford. … There were old friends there to welcome him, and there seemed abundance of ... facilities for scientific work" [2]. Being independently wealthy, Romanes did not need a formal position. He took up residence close to Christ Church College in 1890. Others with Oxford ambitions needed first to secure a position. Any candidate from Cambridge would have to deal with those who considered medicine an art, so any attempt to "Cambridgise" medicine (i.e., add more science) would be resisted [11].

At that time, the post of Deputy to the Linacre Professor at Oxford (to supplement Professor Moseley, who was in failing health) became available. Weldon had earlier promoted a Lankester "memorial" signed by Bateson among others, which should have greatly increased Lankester's chances. But Bateson wrote to Weldon (May 19) noting that: "From the samples of gossip ... that reach me it seems ... unlikely that Lankester will be elected. Now it would be most vexatious to see a post like that go by default to a really inferior man like Jackson, as Edinburgh did to Ewart, yet if no one better applied for it, this must happen. ... Under these circumstances why shouldn't we all go for it?"

Weldon and Bateson independently sought Foster's advice, which was against either applying. Weldon took the advice. Bateson did not. But at one point (June 9), he informed Foster that he had. Then, he changed his mind and told Foster that he was applying (June 15). Indeed, as the possibility of more candidates emerged, Bateson suggested (June 12) to Weldon that Foster had been bluffing. He sent a copy of his application to Lankester who replied (June 15):

> I think you are quite right to submit your statement to the Linacre Electoral Board and nothing could be kinder than your reference to myself. It is the best testimonial sent in for me. I think you are mistaken in looking at the Chair as a purely zoological one. I fear it is really incumbent on its holder to teach Human and Comparative Anatomy. *I* agree with what you say as to the importance of the study of what I call Bionomics, but I don't think you are right in declaring that Darwinism led to its neglect. A few of you young men at Cambridge and of course some Germans, have neglected it for a few years – but you are all coming into the true path. Men of *my* age have all been *naturalists* rather than section-cutters – but section cutting is a grand thing – and not to be despised. I look forward to your promised 'collections' and take the deepest interest in all your views and studies.

Lankester's appointment was announced in *The Times* (July 2), and Weldon soon moved to the Chair he vacated at University College (see below). Bateson's application contained a clear statement of his research agenda. Recognizing its importance, Beatrice reproduced it in her *Memoir*. The parts to be considered here are best appreciated in the context of the prevailing biological paradigm as set out by Lankester in an article on "The History and Scope of Zoology" in the ninth edition of the *Encyclopaedia Britannica* [12].

Lankester distinguished five main branches of biological study which he named morphography, bionomics, zoo-dynamics, plasmology, and philosophical zoology. These are recognizable today as systematics (classification), genetics, physiology, biochemistry, and theoretical biology. Morphography was "the work of the collector and systematist," as exemplified by Linnaeus and Haeckel. Bionomics was "the lore of the farmer, gardener, sportsman, fancier, and field naturalist, including ... the science of breeding, and the allied Teleology, or science of organic adaptations." Zoo-dynamics was "the pursuit of the learned physician" (i.e., biomedical research) as "exemplified by … the modern school of experimental physiology." Plasmology was "the study of the ultimate corpuscles of living matter, their structure, development, and properties." However, Lankester specified that this would be by the "aid of the microscope," rather than by chemistry, which he regarded as a branch of physiology. Philosophical zoology referred to "general conceptions with regard to the relations of living things … to the universe, to man, and to the Creator, their origin and significance." Here, besides "the philosophers of classical antiquity," he referred to "Linnaeus, Goethe, Lamarck, Cuvier, Lyell, Spencer, and Darwin." Through Darwinism had come recognition:

> That the natural classification of animals, after which collectors and anatomists had been so long striving, was nothing more nor less than a genealogical tree, with breaks and gaps of various extent in its record. … The facts of the relationships of animals one to another, which had been treated as the outcome of an inscrutable law by most zoologists, … were among the most powerful arguments in support of Darwin's theory, since they, together with all other vital phenomena, received a sufficient explanation through it.

Finally, Lankester turned to Ernst Haeckel's *Generelle Morphologie,* published in 1866, as putting "into practical form the consequences of the new theory," so developing a logical scheme of classification. This had led to "an extraordinary activity in the study of Embryology," inspired by "Fritz Muller's law of recapitulation which gave the chief stimulus to recent embryological investigations." However, he cautioned that "it is now recognized that 'recapitulation' is vastly and bewilderingly modified by special adaptations in every case, yet the principle has served, and still serves, as a guide of great value."

An application for a position is, of course, a marketing device. But so, to a lesser degree, can be an article in the *Encyclopaedia Britannica*, especially when there is an Oxford professorship on the horizon. From Bateson's other writings, and from his subsequent deeds, it appears that his application "To the Electors to the Linacre Professorship" was a forthright account of his current thinking, fueled by his optimistic, but mistaken, conviction that the scientific world was veering in his direction:

> As I have decided to offer myself as a Candidate, and as my views of the proper development of the study of Zoology differ somewhat from those held by others, I cannot avoid giving at least an outline of the method by which I think that development can best be brought about. … When the theory of evolution first gained a hearing it was felt that it was of primary importance to know first, whether it was true that forms of life had been evolving from each other; and secondly, if evolved, on what lines had this been effected and what was the ancestry of each. All other problems sank into insignificance in comparison with these. Now the readiest method of answering these questions seemed to be the embryological method. By using the so-called Law of von Baer, that the history of the individual is the history of the species, it should be possible to determine the pedigree of the form by examining the manner of development. Furthermore, if stages in the development of different types should be found common to several forms, this would be strong evidence of the common descent of these forms. To find such evidence has been the chief aim of zoological science since Darwin.

Bateson then pointed out difficulties of this approach, concluding that:

> Embryology has provided us with a magnificent body of facts, but the significance of the facts is still to seek. Knowledge, however complete, of the anatomy and embryology of animals

> can at best show us what may be called the formal relations of their structures to each other; it can never show us their genetic relations nor can it forecast their future. The question, then, which Zoology proposes to solve is this: what have been the steps by which animals have acquired the forms which they present, and what will be the future of their development?

Variation he saw as the most important next step:

> Whether we believe with Lamarck that adaptations are the direct result of environmental action, or with Darwin that they have been brought about by natural selection, it is admitted by all that the progression has come to pass through the occurrence of variations. This is common ground. Hence, if we seek to know the steps in the sequence of animal forms, we must seek by studying the variations which are now occurring in them, and ... the laws which limit them. When we shall know the nature of the variations which are now occurring in animals, and the steps by which they are now progressing before our eyes, we shall be in a position to surmise what their past has been; for we shall then know what changes are possible to them and what are not. Until the modes of Variation are known and classified no real advance can be made in the study of Zoology.

He then tried to be more specific about his approach, conveying a hint, but only a hint, of his Mendelian future:

> This study it is my ambition to pursue by the methods employed by Darwin himself; by studying the Variation of the animals and plants with which we are most familiar and especially by encouraging the study of domestic animals; by promoting experiments in breeding and cultivation; by the comparison of local varieties and generally, by amassing material which should serve to illustrate the nature and especially the modes of Variation.

There then followed an account of his past studies on adaptation aimed at determining the utility of structures as predicted by the theory of natural selection. These studies included the "collection of *Crustacea* and *Mollusca* from brackish and other waters in order to determine whether or not these diversities of environment produce specific alterations of structure." A more political application might have then gone directly to his work on cockles (Chap. 2) and glossed over his major results that were not in line with the conventional wisdom—that organisms should be closely adapted to their environments. But he continued: "The result of this investigation goes to show that, while such variations do occur in certain species, in the majority they do not."

Bateson listed four main areas of study: 1: "The Variation of Multiple Parts and of Symmetry." 2: "Variations in the size, number and fertility of animals." 3: "Variations apparently due to change of Conditions." 4: "Local Varieties." He planned a series of major publications under each heading. Each would be primarily a collection of facts. He saw himself as following in Darwin's sure footsteps: "Since Darwin's time no considerable collection of facts of this kind has been made. Each year sees the birth of new theories in this country and elsewhere, but each theory in turn is established by the facts of Darwin and Wallace." And where could all this lead?

> The first of this proposed series of collections will deal with the Variation of Repeated Parts. It will be, as far as possible, a complete account of all observations bearing on the subject. ... The importance of these facts lies in their value as evidence of the magnitude of the integral steps by which Variation proceeds and of the control which the symmetry of the body exercises over these variations.

There is here a glimmer of recognition of conflict between evolutionary forces. Does the pressure for symmetry so dominate that an animal cannot have an extra finger on one hand without also having an extra finger on the other? Alternatively, can symmetry be subverted with an animal having one normal hand and one abnormal hand? If both circumstances can arise, which is the most frequent? At that time, Bateson was hesitant to go into such detail:

> It is impossible to give in abstract an intelligible account of the results to which these facts point, but taken together they will go far to suggest certain laws as to the variation of repeated parts: these laws, if established, will have an important bearing on the conception of the modes of Variation and would lead to a modification of the views now current as to the nature of Homology.

He concluded with "a personal statement":

> I wish it to be distinctly understood that my Candidature is not in opposition to that of Professor Lankester. I recognize in the fullest manner the pre-eminence of Professor Lankester's claim. I have subscribed my name to the Memorial giving expression to this view and I do not retire from the position thus taken; for I hold that in the interests of Science and of the University the choice should fall on Professor Lankester and that the claims of no other candidate should be considered.

Bateson sent "warmest congratulations" to Lankester on his appointment (July 2). As Lankester had implied, the Oxford Electors had looked for an administrator who would organize the teaching, with research much in the background.

Variation

Living or not, objects grouped together because of common characteristics, tend to vary. Peas in a pod vary. Matches in a matchbox vary. Variation can only occur in that which already exists and the nature of that which already exists limits the scope of its variation. Just as the scope of a dice is limited by its structure, so the scope of variation of a living form is limited. From time to time, a human is born with six fingers, but, as far as we know, no human has ever been born with a hundred fingers, or with feathers. We know this with some certainty because of the work Bateson initiated. In his words (*Materials*): "The facts of variation must therefore be the test of phylogenetic possibility."

The Victorians were familiar with games of chance. The variation in landing position of a balanced dice is random. The variation in landing position of an unbalanced dice is non-random. So, from first principles, they supposed that variation in living forms would be either a chance event, or biased. Whether unbiased or biased, a variation might be helpful, unhelpful, or somewhere in between. Within its scope, the dice of life might be multifaceted allowing fine gradations in variation that might appear continuous. Alternatively, variations might be quite discrete (e.g., the usual six-sided dice), so appearing discontinuous. In 1859, Huxley tried to impress on Darwin the possibility that, regarding the type of variation that would bring about a new species, Nature was more likely to work discontinuously—in jumps, (saltations). Two varieties—Ancon sheep and six-fingered humans—seemed to have arisen in this way. Perhaps species arose similarly. Considering two species A and B that had arisen from a common ancestral species, Huxley wrote [13]:

> I know of no evidence to show that the interval between the two species must necessarily be bridged over by a series of forms; each of which shall occupy, as it occurs, a fraction of the distance between A and B. On the contrary, in the history of the Ancon sheep, and of the six-fingered Maltese family given by Réaumur, it appears that the new form appeared at once in full perfection. I may illustrate what I mean by a chemical example. In an organic compound, having a precise and definite composition, you may effect all sorts of transmutations by substituting an atom of one element for an atom of another element. You may in this way produce a vast series of modifications – but each modification is definite in its composition, and there are no transitional or intermediate steps between one definite compound and another. I have a sort of notion that similar laws of definite combination, rule over the modification of organic bodies, and that in passing from species to species 'Natura fecit saltum' [Nature changes in jumps].

Multifaceted or not, would the dice of life be thrown? If so, by what agency? If not thrown, then any variation that occurred might be termed "spontaneous." If the dice of life were thrown, then there would be a thrower. Would this agency be internal or external to the organism? In either case, could the agency direct the bias to the adaptive advantage of the organism? Could the agency "design" the organism? Among these alternatives, current evidence shows that, as Hooker impressed on Darwin, variation is a fundamental property of matter, like gravity. Left by themselves, things tend to get untidy (i.e., to vary). While usually confined in its scope, variation is largely unbiased as to direction. There is no design by means of variation, either internal or external. Variation is not directed. If an organism is closely adapted to its environment, then a variation is unlikely to be helpful. If the environment has recently changed, then a variation may more likely be helpful. But variations, per se, do not occur "for the good of the organism" (teleology).

Stored Information "Made again New"

In the context of variation, Bateson referred to "genetic relations" (see above). But he did not see this in terms of information. Genetic information, like most forms of information, is more than just information—it is *stored* information. At some point in time, information is selected and assigned to a store. At a later point, it may be retrieved from the store. When referring to mental information, we call the retrieval process, memory. But storage of information is not enough—information must be *safely* stored. To select and to store is not to preserve. Stored information must also be *preserved* information. When preservation is not perfect, there is *variation*. When there is variation, there can be evolution.

Most Victorians, like Bateson, were not thinking in informational terms. If they had, they might have considered, from knowledge that the copying of written information is error-prone, that hereditary information might be similarly prone. As we shall see in Chap. 19, Hering and Butler *were* thinking – and very clearly thinking—in informational terms, and they drew strong parallels between mental information and hereditary information. Bateson later came to quote Butler's remark (1878) that "the 'Origin of Variation,' whatever it is, is the only true 'Origin of Species'." But, neither Bateson nor Romanes was able to take information to Butler's conceptual level.

Yet they were able to distinguish the variation of something that had been manufactured using a complementary mold—as when identical decorative garden pieces are cast in clay—from the variation of something that had been manufactured following instructions, as when a clock is assembled from parts. Following the anatomist Richard Owen who recognized "a new and distinct developmental process" [14], Bateson wrote (*Materials*):

> If a man were asked to make a wax model of the skeleton of an animal from a wax model of the skeleton of another, he would perhaps set about it by making small additions to and subtractions from its several parts; but the natural process differs in one great essential from this. For in Nature the body of one individual has never *been* the body of its parent, and is not formed by a plastic operation from it; but the new body is made again new from the beginning, just as the wax model has gone back into the melting pot before the new model was begun.

We now recognize that the "new model is made again new" from a "pot" of chemical parts following instructions that have *themselves* been duplicated "by a plastic operation" using a mold (as when type set in a printer's block acts as a mold or template to duplicate printed information). When this process—the duplication of instructions—is inaccurate, *there is variation in whatever those instructions refer to*. When the instructions vary, what they refer to varies. The "new model" is then different from the old. Over the years, the word variation tended to be replaced by the word "mutation." Bateson noted in his Leidy Lecture in Philadelphia in 1922 (Chap. 17): "The term is commonly employed to give an importance, even an evolutionary significance, to a change for which the common word variation is felt to be an insufficient description."

Materials

There was no great enthusiasm for Bateson's project. Surely, Bateson's work could only be a poor extension of Darwin's *Variation of Animals and Plants under Domestication*? However, negotiations with Macmillan were successful and Bateson's contract secured him "half profits." The publisher was not committed to further volumes and, since the first volume sold poorly, these never appeared.

The literature of the day was replete with observations of the strange—supernumerary nipples in humans, extra horns in goats, changes in eye spots (ocellae) on butterfly wings, and monstrosities of all sorts. Bateson saw his task as the collection, organization, and analysis of this vast literature, to which he added many of his own observations: "To collect and codify the facts of Variation is, I submit, the first duty of the naturalist." Even today, with all the benefits of modern technology, the book would be viewed as an ambitious undertaking. Apart from an eighty-page Introduction, and nine pages of "Concluding Reflections," the 600-page volume contained numerous tables and 209 figures, many in several parts.

In Beatrice's words (*Memoir*), he exhaustively "ransacked museums, libraries and private collections." He wrote to physicians, surgeons, and biologists, and his enquiries were not limited to academic centers: "He attended every sort of 'Show,' mixed freely with gardeners, shepherds and drovers, learned all they had to teach him." Whether at the time he appreciated it or not, he was establishing a wide range of professional contacts that would be of much value in the years ahead. In the Preface (dated Dec. 29, 1893), he acknowledged help from many quarters, including Francis Darwin, Walter Heape, David Sharp, and his St. John's colleague Joseph Lister. The Introduction expanded on the various points he had made in his application for the Linacre Professorship. As was the custom, he paid due homage to Darwin. But then, as was not the custom, he proceeded to undermine Darwin, questioning the all-encompassing role of natural selection in evolution:

> We knew all along that Species are *approximately* adapted to their circumstances; but the difficulty is that whereas the differences in adaptation seem to us to be approximate, the differences between the structures of species are frequently precise. In the early days of the Theory of Natural Selection it was hoped that with searching the direct utility of such small differences would be found, but ... the hope is unfulfilled.

Like Darwin, he saw the problem of the origin of species, which he termed "the problem of Specific Difference," as fundamental: "If the Study of Variation can serve no other end it may make us remember that we are still at the beginning, that the complexity of the problem of Specific Difference is hardly less now than it was when Darwin first showed that Natural History is a problem and no vain riddle." This theme was sustained to the last page: "For though many things spoken of in the course of this work are matters of doubt or of controversy, of this one thing there is no doubt, that if the problem of Species is to be solved at all it must be by the Study of Variation." This was encapsulated in his title—*Materials for the Study of Variation Treated with Especial Regard to Discontinuity in the Origin of Species.*

While organizing his catalog, Bateson realized that for some concepts there were no existing words. Thus, variations in the *number* or *arrangement* of body parts, he deemed "**meristic**" and "**homeotic**," respectively. Variations in the *substances* of which parts were formed, he deemed "**substantive**." A variant organism with an extra, but otherwise normal, finger would be a meristic variant (Greek: *meros* = part). A variant organism with rearrangements of body parts (e.g., an insect's leg where its antenna should be, or a flower's stamen where a petal should be) would be a homeotic variant (Greek: *homeo* = same). In Bateson's words: "The essential phenomenon is not that there has merely been a change, but that something has been changed into the *likeness* of something else." A variant organism with loss of normal coloration (e.g., an albino) would be a substantive variant.

Bateson sought "the *nature* of the series by which forms are evolved," which was the "differentiation" arising between successive terms in an ancestral series leading to today's forms:

> The first questions that we shall seek to answer refer to the manner in which differentiation is introduced in these series. All that we know is the last term of the series [today's form]. By the postulate of Common Descent we take it that the first term differed widely from the last, which nevertheless is its lineal descendent: how then was the transition from the first term to the last effected? If the whole series were before us, should we find this transition had been brought about by very minute and insensible differences between successive terms in the series, or should we find distinct and palpable gaps in the series? In proportion as the transition from term to term is nominal and imperceptible we may speak of the series as being Continuous, while in proportion as there appear in it lacunae, filled in by no transitional form, we may describe it as Discontinuous. … To decide which of these agrees most with the observed phenomena of Variation is the first question which we hope by the Study of Variation to answer.

To understand Bateson's "discontinuity," it is best first to understand continuity. If a road is continuous, then *one process*—walking, running, or riding—should suffice to reach a destination. If a road is discontinuous, perhaps because there is an unbridged river or ravine, then *another process*—wading, swimming, or climbing—must intervene. By the same token, if variation is continuous then, in principle, every possible intermediate between two forms can now exist, and *one process* that has removed intermediates that do not match their environment, suffices to explain the distinctness of the diverse types we see around us. In short, Darwin's natural selection reigns. If variation is discontinuous, then each of two consecutive forms is limited in the extent to which it can vary. Every possible intermediate between the two forms cannot now exist and, furthermore, may never have existed. So, in addition to natural selection, *another process*, that has something to do with the gap between the forms, is needed: "The two classes of phenomena should be recognized as distinct, for there is reason to think that they are distinct essentially, and that though both may occur simultaneously and in conjunction, yet they are manifestations of distinct processes. The attempt to distinguish these two kinds of Variation from each other constitutes one of the chief parts of this study."

He equated discontinuous variation with "organic stability." Thus, "to employ the metaphor which Galton has used so well – and which may prove hereafter to be more than a metaphor – we are concerned with the question of the position of Organic Stability; and in so far as the intermediate forms are not, or have not, been positions of Organic stability, in so far is the variation discontinuous." In his "Concluding Reflections," Bateson summarized:

> Upon the accepted view it is held that the Discontinuity of Species has been brought about by a Natural Selection of particular terms in a continuous series of variations. Of the difficulties besetting this doctrine enough was said in the introductory pages. … For since all the difficulties grew out of the assumption that the course of Variation is continuous, with evidence that Variation may be discontinuous, for the present the course is clear. Such evidence as to certain selected forms of variation has, I submit, been given in these chapters, and … a presumption is created that the Discontinuity of which Species is an expression has its origin not in the environment, nor in any phenomenon of adaptation [to that environment],

> but in the intrinsic nature of organisms themselves, manifested in the original Discontinuity of Variation.

The evidence he was referring to was that an organism, normal in all respects, could produce an otherwise normal offspring except that it had, say, an extra finger, which was a *perfect* finger from the start. Since this occurred in one generation, there was no possibility of the extra digit being gradually formed by a continuous series of successive (generation by generation) small steps. Natural selection could not have been involved. Hence, this type of variation between organisms was discontinuous (quantal): "Meristic Variation in number of parts is often integral, and thus discontinuous." The offspring with the extra digit would be referred as a "variety" or "race" if it were the first of a line of offspring with the extra digit (i.e., if the new character was preserved when the new form was bred with its own type or sometimes even when bred with the normal form).

Today, we know that there are meristic genes controlling finger development and homeotic genes controlling finger location, so an extra digit, or a displaced digit, can result when a mutation occurs in one of these master controlling genes. A meristic gene controlling repetition of parts is an agency comparable to a child controlling the folding of a paper before cutting it to make a series of identical figures ("origami"). Bateson had some sense of this:

> Everyone knows the rows of figures which children cut out from folded paper. There are as many figures as there are folds, each figure being alike if the folds coincide. If the paper is pink, all the figures are pink; if the paper is white, all the figures are white, and so on. If blotting paper is used, and one blot is dropped on the folded edges, the blot appears symmetrically in all the figures. So also any deviation in the line of cutting appears in all the figures; a whole row of soldiers in bearskins may be put into helmets by one stroke of the scissors.

But, in principle, although outcomes differ, we now see that there is *no difference* between the mutational *processes* by which meristic, homeotic, and substantive variants arise. Bateson did not know this. He was following an interesting, but misleading trail with respect to his goal—the origin of species. Decades later, Richard Goldschmidt fell into a similar trap with his doctrine of "hopeful monsters" [15]. Bateson thought that if new varieties and new species both came into existence in one step, then understanding the way one (varieties) arose might help our understanding of how the other (species) arose:

> For if distinct and 'perfect' varieties may come into existence discontinuously, may not the Discontinuity of Species have had a similar origin? If we accept the postulate of Common Descent this expectation is hard to resist. In accepting that postulate it was admitted that the definiteness of Discontinuity of Species depends on the greater permanence or stability of certain terms in the series of Descent. The evidence of Variation suggests that this greater stability depends not on a relation between organism and environment, not, that is to say, on Adaptation, but on the Discontinuity of Variation. It suggests in brief *that the Discontinuity of Species results from the Discontinuity of Variation.*

So he attributed "permanence or stability of certain terms in the series of Descent" (i.e., the preservation of certain newly appearing forms), not to their adaptation in the face of natural selection, but to another *process*, which, with some hand-waving, was referred to as "Discontinuity of Variation." Bateson considered that this latter variational process, like the quantal variations in somatic form and function that people were familiar with in the varieties they saw around them, was "in the intrinsic nature of organisms themselves." But as for the ultimate nature of discontinuity of variation, he was hesitant: "My aim will be rather to describe rather than define the meaning of Continuity as applied to Variation."

For guidance, he turned to Galton's description in *Natural Inheritance* of progression from a monomorphic state, where most individuals were of average disposition regarding a character such as height, to a dimorphic state where there was assortment into two groups (Chap. 3): "The terms Continuous or Discontinuous are applicable to the process of transition from the monomorphic to the dimorphic state according as the steps by which this change was effected are small or large." Galton himself had used as example the rolling of a rough stone, analogous to the dice above. Numerous gentle small steps would suffice for an appreciable change in position of a multifaceted dice, whereas a single large step would be needed to change the face of a conventional six-sided dice. Bateson was not so much thinking of the monomorphic-dimorphic characteristics themselves, but the *process* of their interconversion. In *Materials,* he noted a dimorphism in forceps length among earwigs:

> We are concerned not with the question whether or no all intermediate gradations are possible or have ever existed, but with the wholly different question whether or no the normal form has passed through each of these intermediate conditions. ... Supposing, then, that the 'high' and 'low' [forms of earwig forceps] ... should become segregated into two species – a highly improbable contingency – these two species would have arisen by Variation which is continuous or discontinuous according to the answer which this question may receive.

Claiming to "have no knowledge of the matrimonial arrangements of earwigs," Galton suggested to Bateson (April 22, 1894) that "if they are easily bred in captivity, it seems that an interesting experiment might be made as to whether by selection you can get a pure high or a pure low race of them."

While working on *Materials,* Bateson's travels were curtailed, but not entirely dispensed with. These were a trip to Leyden in August 1891, outings to the opera, and the usual round of chosen, or obligatory, social activities. A letter to Mary (June 23, 1893) indicated attendance at Wagner operas in London with the cryptic comment: "I have still no answer from Mrs. Fields. On Wednesday next, June 28, I shall be stopping with Herringham." And he was not above asking Mary, as he did Anna, to take advantage of her London location during the "long" [summer vacation] to help in his work (July 31, 1893): "Could you manage one day to go to the Zoo and carefully count the number of toe nails on the fore and hind feet of the Indian and African elephants? It will want some care, and if you are not *quite* certain say so. Any one observation of any one foot for absolute certain will be of use." Later postcards indicated the observations were not of use. There was also a commitment to stay with Aunt Honora (his father's sister) at "The Mount" for a week at the end of August (1893), where he hoped that Mary would join him. Following publication of *Materials*, there was a spring trip to Andalusia. He returned with some live beetles that fed on a rare plant and had to appeal to Thiselton-Dyer for samples from Kew Gardens (April 25, 1894).

The Reviews

Materials was published in February 1894. Anna declared (Feb. 12) that my "heart swelled with pride when I saw its handsome proportions." Adam Sedgwick praised the book highly (Feb. 4) but pointed out that embryogenesis was a process subject to variation. To study variations in an embryo's development, it was first necessary to "know its more constant features." Bateson's disparagement of ongoing morphological studies was misplaced. Sedgwick added: "I should also have liked to have seen the words 'formerly Balfour Student' following your name. After all we are only the outcome of what has gone before, and Balfour, who was interposed between us and Darwin, must have had a potent influence in shaping your course." Bateson replied (Feb. 5) affirming his "reverence" for Balfour's name, and lamely claiming his family had talked him out of mentioning it (perhaps indicative of one of his mother's feuds).

In June, Bateson was elected a FRS. Reviews of the book appeared as far afield as the *Sydney Daily Telegraph*, *The American Journal of Science*, *Revue Scientifique*, *Biologisches Centralblatt*, and *Revista de Filosofia Scientifica*. Many reviews were merely descriptive. The biblical journal *The Expositor* noted [16] that "since the publication of the *Origin of Species*, there has scarcely appeared such another monument of individual labour," and the *Manchester Guardian* declared (Sept. 1894): "The work is one of the most brilliant and suggestive which have been added to the biological literature of this country within recent years." Particularly, encouraging must have been the review in *Mind* by Galton [17]. He began by restating his own work on fingerprints (Chap. 3):

> It was shown on ample evidence that they are the most persistent of all the external characters that have yet been noted, and consequently [are] not unimportant in spite of their minute character. … It was also shown that notwithstanding the early appearance of the patterns in foetal life and their apparent importance, they are totally independent of any quality upon which either natural selection or marriage selection can be conceived to depend. … I have failed to observe the slightest correlation between the patterns and any single personal quality whether physical or mental. They are therefore to be looked upon as purely local peculiarities with a slight tendency towards transmission by inheritance. Yet, notwithstanding their immunity from the influence of selection, they fall into three distinct classes, each of which is a true race in the sense in which that word was defined, transitional forms between them being rare and the typical forms being frequent.

Having considered the inutility of certain characters, Galton then turned to the sudden appearance of exceptional mental gifts:

> Can anybody believe that the modern appearance, in a family, of a great musician is other than a sport? [mutation] … No variation can establish itself unless it be of the character of a sport, that is, by a leap from one position of organic stability to another … through '*transilient*' variation. If there be no such leap the variation is … a mere bend or divergence from

> the parent form, towards which the offspring in the next generation will tend to regress; it may therefore be called a '*divergent*' variation. ... The interval leapt over in a transilience may be at least as large as it has been in any hitherto observed instance, and it may be smaller in any lesser degree. Still, whether it has been large or small, a leap has taken place into a new position of stability. I am unable to conceive the possibility of evolutionary progress except by transiliences, for, if they were mere divergences, each subsequent generation would tend to regress backwards towards the typical centre, and the advance that had been made would be temporary and would not be maintained.

He had previously written of "variations proper" or "mere variations," which he had distinguished from "sports." Now, he had introduced the fresh terms "divergent variation," and "transilient variation." As if this had not served to thoroughly confuse his readers, Galton returned to what can be referred to as his "selfish gene" mode which, while comprehensible to modern readers, must have led the Victorians to throw up their hands in despair:

> But what is transilience and what is divergence, physiologically speaking? As we know nothing about the arrangements and movements of the ultimate living units or germs, we can answer only by analogies. The exact answer would require a knowledge of the cause of what, in the nomenclature of Weismann, would be called the architecture of the *id* [structure of the gene], and of which he assumes the existence, but does not attempt to account for. We know that the germ contains the seeds of a vast number of ancestral potentialities, only a very few of which can be simultaneously developed, being to a great extent mutually exclusive. It may therefore be inferred with confidence, that organisation is reached through a succession of struggles for place among competing elements, the successful ones owing their success through position, through superiority in vigour, and so on.

As before, his argument was broadly framed in abstract terms:

> However vague such an explanation may be, it is far from being an inefficient one, for it defines the general character of a process though avowedly incapable of dealing with the details. It applies, moreover, to every theory of heredity which is of a '*particulate*' character; – that is to say, wherever the theory is based on the supposition of a vast number of partly independent biological particles, whose mutual attractions or repulsions, as they successively ripen, result in organisation. Theories that have this general idea for their foundation seem to be the only ones that are in any way defensible, and to all of these the idea of positions of organic stability is applicable. ...It was, therefore, with the utmost pleasure that I read Mr. Bateson's work bearing the happy phrase in its title of 'discontinuous variation,' and rich with many original remarks and not a few trenchant expressions.

However, Peter Chalmers-Mitchell (1864–1945), then a Lecturer in Zoology at Charing Cross Hospital, held that Bateson "overestimates the value" of the discontinuity of species being explained by the discontinuity of variation (*Natural Science*, May 1894):

> The present occurrence of species is absolutely no guide to their original habitats. Temporary geographical isolations of parts of species repeatedly occur, and as repeatedly disappear. ... We have good reason to suppose that groups that have become divergent when separated, have been thrown together again. ... This knowledge may not be, and probably is not, enough to solve the problem of the co-existence of different forms closely allied; but it is enough to give pause before finding co-existence of such species a problem that can be solved only by the assumption of a discontinuous origin.

A more outspoken (anonymous) reviewer for *The Atheneum* (Aug. 18, 1894) found "amazing" Bateson's statement that "the origin of a variety must necessarily be studied first, while the question of the perpetuation of a variety properly forms a distinct subject." The reviewer asked "What then, is congenital variation, if it be not heredity gone wrong? Where can we look for the 'origin of variety' if not in ever-differing combinations of some material carriers of the life-principle? ... Mr. Bateson has the iconoclastic *furor* which befits the leader of a revolt against constituted beliefs; we hope that it may attract followers to his standard, even from the ranks of those poor laboratory homologists from whom his tolerant contempt is so thinly veiled."

The review by H. W. Conn in the American journal *Science* (Jan 4, 1895) wondered "if most of his material does not savour too strongly of abnormal, and, indeed, almost pathological variations, to fairly serve as a basis for a theory of the origin of species." Wallace made the same point in two successive issues of *Fortnightly Review* (March 1895), where he took on not only Bateson, but also those aspects of Galton's work that Bateson had supported:

> Some influential writers are introducing the conception of there being definite positions of *organic stability*, quite independent of utility and therefore of natural selection; and that those positions are often reached by *discontinuous variation*, that is, by spurts or sudden leaps of considerable amounts .... These views have been recently advocated in an important work on variation, which seems likely to have much influence among certain classes of naturalists; and it is because I believe such views to be wholly erroneous and to constitute a backward step in the study of evolution that I take this opportunity of setting forth the reasons for my adverse opinion.

Wallace backed his own authority with that of Charles Darwin:

> Darwin distinguished two classes of variations, which he termed 'individual differences' and 'sports.' The former are small but exceedingly numerous, and the latter large but comparatively rare, and these last are the 'discontinuous variations' of Mr. Bateson. ... Darwin, while always believing that individual differences played the most important part in the origin of species, did not altogether exclude sports or discontinuous variations, but he soon became convinced that these latter were quite unimportant, and that they rarely, if ever, served to originate new species .... Mr. Bateson, however, seems to believe that the exact contrary is the fact, and that sports or discontinuous variations are the all-important, if not the exclusive, means by which the organic world has been modified.

Next, he attacked Bateson's contention that an environment could be continuous:

> The author's main point, that ... specific differences cannot ... have been produced by any action of the environment, because the environment is continuous ... rests wholly upon the obvious fallacies that in each single locality the environment of every species found there is the same, ... [but] the environment as a whole is made up of an unlimited number of sub-environments, each of which alone, or nearly alone, affects a single species, ... The mole and the hedgehog may live together in the same general [geographical] environment, yet their actual environments are very different owing to the different kinds of food, habits, and enemies.

As for the examples, Bateson had used to support his case:

> These can only be classed as malformations or monstrosities which are entirely without any direct bearing on the problem of the 'origin of species.'... All these irregularities and mon-

> strosities are in a high degree disadvantageous, since when subject to free competition with the normal form in a state of nature they *never* survive, even for a few generations. … Rarely in the history of scientific progress has so large a claim been made, and been presented to the world with so much confidence in its being an epoch-making discovery as Mr. Bateson's idea of discontinuous variation corresponding to and explaining the discontinuity of species.

In Wallace's view, natural selection was itself sufficient to confer reproductive isolation between two evolving lines (see Chap. 5).

Typewriters were then coming into use, and unlike those from most other correspondents, a private letter from the wealthy Henry Fairfield Osborn of New York (March 29, 1894) was typed. While generally positive, he could "only partly subscribe to the constructive portion of your work, interesting and forcible as are your arguments and evidences." There was a hand-written letter from marine biologist Frank Cramer of Palo Alto, California (April 10, 1896). Although presented in the form of an enquiry about the feasibility of an experimental approach, it was indirectly critical of the approach Bateson had so far employed, which had yielded no "actual facts" on variations in reproductive capacity:

> It has seemed to me that there is a great need of clear, cold light on the reproductive functions – breeding of species on a large scale and *continuous* observation of the offspring in successive generations. The intricate subject of the sterility of first crosses, and of hybrids, is no nearer being understood than it was when Darwin quit. His own experiments on pigeons, etc., seemed to be directed toward finding out to what extent variation, reversion, etc. were produced by crossing. Nearly all experiments which have come to my notice have had to do with finding out what are the actually existing degrees of sterility and fertility when species are crossed. None worthy of serious consideration have been made, as far as I know, to study the possibilities of producing mutual sterility or the conditions under which it could arise. Doubtless the immense difficulty of securing any results deters men from the experiments. If they were made at all it would seem as if they must be made on domestic varieties, which are notoriously fertile when crossed.

Cramer then turned to possible breeding experiments, again emphasizing the need for large numbers of individuals, and stressing the importance of distinguishing between *sporadic* breeding incapacities involving individuals, and breeding incapacities that were *consistently* observed whenever a cross was attempted between members of particular lines:

> I have no experience of breeding birds or other animals. But I am now so situated … that I could take care of a large number of say, pigeons, and give almost daily careful attention to them. I should feel inclined to undertake the task of keeping 400 or 500 pigeons of two or more varieties and crossing them and giving them every possible opportunity to reveal variation in reproductive capacity both in single varieties and in crosses. I have thought that something might be made out by making mutual crosses by reciprocal pairing of large numbers of individuals of two of the most distinct varieties, carefully studying the individual pairs, their eggs, their rate of breeding, death of embryos, young, etc., and from such data if any differences should appear, select birds for re-pairing [crossing again] in order to determine whether the impotence, if any should arise, is due to individual or varietal causes. By pairing those birds which show any kind of aversion or impotence toward the other variety, with birds of their own variety which have also shown [the] same such aversion or weakness, might it not be possible to learn many interesting things, even though the complete production of mutually sterile varieties were an impossibility or even not expected.

There was here appreciation that success in terms of the production of non-viable or sterile hybrids must be short-lived, since non-viable or sterile organisms, by definition, did not produce further offspring. Thus, the scope of breeding experiments was reduced.

Weldon

Of all the reviews, that which must have hurt the most was Raphael Weldon's. He was the son of Walter Weldon, a journalist and self-taught industrial chemist, who had acquired a modest fortune by developing a process for the manufacture of chlorine. At the age of sixteen, Weldon had gone to University College, London, where he had attended Lankester's zoology lectures. In 1877, with the intention of studying medicine, he switched to King's College, London, but then switched to zoology at Cambridge in 1878, graduating in 1881. Never in robust health, both as an undergraduate and as a graduate he suffered from some form of nervous exhaustion, a state unlikely to have been helped by the successive deaths in the early 1880s of his brother, his mother, and, finally in 1885, his father. Inspired by Francis Balfour, Weldon's early work was in comparative embryology, but, like Bateson, he became increasingly disaffected. Under Galton's influence, in the late 1880s he turned to variation, emphasizing mathematical aspects.

Unlike Bateson, Weldon's life and career became relatively settled at an early stage. After a period at Anton Dohrn's Marine Biological Laboratory at Naples, he gained the approval of Balfour's successor, Adam Sedgwick, and was appointed Lecturer in Invertebrate Morphology in 1884. He married a Girton College graduate, Florence Tebb, in 1883 and, apart from foreign travels and collecting trips (e.g., Bahamas, 1886), was securely based in Cambridge [18]. In 1885, he gave a public lecture on "Adaptation to Surroundings as a Factor in Animal Development" at the Royal Institution in London. Two gentlemen in the audience had arrived early and sat reading "perhaps the two most uncultured papers in London," *The Sporting Times* and *The Bird o' Freedom* [19]. The gentlemen were Samuel Butler and his friend Henry Festing Jones (Chap. 19). Butler later commented that the lecture "was very dull; we thought it would be, but I thought I rather ought to go." Two years later, Butler lectured at the Working Men's College on "The Principle Underlying the Subdivision of the Organic World into Animal and Vegetable." We can be sure that Weldon was not in the audience.

Weldon took leave from Cambridge in 1889–1890 to study at the Marine Biological Laboratory in Plymouth. His first "biometry" paper (1890) in the *Proceedings of the Royal Society* was entitled "The Variations Occurring in Certain Decapod Crustacea" [20]. In a second paper, he took individuals from local races of shrimps and from measurements of various characters ascertained the degrees of correlation following the methods of Galton [21]. The papers were well received, and in 1891, he succeeded Lankester as Jodrell Professor in Zoology at University College and was elected FRS. In London, he began a collaboration with Karl

Pearson, who was soon to publish the first of a series of papers with the general title "Mathematical Contributions to the Theory of Evolution." An informal meeting in December 1893 at the Savile Club with Galton and the chemist Raphael Meldola (1849–1915), led to the establishment of a new RS Committee—The Committee for Conducting Statistical Enquiries into the Measurable Characteristics of Plants and Animals (Chap. 6). The Lankester-chase did not end in London. In 1899, Weldon was to succeed Lankester again, becoming the Linacre Professor in Comparative Anatomy at Oxford. Lankester then moved back to London as Director of Natural History at the British Museum, a position from which he was "retired" by the sorely tried trustees in 1907 at the age of 60.

Weldon's review of *Materials* heightened an antagonism that later broadened into the major dispute between Mendelians and Biometricians. Beatrice excluded from her *Memoir* and *Letters* the early indications of a deteriorating relationship. The two were corresponding amicably, but to his family Bateson wrote (January 1887): "Weldon seems to have done fairly – getting some new beasts, but I greatly fear that he is lost to any work that I think anything of now." Nevertheless, a month later there was a reference to Weldon's generous nature (Feb. 2, 1887): "I have told Weldon to buy me a pipe for which please pay – you will have to ask him about it as there will be difficulty in getting him to 'send in his bill'." In June, Bateson declared that: "I have decided not to go in for the Balfour Studentship. Please don't discuss this with Weldon, or anybody else. I wrote him my feelings about it. He doesn't agree with me and I did not expect he would. Our views of life differ more and more widely as time goes by."

When acknowledging receipt of a copy of *Materials* (Feb. 13, 1894), Weldon mentioned that one of his students had determined the percentage of abnormal forms among over a hundred members of the marine species, *Aurelia ephyrae*. Weldon's studies of relatively *large* numbers of *few* forms (thus facilitating statistical analysis) contrasted with the study of relatively *few* numbers of *many* forms as in *Materials*. Bateson had admitted that "for our purpose we require actual cases of Variations occurring as far as possible in offspring of known parentage; and if, failing this, we make use of cases occurring in the midst of normal individuals of known structure, it must in such cases be always remembered that we cannot properly assume that the varying form is the offspring of such individuals." Indeed, Weldon's own abnormal forms appeared to be "in the midst of normal individuals." On the other hand, the parentage of Bateson's abnormal forms—sometimes isolated specimens in museum bottles—was often far from clear. A more extensive letter from Weldon followed two days later (Feb. 15, 1894). While warm in tone, there were hints at impending reservations:

> I don't want to talk much about your book until I have read it (which I hope to do in the Easter vac.) – But so far I have read the introduction, and some few of the cases about vertebrae – I put selected portions of your sword into the hands of certain selected children as soon as I had read them myself, and we all shouted for joy. I want to tell you simply the frame of mind in which I begin to read the cases. First of all, your *fin de siecle* Arianisms I am ready for: and I accept the ὁμοίουδιον [homoeosis] without a murmur, although page 32 may seem to the irreverent scoffer a travesty of the most solemn enunciation of one problem

> that our language knows, namely that embodied in the opening pages of The Nicene Creed. Let me here remark once for all that if you will write sentences which stick in my memory, I shall have to misquote them on occasion. In all seriousness, Homoeosis seems to me easily to be believed, after simply cutting [the pages of] your book, and remembering one or two of the cases of which you have told me. And of course, if it is to be believed, it is a phenomenon of quite the first order of importance, and worthy of your best epigrams, and of Mr. Clay's largest capitals and blackest 'Clarendon.'

A good start, replete with insider homilies. Weldon then tried to show how he was struggling to understand the variations seen among a group of organisms in terms of the throwing of dice (i.e., a single process):

> At present, I do not quite grasp what you mean by *discontinuity*, by *regression*, and by *oscillation about a position of organic stability*. About *discontinuity* I don't want to bother you with questions, which will possibly answer themselves as I read your book. But I can show you my present frame of mind, about this and about *oscillations etc.*, in a short space, and I should like to do so.
>
> First, Galton's own remarks about '*oscillations etc*' have always seemed to me to be based upon a too exclusive consideration of one aspect of Probability. It is true that, if you shoot at a mark, or if you measure the same thing with a fallible instrument a great many times, your results will cluster themselves about the mark, or about the true measure of the object, with the frequency indicated by the Probability Integral, or, as Galton prefers to phrase it, by the 'Curve of Error.' But so will the results of a game of pitch and toss: and it is hard to see here what is the 'object aimed at,' answering to the mark, or to the true length of the object measured. If you toss pennies, so that the chances of 'head' or of 'tail' are equal, the frequency with which 'runs of luck' occur is given by a symmetrical 'curve of error.' – If you take the more general case, in which the odds are unequal, you get an asymmetrical 'curve of error,' fitted by the same general equation.
>
> I cannot myself conceive, at present, what is the analogue, in the case of an animal, of the mark aimed at. I am inclined to conceive of hereditary causes, at least as acting in the same way as the causes which lead to a given result in throwing dice – abnormality of a parent producing the same sort of result in a single generation as that produced by loading one of the dice. This is one of those vague analogies at which you justly scoff: I only use it to show you how vaguely I think of these things at present. But it will serve to show you what I mean.
>
> If you imagine every brood of animals to be in some way driven towards a definite condition, as a series of bullets from a rifle are driven towards a mark, and if you imagine the differences between the individuals of the same generation to be normally only such as are produced by wind and so on in the case of a bullet, then a change in the general direction of the whole system may well proceed by integral steps, analogous to a change of target by a rifleman on a range. But I cannot myself see that such an analogy fits: and it seems to me purely as a matter for experimental enquiry.

Here, Weldon stated that variation among offspring from the average type (the "mark") might be due to environmental or intrinsic causes (e.g., the wind or a shaky hand), or might be quantal, as if the marksman had changed the target. The latter, in Weldon's mind, related to Bateson's "discontinuity." Weldon seemed to think that one process, tossing a dice, might be a valid way of looking at this without invoking another process—a deliberative change in target:

> Your contention will, I imagine, be that experimentally the evidence shows such sudden changes of 'aims' or of 'positions of stability.' Now that is a point which I have lately had occasion to check with dice. I gave a popular lecture (if you will forgive me for mentioning it!) at the Royal Institution last week, to which I knew that Flower, and Horsley, and some others who might be got to work at variation, might come. I wanted to show the importance of examining large numbers of individuals if the 'Curve of Error' business was to be tested. To make a typical 'Curve of Error' out of a 'chance' result, I made with the help of my wife and a clerk, 26,306 tosses of groups of 12 dice, – recording after every toss the number of dice with 5 or 6 points on the uppermost faces.

In this case, each dice having six faces, then on average one third of the throws should result in a five or six. This would apply whether one threw a single dice twelve times, or a group of twelve dice. On average, four dice out of twelve should show a five or six. Of course, instead of five and six, Weldon could have chosen any two of the numbers from one to six, and still on average four dice would have shown those two numbers. For many throws, four dice landed to show five or six. But quite often three or five dice showed the two numbers. Less often two or six dice showed the two numbers…and so on. Assuming Weldon and his helpers could have thrown, recorded, and collected up the dice in one minute, then, working an eight-hour day, this would have taken them 55 days! Spousal dedication of a high order! They were then able to see to what extent their observations corresponded with the expectations calculated, taking into account that their particular set of dice had a slight imbalance in favor of five and six, as he next explained (see Table 4.1):

> The dice used were not quite true, so that the 'chance' of tossing a 5 or 6 in one throw of one dice was about 0.3377 instead of 0.3333. I enclose the result, compared with that indicated by a 'Curve of Error,' and you will see how closely the two agree. This result approximates to a 'continuous' result; and by using more dice the approximation could have been made closer. You see the regularity with which the frequency of any 'abnormal'

Table 4.1 Weldon's "Table showing the frequency with which a given number of dice, out of a group of 12, turned 5 or 6 points, in 26,306 tosses of 12 dice in each toss"

Number of dice with 5 or 6 points in each toss	Frequency with which this result occurred	Theoretical frequency of the result
0	185	187
1	1149	1147
2	3265	3209
3	5475	5465
4	6114	6263
5	5194	5114
6	3067	3038
7	1331	1332
8	403	426
9	105	101
10	14	16
11	4	1
12	0	0

> result diminishes as the difference from the 'normal' result of 4 dice with the required number of points per throw, increases.

This "continuous" result is plotted in Fig. 4.2a. Weldon next achieved the semblance of a "discontinuous" result by considering a *smaller* number of tosses, where in addition to the main peak of the distribution at 4, there was *sometimes* a small sub-peak. Figure 4.2b shows this at 11, separated from 9 by zero values at 10.

> Now these results were recorded on different sheets of paper, so that I was able to compare the result of measuring some 1000 or 1200 animals with the result of tossing the same

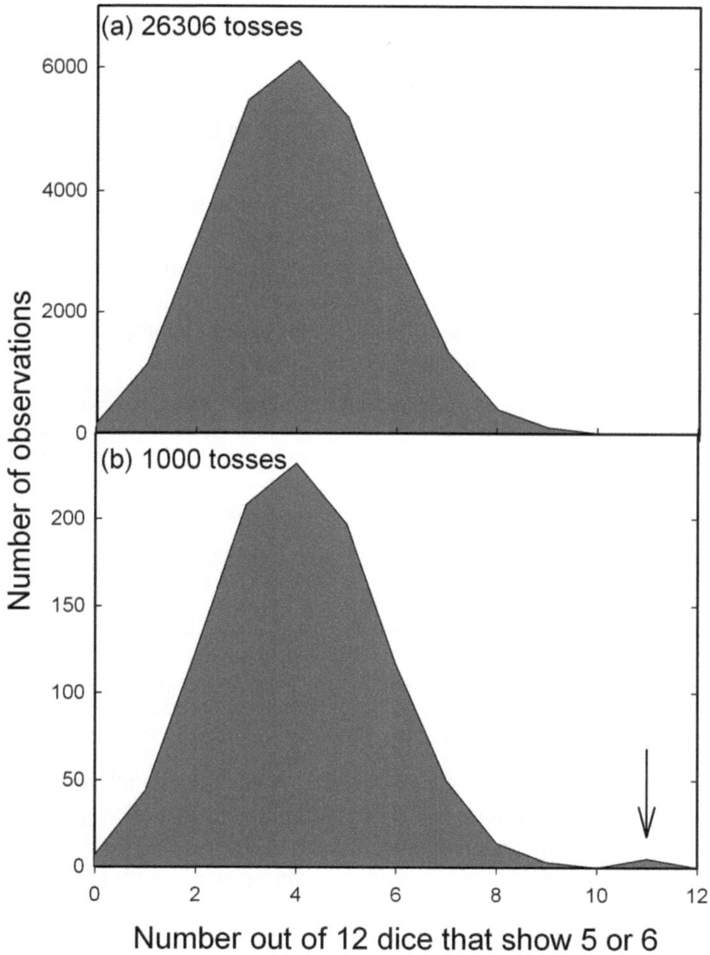

Fig. 4.2 Plot of Weldon's 1894 data from dice tossing as shown in Table 4.1. In (**a**), data for all 26,306 tosses are plotted to reveal a "continuous" distribution (bell curve). In (**b**), there is a simulation of Weldon's occasional result where, in zero tosses out of a thousand, ten dice show 5 or 6, and, in five tosses out of a thousand, eleven dice show 5 or 6. Thus, a "discontinuity" can appear (arrow) as part of what may seem to be one process (dice tossing)

> number of dice. For example, I have such a group in which the result *Eleven* dice with 5 or 6 points occurs, while *ten* dice [with 5 or 6 points] do not occur. Here is a discontinuity of a small kind.

Apparently unaware of the earlier work of Joseph Delboeuf (1831–1896) in Belgium [22], and of John Gulick (Chap. 5), Weldon was close to the concept referred to as "random drift," that was to be given much attention by Sewall Wright in the twentieth century [23]. But Weldon did not see a parallel between what he *himself* had done – selected a small group (1000 tosses) – and what "Nature" might do in the real world (e.g., maroon a small group on an island). Weldon continued:

> Frequently, in small groups of the kind (for this purpose 1000 is a small group) the extreme result occurs more often than it should. The question arises, therefore, how many individual cases you must examine, before you may assert that an extreme deviation occurs more often than it should, in a system which is supposed to be 'oscillating about a mean.' My dice results 'oscillate' about the mean result that when 12 dice are tossed 4 of them turn up either 5 or 6 points. – I have not obtained a single toss in which all 12 dice turned up either 5 or 6 points: the odds against doing so are about 540,000 to one.

Weldon was here treating "discontinuity" in terms of a chance event about a norm, determined by his hypothetical marksman who *all along* had been aiming at 4. He was not interpreting "discontinuity" in terms of a marksman changing the target away from 4, which was more in keeping with Bateson's thinking. From under this statistical smokescreen, Weldon admonished Bateson on the importance of having a large sample as a basis for comparison, a point of which Bateson was aware.

> Now in all such cases, when an *extremely* unlikely event happens, it is recorded [by the investigator]. When a *slightly* unlikely event happens, it is not as a rule recorded. In this way, your examination of 1700 medusae weighs more with me, as an example of your thesis of discontinuity, than the fact that a frog is recorded with a certain well-marked deviation of its vertebral column, and that intermediate cases are not recorded.

Again, this was not new to Bateson. In the Introduction to *Materials,* he had written:

> On the hypothesis of Common Descent, the forms of living things are succeeding each other, passing across the stage of the earth in a constant procession.... To study Variation it must be seen at the moment of its beginning. For comparison we require the parent and the varying offspring together. To find out the nature of the progression we require, simultaneously, at least two consecutive terms of the progression. Evidence of this kind can be obtained in no other way than by the study of actual and contemporary cases of Variation.

Weldon understood that Bateson believed that homeotic mutations—often major changes within one generation—might be examples of the marksman's discontinuity, but he called for evidence on the mechanism of this before he could agree it was not a chance fluctuation that was part of a normal process with random elements. Weldon continued:

> You may be getting bored by this time, so I will leave 'regression,' about which I want to say a great deal. Please understand, that I only want to show you how I feel *before* reading your book. I am quite prepared already to believe that the *direction* of such a variation as that of the frog's vertebral column, or of the excellent lady No. 18 [an insider comment], is determined by *homeosis*: but I am not yet clear as to how much evidence is required before I give up a prejudice in favour of the view that the *amount* of it is determined by 'chance' – i.e., by the same law as that determining the stature of a 'normal' population.

He concluded with a word on regression to the mean in the context of characters that do not blend in offspring, and his own belief that the way ahead lay through mathematics:

> About 'regression,' I will only say this, that Galton was himself a good deal mixed, at least in his exposition, when he wrote 'Natural Inheritance': and that I cannot conceive that characters 'which do not mix' are thereby rendered independent of the phenomenon of regression. One reason why I have done so little work lately is that I have set myself to learn enough mathematics to read Probability seriously: and I think I understand 'regression,' so far as normal variations are concerned. Can you come here some Saturday during the next few weeks, and stay over Sunday? I am myself in travail with a book, though of far less magnitude than yours: but I should very much like to have a long talk about the possible connection between such 'normal' systems of variations as those which I have been examining, and your things.

In a later letter (March 1), Weldon brought forward stronger and more specific criticisms of the chapter on teeth, but this can hardly have prepared Bateson for the generally unfavorable tone of Weldon's review in *Nature* [24]. The review began by praising Bateson's cataloguing, but then turned to Bateson's analysis: "If the criticism and enunciation of opinions had been performed with the same care as the collection of facts, the commentary which runs through the book would have gained in value, and several inaccuracies, due partly to want of acquaintance with the history of the subject, would have been avoided." Weldon noted that Bateson had, but only in a footnote, drawn attention to the fact that the environment of an organism includes other organisms that can interact with the organism under study. Since these other organisms are discrete, they themselves constitute an environmental discontinuity. In this respect, the environment did not have the continuity that was fundamental to Bateson's argument. Weldon thought Bateson's treatment of this aspect dismissive. Furthermore, although Bateson had stated that he was going "to describe rather than define" discontinuity of variation, Weldon was dissatisfied again. Given the emphasis placed on discontinuity of variation, there should have been an unambiguous definition. Finally, Weldon repeated the point that he had made in his earlier letter. Museum specimens had been much employed by Bateson, but museum curators tended to "regard a slight abnormality as not worth bottle, spirit, and a place on the shelf." Thus, Bateson's survey was intrinsically biased against organisms with small variations:

> The only way in which the question [of whether continuous or discontinuous] can be settled for a given variation seems to be by taking large numbers of animals, in which the variation is known to occur, at random, and making a careful examination and record of each. ... A careful histological account of the jaws of five hundred dogs would have done more to show the least possible size of a tooth in dogs than all the information so painfully collected. And so in many other cases.

This was but the first round in an ongoing battle. They both were convinced of the need to study variation—but differed as to the best way to proceed. Both saw the need for increased quantitation of biological data, but Weldon wished to take the quantitative approach much deeper. This had its dangers. However splendid a mathematical analysis, it had to be viewed in a relevant context, which narrow specialization could obscure. We will return to this in the pages ahead.

Bateson and Weldon thought of themselves as "naturalists," not as physiologists. In *Materials,* there was little on the relation of gonads and gametes to reproductive success or failure. Neither Bateson nor Weldon considered that the "discontinuity of which species is an expression" might be based on a type of variation that would create a barrier which would itself define species to the extent that two groups of organisms would be reproductively isolated from each other because of that barrier. The term "reproductive isolation" was one they seldom used. For this, we must turn to Romanes and his ally Gulick.

Summary

The stigma of a second-class degree may have driven George Romanes and Anna Bateson from Cambridge—Romanes, after a period working in Foster's physiology department, to self-sponsored research as Charles Darwin's research associate, and Anna, after a period working with Francis Darwin, back to the "trade" mode of her ancestors. Yet, for a while the achievements of his sisters—Anna in botany, and Mary in history—surpassed those of Bateson himself. While, with the support of Sedgwick and Lankester, Weldon went from strength to strength, Bateson attained only the stewardship of his college kitchens. In 1894, in *Materials for the Study of Variation,* he distinguished variations in the number or arrangement of body parts, from variations in the substances from which those parts were made, but he could not imagine that common mutational processes were involved. Whereas under Darwin's natural selection theory progressive changes over many generations (i.e., many steps) were required to generate new and perfect types, such perfect types (e.g., a type with one perfect extra finger) could actually appear in one generation (one step). Appearance in one generation was a "discontinuity of variation," perhaps relevant to the question of the "discontinuity of species"—how do organisms become divided into discrete groups (species) which share common characters and vary about the mean type? Critics such as Wallace and Weldon dismissed Bateson's book as a mere catalog of monstrosities of little biological relevance. However, Bateson's point was that variation, be it productive of a new variety or of a new species, was something intrinsic to organisms themselves. Furthermore, each generation a new organism must be made afresh—the parental "wax model" must be returned to the melting pot. But this was not seen in informational terms. While embracing Galton's "organic stability," at that time Bateson did not relate this to "reproductive isolation."

Chapter 5
Romanes

> *'We accept natural selection,' ... [Weismann] says, 'because we must—because it is the only possible explanation that we can conceive.' As a politician, I know that argument very well. ... But such a line of reasoning is utterly out of place in science. We are under no obligation to find a theory if the facts will not provide a sound one.*
>
> *Marquis of Salisbury (1894)*

In the 1870s and 1880s, Romanes, and an American missionary in Hawaii, John Gulick, independently questioned Darwin's emphasis on natural selection. As set out in my earlier book [1], they began corresponding with each other in the 1880s and formed an alliance against the "ultra-Darwinians"—Alfred Wallace and August Weismann. The issue was not so much that adaptation to environment as required by Darwin did not seem to be universally present, but that evolutionary mechanisms other than natural selection needed to be identified. The existence of such mechanisms was postulated in 1886 by Romanes in his theory of "physiological selection" [2] and was explored in detail in his final work—volume 3 of *Darwin, And After Darwin*, which was published posthumously in 1897 by his friend Conway Lloyd Morgan with input from Galton [3].

Panmixia—Reproduction Unfettered

Physiologists who wished to understand the function of something eliminated it and observed the result. To know what a nerve was doing, they cut it and then saw if and how the unfortunate creature was impaired. This was not popular with the animal rights activists, against whom physiologists came to battle. Applying this approach to investigate natural selection, Romanes asked, as had Darwin before him, what

A. G. Cock, D. R. Forsdyke, *Treasure Your Exceptions*,
https://doi.org/10.1007/978-3-030-92099-9_5

would happen if natural selection were eliminated? The question was easier to ask than to answer.

Without natural selection, all organisms of a particular species that were born at a particular time, even though varying from each other, would be expected to grow to adulthood. For any particular character, the distribution in the population at birth (the distribution about the "birth-mean") might equal the distribution in the population in adult life (the distribution about the "survival-mean"). For example, if progressive melanization (blackening) were not a feature of moth maturation, the distribution of melanization among young and adult moths would be the same. Without natural selection, all moths, whatever their degrees of melanization, would have an equal opportunity of reproducing with others. Cessation of selection would allow unfettered reproduction ("panmixia"). Each moth would be able to pass on its degree of melanization to the next generation.

On the other hand, continuation of selection, by eliminating certain organisms, would fetter reproduction, since those that had not survived to adulthood would have been eliminated as potential mates. Thus, for characters that might influence survival, the birth mean might not equal the survival mean. If pale moths were selectively eaten by birds, then the distribution of melanization in adults would be different from the distribution in the newborn. There would be fewer opportunities for poorly melanized moths to pass on their paleness to the next generation. The moth population would become progressively darker. Characters that are expected to change during development, such as secondary sexual characters, would not fit this scheme.

How was natural selection to be decreased or eliminated? The domestication of animals and plants in protected fields or greenhouses might achieve this. And "experiments of Nature" were also available. Among these, organs deemed degenerate or "vestigial" appeared as a possible result of *cessation of selection*. Eyes are often degenerate in species that now live in dark caves, but whose ancestors once lived in the open. Hindlimbs degenerate in mammals that now live in the sea, but whose ancestors once lived on land (e.g., whales). However, sometimes these are explained by the continuation of natural selection in different form. Fish need to be stream-lined for swimming. Fish retaining projecting limbs would be selected against. In this case, to the extent that the original evolution of limbs was due to selection, the degeneration of limbs, while still due to selection, could be referred to as due to "counter-selection," rather than to "cessation of selection." Similarly, organisms with eyes whose lids are sealed are likely to get less eye infections. When eyes are not needed, this selective influence trumps open lids. Thus, the onus was on those claiming a cessation of selection to disprove that counter-selection was not operating.

Another factor was what Darwin called "economy of growth." Whatever its function, an organ draws upon the metabolic resources of its body. Supposing the supply of nutrients to be limited, a decrease in size of a superfluous organ would decrease its metabolic demand, leaving more nutrients available for functional organs.

Among members of a population, there would be variants where the size of a superfluous organ had increased, and variants where the size of the organ had decreased. While such types might appear initially in equal numbers, because of "economy of growth" those with larger organs would be at a selective disadvantage and would decrease in number.

The economy achieved might not directly relate to the organ under consideration and might not necessarily relate to nutritional needs. For example, if it were for some reason superfluous, a decrease in size of a liver would decrease pressure on weight-bearing organs—skeleton and muscles—which could then decrease proportionately. A negative variation in liver size could be advantageous in an organism with correlated negative muscular or skeletal variations. In the absence of the negative liver variation, organisms with negative musculo-skeletal variations would be selected against. Whether due to competition for nutrients or otherwise, as noted by Bateson (Chap. 2), the fact that an organism operates as a functional whole implies that a variation in organ A may be accompanied by correlated variations in organs B, C, and D. And since each of the latter influences other organs, an initial variation can be seen as initiating a wave of variations that can affect the entire organism. Thus, when D varies, there are correlated variations in organs E, F, and G.

Where natural selection could not easily be invoked to explain a degenerate organ, those of a strong Lamarckist persuasion (or of a moderate Lamarckist persuasion, as were Darwin and Romanes) would invoke the inheritance/disinheritance of acquired characters (acquired or lost through "use or disuse"). So, it was argued that the first organisms of a species that inhabited dark caves had suffered partial degeneration of their eyes "through disuse," and this trait was passed on to their offspring, perhaps as part of the process Darwin referred to as pangenesis. Thus, understanding the degeneration (atrophy) accompanying cessation of selection might be highly pertinent to the long-standing issue as to what extent, if any, acquired characters could be passed on to immediate offspring.

In 1874, Romanes saw that "whether or not disuse is the principle *cause* of atrophy in species, there is no doubt that atrophy accompanies disuse." Just as *natural selection* was regarded as an agency, or principle, bringing about changes that make organisms better match their environments, so Romanes regarded *cessation of selection* (panmixia) as an agency, or principle, bringing about changes that make organisms better match their environments [4]. Like natural selection, "the cessation of selection depends upon variations being supplied to it; and so, if from any reason a specific type does not vary, this principle cannot act." Here, we see the glimmer of an idea Romanes was later to develop—variation, whatever its cause, must *precede* selection. But the confounding factors alluded to above did not, as Bateson had correctly implied (Chap. 3), allow that "flower of speech," panmixia, to lead to much enlightenment. In this case, physiological thinking was fine in principle but not in practice (Fig. 5.1).

Fig. 5.1 George John Romanes

An Unnoticed Factor

The University Correspondence College offered a way to defeat labyrinthine Victorian examination systems. Described by H. G. Wells, as "one of the queerest outgrowths of the disorderly educational fermentations of that time," he noted that it's "list of tutors displayed an almost unbroken front of Cambridge, Oxford and London 'firsts'" [5]. Among these were Wells, a future writer of science fiction, and Edmund Catchpool, a physicist.

In 1884, a letter from Catchpool on "An Unnoticed Factor in Evolution" was published in *Nature* [6]. It had long been recognized that members of closely related biological species, usually distinguished from each other on anatomical (morphological) grounds, were best distinguished *as species* on physiological grounds—they were unable to interbreed to the extent of producing fertile offspring. Catchpool proposed that the inability of allied species to reproduce with each other was not the *result* of their divergence from a common ancestor into independent species as Darwin had supposed, but was the facilitating *cause* of that divergence (Fig. 5.2).

Apparently, the letter went unnoticed, even by the 36-year-old Romanes who was to spend the last decade of a sadly abbreviated life making the same case [7]. Catchpool's letter began with two facts admitted as "great difficulties" by evolutionists. The first was that which had most concerned Huxley:

> The fact that varieties produced by artificial selection, however divergent, are always fertile among themselves, while species supposed to have been produced naturally by an analogous process are often not mutually fertile even when slightly divergent. ... A period much greater than that of artificial selection should be necessary to produce sterility between

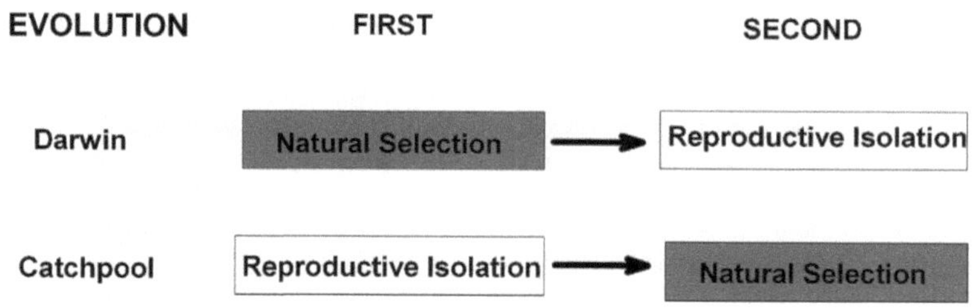

Fig. 5.2 In the general case, for the origin of a species by branching from an ancestral species, reproductive isolation must *precede* natural selection, not the converse. "Spontaneous variations in the generative elements" permit the branching into incipient and then independent species by random drift and/or a natural selection allowed free play unfettered by blending effects

> descendents from the same ancestor; a supposition which would require an almost incredible period for evolution as a whole.

This was something most people could recognize. Artificial selection over relatively short evolutionary periods had produced varieties of dogs that showed extreme anatomical differences yet were still able to reproduce with each other to give mongrel offspring. Catchpool's second fact was more subtle. This was the "blending" difficulty noted by Jenkin (Chap. 3).

> The fact that species evidently derived from a common ancestor, and differing only in small points of marking, though not fertile with one another, are often found side by side in places where it would seem that cross-breeding must prevent any division of the ancestral species into divergent branches, … seems to require that many species now intermixed should once have been geographically separated, sometimes in cases where this is very difficult to imagine.

To overcome these difficulties, Catchpool first pointed to the observable variations between individuals that most people were familiar with. Sometimes these variations were small, such as small differences in eye color, sometimes the variations were large, such as dwarfism or an extra finger. If variation could affect parts of the body which we can directly observe, then it might also affect parts of the body which we cannot directly observe, such as the gonads (testes and ovary). Catchpool postulated that among variations affecting the gonads, some would result in *selective* infertility (sterility) with most other members of the species (i.e., reproductive isolation from those members), but not with a few that had the same type of variation. The fact that a variation had occurred would only be detected if the selective sterility was detected. Until manifest, the variations would be latent rather than patent. Catchpool did not specify whether the variations would be (a) large and so likely to be immediately manifest, or (b) small and cumulative up to a threshold when suddenly manifest, or (c) small and cumulative leading to progressive degrees of sterility. Thus, the variations might accumulate in the germ line from generation to generation until circumstances dictated that sterility be manifest:

> These difficulties are completely removed if we suppose mutual sterility to be not the *result*, but the *cause*, of divergence. As far as can be judged, 'sports' [by this Catchpool meant variations in general, not necessarily large variations] are as likely to occur in the generative

> elements (ova and spermatozoa) as in other parts of the body, and from their similarity in widely unlike groups it seems certain that a very slight variation in these elements would render their owner infertile with the rest of its species. Such a variation occurring in a small group (say the offspring of one pair) would render them as completely separate from the rest of their species as they would be on an island, and divergence (as Wallace has sufficiently shown) would begin. This divergence might progress to a great or a small extent, or even be imperceptible, but in any case the new species would be infertile with the species it sprang from. If this theory be admitted, we must distinguish between varieties and species by saying that the former arise by spontaneous variations in various parts of the body, and only gradually become mutually infertile (thus becoming species), while the latter arise sometimes in this way, but sometimes by spontaneous variations in the generative elements, and are in this case originally mutually infertile, but only gradually become otherwise divergent.

Catchpool made two predictions. First, "we ought to find [pairs of] species (incipient) mutually infertile, but not otherwise distinguishable." Such "sibling species" have since been described. Second, "we ought to find that island and other isolated species, which have arisen, not by limited fertility, but by geographical instead of physiological separation, are often mutually fertile even when as widely divergent as the artificial varieties of dogs and pigeons." Here, Catchpool supposed that migrant members of a species separated from the main species, say on an island, would usually vary from the mainland species because of "spontaneous variations in various parts of the body," but not because of "spontaneous variations in the generative elements." Thus, if the island types were returned to the mainland, they would still be able to reproduce with the mainland types (just as a poodle can reproduce with a greyhound, and a fantail pigeon can reproduce with a pouter pigeon).

Branching

Fundamental to Catchpool's thinking was divergent or branching evolution. This is not so simple as it appears. It is best approached through consideration of unbranching, linear, evolution. In Fig. 5.3a, the linear vertical arrow shows the evolution of form B from ancestral form A. Sampling at various points down the arrow would reveal intermediate forms that, in the general case, had come to appear, and/or function, less like A and more like B. Paleontologists sometimes find such forms in the fossil record. Indeed, from such fossils, it is sometimes possible to isolate fragments of DNA, which can be sequenced. Through careful superimposition of the base sequences of neighboring fragments, it is then possible to reconstruct the entire DNA "blueprint" of the organism as it had once existed. As portrayed by Crichton in *Jurassic Park* [8], it then becomes theoretically possible to reconstruct the entire organism. This raises the question, if we were to reconstruct ancient form A, would it be able to reproduce with modern form B? In other words, having eliminated the temporal and any geographical barrier between the two forms (by placing them together in our "Jurassic Zoo") would they produce offspring? Almost certainly not. This would be due to some disparity in their genes that would prevent effective cooperation.

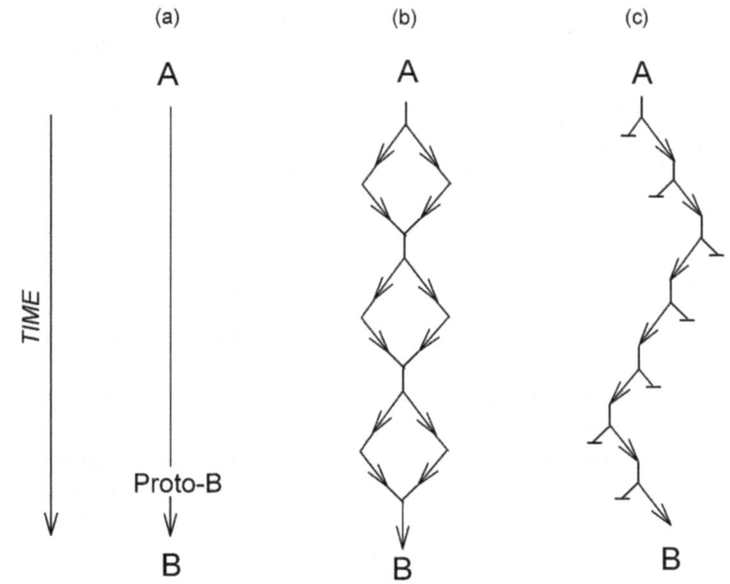

Fig. 5.3 Linear, non-branching, evolution from ancestral form A to modern form B. (**a**) Through time, intermediate forms ("proto-B") would become less like A in anatomy and physiology and more like B, due to changes in genes. Branching into varieties, due to changes in genes, would be suppressed either due to blending effects (**b**), or to natural selection eliminating one variety (**c**)

Anatomical and physiological features (including mental states), and gametic characteristics, are all affected by genes. Even if not emotionally disparate, two forms might differ anatomically or physiologically to such an extent that copulation would not be possible. And even if copulation were possible, gametic characteristics might have differed so preventing their fusion to form the zygote (the fertilized egg). The two forms would be reproductively incompatible because of a failure, at some level, in the transmission of gametes. There would be a *transmission* barrier between them.

But what if we were to reconstruct, not the most ancient form (A), but some intermediate form closer to B (let's call it "proto-B") that would correspond to a time point on the shaft of the arrow? At some point, as we move progressively down the arrow, we would arrive at a proto-B form that was so anatomically and/or functionally like B that the transmission barrier would be overcome. The two forms would then be able to copulate to generate a zygote. However, almost certainly, even if able to complete some cell divisions, that zygote would not be able to develop into an embryo and thence into a mature adult organism (hybrid inviability). In other words, the transmission barrier would now have been replaced by a *developmental* barrier, due to disparity between the parental genes required for development. The genes would be unable to cooperate.

Nevertheless, moving progressively down the arrow, we would eventually arrive at a form of proto-B that was so close to B that the developmental barrier would fall.

No longer would the two forms be "reproductively isolated" in the sense of not being able to produce a child. But would they still be reproductively isolated in the evolutionary sense—namely, would the seemingly healthy child be able to continue the line by reproducing with a B, or with another proto-B, of the opposite sex. Almost certainly yes. And, almost certainly the offspring would be fertile. The developmental barrier would *not* have been superseded by a *sterility* barrier (hybrid sterility).

If, unlike Darwin who remained vague in his distinction between species and varieties, we define a species as a group of organisms that are fully reproductively compatible (i.e., there are no barriers), then we should note that, although we may think of the vertical arrow (Fig. 5.3a) as illustrating progressive *within species* evolution, in fact it is evolution within a *line* of organisms that progresses from species to species. It just happens we have never been able to experimentally cross any A or proto-B with a B to demonstrate their reproductive incompatibility.

The only illustration in Darwin's great book (1859) was the branching from early forms to produce a multiplicity of modern forms. How does the linear form of evolution (Fig. 5.3a) relate to Darwin's branching evolution? Before addressing this, we should first note that the vertical arrow is better drawn as in Fig. 5.3b to demonstrate the short-lived branching that we would expect to have occurred. From time-to-time form A and its descendants would have differentiated into two contemporaneous *varieties* of approximately equal fitness, but the members of one variety would be reproductively compatible with members of the other variety, so there would have been a tendency for their characters to blend, thus eliminating their differentiation into the two varieties (Chap. 3). Alternatively, before blending could have occurred, classical Darwinian natural selection could have operated so favoring members of one variety ("fit") over the other ("less fit"); again, this would result in the essentially linear pattern of evolution (Fig. 5.3c). All that would be needed for this mode of evolution would be the classical Darwinian triad—variation, natural selection, and inheritance.

In branching evolution as proposed by Catchpool, types of approximately equal fitness would initially be produced since there would be no phenotypic differences for natural selection to act upon. To separate the "fit" from the "fit," there would need to be a barrier to reproduction to prevent blending. *Another process would be required.* The triad—variation, natural selection, and inheritance—would have to become a tetrad—variation, reproductive isolation, natural selection, and inheritance. Any shape or form of barrier would suffice for reproductive isolation, but, in the general case Catchpool supposed a sterility barrier.

At the bottom of the inverted Y in Fig. 5.4 are two modern forms B and C that arose from ancestral form A. The two forms are reproductively isolated by virtue of differences in genes that confer psychological, anatomical, and physiological differences. However, as we move back through time, up along the limbs of the inverted Y, one by one, these genic differences would disappear. Eventually, a stage would be reached where copulation, and even fertilization to produce a zygote, could occur between a proto-B form and a proto-C form. The transmission barrier would have fallen. Nevertheless, the parental gene products would not be able to cooperate

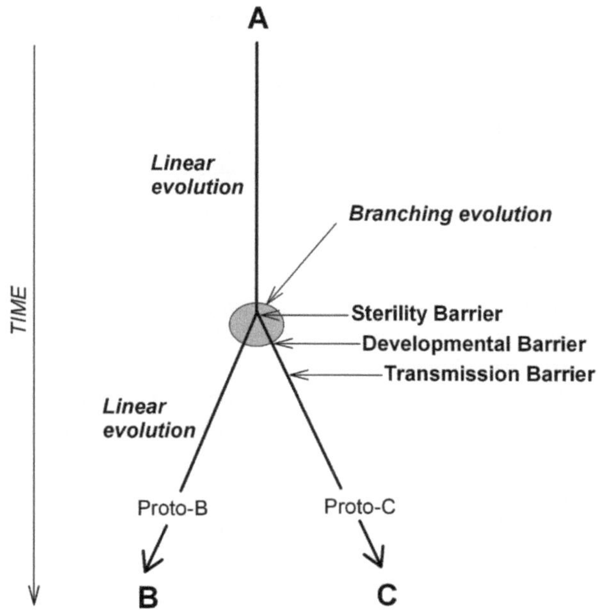

Fig. 5.4 Branching evolution from ancestral form A into modern forms B and C requires reproductive isolation. This is achieved by three sequential barriers—the sterility barrier, the developmental barrier, and the transmission barrier. These reinforce and replace each other as the two forms (proto-B and proto-C) diverge

successfully in the embryo and development would fail. The hybrid embryo would be inviable. Moving closer to the fork of the Y, the parental genes required for development would progressively loose their differences (become able to work together) and the developmental barrier would fall. If two contemporaneous forms (proto-B and proto-C) were then crossed, a child would be born that could grow into an adult. Only one barrier would then remain—the *first* barrier that had arisen between the two forms—the sterility barrier.

Physiological Selection

Romanes made essentially the same points as Catchpool in his theory of the origin of species by "physiological selection" presented to the Linnean Society in May 1886 [2]. Romanes had wealth, leisure, and most important of all, the tutelage of Charles Darwin from 1874 to 1882. Romanes saw that Darwin's natural selection was *itself* an isolating agent of major importance. Natural selection, by culling off the disadvantaged, worked by isolating the advantaged from the disadvantaged, thus preventing panmixia. Natural selection isolated the *fit* from the *unfit*. This was the basis of linear evolution—the modification of forms to better reproduce their own kind *without*, at least in the short term, a transformation into a new species (Fig. 5.3c). For a species to branch into two, there would have to be some mechanism for isolating the *fit* from the *fit*. This could be brought about by geographical isolation so that members of one group of fit organisms would be unable to reproduce with members of another group of fit organisms. There would then be random differentiation of the fit in one location from the fit in another location due to Catchpool's "spontaneous variations in various parts of the body." Despite an absence of "spontaneous variations in the generative elements," the differentiation could eventually attain an extent such that the two types might no longer be able to reproduce with each other when and if the geographic barrier were lifted. For example, one type might have grown larger, so making copulation difficult with the smaller type (transmission barrier).

As we saw with our Jurassic Park example, the *fit* can also be differentiated from the *fit*, so impairing reproduction, when there is a time barrier. Another example on a shorter time scale would be that a "spontaneous variation in various parts of the body" of a plant might lead it to flower earlier than most members of the main species. Such a type would only exchange pollen with another early flowering type. So early flowering forms would be reproductively isolated from later flowering forms by a temporal barrier. At an early stage, pollen kept from the early flowering form would still be fully capable of fertilizing flowers of late flowering forms. But over time, incompatibilities between early pollen and late ovules would randomly arise, and this mutual incompatibility would eventually define the two forms as distinct species.

However, Catchpool and Romanes had in mind a more fundamental type of isolation of the *fit* from the *fit*, which was not primarily based on geographical or temporal differences. This type of isolation was an abstraction for which there was no known biochemical or cytological basis. All they could postulate was that it reflected a primary variation that was *intrinsic* to each organism. In Romanes' words [2]:

> In many cases, no doubt, this particular … variation, has been caused by the season of flowering or of pairing having been either advanced or retarded in a section of a species, or to sundry other influences of an *extrinsic* kind; but probably in a *still greater number of cases* it has been due to what I have called *intrinsic* causes, or to the 'spontaneous' variability of the reproductive system itself. (My italics)

Romanes had acquired the nickname "The Philosopher," and indicating his approval of this, he sometimes signed his more intimate correspondence in this way. However, "philosophical thinking" tended to be disparaged by the hard-nosed "men of science," and the name cannot have helped his cause. Anticipating some criticism, Romanes concluded his Linnean Society address by conceding that the physiological basis of his abstraction was unknown [2]:

> My suggested explanation of the origin of species opens up another and a more ultimate problem—namely, granting that species have originated in the way supposed, what have been the causes of the particular kind of variation in the reproductive system which the theory requires? This, of course, is a perfectly intelligible question, and one that must immediately suggest itself to the mind: my failure to meet it is therefore apt to give rise to the impression that my theory is imperfect. But … this question is really not one with which

> the theory of physiological selection can properly be regarded as having anything to do. This theory has only to take the facts of variation in general as granted, and then to construct out of them its suggested explanation of the origin of species. No doubt it would be most interesting to discover the causes of every variation that constitutes the beginning of a new specific character; but our inability to do this does not invalidate the theory of physiological selection, any more than it does the theory of natural selection. ... It is enough for the explanation which is furnished by Mr. Darwin's theory of the evolution of adaptive structures by natural selection, that the variations in question take place; and similarly as to the present theory of the evolution of species by physiological selection.

In August 1886, Romanes was hailed by *The Times* as "the biological investigator upon whom in England the mantle of Mr. Darwin has most conspicuously descended." Nevertheless, the theory drew a storm of criticism from Francis Darwin, Lankester, Wallace, Huxley, Thiselton-Dyer, and others. The letter in *Nature* by Catchpool was brought to Romanes' attention, and he promptly acknowledged Catchpool's priority, to which Catchpool responded in *Nature* with modest qualifications. Despite their negativity, some critics provided useful guidance regarding terminology. Thiselton-Dyer attacked physiological selection in his Presidential Address to the BA in 1888 [9], and there followed an extensive series of letters in *Nature* [1]. He used the term "reproductive isolation," which has since come into wide usage:

> In point of fact, what Mr. Romanes calls physiological selection may be more accurately described as reproductive isolation. He supposes that individuals of a particular species arise which, from some cause or other, are incapable of breeding with other conspecific individuals. They are therefore in one aspect isolated as if they were on an oceanic island. This being so, any casual variations which they exhibit will be perpetuated, he thinks, whether adaptive or not. And in this way he also thinks that species distinguished by non-adaptive characters have arisen. The idea is interesting ... however ... I have arrived at the conclusion that it is not a principle of very much value.

Wallace in his criticism introduced the term "physiological complements" for two individuals who had become reproductively isolated from most members of the main species but would produce fertile offspring when mutually crossed (i.e., they complemented in not being reproductively isolated from each other).

For present purposes, the physiological basis of Romanes' "spontaneous variability of the reproductive system" can be referred to as something determining an organism's reproductive "accent." This conveys a sense of compatibility or incompatibility. Just as when accents are compatible two humans can communicate and hence are more likely to reproduce together, so when accents are incompatible (i.e., there is a linguistic barrier), communication, and hence reproductive potential, is impaired. The probable biochemical basis of physiological selection—the "accent of DNA"—was the subject of earlier books and is outlined here in the Part VI.

The distribution of some quantitative measure of accent (say, the extent to which "h's" are dropped, as in the cockney accent) is shown in Fig. 5.5. The bell curve shows that most individuals are of accent type B. They are the majority type and therefore are most likely to encounter each other. So, B-with-B reproductive interactions to produce fertile B-type offspring are the most frequent. Much rarer are individuals of type A on one limb of the bell curve. It is most likely that A types will

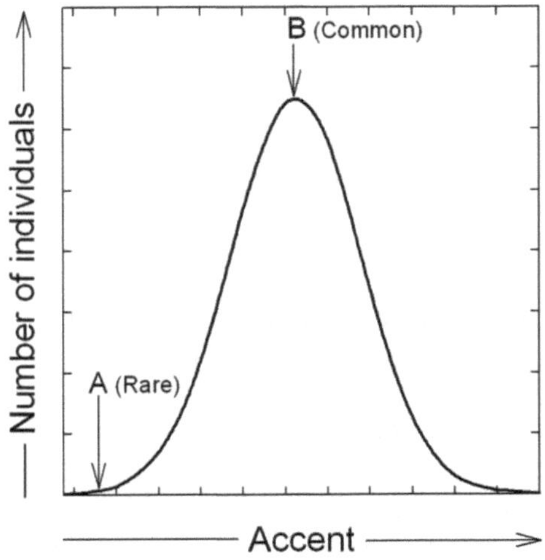

Fig. 5.5 Distribution of an abstract entity—"accent"—that results from "spontaneous variations in the generative elements." **B** indicates the population norm. **A** indicates a rare type. For simplicity, we consider just these two types which can cross but the offspring is sterile (a "mule"). However, in practice there might be many intermediates with varying degrees of sterility in offspring

meet only the abundant majority type—type B—with which, by virtue of the differences in accent, they are likely to be reproductively incompatible (if offspring are produced, they are sterile). Very rarely, an A type will meet an A type of the opposite sex and reproduction to produce fertile A-type offspring can occur. A scheme for this is set out in Fig. 5.6.

The main line of organisms (long vertical gray rectangle) proceeds in linear, unbranching fashion. On rare occasions, a potential new line can for a brief period maintain itself by reproduction and hence expand in numbers (short vertical gray rectangle). However, under usual conditions, the maintenance of A types in the population depends largely on their continuous formation through "spontaneous variability of the reproductive system," rather than because, by chance, they succeed in reproducing themselves. Reproduction of the A types will best occur when there are changes in their genes that confer characters that are favored by natural selection, so providing selective footholds that, by virtue of the *pre-existing* reproductive isolation, cannot be thwarted from blending with B types. In other words, reproductive isolation empowers natural selection (Fig. 5.7).

As will be considered in Part VI, language has both content and accent. You cannot have intelligible content without an accent, but you can have an accent without an intelligible content. Remove intelligible content and all you have left is the accent. For example, you could ask a Frenchman, an Englishman and an American to read a gobbledegook sentence and most listeners would be able to tell which accent was which. To that extent, accent is *more fundamental* than content. It can be seen as a "residue" (secondary information) upon which content (primary information) is rests. The usage "residue" is here somewhat different from Galton's (Chap. 3).

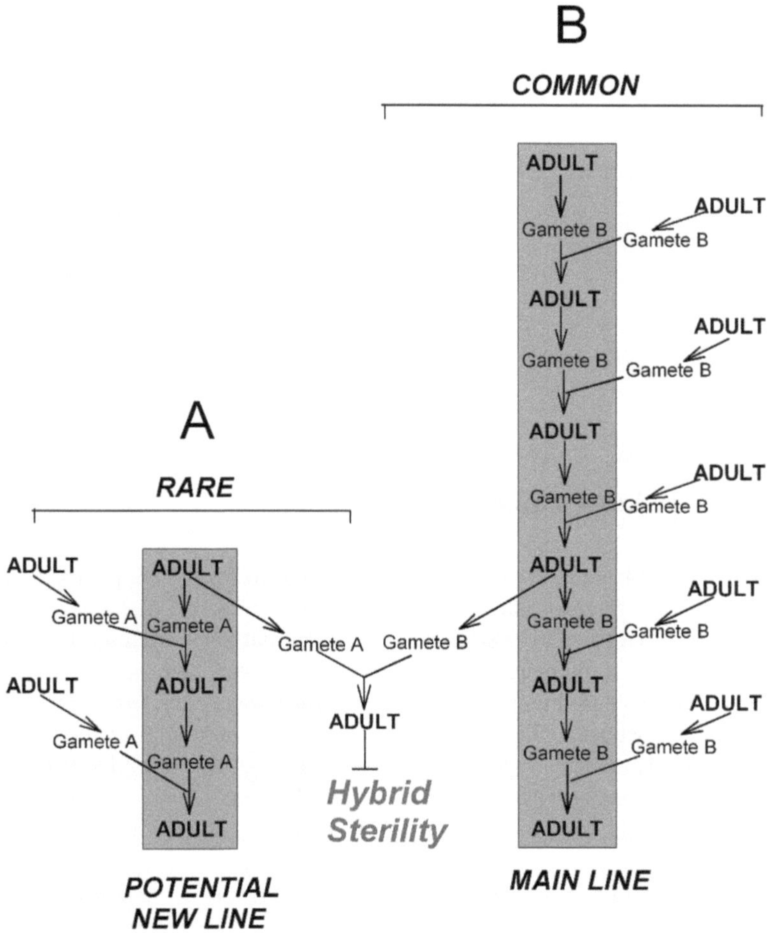

Fig. 5.6 Reproductive isolation (due to hybrid sterility) between the population norm (type **B**) and a rare type (**A**). In the course of normal linear unbranching evolution type **B** meets type **B** and produce **B** type gametes which convey the **B** accent through the generations. On rare occasions, common type **B** meets rare type **A**. The resulting child grows into an adult, but the adult is sterile, so the line ends. Thus, parental types **A** and **B** are *reproductively isolated* from each other. On *very* rare occasions, there is normal linear unbranching evolution when rare type **A** meets another rare type **A**. This will usually not recur because of the rarity of type **A**. If however there is a concomitant "spontaneous variation in various parts of the body" of an **A** type, leading to a neutral or advantageous adaptation, that adaptation will be protected from blending effects because of reproductive isolation from the main population of type **B**. If there is an advantage, **A** will become less rare (i.e., it will be positively selected) and so the probability of meeting more **A** types to perpetuate the new line is increased. The conditions are then right for non-linear, branching, evolution. An "incipient" species can arise

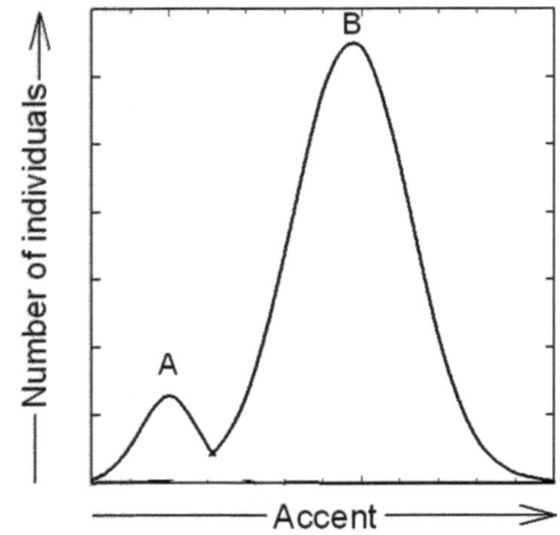

Fig. 5.7 When rare types A reproduce with their like in isolation from B, then their frequency in the population can increase. A discontinuity arises and a separate bell curve appears

Francis Darwin's Criticism

Writing from Cambridge, Francis Darwin in a letter to *Nature* (Sept. 1886) quoted from *The Origin of Species* and said that in the 1860s his father had toyed with what Romanes had come to call physiological selection, but had ended up dismissing it [10]. Furthermore, in the margin of Thomas Belt's book *A Naturalist in Nicaragua* [11], where the same argument had been made *a decade earlier*, his father had written: "No, No." Instead of Catchpool's "spontaneous variations in the generative elements," and Romanes' "'spontaneous' variability of the reproductive system," Belt (1832–1878) had ascribed variation to "the elements of reproduction":

> The varieties that arise can seldom be separated from the parent form and from other varieties until they vary also in the elements of reproduction. … As long as varieties interbreed together and with the parent form, it does not seem possible that a new species could be formed by natural selection, excepting in cases of geographical isolation. All the individuals might vary in some direction, but they could not split up into distinct species whilst they occupied the same area and interbred without difficulty. Before a variety can become permanent, it must either be separated from the others or have acquired some disinclination or inability to interbreed with them. As long as they interbreed together, the possible divergence is kept within narrow limits, but whenever a variety is produced the individuals of which have a partiality for interbreeding, and some amount of sterility when crossed with another form, the tie that bound it to the central stock is loosened, and the foundation is laid for the formation of a new species. Further divergence would be unchecked, or only slightly checked, and the elements of reproduction having begun to vary, would probably continue to diverge from the parent form, …. Thus, one of the best tests of the specific difference of two allied forms living together is their sterility when crossed, and nearly allied species separated by geographical barriers are more likely to [retain the ability to] interbreed [if the barriers are lifted] than those inhabiting the same area.

Romanes replied to Darwin's letter a week later expressing delight that "so distinguished a naturalist" as Belt had anticipated the theory of physiological selection. However, Romanes noted that, from the evidence Darwin had produced, it did not seem "that the theory of physiological selection was ever present in the mind of" Charles Darwin. A week later Francis Darwin, having moved to the "Golf Club, Felixstowe," wrote a short, exasperated note: "I am sorry that I have not succeeded in making my meaning clear to Mr. Romanes. I had hoped that my former letter … would have given some indication as to my father's views. With regard to the sentence quoted from the 'Origin of Species,' our views seem to differ so much that it seems useless to prolong the discussion." Romanes replied:

> If any one will turn to the sentence in question (p. 247, 6th ed.) he will find that it constitutes an integral part of an argument showing that sterility between species cannot have been brought about by natural selection. The argument is that, *even supposing sterility with parent forms to be an advantage*, it is an advantage which could not be seized upon by natural selection, and hence that some other explanation of such sterility must be found. Now, so far as I can see, there is here not only no shadow of the theory of physiological selection, but the whole argument is proceeding on totally different lines. For the very essence of this theory [of physiological selection] is that the sterility in question *need not be supposed to be an advantage*, and therefore that any variation in the way of such sterility *does not require* to be selected through the struggle for existence, being of its own nature a variation which survives.

Francis Darwin then deeply engaged in editing the letters of his father, to be published in 1887, did not reply. Why should he? Huxley and his acolytes were baying for blood.

Lankester's Criticism

Lankester was one of those who had assisted Huxley in the 1870s when the government's belated recognition of science (the 1870 Elementary Education Bill) had led to the funding of a laboratory-based lecture course for science teachers in London. Lankester's father had founded the *Quarterly Journal of Microscopical Science*, which Lankester coedited from 1869 onwards [12]. With Huxley and others, he had exposed fraudulent "spirit writing" by the notorious Mr. Slade. A letter from Lankester formally opposing Romanes was declined by the editor of *Nature*, Romanes' friend Norman Lockyer, who told Lankester that he had shown the letter to Romanes [13]. Lankester replied: "You are quite right not to print my letter to Romanes, as it is not argumentative but purely denunciatory. I am glad he has seen it, as he will now know what a humbugging piece of foolery I consider his attempt to say 'Darwin-and-I' and 'the Darwin-Romanes theory', is. It is time that he knew that I consider him a wind-bag."

However, Lankester later took an invitation to review for *Nature* Wallace's book *Darwinism* as an opportunity to renew the attack [14]. Romanes, whose election to Fellowships of the Linnean Society (1876), and of the RS (1879), recognized his extensive laboratory studies of the nervous system in jellyfish, was declared to be no laboratory man: "Who are those who seek to minimize natural selection and to set up the false gods of variation, use and disuse, etc.? Certainly not laboratory men. Is the Duke of Argyll a laboratory naturalist? Is Dr. George Romanes? [Actually, Romanes had no doctorate] Is Prof. Cope? Are Mr. Herbert Spencer and Professor Patrick Geddes? I venture to say they are not."

Having assigned Romanes some strange bedfellows, indicative of his disdain for Romanes' non-laboratory speculations, Lankester then classified himself, along with Weismann, as a laboratory man (which was true). He also took the opportunity to emphasize the value of non-laboratory field studies noting "Bateson's researches in Tartary, Caldwell's in Australia, Poulton's experiments on insects, and Moseley's 'Notes of a Naturalist on the Challenger'." Finally, he praised Wallace's treatment of the key issue, hybrid sterility:

> In his chapter on the infertility of crosses, Mr. Wallace treats at length and [with] admirable effect a very important subject, as to which he is full of ingenious novel suggestions and apposite facts. His criticism of Mr. Romanes's essay entitled 'Physiological Selection,' appears to me to be entirely destructive of what was novel in that laborious attack upon Darwin's theory of the origin of species.

There followed a series of open letters in *Nature* between Lankester on one side, and Romanes and his ally, Gulick, on the other [1]. Perhaps Lankester reconsidered the matter in 1905 after giving the annual lecture in a series Romanes had endowed at Oxford [15]. A year later in an address to the BA (Chap. 9), he observed that: "The principle of physiological selection advocated by Dr. Romanes does not seem to have caused much discussion and has been unduly neglected by subsequent workers. It was ingenious, and was based on some interesting observations, but has failed to gain support." Lankester's skills at rooting out hypocrisy in science were later found wanting in the Piltdown hoax case [12].

Huxley's Criticism

Meanwhile, the elderly Huxley used a RS obituary notice on Charles Darwin to continue the attack. While preparing the obituary, he reread *The Origin of Species*. A letter to Foster noted (Feb 14, 1888):

> I have been reading the *Origin* slowly again for the nth time, with the view of picking out the essentials of the argument, for the obituary notice. Nothing entertains me more than to hear people call it easy reading. Exposition was not Darwin's *forte*—and his English is sometimes wonderful. But there is a marvellous dumb sagacity about him—like that of a sort of miraculous dog—and he gets to the truth by ways as dark as those of the Heathen Chinee. I am getting quite sick of all the "paper philosophers," as old Galileo called them, who are trying to stand upon Darwin's shoulders and look bigger than he, when in point of real knowledge they are not fit to black his shoes.

He wrote on similar lines to his close friend Joseph Hooker (March 9, 1888). This time there was no ambiguity regarding his target:

> What little faculty I have, has been bestowed on the obituary of Darwin for R. S. lately. I have been trying to make it an account of his intellectual progress, and I hope it will have some interest. Among other things I have been trying to set out the argument of the "Origin

> of Species," and reading the book for the nth time for that purpose. It is one of the hardest books to understand thoroughly that I know of, and I suppose that is the reason why even people like Romanes get so hopelessly wrong.

The obituary was published later that year [16]. While Romanes was not mentioned directly, there can be little doubt as to Huxley's intent:

> Although … the present occasion is not suitable for any detailed criticism of the theory [of Darwin], or of the objections which have been brought against it, it may not be out of place to endeavour to … show that a variety not only of hostile comments, but of friendly would-be improvements lose their *raison d'être* to the careful student. Observation proves the existence among all living beings of phenomena of three kinds, denoted by the terms heredity, variation, and multiplication. Progeny tend to resemble their parents; nevertheless all their organs and functions are susceptible of departing more or less from the average parental character; and their number [the number of progeny] is in excess of that of their parents. Severe competition for the means of living, or the struggle for existence, is a necessary consequence of unlimited multiplication; while selection, or the preservation of favourable variations and the extinction of others, is a necessary consequence of severe competition. 'Favourable variations' are those which are better adapted to surrounding conditions. It follows, therefore, that every variety which is selected into a species is so favoured and preserved in consequence of being, in some one or more respects, better adapted to its surroundings than its rivals. In other words, every species which exists, exists in virtue of adaptation, and whatever accounts for that adaptation accounts for the existence of the species. To say that Darwin has put forward a theory of the adaptation of species, but not of their origin, is therefore to misunderstand the first principles of the theory. For, as has been pointed out, it is a necessary consequence of the theory of selection that every species must have some one or more structural or functional peculiarities, in virtue of the advantage conferred by which, it has fought through the crowd of its competitors and achieved a certain duration. In this sense, it is true that every species has been 'originated' by selection.

In a reprinting of his earlier *Darwiniana* lectures (1893), Huxley pointed to four problems that remained—the logical basis of natural selection, hybrid sterility, the cause of variation, and whether acquired characters could be inherited [16]. He also hinted at a struggle between different forms of inheritable information for passage through the generations (a preview of modern "selfish gene" concepts) and concluded with a swipe at Romanes:

> I remain of the opinion … that until selective breeding is definitely proved to give rise to varieties infertile with one another, the logical foundation of the theory of natural selection is incomplete. We still remain very much in the dark about the causes of variation; the apparent inheritance of acquired characters in some cases; and the struggle for existence within the organism, which probably lies at the bottom of both of these phenomena. … I have reprinted the lectures as they stand, with all their imperfections on their heads. It would seem that many people must have found them useful thirty years ago; and though the sixties appear now to be reckoned by many of the rising generation as a part of the dark ages, I am not without grounds for suspecting that there yet remains a fair sprinkling even of 'philosophic thinkers' to whom it may be a profitable, perhaps even a novel, task to descend from the heights of speculation and go over the A B C of the great biological problem as it was set before a body of shrewd artisans at that remote epoch.

Indeed, it had been the loftiest speculator and shrewdest artisan of all, Charles Darwin, who had implied a struggle for existence within the organism in his 1868 hypothesis of pangenesis, where new "gemmules derived from the same [character] unit after it has been modified go on multiplying … until at last they become sufficiently numerous to overpower and supplant the old gemmules"

Bateson's Support

Bateson sent *Materials* to Huxley, who replied (Feb. 1894): "How glad I am to see … that we are getting back from the region of speculation into fact again. There have been threatenings of late that the field of battle of Evolution was being transferred to Nephelococcygia [Cloud Cuckoo Land; [17]]. I see you are inclined to advocate the possibility of considerable 'saltus' on the part of Dame Nature in her variations. I always took the same view, much to Mr. Darwin's disgust, and we often used to debate it."

This appears as a reference to Romanes that Bateson would likely have understood. Yet, Bateson's first encounter with Romanes' physiological selection had been several years earlier during his travels on the Steppes (Chap. 2). Here far from the pens and, more importantly, tongues of the Huxley clan, he had been able to form a relatively unprejudiced opinion. His sister Anna had sent the three consecutive issues of *Nature* in which Romanes' lecture at the Linnean Society had been serialized [2]. She was then working with Francis Darwin and was probably fully aware of the storm of criticism the lecture had aroused. From the following, we can infer that, perhaps influenced by Darwin, Anna had disparaged Romanes' paper, both with respect to quality and novelty. However, the twenty-five-year-old Bateson strongly differed from his seniors back in England (Dec. 4, 1886; Kasalinsk):

> Thanks for those 'Natures.' I don't agree with you that Romanes' paper is poor. It seems a fair contribution and at all events does, as he says, put the whole view on a much more logical basis. The scheme thus put will at least work logically while the other, as left by Darwin, would not. Of course, as to the novelty of the suggestion I know nothing, and I don't much care. I did not suppose Romanes would ever write as good a paper. Of course there is nothing "genial" (adj. derived from "genius" in continual use by Tarnovski) in it, but it is a straight forward, common-sense suggestion.

A letter from Weldon indicates how another young scientist viewed Romanes. Bateson had cited Romanes in *Materials* only in the context of the latter's experimental work. Interlacing humor with sarcasm, Weldon noted that observations on jelly fish by Romanes (who would have considered himself a physiologist, not a naturalist) were not in accord with those of one of Weldon's students (Feb. 13, 1894):

> Many thanks for your great piece of work—I will not insult you by trying to make any remarks about it for a month or so, because it is evidently one of those abominable works which have to be properly read. Curiously enough, in cutting it I notice a point on which a student of mine has been working. … This man has also drawn quantities of *Luceruaria* and of hydromedusae …. He has many abnormal hydroids and I fancy that the 'many thousands examined by Romanes G. J.' [a quote from *Materials*] were examined in the superficial manner which is characteristic of that eminent naturalist.

We can now note that many of Romanes' experimental observations have stood the test of time. For example, to acknowledge his "fundamental contributions" to neuromuscular physiology, it was proposed in 1944 that "the recovery of excitability of tissue upon reversal of polarity of a stimulating current be called the 'Romanes effect'" [18].

Cunningham's Support

As we have seen (Chap. 4), when Weldon finally expressed an opinion on *Materials*, it was quite negative. One of Weldon's colleagues at Plymouth was Joseph Cunningham (1859–1935), a Lamarckist with an interest in the history of science, who had recently translated a key work of Eimer (19). Cunningham wrote (Feb. 16, 1894) congratulating Bateson on "your big book" but lamented that "you have scarcely taken a right view of my work on flat fishes." Cunningham had observed increased pigmentation following irradiation of their white undersides. Bateson replied (Feb. 24) pointing out some of the problems with fish-tank experiments and adding: "I see clearly the interest of your results, but I do not feel wholly convinced that the cause of the pigmentation was the illumination from below. Nevertheless, I have no *a priori* objection to such a view, and I admit that your case is strong." Romanes wrote approvingly of Cunningham's experiments in volume two of his *Darwin, and After Darwin*, which was published posthumously in 1895 by Lloyd Morgan. That year, in the journal *Natural Science* Cunningham expounded at length on Bateson's failure to mention Romanes' theoretical work [20]:

> The most recent important contribution to the study of evolution in this country is Bateson's 'Materials for the Study of Variation, treated with especial regard to Discontinuity in the Origin of Species.' … Valuable as the work is, and important as are the aspects of variation to which the author draws attention, one cannot help being astonished at the fact that Mr. Bateson writes as though, apart from adaptation to the environment, no explanation of the discontinuity of specific forms had been offered before he himself took up the problem. He does not mention that, so long ago as 1886, Romanes maintained that Natural Selection was a theory of the origin of adaptations, by no means a theory of the origin of species; or that two years later Gulick published most important evidence of the origin of varieties by mere divergent variation through isolation, *without any adaptation at all*. In fact, the modes in which variations, instead of being individual peculiarities within a species or other taxonomic group, are to become constant and characteristic of a distinct group, are never discussed by Bateson. … There is some logic in his contention that, since environments are often continuous, their influence cannot always explain the discontinuity of species. But it is clear that, if progressive modification in different directions goes on in two groups of a single species which are isolated so that no interbreeding takes place, then two species will be formed, and the discontinuity between them will become greater at every generation. This result will follow, however gradual and continuous may be the modification in each group. Mr. Bateson's argument is defective, therefore, in two respects: first, there is no necessity for discontinuity of [observable] variation to explain discontinuity of species; secondly, he has not attempted to show where discontinuity of modification probably did occur historically in the evolution of any particular group.

Cunningham then considered evolution of three species of flatfish:

> The relations of these three species, on the view that their differences are not adaptive, illustrate certain principles which have been elaborated by Romanes and Eimer. They exemplify the class of facts which Romanes intended to explain by his theory of physiological selection, which, reduced to its simplest terms, states that species whose geographical ranges overlap or to a large extent coincide, could not become distinct if they were constantly inbreeding. … The main point is that so long as intercrossing took place, variations that occurred in one group of individuals would sooner or later become the common property of all the members of the parent form; and that when the three groups were kept apart, a variation that arose in one would be confined to that one.

Furthermore, variation was the name given to a phenomenon which, at that point in time, we should be content to simply observe:

> We may take it as settled by observation that when groups of species are isolated they will diverge by variation. It may be asked, why should two groups of individuals, having originally the same habits and spread over the same area, enter upon divergent lines of modification, simply because they are separated? We do not know, but we can safely say that they will do so. We have first to ascertain what does take place before we can find out why it takes place; and when we find that differences between species are *not adaptations to different modes of life*, we have simply to study these differences and their history as structural features.

Given some form of isolation, the rate of observable variation was not of major relevance:

> I think, however, that the discontinuity of species is sufficiently explained by the view of Romanes and others concerning the effects of the isolation of strains or races, whether by geographical, environmental, or physiological barriers to inbreeding. In my opinion, therefore, discontinuity of [observable] variation is not necessary, as Bateson maintains, to explain the discontinuity of species. However gradual may have been the process of modification by which two allied species have been derived from a common ancestral species, the existing discontinuity is a necessary consequence of the divergence of the two lines of modification, provided that no intermixture of the variations in the two lines has taken place.

Cunningham hereby undermined Bateson's arguments based on large observable variations. While Bateson's focus was on the anatomical and physiological variations that were concomitant to species formation, Romanes' focus was the "reproductive system" which was as liable to variation as any other body system. He held that it was *this* variation that created reproductive incompatibilities between members of a species, thus *preserving* anatomical and physiological variations, *small or large*, so that morphologically distinct species might *then* appear to emerge, perhaps discontinuously. Romanes probably regarded the physiological variation of the reproductive system as generally *slow and progressive* because it was "collective" in nature. For example, he listed as one of the misconceptions of his opponents the notion [3]:

> That I imagine physiological varieties always to arise 'sporadically,' or as merely individual 'sports' of the reproductive system. On the contrary, I expressly stated that this is *not* the way in which I suppose the 'physiological variation' to arise, when giving origin to a new species; but that it arises, whenever it is effectual, as a 'collective variation' [of the

> reproductive system] affecting a number of individuals simultaneously, and therefore characterizing 'a whole race or strain'.

Such a collective variation (in what we may now interpret as the "accent" of DNA) could build up on a group of related individuals over many generations, remaining latent until conditions favored the emergence of recognizably new species (see Part VI). To this extent, we can equate "accent" with Bateson's "residue."

A year later, Lankester referred in *Nature* to "the independent and anti-Darwinian theories of Mr. Romanes and Mr. Bateson" [21], thus recognizing some similarities, but perhaps also some dissimilarities, in their lines of thought. When reviewing Bateson's *Problems* (1913), Cunningham again charged him with neglecting Romanes' (Chap. 14). However, in 1922 Cunningham dismissed as "false metaphysics" Bateson's further development of the ideas of Romanes, who was by then quite forgotten (Chap. 17).

Bateson's Lectures

Bateson considered *Materials* the first volume of a series. But in research there is always the temptation to make one more observation, to do one more experiment, before putting pen to paper. With the poor sales of *Materials,* it was easy to yield to this. Although diverted from writing, he was still giving lectures, and his archives contain notes for these as well as for the intended future volumes. His lecture notes are probably more reliable than his "begging letters" (grant applications) and formal reports, where political factors might have been more in play. Although somewhat garbled, the notes reveal the development of his ideas through the 1890s and the Mendel era, while hinting at deeper philosophical goals.

From a file marked "Various" begun in 1889, it seems that Bateson then saw sterility as an example of a variation, just as more apparent anatomical or physiological differences could be considered as variations. While averring the goal to be the "absolute truth," he began by cautioning that the researcher should proceed a step at a time, beginning with the most logical step, which, in his view, was the description of variation as a phenomenon:

> You must understand that I ask your attention to this subject, not because there is in it anything especially new to tell and which you ought to know as knowledge, but rather because I wish to put before you the extreme meagerness of our knowledge of these things on the one hand, and the extreme importance on the other. It is not the things that we *know* about variation which are important; it is the fact that we don't know what we ought, and that we might, if we tried. *We are all students of Biology, and I take it that we are all determined to do some piece of original work*—to do it as soon as we may and to do it in the best field. We are all determined when the chance comes, to find out some secret; and we all want to find out the secret that shall mean most, that shall give us the answer to the questions that we most want answered, that shall give us the Truth which will cover most ground and that shall help us on more towards Absolute Truth than the same effort put into another field.

It is much to treat the subject of *Evolution* as a whole, and to introduce only such facts, and to appeal only to such facts as to the mode of occurrence of variations, as may seem necessary to illustrate or support the view of Evolution which is being put forward. In this way I believe a logical error and a serious error in method is made. I propose rather to study variation as a phenomenon for its own sake. When this is done, if it be found possible to obtain any reasoned scheme of the occurrence of variations, *then* perhaps it *might* be fitting to apply the knowledge thus gained to trying to construct a scheme of Evolution.

Under the heading "Introduction," the discourse continued in note form:

The Study of Variations. Importance of the study of variation in any attempt to obtain a just view of Evolution or even Phylogeny. Since Evolution, however effected, consists solely of the accumulation and integration of variations, a knowledge of their nature, magnitude, modes of occurrence and causes, must be the first step ….

Methods of Studying Variations. Sources of error. Possible differences in kind between variation among domesticated forms and wild ones.

The Nature of Variation. Especially of:

(1) Variations *directly affecting Fertility.* (e.g., variation in number of offspring; length of life; age of maturation; period of germination of seeds; period of gestation, etc.; the age of parent as affecting number and quality of offspring). Sterility as a variation.
(2) Variations which consist in a *Repetition of Parts.*
(3) Variations in the degree to which the sexes are distinguished. (Specific variations in number of each sex among offspring produced by certain forms.) Hermaphrodite forms. Multisexual forms. Castration. Changes of sex in [the] same individual. Question as to vegetative and sexual reproduction being complementary.
(4) Variation in Symmetry and the control exercised by Symmetry.

Influence of Conditions. Variations directly consequent upon conditions such as: Amount of nutrition. Climate. Soil. Salinity. etc. (Poulton's experiments—Buckman—Schmankewitsch—Kerner—Semper—Hoffman, etc.). Local varieties (of plants—salmon—birds—mollusks, etc.). Grafts and graft hybrids (complementary grafts). ... Bud variation and reversion of buds. Types which are not visibly affected by conditions, and limits of conditional influence. *Inheritance of acquired characters.*

Hybrids and cross-fertilization. Variations to ensure cross-fertilization and variations to prevent it. List of experiments and observations *to be made.*

Specific Prepotency [dominance] *of Certain Variations.* List of results of complementary crosses among plants. Prepotency of polydactyl forms (e.g., Silky fowls, Polled cattle, Wild type over domesticated, etc.).

Do animals ever revert to a more complicated ancestral type …? That is to say, does an animal whose normal adult state is one of less complexity than that of its ancestors, ever reproduce in itself any very complicated structure which its ancestors possessed, but which does not ordinarily occur in the larval state? If it be supposed that an animal is built up of a number of various … organic elements and that a new variation is due to the entrance of a new element of this kind, then if a variation disappears it must be supposed that the element which caused it is gone; if then the variation reappeared, is it to be supposed that the element has passed on in a suspended state [latency], and has suddenly sprung into action, or is it more likely that it has reentered afresh?

It may be as well to point out that the doctrine of the Evolution of living things is arrived at empirically and it is not suggested that we have as yet any understanding of the ultimate physical processes by which Evolution is brought about. The fact of evolution is perceived and recognized in its occurrence and thus only. It is known by its results, as for example gravitation is, and not in its essence. All processes are thus known to us and thus only. Let it not then be a reproach to the views of Nature that will be here set forth, that if

> adopted it must tend to an empirical rather than a 'causal' solution of the problems of life: for in so far as this knowledge when we shall have attained it will be empirical, it will but have a quality common to all knowledge and in proportion as it is empirical, so much the ... [the rest is lost, or was not completed].

The Bateson archives contain notes made from journals and books under many headings. Notes on separate pieces of paper were labeled on the top right-hand corner (underlined twice) as to the category of subject matter: "conditions" (e.g., soil effects, fertilizers, color of hydrangeas), "sex," "fertility," "size," "duration of life," "maturation," "hybrids," "regeneration," "grafting." He also added clippings from magazines and newspapers. Usually, there was no discrimination between plants and animals. That his reading of Romanes might have had some influence is indicated by occasional musings: "Does mutual sterility ever exist (in otherwise fertile forms) apart from other variations?" The word "isolation" was used, but not often.

In the "hybrid" file, there are replies from Barr of Covent Garden (Jan. 1890), which indicate the wide net cast in search of information. The replies referred to some principles of hybridization laid down by the Manchester horticulturalist Dean Herbert, who had been much cited by Darwin. There is a list of eight questions in Bateson's handwriting:

> In a case where a cross between two species is possible the following questions may be suggested: 1. Is the result of the union always or often the same? Or is it indefinitely variable? 2. Is a union possible between the two forms, whichever be taken as the father? If such '*reciprocal*' crossing is possible, are the results the same whichever be taken as the father? Or are they constantly different? 3. In cases where a *constant* result is produced, what characters can be referred to the father and what to the mother? 4. In cases where there is variability among the results, is there any marked '*distribution*' in the inheritance? (e.g. do some resemble one parent especially and others the other, or are the characters blended?). 5. Cases of sexual fertility of hybrids (i) with each other, (ii) with either parent forms, (iii) with other varieties or species. 6. Cases of partial or complete sterility between varieties known to have been produced ... from the same species. 7. In cases of hybrids (whether sexually fertile or not) is asexual reproduction more or less freer than in the parent forms? 8. Herbert believed that the sterility of hybrids depended rather on *the degree to which the conditions of life* of the two parents differed, rather than on the degree to which their structures differed. Cases wanted supporting or tending to refute this view.

In later lectures, Bateson seemed to be reaching for a duality, one element of which related to natural selection, the other to some intrinsic stabilizing factor (perhaps related to Galton's "organic stability"). The first lecture of a term (I. 1897) read: "The modes in which new forms may become established without intermediates becoming established. ... The fixity and capacity for survival of each may be due to *fitness* of each to [environmental] conditions, or to a '*stability*' belonging to each, apart from its direct relation to [the] external world." The systematists were disparaged: "To arrange forms in [a] logical order is not to show the manner of their evolution." A "recapitulation of main points," and "some salient deductions," were set out in the last lecture (VIII. 1897):

> There *are* those *separate* varieties—varying individuals—capable of being grouped just as species can be. Why are they not incipient species? The objection is simply that the evolutionists had begun to believe everything could be explained by *adaptation*—a definiteness of form was to be due to *fitness*. If this definiteness—i.e. specificity—is *not* due to adaptation, it is still to be explained—but it is still to be explained in any case, for there *is* definite-

ness of variation and no one has suggested that this is due to adaptation. [It] has been suggested [that it is] due to *Reversion* [to an ancestral form]. [Let us] examine this for a moment ….

After dismissing reversion, he concluded: "In the present state of knowledge, everything in my judgement points to evolution proceeding in great measure by means of the *large* variations. But if large variations [are] due to Stability, what *is* that? [We are] thrown back to empiricism."

Lectures once given may be repeated from the same notes in future years, sometimes with additions inserted. Some lectures dated 1899 have the further date 1900 added:

Problem of species is *not* merely how do species arise from each other; but how does it happen that there are such things as species at all? … In all discussions of [the] origin [of species], remember [that] two essentially distinct questions are presented.

(1) How do the variations arise as an objective fact—what are the causes of variations—what is their nature?
(2) Perpetuation. And in addition, for [a] complete solution [we should add]:
(3) Adaptation, though this is wrapped in (2) to some extent.

As for the origin of variation: "The *Causes* of variation are thus conceived as something distinct from the causes to which the varying characters are due when established …. Think of organisms as substances—complex if you will—but still substances—and you will see how utterly these considerations are inapplicable. We can scarcely speak of a 'cause' of variation, rather than of a 'cause' of normality." Lecture V given in the October term of 1899 read:

It is not true that the greater the variation the more rare it is. The word which seems best to express this is *Discontinuity*—not an absolute but a comparative term. … But if discontinuity is not *itself* of direct utility, then it is an indication of [the] determinateness of *Stability* in its various manifestations. Remember, the importance to us of these variations rests on the intention to build species out of them. We *hope* to find variations out of which species can be built. Are not these they? Species is discontinuous—admittedly in the main. Differences in environment—apart from chemical and organic constituents—are largely continuous. Yet rarely do we find species merging in the way that this would lead us to expect. If we can see indications of Stability in certain degrees of variation rather than in others—by so much the easier does [the problem of the] origin of species become. Darwin's horror of this idea [was] probably founded on the feeling that it was only deferring the evil hour—it was merely throwing back on the 'nature of the organism' what he hoped to explain more directly.

But has the organism then no natural limitations to its power to change? To many, all idea of discontinuity is repellent—states of matter—chemical properties—to be ultimately explained away—[statements such as] 'the obvious visible discontinuity masks a real continuity!' abound. The theory of gradual descent implies belief that all intermediate states are equally possible and have existed as normals. Difficulties thereby [are] vastly increased.—an intermediate *population* [is] rare—No real necessity for such [an] assumption. … Think of what organisms are—complex systems of chemical and molecular actions. We are asked to suppose they have no natural or inherent order—that without Selection and environmental control they would be *Chaotic*. But when [we] turn to [the] facts of variation [it is] not necessary to suppose any such thing.

The lectures at this time were largely a rehash of *Materials*, with frequent reference to Galton. However, he did turn to differential fertility:

> Lastly comes the physiological test of fertility. By it we dispose at once of most domestic varieties of animals and plants to some extent. When we come to consider hybrids [we] shall find structural differences [are] no *certain* criterion of inability to produce offspring. Nevertheless, sterility [is] very usual—not total but to some degree.

There exists a printed syllabus of Bateson's Cambridge Lectures on the "Practical Study of Evolution" for the 1899–1900 academic year. For the Lent term, this included: "Fertility and Sterility; general considerations. 'Physiological Selection'." Lecture V read:

> We have seen how a great diversity of animals may exist simultaneously in [the] same location under a variety of forms. We cannot suppose that all are the most fit and we are forced to the conclusion that ... in each of these cases, and in respect to their varying characters, *environmental exigencies are not precise*. And yet, though in all these cases the ordinary reasoning by which the precision of colour is referred to environmental exigency, is out of court, yet the colours may be precise.

Lecture X read: "Heredity. No glimmering of an idea what kind of process it is. [We know that] Characters [can be] separately transmitted. No *a priori* reason for imagining any two characters of a species [to be] correlated." Lecture XI tended to imply that Romanes had merely elaborated the works of others: "Fertility. Enormous importance of small changes in [fertility] as influencing prospects of perpetuation .... Idea that differential fertility may operate is an old one. Older anthropologists. Broca. Pouchet ... Romanes' Physiological Selection."

In Lecture XV, Bateson made the important distinction that, whereas in higher animals the germinative tissues (ovary, testis) were separated from the other tissues (soma) at an early stage, this did not apply to higher plants: "As to Differentiation of Germ Plasm. We must reply that in flowering plants undifferentiated tissue may be germinative. The power to make the response is transmissible, not the response."

In 1900, Mendelism burst on the scene (Chap. 8). The first mention of Mendel was in Bateson's second lecture on "Heredity" in the Lent Term of 1901, and the contrast with Galton was noted: "Galton brings *each* ancestor into account. [He] also supposes that all characters *may* be apparent in any generation." In November 1901 in his second lecture, Bateson spoke of "Mendel, de Vries, Correns and Naudin" together and mentioned "purity of germ cells, equal number of each," and "dominance not essential." As for Galton, he was declared to have "failed from the fact that his original conception of the problem [was] wrong. ... He came to it with the mind of a statistician, whereas we want [the] mind of a chemist."

Lecture after lecture through November and December 1901, Bateson posed the fundamental Mendelian questions: "Mutation. Detection of germ cells. Where do the characters divide? ... Unit characters. ... How many of them? Is there any residue? When do [the] characters divide? May sterility be incapacity of division? De Vries mutations. ... No hint of origin of interracial sterility. Power of Selection limited."

Lectures dated 1903 under the heading "Fertility" noted the "immense importance of sterility on crossing when it occurs," and contained the heading: "Romanes—Physiol. Sel.". Intriguingly, the same notes contained the word "Drosophila" in large lettering with a circle round it. This could have been added when preparing a later lecture. Despite his preoccupation with the details of Mendelian analysis, lecture notes dated 1904 suggested that the variation problem remained a major concern and that he thought its solution could lead to answers to ultimate questions:

> *Next* far more important—[variation is studied] as a contribution to the *final settlement of things*. Need not say much on this. To anyone who has thought of these things it is familiar that everything else [is] secondary to this. That all discussion of political, social and other such questions is frivolous, since the data [are] unknown. ... This much being supposed, note that [the] problem has 2 parts. [There are] 2 things to explain—(a) Causes that make them vary. (b) Causes that make them stop in fixed types. Here all theories differ.

In October 1905, there were six weekly "Advanced Lectures on Botany" on "The Facts of Heredity," at University College, London. The three-page formal syllabus began by referring to "Heredity a branch of physiology," and "Study of heredity long neglected. First general recognition that the phenomena are orderly due to work of F. Galton. Weismann's contribution." The syllabus ended with "The bearing of the facts on the problem of Species." However, this, and later lectures, revealed an increasing engrossment with problems of Mendelism, and physiological selection was no longer mentioned (notes dated as late as 1909). The notes for January 1907 began with: "Fertility and sterility. Drosophila. ... Differential fertility—importance of." The notes went on to describe crosses between Drosophila lines of low and high fertility, and the transmission of the fertility trait.

Bateson and Romanes

The best that can be found concerning Bateson's final view of Romanes—and here the linkage is far from explicit—comes from his BA Address at Cambridge in 1904 (Chap. 7). Here, he made a pitch for returning to the observation of natural forms both in the field and under cultivation, and disparaged those whom Huxley would have called "paper philosophers":

> The evidence of the collector, the horticulturalist, the breeder, the fancier, has been treated with neglect, and sometimes, I fear, with contempt. That wide field whence Darwin drew his wonderful store of facts has been some forty years untouched. Speak to professional zoologists of any breeder's matter, and how many will not intimate to you politely that fanciers are unscientific persons, and their concerns beneath notice? For the concrete in Evolution we are offered the abstract. Our philosophers debate with great fluency whether between imaginary races sterility could grow up by an imaginary Selection; whether Selection working upon hypothetical materials could produce sexual differentiation; how under a system of Natural Selection bodily symmetry may have been impressed upon a formless protoplasm. ... Enthusiasm for these topics is sometimes fully correlated with indifference even to the classical illustrations; and for many whose minds are attracted by the abstract problem of inter-racial sterility there are few who can name for certain ten cases

> in which it has been already observed. And yet in the natural world, in the collecting box, the seed bed, the poultry yard, the places where Variation, Heredity, Selection may be seen in operation and their properties tested, answers to these questions meet us at every turn—fragmentary answers, it is true, but each direct to the point.

But Bateson was not allowed to forget Romanes. Daniel MacDougal of the New York Botanical Garden congratulated Bateson on being awarded the Darwin Medal (Dec. 5, 1904): "We are now reading the pages of the proof of the book of de Vries's lectures given over here this summer. It makes a book about the size of 'Darwin and After Darwin' [Romanes' title] and will be easier reading … than 'Die Mutationtheorien'."

Shortly thereafter, Bateson wrote "Evolution for Amateurs" for *The Speaker* (June 24, 1905). This was a review of an English translation by J. Arthur Thomson and Margaret Thomson of Weismann's *The Evolution Theory*, which Bateson thought was dated ("it should have appeared thirty years ago") in its emphasis of the "simple creed" of natural selection. "The Westphalian sage" had rested "content to treat the discoveries of recent years as non-existent" and, in particular, had criticized de Vries as overrating "the value of his facts" on Mendelism. Weismann's "gospel of selection" led Bateson to "think of that other philosopher [Voltaire] who, observing how the nose fits the spectacles and the feet the shoes, deduced with the same confidence and more brevity that all was for the best." Weismann (but not Galton) was praised for having "first taught us to distinguish the 'soma,' or body, from the germ."

The extent of Bateson's divorce from his Romanesian roots is evident from a brief note to Hogben written in 1924 (Feb. 19), when the possibility arose of merging the Genetical Society with one concerned with the physiology of reproduction (Chap. 22): "Genetics cannot be 'extended' to include the physiology of reproduction. It *is* the physiology of reproduction. For historical reasons which I sometimes think I will set out, the official physiologists some 40–50 years ago abdicated from this—the most vital part of their science. They now, as you have found, affect to ignore it." As Secretary to the Physiological Society in the 1880s, Romanes would clearly fit the category of "official physiologist."

Thus, it appears that Bateson was not aware of any intellectual debt to Romanes, and his 1904 address (above) suggests some antagonism. What of the converse? There are few indications that Romanes was aware of Bateson's works, but then Bateson only began publishing in the mid-1880s. In the 1890s, Bateson's letters in *Nature* indicated that he was replacing the failing Romanes as "official" gadfly to the evolution establishment (Butler had been the "unofficial"). The first was a controversy over "aggressive mimicry" with Poulton of Oxford (Chap. 6). In 1892, when writing (Dec. 2) to Poulton from Madeira where he was hoping to mend his health, Romanes referred to Bateson's aggressive tone. It so happened that, a week before publishing Weldon's scathing review of *Materials* in 1894 (Chap. 4), *Nature* had published a letter from Weldon that made it difficult for Romanes not to acknowledge Bateson. Weldon's letter had been written in response to a letter from Romanes published the previous week. Here, recapitulating his letters to *Nature* two decades earlier on "cessation of selection," Romanes had considered how organs come to degenerate when the selective forces presumed to have brought them to their mature forms, were no longer present. Weldon wrote:

> With regard to the extreme statement [of Romanes] that '*any* failure in the perfection of hereditary transmission will be weeded out' by selection in a wild state, I would urge the need, which has lately been well pointed out by Bateson, of a *quantitative* measure of the efficiency of selection. The frequency of even considerable abnormalities in specialized organs of wild adult animals, of which so many admirable examples are described in Mr. Bateson's recent work on variation, show, if it needed showing, that natural selection is in most cases an imperfect agent in the adjustment of organisms.

A week later, and only a few weeks before his death, Romanes replied:

> If Prof. Weldon will read what I wrote last year in the *Contemporary Review* … he will find that in this matter I am quite on the side of Mr. Bateson and himself. It has always been my endeavour to argue that the ultra-Darwinian school of Wallace and Weismann are pushing deductive speculation much too far in maintaining 'The All-Sufficiency of Natural Selection.' I shall never believe—any more than Darwin believed—that what I have called 'selection value' is unlimited. But this is not incompatible with the belief that *in whatever degree* natural selection may have been instrumental in the construction of an adjustment, *in some degree* must its subsequent cessation tend to the degeneration of this adjustment.

Romanes' passing "in the outset and full promise of a splendid scientific career." was lamented by the Marquis of Salisbury in his BA address at Oxford later in the year (Chap. 19).

Gulick

But Romanes' ally, Gulick, remained very much alive. Although Bateson had not mentioned him in *Materials*, his notes for the never-realized second volume contained references to Gulick's studies on tree snails in the Sandwich Islands (now Hawaii). In earlier studies of Canary Island fauna, Leopold von Buch (1774–1853) had noted [22]:

> Upon the continent the individuals of the genera by spreading far, form, through differences of the locality, food and soil, varieties which finally become constant as new species, since owing to the distances they could never be crossed with other varieties and thus be brought back to the main type. Next they may again, perhaps upon different roads, return to the old home where they find the old type likewise changed, both having become so different that they can interbreed no longer. Not so upon islands, where the individuals shut up in narrow valleys or within narrow districts, can always meet one another and thereby destroy every new attempt towards the fixing of a new variety.

However, in the case of relatively immobile snails, Gulick found that individuals shut up in narrow island valleys could very easily fail to meet each other. Each valley was a sub-island. By moving away from its peers, a migratory snail would tend to isolate itself and its kin, thus founding a new type ("founder effect") not through selection because of adaptation, but by random variation, a process later known as "random drift." In 1905, Gulick produced a major monograph on evolution that referred to Bateson quite positively [23]:

> In his volume entitled 'Materials for the Study of Variation,' ... Bateson points to the lack of correspondence between the diversity of physical environments and the diversity of specific forms as a feature unexplained by the theories either of Lamarck or Darwin .... He suggests that the explanation for this lack of correspondence must be sought in the organic group [i.e. internal to the organism], and not in its environment; and that the study of variation is the chief hope, though even that may fail. I entirely agree with Mr. Bateson in regard to the importance of variation and of the factors in the species that control variation; and prominent among these factors I find either the power of free communication and [consequent] intergeneration, or the lack of that power. ... Isolation, therefore, ... goes far toward explaining the phenomena which we have been considering, and which are essentially the same as those which Mr. Bateson has cited as being in pressing need of explanation.

In *Problems* (1913), Bateson reciprocated, citing Gulick's monograph: "I incline far more to agree with Gulick who, after years of study of the local variations of the Achatinellidae, came to the conclusion that it was useless to expect that such local differentiation can be referred to adaptation in any sense."

In his evaluation of Romanes, Bateson was probably much influenced by the "paper philosopher" image imputed by Huxley. However, Romanes was a committed experimentalist who, on the advice of Darwin, undertook extensive studies, none of which was formally published, in an attempt to demonstrate transfer of information by way of "gemmules" from soma to germ line (Chap. 3). He continued to bark up this Lamarckian tree long after the death of Darwin.

Memes

When it came to viewing hereditary transmission as the transmission of information *by means of gametes*, Romanes, like Bateson, would not heed the admonitions of Hering and Butler (Chap. 19). However, Romanes could see the transmission of information from generation to generation *by means of culture* as a new form of evolution. In that it involved a "non-physical natural selection" in a "psychological environment," it is reflective on the "mneme" or "meme" concepts developed by Butler, Semon, and Dawkins. In the second volume of *Darwin, and After Darwin* Romanes wrote [24]:

> Civilized man enjoys an advantage over savage man far in advance of those which arise from a settled state of society, incentives to intellectual training, and so on. This inestimable advantage consists in the art of writing, *and the consequent transmission of the effects of culture from generation to generation.* Quite apart from any question as to the hereditary transmission of acquired characters, we have in this *intellectual* transmission of acquired *experience* a means of accumulative cultivation quite beyond our powers to estimate. For, unlike all other cases where we recognize the great influence of individual use or practice in augmenting congenital 'faculties' (such as in the athlete, pianist, etc.), in this case the effects of special cultivation do not end up with the individual life, but are carried on and on through successive generations *ad infinitum*. Hence a civilized man inherits mentally, if not physically, the effects of culture for ages past, and this in whatever direction he may choose

to profit therefrom. Moreover—and I deem this an immensely important addition—in this unique department of purely intellectual transmission, a kind of non-physical natural selection is perpetually engaged in producing the best results. For here a struggle for existence is constantly taking place among 'ideas,' 'methods,' and so forth, in what may be termed a psychological environment. The less fit [ideas] are superseded by the more fit, and this not only in the mind of the individual, but, through language and literature, still more in the mind of the race.

Summary

Contrasting with Darwin's view that reproductive isolation was a *consequence* of divergence due to natural selection, Belt, Catchpool, and Romanes proposed that reproductive isolation was a *facilitating cause* of divergence, which might, or might not, involve natural selection. William Bateson, unlike those allied to Huxley, endorsed Romanes' viewpoint in a letter to his sister, and probably in his lectures, but not in his scientific writings. Cunningham drew attention to the resemblance between the evolutionary writings of Romanes and Bateson, who seems to have dismissed Romanes as a mere "paper philosopher." Bateson's "discontinuous variation" as an originator of species can now be seen as reflecting the *empowerment* of natural selection by a new type of variation that would confer reproductive isolation through a process referred to by Romanes as "physiological selection." This would allow shifts from one position of "organic stability" to another, thus preventing the blending of characters that would occur when breeding was unfettered ("panmixia"). For present purposes, the basis of physiological selection is referred to abstractly as the "accent" of whatever in the "generative elements" conveys hereditary information.

Chapter 6
Reorientation and Controversy (1895–1899)

The term controversial is conveniently used, by those who are wrong, to apply to persons who correct them.

Bateson (Feb. 1907).

This was a time for reorientation—a lull before the Mendelian storm. Yet the *Materials* mode of thinking had a momentum from which it was difficult to break. Between 1895 and 1899, Bateson shifted from the role of wandering lone naturalist—observing nature in the wild—to that of a home-based, team experimentalist, collaborating with others in the selection of organisms to breed and in the recording of patterns of inheritance among their offspring. He was no longer an observer and collector of experiments *of* Nature. Now, he experimented *on* Nature. While a time for reorientation, it was also a time to stop standing aside from the controversies, scientific and otherwise, that sent the gossipy Victorians scurrying to the pages of the daily *Times*, the weekly *Nature*, the *Fortnightly*, and the monthly *Athenaeum* and *Nineteenth Century*. Bateson was no longer in the wings. He was on-stage! While today the disputes may appear of little consequence, the professional relationships that were cemented or broken in their course would prove critical in the battles ahead.

Aggressive Mimicry

In warfare, sailing under false colors (mimicry) can be a way of escaping attack. It can also facilitate a surprise attack (aggressive mimicry). Bateson's first public controversy was with Edward Poulton (1856–1943), who had been at Oxford since 1873 and had recently published *The Colours of Animals: Their Meaning and Use, Especially Considered in the Case of Insects* [1]. He had also coedited a translation of Weismann's *Essays upon Heredity and Kindred Biological Problems*.

A. G. Cock, D. R. Forsdyke, *Treasure Your Exceptions*,
https://doi.org/10.1007/978-3-030-92099-9_6

Poulton was afforded four pages in *Nature* (Oct. 6, 1892) to review *Animal Colouration* by F. E. Beddard, who had challenged Poulton's endorsement of the power of natural selection. Poulton first referred to "an ignorant assumption" of one German author, Theodor Eimer, and then attacked the work of another (*Animal Life*, by Karl Semper). Whereas Semper appeared "to have written his preface before he considered the materials from which he proposed to write the book," the "contradictory statements" and "inadequate conceptions" of Beddard were "clearly due to haste and want of sufficient reflection." Poulton called for more "exact numerical investigations" giving as examples "those published by Galton upon man, by Wallace upon various animals, by Weldon upon *Cragnon*, and by Lloyd Morgan upon bats."

Whether Beddard attempted a reply is not known. But Bateson did. He chose to take up Poulton on the mimicry by certain flies (*Volucella*) of the bodily appearances of certain species of bees and wasps (Hymenoptera). The naturalist and explorer Henry Bates, a friend of Wallace, had described "protective" mimicry as a phenomenon by which certain non-stinging insects had adapted to appear like stinging insects (Batesian mimicry). The Darwinian explanation was that this warned away would-be predators, so that a fly which looked more like a bee than its fellows would have survived preferentially. This was a non-aggressive mimicry in that it favored an escape, rather than attack, strategy, on the part of the mimicker. Of the three parties—fly, predator, and bee—the role of the bee itself was minimal. However, the case Beddard considered appeared to be one of aggressive mimicry, in that the adaptation involved parasitism by the fly on the bee. This involved just two parties—fly and bee.

Poulton had proposed that the disguise permitted the fly to enter a bees' nest with impunity, where, like a cuckoo, it would lay its eggs [1]. Whereas a baby cuckoo might push its surrogate parents' offspring out of the nest, a newly hatched fly larva would actually feed on the bee larvae. To support his case, Poulton had noted that two varieties of fly with different coloring parasitized the nests of two bee species with respective coloring. Bateson pointed out that the evidence for a match between the coloring of a fly and the bee species it preyed upon was weak. Indeed, larvae of the two varieties could sometimes be found together in the nest of one bee species. Bateson concluded: "The publication of statements like this of Mr. Poulton's, omitting most salient facts – facts, besides, which, though adverse to his speculations, add a ten-fold interest to the subject – is surely unfortunate.... It is well to remember that the value of facts is not to be measured by the ease with which they may be momentarily fitted to the sustenance of a facile hypothesis."

Poulton replied two weeks later (Nov. 10) stating that aggressive mimicry was so well known as an example of natural selection that, given the pressures of space, he had avoided going into the history. His sources had included various museum displays. Lloyd Morgan in his book *Animal Life and Intelligence* had come to the same conclusions having obtained primary data from a museum display. Poulton pointed out that, although he might indeed have been mistaken, the fact that two varieties of different color might emerged in the same nest was "only to be expected." Being varieties (not species), they should be able to breed with each other. What was important was "not the colour of the offspring that emerged, although this is of high interest on other accounts, *but the colour of the parents which enter.* ... It might be supposed that Mr. Bateson would have understood this, but it is perhaps too much to expect from a critic who is so aggressively uninterested."

In his reply (Nov. 24), Bateson, having inspected the museum specimens, concluded that they had been mislabeled and thus extended the scope of his criticism to the museum authorities and to Lloyd Morgan. He concluded that "In the absence of any direct evidence in its favour, and inasmuch as it is inconsistent with many ascertained facts ..., the hypothesis of 'Aggressive Mimicry' should surely be withdrawn." As mentioned in Chap. 5, Romanes wrote (Dec. 2) to Poulton: "I have now read the correspondence in 'Nature.' It seems to me that Bateson is quite absurdly 'aggressive,' even supposing that he proves to be right." Poulson replied to Bateson in *Nature* (Dec. 8) noting that, in addition to Lloyd Morgan and Beddard, another authority, Romanes, had also availed himself of the museum specimens in his published work.

In this, his first major public controversy, Bateson made clear that scientific disagreement relating to professional competence could not easily be separated from other matters. Mischievously, Newton had suggested to Arthur Shipley (who had collaborated with Poulton in the Weismann translation) that Bateson might like to sign a letter supporting Poulton's application for an Oxford Chair. Bateson replied to Shipley (Jan. 19, 1893): "I have to thank you for your note inviting me to sign the circular letter (enclosed) to Poulton supporting his candidature for the Hope Professorship. I regret that I find myself unable to join in it as I do not share the views expressed." Shortly thereafter Poulton become Hope Professor of Zoology, a position he held for four decades. From this high position, Poulton returned to the attack in a later book (Chap. 11). This rallied Punnett to the defense. There was a visit to Ceylon to examine the phenomenon at first hand (Chap. 14) and a book entitled *Mimicry in Butterflies* (1915).

It might be hoped that, a century later, the issue of aggressive mimicry between flies and bees would have been resolved. In a review it was noted [2] that this was a field full of "speculation and observations of strange phenomena, but no really good experiments." Thus, as far we can determine, the dispute remains unresolved. The passage of time was insufficient for Bateson's contemporaries to see the error of their ways if such was the case (Fig. 6.1).

Cultivated *Cineraria*

In 1895, instead of the usual readings of papers, a "new method" was instituted at the RS (Feb. 28), as an "experiment." Henceforth, there would be an open discussion of some topic of general interest, to which non-members would occasionally be invited. On this occasion, Weldon, Thiselton-Dyer, Lankester, and Bateson contributed. Weldon gave a preliminary account of a *Nature* paper [3] that would shortly

Fig. 6.1 W. F. Raphael Weldon

appear (March 7). While not mentioning Bateson by name, it was a renewal of his earlier attack (Chap. 4):

> There will commonly be found a few individuals [organisms] which differ so remarkably from their fellows as to catch the eye .... *Some naturalists* have been led, from the striking character of such variations, to assume for them a preponderant share in the modification of specific character [species-specific character] .... *They* regard change in specific character as an event which occurs, not slowly and continuously, but occasionally and by steps of considerable magnitude, as a consequence of the capricious appearances of 'sports' (our italics).

Weldon then went on to report investigations of the dimensions of the female crab (*Carcinus moenas*) carried out with the support of a RS Committee (of which he was Secretary). He declared that: "The questions raised by the Darwinian hypothesis are purely statistical, and the statistical method is the only one at present obvious by which that hypothesis can be experimentally checked." The method for directly assessing natural selection appeared relatively unambiguous:

> In order to estimate the effect of small variations upon the chance of survival in a given species, it is necessary to measure *first*, the percentage of young animals exhibiting the variation; *secondly*, the percentage of adults in which it is present. If the percentage of adults exhibiting the variation is less than the percentage of young, then a certain percentage of young animals has either lost the character during growth or been destroyed. The law of growth having been ascertained, the rate of destruction may be measured; and in this way an estimate of the advantage or disadvantage of the variation may be obtained.

The following week in *Nature* Thiselton-Dyer gave Weldon's studies high praise [4]:

> An important light is thrown upon what I may be allowed to term the *stability problem* by the remarkable investigation recently presented to the Royal Society by Professor Weldon. It is one which I am persuaded would have given Mr. Darwin much pleasure. ... The result was to show that 'selective destruction' takes place in early life amongst individuals, which deviate from the 'mean specific form.' Professor Weldon arrives at the conclusion 'that the position of minimum destruction should be sensibly coincident with the mean of the whole system,' ... which 'may be expected to hold for a large number of species which are sensibly in equilibrium with their present surroundings, so that their mean character is sensibly the best.' The actual statistical demonstration of this fact, in my opinion, deserves to rank amongst the most remarkable achievements in connection with the theory of evolution.

Dyer went on to "entirely agree" with Weldon "in minimizing the value of 'sports' in evolution," which do no more than reveal "the possibilities of variation and the fact that it may be discontinuous." Then, as an example of the power of the accumulation of small variations in evolution, Dyer exhibited specimens of *Cineraria* (or *Senecio*) *cruenta* (the supposed wild ancestor) alongside the modern cultivated form, as an example of what could be done by "the gradual accumulation of small variations," without the incorporation of "sports."

Bateson at that time was away on field studies in Spain, but Cunningham replied (March 28) noting that Weldon had not carried out an "actual statistical demonstration," and that "Prof. Weldon's results, on his own showing, have done more against selective destruction that for it." On his return, Bateson came across this *Nature* correspondence. He chose to reply first to Dyer and to limit himself to just one point—the origin of *Cineraria*. Bateson stated (April 25) that the general opinion among plant breeders and fanciers, back to the time of first cultivation (about 1800), was that the cultivated form had originated by hybridization of several wild species; there was, moreover, ample evidence of the selection of some mutant forms (sports). In a letter than ran to nearly four columns, Bateson cited authorities for this in extenso.

Dyer replied, Bateson in turn replied, and then Weldon joined the controversy—on Dyer's side (May 16). He began by claiming that "Mr. Bateson has omitted from his account of these records some passages which materially weaken his case." In other words, Bateson, while displaying an erudite search of the original literature, had actually selected items that supported his case. Weldon ended even more cuttingly: "His emphatic statements are simply evidence of want of care in consulting and quoting the authorities referred to." Bateson took offense at this, as casting doubt on his good faith, and Weldon published a note of disclaimer, expressing pained surprise that his words should have been taken in that sense. Bateson wrote privately to Weldon (May 20) suggesting a meeting:

> Last night I sent to *Nature* a letter in answer to you .... There are facts which, if I read them rightly, put your charges in a very unpleasant light. On thinking it all over, and after consulting my friends here, I should like to give you an opportunity of hearing my defence. I am conscious that I owe a great deal in life to you and I am unwilling to treat you as I would Dyer and his like. Will you see me tomorrow? I could leave at 10. a. m., so that I could call at University College at 12 noon or later; but I must catch the 3 p.m. train home. Please wire.

The meeting took place the following day, and Bateson made a note to himself on what was said at the meeting. The note, obscure in parts, began on May 21 and concluded with the date May 23.

> Called and saw Weldon at U.C. Read him substance of my letter to *Nature* – also letter to Editor as to publication with Dyer's. He [Weldon] defended his letter and declined to withdraw, but said he regretted that it should be thought he meant to suggest that I had suppressed evidence because [it was] damaging [to my case]. ... Dyer is 'bluffing'. He [Weldon] told me of passage in "*Cross-Fertn*" p. 335 [Charles Darwin's book] showing that Dyer had been in communication with Darwin and that the latter believed Cinerarias to be probably hybrids (between herbaceous x woody species) – says [he, Weldon] asked Dyer who was 'Moore'? and Dyer 'went perfectly scarlet.' I suggested he should say he had been mistaken – he declined. But said he should say he was sorry for misinterpretation of his reflections on me. (Weldon's position in writing is therefore that of the accomplice who creates a diversion to help a Charlatan. I cannot at all understand his motives or how he can bring himself to play this part.)

Shortly thereafter, Bateson wrote to Weldon:

> I have to thank you for your letter and enclosure. The assurance that you did not intend to suggest that I had tampered with the evidence is of course accepted, and the expression in my letter to *Nature* relating to this shall not appear. – In fact the letter will now be largely recast. You will not I suppose think that I take a liberty if I refer to the passage in "Cross Fertilization" showing Darwin's opinion, and establishing a presumption of self-sterility in the Cineraria. I shall of course not mention the occurrence of Dyer's name in the passage, and I shall acknowledge that I have seen it through the courtesy of an opponent – not of course introducing the name of my informant.
>
> Having regard to the relations in which we once stood to each other, it would not I think be right that I should be less than explicit in speaking of the impression that this incident has left with me. Had I known nothing of your mind beyond what your published letter told, I should scarcely have thought your intervention either wise or good-natured, but that would have been all. I have now learnt from you that it was undertaken to support a man whom you believed to be insincere – "bluffing" as you said. To what good end such an action may lead I do not know, nor indeed would I willingly think of it any more. It is through things like this that I have come to doubt whether the world of science is a school of truth or of chivalry.

Weldon replied in exasperation (May 24):

> I can do no more. First, you accuse me of attacking your personal character; and when I disclaim this, you charge me with a dishonest defence of some one else. I have throughout discussed only what appeared to me to be facts, relating to a question of scientific importance. If you insist upon regarding any opposition to your opinions concerning such matters as a personal attack upon yourself, I may regret your attitude, but I can do nothing to change it.

The meeting on the May 21st probably did more than any other single incident to sour their relations. The crucial point was that Weldon let slip that he thought Dyer was "bluffing." Dyer was representing himself as an orthodox defender of Darwinism and yet, in his book *Cross and Self Fertilization*, Darwin himself had stated that cultivated *Cineraria* were "probably derived from several fructicose or herbaceous species, much intercrossed." Moreover, Dyer could hardly be unaware of this, since, in a footnote Darwin acknowledged Dyer's help in giving information about the varieties [5]: "I am much obliged to Mr. Moore and to Mr. Thiselton-Dyer for giving me information with respect to the varieties on which I experimented. Mr. Moore believes that *Senecio cruentus, tussilaginis*, and perhaps *heritieri, maderensis* and *populifolius* have all been more or less blended together in our cinerarias."

The *Cineraria* controversy, which was also reported in the *Gardeners' Chronicle* (May 11, 1895), was rich in imputations of bad faith. The only written authority supporting Dyer's view of a single origin was an article by one of Dyer's staff at Kew—R. Allan Rolfe (*Gardeners' Chronicle*, 1888). Bateson later received a letter from J. H. Burkill of the Botany School at Cambridge (Oct. 17, 1895). During a visit to Kew Rolfe had asked him to tell Bateson his view of the controversy. The 1888 article had, for some reason, been altered by the editor, Maxwell Masters. Rolfe was himself in favor of a hybrid origin. When the controversy arose Rolfe had been advised by a colleague not to point out the mixed origin of the article to Dyer. Bateson was to treat all this as confidential, as Rolfe feared for his position.

It would be interesting to know who was right about *Cineraria*. One relevant point would be chromosome numbers. However, the genus is now merged with *Senecio*, with over 500 species. Darlington and La Cour's *Chromosome Atlas of Flowering Plants* shows that there is a great deal of polyploidy in the genus, and many specific names have changed since Bateson's day. So, once again, no immediate truth emerges. The Cineraria controversy overtly died when the Editor of *Nature* refused to publish more. But Bateson would not let the bone go. In 1895, in the *Proceedings of the Cambridge Philosophical Society,* he published "Notes on Hybrid Cinerarias Produced by Mr. Lynch and Miss Pertz." And there was further comment in 1900 in a paper on "Hybridization and Cross-Breeding."

The episode brings to light Francis Darwin's role in trying to sooth troubled waters. At a meeting of the RS in 1896, Weldon thought that Bateson was deliberately trying to avoid him and mentioned this to Darwin. On hearing of this, Bateson wrote in denial to Weldon (Dec. 6), who replied (Dec. 8): "I am sincerely sorry that I have done you an injustice. The whole thing has been very painful to me; and I am glad to hope that we may now meet on better terms. May I now do what I did not venture to do before, and offer you all good wishes and congratulations on your marriage. Yours very truly, W. F. R. Weldon." Bateson replied curtly (Dec. 10) thanking him for the good wishes and concluded: "I too shall be glad if it is possible for us in the future to meet on better terms. W. B."

As for Dyer, it must have been an uphill battle for Francis Darwin. In his dotage, Dyer seemed to mollify. He wrote to Bateson in 1914 (May 24): "I gather that you are supposed to resist the chromosomes being the carriers of characters. I have long thought that the cytoplasm must play some role in the matter. … F. Darwin was here the other day and seemed doubtful about the efficiency of natural selection. … Strict skepticism as to old beliefs is an essential condition to any progress."

Crabs

But Bateson had not forgotten other issues raised at the portentous RS meeting (Feb. 28). Weldon's crab studies were criticized in *Nature* not only by Cunningham and Bateson, but also by Lankester. In a paper in the *Proceedings of the Royal Society* (1895) on variation among crabs [6], Weldon had attempted to demonstrate the operation of natural selection through the decreased variability of various measurements in older crabs (implying that crabs that had deviated from the average had been eliminated by natural selection, so that variability had decreased in the population).

Under the title "The Utility of Specific Characters," Lankester in *Nature* (Aug. 20, 1896) questioned these studies. Darwin had held that a character might in some way be correlated with another, but only one would be the direct target of natural selection. In some cases, this was obvious—for example, natural selection might increase arm muscle mass and correlated with this there would be an increase in body weight, which itself, in the range observed, might be little impacted by natural selection. More often, the correlation would not be obvious, in which case the question arose as to which of two characters was target, and which the correlate. Weldon had ascribed to Lankester the view that it was "legitimate" to pick one of these characters ("antecedent phenomena") as "the only effective cause of change in death rate," the other being a "merely unimportant concomitant … of this one essential change." Lankester demurred:

> If the study of the antecedent or associated phenomena by means of hypothesis and test-experiment had not been and cannot reasonably be carried out, the naturalist cannot … reasonably either 'pick out' one of them and assert that it is the cause of increased or decreased death-rate. … Surely in the case of Professor Weldon's crabs, most naturalists would take the view that the frontal measurements may *possibly* be operative in saving the life of the crab, or *may* be only a correlative of some other life-preserving structure; that its quality in this respect should be inquired into by means of hypothesis and experiment; and that, until this is done, it is *premature* to speak of a particular frontal proportion as having for its *effect* the survival of those crabs distinguished by its possession. The chief task of the student of living things seems to lie in the search for such explanations. … You cannot … reduce natural selection, as Prof. Weldon proposes, to an unimaginative statistical form, without either ignoring or abandoning its most interesting problems.

In October 1896, Bateson wrote to Francis Galton, as chairman of the RS Committee supporting Weldon's work, extending these criticisms. A decade later Pearson retrospectively summarized the difficulties that had confronted Weldon's study [7]:

> How is the … deviation from type to be measured at each stage of growth? What is to determine 'adult' life? What measure is there of the time during which the individual adult crab has been exposed to the selective destruction? Weldon undoubtedly chose the crab because of the facilities it offers for measurement. But its age then becomes an appreciation based merely on the obviously close, but probably imperfect correlation between age and size. Further, the law of growth, complicated rather than simplified by the moults, and the question as to how far the variability of the characters dealt with is affected by growth, combine, in the case of crabs, to form an exceedingly difficult problem. It is practically impossible to keep a large series of crabs through the whole period of adolescence, and if it were possible,

> it is far from certain that the claustral environment necessary would not sensibly affect their law of growth.

Bateson's arguments, a decade earlier, had led to a voluminous correspondence, lasting into 1897; no punches were pulled on either side, but on the other hand, there was no personal vituperation. Much of the correspondence was carried out privately by way of Galton, which might suggest that relations between Weldon and Bateson were by now so strained that direct communication was impossible. This seems unlikely since from time to time during 1897 dinner invitations were exchanged.

Weldon's paper had been published, not under his own name, but as a Report *from* the Committee (not *to* the Committee, as in Bateson's own later series of Reports; Chap. 8). There was an appendix under Weldon's name, but this served only to heighten the illusion that the body of the paper was the joint responsibility of the Committee. So, it was quite natural that Bateson should write in the first instance to Galton as chairman of the Committee, and it would have been up to Galton to suggest that Bateson continue the correspondence directly with Weldon—as, eventually, Galton did. Galton wrote to Bateson (Nov 20, 1896) concerning the next meeting of the Committee: "At it, I intend to propose a resolution … to the effect that the responsibility for statements in such papers as they present to the R. Soc. rests with the authors of those papers."

At a committee meeting (Jan. 14, 1897), Weldon, seeking Bateson's input, handed him a draft of his latest paper on crabs and shortly thereafter (Jan 16) noted: "I propose to finish it, and to rewrite it and publish without consulting the Committee further." Bateson replied two days later:

> As to this new work … I scarcely know how to say what I feel. Yet if I say nothing I may hereafter be reproached for my silence. You have given vast labour to it, and I cannot doubt that you have seriously thought yourself justified in the conclusions you put forward. Perhaps too you have thought a little of the responsibility you take in using your authority to publish such work – for as you know, though many will accept your conclusions, few will read a paper like this and fewer still will even make an attempt to follow it or to judge the matter for themselves. Nevertheless to me the evidence seems [so] wholly inadequate and superficial, that I do not understand how a responsible person can entertain the question of accepting it. I very truly regret that you should give your countenance to such a production.

Weldon replied (Jan. 24, 1897): "Thanks for your very plain speaking. Since your opinion of work which has cost me some trouble is so poor, it is certainly better to say so frankly. I do not agree with you, as I suppose you expected, but that cannot be helped." Although not stated as such, it also could not be helped that, through his unremitting attacks, Bateson had so shaken Galton's confidence in his Committee's agenda that, de facto power had been wrested away from Weldon and his biometrical supporters and was now in the hands of Bateson, who from the outset had been excluded from its deliberations. It was a coup of major proportions. Henceforth, with reformulated goals, the Committee was to be known as "The Evolution Committee."

The Evolution Committee

A RS "Committee for Conducting Statistical Enquiries into the Measurable Characteristics of Plants and Animals" was formed in 1894 with Galton as Chairman and Weldon as secretary [8]. Other members were Francis Darwin and A. Macalister of Cambridge, Meldola of London, and Poulton of Oxford. Macalister attended no meetings. Most of the early meetings involved the approval of funds for Weldon and his reporting of his preliminary results. In 1896, at Galton's suggestion, a mathematician (S. H. Burbury) was added. Pearson was also added, but he attended no meetings because, as he noted in a later letter of resignation, they "have always taken place at a time I cannot attend."

There was an informal meeting (Dec. 4, 1896), with two non-members present (O. Salvin, Heape). Bateson was invited but was unable to attend. Under consideration was the possibility of setting up an Experimental Farm (possibly centered at Darwin's Down House), which would have a resident caretaker and would provide research materials and opportunities for breeding studies. An early push in this direction had been given by Wallace who, after a vigorous public debate with Romanes and Gulick, was inclined to test their ideas, if only to disprove them. Wallace wrote to Galton (Feb. 3, 1891):

> Don't you think the time has come for some combined and systematic effort to carry out experiments for the purpose of deciding the two great fundamental but disputed points in organic evolution: – (i) Whether individually acquired external characters are inherited, and thus form an important factor in the evolution of species, – or whether, as you and Weismann argue, and many of us now believe, they are not so, and we are thus left to depend almost wholly on variation and natural selection. (ii) What is the amount and character of the *sterility* that arises when closely allied but permanently distinct species are crossed, and then 'hybrid' offspring bred together. Whether the amount of infertility [is the same] between the hybrids of species that have presumably arisen in the *same area*, and those which seem to have arisen in very *distinct* or *distant areas* – as oceanic or other islands. ... To be really good however the hybridity experiments (and the others too) would have to be carried out with large numbers of animals, and thus some sort of experimental farm would be required. ... Then, using small animals such as *Lepus* and *Mus* among mammalian, some gallinaceous birds and ducks, and also insects, a good deal could be done ... at a small cost.

In his capacity as Chairman Galton wrote to Bateson (Jan. 1, 1897): "Both Weldon and myself are determined that you should join us. Would it be agreeable to you that we should propose your name?" Bateson refused (Jan. 3): "I had better not. I am not convinced that the present lines of enquiry of the Committee are fruitful, and I do not think it is likely that the results will be at all proportionate to the labour expended." However, after a talk with Galton (Jan. 6) he relented, and Galton (Jan. 8) asked him to attend a meeting (Jan. 14) where his membership would be proposed: "Please consider this an invitation to attend, and to suppose yourself invisible until formally elected." At that meeting, it was resolved:

1. To institute enquiries to the extent to which existing establishments, whether private or public, might be induced to co-operate in collecting data, having

reference to Heredity, Variation, Hybridism, and other biological phenomena requiring continuous observations.
2. To institute enquiries as to the feasibility of establishing one or more experimental stations where data of the above and similar character might be collected and discussed.
3. To establish a Sub-Committee for the purpose of conferring with Breeders.

Moreover, the scope of the Evolution Committee was stated to include: Inbreeding; Prepotency (dominance); Variation (its amount and transmission); Hybridism; Fertility and Sterility; Coefficients of heredity, including fraternal variation; Panmixia; Telegony (an interest of Cossar Ewart); Transmission of acquired modifications; Mental state of mother affecting the offspring (as was believed by many breeders). In that and subsequent meetings, several others were co-opted, as either permanent or "accessory" members—Beddard, E. Clark, J. Evans, Cossar Ewart (zoologist; 1851–1933), Frederick D. Godman (etymologist; 1834–1919), Heape, Lankester, E. J. Lowe, Maxwell Masters of the RHS, Osbert Salvin, and Thiselton-Dyer. Bateson submitted a thirteen-page brief to Galton (Feb. 1, 1897) indicating the lines of research he personally hoped to undertake:

> The subjects … on which we want evidence are these. 1. The facts as to the transmission of parental characters and the conditions determining the transmission (including the questions of the blending of characters, the fixation of characters, etc.). 2. The conditions (other than those included in 1) determining (a) number of offspring; viz. fertility and sterility. (b) sex of offspring.

At that time, he envisaged studies with fowl, canaries, moths, and butterflies. At the next meeting (Feb. 11), a change of name to "The Evolution (Animals and Plants) Committee of the Royal Society" was approved. The task of the Committee was seen as the "accurate investigation of Variation, Heredity, Selection, and other phenomena relating to Evolution." Weldon, who remained as Secretary, was not happy. A day after the meeting, he wrote to Galton: "The Committee you have got together is entirely unsuited. … It is far too large, contains far too many of the old biological type, and is far too unconscious of the fact that the solutions to these problems are in the first place statistical, and in the second place statistical, and only in the third place biological."

Late in 1897 (Oct. 29), Weldon wrote telling Bateson about his recent researches on survival of crabs of differing carapace sizes that had been held in different tank environments, and advising him that there was some spare money in the Committee kitty:

> We are both very sorry that we are engaged on Sunday and so cannot accept your invitation. Will you come and dine on Wednesday if you are free in the evening, or if not, will you come and fetch me from University College at any time after two? … I wish you joy in your allotment. It should do a lot of things. There is rather more than £100 unspent. Galton and Heape want some, I fancy, and I do not think anyone else has any plans, or asks for any money.

Within a few weeks, Bateson was submitting an "application for a grant in aid" of research to Galton (Nov. 19, 1897). He began by describing where the work would be done and who would do it:

> I hear from Weldon that the Committee will probably meet on Tuesday next and I am therefore making application for a grant in aid of experiments on heredity. As I told you the other day, I have taken an allotment (about 3/8 acre) on ground belonging to the University adjoining the experimental ground of the Botanical Gardens. ... The experiments will be carried out partly by myself and partly by Miss Saunders, whose paper on *Biscutella* was lately published in *Proc. Roy. Soc.* If any grant is made to me it is understood that it will be applied to the maintenance of our experiments jointly.

Next, he turned to the nature of the work:

> The general object of our investigation is to determine in suitable cases the degree to which varietal characters do or do not blend in offspring of mixed parentage. For the past three seasons I have been trying to do this in the case of local varieties inhabiting adjoining areas, taking for subjects (i) *Pieris napi* and the alpine variety *bryoniae;* (ii) *Pararge aegeria* and the northern variety *egerides*. ... On the botanical side, Miss Saunders has begun experiments of the same kind on *Biscutella laevigata* (hairy and smooth varieties), several species of *Papaver*, and *Atropa belladonna* (black and yellow fruits). She has also collected material for beginning experiments on *Lychnis verpertina* and *diurna*. ... I wish to begin some experiments of the same kind on two breeds of poultry. Some simple experiments on poultry would not be very costly; but as I wrote to you in my letter of Feb. 1, it appears to me desirable that special experiments should be made on the degree to which the transmission of parental characters depends on the degree to which either parent is closely bred. The whole question of the consequences of in-breeding is obscure, but if any evidence could be got on this particular point it would be most important.

Finally, he made a case for funds which happily matched the amount he knew to be available:

> The first need however is an assistant, and I should regard any grant received as intended primarily for this object and for the general expenses of the experiments on plants and lepidoptera; but if I could get as much as £75 I should start the poultry experiments next spring. As I do not know what other applications will be made I hardly like to apply definitely for so much, because of course there must be equitable division. I have been corresponding with my sister [Anna] as to the continuance of her work on hybrid Cypripediums. She says she is now so much tied up to her own place that she must give up the idea of the traveling which would be involved in frequent visits to the growers. She hopes however that some of them may be induced to send her material. ... I hope before long to have an opportunity of consulting with Dr. Masters about this. Will you kindly send this letter to Weldon together with my letter of Feb. 1 which I think you still have?

Bateson seems to have encouraged Anna to submit a proposal on orchids (Feb. 17, 1897):

> Your notes are much to the point. ... A circular is to be issued asking persons willing to undertake work to specify what it is and how the Committee can help them. You should have a copy of this. ... I think yours quite the most promising scheme yet submitted! ... By the way Masters said he knew a man who was measuring lengths of petals in orchids – but I did not get any talk with him afterwards, so I do not know who it is. Anyhow, it needn't interfere with you. May I show him your scheme when our circular goes out?

It is possible that the reference to orchids represents the first wind of one who was to become Bateson's greatest ally, Charles Hurst. At that time, Hurst was, in a small way, collaborating with Pearson. It seems that Anna did not pursue the orchid work.

At an early-stage Bateson formed an alliance with the Scientific Committee of the RHS. In January 1898, the Evolution Committee appointed a Sub-Committee "to confer with the Royal Horticultural Society," and at a meeting with the Committee Chairman, Masters (Feb. 9, 1898), Bateson asked that RHS members carefully record details of their crosses. Questionnaires would be circulated to those likely to help. Although supportive in principle of the idea of an Experimental Farm, Bateson opposed it (Chap. 7), perhaps because he saw it as drawing funds away from other activities. The idea was dropped. Weldon described his continuing crab studies in a Presidential Address to the Zoological Section of the BA in Bristol in 1898. He was still managing Committee affairs and in June 1899 gave Bateson bad news:

> The Evolution Committee is in rather a mess. The Government Grant Committee has only given us £50. We must meet and consider. … I think, as I have thought for some time, that you, or anyone else working at definite problems, would do better to deal directly with the Government Grant Committee and that the Evolution Committee is a mistake.

The storm was weathered. In January 1900, Weldon sent Bateson a £75 cheque. However, at about that time Pearson, Weldon, Galton, Dyer, and Meldola brought into question the future of the Committee by resigning *en masse*. Meldola wrote to the RS (Jan. 22, 1900):

> It was … our hope that we might have been able to start something in the nature of a zoological (inland) station for systematic experimental and observational work requiring constant supervision over long periods of time. This hope has not been realised and the work of the Committee seems now to have resolved itself into applications for grants from the Govt. Grant Committee for particular members of the Committee to carry on researches. But such applications could be quite as well made by the individual members directly in the ordinary way, and I do not see any reason for multiplying the already overwhelming number of Committee meetings that scientific men resident in London are being constantly called upon to attend.

Responding to a request from Foster (for the RS) concerning the Committee's future, Bateson agreed to make a case for its continuation (Jan. 28, 1900). He added: "We all very genuinely regret Galton withdrawing his name from the Committee, but I am not alone in feeling that so long as he acted as Chairman progress was almost impossible. The loss of the other members does not affect us in any way. I have no doubt that Godman who has generally attended our meetings will be willing to become Chairman." Heape was more combative, writing to Bateson (Feb. 1, 1900):

> I have been thinking of our conversation about the Evolution Committee and Foster's request for a letter. It seems to me that the course we are asked to take is a most unusual one. We are in point of fact appointed to serve for 12 months [they had just been reappointed], and in spite of that fact we are requested to show cause why we should exist. I do not hesitate to say that were I in your place I should refuse to reply to the question unless it was made officially.

Bateson's reply to Foster worked. The following December the Committee was reappointed (Bateson, Burbury, Darwin, Ewart, Godman, Lankester, Macalister, McLachlan, Masters, and Poulton, with "accessory members" Clarke and Somerville). From 1900 until 1910, with Bateson as Secretary and Godman as Chairman, the committee was Mendelian-dominated, with four members regularly in attendance (Bateson, Darwin, Godman and Clarke). In 1906, the horticulturalist L. Sutton and biologist Leonard Doncaster were co-opted. The Committee was probably more useful as a talking-shop than as a direct promoter of research. Only one smallish piece of work (a collective enquiry into melanism) arose from its activities, although other collective enquiries were started. To Bateson, the Committee was useful in several ways; as a source of moral support at a time when his position in Cambridge was isolated; financially, through research grants of up to £75 per annum from 1898 to 1903; as a vehicle for the publications of his new Mendelian school after 1900. The five long *Reports to the Evolution Committee* (1902–9) by Bateson and associates might have been difficult to place, unless in much curtailed form, in any ordinary publication.

Melanism in Moths

Bateson had been breeding melanic (black) forms of butterflies in the Department of Animal Morphology, but without much success due to infection of his colonies (Chap. 12). Since the first observation of a melanic form of the peppered moth in Manchester in 1848 [9], there had been numerous sightings, suggesting that some emergent selective pressure was favoring the survival of these rarities (*Amphidasys doubledayaria*, now known as *Amphidasys carbonaria*) over the more usual form (*Amphidasys betularia*, now known as *Biston betularia*). In 1896, an authority on moths, J. W. Tutt, had suggested that in smoky industrial areas the darker color would more effectively camouflage the moths against avian predators—a clear example of Darwinian natural selection in action [10].

Of course, Bateson was skeptical. By May 1900, he had persuaded the Evolution Committee to allow him to solicit the assistance of amateur naturalists on a "Collective Inquiry as to Progressive Melanism in Moths." A circular noted that "dark forms of several species of moths have recently appeared and become increasingly abundant. … It is to be regretted that no systematic or statistical records of these phenomena have been kept, and it appears to the Committee that if such a record be now instituted and continued for a period of years it cannot fail to have considerable scientific importance." The results of the study were collated by Doncaster in 1906 [11].

Bateson concluded in 1913 (*Problems*) that "dark specimens had appeared sporadically" and that "strains derived from these dark individuals have gradually superseded the normal type more or less completely," an apparent "instance of evolution proceeding much in the way which Darwin contemplated." But would black really camouflage against predators?

Fig. 6.2 Karl Pearson

> To my mind there is a serious objection to the theory of protective resemblance in application to such a case as the *betularia* forms, which arises from the fact that the black *doubledayaria* is a fairly conspicuous insect anywhere except perhaps on actually black materials, which are not common in any locality. Tree trunks and walls are dirty in smoky districts but they are not often black, and I doubt whether in the neighbourhood of Rotherham, for instance, which is one of the great melanic centres, *doubledayaria* can be harder to find for a bird than *betularia* would be. ... Those who see in such cases examples of the omnipotence of Selection must frequently find themselves in this dilemma. Taking the evidence as a whole, we may say that it fairly suggests the existence of some connection between modern urban developments and the appearance and rise of the melanic varieties. More than that we cannot affirm.

The issue became a *cause célèbre* for the selectionists [12]. Bateson may have been near the truth (see Chap. 24) (Fig. 6.2).

Homotyposis

Bateson's first scientific clash with Pearson came at the end of 1900. By this time, Bateson was into Mendelism (Chap. 8), but it seems that Pearson's attention was not drawn to Mendelism until October 16th when Weldon wrote: "I have heard of and read a paper by one, Mendel, on the results of crossing peas, which I think you would like to read. ... I have the R. S. copy here, but I will send it to you if you want it." The phenomenon which Pearson called "homotyposis" relates more to the pre-Mendelian stage of Bateson's thinking and is best considered here.

In *Materials,* Bateson had referred to the repetition of similar parts as "merism." Digits, vertebrae, petals, were all members of "meristic series." The resemblance between two sides of a body—bilateral symmetry—was a special case of this, as were the limbs of starfish and daisy petals (radial symmetry). He had proposed that: "If … there is a true analogy between the process by which new organisms may arise *asexually* by Division, and the process by which ordinary meristic Series are produced, it follows that Variation, in the sense of difference between offspring and parent, should find an analogy in Differentiation between the members of a Meristic Series" (our italics). In 1894, his hopes for this approach were high:

> If reason shall appear hereafter for holding any such view as this, the result to the Study of Biology will be profound. For if it shall ever be possible to solve the problem of Symmetry, which may well be a mechanical one, we shall thus have laid a sure foundation from which to attack the higher problem of Variation, and the road through the mystery of Species may thus be found in the facts of Symmetry.

Pearson saw an opportunity for biometric analysis. On October sixth 1900, the RS received his paper on "homotyposis," meaning the degree of similarity between repeated parts, which he referred to as "undifferentiated like organs." Under this heading, he considered both single cells, such as blood corpuscles, and multicellular organs, such as plant leaves. Thus, two leaves from a tree were "homotypes" or "homotypic organs." His coefficient of homotyposis, measured by comparing the variabilities of a measurement made on such repeated parts, was claimed to relate to the correlation between brethren for the same measurement. Coming from the Professor of Applied Mathematics at University College, and replete with fancy equations and a new buzz word, the two RS reviewers responsible for deciding suitability for publication would normally have accepted it. But Bateson was one of the reviewers. He sent in an adverse report. Nevertheless, the paper went forward for reading at the next meeting (Nov. 15), and Pearson's five-page abstract, that had been pre-circulated to the Fellows, was published in the *Proceedings of the Royal Society* (Jan. 17, 1901). Eventually, the ninety-five-page full paper was published (Dec. 1901) in the *Philosophical Transactions of the Royal Society* [13].

When the November reading was announced, Bateson wrote to Foster asking if he might have ten minutes during the discussion. He also gained agreement that it would be acceptable for him to base his remarks on the full paper, and not just on the five-page abstract, which was all that other Fellows present at the meeting would have seen. Bateson duly delivered his remarks and afterward Pearson had the opportunity to amend the paper (the published version of which contained a footnote dated July 1901). Foster invited Bateson to write up his comments as a paper, to be published in the *Proceedings*—which Bateson did (received by the RS Jan. 25, 1901). This was published in the April 1902 issue, but it was dated 1901, and contained an "added note" indicating that Bateson had been able to amend it as late as November 1901.

In his initial abstract, Pearson wrote: "If an individual produces a number of like organs, which so far as we can ascertain are not differentiated, what is the degree of diversity and of likeness among them? Such organs may be blood-corpuscles, hairs, scales, spermatozoa, ova, buds, leaves, flowers, seed-vessels, etc., etc. Such organs I term *homotypes* when there is no trace to be found between one and another of differentiation in function." By emphasizing an absence of differentiation, he excluded some of Bateson's meristic series, such as fingers and vertebrae. Indeed, he later introduced the idea of replaceability—that one leaf could perfectly replace another—something that fingers and vertebrae could not achieve, since they were related in an "organic" sense, meaning that "they have a function to perform in common." The abstract set out "the problem":

> Is there a greater degree of resemblance between homotypes from the same individual than between homotypes from separate individuals? If fifty leaves are gathered at random from the same tree and from twenty five different trees, shall we be able to determine from an examination of them what has been their probable source? Are homotypes from the individual only, a random sampling, as it were, of the homotypes of the race?

As it stands, even today this would not be without interest, but Pearson's abstract went further:

> A theory of fraternal hereditary resemblance is given on the basis of the likeness of brothers being due to homotyposis in the characters of spermatozoa and ova put forth by the same two individuals [their parents] and uniting for the zygotes whence the brothers arise. It is found that the mean value of fraternal correlation ought to be equal to the mean intensity of homotypic correlation. We have so far worked out nineteen cases of fraternal correlation in the animal kingdom, and their mean value = 0.4479, *i.e.,* is sensibly equal to the intensity of homotyposis in the vegetable kingdom. It is, therefore, very probable that heredity is but a phase of homotyposis, and that the latter approximates to a certain value throughout living forms. ...The results of our first investigation in this field seem to support the view just expressed, and to indicate that the Principle of Homotyposis ... is a fundamental law of nature, which will enable us to sum up in a brief formula a great variety of vital phenomena.

Bateson was not happy with this. The large number of "competent helpers" and "collaborators" acknowledged by Pearson indicated that he had financial resources far greater than those available to Bateson for his own work. Yet the non-confrontational manner of Bateson's opening remarks, typical of his entire paper, suggests more a dissatisfaction with the underlying science than envy, or anger at the possibly unwise dissipation of resources:

> Professor Pearson raises an issue of extraordinary importance. In any attempt to perceive the true relation of variation to differentiation, and to analyse the essential similitude existing between Heredity [repetition of individuals] and Repetition of Parts, we reach a fundamental problem of biology. ... At the outset I wish to express the conviction that the leading idea which inspired and runs through the work is a true one. Prof. Pearson suggests that the relationship and likeness between two brothers is an expression of the same phenomenon as the relationship and likeness between two leaves on the same tree, between the scales on a moth's wing, the petals of a flower, and between repeated parts generally. The conception of heredity is thus greatly simplified, and that phenomenon is seen in its true relation to the other phenomena of life, becoming merely a special case of the phenomenon of Division and the repetition of parts.

He went on, however, to point out a problem with the definition of two organs as homotypically related. To produce organs, the original zygote had to divide and differentiate into different cell types. Homotypic and heterotypic organs resulted.

Pearson's study concerned comparisons between homotypic organs where differentiation had resulted in similar forms—scale versus scale, leaf versus leaf, corpuscle versus corpuscle—not comparisons between heterotypic organs where differentiation had resulted in dissimilar forms—stem versus root, wing versus leg, liver versus brain. After their original differentiation (i.e., regulated variation), homotypic organs were assumed to vary from each other only indiscriminately (unregulated variation). By dealing only with "undifferentiated like organs," Pearson had ruled further regulated variation (differentiation) between like organs as beyond the scope of his study.

Bateson's point was that it was not possible to draw a hard line between "specific" differentiation and "normal" unregulated variation. Pearson could not exclude the possibility that some degree of differentiation (regulated variation) had occurred between the organs he had classed as homotypic. In particular, when classifying gametes as homotypes, Pearson had written:

> Turning to the process of reproduction, the offspring depend upon the parental germs, and it would thus seem that the degree of resemblance between offspring must depend on the variability of the germ cells and the ova which may each be fairly considered as 'undifferentiated like organs.' Here again we are not compelled to assert that much or little is due to environment and little or much is due to inherent ancestral influence. All we assume is that such causes as produce the likeness between leaves of the same tree, or florets on the same flower, produce the likeness between spermatozoa or ova of the same individual, and that on this likeness the ultimate resemblance of offspring from the same parent depends. We have then to investigate how the quantitative resemblance between offspring of the same parents is related to the quantitative relationship between the undifferentiated like organs in the individual. … Now the reader will perceive at once that if we can throw back the resemblance of offspring of the same parents upon the resemblance between the undifferentiated like organs of the individual, we shall have largely simplified the whole problem of inheritance. Inheritance will not be a peculiar feature of the reproductive cells.

Emphasizing the latter point, Pearson went on to affirm that "the problem of inheritance is to a large extent the same as the problem of variability in the individual …. It is part of a much more general problem having nothing to do with sexual reproduction …. Heredity is only a special case of homotyposis." However, gametes were known to be highly differentiated (Chap. 8). Bateson wrote:

> Does he recognize that variation between brothers is comparable not merely with variation between repeated parts, *but also with differentiation*, and with predominantly orderly variation among such parts? … Ova and spermatozoa can be treated as 'undifferentiated like parts' so long as their variations, judging by the resulting offspring, are sensibly irregular. Can we recognize differentiation among them as distinct from variation? Certainly we *sometimes* can. In determining the correlation of confraternities, the parentage enables us to distinguish the fraternal groups correctly, and consequently a fraternal correlation may be truly determined. *For to do so we are not compelled to distinguish differentiation from variation.*

A problem arose when Bateson courteously sent a copy of the proofs of his paper to Pearson, who replied indignantly enclosing the proof "unexamined." It was, he maintained with some justice, highly improper that Bateson's paper should be printed (and, he suggested, circulated to Fellows, though this seems not to have happened) before his own (which it criticized) had even been put in proof, let alone published. This was an abuse of the privileged position of a reviewer. Bateson immediately withdrew the paper, which had been submitted to the RS on the understanding that it would not be published until after Pearson's. He also hastened to reassure Pearson of his own good faith. Foster's assistant wrote to say that it had never been intended to publish Bateson's paper before Pearson's: it had been put into type just "to get it out of the way."

The exchange of letters between Pearson and Bateson came at a time when it was inevitable that the homotyposis controversy should become entangled with the one on Mendelism. Implicit in Bateson's criticisms—not of Pearson's mathematics, but of the underlying biology—was a criticism of Pearson's biological advisor, Weldon. In fact, Bateson sent an impassioned plea (Feb. 14, 1902) to Pearson for peace and friendship. The reply was noncommittal—even an outright rejection (Feb. 15, 1902). Pearson made clear the main reason was his own friendship with Weldon. Furthermore, Pearson thought that the RS *Proceedings* ought to be reserved for original contributions to knowledge and closed to controversial matters of this kind.

Biometrika

The day after the sour reception of homotyposis at the November RS meeting, Weldon wrote to Pearson: "The contention 'that numbers mean nothing and do not exist in Nature' will have to be fought .... Do you think it would be hopelessly expensive to start a journal of some kind?" Thus *Biometrika* was created, with Galton as "consulting editor" and including as "coeditor" Charles Davenport (1866–1944) of the University of Chicago, who had recently authored a short text on biostatistics [14]. "A guarantee fund sufficing to carry the Journal for a number of years was raised at once; good friends of Biometry coming forward to aid the Editors" [7]. Weldon began the first issue with an Editorial:

> The first condition necessary, in order that any process of Natural Selection may begin among a race, or species, is the existence of differences among its members; and the first step in an enquiry into the possible effect of a selective process upon any character of a race must be an estimate of the frequency with which individuals, exhibiting any degree of abnormality with respect to that character, occur. The unit, with which such an enquiry must deal, is not an individual but a race, or a statistically representative sample of a race; and the result must take the form of a numerical statement, showing the relative frequency with which various kinds of individuals composing the race occur.

Supported by "good friends" with deep pockets, there was now no impediment to publication of their views. In the April 1902 issue of *Biometrika* [15], there appeared Pearson's formal *twenty-five*-page response to Bateson, entitled "On the Fundamental Conceptions in Biology." As Editor, he offered Bateson space for a reply. Bateson produced a twelve-page paper, which Pearson said was too long, but he would publish a version cut to six pages. This Bateson was "unable to accept." So he arranged for Cambridge University Press to print it privately in the same format as *Biometrika* (of which the Press was the publisher). The paper was expanded from the original, and doubtless some of the barbs were sharpened. Only a hundred copies of this brilliant, twenty-three-page, polemical squib, *Variation and Differentiation in Parts and Brethren*, were printed.

In his starkly confrontational reply of April 1902, Pearson had attacked not only Bateson but also the "obscure" ideas of de Vries, and the Zoological Committee of the RS—the "old school of biologists" under whose aegis Bateson's "confused and undefined notions" had been published. Pearson lamented Bateson's employment of "the sort of language we know so well in mediaeval works of physics," and struggled to understand "the sense in which Mr. Bateson uses his terms." Since Bateson had introduced ideas from his "bulky volume" *Materials*, Pearson considered a critique of the latter was fair game. Here, he was on strong ground:

> If Mr. Bateson wishes to attack the problem of evolution by what he terms discontinuous variation, he must go further than forming a useful catalogue of museum and collectors' deviations from 'type.' He must trace first whether in any given case they are or are not inherited, secondly he must discover whether or not the individuals who possess them are more fertile than the 'type,' thirdly whether the death rate is with regard to them selective or non-selective. … Starting with a race having among its members a few with a recognisably discontinuous variation, he must show how its descendents at a later period have the discontinuous variation of the earlier period as a dominant character. In other words he must deal with the vital statistics of populations, or proceed *biometrically*.

It was not until page seventeen of his reply that Pearson turned to "the problem of homotyposis," to set out the "general tenor of my reasoning":

> The production of gametes seems a process analogous to that of the production of any like organs by an individual, and the average value of the correlation of such organs ought to give us a value approximating to that of the average correlation between the scarcely measurable organs and characters of the gametes themselves. … Absence ['of observable *differentiation in the gametes themselves*'] was the very sufficient reason for comparing the correlation of characters in the gametes with the correlation of *undifferentiated* like organs. Hence the source of my definition of homotypes as 'undifferentiated like organs.' It will be seen at once that the whole of Mr. Bateson's argumentation is purely idle and … largely captious.

In his reply, Bateson noted that Pearson's postulate of interchangeability, as a characteristic of organs deemed homotopic, was a late addition, apparently provoked by a newly acquired perception, regarding "the subject of differentiation," that "the greatest difficulty … is … that of … ensuring that … there is not remaining an organic correlation due to the necessity of adjacent parts 'fitting'." In such circumstances, Bateson hoped that the "'idleness' of my objections will be less manifest than at first."

Referring to the emergent Mendelism, Bateson noted that "simpler methods have already penetrated far below that surface which the instrument of the biometrician explores with such minuteness." Yet, "we can feel nothing but admiration for those statistical methods which, as perfected by Prof. Pearson, are yielding many useful results not otherwise attainable, yet their limitations must be constantly remembered." He felt that Pearson had over-reached himself in his claims for the power of the biometric approach. There was no exclusive virtue in one approach, namely biometrics, to fundamental problems. Indeed, Bateson looked forward to microscopic studies which might reveal "at which of the gameto-genetic divisions [at which stage of meiosis] segregation of characters takes place .... There are ... several cases where the germ of one organ arises by a definite differentiant division from a single cell or group of cells, and I am not without hope that the study of the minute histology of such divisions may give an indication as to the nature of the differentiant divisions in gameto-genesis also." The modern reader may see as Pearson's mistake that he was not content to let the mathematical procedures themselves define abstract entities. Instead, he attempted, at too early a stage, to flesh out his abstractions in strict biological terms. In this circumstance, he could not fail to meet with opposition from such as Bateson.

In response to Pearson's request for definitions, Bateson declared he had never "pretended to define conceptions [that were at the time] half-formed." The omission was "due, not to carelessness, but to design." The continuing respectful tone of Bateson's communications indicates that he was more sensitive than Pearson to the need to tread cautiously when fields are opening up and investigators are struggling to frame new concepts. Nevertheless, despite Bateson's references to Mendel, both were mainly thinking in terms of bodily appearance (phenotype) rather than genotype (information encoding that bodily appearance). And Bateson, although claiming that "discontinuity in variation is a term in degree" (*Science Progress* 1897), was still tending to think in terms of discontinuous variations of a type which, in one generation, were manifest as extreme morphological changes—e.g., an extra finger.

Times Spawns *Daily Mail*

In the Bateson archive, there is a short undated manuscript in Bateson's hand, headed "Discontinuity," which is likely to have been written no later than 1903, and probably earlier. It is of particular interest since it displays Bateson's mind, not only in confrontational mode, but also in informational mode, albeit metaphorically and without the deep insight of Butler and Hering (Chap. 19). Those who followed the drastic "mutations" in *The Times* of London in the 1990s may have cause to chuckle:

> The doctrine of Discontinuity is that new types may arise not by the gradual transformation of an existing population as a whole into another type, but by the successful self-assertion of a form possessing some specific novelty in or beside the old one, which it may or may not ultimately replace. The process of gradual transformation may occur too, and no doubt sometimes does, but I incline to the view that in the production of significant changes discontinuity is the rule. So for example, we could conceive the tone and price of the *Times* newspaper changing by continuous variation till it would need a new name, and become the *Daily Mail*; but it would be more in accordance with experience if we found the successful competitor arising as a recognizable novelty from the first. In the ordinary life of many species these potential competitors are often being born, but they find no niche and perish. Given the opportunity they persist. In the output of *gametes*, probably even stable races are constantly throwing off a whole series of varieties which, either because they meet no compatible counterpart, or from some other ineptitude, fail either to be represented in the zygote form, or to reach maturity.

Summary

After the publication of *Materials,* Bateson remained very much in *Materials* mode as a "naturalist" collecting insect and plant specimens. It was from this perspective that he was drawn into public controversies with major establishment figures—Poulton (Professor of Zoology at Oxford), Thiselton-Dyer (Director of Kew Garden's), Weldon (Professor of Zoology at University College, London), and Pearson (Professor of Applied Mathematics at University College, London). These men were either on, or could strongly influence, Galton's RS Committee for Conducting Statistical enquiries into the Measurable Characteristics of Plants and Animals. Bateson's success in the controversies—particularly that concerning measurements of crabs where his ally was Lankester—led in 1897 to the addition of several biologists, himself included, to Galton's Committee, which was renamed the RS Evolution Committee. Although Galton remained as Chairman and Weldon as Secretary, Bateson had captured the Committee. Its agenda became more biological and through it Bateson received RS funding. He opened up lines of communication with the RHS, squelched the move to open an Experimental Farm, and pushed for the distribution of a circular soliciting the help of amateur naturalists in collecting data on melanism in moths. By 1900, the capture of the Committee was complete. The biometricians resigned *en masse* and Bateson became its Secretary. A subsequent controversy over "homotyposis" led Pearson and Weldon to establish a new journal, *Biometrika.*

Chapter 7
What Life May Be

In June 1896 we were married, and I began to learn what life may be.

Beatrice Bateson. Memoir

Between the years 1895 and 1899, the reorientation of Bateson's life took many paths apart from the controversies of the previous chapter. While firing off missives to *Nature* and shuttling to London for committee and lobbying activities, contact with Beatrice was re-established, a residence was found both for their own offspring and those of numerous moths, butterflies, and fowl, and there were the unending controversies associated with college and university politics. Moreover, family responsibilities made necessary a better income. Should Bateson relinquish Cambridge for the Secretaryship of the Zoological Society?

Mrs. Herringham

While his anti-Darwinian position won little favor among scientists, *Materials*, Bateson's catalogue of deformities and monstrosities, greatly interested physicians and horticulturalists who were to prove staunch allies in the years ahead. Notable among these were the physician Wilmot Herringham (1855–1936) and, more importantly, Mrs. Christiana Herringham (1852–1929). How the first contact was made is uncertain. She was a friend of Mary Bateson. In 1888, Bateson may have noticed a paper by Herringham in *Brain* on the high familial incidence of peroneal muscular atrophy. In any event, in 1892 (Oct. 17), Herringham replied from his Upper Wimpole Street address to an enquiry from Bateson on variation in human ribs, and in 1893, Bateson visited them in London.

In 1894, his *magnum opus* behind him, Bateson felt more able to go on collecting trips. In the spring, he spent a month in Spain, sailing (March 20) by way of

A. G. Cock, D. R. Forsdyke, *Treasure Your Exceptions*,
https://doi.org/10.1007/978-3-030-92099-9_7

Gibraltar, and carrying with him, as during his Steppes travels, a field notebook. The trip was a solitary one, although his notebook described a few encounters along the way. In July, he set off again traveling through Switzerland, Italy, and Austria. This time he had a traveling companion – Mrs. Herringham. His field notebook recorded: "1 July left London with Mrs. Herringham. 2 July arrived Aigle. Hotel Beau. Site good. Set out collecting 2 p.m. Mrs. H. went sketching doesn't seem very tired. Caught vast number *Galathea* variable in oculii." The notebook continued with reports of plants and insects seen at different locations. The two seem to have separated and then joined up again, since a further note stated: "10 July. Showery. Wired and wrote to Mrs. H. to tell her to come up here. Wrote to M. B. and to W. P. H.. Girls school in this house. *Delius* here. Vars. [varieties] of *urticoe*. Mrs. H. came." And a few days later, there was the entry: "12 July Coll. [collected] at back of Rue de la Vache. Long walk with Mrs. H. lost our way." He wrote to Anna Bateson (July 26, 1894) from Pie di Mulera:

> Here at length I am in true Italy, trellised vines and all the rest, and I wonder why anybody lives anywhere else. … It has been a somewhat singular tour. You will no doubt have heard how I came to start out with Mrs. Herringham. I still have her in tow and she seems grateful. The fact was that I saw early on that she was quite unfit to be left by herself, and as we get on well, we have proceeded together. She is a very strange person, but in many ways interesting and quite superior to bodily fatigue, which is an undisputable quality in a fellow traveler. She came over Monte Moro (including a stiffish bit of snow) without any distress. Then she is considerable as an artist and has good natural history instincts, making herself very useful in catching and folding up. Mentally, I don't know that I have quite got to the bottom of her, but now that she is in better spirits she is sufficiently companionable. The people in the inns are naturally puzzled to discover our relationship. I see them 'exercising of their brains' endlessly on this score. At first she sat mute at tables d'hote, and then it was odd enough, but now that she begins 'to take notice' it is still stranger. If she were a person of youthful and attractive exterior it would be a difficult position, but as she is not, but rather a 'bekümmertes Weib' [grieving woman], it is easier.

The agenda was not entirely set by Bateson. A day later, he wrote:

> We have moved on at Mrs. H's special desire for two final days at an Italian Lake – Orta. – It is beautiful beyond words. I am sitting this afternoon writing on the inn balcony simply because I can't leave this view. The colour of the lake is quite new to me – a sort of deep colour between Prussian blue and ultramarine shot with pale bright green. In the middle is a tiny island covered with white buildings and reddish roofs. Add the vines and the flowers and the whirr of the insects, and the glorious burning sun 'filling the heart with food and gladness.' The house is almost empty. This noble sun has driven them away, thank heaven. Another of Mrs. H's good points is that she fears no heat of the sun. She is just now sitting I believe in the full blaze sketching.

That the relationship bloomed, in an entirely platonic fashion, will become apparent in the pages ahead. Suffice it to say that, on hearing of Bateson being awarded the Darwin medal of the Royal Society in 1904, Wilmot Herringham wrote: "I am extremely delighted at this. You could not have a higher acknowledgement of your work. I shall be delighted to dine. Of course you'll put up here."

While little trace of it can be found in his correspondence, it is likely that Bateson's fellow traveler had more need of his company than he of hers. Christiana Herringham was the eldest daughter of wealthy London stockbroker and art collector, Thomas Powell, who, following the passage of the Married Women's Property Act in 1882, settled the sum of £31,000 on her. At an early age, she developed a passion for the arts and, following Cennini's manual published in 1390, taught herself to copy the Old Masters who painted in tempera. In 1878 at a tennis party in Blackheath, she met Wilmot, son of a Somerset clergyman, and then a relatively impoverished medical student at St. Bartholomew's Hospital. After a stormy engagement, largely reflecting her unwillingness to discuss religion in which she had no interest, they were married in 1880. Two sons were born in 1882 and 1883. In 1891, the eldest son developed joint pains following what may have been scarlet fever. After a protracted illness, he died of kidney failure in August 1893. Christiana asked that no one wear mourning, and Wilmot asked his parents to respect her wishes. She was exhausted and distraught. She needed a break. Enter Bateson – a man of high purpose and culture who shared her interest in art and her religious skepticism. The surviving photographs reveal a lady whose charms were not quite as dour as Bateson reported [1].

That the trip was as therapeutic as Wilmot might have anticipated is suggested by subsequent events. Christiana returned and set out to translate Cennini. Her book was published in 1899. Her fortune expanded £12,000 on the death of her father in 1897. His parting injunction was that she use her inheritance with "kindness and thoughtful liberality." Christiana touched many lives, especially in the art world and in the women's liberation movement. Her friends included the artist William Rothenstein (1872–1945) who in 1910 traveled with her to India where she copied ancient Buddhist wall paintings in caves near Hyderabad (Chap. 14). He later noted [2]:

> When she singled me out, insisting I had more to say in my work than other painters, I found it hard to maintain the standard she expected. Having gained a rare knowledge through copying – her copies were unique – she looked on contemporary painting for that combination of intense observation of particular form coupled with a worthy subject matter, which was the glory of Florentine artists. If I fell short of one or other quality she was sternly critical. 'Why, that is only *genre* painting, which many artists can produce;' and since she was one of the few who encouraged me in my aims, and who understood them, Mrs. Herringham's friendship was an asset in my life.

"Very Much Alone"

Apart from academic work and university politics, there was a major source of income to attend to, the Stewardship of the College kitchens and farm. A letter to Mary (August 14, 1891) indicated the scope of Bateson's activities:

> I am overdone with work. … I am totting up eggs at 7/6. Butter at ¼ (how much for 5 doz. and 11 lbs?), or at 1/1 (how much for a pail of 1 cwt. 3. 14?) and so on. I am cooperating with the Regius Profs. of Divinity, Latin and Greek, the Master of S. J. C., the Profs. of Zoology and Arabic and Archeology and Neil, Larmor, and Ardlison to keep Greek for the Little-Go [examination]. I am (privately) Secretary to the Ctee., though I stipulate that it should not appear as such.

Beatrice observed (*Memoir*) that he carried out his duties as College Steward "with zeal and zest," and "when in 1902 the scandals of College kitchens were entertaining every household in Cambridge, it was a matter of self-congratulation to him that he had steered the Johnian kitchen staff safely through those seas of temptation wherein so many at that time foundered." Regarding his Stewardship, a St. John's colleague, F. F. Blackman, noted [3]:

> His was an impatient spirit and could not in all things keep the common touch. This failure, at its lowest level, was voiced by his confession 'before a barmaid I am dumb;' while ... he admitted that he found difficulty, when he was College Steward, in keeping in touch with the undergraduates' 'dinner committee' and their discussion of dietary details. As it was one of the duties of the Steward to consider all their suggestions, he ended by selecting the most congenial spirit amongst them to act alone as a go-between.

New undergraduates at St. John's soon became familiar with his exploits:

> Round the name ... legends soon accumulated; that his views on evolution were heterodox; that he disapproved of attempts to reconcile science with religion; that he had shot a man on his Eastern scientific travels; that he had been the proud owner of a bulldog and then given him away to a porter at Waterloo Station. Clearly a man of more heroic mould than we expected to find among the dons of our College! A few years later I came to know the actual William Bateson and was nowise disappointed.

It appears that, sometime around September 1892, he did indeed acquire a dog, but the responsibilities of ownership were such that he and the dog soon parted company.

The urge to travel remained strong. In the spring of 1895 (March 15), there was a solo collecting trip to the Spanish Pyrenees, traveling by train by way of Paris to Madrid, where he was "sickened" by a bull-fight. He took in the cultural delights of El Escorial (where it snowed), Granada, and Cordoba, arriving back by way of Bordeaux, Tours, and Paris (April 12) to find that the Cineraria issue had taken off in the pages of *Nature* (Chap. 6). Then in June and July, there was a "collecting" and "sight-seeing" trip to Italy, the first part of which was with Mary. In September, there was a letter to Mary from France (September 7, 1995; Challes les Eaux).

> I am doing fairly, but no great success yet. Surrounded by a bubbling world I maintain myself in a solitude complete. It is a rather semi-fashionable Baths here, which suits me fairly well as regards lowland varieties: in a week or less I go to some Alpine place. Though late, the insects are still on the wing. ... By the way *tout le monde va à bicyclette* and all the little ladies have most striking small clothes for that purpose. If I dare to know where to get them, I would bring you a pair. They combine comfort, I suppose – with – may I say it? – not only respectability but even decency.

Still in France, he wrote to Anna (September 17, 1995; Grenoble):

> From a business point of view, I have not done much this trip. I have a few larvae, but not of much account. The season is over. ... So I came down here to do a little sight-seeing before turning homewards. At an inn at La Grave near here I met the A. Marshalls [Cambridge economist] who had been there 7 weeks. He travels with 5 cases of books! For so novel a mind he has an astounding power of boring one. I really felt quite glad to be quit of him, though I have been very much alone this tour – a condition that becomes less pleasant as one grows older ... . In a 'Times' at this house I see Dyer (at the Brit. Ass.) has been digging at me – not by name, *bien sûr*, but very evidently.

University Politics – Gender Issues

Bateson was much engaged in controversies at Cambridge, both at the college and university level (as were Weldon and Pearson at the University of London, where their opponents included Thomas Huxley). One day in 1891, St. John's awoke to find a red-pillar box near the gate. Bateson persuaded 15 Fellows to petition the Master (February 27): "We feel that the presence of the pillar-box in any form must take away the charm and the stateliness of the gateway. The beauty of our ancient buildings is still the great and singular glory of the College .... We are confident that if the circumstances are represented to the Post Office, no objection to the removal will be raised." It was soon removed.

However, Bateson's major concerns were at the university level, where, goaded by his parent and siblings, he fought for the rights of women to have access to study and research facilities, to be issued with degrees, to take part in university business, and to be eligible for academic positions. As will be described in Chap. 23, the opposition was formidable. As so often in human affairs, disagreement in one area could lead to disagreements in others. Those who disagreed on women's rights might be less inclined to agree on other issues. On the other hand, an alliance in one area might propitiate alliances in others. Many men – and, indeed, some women – held strongly that women were constitutionally incapable of academic work:

> Seeing that the average brain-weight of women is about five ounces less than that of man, on merely anatomical grounds we should be prepared to expect a marked inferiority of intellectual power in the former. ... In actual fact we find the inferiority displays itself most conspicuously in a comparative absence of originality, and this more especially in the higher levels of intellectual work. ... The disabilities under which women have laboured with regard to education, social opinion, and so forth, have certainly not been sufficient to explain this general dearth among them of the products of creative genius.

These are the remarks – given in 1887 with "almost brutal frankness" – of one of England's leading neurophysiologists, George Romanes [4]. Romanes appeared undaunted by the fact that when he had, under somewhat controlled conditions, tested the relative abilities of men and women quickly to read a text and then answer questions on it, the women had come out well ahead. Nevertheless, he confessed:

> When I was at Cambridge, the then newly-established foundations at Girton and Newnham were to nearly all of us matters of amusement. But we have lived to alter our views, for we ... see how that was but the beginning of a great social change, which has since spread, and is still spreading, at so extraordinary a rate that we are now within measurable distance of the time when no English lady will be found to have escaped its influence. ... But, while we may hope that social opinion may ever continue opposed to the woman's movement in its more extravagant forms – or to those forms which endeavour to set up an unnatural, and therefore impossible, rivalry with men in the struggles of practical life – we may also hope that social opinion will soon become unanimous in its encouragement of the higher education of women.

Whether Miss Philippa Fawcett read this we do not know. But in 1890, she scaled the much-coveted "senior wrangler" peak of Cambridge mathematics. Regarding degrees for women, Cambridge was behind many comparable institutions, a point Pearson made forcibly when, as a Cambridge graduate, he was approached by Bateson for support (December 30, 1895):

> I beg to acknowledge your circular as to degrees for women at Cambridge, but I cannot sign it as I am not a member of the Senate. I am glad to see that something is being done *inside* the University in the matter, although, I think, it is only a couple of years since Dr. [Henry] Sidgwick thought a petition in favour of women being allowed degrees and largely signed by outsiders like myself 'very ill-timed.' The wording of your present circular seems to me misleading. I cannot think that any one acquainted with the facts would speak of Cambridge as a 'pioneer in the movement for extending education to women.' The matter had already been fought out and settled in the London Colleges, before the Women's Colleges at Cambridge were thrust on a very unwilling University. What is 'The example set by Cambridge, which has been followed by most of the other Universities of Great Britain?' … By all means, throw open your degrees to women, but do it quietly and without any tall talk of being pioneers forsooth in woman's academic freedom.

The Principle of Newnham College, Eleanor Sidgwick, had declared in 1890 that in the female colleges, there were those "who study for study's sake; who desire to devote their time to learning or science, and hope, however humbly, to help carry knowledge a little further, or at least to help others to care for such things." Yet, "there are fewer facilities for them in later life; fewer advanced studentships, fellowships and professorships." Nevertheless, progress was made. As mentioned in Chap. 4, Anna Bateson and Dora Pertz both graduated from Newnham College in botany and were able to pursue research, albeit as assistants.

Also among the Newnham botanists was Edith Rebecca (Becky) Saunders (1865–1945) who in 1888 obtained first-class honors in the Natural Sciences Tripos and, like Anna, won a Bathurst Studentship. She became a botanical demonstrator at the Balfour Biological Laboratory for Women, which had been established in 1884, and later she became its director [5]. Saunders was one of Bateson's most reliable long-term collaborators. In August 1895, she began studies of crosses between varieties of plants with smooth (glabrous) leaf surfaces and varieties with hairy leaf surfaces. These characters proved to be passed discretely in an all-or-none, nonblending, manner from parents to offspring. However, she did not go beyond the first generation (no further crosses between the mature forms of the child plants). Her first paper "On Discontinuous Variation Occurring in *Biscutella laevigata*" appeared in the *Proceedings of the Royal Society* in 1897.

Another issue of concern was compulsory Greek (Chap. 23). The requirement for proficiency in Greek had long barred persons without a classical education from entering the University. Bateson issued a "fly sheet" entitled "For Greek" (October 23, 1891). On this occasion, he won. The "grace" to consider making Greek non-compulsory was rejected in a vote of the University Senate with 525 votes against ("non placet"), and 185 votes for ("placet"). The issue was to reemerge time and again in the years ahead. In 1905, one of Bateson's students, J. Stanley Gardiner, writing from Mauritius, considered the possibility of defeat (August 17):

> If Greek goes I would like to see in the *same* Grace a stiff examination in the English language to take its place. You were very nasty, as it seemed to me at the time, in making me rewrite my first paper 4 times before you would send it even to the Proc. Camb. Phil. Soc., but I have never ceased to bless you since, as my lack of knowledge of English (regarding it from a grammatical point of view) has always been a great trouble to me.

In the early 1900s, one of Bateson's collaborators, R. C. Punnett, served with the mathematician, Godfrey Hardy (a fellow cricketer), as joint secretary to the Committee for the retention of Greek in what was known as the "Previous Examination." This served Punnett well when pressed to analyze inheritance more mathematically (Chap. 10).

Marriage

In June 1896, Beatrice "began to learn what life may be" – a decidedly reticent one liner in her *Memoir*. As mentioned in our Prologue, in 1895, shortly after her father's death, Beatrice published a short story in *The English Illustrated Magazine* which was clearly modeled on her own experience [6]. The story represented what the future held for her – a lonely spinsterhood, enlivened or haunted by memories of what almost came to pass. It seems likely that the story was not just an outpouring of her heart but was intended as a signal to Bateson: a message that it was not by her choice, but through her family's coercion, that they had been separated. At any rate, several months after its publication, the message reached him, by way of the wife of the philosopher, Alfred North Whitehead. It appears that Mrs. Herringham then became involved. She wrote (January 29, 1896):

> I will certainly do what you ask if I can, but I don't read much monthly literature. I did not think it was easy for you to meet, but you make me feel a little as if you had made me arbiter of your fate. I certainly wish that if you are to meet again it should be in such a way that you could look afresh with eyes seven years older. I shall not be able to help thinking about this very often. If it were possible to be the means of your accidentally meeting I should be glad. Whether I should be glad years after is another thing. You say you have often wished to tell me this. You had told me a great deal of it in scraps – but I did think your parting had been more of a hasty misunderstanding – I know how much you cared – I don't want to say any more.
>
> I liked your Greek paper. I restored our young man to the study of that language. But I suppose one gets at a great deal of it through art. I hope so for I am afraid I shall never learn Greek. You should see the Velesquez again. I went with Miss Pridranx and we idled around and real[ly] took in the portraits a bit – there are more than I realized when I went with you. The best one of the Spanish Princess is really great. I mean to go abroad this spring as soon as I think it is light and warm enough to work. I think I shall start with some copying in Florence.
>
> Dear me. I wish I could help you. We are all so [im]possibly alone. We try to aid each other but we don't get out of our own hedges much.

So, at some point, Mrs. Herringham may have acted as an intermediary. Just how she brought about this delicate task, whether by a direct approach to Beatrice, following by incitement to rebellion, or by persuading Mrs. Durham that Bateson had by now proved himself to be sober, worthy, and respectable, is unclear. In any case, soon there was a letter from Bateson to Beatrice (April 24, 1896):

> From things that have lately reached me I have been led to think it possible that you may be willing to see me again. If it is not so, you tell me and that will be all; but if it is so, will you some day meet me? If we meet, it must, I think, be as friends simply, that we may see each other after the changes seven years have worked, freely and without committal of any kind on either side. I think that would be the right way. Unless you choose otherwise, I would suggest we should meet in some open place and that we should have a walk together. It would be less trying so. I will not say any more now, save that it has been for a long time my earnest desire to meet you again, if only as one who was once my dear friend, without regard to the future at all.

The wedding was held in London in June 1896. It was held in a *church* – St Paul's, South Hampstead – with the Reverend Tom Fields officiating.

Beatrice, aged 28 at the time of her marriage, was the only one of the six Durham sisters to marry. She and two of her sisters (Florence Margaret and Francis Hermia) were listed among those attending Bedford College, in London, in the early 1880s. In the 1891 census, she was a "student of music." Florence attended Girton College and took part 2 of the Natural Science Tripos in physiology in 1892. She then lectured in London (Royal Holloway College) until 1899 when she became a demonstrator in physiology at the Balfour Biological Laboratory. In her later years, she studied the genetics of alcoholism. Hermia also went to Girton College and became a historian. Mary Edith (Dick) Durham (the eldest sister) traveled and wrote about the Balkans. Edith Durham had disapproved of Bateson as a future brother-in-law. She had a slight limp, and Gregory Bateson later related to Alan Cock that his parents disliked "Dickie." One day in a hatch of experimental chicks, there was a half-crippled individual for whom the name "Dickie" was duly entered into their research ledger (Fig. 7.1).

Life

Beatrice was soon to learn about life. For a start, where were they to live? Bateson knew the properties held by St. John's College's very well. His papers contain his "Preliminary Report as to Various Ways for Providing Accommodation for Married College Officers" (February 21, 1896). The fate of the Report is not clear. At first, Bateson and Beatrice lived at Norwich House, Panton Street, near to the Cambridge Botanical Garden, where in their nearby allotment Bateson bred poultry and Miss Saunders crossed varieties of plants. However, the scale of the work expanded to such an extent that they had to seek a second allotment.

With Bateson working at or close to home, he was seldom far from Beatrice. Several hours after his departure on even a brief trip, the reliable Victorian mail

Fig. 7.1 Beatrice Bateson

service would be delivering a letter to her. One from the Midland Grand Hotel, London (January 6, 1897), is not untypical: "Meet me at 3 Hanover Square at 12-30. You can go in and ask the clerk at desk for me. I am to see Galton this afternoon, but we could do a Gallery first." In the spring of 1897, it is likely that Mary accompanied them on a trip to Touraine, in the Loire valley (Letter to Anna; February 17, 1897). While Beatrice was in London visiting her mother at 77 Avenue Road, St. John's Wood, Bateson wrote (May 9, 1897): "I have accepted an invitation for the Kennedy's Friday 14 May, dinner. This is going to be the first dull evening I have had for ever so long!"

In July, there was another trip to the Continent, with Bateson departing first. Regarding an impending visit from his literary sister, he wrote from London to Beatrice at Norwich House (July 2, 1897): "Get plenty of A strawberries for Margaret." Later, on the same day, he was in Dover, from which a postcard was received. The following day he was writing from the Hotel Nationale, Bâle (Basel), in Switzerland: "Take care of yourself and caterpillars. I shall expect to see you at Airolo on Wednesday." For some part of the trip, they may have been accompanied by Saunders since Beatrice annotated a later letter: "In 1897 we were at Tosa, etc., also collecting. We took Edith Saunders with us to study Biscutella hairy and smooth." Apparently, when sheltering from the rain, Saunders drank a little too much "Asti," and she departed "from the strict demeanor of the student."

Bateson reported (October 24) to Beatrice, who was again staying with her mother, that he had been to his mother's flat at Oxford and Cambridge Mansions, St. John's Wood, where there had recently been a "gathering of Aunts and Dowagers" including Beatrice's mother, whom Bateson considered to have been "delightful – quite the best horse of the party." By this time, Beatrice was pregnant. In December,

Bateson was writing to her from Brighton where he had "had a splendid walk along the crest of the Downs and saw some butterflies at Lewes with my tea." Beatrice became depressed and again went home to mother. Bateson wrote (January 4, 1898):

> I don't know what to say, beyond that I am sure the more you can interest yourself in other things the better. That is rather empty counsel, nevertheless, on the whole I daresay if you get the little one's kit all in readiness, it will ease your mind as much as anything. … I don't quite understand your sentence about not feeling 'happy about it any more.' Try to keep your heart up, poor girl!

The following day Bateson wrote: "Glad to get your note and hear you are a bit brighter."

So, the family became three. John was born on April 22, 1898, at Norwich House. To celebrate, Bateson purchased an original edition of William Blake's *The Book of Job* (1824). For a while, Beatrice kept a diary noting John's progress. Breast nursing stopped at 2.5 months. Vaccination was in October 1898 (twice because the first did not take), and he was circumcised under ether anesthesia in June 1899. In July, Beatrice took a few days off at her mother's without the baby. Bateson reported (July 14, 1898): "No new cases of 'disease' in any of the several nurseries. The 3 isolates are doing very well. Your own young is rather extra well, I believe. No noise by night. At least none reached me. As bidden, I looked in in the morning to find only a sleeping mass, but I was honoured with an unsteady stare …. The nurse said 'Yes, I was just thinking,' when I told her I was writing you a bulletin. So it is just as well I did." A few days later, there were: "New troubles. The disease among the chickens has broken out with a fresh virulence."

Beatrice came back (July 18) and immediately departed to Edith Bateson's cottage at Much Hadham with John. Bateson wrote (July 22, 1898):

> Your note very welcome. The rush is awful. Butterflies coming out in the pots in spite of Frederick boxing off. The 2nd boy is useless and only makes F. waste time. The last of the young chicks died today. It is disappointing. I had quite thought it saved. … Enclosed from DS I have had some talk with him about it, but on the whole I think had best not 'think at all about the place.' Few I fear would take Sharp's very flattering estimate of my capacity and I doubt whether I should really like the work or do well in it.

The cryptic remarks about David Sharp (1840–1922), the Curator of Insects at the Cambridge Museum of Zoology, indicated that new family responsibilities were pressing him to attain a more substantial position. Bateson begged Beatrice to return (July 24, 1898): "Am getting dreadfully behind. Is there any chance you could conveniently come back here from town and give me a day's work on Tuesday? Don't unless you feel thoroughly easy about leaving the cottage." It seems that she did, but then went off leaving him holding the baby again. Bateson wrote (July 28, 1898): "My own parent will have a good deal of criticism and remark to offer regarding J. B. when she sees him. It is a pity he can't keep asleep between 4 and 7 a.m. … Mrs. Ward [wife of philosopher James Ward] and friend called yesterday and we had a nice chat. She wanted to see the infant. I said there was one handy in the street she could look at – which indeed there was – but she insisted on the veritable J. B." There was a long reprieve when (August 2) Beatrice took John to the seaside (Aldeburgh) with her mother for four weeks.

Bateson considered future plans. A move of Lankester from the Oxford Chair (to which Bateson had aspired; Chap. 4) to the Directorship of the Natural History Museum had led Bateson to consider an application either for his Chair or the associated Prosectorship. However, by now he was more interested in the possibility of becoming Secretary of the Zoological Society, with responsibility for the operation of the Regent's Park Zoo. Having been a member of the Council of the Society, he was quite familiar with its operations. In a candid letter (August 3, 1898), Bateson wrote to the aging incumbent, P. L. Sclater, enquiring whether he might be retiring in the near future:

> I realize fully how various are the duties of the post, how difficult it must be to discharge them efficiently, especially for the man who follows so distinguished a predecessor, but I am bold enough to think I might be successful. Apart from a love for Zoological Science – which doubtless many have no less strongly – I have had some experience of business, which among zoologists is a less common qualification. In 1892 I was appointed Steward of my College (which is a large one) at a time of difficulty when, through various causes, especially certain dairy and farming operations, its domestic affairs had got into some confusion. With these difficulties I have fortunately been able to cope successfully and I have the satisfaction of knowing that my services have been appreciated by the college. In the capacity of Steward I have had the financial management of a considerable office with an expenditure of about £12,000 a year, and the personal control of a large staff of servants. I have thus had opportunities of acquiring a knowledge of business in some degree comparable with that of the Zoological Gardens.

Bateson wrote to Beatrice (August 3, 1898): "This morning I wrote to Sclater as proposed. Newton sent me a draft which seemed to me in most respects excellent. So that step is taken."

With Beatrice away, with problems with "the 2nd boy," and with an impending International Zoological Congress at Cambridge, one would imagine Bateson to have been firmly anchored. Not so. He asked Anna Bateson to accompany him on a short group tour abroad but reported (August 3): "Anna declined the tour, so I have bought my own ticket." He was off the following day, writing from All Acqua, Airolo (August 6): "With what you say about the Oxford post I largely agree, but I fear there is very little chance of an outsider being chosen. I am awaiting Sclater's answer in some anxiety." Two days later, he wrote, still from All Acqua: "I hope John doesn't blow his horn all night. Thirty years hence I daresay he will read our names in these Inn [visitors'] books – unless, which is more likely, he is given up to County Cricket and 'Surrey's Huge Effort' – the last news bill I saw."

Norwich House was in disarray. He wrote to Beatrice (August 13) "Got back at 11.30 to find widespread troubles … Larvae starving terribly. Butterflies coming out and hopelessly battered in their pots." Then there were the Congress bookings:

> Congress and Trust business in confusion. Sclater's answer herewith. It is much more satisfactory than I expected. Send it back to me. The Oberthürs cry off. So perhaps you will not come [back] for the Congress. If you would it would be a help of course, for there will be more than 1 pair of hands can do. … Unless you are especially bent on stopping, do come back at the time appointed. I begin to be a good bit bored. Travelling alone is poor form nowadays. Several incidents to tell, but nothing of magnitude. … As to J., – I daresay you will now bring him back. I am very sorry Oberthür is not coming. On several hands we hear of parties crying off. Some say we shall only have about 350.

It turned out that Sclater was not contemplating retirement at that time, and his reputation untarnished by the controversies (Chap. 6), Weldon moved to Lankester's position at Oxford. Later in the year, Bateson entered "begging letter" mode. The Duke of Bedford had established an experimental fruit farm at Woburn Abbey in 1894 [7] and had expressed interest in Bateson's work. Bateson wrote (December 5, 1898) to "My Lord Duke" proposing to help set up an Experimental Station at Woburn that he would supervise from Cambridge. He enclosed documentation relating to the Evolution Committee and its agenda. The Duke declined (December 9).

Another hope was that his eye for fine art might prove of speculative advantage. Beatrice later reminisced (note circa 1928 attached to letter of March 23, 1900): "There was always a dream of a wonderful treasure of art that should lift us out of our financial squalor." Thus, while glad that the family had escaped from "trade," when it came to art, Bateson was not above a higher class of trading. In 1899, Beatrice was pregnant again, and the only letter from Bateson for the first six months was one from Bath (January 29) where he was collecting art. However, soon their finances were to become more stable. Bateson obtained his first university appointment as Deputy to Newton in the Department of Animal Morphology.

For most of July 1899, Beatrice and John were away at Edith Bateson's. During their absence, there was an important meeting of the RHS at Chiswick, London (Chaps. 8 and 9). Visitors included de Vries, who prior to the meeting stayed at Norwich House. Bateson wrote (July 7): "de Vries writes he must 'take great care of his health and be very sober.' I suppose F. D. intoxicated him on the last visit." The first impression was mixed (July 8):

> De Vries is a really nice person, very simple and rather rough in style. It is delightful to have a person with whom one can change a word. He is an enthusiastic discontinuitarian and holds the new mathematical school in contempt. So we hit it off to admiration. Saunders talked and chattered as I never saw her so before. De Vries' English is sufficient but very incorrect. He is still in the stage of 'becoming' [rank in society] – his hat and so forth. I am so sorry you should not have met him. He has edited a garden paper and served on horticultural juries in order to learn the tricks of the horticulturalists. He is to be on the fruit 'Jury' of this conference. I fear he didn't use 'is 'ot water. He declined a cold bath before dinner yesterday with an emphasis that might have been put on the stage. He has I think no sponge. His linen is foul. I daresay he puts on a clean shirt once a week 'whether he requires it or not' as I have heard said. Art doesn't reach his soul. … Today being the hottest day of the year he asked to have the drawing room window shut. 'Tank you. Yes. It is better so, I zink.' Tonight we are to have [to dinner] F. D., Sedgwick, Gardiner, Lister – six in all.

Bateson's notion of de Vries' social standing did not match his pedigree – his father was a parliamentarian who became Minister of Justice, and there were academics among close relatives. Later in the day, Bateson wrote:

> It is just as I feared. They have put me down to propose 'Hybridists' at the dinner on Wednesday. So my work is cut out – and the paper not half written yet. … I don't see how I can come over [to Much Hadham] on Monday. It is going to be an awful rush – paper still undone and all Miss Saunder's specimens to be got ready and labels written for them. … If you can think of some bright telling crisp bits for my after dinner speech please send them on.

De Vries left the following day for London, and Bateson noted: "de V. did *not* take his bath. I fear also he did not tip Lena." At the conclusion of the meeting, Bateson wrote from London (July 13):

> Now the whirl is over. … de Vries has been a great stand-by. We were the only two of the recognized scientific people who attended after all. Foster was to have come but he was ill. My paper did not do much good I think, but I believe I was fairly successful post-prandially. We had a distinguished setting of 2nd class notables – Belgians and Dutch Ministers, and a few Judges together with brethren of the Apron and potting bench. Two days of it are enough and today's function will be a little superfluous. Anna's hands are like other people's pumice stone. … At the Horticultural dinner … at the top of the table for lunch, Anna, Dora Pertz, and Saunders, with one other, were the chief ladies. They were set at equal distances like jewels in a crown. Saunders is so terribly bent on business. The Conference struck her as frivolous, I fear. Vilmorin (whom she always calls 'Vilmórrin') looks as if he has walked straight out of Balzac, with smooth cheeks and large black *favoris de notaire*. I am afraid business is first with all these persons. Rolfe is a wretched little creature. Dr. Masters was decorated with the 'Insignia of Leopold," which as he looked the veriest greengrocer, seemed unnecessary.

Again, chaos had broken out in his absence (July 16): "We shall be glad to have you back. There is no luck about the house! … Moth and chick breeding had gone wrong. F misunderstood my directions and they [a batch of chicks] had to be thrown away. N. B. Mevrouw de Vries does *his* plants in his absence and knows them every one – so he said. All the same, he was dreadfully anxious to get back."

The reference to the house relates to their quest for a new house with a large garden. Beatrice (circa 1928) added a note to the correspondence: "Riversdale, [was] a house we bought in Grantchester. We then found it very difficult to hire land adjoining. King's college demanded too high a price for us. In the meanwhile we brought Merton House, Grantchester and after short anxiety sold Riversdale … at a small profit – to our intense relief." On July 19, Bateson reported that he had made an offer to buy Merton House from Dr. and Mrs. Keene. It was accepted, but due to Dr. Keene's illness and to its state of disrepair, they could not move in immediately. So their second son, Martin, was born (September 1, 1899) at Norwich House (a few days before the Jameson raid that initiated the Boer War).

Beatrice noted in her dairy that from September 23rd, Martin was with a nurse at 64 Jesus Lane, presumably to facilitate the move to Grantchester. Like John, he was circumcised (October 4, 1899). Beatrice offers no comment on this – given the Batesons' atheism, it was presumably for medical rather than ritual reasons. Circumcision was then advised in some medical circles to control "deviant" sexuality – specifically masturbation [8]. Beatrice's baby diary tailed off around 1900, reopening briefly in 1904 for their third son (born May 9 at Grantchester). By this time, Bateson had become aware of Gregor Mendel, and the boy was named Gregory. When they moved from Merton House in 1910 (Chap. 14), it was purchased by Lister of St. John's, who in 1911 married one of the Bateson school, Dorothea Marryat.

By now their capital was likely exhausted, so that there would have been little available for research. The animal experiments were continued at Grantchester, freeing up ground for Saunders, so the second allotment near the Botanical Gardens could be given up. Despite the labor, Beatrice in her *Memoir* conveys a sense of intense happiness:

> Merton House, Grantchester was most admirably adapted to his needs. It was within University bounds, it had a good paddock, protected by a belt of shrubbery, good outhouses, and a very good garden and orchard. ... We had now ground; we had poultry-pens well stocked; we had row upon row of peas, poppies, lychnis; the garden was full of big and little experiments – some, tentative trials of subjects; some, serious undertakings. Our 'gardener' Blogg, was a capital man and a delightful person, but by profession a coachman and already ripe in years. (He made all the proper stable noises with his mouth while cleaning the hen-houses.) Besides him we had our own hands. When lecturing and College work were done, the rest of the day was spent in hard manual labour; from the merest menial drudgery to high flights of scientific speculation, hand and brain were hard at work. There was all the sorting, sowing and gathering of seed to be done personally; the fertilizing and recording; most of the digging, hoeing, weeding, staking and watering; the five incubators, each 100 egg power, and as many rearers (all run with oil lamps); the tiny chicks; and at times hundreds of larvae to be attended to. All writing (not reckoning the ordinary post, which was often heavy) was done at night.

Down House

The value of experimental stations had been recognized by Galton's Committee (Chap. 6). Prior to Bateson joining the Committee, Frances Darwin proposed using his father's house at Downe for this purpose. In July 1899, Foster sought a meeting with Bateson to discuss the matter. Of more significance, George Darwin (the mathematical Darwin) also sought his opinion. Bateson had opposed the proposal earlier largely because of the absence of a library and, in those premotor days, its remoteness from other centers. Bateson's reply (July 21, 1899) left Darwin in no doubt about his agenda, and his interest in statistical analyses and, although not explicitly stated, in the use of animals and plants that were of direct interest to man (and thus might attract support):

> If Down were preserved as a biological station it would, I take it, be devoted to work having a direct bearing on Evolution. The primary object of such an institution should be the maintenance of investigations which require to be continued for long periods of time. Of these the most important would certainly be an attempt to determine accurately, by experimental breeding, the laws of inheritance in animals and plants. Such experiments should be so designed as to throw light on many points:

1. The magnitude of variations.
2. Modes by which variations may be perpetuated; the degree to which different varietal characters are, or are not, capable of blending by intercrossing.
3. The influence of intercrossing on the fertility of races – fertility being used in a wide sense.
4. The conditions of transmissibility; the nature of prepotency [dominance]; the possibility of altering the power of transmission by selective mating – notably by consanguineous breeding.

> Work of the kind contemplated should, of course, be begun simultaneously in both animals and plants. To have permanent value, however, such experiments must be performed on a scale sufficient for the application of Statistical methods. By this condition the larger animals are necessarily excluded; but in addition to the Botanical work, much might be done with several of the smaller animals, especially the fixed breeds of poultry, pigeons and perhaps canaries; several species, and fixed varieties of insects, especially Lepidoptera, and perhaps others. The beginning would be on a very small scale, but the work which could be carried on is limited almost entirely by the amount available for labour, as simple appliances would suffice. Both in the case of animals and plants, the chief expense would be under this head.
>
> The scientific importance of this class of work cannot be over-estimated. The field is almost untouched. It is for want of continued observations on inheritance in animals and plants that so little progress has been made in the science of Evolution since Darwin's work. From the practical side also, though the gain may be remote, it cannot be doubted that, sooner or later, considerable results will be reached. What the precise form of these results may be cannot be foretold, but it is certain that if only an outline of the laws of Inheritance could be obtained, the influence of such knowledge on the art of the breeder must be immediately felt.

The issue reemerged in 1903 when the RHS sounded out Bateson on a field station at Wisley (Chap. 8). Decades later, Thomas Morgan, having been awarded the Darwin Medal of the RS, notified the Secretary (W. B. Hardy) that he wished the prize money of £100 be given to "some good cause in England." He enquired informally of Bateson (December 29, 1924) on the suitability of putting it toward "a nest egg for purchasing the Darwin house, to be given to the British Nation." Bateson replied (January 12, 1925): "A good while ago – more or less in connection with the Darwin Celebration of 1909 if I remember rightly – there was some talk of buying Down. I never heard that the suggestion got much support. It was however mixed up with the notion that the place could be used as a research station, for which it is not very suitable. Whether there would be any disposition now to buy the house as a show place I cannot say. Personally I think that the cult of souvenirs is overdone, but I may be alone in that opinion." He communicated similar thoughts to Hardy. Despite this, the Down House heritage was eventually recognized, and it was indeed acquired, not only for "the British Nation," but for the world.

Bernard's Symposium

Henry Bernard (1853–1909) graduated in mathematics from Cambridge in 1876, was ordained priest in 1878, and traveled extensively serving for six years as Assistant Chaplain in Moscow. He developed a strong interest in zoology and at some point may have studied with his "friend and teacher" Ernst Haeckel in Jena [9]. In 1894, he was appointed to the cataloguing department of the British Museum, where he worked on corals.

Our present interest is that in 1900, arising from the difficulties he had encountered in classifying corals, he circulated an essay on the species concept to several leading biologists and invited their comments. The correspondence was chain-like. Each respondent returned the essay and his letter to Bernard, who forwarded it together with all previous letters to the next respondent. When he had finished with them, Bernard sent the letters to Bateson. They were found among his papers in an envelope inscribed "Bernard's Symposium." The letters reveal how the species problem was viewed at the turn of the century by a small group of British biologists of diverse backgrounds and viewpoints. Furthermore, it reveals the confusion Bateson had to contend with. Bernard's original essay has not come to light, but some sense of it can be derived from a brief abstract of a later address given at the Linnean Society in 1901 (February 7), which was attended by some of the respondents. Bernard's major point was that:

> The 'natural order' can only be based upon an exhaustive study of all the discoverable variation, and only then will it be possible to arrange these variations into natural groups or 'species.' ... The present exclusive adherence ... to the Linnean binomial system, which implies classification when classification can only be attained as the end and crown of our work, is philosophically absurd and practically disastrous. The absurdity of starting by assuming what it is the object of our researches to find is self-evident.

As a practical measure, Bernard proposed a numerical system to replace the classical system of Linnaeus. The first recipient of the essay was Lankester, now back in London. He side-stepped the issue himself and urged Bernard (October 17) to consult with others, naming Weldon, Bentham, and Thiselton-Dyer. The latter replied from Kew (October 20) confessing that he had "never been able to clearly comprehend the principles of Zoological nomenclature," but he was less troubled with plants. He held that "the evidence is overwhelming that 'natural sterility' is not a specific criterion," and "the best botanical taxonomists now regard 'species' as abstractions representing the average characters of a variable group. ... The essence of the idea of species is *discontinuity*. Two allied species are reckoned valid if the discontinuity they exhibit cannot be explained away by possible variation."

The next respondent was Weldon from Oxford (November 4, 1900), who was "glad to find that you give up the idea of mutual sterility as a test of specific distinction. ... Such a test would sometimes oblige us to subdivide groups morphologically similar, while it would at other times oblige us to unite groups morphologically distinct. ... Morphological discontinuity must be the test of specific distinction." On the other hand, to the definition one ought to add "some statement of the characters which must be discontinuous in order to involve specific distinction. As it stands, the statement is incomplete, and I imagine that it cannot be formally and generally completed further than by saying that the character must be common to the sexes." Weldon then turned to the problem of characters that exhibited "apparent discontinuity in one sex and apparent uniformity in the other," noting that "except by using mutual fertility as a test ... it is impossible to justify many of these species, and at the same time admit the 'dimorphism' of others." He then noted "that two distinct physiological species [i.e. mutually infertile] may be morphologically indistinguishable in one sex, I am quite prepared to admit; but I do not think that such species need to be morphologically recognized."

Bateson replied to Bernard from Grantchester (November 19) noting that "with your essay I have read the accompanying letters of Lankester, Weldon and Thiselton-Dyer." Regarding Bernard's concerns:

> What *is* species? ... How may living things best be catalogued? I believe you are on the right track in insisting that these two questions be dealt with separately, and that the last should be taken first. The first we cannot answer, the second we must. Logically I see no objection to your numerical system. ... It is a question I cannot possible judge, being without experience in these matters. But the problem of *Species*, its nature and significance, interests me more than any other and I must thank you for asking me to take part in your symposium.

He was unable to resist commenting on the remarks of others, beginning first with the Director of Kew Gardens, who occupied a position of dominance in British Botany rather like that held by Lankester in Zoology:

> Thiselton-Dyer, with characteristic vigour, thinks that all Nature wants is a firm hand. His encyclical is filling my breast with food and gladness. I have read it again and again. He tells us all is as it ought to be. Each fulfils his destiny. *Species* are discontinuously separated from each other and are careful to vary within their own range. Obliging *Variations* are from time to time permitted to show us how it all came about. These are *Reversions* and need give us no concern. Knowing they are reversions they will abstain from imitating existing species, which is strictly forbidden. The business of *botanists* is to name the plants, and the business of *Kew* is to keep botanists in their place. So all is for the best. Pangloss himself was a sceptic in comparison with this. And they say it is an age of unbelief!

After a few, much milder, comments on Weldon, he then proposed his own compromise:

> Why should we not frankly recognize that the phenomenon of specific difference may present itself in many forms, different in different groups, just as the phenomena of chemical differences do? Some of them are structural, for example size, colour, number, proportions, *facies*, pattern, hairiness and so on; others for want of more precise analysis, we call physiological (fertility, etc.). Of these latter the cataloguer may by the exigencies of the case, have a very imperfect knowledge, but such knowledge as he has he should exhibit. These are not the least, but the most important facts he can tell us.

Then Bateson turned to the species question:

> Pearson and his friends present us with a statistical picture of *normality*, and thence are feeling after a statistical conception of species. I was a good deal taken with this idea at one time, and to my Evolution class here I teach it as an approach to truth. But as I see more of the facts I begin to doubt if this is the right road. I am gradually inclining to the belief that *specific differences* have severally a more tangible basis, and that some much more concrete phenomena lie at the root of at least a large part of the manifestations of the Species. Whatever change is made as the result of your agitation, I trust ... that the question will in no wise be prejudged.

Sharp wrote (November 21) urging, as had Bateson, that, as a practical matter, the nomenclature issue should not be delayed until the species question was settled. And Alfred Wallace wrote (November 29) noting that:

> I can quite understand that in the lower forms of life you are working at [i.e. corals] … the attempt to *define* species may be hopeless, and it may be because *nature* has in most cases *not yet defined them*. But with the higher vertebrates and insects with which alone I have any intimate acquaintance the difficulty rarely arises. … Neither do I see much difficulty in giving a clear definition of 'species' on the principles of evolution. … The numbers of well marked, and at first sight *good* species, which yet show a *complete transition* to other species, are very few indeed compared with the very large number which show no such [morphological] transition."

Wallace then as "mainly a statement of fact, founded on the theory of evolution by natural selection," gave his definition of species: "A species is a group of individuals which reproduce their like within *definite* limits of variation, and which are *not* connected with their nearest allied species by insensible variations."

Summary

In 1894, his magnum opus behind him, Bateson went on a collecting trip with the wealthy Christiana Herringham, who was grieving the death of a child. While she returned invigorated, he remained "in a solitude complete" and "very much alone" despite his duties as Steward of the College kitchens and farms, his active involvement in university affairs, and numerous collecting and sight-seeing trips abroad, sometimes accompanied by a sister. In 1895, a Demonstrator in the Balfour laboratory for women, Rebecca Saunders, initiated breeding studies in plants, and became Bateson's most reliable long-term research associate. Meanwhile, in 1896, with the help of the wife of the philosopher, Alfred North Whitehead, and of Mrs. Herringham, Bateson was able to re-establish contact with Beatrice Durham. Thus, he was able to cement three important relationships: with Mrs. Herringham, with Miss Saunders, and with Beatrice. They were married in a church by the clergyman whose malchaperoning had facilitated their fleeting engagement in 1889. Shortly after the birth of their second child, they moved to a house with a large garden at Grantchester. Attempts to expand the support of their research beyond that provided by the RS ("begging letters" and speculations in art) were unsuccessful. Feelers regarding the Secretaryship of the Zoological Society led nowhere. However, in 1899, Bateson obtained his first university teaching position – as deputy to Newton, the Professor of Zoology and Comparative Anatomy. His links with the Darwin's may have been further cemented by discussion with George Darwin on Down House as a future Biological Station. The establishment views emerging during a Symposium on the classification of species indicated that there were likely to be difficulties ahead.

Part II
Mendelism

Chapter 8
Rediscovery (1900–1901)

He was over the stepping-stones and away, scrambling up the further bank whilst the Biometricians, chiding, were still negotiating the difficulties of the first step.

Beatrice, Memoir.

The "rediscovery" of the work of Gregor Mendel has often been related. There is no unanimous agreement on the exact chain of events. It appears that around 1900 a friend of de Vries, knowing that he had been crossing varieties of plants and observing the distribution of characters among offspring, sent him a copy of Mendel's 1866 paper [1–3]. Before the era of easy copying, the paper may have been one of the original offprints the publisher had made available to Mendel to distribute as he chose [4]. Two other botanists, Correns in Leipzig and Tschermak in Vienna, independently came across the paper. Mendel's work was mentioned by de Vries in his "Sur la Loi de Disjunction des Hybrides," published in the French journal *Comptes Rendus*. There was also a paper in *Berichte der Deutsche Botanische Gesellshaft* (April 25, 1900). In those days, offprints of their papers were often sent to authors simultaneously with the publication of a journal, so, given the rapid mail transmission times then in effect, Bateson could have received an offprint from de Vries by about April 30th. Bateson read it while on the train from Cambridge to London to give a lecture at a meeting of the RHS in May 1900. In her *Memoir*, Beatrice tells us that the paper so excited him that he changed the content of his intended lecture. Thus began a struggle to make the work of Mendel known and accepted by the English-speaking world. Just as Huxley had proclaimed "I am Darwin's bulldog," it seems not inappropriate to regard Bateson similarly with respect to Mendel.

A. G. Cock, D. R. Forsdyke, *Treasure Your Exceptions*,
https://doi.org/10.1007/978-3-030-92099-9_8

Bateson's Anticipations

Bateson told Newton (January 14, 1890) of his plans for breeding studies that would begin with a "single breed or species and [be] extended as opportunity occurred." The fanciers of various types of animals constituted a reservoir of breeding expertise that he would attempt to tap. As an administrator, Newton (January 22) saw the problem in terms of attracting funding and gaining an imprint of respectability through association with a society, such as the Zoological Society. Bateson replied (January 22):

> The function of any society in connecting itself with such an enquiry would be chiefly to provide a status and funds .... My feeling is that a man working alone would be able to do as much as a Committee and perhaps more, if he overcame the initial difficulties of getting into relation with a sufficient number of the right people [e.g. fanciers], and succeeded in impressing them with his fitness to make good use of the facts supplied by them. But ... it would be a harder and much longer process for him to get a start than if he were backed by a recognized body. Of course, Darwin had to invent the subject itself *ab initio*, but it has often struck me as a remarkable fact that he did not succeed in getting much *systematized* or *statistical* information .... My original intention was to try breeding on my own account, but the other way seems much better .... As regards funds I have something to go on with, and should start in a small way.

In *Materials*, Bateson stressed the importance of following variation from generation to generation: "To study variation it must be seen at the moment of its beginning. For comparison we require the parent and the varying offspring together. To find out the nature of the progression we require, simultaneously, at least two consecutive terms." Cramer's letter (Chap. 3) and the studies of de Vries, Correns and Tschermak, reveal he was not alone in this view. Also in *Materials*, he offered "concluding reflections" regarding *two classes* of observable characters: those that were inherited in an either-or fashion without blending ("alternative inheritance"), and those that blended in offspring ("blending inheritance"). If those who came to oppose him had fully understood this distinction, much controversy might have been prevented:

> An error more far-reaching and mischievous is the doctrine that a new variation must immediately be swamped, if I may use the term that authors have thought fit to employ. This doctrine would come with more force were it the fact that as a matter of experience the offspring of two varieties, or of variety and normal, does usually represent a mean between the characteristics of the parent. Such a simple result is, I believe, rarely found among the facts of inheritance. It is true that ... there is as yet little solid evidence to which we may appeal, but in so far as common knowledge is a guide, the balance of experience is, I believe, the other way. Though it is obvious that there are certain classes of characters that are often evenly blended in the offspring, it is equally certain that there are others that are not.
>
> In all this we are still able only to quote case against case. No one has found general expressions differentiating the two classes of characters, nor is it easy to point to any one character that uniformly follows either rule. Perhaps we are justified in the impression that among characters which blend or may blend evenly, are especially certain quantitative characters, such as stature; while characters depending on differences in number, or upon qualitative differences, as for example colour, are more often alternative [non-blending] in their

inheritance. … Nevertheless, it may be remembered that it is especially by differences of number and by qualitative differences that species are commonly distinguished. Specific differences [differences between species] are less often quantitative only.

But however this may be, whatever may be the meaning of alternative inheritance and the physical facts from which it results, and though it may not be possible to find general expressions to distinguish characters so inherited from characters that may blend, it is quite certain that the distinctness and Discontinuity of many characters is in some unknown way a part of their nature, and is not directly dependent upon Natural Selection at all. The belief that all distinctness is due to Natural Selection, and the expectation that apart from Natural Selection there would be a general level of confusion, agrees ill with the facts of Variation. … But beyond general impression, in this, the most fascinating part of the whole problem, there is still no guide. The only way in which we may hope to get at the truth is by the organization of systematic experiments in breeding, a class of research that calls perhaps for more patience and more resources than any other form of biological enquiry.

In the late 1890s, Bateson's formal agenda was set out in letters to Galton as chairman of the Evolution Committee, and to George Darwin (Chap. 6). He elaborated on this at the July 1899 RHS conference (Chap. 9). In his address – "Hybridization and Cross-breeding as a Method of Scientific Investigation" – he declared the ultimate problem to be "the *problem of species*," and, given "the doctrine of Descent," it then followed that:

Wherever … two closely allied varieties exist, the problem of species is presented in a concrete form: How did variety A arise from variety B, or else B from A, or both from something else? This question involves two further questions: 1. By what steps – by integral changes of what size – did the new form come into being? 2. How did the new form persist? How was it perpetuated when the varying individual or individuals mated with their fellows? Why did it not regress to the form from which it sprang, or to an intermediate form?

When only one step was involved, so that "the variation is found already at its beginning in some degree of perfection," then Bateson applied the term *discontinuous*. Thus, "it is no longer necessary to suppose that … long generations of selection and gradual accumulation of differences are needed." Bateson also called for recognition that many characters were passed intact, without change, from generation to generation:

If instead of abstract ideas the facts of cross-breeding are appealed to, it is found that so far from this blending and gradual obliteration of character being the rule, it is nothing of the kind. In many characters, on the contrary, it is at once found on crossing that the varying character may be transmitted in as perfect a degree as that in which it was found in the parent. It need scarcely be said that there are many structures which do not thus retain any integrity when crossed, but there are many that do. Which characters are thus unblending, and which blend, must be determined by careful cross-breeding …. The recognition of the existence of discontinuity in variation, and the possibility of complete or integral inheritance when the variety is crossed with the type [typical form], is, I believe, destined to simplify to us the phenomenon of evolution, perhaps beyond anything that we can foresee.

Given this recognition, the path ahead was clear:

At this time we need no more *general* ideas about evolution. We need *particular* knowledge of the evolution of *particular* forms. What we first require is to know what happens when a variety is first crossed with its *nearest allies*. If the result is to have any scientific value, it is

> almost absolutely necessary that the offspring of such a crossing should then be examined *statistically*. It must be recorded how many of the offspring resembled each parent and how many showed characters intermediate between those of the parents. If the parents differ in several characters, the offspring must be examined statistically, and marshalled, as it is called, in respect of each of those characters separately.

While Bateson discussed only what could be seen – the phenotype – he noted that: "We are far from knowing which kinds of variations may … be definite and palpable, and which are not." Thus, he acknowledged the possible existence of variations that were latent. The conference was the first of a series of what became known as the International Congresses of Genetics, to which basic and applied researchers increasingly flocked. Among those in attendance was the 51-year-old de Vries, who spoke on "Hybridisation as a Means of Pangenetic Infection."

Romanes Anticipations

The phenomenon of the "integral" inheritance of characters was known. Like the Hursts in England (Chap. 10), the Vilmorin family in France had long been involved in commercial aspects of horticulture, and in 1856, Louis de Vilmorin began breeding studies with lupins and attained close to 3:1 ratios (see below) in offspring. He noted that plants that were indistinguishable by eye might differ greatly in the characters transferred to offspring. Darwin in his correspondence in the 1860s showed an awareness of similar studies by Charles Naudin (1815–1899) in *Les Jardins des Plantes* in Paris [5].

In 1868, in his *Variation of Animals and Plants under Domestication*, Darwin observed that "crossed forms of the first generation are generally nearly intermediate in character between their two parents; but in the next generation the offspring commonly revert to one or both of their grandparents, and occasionally to more remote ancestors." Furthermore, he toyed with ideas that the gonads might contain both "pure" and "hybridized" character units ("gemmules"; Fig. 3.2 in Chap. 3), and that "gemmules of the same nature would be especially apt to combine." Thus, "when two hybrids pair, the combination of pure gemmules derived from the one hybrid with the pure gemmules of the same parts derived from the other, would necessarily lead to complete reversion of character." Arguing more quantitatively, Galton wrote to Darwin in 1875 showing how such integral inheritance might work (see below). In 1876, Darwin published *The Effects of Cross- and Self-Fertilization in the Vegetable Kingdom*. This was the product of a decade of experimental breeding studies that seem to have been motivated more by a desire to understand the increased vigor sometimes displayed by hybrids (itself a topic of fundamental importance), than to understand reversion to grandparental types.

It was Mendel who had taken the critical step, moving from qualitative to quantitative, with a knowledge that statistical variations should be taken into account. Thus, when seeking generalizations, the results from numerous experiments were collectively compared following the statistical guidelines of the day. In November

1880, to help Romanes prepare an article on "Hybridism" for the *Encyclopedia Britannica* [6], Darwin loaned him his new copy of *Die Pflanzenmischlinge*, a book on the history of hybrid breeding by Wilhelm Focke (1834–1922). The associated letter is of interest in revealing how a distinguished authority can set a junior off on the wrong track [7]:

> Kölreuter, Gärtner and Herbert are certainly far the most trustworthy authorities. There was also a German, whose name I mention in 'Origin,' who wrote on Hybrid Willows [Max Wichura]. Naudin, who is often quoted, I have much less confidence in. By the way, Nägeli (whom many think the greatest botanist in Germany) wrote a few years ago on Hybridism .... The title will be sure to be in Focke.

Naudin had taken a position very close to Mendel's, and "the greatest botanist" had been Mendel's main scientific correspondent yet had failed to understand him. Focke's book referred to Mendel, and Romanes added the citation to his article and returned the book to Darwin. Whether he later acquired his own copy we do not know, but he later referred Thiselton-Dyer to certain pages in Focke's book [7]. In the same correspondence, Romanes referred to Naudin's *Datura* (thorn apple) crosses, and to "Focke, with whom I have been in correspondence from the first." Indeed, by 1891, he was actively promoting an "*Institute Transformiste*" – a large "garden or farm for the purposes of studying questions of hereditary transmission" [7]. A letter written to Schaffer a few days before Romanes' death in 1894 indicates that he was probably well ahead of Bateson in his Mendelian mode of thought [8]:

> I have found after several years experimenting with rabbits, rats and [undecipherable], that one may breed scores and hundreds of first crosses between different varieties, and never get a single mongrel throwing intermediate characters – or indeed any resemblance to one side of the house. Yet, if the younger are subsequently crossed *inter se* (i.e. brothers and sisters, or first crossings) the crossed parentage at once repeats itself. *Ergo*, even if the pups which are to be born appear to give a negative result, keep them to breed from with one another.

Bateson's research notebooks provide no evidence for his conducting such brother–sister matings in birds until 1901. However, by July 1899, Saunders had done the equivalent in plants (i.e., had produced an F_2 generation in *Lychnis*). The characters under study (hairy versus smooth, otherwise known as "glabrous") showed different patterns of inheritance in different groups of plants, and it seemed that each species and character followed its own rules of inheritance [9].

Rediscovery

Life was going on much as usual. In the spring of 1900, Bateson sent letters to Beatrice from various locations in England where he was visiting art dealers or horticulturalists. On April 24th, he was the RS delegate to the Scientific Committee of the RHS. Then one day in May, he took the train to London.

Bateson gave his RHS lecture on "Problems of Heredity as a Subject for Horticultural Investigation" on May 8, 1900. The version published in Beatrice's

Memoir was updated from the version actually delivered, for it refers to papers by Correns and Tschermak that had not been published on the day of the lecture. As a result of his reading of the de Vries paper, he modified his lecture to incorporate brief accounts of Mendel's pea experiments, which had been confirmed by de Vries (in several genera), by Correns (in maize), and by Tschermak (in peas): "That we are in the presence of a new principle of the highest importance is, I think, manifest." Bateson's "conversion" to Mendelism was prompt and complete. His temperament was cautious and skeptical, but he did not react in this fashion to Mendelism. His own theoretical and experimental studies had prepared his mind. The impression was deep and lasting.

The approach taken in his lecture was to honor the great contributions Galton had made (Chap. 3), while pointing out that the latter's predictions – that each parent contributed, on average, one half to the whole heritage, the grandparents, one quarter, etc. – had been found wanting in the cases studied by Mendel. Here nonblending (alternative) inheritance applied and some characters were "dominant" to others (which were accordingly deemed "recessive"). Indeed, Mendel had deliberately selected certain characters because their inheritance displayed the dominance–recessive property. Thus, in these cases, character-display from generation to generation was discontinuous. Mendel's work had revealed laws of discontinuous inheritance. Since the work had been confirmed by the independent studies of others, there could "be no doubt that Mendel's law is a substantial reality."

However, Bateson pointed out that "the numbers with which Mendel worked, though large, were not large enough to give really smooth results." There were "a few rather marked exceptions," and he wondered "whether some of the cases that depart most widely" could "be brought within the terms of the same principle or not." Furthermore, although Mendel had carefully selected his characters for their dominance–recessive relationships, in the offspring of crosses, "the degrees to which the characters appeared did vary, and it is not easy to see how the hypothesis of perfect purity in the reproductive cells can be supported."

Bateson admitted that "we still have no glimmering of an idea as to what constitutes the essential process by which the likeness of the parent is transmitted to the offspring, … of the physical basis of heredity we have no conception at all. … We do not know what is the essential agent in the transmission of parental characters, not even whether it is a material agent or not. … But apart from any conception of the essential modes of the transmission of characters, we *can* study the outward facts of transmission." It might then be possible to show how widely Mendel's results applied and to determine the basis of any anomalies.

For this, funds were needed. Ever mindful that his passion for art might be turned to their financial advantage, that summer Bateson went art collecting in Kandersteg, Germany (August 9, 1900), and in October, he and Beatrice were in Paris for a week where there was a big exhibition. Half-way home, they realized that one of their Grantchester paintings (a Boucher that had been purchased for a few shillings) was highly priced in Paris. Bateson returned to Paris, and Beatrice went home and sent him the painting. He sold it (dare we say "traded it") for £40.

A translation of Mendel's paper (the first draft was by Charles Druery) was published in the *Journal of the Royal Horticultural Society* (1902), and Bateson wrote introductory remarks:

> The conclusion which stands out as the chief result of Mendel's admirable experiments is of course the proof that in respect of certain pairs of differentiating characters the germ cells of a hybrid, or cross-bred, are *pure*, being both carriers and transmitters of either the one character or the other, not both. That he succeeded in demonstrating this law for the simple cases with which he worked is scarcely possible to doubt. In so far as Mendel's law applies, therefore, the conclusion is forced upon us that a living organism is a complex of characters, of which some, at least, are dissociable and are capable of being replaced by others. We thus reach the conception of *unit-characters*, which may be rearranged in the formation of reproductive cells. It is hardly too much to say that the experiments which led to this advance in knowledge are worthy to rank with those that laid the foundation of the Atomic laws of Chemistry.

Gametes were here described as "carriers and transmitters" of the characters *themselves*, not of something that in some way *represented* the characters. Although the distinction was present in his mind, Bateson did not clearly distinguish what subsequently became known as phenotype and genotype – a recurring feature of his writings. Mendel did not mislead him in this respect since he had written (1866) of "elements ... which determine" the differences between gametes, and noted that "the production would be made possible of as many sorts of egg and pollen cells as there are combinations of the *formative elements*" (i.e., elements which *form* – dare we say *inform* – the phenotype). For Bateson, the word "characters" seems to have been short-hand for "elements which determine the characters," a leap many of his readers would have been incapable of taking. At the outset he had invited readers to imagine: "For it will be seen that the results are such as we might expect if it is imagined that the cross-bred plant produced pollen grains and ovules, each of which bears only *one* of the alternative varietal characters and not both." He went on to consider why it had taken so long for Mendel's work to be recognized:

> The cause is unquestionably to be found in the neglect of the experimental study of the problem of Species which supervened on the general acceptance of Darwin's doctrines. The problem of Species, as Gärtner, Kölreuter, Naudin, Mendel, and the other hybridists of the first half of the nineteenth century conceived it, attracted thenceforth no workers. The question, it was imagined, had been answered and the debate ended. No one felt any interest in the matter. A host of other lines of work were suddenly opened up, and in 1865 the more vigorous investigators naturally found these new methods of research more attractive than the tedious observations of the hybridisers .... We must go back and take up the thread of the inquiry exactly where Mendel dropped it.

Weldon's Analysis

Weldon submitted (December 9, 1901) a paper entitled "Mendel's Laws of Alternative Inheritance in Peas" to *Biometrika*, of which he was a coeditor [10]. It appeared in the January 1902 issue and was read by Bateson on February eighth, although an abstract had been "most courteously sent ... by an editor of *Biometrika*" at some earlier date. Before considering Weldon's paper, there will be a brief digression on plants.

Because of fundamental differences between their life cycles, plants are often more suitable for breeding studies than animals. Plants of one type can be crossed with plants of another type by placing the male pollen from one on the female stigma of the other. The resulting zygote, instead of thenceforth developing to produce a mature form, generally arrests at the embryo (seed) stage. If it did not, there would be no apples, oranges, etc. In this respect, plants resemble butterflies in that there is an intermediate caterpillar stage, which can remain quiescent before proceeding to the adult form. When circumstances are propitious, the seed germinates to complete the developmental process. The importance of this is that, apart from the seed coat furnished by the mother plant, the seed is an embryo, an early manifestation of a new individual contributed to by both paternal (pollen) and maternal (ovule) elements. For example, both may contribute to give the first leaves (cotyledons) their appropriate color.

Another characteristic of plants is that they are immobile. They do not go looking for a mate. So, they attract insects that will transfer their pollen. But in the natural state, surrounded by many competing species or in a sparse terrain, it is possible that a plant will not succeed in attracting an insect bearing appropriate pollen. Fertilization would then not occur, so the line would die out. To counter this possibility, many plants develop both male and female parts, thus retaining the option of self-fertilization. In other words, plants that did not so evolve (as hermaphrodites) may have perished by natural selection. Indeed, in many cases, including peas, self-fertilization is the rule, rather than the exception.

Weldon's paper began with a description of Mendel's main observations. When Mendel crossed a line of plants which always produced seeds with yellow cotyledons with a line that always produced seeds with green cotyledons, the resulting seeds (the "first generation"; F_1) *all* had yellow cotyledons. It did not matter whether a yellow-seeded plant was the pollen parent, or whether a green-seeded plant was the pollen parent. The yellow seed character appeared to be dominant over the green seed character. Had the element determining the green character entirely disappeared, or was it merely latent? To find out, these F_1 seeds were allowed to germinate to produce new plants, from which a "second generation" (F_2) of seeds was derived following self-fertilization. Among a total of 8023 seeds, 6022 were yellow and 2001 were green. The green character had reappeared, and thus the element determining it must have been latent in the F_1 plants. For every green seed, there were approximately three yellow seeds (3:1 ratio). This frequency of green seeds was too high to be accounted for by new sporadic variations.

Mendel's famous interpretation was that the element determining each character was inherited in an integral fashion. An adult plant could transmit two elements corresponding to a particular character, but a gamete (male pollen or female ovule) could only transmit one element. With respect to a particular character, each gamete was pure. The elements determining the characters were alternatives. Individuals of the F_1 generation were all hybrids with respect to the elements determining yellow and green, and so the resulting gametes corresponded either to yellow or to green. There would then be four possible combinations of gametes from which the F_2 generation of seeds would be derived:

OVULE		POLLEN	SEED	TYPE
♀ yellow	x	♂ yellow	Yellow	Pure
♀ yellow	x	♂ green	Yellow	Hybrid
♀ green	x	♂ yellow	Yellow	Hybrid
♀ green	x	♂ green	Green	Pure

If yellow were dominant, then three of these would still produce only yellow seeds, but when both gametes conveyed the green character, the resulting adult plants would all produce green seeds – thus the three to one ratio.

Weldon also made the point that if the second generation were allowed to self-fertilize to produce a third generation, etc., then the proportion of hybrids would progressively decrease. He tabulated from Mendel's paper the likely proportions in each generation:

GENERATION	YELLOW	HYBRID	GREEN
F_1	0	1	0
F_2	1	2	1
F_3	6	4	6
F_4	28	8	28

Although he did not further simplify the ratios in the F_3 and F_4 generations, it seemed that Weldon had fully understood Mendel's message. Given Weldon's penchant for numerical analysis, it would be surprising if he had not. Yet, while having "no wish to belittle the importance of Mendel's achievement," he proceeded to elaborate more extensively, and more numerically, on Bateson's own caveats, quoting the above remark that Mendel's numbers "are not large enough to get really smooth results." Weldon held that there was "a grave discrepancy between the evidence afforded by Mendel's own experiments, and that obtained by other observers, equally competent and trustworthy." There were Mendel's strict distinctions between yellow and green peas, or between round and wrinkled peas. These were not valid. There were intermediates between the forms. In the case of a pea type known as "Telephone," for example, "hardly any two seeds are alike. So that both the category 'round' … and the category 'wrinkled' … include a considerable range of varieties." He was to expand on this in a later paper entitled: "On the Ambiguity of Mendel's Categories" (Chap. 9). What was the explanation for this ambiguity? "Mendel does not take the effect of differences of ancestry into account, but considers that any yellow-seeded Pea, crossed with any green-seeded Pea, will behave in a definitive way, whatever the ancestry of the green and yellow peas may have been." This was "a fundamental mistake which vitiates all work based upon Mendel's method."

After he learned of Mendel's work, Bateson's research notebooks displayed a heightened pace, with extension of breeding to the second (F_2) generation [9]. For each year of RS funding (1898 onward), progress reports to the Evolution Committee had been required. Bateson reported on the animal work (insects, fowls), and Saunders reported on the plant work. These reports were not formally published. In 1901, they completed a more extensive report and submitted this to the RS Evolution Committee (December 17). Prior to publication, there was an opportunity in March 1902 to include an "added note." Nevertheless, the bulk of the writing was completed in December and, despite many last-minute proof corrections, Bateson was at last free to move back to research mode, and away from writing mode. But Weldon's activities ensured that this was not to be.

It was neither in Bateson's nature to – nor given that it came from an eminent source who was in a position to influence career choices, research priorities, and funding decisions, was it appropriate that he did – ignore Weldon's attack on Mendel. Writing "in consultation with" Saunders, the incensed Bateson within a few months cobbled together a 208-page polemic, *Mendel's Principles of Heredity: a Defence* (hereafter "*Defence*"), which was promptly published by Cambridge University Press. Thus, two years after the "rediscovery," in 1902, two major works emerged from the Bateson "school" at Cambridge. First, there was the more technical RS report (*Reports to the Evolution Committee - Report 1*; hereafter referred to as "*RS-Report 1*") coauthored with Saunders, which will be considered next. Second, there was the more general *Defence* to be considered in the next chapter.

Royal Society Report

RS-Report 1 described their studies in plants and fowl that, taken with those of de Vries, Tschermak and Correns, provided strong support for Mendel. There was now a new biological science centered on "character units, the *sensible manifestations of* physiological units of as yet unknown nature" (our italics). No longer would "various species, breeds, varieties, and casual fluctuations" be regarded "as all comparable expressions of one phenomenon, similar in kind." No longer would there be "insufficient recognition of the possibility that variation may be, in its essence, specific (Fig. 8.1)."

The new science was not named at that time, but it came with new concepts for which an immediate coining of terms was imperative. The difficulties encountered in the struggle to bring order to the new field exasperated the RS secretariat that had to cope with multiple last-minute revisions. The phrase "sensible manifestations of" implied recognition of a phenotype–genotype distinction, but there were no equivalents of "phenotype" and "genotype" – words that were coined several years later (Chap. 10). However, for the first time, the now familiar words, *homozygote*, *heterozygote*, and *allelomorph* (later shortened to *allele*), appeared. A cross between two

Fig. 8.1 Edith Rebecca Saunders

parents would produce the *first filial* generation (the F_1 generation). The offspring of successive crosses of members of each filial generation with themselves were referred to as the F_2 generation, the F_3 generation, etc.

Organisms that were homozygous with respect to a character were allelomorphically "pure," generating uniform gametes, so they would "breed true" when crossed with similar homozygotes (i.e., there was "fixity of form"). On the other hand, organisms that were heterozygous with respect to a character were not allelomorphically pure and so would not normally "breed true." The "want of fixity of certain forms, though continually selected [bred with like forms], may at once be explained by the hypothesis that they are heterozygous only, and have no gametes corresponding to them" (i.e., there were no individual gametes corresponding to their phenotype). In other words, there were *two* kinds of gametes (say A and a), but *three* kinds of individuals in the population (say AA, Aa, aa). Two of these, AA and aa, were homozygous. One, Aa, was heterozygous. Should dominance be incomplete, then the heterozygote could appear as a distinct form.

The occasional appearance of a particular "rogue" form (considered a "reversion") might often be explained on "the hypothesis that such a 'rogue' is a recessive form." Sometimes, a homozygous recessive form might not reach adulthood: "A recessive allelomorph may even persist as a gamete *without the corresponding homozygote having ever reached maturity in the history of the species*." Thus, when two gametes transmitting recessive alleles united as a zygote, the combination might happen to be lethal.

The word "mutation" was now used, and there was a clear distinction between two kinds of variation, one *common* involving "type-gametes," and one *rare* involving "aberrant gametes." The common variation would be due to the chance union by the crossing of dissimilar "type-gametes" to produce a heterozygote. This "crossing may be truly spoken of as a 'cause' of variability" (i.e., production of heterozygous forms that differ from the homozygous forms), but it was created by a *reassortment* (through crossing) of what was *already present* in the species (i.e., a rearrangement). On the other hand, the rare variation would be "an *originating* variation – the 'mutation' of de Vries," so that "the output of a certain [small] number of such aberrant gametes is normally incidental to the development of type-gametes." Such sporadic "sports ... may be exceedingly rare, and therefore produced by few individuals only."

While from most of Saunder's plant crosses, the ratios of the various F_2 forms were in keeping with Mendelian expectations, some of her results with stocks (*Matthiola*) were strange. Two pairs of characters that seemed to be allelic alternatives were hoariness (hairy surface) versus glabrousness (smooth surface), and green seed color versus brown seed color. It was noted that "in certain combinations there was close correlation between (a) green color of seed and hoariness, (b) brown color of seed and glabrousness." This was later referred to as the "coupling" of characters (i.e., linkage).

That which *represented* a character in the gamete was at first given the name "element." It was "possible to conceive of the elements contributed by the two gametes as engaged in a conflict," which would somehow be resolved in the adult (zygote). Thus, "we do not *know* whether the character exhibited by such [a] zygote is really the product of the allelomorphs of *both* gametes, or is due to the exclusive development of that of one gamete only." Yet, "from what is known of discontinuous variation in general, we incline to the view that even though the figures point to a sharp discontinuity between dominant and recessive elements, we shall ultimately recognize that the discontinuity between these elements need not be *universally* absolute." The modern reader may see here a hint – just a hint – at recombination occurring at the DNA level.

There was also the phenomenon of mosaicism to be explained. Bateson and Saunders considered the fruit of *Datura* where the character "thorny" was dominant over the character "thornless." Usually, the fruit were either entirely thorny or entirely thornless. However, sometimes thornless fruit displayed thorny patches, hence appearing as a "mosaic." There appeared to have been variation within an individual adult diploid organism. How did this come about? Later seen as an example of somatically-derived mosaicism (Chap. 13), at that time they struggled since they could not exclude an explanation at the level of the gametes (i.e., germline-derived mosaicism).

The Residue

In correspondence with Darwin (December 19, 1875), Galton had considered the distribution of gemmules corresponding to the characters for blackness and whiteness and arrived at ratios identical to Mendel's:

> If there were 2 gemmules only, each of which might be either white or black, then in a large number of cases one-quarter would always be quite white, one quarter quite black, and one half would be grey. If there were 3 molecules, we should have four grades of colour (1 quite white, 3 light grey, 3 dark grey, 1 quite black and so on according to the successive lines of 'Pascal's triangle').

At that time Galton had interpreted the occasional appearance of so-called ancestral characters among offspring as indicating that the germline contained latent elements (Chap. 3). There were "primary elements," which would include the patent "embryonic elements" that "contain materials out of which structure is evolved." Galton proposed that if these were, in some way, subtracted from the totality of the primary elements, there would be left behind a "residue" of the latent elements. The extent to which Bateson and Saunders were influenced by this is not clear, but in the concluding pages of *RS-Report 1,* the word "residue" emerged: "Has a given organism a fixed number of unit characters? Can we rightly conceive of the whole organism as composed of such unit-characters, or is there some residue – a basis – upon which the unit characters are imposed?" This was expressed in slightly different terms a few months later in *Defence*: "From the fact of the existence of interchangeable [allelic] characters we must, for purposes of treatment, and to complete the possibilities, necessarily form the conception of an *irresoluble base*, though whether such a conception has any objective reality we have no means as yet of determining."

The discussion in *RS-Report 1* indicated they were taking the residue concept much further than Galton, and were relating it to Darwin's fundamental question – the nature of the origin of species – a "phenomenon" that did *not* "attach to" the elements (now known as genes), which determined allelomorphic characters:

> We know, of course, that we cannot isolate this residue from the unit characters. We cannot conceive a pea, for example, that has no height, no colour, and so on; if all these were removed there would be no living organism left. But while we know that all these characters can be interchanged, we are bound to ask is there something not thus interchangeable? And if so, what is it? We are thus brought to face the further question of the bearing of Mendelian facts on the nature of Species. The conception of Species, however we may formulate it, can hardly be supposed to attach to allelomorphic or analytical varieties. We may be driven to conceive 'Species' as a phenomenon belonging to that 'residue' spoken of above, but on the other hand we get a clearer conception of the nature of sterility on crossing.

Why a "clearer conception?" There are various types and causes of sterility. The sterility to which they referred in *RS-Report 1* was that which was *consistently* obtained from a cross between members from allied species that were so closely related that offspring could be produced. The hybrid offspring was *consistently*, to some degree, sterile, and this "hybrid sterility" was associated with maldevelopment of the gamete-producing organs (the gonads), so that fewer and/or malformed gametes were produced. Since the species were closely related, it was inferred that they had derived from a common ancestral species. Since the offspring of crosses between members of that ancestral species could be assumed not to have displayed consistent sterility (otherwise, there would have been no descendants), then the hybrid sterility revealed the emergence of a reproductive barrier between the descendent species, which could have been an *originating* barrier leading to the state of *reproductive isolation* that *defined* them as distinct species (Chap. 5). Thus, Bateson and Saunders continued:

> Though some degree of sterility on crossing is only one of the diverse properties which may be associated with Specific difference [difference between species], the relation of such sterility to Mendelian phenomena must be a subject for most careful enquiry. So far as we yet know, it seems to be an essential condition that in these [Mendelian] cases the fertility of the cross-bred [hybrid] should be complete. We know of no Mendelian case in which fertility is impaired. We may, perhaps, take this as an indication that the sterility of certain crosses is merely an indication that *they cannot divide up the characters among their gametes*. If the parental characters, however dissimilar, can be split up [segregated], the gametes can be formed. ... The inability to form gametes may mean that the process of resolution cannot be carried out. ... That the sterility of hybrids is generally connected in some way with inability to form germ-cells correctly, especially those of the male, is fairly clear, and there is in some cases actual evidence that this deformity of pollen grains of hybrids is due to irregularity or imperfection in the processes of division from which they result. It is a common observation that the grains of hybrid pollen are too large or too small, or imperfectly divided from each other.

More significantly, they cited studies of Michael Guyer on chromosome behavior during meiosis in hybrid pigeons as reported in the journal *Science* in 1900 [11]. Here, Guyer, "apparently in ignorance of Mendel's work," had come to similar conclusions based on direct observations of the chromosomes (Chap. 13). Bateson and Saunders concluded:

> As soon as all means shall have been found of making visible that differentiation which we now know must exist between the germ-cells of the same heterozygote, a vast field of research will be opened up. Till then, the microscopical appearances accompanying the segregation of the characters must remain unknown, and we are obliged to resort to the cumbrous and protracted method of deduction from the statistical study of the zygotes [organisms] formed by the union of the several kinds of gamete.

In other words, they had to engage in what came to be known as "classical genetics." Indeed, anticipating Hardy-Weinberg (Chap. 10), they opined:

> It will be of great interest to study the statistics of such a population in nature. If the degree of dominance can be experimentally determined, or the heterozygote recognised, and we can suppose that all forms mate together with equal freedom and fertility, and that there is no natural selection in respect of the allelomorphs, it should be possible to predict the proportions of the several components of the population with some accuracy. Conversely, departures from the calculated result would then throw no little light on the influence of disturbing factors, selection, and the like.

The Bateson School

After completing his writing marathon in March 1902, there was time to catch up and make long-intended visits. Surviving correspondence with Charles Hurst dates from March 24, 1902 (Chap. 10). There was a first visit to Burbage where Bateson met old Benjamin Hurst, a fellow admirer of the works of Balzac. Hurst's biographer, Rona Hurst, records that "Uncle Ben ...was ... a man much to Bateson's taste and they got on famously together" [12]. There was also time to update his lectures.

Interviewed years later by Lipset, Nora Barlow (a granddaughter of Charles Darwin) recalled the excitement of a lecture by Bateson [13]:

> My first introduction to the whole subject ... was when William Bateson was giving what we called his bible class, in a remote lecture room, in the back of one of the colleges. It was outside the ordinary curriculum. It was a five or six o'clock lecture. And there he introduced a small set of people into the elements of the new Genetics. Mendelism was just coming in. ... He was a brilliant lecturer and, of course, he had an entirely new view of ordinary heredity. ... It was very inspiring indeed.

Another member of the bible class was Geoffrey Keynes, who in 1907 was elected to "the select Natural Science Club" with which Bateson was associated [14]:

> The latter's lectures were not part of the syllabus for students of Zoology, but I was excited by the new ideas just then coming to the surface as 'Mendelism,' of which Bateson was the leading exponent in England, and so I went to hear him. ... Soon I was a regular visitor at the Batesons' house in Grantchester. Bateson was generally regarded as a somewhat formidable character, but I found him to be full of intellectual interests, great fun in conversation and, to my delight, a collector of the works of William Blake – indeed he was the possessor of one of Blake's most startling colour prints showing 'Satan Exulting over Eve.' It hung in his dining room, so that I could sit opposite and gaze at it throughout the meal.

Decades later, Keynes was to reproduce it in his biography of Blake, which was dedicated to Bateson. Commenting on Bateson's lectures, another student, A. J. Willmott, reminisced [15]:

> I do not think that as a lecturer he was lucid. His tables on the blackboard frequently were rubbed out and started with fresh gusto a second, even a third, time – but we always arrived. ... His main strength lay in his capacity to stimulate. He held 5 o'clock classes, during which we read, in one year the *Origin of Species*, and another, *Animals and Plants under Domestication*. We read one or two chapters during the week, and met for discussion of points of interest. ... I remember an excursion to Oxshott with his sons, who wanted some of the beetle larvae kept as cows by the wood-ants. We started prodding the ant-hill with sticks. But Bateson went down on his knees, swept the top off with one wave of his arm, dug a pit into the centre and shoveled out heaps from its base with his hands, which at the end he casually rubbed together.

Initially, all Bateson's research associates were women, most of whom did not have the security of Saunders. These included Nora Darwin, Beatrice's sister Florence Durham (1869–1949) of Girton College, and four graduates of Newnham college – Hilda Killby (1877–1962), Muriel Wheldale (1880–1932), Dorothea C. E. Marryat (1880–1928), and Igerna B. J. Sollas (1877–1965). There was room for considerable initiative in acquiring space and resources. Saunders with the help of Marryat also carried out plant breeding studies in Newnham College gardens. Sollas and Killby reared guinea pigs and goats in a field near the College grounds. Durham used a room in the Museum site to breed mice. With this embryonic Mendelian school, a number of men became, to various degrees, associated – Reginald Punnett, Robert Lock, Reginald Gregory, Leonard Doncaster, Wilfred Agar, William Balls, Henry Biffen, Thomas Wood, and Charles Hurst (Chap. 10).

Despite the availability of much voluntary labor, financial support was necessary. Bateson's frankness about his goals may not have served him well in this respect. When a large tract of land became available to the RHS at Wisley, Bateson was asked his opinion on the setting up of a research station. He replied to the Secretary, William Wilks (September 25, 1903), noting the great need for "a properly equipped station for the study of breeder's problems by scientific methods," thus suggesting future results of immediate utility. However, he then declared that: "None of the American or Continental stations do what is wanted. They are all *bound* to try for immediate results of utility, and the permanent importance of their work suffers greatly thereby, as several of them admit. … To lay the foundation of a comprehensive and precise knowledge of heredity and variation would be a work worthy of the Society, and in the end gain the approval both of naturalists and practical breeders of all nations." It seems that they took his advice, but there were other pressures on the RHS budget, and it was not until 1908 that the Wisley Laboratory and Research Station was formally opened by Lord Avebury.

In 1903, Cambridge University received a bequest from Mr. F. J. Quick to promote "study and research in the sciences of vegetable and animal biology." Among the Trustees of the Quick Bequest were Lankester and Newton. Through the latter, Bateson submitted a proposal for a research station (November 1903). With continuing input from Bateson, the matter ground on until 1906 when the Quick Professorship was announced – in *Protozoology*! In high dudgeon, his senior colleague in Newton's Department of Animal Morphology, Adam Sedgwick, put the blame squarely on Bateson himself who, for some reason, had declined to serve on a Board of Managers (October 16, 1906): "The Chair is thus lost to Science and that it has been so lost is entirely due to your action in refusing to serve as a Manager." Concerning the "Quick scandal," Bateson replied the following day: "I am merely amazed at the want of scientific judgement which led to such a choice. Protozoology is a backwater of biology – an interesting one, of course – but not part of the mainstream of discovery. … I can only say I have done what I could. Nothing is further from my thought than any desire to quarrel with you about this affair, and I hope we shall not feel it necessary to refer to the matter again." Of course, the matter was referred to again, and the correspondence ground on for some months. To Mrs. Herringham, he was less restrained (undated letter): "It has been a great disappointment that no part of the Quick Fund has come to us. At one time it seemed almost certain that we should get part, and I quite hoped to have got the work on to a permanent footing, but, mainly owing to Lankester's perversity – as I believe – that hope is gone."

It was reported that at a meeting of the Trustees (June 27, 1905), the fact that Sir Michael Foster had considered protozoology as "likely to be fruitful of important results" was given much weight. However, many years later, Bateson still considered Lankester's role to have been pivotal (Letter to Hogben, February 19, 1924): "Adam Sedgwick, before Lankester got at him, wished the Quick foundation to be allotted to the subjects on which I was at that time lecturing, including regeneration and as much *Entwicklungsmechanik* [mechanism of development] as I could manage – a most essential part of Genetics, with which biologists and physiologists alike in Cambridge would have nothing to do."

Reviewing the matter in her *Memoir*, Beatrice wrote that "the shortage today of well-trained men, equipped in sound scientific method, to fill the many Agricultural posts at home and abroad is an interesting commentary on the disregard and neglect which his efforts met with at this time." From Bateson's point of view, the rejection of his proposal hit doubly hard because, spurred on by his unfailing optimism, he had wasted much effort at a critical point in the development of Mendelism. At about this time, Francis Darwin organized a "loyal address" for Foster, who was Member of Parliament for London University from 1900 to 1906. Since the signatories included Sedgwick and Shipley, elementary politics might have dictated that Bateson join. He replied to Darwin (undated letter):

> Thank you for sending the loyal address to Foster. I think a lot of the things you say, but 'Amen sticks in my throat.' The language seems to me to pass the bounds of reasonable adulation. You could hardly say more if he were, say, Pasteur, or some other of the great. He is a fine teacher to whom we are all beholden. I am proud to have been one of his pupils. He has done plenty of good work for Cambridge, but he has done his full share of mischief too, and in these later years his influence has been by no means always used for good. It is disagreeable to have to say so to a thing like this. It is curmudgeonly and to do so perhaps marks the weakly minded, but one must have the courage of one's weakness.

In 1903, Beatrice was pregnant again, and further help became essential. There came an unexpected offer from Mrs. Herringham (December 16): "If I died and left you a legacy for your scientific work you would be very pleased – with the legacy. If a person doesn't intend to leave legacies, may that person give gifts for similar sorts of purposes, – which means, could you allow your work to be helped in this way? Possibly you know some one to whom it would be a lift to take a salary to work for you."

In 1901, with Roger Fry (1866–1934), who had begun to study science at Cambridge but then had switched to art, she had helped form The Society of Painters in Tempera. In July 1903, also with support from Fry – the man who was to become "the evangelist of post-impressionism" – she donated funds to launch the National Arts Collections Fund aimed at combating those such as J. P. Morgan, who considered the art collections of Europe fair game. That the Americans might also want to buy Bateson was at the back of her mind, for her letter went on: "I have been thinking of this for a long time since I asked you last summer what sort of help you wanted. This is *private*. It seems a pity that you should not be able to get on fast with this work, and that you should hanker after America. If you are not pleased with my effort at patriotism, just forget. What I meant is about £100 a year for two or three years. I know that isn't as much as you want. Could it help?" To Bateson's affirmative she responded (December 19, 1903):

> Your letter gave me a warm feeling too. *Thought* you would like it, but one wants to know. I have actually £150 available at the present time, and it would be a pity that you should not have satisfactory help if you have it at all. If what I said will not secure you effective help you had better tell me, because I should not have *less* available next year than now and as far as I can see. ... As for 'private' you could keep it so till we can have a talk. I think people ought to have 3 children and may have 4. Some people can't, which has been my fate: and when they are like Martin and John you need not grumble. *Private* did not mean from Beatrice.

Her other projects included the establishment in 1903 of *The Burlington Magazine*, a forum for "connoisseurs of the arts," and in 1909 of *The Englishwoman*, the magazine of the Women's League of Suffrage Societies. The Herringham gift to Bateson had no strings attached. In her donations to the Arts, she was more involved. When the male-dominated art-elite frowned at this, she stood aside with some reluctance: "I threw the stone in, so you must forgive me watching the eddy" [16].

The funds were limited in duration, but it was a start. In London, Pearson was more fortunate. In 1903, he received £500 per annum from the Drapers' Company, and the funds continued to roll in until 1932. Nevertheless, at last, Bateson was able to consider a partner. Writing on Christmas day to Reginald Punnett (1875–1967), who since 1902 had been Demonstrator in Morphology in Newton's Department, Bateson noted that: "My wife has hitherto done a large part both of the recording and of the many menial operations that such work involves, but I am sorry to say she will be more or less incapacitated this next season, so that help in some form or other I must get." He gave an outline of the financial basis of the operation and offered the position of "partner" to Punnett, with a stipend of £80 a year, noting that "it should be understood that for both of us there will be a good deal of merely *menial* work to do. This menial work is nevertheless sufficiently responsible, and while a responsible paid assistant would cost too much, such work, on the other hand cannot be trusted to a lad." There should be no worries about "'priority' – and all that." They would work and publish in common. He then sketched out his ideas as to how "this will grow into a big thing," adding that "I have talked the matter over with Sedgwick [Punnett's immediate supervisor], and he approves my scheme with some enthusiasm."

However, Punnett was not Bateson's first choice. On receiving the Herringham offer, he immediately wrote (December 18) to Leonard Doncaster (1877–1920), who was then working at Cambridge under Harmer in the Zoological Museum. Doncaster replied the next day expressing interest but pointed to problems and sought further information. Eventually, he declined. Unlike Punnett, he was unwilling to abandon his own research. In the course of a rapid series of letters, Bateson expressed the wish that "if you joined us you would, I hope, rather throw yourself into it and not look on it as so many hours a day." Nevertheless, Bateson promised that "you could certainly go on with the sawfly or any kindred enquiry. You could have a microscope here and a room to yourself as a rule, though we couldn't manage the section cutting here." Furthermore, Bateson did not see any difficulty about his retaining the Museum appointment.

Although not spelled out, underlying Doncaster's formal refusal (December 23) was probably the belief that the partnership would not work, and, even if it did, that it would not assist his career. In other words, it seems he had no conception of the magnitude of the great opportunity Bateson was offering, and/or he had a higher conception of the importance of his own research program:

> I have planned out and started a pretty extensive scheme of work for myself; apart from my pigeons here, which my sisters look after, but which need some supervision, I have (1) the sawflies, (2) the moths, (3) the rats. The sawflies take … a lot of time in cutting and searching sections, but in the summer they will take much more, for they must be under almost

continual observation in order that the eggs may be pickled at just the right times, and when I want fertilized eggs I shall have to keep a close watch on them to see that copulation takes place. … If I could afford to disregard the future altogether, I should probably be willing to put off some of these things for a year or two, and then with added experience I should doubtless be much more successful. But I feel that unless I am to remain all my life in my present not very satisfactory position at Cambridge, there is an unpleasant necessity that I should produce independent work, that not only satisfies myself but also will convince others that I know what I am about. I know you disapprove of this attitude [of expediency], and I believe I do so no less than you, but I do not see how it is to be avoided. I am also very keen on the work I have started for its own sake, and should regret giving it up, and I feel that I could not do both it and the work you propose *well*. It is absolutely necessary that to do either well I should put my whole strength into it and I feel that I should have neither time nor energy for both.

When later reviewing the correspondence for her *Memoir*, Beatrice noted: "The whole biological school being in arms against Will was not likely to be a help to D. in post-hunting. He was a true Quaker with an eye on the world and very cold. The partnership would have been a dismal failure and it was a great relief to find at the end of his [letter of] self-revelation that he still refused." Thus, Beatrice was pleased when Doncaster, described by Bateson in 1921 as one whose "development and bearing towards the world" had throughout his life been governed by "Quaker principles," turned down the offer. Nevertheless, the refusal may have shaken Bateson, and the words in his initial letter to Punnett appear more carefully chosen than in his initial letter to Doncaster. Punnett took a few days to respond, but his response when given was forthright and unconditional:

Very many thanks for your welcome letter. There is nothing I should like better. I am keen on the problems and have a stock of patience above the average, so that I am not at all frightened of menial work. My only misgiving is that you may be over-rating me. More than a year ago I had settled in my mind to get into something of the kind and I had decided this year to clear such work as I have in hand by midsummer, whilst starting some mice experiments as preliminary canter before settling down to them in September. I had intended to spend a couple of months in America this summer, but that can wait. I shall be in Cambridge about the 8th or 9th and will then come out to see you.

Since the experiments were carried out in the gardens of the Batesons' house, and since Beatrice herself would continue to help, she would have had a lot of contact with any partner. Perhaps Beatrice was right in her premonition that a partnership between Bateson and Doncaster would not have worked. In early January, Beatrice went to stay with the Herringham's in London, and a letter to her from Bateson (January 10, 1904) shows tempered approval of the new recruit:

I had a morning with Punnett yesterday. He talks vaguely of not taking his stipend, and putting it into the work; but I did not encourage extravagance as yet. He is of course very raw and did not find fertilizing Primula very easy. But I think he means to throw himself into it. Nothing of [R. P.] Gregory yet, and Punnett had to leave early to play golf. This may become rather irritating by repetition. … Long spell of Saunders today. Her intelligence is certainly much quickened from two years ago: but trying moments come. Her voice rings in my head.

So, while accepting the offer, Punnett declined the stipend. He had income from a fellowship at Gonville and Caius College. Shortly thereafter, against Sedgwick's advice, he resigned his Demonstratorship on being awarded the Balfour Studentship (£200 per annum) that Bateson had coveted decades before. The close relationship that developed between Bateson and Punnett was reflected upon by Gregory Bateson years later in an interview with Lipset [13]. Visitors to the Grantchester house were frequent, and often international. Young Gregory saw Punnett very much as playing second fiddle to his father. One day, one of two visitors seemed to defer to the other. Gregory was said to have asked: "Is he that man's Punnett?"

Beatrice played numerous roles. Sometimes, she acted as clerk, and her initials "CBB" are in the notebooks that are now part of the Cambridge University Library collection. On joining Bateson, Punnett took over the research notes ("ledgers"), and they became much more systematic. By the time of his retirement from Cambridge, 100 volumes had accumulated. Years later Punnett recollected [17]:

> The great majority of the eggs were never allowed to hatch, for some of the characters on which we were working were sufficiently developed for determination at about the eighteenth day. So we had periodical 'openings' which were recorded in a separate notebook known as the 'Dead Book.' On the day of an 'opening' we adjourned to the outhouse … having previously collected Mrs. Bateson to clerk for us, a function which she performed with the greatest efficiency and devotion. Having settled her in a chair at the trestle table with the 'Dead Book' and a large bowl, Bateson took off his coat and produced his knife with the big, blunt blade, while I stood by with a pair of scissors. He then took up an egg, read off the numbers of the pen, the hen and date of laying, and after 'Have you got that, Beatrice?' proceeded to stab and peel off the shell into the aforesaid bowl, and to call out the peculiarities of that particular embryo such as lt., nts., r.c., n.e., f.1., which was to be interpreted as 'light down, no coloured ticks seen, rose comb, no extra toe, feathering on one leg.' After which the chick was handed to me, who slit it so as to expose the sex glands and give Mrs. Bateson the sex to complete the entry.

Despite being so tied down, Beatrice in her *Memoir* recalled that there were outings:

> During these years, from early spring to late autumn we never left home together, except on 'Flower Show' day. Then we rose early, 'did' the incubators, and bicycled to the station for an 8-0 clock morning train …. We tried once or twice to finish with a theatre, but the midnight ride back to Grantchester from Cambridge station – the eggs still to be turned, the lamps adjusted – taught us to be content without this extra pleasure.

Summary

Mendel showed that the "unit-characters" seen in seeds or mature plants depended on the "elements … which determine" them in the gametes that had united to form the zygote from which the seeds or mature plants had derived. A plant could be pure-bred with respect to a given character even though its parents were cross-bred ("hybrids"). Purity depended on the chance meeting of two gametes bearing similar elements (now known as genes). The view long held by horticulturalists that purity of type required continued selection was shown to be incorrect. Indeed, an individual could be pure with respect to one character and cross-bred or impure with respect to others. The Biometricians viewed the work as fundamentally flawed.

Bateson's battle to promote, verify, and extend Mendelism began at a meeting of the RHS in May 1900. In their 1902 *Report to the Evolution Committee,* Bateson and Saunders recognized that Mendel's unit characters were "the *sensible manifestations of* physiological units of as yet unknown nature." New terms were coined, but the distinction between character units and the corresponding physiological units (elements) remained vague. While Mendel's work had revealed the laws of the discontinuous inheritance of patent characters, Bateson and Saunders considered that initiation of the discontinuity between species (i.e., the origin of species) would usually require something more – perhaps Galton's "residue" of latent elements acting as a "base" that was "irresolvable" by Mendelian analysis. Aid from the school of largely voluntary female workers that had begun to build around them was supplemented in 1904 by a partnership with Punnett, funded by Mrs. Herringham.

Chapter 9
Mendel's Bulldog (1902–1906)

Karl Pearson is a biometrician,
And this, I think, is his position,
Bateson and co., hope they may go
To monosyllabic perdition.

Naomi Mitchison [1]

By May 1900, relations between Bateson and the Biometricians were thoroughly and probably irremediably soured. It only needed a fresh issue for quarrels to break out again. Mendelism was admirably suited to serve this purpose, seeming initially to come down on the side of discontinuity, although ways were found of fitting the facts of continuous, as well as discontinuous, variation within a Mendelian framework. Fundamentally at issue was the proper place of statistics in biology. For the Biometricians, statistical analysis was something central to their philosophy – the means whereby biology could be rescued from the status of a second-class and mainly descriptive science and elevated to that of a first-class and analytical science. The statistical apparatus which they developed was above any particular theory, a point Pearson was fond of emphasizing. Regressions and correlation coefficients were no more than a summarization of the facts, free of any theoretical content. For the Mendelians, statistics had no central role. It was a tool – one among many – to be used where appropriate, but in no way having a special place above others.

This prior souring of relations and strong and argumentative personalities on both sides are perhaps responsible for the ensuing dispute between Mendelians and Biometricians (or Ancestrians as the latter were sometimes called because of the importance they attached to Galton's Law of Ancestral Heredity) being almost exclusively a British phenomenon. This is not to say that in other countries Mendelism was accepted instantly. It did not suffice for de Vries, Correns, and Tschermak to blow the trumpets to bring about the collapse of the citadel of traditional biology. As in Britain, there were many biologists of conservative outlook who were converted only belatedly and reluctantly, if at all. But their opposition had

A. G. Cock, D. R. Forsdyke, *Treasure Your Exceptions*,
https://doi.org/10.1007/978-3-030-92099-9_9

a quite different basis from that of Weldon and Pearson, who regarded themselves, not as conservatives, but as being in the van of biological progress. Their opposition was based not on traditionalist inertia, but on their perception that Mendelism constituted a threat to their own newly founded branch of biology.

Defence

The first 104 pages of *Defence* consisted of a "brief historical notice" by Bateson and a translation of Mendel's paper. The rest dealt, point-by-point, with Weldon's arguments, not failing to include within its scope various pronouncements of Pearson, whose response was prompt (June 5, 1902): "I think I ought at once to inform you that I could not possibly pass for the pages of *Biometrika* anything in the tone of your recently published book on Mendel. If you like to keep within the bounds of courteous scientific controversy our pages are open to you, but not to vituperation which suggests nothing but the days of Scalinger." Likewise, Galton thanked Bateson for the book, but added that he regretted "the tone of your controversial remarks."

Bateson's usual allies were effusive. Walter Heape wrote (June 10): "As for your preface – well, I take it as a declaration of war! And I am with you to the best of my abilities. In the role of 'knocker up' I am not quite sure you would receive unstinted praise or continuous support from clients, but as a knocker-down I venture to think you will have a lot of backers." Wilmot Herringham wrote (July 6): "A most fascinating and exciting thing. … The main points are clear and intelligible, and I think you do not in the least exaggerate their immense importance. I could read no other book till I had finished it." The book soon sold out. Bateson decided that it had done its work and did not call for a reprinting. Having referred to those who "were content supinely to rest on the great clearing Darwin made long since," the Preface attempted to criticize Weldon while not unduly antagonizing the other Biometricians:

> Were such a piece [of writing] from the hand of a junior it might safely be neglected; but coming from Professor Weldon there was the danger – almost the certainty – that the small band of younger men who are thinking of research in this field would take it they had learned the gist of Mendel, would imagine his teaching exposed by Professor Weldon, and look elsewhere for lines of work. … The reader who has the patience to examine Professor Weldon's array of objections will find that almost all are dispelled by no more elaborate process than a reference to the original records. With sorrow I find such an article sent out to the world by a Journal bearing, in any association, the revered name of Francis Galton, or under the high sponsorship of Karl Pearson. I yield to no one in admiration of the genius of these men. Never can we sufficiently regret that those great intellects were not trained in the profession of the naturalist.

In *Materials*, Bateson had considered the legal origin of the words "heredity" and "inheritance," which implied that a material object had *itself* been passed from generation to generation:

> Heredity and Inheritance ... are of course metaphors from the descent of property. ... The metaphor of Heredity misrepresents the essential phenomenon of reproduction. In the light of modern investigations, and especially those of Weismann on the continuity of the germ cells, it is likely that the relation of parent to offspring, if it has any analogy with the succession of property, is rather that of trustee [the person who supervises a transfer] than of testator [the person who, at his death, transfers something material to another].

This was elaborated on in *Defence*: "On a previous occasion I pointed out that the terms "Heredity" and "Inheritance" are founded on a misapplication of a metaphor, and in the light of the present knowledge it is becoming clearer that the ideas of 'transmission' of a character by a parent to offspring, or of there being any 'contribution' made by an ancestor to its posterity, must only be admitted under the strictest reserve, and merely as descriptive terms."

Such circumlocutions – for example, denying *any* contribution by an ancestor – have to be understood in context and must surely have confused rather than informed. Nevertheless, in general, his arguments were clear. Furthermore, Weldon, having mainly studied animals, was not conversant with the technical details of plant breeding. *Defence* pointed out the need to distinguish seed *cotyledon* color (an embryo-derived character) from seed *coat* color (a maternally-supplied character), and the need to distinguish varieties with opaque and transparent coats – the former masking the cotyledon colors, the latter showing them. Weldon did not appreciate that characters, such as height, that appeared to vary continuously (not step-wise) among members of some species, could nevertheless be explained in Mendelian terms. In *Natural Inheritance,* Galton had written:

> Human stature is not a simple element, but a sum of the accumulated lengths or thicknesses of more than a hundred bodily parts, each so distinct from the rest as to have earned a name by which it can be specified .... This multiplicity of the elements, whose variations are to some degree independent of one another, ... the larger [their] number ... the more nearly does the variability of their sum assume a 'Normal' character.

Unlike humans where many factors could affect height, in peas there appeared to be one preponderant factor – the internodal length (so pea plants were either short or tall, with no intermediates). By a "normal character" in humans, Galton was referring to a bell-curve that displayed the distribution of heights in a population (i.e., the character displayed a "normal" distribution). Bateson, like Mendel before him, saw that if the expression of a character required multiple pairs of alleles, neither of each pair being dominant, then "each heterozygous combination of any two may have is own appropriate stature," so that "the mere fact that the observed curves of stature give 'chance distributions' is not surprising." In this case, "the purity of the gametes ... might not ... be capable of detection." This was briefly alluded to in *Defence*, and more extensively in *RS-Report 1*:

> In, for example, the stature of a civilized race of man, a typically continuous character, there must certainly be on any hypothesis more than one pair of possible allelomorphs. There may be many such pairs, but we have no certainty that the number of such pairs and consequently of the different kinds of gametes are altogether *unlimited* even in regard to stature. If there were even so few as, say, four or five pairs of possible allelomorphs, the various

> homo- and hetero-zygous combinations might, in seriation, give so near an approach to a continuous [normal] curve, that the purity of the elements would be unsuspected.

Yule touched on this in 1906 [2], and in 1918, a recruit to the biometric ranks, Ronald Fisher, was to further elaborate in a famous paper entitled "The Correlation between Relatives on the Supposition of Mendelian Inheritance" [3].

Boom Begins

The impact of *Defence* became clearer when Bateson attended what was billed by the Horticultural Society of New York as the Second International Conference on Plant Breeding and Hybridization (the first being the 1899 conference in London). Bateson represented the RHS. He sailed from Liverpool on the Majestic (September 17, 1902). Notable fellow travelers included a "dry Edinburgh Scot" named Simpson, who was "distinctly addicted to religion – a fatal blemish in a biologist." Bateson learned that Simpson "was not present at the meeting when W[eldon] attacked Ewart" (presumably in Edinburgh where Cossar Ewart, who had been a friend of Romanes, was Professor of Natural History). However, Simpson "knows practically nothing of Mendel and the Controversy. But with such a cunning type one doesn't quite know where the truth lies." Simpson introduced Bateson to the "too highly respectable" Professor Daniel Gilman and his lady, the latter being found more congenial than her husband who, by virtue of his involvement with the Carnegie Institute in Washington, was "rather tiresomely conscious of his great positions and powers to give or withhold patronage." They docked in New York (September 24), and Bateson went to the Hotel Lafayette, a French hotel he came to like and was to return to on future visits.

He was taken (September 28) to Buffalo to see Niagara Falls by the Curator for Natural Sciences at the Museum of the Brooklyn Institute of Arts and Sciences, Alfred Mayer. The Conference opened (September 30) with an address by Bateson, which will be considered in the next chapter. He wrote home excitedly (October 3): "At the train yesterday, many of the party arrived with their 'Mendel's Principles' in their hands! It has been 'Mendel, Mendel all the way,' and I think a boom is beginning at last." When he gave a lecture on "Heredity" at the Brooklyn Institute, he was introduced as the man who had become a "household word." At Baltimore (October 6), he met with "old Brooks" (Chap. 2). After visiting the Department of Agriculture and the Smithsonian Institute in Washington, he sailed home (October 8) on the Germanic from New York.

Shortly after his return, there was an enquiry from Mayer (October 20) as to his interest in a position at the Brooklyn Institute, the grounds of which "afford ample space for raising plants, etc." Bateson respectfully declined, noting: "I am not altogether hopeless of starting an Experimental Station in England." The time seemed propitious for an application to America for funds to support this. Apparently prompted by Herbert Spencer's interest in the possibility of the inheritance of acquired characters (Lamarckism), Mr. Andrew Carnegie, a financially successful Scottish emigrant, had written in 1900 to Francis Darwin indicating he might support research. Another explanation might be that Carnegie had received an application from Davenport (see below) and wanted another to compare it with. A brief expression of interest from Bateson (November 21, 1900) had come to nothing, but now the situation was different. Bateson submitted (October 1902) a formal application to Gilman, ex-President of Johns Hopkins University, who now managed the affairs of the Trustees of the Carnegie Institution. Gilman advised (November 4) that currently they had a lot of business in hand and would not get round to a decision until the new year. To Bateson's offer (November 16) to name people to contact for testimonials, Gilman replied (December 1) that this would not be necessary. Needless to say, like many of the other "begging letters" Bateson submitted in those years, the application was not successful. But about that time Carnegie agreed to fund Davenport's proposed "Station for Experimental Evolution," which formally opened in 1904 with de Vries giving an inaugural address (where he suggested that radiation might be mutagenic). Not only did Davenport have Carnegie, he also had a "Mrs. Herringham." She was Mary Harriman, daughter of a railroad magnate. She helped him establish a Eugenics Record Office at Cold Spring Harbor in 1910.

Despite suggestions that he might succeed Foster as Secretary of the RS (upon whose Council he served from 1901 to 1903), and that the prospects were good for becoming Secretary of the Zoological Society when Sclater eventually resigned in October 1902, Bateson did not pursue another appointment. After a stormy contest, the latter post went to Chalmers-Mitchell. To Beatrice, Bateson wrote (November 17, 1902):

> Looking at the matter fairly I don't believe I could do good work, and the Zoo too. As I told you, Lankester thought the Z. S. would expect their man to give his whole time, and I believe he would have to, if he was to be a success. *It would never do to grudge every day given to the Zoo*, and that is just what I should face. ... I have a feeling we shall get the Carnegie money, and if not that, some other, and day by day I feel clearer that my proper place is on the land. I feel sure we are on a splendid line and the next few years are *the* years. If we are to get the first crop off this work it must be done then. ... All things considered, I am in an exceptionally good position now – enough to live on, and plenty of leisure – and there is always the chance of an endowment for the work. ... In my present view, I wonder we have taken the question of the Zoo so seriously!

Meanwhile, people of diverse skills and expertise were becoming interested in Mendelian investigations, either independently or in association with the Bateson school. In early 1903 (January 21), Bateson included R. Staples Browne, an expert in pigeon breeding, in his grant application to the RS. An important ally in the USA was William Castle (1867–1962), a student of Davenport. When Castle wrote in the *Proceedings of the American Academy of Arts and Sciences* supporting Mendelism [4], Pearson responded questioning his mathematical competence. In 1903, Bateson joined the Ad Eundem Club – an exclusive Oxford-Cambridge dining club – where he doubtless spread the word. And at St. John's College, the physicist Joseph Larmor leant a sympathetic ear. Bateson wrote to him (May 28, 1903):

> Jokingly no doubt, you remarked last night that a jury of good men and true should be empanelled to try the Mendelian issues. Though the remark was not meant seriously, it occurs to me to say that if 3 or even 2 such good and true men could be found willing to take the trouble, I ask no better. I don't suppose they can be found, but let the suggestion be borne in mind. I believe in differences of this kind such private commissions have in the past been of use – notably in Pasteur's case. The Mendelians claim that the new analysis is an instrument of the highest possible value – but while their will is 'technically' opposed, the biological world will not bother itself to look into the facts, and both sides are blocked. We feel perfectly assured that no one who sees the phenomena can feel any doubt about them, but it is impossible to *make* people look at them. I heartily wish there were some mechanism of getting your suggestion carried out.

Beatrice accompanied Bateson to a BA meeting at Southport (May 26, 1903), and in August, she was away with the children at Felixstowe. By Christmas, it was apparent that she was again pregnant. On Christmas day, Bateson departed. On her typed transcript of his letters, Beatrice later recalled: "This was one of his sudden dashes for a complete change. A restless craving that was irresistible even though the result was disappointing." A day later there was a letter from Rouen. The commentary was typical: the food (usually bad), strangers he had bumped into (usually like someone they knew or a character in a novel), churches he had seen (usually uplifting), concerts and plays he had been to (usually disparaging), and novels he had read (usually with implausible plots). He toyed with the idea of moving on to Paris, but the Herringhams were there, and he did not want to bump into them. Beatrice added to her typed transcript: "Economy was his motive in avoiding friends. He took his holiday but parsimoniously cheaply. Money was for research not for pleasures. Shops meant antiquities. Partly amusement, interest, and pleasure in handling pretty things, partly an unrealized hope of finding a treasure which should endow his work." Nevertheless, by December 28th, he was so bored with Rouen that he decided to risk the Herringhams. From Paris, he wrote: "As today it is Monday and Museums are shut I am once more rather loose ended. … I go in fear of running into the Herringhams. … I have much gained by two days absolute solitude, which I always enjoy in my fashion, but I expect tomorrow will find me ready for home."

Yule

In November 1902, Weldon published a paper in *Biometrika* "On the Ambiguity of Mendel's Categories" [5]. Bateson had "been misled" by "Mendel's very imperfect system of units." Bateson had swallowed uncritically "Mendel's system of dividing a set of variable characters into two categories, and of using these categories as statistical units." Thus, the "yellow" and "green" cotyledon races of peas needed to be more carefully distinguished in terms of color variability among members of the race, and color variability among seeds of an individual plant. This was classical Galton-Pearson thinking, and Weldon made clear that having examined the recent works of Bateson and Saunders and noted their "confusion between resemblance to a race and resemblance to an individual," he considered that there was "not reason to modify" the position he had taken – namely, "that the characters of cross-bred [hybrid] individuals, derived … from … 'pure' parents, can be regarded as depending upon the characters of [all] the ancestors from which the 'pure' parents are descended." He concluded that:

> It is deeply to be regretted that so many interesting experiments, involving so much time and labour, should be recorded in a form which makes it impossible to understand the results actually obtained, and so gives rise to misconceptions both in the minds of the recorders and in others. … The accumulation of records, in which results are massed together in ill-defined categories of variable and uncertain extent, can only result in harm.

At this point, G. Udny Yule (1871–1952), who had been Demonstrator and Assistant Professor in Applied Mathematics under Pearson from 1893 to 1899, tried to reconcile the divergent views in *The New Phytologist* [6]. While declaring that "I am inclined to agree with Mr. Bateson as to the *possibly* very high importance in practice and theory of Mendelian phenomena," he regretted Bateson's "style and manner of treatment," for surely "the language of unbridled enthusiasm and lavish abuse creates nothing but mistrust." As set out in Chap. 3, Yule first "cleared the ground" by outlining the basis of the Galton-Pearson regression approach. He concluded that:

> It is essential, if progress is to be made, that biologists – statistical or otherwise – should recognize that Mendel's Laws and the Law of Ancestral Heredity are not necessarily contradictory statements, one or another of which must be mythical in character, but are perfectly consistent the one with the other and may quite well form parts of one homogeneous theory of heredity.

At the same time, he veered toward Bateson noting: "I cannot include under the same heading the special laws as to the operation of Ancestral Heredity which were formulated by Galton and Pearson. These laws have, beyond question, been of service in suggesting lines of research, … but the *fixity* of the numerical constants involved … has not stood the test of time." In 1912, Yule became Reader in Statistics at Cambridge and a Fellow of St. John's College.

Darbishire

Meanwhile, Weldon, aware that much of the evidence derived from plant breeding, had decided to examine Mendelism in animals, his own domain. With Arthur Darbishire (1879–1915), he had begun crossing Japanese waltzing mice with albino mice [7]. Darbishire was a demonstrator in zoology and comparative anatomy, but the situation became complicated when he moved (October 1903) to a similar demonstratorship at Owens College, Manchester, since the mice remained at Oxford. In a series of reports, Darbishire and Weldon interpreted their results as contradicting Mendel. Darbishire began corresponding privately with Bateson in December 1902. Bateson, with his sister-in-law Florence Durham, had also begun experiments with mice and was familiar with technical aspects of their breeding. From March to May 1903, there was a flurry of public correspondence (*Nature*) between Weldon and

Bateson, part of which (from Bateson April 7) was declined by the Editor on the basis of its length. Eventually (May 14), Weldon called for a postponement:

> By modifying first one and then another of Mendel's statements, his name is made to shelter almost any hypothesis, and almost any experimental test is evaded. In the next number of *Biometrika* Mr. Darbishire will publish a series of new results, which have an important bearing on the application of Mendel's 'principles' to his mice. Until these new facts are available, I do not think further discussion will be profitable, and therefore I do not propose to continue this correspondence.

The Editor agreed: "This correspondence must now cease." All eyes were now on Darbishire. At the Southport BA Meeting, there was a confrontation in the Zoology section. Darbishire presented some of his mice results. Bateson presented a review of "The Present State of Knowledge of Colour-Heredity in Rats and Mice," which was published in the *Proceedings of the Zoological Society*. He pointed to problems such as the selective eating of newborn by adults, and the need to accurately mark individuals and to score colors in individuals of the same age. Above all, there was the need to be certain of the homozygosity of the starting lines. He cited with approval the pioneering studies of Lucien Cuénot, who was promulgating Mendelism in France, and disparaged those of von Guaita (whom Darbishire and Weldon had followed). He concluded: "The majority of the observations are in accord with the Mendelian hypothesis in a simple form." Hurst wrote to Bateson (May 27, 1903):

> I was delighted with the Zoology Society Meeting last night, ... because it gave me an opportunity of personally measuring the weaknesses of the Oxford people. Frankly, I was very disappointed with Weldon, true he was smart and acute, but very narrow and not at all philosophical in either his manner or his matter. I think he 'reads' much better. With regard to Darbishire, he is of course secondary, though no doubt useful to carry out the experiments. His exposition of the experiments was rather vague and muddled last night. I sincerely hope his mind is clearer when mating his mice and recording the results: one thing is certain, Cambridge has nothing to fear from Oxford in regard to this question. I was glad to hear Weldon accept the Mendelian principles in regard to albinism. This is something at any rate, and it is quite possible now that our good friend will evolve gradually into a Mendelian.

Armed with the much-heralded data of Darbishire, in June Weldon extended his criticisms of "Mr. Bateson's Revisions of Mendel's Theory of Heredity" [8]. There was an impasse. What Weldon saw as revealing a contribution from remote ancestry, Bateson saw as due to impurity of the initial lines chosen for crossbreeding. Communication between Darbishire and Bateson continued, including a trip to Oxford (September 1903) where Bateson met another of Weldon's students, Edgar Schuster, a nephew of Arthur Schuster (Chap. 22). Darbishire's work was a major source of contention at the 1904 BA meeting (see below).

Garrod

Besides the horticulturalists, Bateson's arguments held sway with many physicians, prominent among whom was Archibald Garrod (1858–1936). In London, Garrod had collaborated with Frederick Gowland Hopkins (1861–1947), a medically qualified "physiological chemist" (as biochemists were then known) at Guy's Hospital.

Bateson had consulted Hopkins (September 1, 1892) concerning a moth pigment that he thought might be a uric acid derivative. Hopkins came to Cambridge in 1898 at the invitation of Foster to develop "physiological chemistry." Building on the work of German organic chemists, and convinced that chemical events within organisms were not as inscrutable as many thought, he had to struggle for recognition of Biochemistry as a discipline, much as Bateson had to struggle for recognition of Genetics. Hopkins supported the founding of the Biochemical Society in 1911 and in 1914 became the first Professor in Biochemistry at Cambridge.

In *RS Report-1*, Bateson and Saunders discussed "marked individual peculiarities" that might not be seen in the following generation. They saw that "the absence of the character in the first [F_1] generation may indicate merely that it is recessive, and its reappearance in the next generation may be due to the heterozygote having bred with another individual also bearing the recessive allelomorph." As examples of this, they pointed to certain rare human diseases, noting recent work of Garrod on the disease alkaptonuria [9]:

> Garrod has noticed that [in] no fewer than five families containing alkaptonuric members, more than a quarter of the recorded cases, are the offspring of unions of *first cousins*. ... Now there may be other accounts possible, but we note that the mating of first cousins gives exactly the conditions most likely to enable a rare and usually recessive character to show itself. If the bearer of such a gamete mate with individuals not bearing it, the character will hardly ever be seen; but first cousins will frequently be bearers of *similar* gametes, which may in such unions meet each other.

Garrod developed this idea in his *Inborn Errors of Metabolism* (1909) – a major synthesis of the biochemistry of genetic diseases – which related each gene to an enzyme (otherwise known as a "ferment") that catalyzed a step in a chemical chain of reactions. If one gene failed, then the corresponding enzyme would fail, and one step in the chain of reactions – a metabolic pathway – would fail. If, for example, the pathway was concerned with the synthesis of a colored pigment, then pigment production would cease; the variant organism with the mutated gene would be detected by the observed color change. It seems that it was Bateson who first wrote to Garrod, since the first letter from Garrod to Bateson (January 11, 1902) began:

> It was a great pleasure to receive your letter and to learn that you are interested in the family occurrence of alkaptonuria. ... I can find no record of its transmission from one generation to another, nor can I hear of any children of alkaptanuric subjects. However, Dr. Osler is enquiring about 2 American brothers who may have children. ... I have for some time been collecting information as to specific and individual differences of metabolism, which seems to be a little explored but promising field in relation to natural selection, and I believe no two individuals are exactly alike chemically, any more than structurally.

A few days later, he wrote in reply to Bateson noting that "Hopkins is an old friend of mine and we have written several papers together." In 1934, when commenting on the praise he had received for *Inborn Errors*, Garrod wrote: "Hogben in his paper on alkaptanuria has, I think, given me more credit than I am entitled to, seeing that it was Bateson who saw the daylight" [10].

Full understanding of the characters they were studying required an understanding of the underlying biochemistry. Florence Durham studied the pigments responsible for hair color (Chap. 10), and Muriel Wheldale was to become a world leader in plant pigments (Chap. 11). In 1903, Reginald Gregory observed in *The New Phytologist* [11] that the difference between round and wrinkled peas was associated with a difference in their starch grains. Concerning Weldon's questioning of the strict classification of different strains of peas into two types, including the irregularly round race known as "Telephone" (Chap. 8), Gregory found no pea that could not be confirmed as either round or wrinkled through microscopic examination of the starch grains. In 1909, Bateson (*Principles*) noted that it was the power to convert sugar into starch that kept seeds round rather than wrinkled: "The round seed of peas, or of maize, is one which contains something possessing the power of turning most of the reserve-materials into starch. If the dominant factor endowed with this power is absent, much of the sugar remains sugar, and the seed wrinkles in ripening. The actual physiological processes involved are doubtless more complex than this, but there is no mistaking the essential nature of the distinction between the round and wrinkled seed."

Linnean Fight

The Report of a meeting of the Linnean Society in early 1904 (February 18) stated that Rowland Biffen read a paper on "Mendel's Laws and their Application to Wheat Hybrids" [12]. He showed that both anatomical characters, such as bristles, and physiological ("constitutional") characters, such as time of ripening and immunity to parasites, followed Mendel's laws. Weldon then "spoke at length on his views of the Mendelian hypothesis, referring in illustration to observations on hybrid albino mice and albino human beings." It seems that the ensuing discussion needed to be interrupted. Thiselton-Dyer "suggested that it would be convenient to adjourn further discussion until after the reading of Mr. Bateson's paper." Bateson then exhibited a series of Primulas (*Primula sinensis*) that had varied in Sutton's nurseries over five seasons. The Report concluded by noting that "Prof. W. F. R. Weldon criticized the paper and was replied to by Mr. Bateson." Hurst was present, and the latter's biographer describes it as a "fight." It seems the Mendelians got the worst of it. In response to Bateson's depressed account, his student Lock replied from Ceylon (March 23, 1904): "I was sorry to hear about the Linnean. You floor him so completely upon paper, that it is a pity the arts of the orator should be allowed to prevail."

On May fourth, Bateson declined an invitation to hear Galton at the Sociological Society, and on May ninth, Beatrice delivered their third child (Gregory). Shortly thereafter (May 14), Hurst received a letter from Bateson: "You must read and digest the new Cuénot. It contains the whole – or nearly the whole – clue to Darbishire's mess." Synthesizing a host of bewildering observations, Cuénot now argued that coat color in mice was a compound character, requiring both a basic color-forming factor and a factor determining whether or not this color-forming factor would operate. Furthermore, Cuénot showed gray (agouti) to be dominant to black, and yellow to be dominant over both gray and black. From crossing yellow mice among themselves, Cuénot found that they were *always* heterozygotes and possessed the element determining the gray color as a recessive. Mendelian segregation would predict that if crossed there would be three yellow to one gray recessive. In fact, the ratio turned out to be two to one, and the two were always heterozygotes. Where was the yellow homozygote? The solution was that two copies of the yellow-determining element in one organism were collectively lethal. Cuénot had not disproved Mendelism, as he had revealed a "synthetic lethal" mutation.

At this time, Bateson had just completed *RS-Report 2* (Chap. 10) and had turned to other matters: an address at a BA Meeting in August (see below), a paper to be read in his absence at the above Sociological Society Meeting (Chap. 15), and an application to the Balfour Fund for greenhouses for studies of cross and self-fertilization in Primulas. The application was rejected on a technicality – that the Fund could only support work carried out within existing buildings. Davenport was having more luck. He wrote (May 17) reporting that the Carnegie Station for Experimental Evolution was now established, and asking that Bateson's Grantchester operation be formally dubbed a "correspondent" station, to foster communication, exchange of papers, etc. Later in the year, Davenport's political skills were manifest in an editorial article in *Science* (November 12). While paying tribute to Pearson and Weldon, and hailing mathematics as "the queen of the sciences," he mentioned Bateson's BA Address (see below) and concluded that: "This conception of species, which has arisen during the present decade, has its germ in the work of Mendel, and in consequence of the stimulating researches of de Vries, Correns, Tschermak, Bateson and others, has developed into a stately doctrine, a doctrine which bids fair to revolutionize biology as the atomic theory did chemistry." A Biometrician was breaking ranks!

BA Address

On August 18th and 19th, 1904, the BA met at Cambridge with the Prime Minister, the Right Honorable Arthur J. Balfour, as President. Years later, Punnett recalled Bateson's preparation for the meeting [13]:

> Bateson was to be President of the zoology section at the coming meeting of the British Association. I had never attended a meeting of this body and had not thought of it as of any importance to the work we had in hand. Bateson, however, had realised that this was a great opportunity. It was constantly in his thoughts and he neglected nothing in order to make his own part a success. When the sweet pea season came in June he used to take a chair and a little table out into a small copse at the back of the chicken pens, leaving the strictest orders that he was on no account to be disturbed by anyone. So each day when I arrived with the *Morning Post* – a paper of which Bateson approved since it contained the best account of art sales, and I don't think he read much else – I set to work cleaning, recording and labeling sweet peas while Bateson, in the seclusion of his copse, struggled with his address.

On the first day of the meeting, Bateson addressed the general assembly in the new Geology Building [14]. As related in Beatrice's *Memoir*, Bateson thought it would be desirable some day to republish the address because "my friends are finding these things out for themselves now and I have a vain desire to tell them I was there first." He began with his usual discourse on method. It was time to return to experimental breeding, an area that had become "thinly tenanted" since Darwin's *Origin of Species* which, while opening up many interesting subjects "deemed more amenable to human enterprise," had diverted attention from "the main line of enquiry," namely the "fundamental nature of living things":

> The successes of descriptive zoology are so palpable and so attractive, that, not unnaturally, these which are the means of progress have been mistaken for the end. By now the survey of terrestrial types by existing methods is happily approaching completion, and we may hope that our science will return to its proper task.

However, until there was a return to "forgotten paths," this would be a lonely task:

> Those whose pursuits have led them far from their companions cannot be exempt from that differentiation which is the fate of isolated groups. The stock of common knowledge and common ideas grows smaller till the difficulty of intercommunication becomes extreme. Not only has our view changed, but our materials are unfamiliar, our methods of inquiry new, and even the results attained accord little with the common expectations of the day.

Apart from this lonely few, and unlike early workers (Kölreuter, Gärtner) who "were informed by the true conception that the properties and behaviour of species were themselves specific," the present generation suffered from a "misconception," a "confounding" and a "fancy." These were: the "misconception of the nature of specific difference [the difference between species] as a thing imposed and not inherent;" the "indiscriminate confounding of all divergences from type into one heterogenous heap under the name 'Variation';" and "the fancy that, but for Selection, the forms of animals and plants would be continuous and indeterminate." Indeed, to challenge "the omnipotence of Selection" was professionally hazardous: "On pain of condemnation as apostate," one would be regarded as "a danger to the dynasty of Selection." Thus, "Selection is a true phenomenon; but its function is to *select*, not to create." For "no theoretical evolutionist doubts that Selection will enable him to fix his character when obtained." But, Bateson stressed, first it had to be *obtained*. "Variation leads; the breeder follows." So, "to prove the reality of Selection as a factor in Evolution is … a work of supererogation. With more profit may experiments be employed in defining the *limits* of what Selection can accomplish. For whenever we can advance no more by Selection, we strike the hard outline fixed by the natural properties of organisms."

Bateson was not disdainful of histological studies, and he named those of Strasburger, Boveri, Wilson, and Farmer. Indeed, he felt "sure that in the near future we shall be operating in common," for "the experience of the breeder is in no way opposed to the facts of the histologist." The point at which the disciplines "unite will be found when it is possible to trace in the maturing germ [gametes] an indication of some character afterwards recognizable in the resulting organism." He wanted a marker in the gametes that could be correlated with some anatomical or physiological characteristic. In this context, he spoke of the "coupling of characters," which segregated together in gametes (i.e., characters that demonstrated what is now called linkage), and pointed to sex-linked characters, such as color-blindness and hemophilia in humans, which indicated "some entanglement between sex and gametically segregable characters." Indeed, "of all … distinctions, none is so universal or so widespread as that of sex: may it not be possible that sex is due to segregation occurring between gametes, either male, female, or both? It will be known to you that several naturalists have been led by various roads to incline to this view." He regarded the latter as "familiar ground," perhaps referring to McClung's observation (Chap. 13) of what we now refer to as a sex chromosome. However, Bateson declined to use the word "chromosome."

By this time it had been verified for many plants and animals that for heterozygotes "at least one cell division in the process of gametogenesis is therefore a differentiating or *segregating* division, out of which each gamete comes sensibly pure in respect of the allelomorph it carries." As for "the actual moment of segregation," while "no quite satisfactory proof" yet exists, "the reduction-division [of meiosis] has naturally been suggested as the critical moment." It was true there were still poorly understood "complexities of segregation," by which "gametes of new types, sometimes very numerous, are produced by the cross-bred." How could this occur? Bateson continued:

> We shall, I think, be compelled to regard these phenomena as produced either by a *resolution* of compound characters introduced by one or both parents, or by some process of *disintegration*, effected by a breaking-up of the integral characters followed by recombinations. It seems impossible to imagine simple recombinations of pre-existing characters as adequate to produce many of these phenomena. Such a view would involve the supposition that the number of characters pre-existing as units was practically infinite – a difficulty that as yet we are not obliged to face. However that may be, we have the fact that resolutions or disintegrations of this kind – or recombinations, if that conception be preferred – are among the common phenomena following crossing, and are the sources of most of the breeder's novelties.

As we shall see in Chap. 13, the coupling of elements determining characters, as reported by Bateson and Saunders, had already been interpreted by Walter Sutton in modern terms – a linear arrangement of the elements (genes) along chromosomes (Fig. 9.1). And Boveri in a 1904 text had envisaged the uncoupling of elements linked *within* a chromosome, namely intrachromosomal recombination (Fig. 9.2). Nevertheless, by 1904, despite some difficulty with the German language, Bateson had become familiar with de Vries' concept of unit characters as outlined in *Intracellular Pangenesis* (Chap. 3):

> The existence of unit-characters had, indeed, long been scarcely doubtful to those practically familiar with the facts of Variation [here de Vries was cited], but it is to the genius of Mendel that we owe the proof. We knew that the characters could behave as units, but we did not know that this unity was a phenomenon of gametogenesis. He has revealed to us the underworld of gametes. Henceforth, whenever we see a preparation of germ cells we shall remember that, though all may look alike, they may in reality be of many and definite kinds, differentiated from each other according to regular systems.

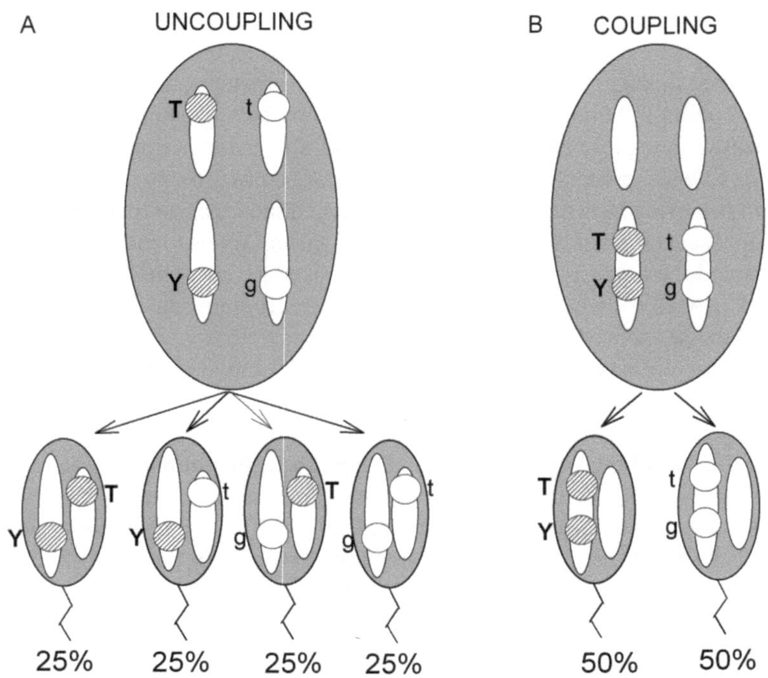

Fig. 9.1 Implications of (**a**) "uncoupling" and (**b**) "coupling" for the distribution of two allelic gene pairs among gametes. The large ovals refer to F_1 gonadal cells from which haploid gametes will be derived. The diploid F_1 cells contain two chromosome pairs (long rods), one of each pair being derived from the organism's father (say a **TY** homozygote) and the other from its mother (say a tg homozygote). Two pairs of allelic genes are shown as small circles, with the dominant alleles of each pair (**T** and **Y**) being cross-hatched, and the recessive alleles of each pair (t and g) being left unhatched. Whether on different chromosomes (**a**) or on the same chromosome (**b**), **T** and **Y** (and t and g) are nonallelic genes in that they are not alternatives – they do not occupy corresponding positions on chromosomes. In A, uncoupling in half the gametes from the combinations of genes that existed in the original parent (i.e., **T** coupled with **Y**, and t coupled with g) reflects the fact that each allelic gene pair belongs to a distinct chromosome pair. In B, coupling remains intact in all gametes due to both allelic gene pairs belonging to the same chromosome pair. Arrows point to the average distribution of the genes among the resulting gametes (ovals with tails to indicate spermatozoa). Each haploid gamete can accept only one member of each chromosome pair. In **a**, there are four possible combinations (four classes of gamete) and each combination will, on average, be 25% of the gamete population. In **b**, due to coupling, there has been no reassortment of genes, and there are only two classes of gamete, which correspond to the two parental combinations of genes. Each combination, on average, will be in 50% of the gamete population unless some *intrachromosomal* recombination has occurred (see Fig. 9.2). Note that the actual characters observed (the phenotype corresponding to each genotype) will depend on (i) the degrees of dominance of the various genes and (ii) whether there are interactions between gene products

Still seeming to use his short-hand term "character" interchangeably with "element," Bateson continued: "It is doubtful whether segregation is rightly represented as the separation of *two* characters, and whether we may not more simply imagine that the distinction between the allelomorphic gametes is one of presence or absence

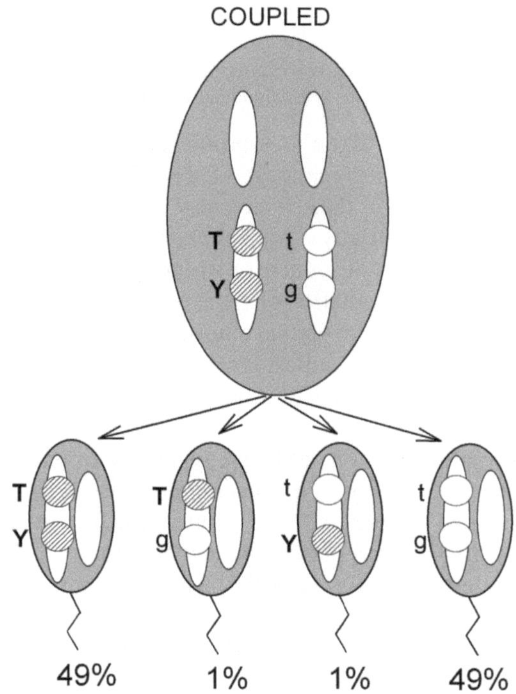

Fig. 9.2 Intrachromosomal recombination results in less than 50% representation of each parental combination in the gamete population. The details are as in Fig. 9.1. We now know that, at low frequency (expressed as 2% in this example), members of a homologous chromosome pair can exchange segments between themselves, so genes in those segments would also be exchanged

of some distinguishing element. De Vries has devoted much attention to this question in its bearings on his theory of Pangenesis." And what was an element corresponding to an integral character? What was its physical nature? Was its size inviolate? Could it be split?

> We do not declare the size of the integer is fixed eternally, as we suppose the size of a chemical unit to be. The integrity of our characters depends on the fact that they *can* be habitually treated as units by gametogenesis. But even where such unity is manifest in its most definite form, we may … generally find a case where the integrity of the character has evidently been impaired in gametogenesis, and where one such individual is found [i.e. a recombinant] the disintegration can generally be propagated [i.e. it is stable and can be transmitted to offspring].

As manifest in the distribution of characters among offspring, Bateson considered a fact, "nearly as significant" as segregation itself, was that "the gametes bearing each member of an [allelic] pair generally are formed in equal numbers by the heterozygote, if an average of cases be taken." He held that "this fact can only be regarded as a consequence of some numerical symmetry in the cell divisions of gametogenesis." Having already alluded to the reduction division of meiosis, it would seem to modern minds that he was thinking of chromosomes, but, again, chromosomes were not mentioned. Bateson thought that the equality was telling us something about the transmitting agency in inheritance – what we now know to be DNA:

> The fact that equality is so usual has a direct bearing on conceptions of the physical nature of Heredity. I have compared our segregation with chemical separation, but the phenomenon of numerically symmetrical disjunction as a feature of so many and such different characters seems scarcely favourable to any close analogy with chemical processes.

He inferred: "If we may profitably seek in the physical world for some parallel to our gametic segregations, we shall, I think, find it more closely in mechanical separations, such as those which may be effected between fluids which do not mix freely, than in any strictly chemical phenomenon."

Next Bateson turned to variation itself, noting that "as regards new characters involving the addition of some new factor to pre-existing stock we are almost where we were. When they have been added by Mutation, we can now study their transmission; but we know not whence they come." The best he could come up with was an aphorism attributed to Virchow, "that every variation from type is due to a pathological accident." He observed that there was a need to distinguish whether a case of variation "was one of heterozygosis or of nascent mutation?" In other words, is an observed variant form "a *mutation* …, or is it simply due to a recombination of pre-existing characters?" He pointed out that mutation was usually accompanied by "statistical irregularity," which "contrasted with the numerical symmetry of the gametes after normal heterozygosis" when, in nondominant cases, the F_1 offspring could appear as relatively uniform anatomical or physiological variants from the parental types. Thus, mutation could be "simply the segregation of a new kind of gamete, bearing one or more characters [elements determining characters] distinct from those of the [varietal or species] type."

Bateson went on to point out that the *continuity* of form from generation to generation (heredity), and the *discontinuity* of form from generation to generation (variation) were not in opposition:

> Just as the normal truth to type, which we call Heredity, is in its simplest elements only an expression of that qualitative symmetry characteristic of all non-differentiating cell divisions, so is genetic variation the expression of a qualitative asymmetry beginning in gametogenesis. Variation is a novel cell division. So soon as this fact is grasped we shall hear no more of Heredity and Variation as opposing 'factors' or 'forces' – a metaphor which has too long plagued us.

By this, he implied that when a cell divides a mutation may arise in such a way that often only one of the child cells conveys the mutant character. Although thinking primarily of gametogenesis, he added as a footnote the idea that "the parallel between the differentiating divisions by which the parts of the normal body are segregated from each other, and the segregating processes of gametogenesis, must be very close. Occasionally, we even see the segregation of Mendelian characters among zygotic [adult] cells." In the latter case, he may have been thinking of mosaicism, such as thorny patches in *Datura* (Chap. 8) and leaf variegation; here yellowing in a leaf sector revealed that a mutation had occurred during leaf development, resulting in a yellow cell lineage. We now know that this is usually different from normal differentiation, say of a leaf from a stem, which involves mechanisms for the differential expression of unmutated genes. In essence, a leaf is a leaf because genes determining leaf characters are expressed and genes determining stem characters, although present in all the cells of a leaf, are not expressed. Likewise, a stem is a stem because genes determining stem characters are expressed, and genes determining leaf characters, although present in all the cells of the stem, are not expressed.

There then came what Bateson referred to as "the inevitable question," which invariably related to the formation of biological species: "What makes the character group split? Crossing we know may do this; but if there be no crossing, what is the *cause* of Variation?" Since he refers here, not to a character, but to a character *group*, it is probable that he is asking, but not answering, the question as to how a new species (consisting of organisms with a group of characters some of which are different from those of the originating species) emerges.

Although the debate with the Ancestrians was to be the following day, in his speech, Bateson hinted at what was to come: "The imposing Correlation Table into which the biometrical Procrustes fits his arrays of unanalyzed data is still no substitute for the common sieve of a trained judgment." Furthermore, "when a pair of gametes unites in fertilization, the characters of the zygote depend directly on the constitution of these gametes, and not on that of the parents from which they came." When inheritance is Mendelian:

> Fixity of type, instead of increasing gradually generation by generation, comes suddenly, and is a phenomenon of individuals. Only by the separate analysis of individuals can this fact be proved. The supposition that progress towards fixity of type was gradual arose from the study of masses of individuals, and the gradual purification witnessed was due in the main to the gradual elimination of impure individuals, whose individual properties were wrongly regarded as distributed throughout the mass.

His next remarks were more pointed:

> While formerly we hoped to determine the offspring by examining the *ancestry* of the parents, we now proceed by investigating the *gametic composition* of the parents. Individuals may have identical ancestry (and sometimes, to all appearance, identical characters), but yet be quite different in gametic composition; and, conversely, individuals may be identical in gametic composition and have very different ancestry. Nevertheless, those that are identical in gametic composition are the same, whatever their ancestry. Therefore, where such cases are concerned, in any considerations of the physiology of heredity, ancestry is misleading and passes out of account.

BA Fight

Writing retrospectively in 1906, Pearson [14] described Bateson's Presidential Address as "chiefly an attack upon biometric work and methods," and noted that "the discussion which followed culminated in the President holding aloft the volumes of this journal [*Biometrika*] as patent evidence of the folly of the school." Perhaps Pearson was referring to the discussion the next day. The word had got round, and the room was packed. Some doubtless thought it might be the biggest BA fight since Huxley-Wilberforce in 1860. *The Times* (August 20, 1904) reported the audience as giving "demonstrations of partisanship so new to the serene abstractions of Section D."

The two sides spoke alternately. First Saunders showed for stocks that standard Mendelian results were sometimes complicated by phenomena such as gametic coupling of distinct characters, and interaction between characters in the zygote so that a second character was only seen in the presence of the first. These might require several generations of breeding to analyze. Then, presumably much to Weldon's consternation, Darbishire showed for mice that the Mendelian ratios were not at variance with the Galton formulae. Then, Hurst described his rabbit studies. Weldon opened the discussion noting that the Mendelians were happy to refer to plants as glabrous (smooth) or hairy (the dominant character), but how hairy were the second-generation forms compared with the original parent? He had observed a great range in values of hair counts, suggesting a need for more precision as to the definitions of glabrous and hairy. According to Punnett, "Weldon spoke with voluminous and impassioned eloquence, beads of sweat dripping from his face." [13].

After lunch, Punnett described their poultry data, and Minot from the USA spoke of his work on guinea pigs. Then Bateson rose to reply to Weldon's points, especially emphasizing the purity of unit characters as the explanation for the reversion to ancestral types (through recessive homozygosity), and the unvarying gametic constitution of heterozygotes as revealed through back-crossing with the parents (homozygotes). Then he turned to the younger members of the audience who, he said, would recall this day as a critical moment in the history of biology. The Ancestrians were likened to the flat-earthers of medieval times who had struggled to reconcile this belief with the observed movements of the heavenly bodies.

Karl Pearson, continuing the ad hominem approach, despaired at the Mendelians' lack of exactitude due to the incompleteness of their mathematical knowledge. He wound up appealing for a three- or four-year truce. The controversy would be "settled by investigation, not by disputation." At that point, the Reverend R. T. Stebbing stood up to proclaim "let them fight it out," and Chairman Hickson closed the debate hoping that biologists still "sitting on the fence" would have gained enlightenment. No immediate report appeared in *Nature*. Bateson commented on this in a letter to Beatrice who was away from Cambridge (August 27, 1904):

> P. agrees with me that the silence of 'Nature' is significant, and I no longer feel much doubt that 'biometry' is damaged heavily. … Tomorrow I dine with Newton, who looked in this afternoon, and on Monday with the Whiteheads. … The sketch of me in 'Zool. Section' (Daily Graphic 23 Aug.) represents an individual in appearance between the Duke of Argyll and Mr. Rutter, so I ought to be content. … The Brit. Med. Journ. prints part of my address, then W's speech, but not my answer.

The meeting was reported in *The Times* (August 20) and, eventually, in *Nature* (August 25 and September 29). Sadly, Pearson turned his fire on Darbishire, asking rhetorically in *Nature* (September 29) whether Mr. A. D. Darbishire of Oxford was the same as Mr. A. D. Darbishire of Manchester: "Which writer shall a member of the inquiring general public trust? If the two writers be the same, must we assume that in Oxford under the influence of some recessive biometer, Mr. Darbishire failed to see that 97 in 555 was a reasonable quarter … but that he has learned in Manchester, or perhaps in Cambridge from some dominant anaesthetist, that these things really are so?" Later in *Nature* (October 27), Pearson took on Lock, who had just published his Maize-breeding studies.

Sudden Wanderings

In September 1904, Bateson and Beatrice with "a 15 day [train] ticket" went on a walking tour. In a letter to his mother (September 5) from Hutton-le-Hole, he thanked her for sending money for a German governess for the boys. "I have always been glad of my early French and greatly regret no fluency in German, which is more use, if less pleasure." Later (September 23), there was a solo visit to a fancier at Barnstable to collect birds for breeding experiments. In November came notification that Bateson had been awarded the Darwin Medal of the RS "for your important contribution to the theory of evolution by researches on variation and heredity." The Medal came with a check for £100. Among letters of congratulation was one from his old governess, Frances Lakin, who had seen the announcement in *The Times*. She had been closely following the careers of the Batesons: "Will you tell Mary from me how greatly I enjoyed Medieval England – not a *word* was skipped – it is most interesting."

In November 1904, Weldon began a series of eight public lectures in London on "Current Theories of the Heredity Process" which Bateson attended. Then, acting on a sudden whim, over the Christmas period he and Beatrice visited Vienna and Brünn (Chap. 20). In January, Saunders attended one of Weldon's lectures that Bateson could not attend, and she sent him a postcard about it the next day. In July, Bateson seems to have had another sudden urge to go off alone, this time by bicycle and train. In a note on her transcript of his letter (July 30, 1905), Beatrice later commented that "this short outing had for motives the search for a suitable holiday resort" for the family:

> Wet at Selby, but I started forth wondering why I left home. In a little while it cleared and I had a good run to Hull, getting in too late for shopping. Hull is a view of Yarmouth given up to business instead of pleasure. If you want to see the seething masses, this is the real place. Yarmouth is the swarm, but this is the hive. Till 11 last night – and how much later I don't know – some *miles* of streets were choked with Sat-night crowds. I went to a play – as the waiter told me, it was a 'drama.' It is so long since I have met the drama except by hearsay that it struck very fresh. 2 acts were enough. In them we had two several murders on the stage, in different scenes with completely different motives; a blond girl abducted and stowed away in a cellar in Paris; a safe broken open ... Revolvers out every few minutes. Detectives in false beards, posted as 'blind' men at Char. X. Station, etc, etc. in one confused mess. It is comfortable in this house and I intend to come back tonight after seeing Spurn Hd. which is so sternly condemned by the Guide [book] that it may be something uncommon. Hope you are feeling better. W.B.

The next day he wrote a rambling letter from Scarborough (August 1, 1905), which ended: "The air is 'splendid' and has set me up a lot. Unless Punnett signals for me I think I shall stop a few days more, perhaps till Sat. or Mon."

At the next RS Meeting (December 7), Bateson, Saunders, and Punnett presented a paper entitled "Further Experiments on Inheritance in Sweet Peas and Stocks. Preliminary Account," and Hurst presented his horse studies. The latter resulted in a dash to France and much searching of studbooks (Chap. 10). There was another RS meeting (January 18), and two weeks later, Bateson was back in London exhibiting Hurst's rabbit skins at the Neurological Society where he gave an address on "Mendelian Heredity and its Application to Man."

Published in *Brain*, the address is likely to have furthered support from the medical profession. He explained the Mendelian principles using black and white draughts as "germ cells" to illustrate the separate paternal and maternal contributions to their offspring. From the outset, he employed his usual short-hand, referring to alleles themselves as "characters," and noting that they were "treated by the cell divisions in which the germs are formed as *units*." Furthermore, "segregating characters or *allelomorphs* as they are called, are ... always constituted in pairs."

Pedigree diagrams, showing the passage of traits through families, had long been available, although not interpreted in Mendelian terms; in America, Farabee had followed the dominant inheritance of digits with two phalanges instead of three (i.e., fingers were like thumbs), and in Britain, Edward Nettleship (1845–1913), who began corresponding with Pearson and Bateson in 1904, had followed the inheritance of congenital cataract [15]. In his address, Bateson used their diagrams to display Mendelian dominance-recessive relationships. He pointed out that physicians could now begin to assign probabilities and appropriately reassure patients. Another disease link, albeit in plants, was that Biffen had shown wheat resistance to rust to be a recessive character. Since Bateson was addressing neurologists, he did not fail to mention Japanese waltzing mice, whose strange movements probably related to "malformations of the semi-circular canals" of the ear.

Shortly thereafter (March 4), there was an enquiry from Lankester who was to give the Presidential Address to the BA at York that summer. He wanted to know, concerning "heredity," the most significant discoveries or theories, those due to British workers, and those that had been assisted by the BA. Bateson replied (March 11) enclosing a copy of his 1904 BA Address with the names of non-British and British workers marked in different colors. He mentioned that the BA had given £35 for his own work in 1903 and 1904.

Since the BA meeting overlapped with a major RHS meeting (Chap. 10), Bateson was unable to attend. But he probably read the Address as reported in *Nature* [16]. True to form, Lankester spoke of the origin of species by Darwinian natural selection as "more firmly established than ever," and disparaged the "many attempts to gravely tamper with essential parts of the fabric" as Darwin had left it, and "even to substitute conceptions for those which he endeavoured to establish, at variance with his conclusions. These attempts must ... be considered as having failed."

The Bateson school was using greenhouse space in the Botanical Gardens, but desperately needed its own greenhouses. At this relatively late hour, funds began to flow from the Darwin brothers. Francis wrote (March 11, 1906): "I am on the track of a small sum for your work." He saw the sum of £200 that they might provide as acting as a nucleus for further donations if there were some formal body for their receipt. Bateson replied, "The bodies which suggest themselves are (1) Seward's Botanical Gardens Committee, (2) The Evolution Committee." It ended up as the

Botanical Gardens Experiments Fund, a timely event since Punnett's Balfour Studentship was soon to expire.

In April 1906, Bateson went to his new writing retreat, the Hill House Inn at Happisburgh in Norfolk, to work on the manuscript of what was to be again called *Mendel's Principles of Heredity* but without *Defence* tagged on (hereafter *Principles*). He was thrilled (April 1) to see a lifeboat launched for rescue and continued by train (April 3) to Yarmouth and Lowestoft, where he visited various art dealers. He toyed (April 4) with the idea of Beatrice joining him at Felixstowe, where his mother was staying: "I would say let us both come here, but I don't like leaving everything to RCP and if you go away I think I ought to be at home."

Then there Was One

Weldon and his wife continued the mouse-breeding. During the Easter vacation of 1906 while staying at a country inn, he was attacked by influenza, which on his return to London (April 11) developed into acute pneumonia. He died in a nursing home two days later. He was buried at Holywell, Oxford. A strange obituary in *The Times* (April 16), instead of reviewing Weldon's life, dwelt on the 1904 debate at Cambridge and named Bateson. On the same day Bateson wrote to Beatrice, who was away at the time:

> I owe a great deal to him. It was through the chance of meeting him that I became a zoologist, and afterwards through him that I got my first start with *Balanoglossus*. Until the time – about 16 years ago – when his mind began to embitter itself against me, I was more intimate with him than I have ever been with any one but you. I rather wish I could write to Mrs. Weldon, imagining ourselves back in earlier days. Perhaps I shall, but I do not know enough of her present state of mind to feel sure that it would be taken in good part. Perhaps not writing may equally be misunderstood. How big a disturbance this will make in our area I hardly yet know. If any man ever set himself to destroy another man's work, that he did to me – and now suddenly to have one of the chief preoccupations of one's mind withdrawn, leaves one rather 'in irons' as sailors say. 'Biometrika' I should think, will come to an end in a year or so. We shall have very seriously to consider whether I am to stand for his professorship. It seems indecent – apart from the fact that our roots are so deep here now. There is much to talk of and I shall be glad to have you back. W. B. I enclose *Times* notice, a disgusting and tactless piece of work.

In her notes, Beatrice records that Bateson received a "very pressing letter" from Oxford regarding his possible interest in the vacant Chair in Zoology. Bateson turned the offer down adding: "I can sympathize with a fraternity sentenced to receive Lankester as a perpetual pensioner, but I don't think I can avert your fate."

On April 17th, Bateson's Aunt Honora died. The following day "a correspondent" supplemented the earlier *Times* notice of Weldon's death with a more obituary-like obituary. Shortly thereafter Bateson received a tactfully worded letter from C. Herbert Fowler, who had worked with Weldon at University College. He was seeking contributions to "a memorial to commemorate Weldon's personality and work," that would "further by donations to a prize" the "line of work," Weldon had been pursuing. Alternatively, or as well, donations were sought toward "a bust of something of that kind." Bateson replied (June 24, 1906):

> First let me thank you for the exceedingly considerate way in which you have written. I had seen the circular as to the Weldon memorial, and had already come to the conclusion that, if I were to receive an invitation, it would not be possible for me to join. Since getting your letter I have thought over the matter again from many points of view, but I have not been able to change this decision. You will not perhaps expect that I should say more, but so kind a letter as yours requires, I think, some fuller reply. To Weldon I owe the chief awakening of my life. It was through him that I first learnt that there was work in the world which I could do. Failure and uselessness had been my accepted destiny before. Such a debt is perhaps the greatest that one man can feel towards another; nor have I been backward in owning it. But this is the personal, private obligation of my own soul. To take part in the public memorial, even though my name might be unpublished, would be to join in approving a scientific career which I cannot think of without disgust and contempt. It would be only through weakness that I could associate myself with those who are about to do him this honour: and if want of magnanimity is imputed to me, I must endure it.

And what of Weldon's associates? In June 1905, further mouse experiments "made at the suggestion and with the advice and help of Professor Weldon" had been published in *Biometrika* by Schuster [17]. He concluded: "It is hardly necessary to point out that with regard to characters, colour-productiveness and albinism, the mice under consideration here behave in complete accordance with Mendel's laws." Another biometrician had strayed from the fold. Darbishire likewise. He had much correspondence with Hurst, and the extent of his conversion is evident from a review he wrote for *Nature* [18]:

> Scant justice has been done to the greatness of Mendel's work and to the conceptions based upon it. … That Hurst can predict the difference between the result of mating two pairs of rabbits externally identical, by means of a knowledge of the difference between their gametic constitution acquired by previous breeding from them, constitutes, it seems to us, the longest stride the study of heredity has made for some time past.

In 1906, in a presentation to the Manchester Literary and Philosophical Society [19], Darbishire observed that: "The true function of the biometrician is to give us statistics of average conduct where we cannot predict individual conduct." He went on to academic positions in London and Edinburgh, was often a reviewer for *Nature* (Chap. 10), and in 1910 published a translation (with Farmer) of de Vries' *Die Mutationstheorie*, followed in 1911 by *Breeding and the Mendelian Discovery*. After Oxford, Schuster joined Pearson in London (Chap. 14), but his conversion to Mendelism was made clear in a book on eugenics published in 1913 [20]:

> The study of inheritance in recent years has been pursued vigorously along two different paths. First, by the statistical summarization of the facts, … and secondly, by the experimental hybridization of animals and plants. The latter method, when carried out on the lines laid down by Gregor Mendel, has alone led to some real understanding of the underlying physiological process, and it appears capable of leading to more.

In 1915, Darbishire enlisted in the First World War and died of meningitis in a military hospital in Scotland [7]. In the same year, his *An Introduction to Biology* had as its "main constructive thesis ... the idea, which we owe to Samuel Butler, that the details of the process of evolution can be studied most minutely in man." As for Darbishire's past role as a biometrician:

> The influence of interpretation upon description is seen in the work of the budding investigator, when he first embarks upon a piece of 'original research.' The subject of inquiry is usually suggested by the lad's teacher, who, in the worst cases, wishes the results of the inquiry to point in a particular direction, and does not conceal his wishes from his pupil, and who in all cases indicates the lines along which, and the methods by which, the research should be carried out. But this is not ... dangerous because the young investigator soon comes to have views of his own, and rebels against his master. The effect becomes dangerous when it is one's own unconsciously entertained interpretations, which, unknown to us, hold our hand and direct the pencil which we fondly believe to be carrying out the pure, unadulterated work of description.

With Weldon's death and Galton fading, there was but one of the original biometricians left – Pearson. Biometrics remained live-and-well in his hands. When Bateson's friend, Lister, espoused Mendelism at the 1906 BA meeting [21], Pearson replied promptly in *Nature* to this "latest critic of biometry" [22]. Lister responded [23] criticizing Pearson's US protégé, Raymond Pearl (1879–1940), who for a while appeared to be acting as "Pearson's bulldog," but later became a Mendelian convert, his coeditorship of *Biometrika* being terminated in 1910. That year Pearl did not visit Bateson but, the next best thing, he visited Johannsen, and in 1911 published a paper reporting something like pure lines in chickens [24]. Under a different name, led by Fisher and J. B. S. Haldane in Britain and by Sewell Wright in the USA, Biometrics was to reemerge, hydra-like, as a dominant force in the biological sciences of the twentieth century. Florence Weldon died in 1936. She endowed a Chair of Biometry at University College, and the first occupant was Haldane.

In Irons

The "boom" that Bateson had proclaimed from New York in 1902 was continuing. Mendelism was gathering momentum. Funds were trickling in from various sources. But regarding his own situation, nothing seemed to change. When you point your sailboat too close to the wind and get stalled, you are "in irons." In 1906, Bateson was stalled. His hopes had been high for the Quick Professorship, but it had gone to protozoology (Chap. 8). So, 1906 became just another business-as-usual year – high business, exciting business, but still business-as-usual.

It was in July 1906 that the new science was publicly baptized "genetics." The word itself had long been in use. In 1901, Galton had used it in quite a modern sense [25]. Bateson had written to Sedgwick in 1905 (April 18):

> If the Quick Fund were used for the foundation of a Professorship relating to Heredity and Variation the best title would … be 'The Quick Professorship of the study of Heredity.' No other word in common use quite gives this meaning. Such a word is badly wanted, and if it were desirable to coin one, 'GENETICS' might do. Either expression clearly includes Variation and the cognate phenomena.

This now became explicit. The occasion was the "Third International Conference on Hybridization and Plant-Breeding" organized by the RHS as a continuation of the conferences of 1899 and 1902, but now including some animal-breeding studies. The title of the conference proceedings, published in 1907, was amended by Wilks from "Hybridization and Plant-Breeding," to "Genetics." Bateson spoke on "The Progress of Genetic Research:"

> I suggest for the consideration of this Congress the term *Genetics*, which sufficiently indicates that our labours are devoted to the elucidation of the phenomena of heredity and variation: in other words, to the physiology of Descent, with implied bearing on the theoretical problems of the evolutionist and the systematist, and application to the practical problems of breeders, whether of animals or plants.

Bateson also more precisely defined terms such as "pure-bred" and "cross-bred," using playing cards as an analogy:

> We have at last a critical appreciation of the physiological meaning of the term 'pure-bred' as applied to a plant or animal. … An individual is *pure-bred* when the two cells [gametes], male and female, from which it develops, are *alike in composition*, containing identical elements or characters. No long line of progenitors is needed to produce a pure-bred plant. … It matters not how the parents are bred. They may be mongrels, as heterogenous in composition as packs of cards; but if from the two packs *similar* cards happen to be dealt, the product of these two cards is pure. And as in cards we may consider their attributes of colour, suit, and number as distinct, so in the living thing we know that the several features or physiological characters may be treated as distinct in the cell-divisions by which the germ cells are formed. From this separability of characters or distinctness of the characters it follows that an organism may be pure-bred in one respect and cross-bred in another.

While speaking here correctly of "an individual" that "develops" from male and female gametes, he referred to such gametes as containing "elements or characters," again implying that the two words were interchangeable in meaning. At the same meeting, Hurst was more precisely referring to gametes as *transmitting* "unit-factors" and zygotes as *displaying* "characters" (Chap. 10). Rather than clarifying this, Bateson was perhaps more focused in converting those of a biometric persuasion. He went on to consider various kinds of "reversion." First, there was "the simplest sort of reversion" due to the "reappearance of a recessive character." Such "recessives never get the chance of appearing until they [the corresponding elements] are introduced into the organism simultaneously from both sides of the parentage."

New characters could arise by other mechanisms, and these were also often labeled as "reversion to an ancestral form, which may be infinitely distant." Under this heading, Bateson included the "reversion in crossing" that could "occur even when types of absolute purity were crossed together." Here, "the reappearance of the ancient characteristic is caused by the meeting together of distinct [non-allelic] elements, long parted. In some unknown way these two factors 'let each other off.' Both factors must be present together in order that the feature in question may be developed." This appears to be an early description of what we now see as gene–product (e.g., protein–protein) interaction which, if it appeared in a positive sense, Bateson was to call an "epistatic" interaction, or in a negative sense, a "hypostatic" interaction. Again, there was no discrimination between "elements" and "factors." The "meeting of long parted factors" had to be distinguished from variation "due to the separation or elimination of factors." There was also variation "due to the addition of *new* factors. Genetic research has thus provided the first indication of the physiological process which results in the birth of variation."

Royal Society Report

At the RHS's Third International Conference on Genetics in 1906 (Chap. 10), silver medals were awarded to Saunders, Hurst, and Biffen, while gold went to Bateson, Johannsen, and Vilmorin. In the course of dinner toasts, Bateson praised those who had made the *RS-Reports* possible: "Had it not been for the work ... done by my friends and pupils – first of all by my colleague, Miss Saunders, whose name has been so deservedly honoured tonight ... I could never have dared ... to have asserted that Mendelian research has been and is of the importance that we now know it must possess."

In 1905, there had appeared the *Second Report to the Evolution Committee of the Royal Society* (*RS-Report 2*; see Chap. 10), and in 1906 the *Third Report* (*RS-Report 3*) by Bateson, Saunders, and Punnett. Here, they more clearly distinguished between characters whose genes were allelic (allelomorphic) to each other, and characters whose genes were nonallelic. The first would include cases of "monohybridism" which gave the familiar 3:1 ratios in the F_2 generation. The latter would include "dihybridism," where, in the simplest cases, the two sets of alleles would appear to independently assort (Fig. 9.1a). Examples were the alleles corresponding to "rose comb" and "pea comb" in chickens, neither of which was dominant to the other. In this case, when one of each allele was present in the offspring there appeared a new character, the "walnut comb" (see Chap. 12). In some cases, however, members of two apparently independent sets of alleles did not independently assort (i.e., "gametic coupling;" Fig. 9.1b). *RS-Report 3* concluded that "this discussion cannot now be carried further," yet "it is clear that gametic coupling, however caused, plays a large part in the phenomena of heredity."

In August 1906, Bateson combined materials from his earlier addresses to summarize "The Progress in Genetics since the Rediscovery of Mendel's Papers" in a new journal, *Progressus rei Botannicae*, edited by the Dutch botanist, Jan Lotsy. The language was still ambiguous, but less than usual – a "type" (with one or more observable characters) being "represented" in gametes. The much-prized blue Andalusian fowl had long baffled breeders, who had never succeeded in getting it to breed true to type: "In certain strictly Mendelian cases F_1 is *normally* a blend-form," and in the case of the Andalusian fowl, "blue is the heterozygous type," so that "we meet the fact that selection is powerless to fix a type which is not represented in the gametes." This emphasized that there was no necessary conjunction between what was observed and what was transmitted in an individual germ line – Andalusian blue appeared as a concoction generated in some way when two characters that were represented by elements in separate gametes were brought together in a new organism. Thus, Andalusian blue in fowl was solely *zygotic* in character; it was not a *gametic* (i.e., gametically-borne) character. There was no individual element representing it in a gamete.

Given that it was "practically impossible to make any *general* statement as to which characters are dominant and which are recessive," nevertheless it was possible "that in the dominant some element is *present* which is *absent* in the recessive type." This was the much misunderstood "presence and absence hypothesis," that Bateson had touched upon in his 1904 BA address. From the preceding context, it might seem that Bateson meant an "element" as *representing* and different from an observable character; such an element could be either present *in gametes* or absent *from gametes*. He then, however, switched from "elements" to "factors" as being either present or absent:

> The difficulty in applying such a generalization lies in the fact that not very rarely characters dominate which to us appear negative. … In Sweet Peas and other plants there are instances of lighter colours being dominant over the more fully pigmented. Consequently we are almost precluded from regarding dominance as merely due to the presence of a factor [element] which is absent from the recessive form. Not impossibly we may have to regard such negative characters as due to the presence of some inhibiting influence, but in our present state of knowledge there is no certain warrant for such an interpretation. As will be evident, especially in cases where colour is concerned, a vast number of the observed facts can be readily represented in terms of the presence and absence of definite factors [elements], and it is quite conceivable that Mendelian inheritance in general may be reducible to such terms.

When one pair of alleles was under consideration (the "monohybrid" case) and one was clearly dominant, then Mendel's 3:1 ratio would appear in the F_2 generation. Figure 9.3 gives, as an example, the case of tallness (T) and smallness (t) in peas. When two pairs of alleles were under consideration (the "dihybrid" case) and one of each was clearly dominant, then the F_2 ratios came out at 9:3:3:1 (Fig. 9.4). Here, "9 members exhibit *both dominant* characters, each of the groups of 3 have *one dominant* character, while the 1 has *both recessive* characters. Similarly, where three pairs of dominant characters were concerned the F_2 ratios came out at 27:9:9:9:3:3:3:1." Thus, as more characters were included, the complexities of analysis greatly increased.

There were also definite, but different, sets of numbers for F_2 progeny when degrees of dominance, or gene–product interactions, were present: "Now so long as the characters of each allelomorphic pair are not only independent in their

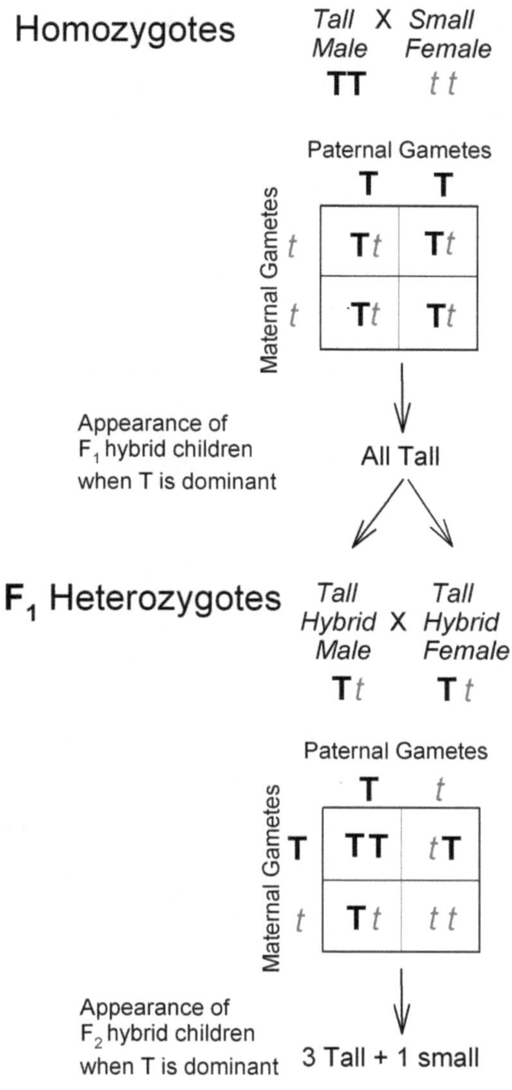

Fig. 9.3 The monohybrid case. A male with two allelic copies of the gene determining tallness (**T**) produces equal numbers of gametes corresponding to those two copies. A female with two allelic copies of the gene determining smallness (***t***) produces equal numbers of gametes corresponding to those two copies. Their children (the F_1 generation, in boxes) each possess one gene determining tallness and one gene determining smallness, but all are tall so revealing **T** to be dominant over its alternative ***t***. When these hybrid F_1 children are crossed (i.e., they are parents for the F_2 generation), each produces equal numbers of gametes corresponding to tallness (**T**) and shortness (***t***). When these gametes meet, on average three children will possess at least one copy of the **T** allele, and so will appear tall, and one child will possess two copies of the ***t*** allele and so will be homozygous and small. Of the three tall phenotypes, one will be homozygous (**TT**), and two will be heterozygous (**T*t***) in that they possess both alternatives

transmission but are without perceptible effect on each other in the zygotes, such simple ratios will be evident. In such cases, types distinct from the parental types will appear, but they will be recognizably due to recombination." Sometimes, new types ("novelties") would appear [e.g., Andalusian blue] which, although not immediately recognizable as due to recombination between alleles, were found on subsequent analysis to have been so formed. But in other cases, the novelties were due to "interaction between characters belonging to distinct allelomorphic pairs."

Citing the work of Cuénot, Hurst, and Durham, Bateson pointed out that in these cases, the F_2 ratios were predictable, although departing from classic Mendelian forms. Among them would be (i) a character that was "imperceptible unless it meets in the zygote another character belonging to another allelomorphic pair," and (ii) "characters which depend for their appearance on the presence of *two* factors *neither* of which can be perceived in the absence of the other." An example of the latter was "the appearance of coloured flowers in F_1 from the cross of two white Sweet Peas." When taking into account degrees of dominance and interactions between allelically different and independent characters, the numbers of each type of offspring in each generation could be calculated:

> When all the factors belonging to all the allelomorphic pairs are distributed independently among the gametes, the number of each kind of gamete and the results of their several unions can be calculated by the simple rules of probability.

Yet sometimes these rules were violated. As had also been noted by Correns, violation sometimes gave "evidence of a linking or *coupling* between distinct characters." Sometimes, the coupling appeared complete, sometimes incomplete. "When such coupling is complete, the two characters of course can be treated as a single allelomorph." In this case, the rare double-recessive would appear with an average frequency of 1/4 (Fig. 9.3), rather than 1/16 (Fig. 9.4). Sometimes, the coupled characters were of the same nature, and sometimes of different nature, so knowing one character did not allow prediction as to what type of characters it might be coupled to.

Hybridization a Decoy

As if these complexities were not enough, miscellaneous decoys remained to trap the unwary. In his 1899 address at the RHS conference, Bateson had noted that "all those present are aware of the great and striking variations which occur in so many orders of plants when hybridization is effected." However, he had cautioned that natural species were unlikely to have arisen in this way. He was both right and wrong. Many species, especially plant species, have arisen in this way (by a process described below). But two species must be in existence *prior* to their intercrossing (hybridization) to produce such new species. Somewhere back down along the evolutionary chain there must have been a single species which had diverged into two species that were sufficiently closely related to breed together, so that future hybrids

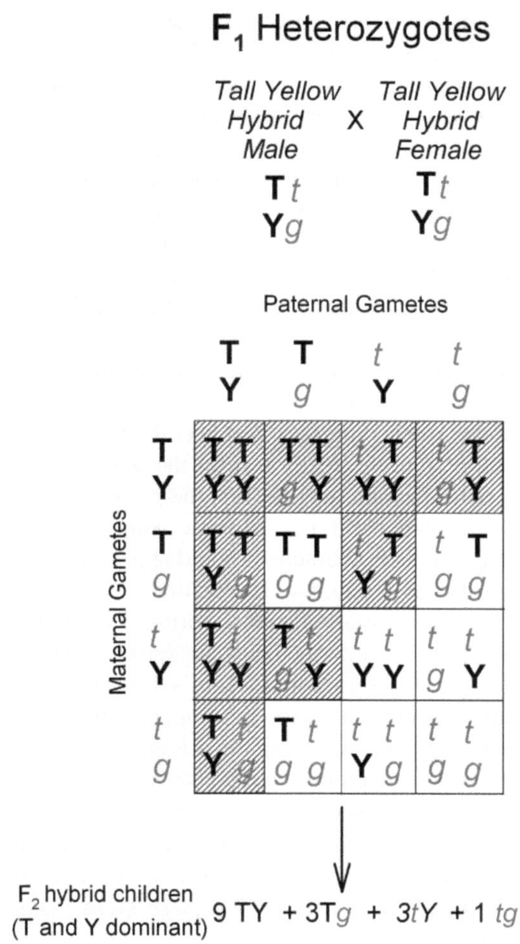

Fig. 9.4 The dihybrid case. Whereas genes determining other characters were ignored in the monohybrid case (Fig. 9.3), in this dihybrid case, the allelic gene pair (**T** and t) that determine tallness and smallness are considered with the allelic gene pair determining seed color characters – yellow (determined by gene **Y**) being dominant over green (determined by gene *g*). That genes determining height are nonallelic with genes determining color is indicated by the vertical separation of their symbols. In this case, there is no coupling, and in the F_1 parents (diploid), both allelic pairs are present, and their gametes can be either **T** or *t*, and either **Y** or *g*, giving four possible combinations in equal numbers (i.e., 25% each). When the four types of paternal gametes meet the four types of maternal gametes, 16 combinations can appear in their F_2 hybrid children (in boxes). If **T** or **Y** are present together, either as single or double copies, then the children display both dominant characters together (i.e., yellow seeds grow into tall plants). The nine children corresponding to this are represented by cross-hatched boxes. The other combinations are in white boxes. The top left cross-hatched box and the bottom right white box correspond to the grandparental pure lines (homozygotes) from which the original parental F_1 heterozygotes might have derived. When two pairs of Mendelian characters are concerned in a cross, the average result can be directly calculated. In the case of hybrid tall yellow peas, the formula is: (3 T + 1 *t*) x (3Y + 1 *g*) = 9TY + 3T*g* + 3*t*Y + 1 *tg*. Here, four phenotypes can be distinguished in ratios 9: 3: 3: 1. Depending on the characters under study (their degrees of dominance and interactions), Bateson and his co-workers were able to find examples of other phenotypic series: 15: 1 (from [9: 3: 3]: 1), 12: 4 (from [9: 3]: [3: 1]), and 9: 7 (from 9: [3: 3: 1])

might be derived by intercrossing. Those two species must have arisen from a common ancestor by some process other than hybridization.

Hybridization can be compared to the shaking of a kaleidoscope. The variously shaped colored pieces reassort at every shake to reveal a variety of patterns. But the pieces themselves do not change. Before shaking can commence, the kaleidoscope must be loaded with a set of colored pieces. Thus, two processes contribute to the patterns that are seen: first there is the preparation of the pieces; then their shuffling.

In a biological context, it is the preparation of the pieces, rather than their shuffling, that seems to be most important where the formation of a new species is concerned. However, we now know that if two sets of pieces from two species can be brought together, their *summation* may result in a new species, which is *instantly* isolated reproductively from the parent species – a phenomenon known as polyploid speciation (to be considered below).

Also speaking at the 1899 conference had been Hurst (Chap. 10), whose orchid breeding studies had led him to greatly respect the power of hybridization. To be deemed successful, hybridization requires first the production of a form with a different character from the parental types. At some point following crossing, a form with new, perhaps intermediate, characters should appear. In the case of the characters Mendel studied in peas – tallness and smallness – there is usually a quantal transfer from generation to generation, and no forms of intermediate height appear. For some characters, however, intermediates are formed. What is then required is that the intermediate character "breed true." This is the criterion of success. But usually, whereas an intermediate character may appear in the F_1 generation (e.g., Andalusian fowl), on further intercrossing the original parental characters return in the F_2 generation. The F_1 character does not breed true. It does not consistently produce its own kind, and only its own kind, when crossed with its own kind.

Another speaker at the 1899 conference was de Vries. He indeed had obtained hybrids from the evening primrose (*Oenothera Lamarckiana*) that bred true. The plants were large and hence were named *Oenothera gigas*. Although de Vries had not known it at the time, the largeness was partly due, not to an increase in the total number of cells in the plant (as in the case of tallness in peas) but to an increase in the size of individual cells. The cells were large because the crossing process had somehow resulted in two genomes (sets of chromosomes) occupying a common nucleus.

Large nuclei are usually accompanied by a large cytoplasm, so the cells are of regular appearance, but large. The pieces had not changed or mixed in the kaleidoscope, but the contents of two kaleidoscopes had come together. Indeed, regarding chromosomes Bateson later noted in an address entitled "Facts Limiting the Theory of Heredity" (1907): "Very closely allied types may show great differences … Miss Lutz has found … *Oenothera gigas* … has 28 [chromosomes], while *Oenothera lata* has 14 [chromosomes]." These chromosomal differences were true of *all* cells *whatever tissue was sampled*. So, instead of being diploid plants (*lata*), they were tetraploid (*gigas*).

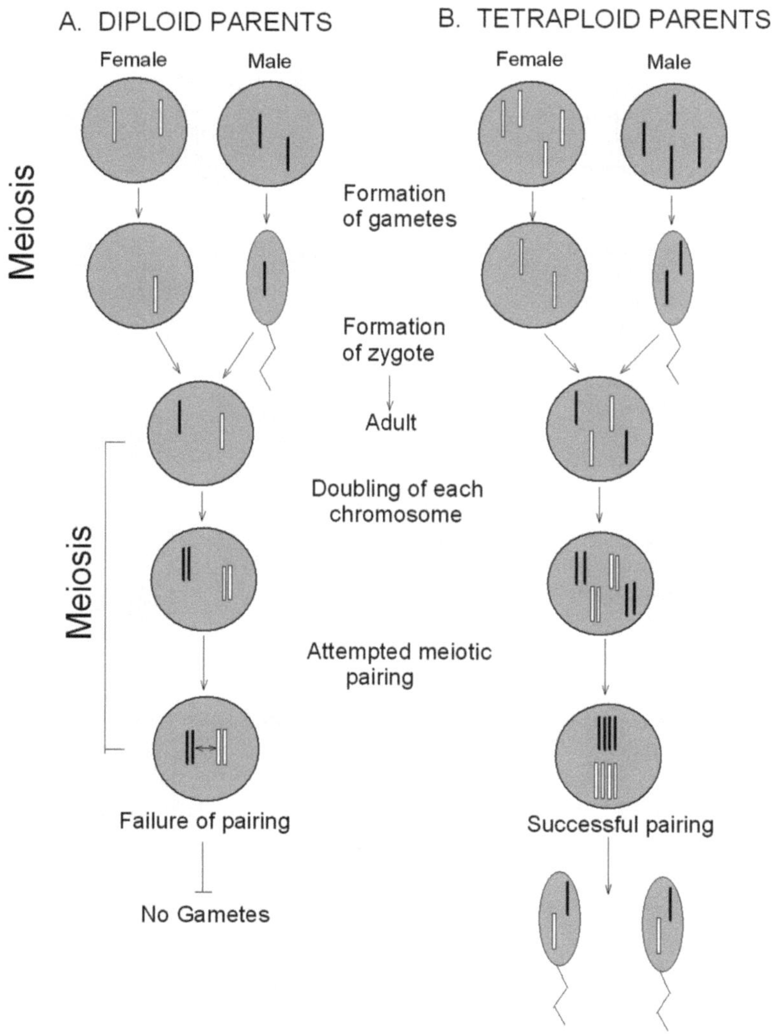

Fig. 9.5 Formation of hybrids between two allied species in the case of (**a**) normal diploid individuals, and (**b**) tetraploid individuals. For simplicity, each parent has only one type of chromosome, represented by paired linear rods. The differences between the two parental species are symbolized by their chromosomes being white and black, respectively. In **a**, each parent generates haploid gametes containing only one chromosome. In **b**, each parent generates diploid gametes containing two chromosomes. In **a**, the hybrid zygote, when it becomes an adult and attempts gametogenesis, is unsuccessful (i.e., hybrid sterility), because the differences between the parental chromosomes are sufficient to impede pairing during meiosis. In **b**, the same two species can be regarded as having generated, through errors in meiosis, individuals with four copies of the chromosome (autotetraploids). When white is crossed with white (not shown), or black with black (not shown), each chromosome does not have a *unique* pairing partner, since there are *four* potentially pairing chromosomes, which get entangled (see Chap. 16). However, a few viable diploid gametes might be produced. When white is crossed with black (shown), the hybrid zygote (allotetraploid) as an adult is able to complete gametogenesis successfully because, although the differences (white and black) remain, each chromosome has a *unique* pairing partner. The gametes (male only are shown) differ from the parental gametes in having one black and one white chromosome. The beginning of a tetraploid line, where the chromosomes of the two founding organisms retain their individuality, is now feasible (see Fig. 9.6)

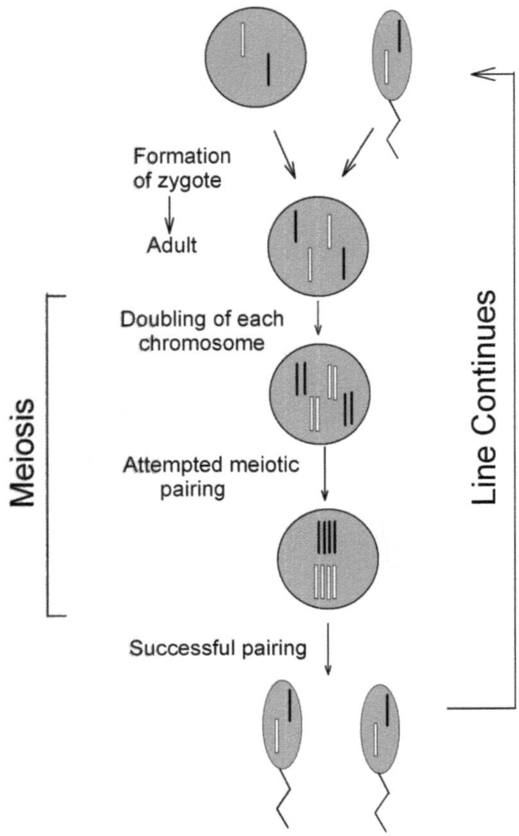

Fig. 9.6 Foundation of a tetraploid line from the two founding species shown in Fig. 9.5. The diploid gametes contain chromosomes originating from each species (symbolized by the white and black fills). When the hybrid zygote becomes an adult and attempts gametogenesis, each chromosome has a *unique* pairing partner, and the line has the potential to continue. The two parental chromosomes may generate a tetraploid phenotype different from those of the parents. It is a "constant hybrid." Since relatively few meiotic generations are required, and the tetraploid is reproductively isolated in being unable to cross successfully with the original diploid type, the phenomenon appears as an example of instant speciation

It is now well-established that certain polyploids can breed true, because each chromosome has a pairing partner (Figs. 9.5 and 9.6). The hybrid sterility seen when allied species are crossed can be "cured" if the diploid parental lines are first made tetraploid. Indeed, this is one of the *strongest* pieces of evidence that the pairing of chromosomes in meiosis must be specific – *like pairing with like* (homologous chromosome pairing with homologous chromosome). What does this mean at the molecular level? Is there something able to cross-link chromosomes if they are homologous, but not if they are not homologous? Or is it something about the structure of the chromosomes themselves that makes them pair specifically? The chapters ahead lead to a possible answer in Part VI.

Nevertheless, although polyploid species are abundant (at least in plants), this is not how species usually arise. Thus, for de Vries, polyploidy was a decoy that tended to side-track him into an area that, although of great interest, was only indirectly informative on the nature of variation among organisms. Bateson did not make this mistake. The early correspondence from de Vries to Bateson revealed an earnest desire to avoid misapprehension. In 1900, de Vries wrote from Amsterdam (October 25):

> I am very much pleased at your remark on my quotation of your book, for there must be no discontinuity between us, not even in the use of the word. I gladly accept the general use, even I have always called in my lectures the chuchled [? poor English of de Vries] variations continuous, and those I now call mutations discontinuous. I hope we will agree on this point also. To answer your questions on hybrids is an easy matter. If I say *l'hybride ne saurait tenir le milieux entre eux*, I mean the monohybrid [where parents differ in only one character], also of discontinuously differing parents. For continuous cases are excluded from my papers; for them I plainly admit *Mr.Galton's* rule of the midparent, and his 'laws of regression'.

In peas, height, despite being a quantitative character, varies discontinuously (i.e., either tall or small). More often, quantitative traits vary continuously. This reflects contributions from several genes, which get kaleidoscopically shuffled to generate a range of combinations in new individuals. In such cases Galton's laws seem not to be violated (see above). In 1901 (July 1), de Vries suggested that one of Bateson's crosses was breeding true like his Oenothera crossings: "I learned [of] your new case of crossing with great interest. Does it not coincide with my Oenothera crossings which give also some Dominants and some Recessives, both of which remain constant in the next generation? It would be a very great thing if you had found another case of this, and from your letter I do not see why it should not be so."

Following the publication of the first volume of his *Die Mutationstheorie*, de Vries wrote (October 31, 1901) suggesting, not that his Oenothera hybrids were exceptions, but that cases of Mendelian inheritance were exceptions:

> I prayed you last time, please don't stop at Mendel. I am now writing the second part of my book which deals of crossing, and it becomes more and more clear to me that Mendelism is an exception to the general rules of crossing. It is in no way *the* rule! It seems to hold good only in derivative cases, such as real variety characters. My Oenothera's, and other Oenothera's too, don't Mendelise at all, with the exception of *O. hemistylis*. I have made a long series of experiments in crossing half-races and middle-races of dicotyledonous plants (such races having as it seems never been studied before). They don't Mendelise at all.

The concept of "mutation" as set out in the first volume of *Die Mutationstheorie* was as a sudden change which would create a potential new species. This was in line with Bateson's *Materials*, but Bateson's thinking was now moving ahead rapidly, and there was to be a parting of the ways. Nevertheless, in a review of de Vries' book in the April 1902 issue of *Biometrika* [26], de Vries and Bateson were treated as like-minded by Weldon. Even in recent times, there is often reference to the "mutationists" de Vries and Bateson [27, 28].

Undeniably, de Vries had obtained new forms in the F_1 generation ("first crosses") that would indeed "breed true," reproducing their phenotypic characters when crossed again with each other (Fig. 9.6). Bateson took the opportunity of a review of the first of a two volume German text on evolution by Lotsy (*Nature* 1906), to express his misgivings:

> To those who know that the mutations of Oenothera are not errors of observation, and hesitate to accept them as the single key to the final mysteries of evolution, the question begins to press: What *are* those mutations? Upon this point the teaching of genetic research is clear. Before we can form a definite view as to the nature of any given mutation we must know its gametic relations to the type from which it sprang, and to the sister-mutations. So

> far, these relations, as expressed by the ratios in which the forms appear, seem to be almost always irregular in the Oenothera cases. … All that can be positively asserted is that these mutations are forms arising discontinuously, and that their distinctions are exactly comparable with those that often appear to characterize species. But now that we understand what a medley of phenomena is included in the term 'specific difference' it become necessary to go further and to ascertain which phenomenon is exemplified in each case. … Meanwhile, however, it must be conceded that there are serious difficulties in the way of a purely Mendelian account of the Oenotheras – more perhaps than Prof. Lotsy indicates.

Bateson's reservations were made explicit in his 1906 *Progressus* review. He considered that many reported examples of "first-crosses breeding true" had not been "satisfactorily observed on a large scale," so that "if genuine examples exist, they must be exceedingly rare." Sometimes the phenomenon had been found due to "parthenogenesis, and consequently the reproduction is really of an asexual nature." Without doubting the importance of de Vries' observations, he noted:

> The intercrossings carried out by de Vries among the wonderful mutational forms raised by him from Oenothera have given rise in many cases to results which are undoubtedly very anomalous, but until these cases have been analysed more fully and studied in the light of recent knowledge of the inter-relations which may subsist between characters, it is difficult to be sure as to the significance of these phenomena. A better knowledge of these curious cases would probably lead to important advances in genetics.

Like Bateson, Spillman (see Chap. 10) sensed something odd with de Vries's demonstration of speciation in one generation in Oenothera. Writing in an US Department of Agriculture *Bulletin* in 1909, Spillman noted [29]: "These fixed forms which occur in the progeny of hybrids are sometimes mistakenly called 'mutations.' They are in no sense mutations of the sort comprehended by that much misused term as it is at present understood. They are simply recombinations of characters which, before the hybridization occurred, existed in different combinations." The discontinuity that grew between de Vries and Bateson endured. But Bateson helped get *Die Mutationstheorie* translated into English. He wrote (November 4, 1907) to the German publisher (Veit and Company) proposing collaboration with Cambridge University Press. The translation (by Farmer and Darbishire) ended up with other publishing houses. In 1923, Bateson negatively reviewed a multiauthor text edited by J. C. Willis, which included an article by de Vries (Chap. 18).

Mary

Bateson's fights did not always involve the tall and powerful, but they added to the many stresses he was under. In an April 1906 issue of *The Speaker*, J. B. Burke's *The Origin of Life* and McCabe's translation of Guenther's *Darwinism and the Problems of Life* were anonymously reviewed by Bateson (negatively). Burke replied (April 28) claiming "atrocious and malicious libel" and demanded that statements on his work in the Cavendish Laboratory at Cambridge "be withdrawn and an apology offered." After consultation with Bateson, the Editor issued a brief apology.

Fig. 9.7 Mary Bateson

A few weeks later (May 10) in *The Cambridge Review*, under the initials N. R. C., there was an even more damning review of Burke's book [30] (Fig. 9.7).

In September 1906, Bateson set off alone by bicycle and train on an art-collecting holiday. His letters to Beatrice began in Birmingham (September 19) and indicated travel to Coventry, Kenilworth, Leamingston, and Warwick. At Birmingham (September 20), he had a pleasant stay with a dealer's family, and replied "Aye, Aye!" when asked whether he was a Liberal. There was much concern over the new Education Bill which was seen as threatening religions (the family were Catholic). His trail led to Wolverhampton, Shifnal, Much Wenlock (where there was some fossil hunting), and Ludlow (September 24). He returned to greet de Vries, who was visiting Grantchester (November 26). Then, on November 30th came the untimely death of his sister Mary, at age 41.

Mary had taught at Newnham from 1888 to 1906, had been honorary secretary of the Cambridge Association for Women's Suffrage from 1892 to 1898, and had recently led a delegation representing women graduates to the Prime Minister. A memorial fund in her name raised £786; the subscribers included Francis and George Darwin, Florence Durham, Mrs. Herringham, Alfred Marshall, Mrs. Sidgwick, and Alfred Whitehead. The proceeds went to an endowment for the Fellowship Fund of Newnham College, one of the fellowships being designated the Mary Bateson Fellowship.

Her close colleague, Frederic Maitland, wrote in sympathy describing her high achievements (December 1), and Bateson thanked him (December 2) acknowledging that "in all her later work you were the inspiring force," and adding: "She has gone in full work: so may we all." He wrote to their mother (December 10) concerning Mary's will, effects, and papers, noting that he had spoken to Mrs. Sidgwick (of Newnham) and adding: "We leave tonight for Braunschweig, Germany, and expect to be back before Christmas." In her *Memoir,* Beatrice wrote: "A little alarmed, I think, at his overwhelming fatigue, he took himself in hand with energy and purpose, and at the end of a week or ten days threw off the physical depression that had almost overmastered him. We walked in the snow of the Harz mountains, and afterwards wandered through the Museums and Galleries of Northern Germany."

Meanwhile, regarding their personal fortunes there was a shift in the wind. Bateson turned down two requests for books, but a third was one he could not refuse. First, early in 1906, a letter from Osborn of the American Museum of Natural History, sent after consultation with Wilson, offered Bateson $1000 for lectures and a book in the "Columbia Biological Series," for which he would be entitled a "large percentage" of the profits. Bateson replied (February 14) hedging about the lectures and especially the book since he was already committed to Cambridge University Press for *Principles*: "I doubt if – so to speak – the cistern would be ready for a really fresh discharge again so soon." The matter was not pursued.

Then, there came a request for a genetics text from Harmer (December 27, 1906), who with Shipley was co-editing the Cambridge Natural History series. Bateson replied: "I am trying to finish the Mendel book for the Press – long overdue. … The relation of Mendelian phenomena to the species problem at large and to de Vries's Mutations in particular, is still obscure, and until this awkward gap in the evidence is at least partially filled it is impossible to deal with the whole field of Genetics in a coordinated fashion."

Finally, another letter arrived from America, this time from Yale (written December 7) where the annual Silliman Memorial Lectures had been endowed in 1883. Bateson was invited for the 1907–1908 academic year, at a fee of $2500, to give 8 to 12 lectures to be published as a book "at the expense of the university." Silliman had wanted the lectures to "be such as will illustrate the presence and wisdom of God in the natural and moral world," but that was not dogmatically interpreted. The fund could be used for a course which was "not positively and avowedly materialistic in its conception." Bateson wrote (January 27, 1907) that his title would be "*The Problems of Genetics*," (hereafter "*Problems*") and that he hoped Cambridge University Press would be entrusted with the sales in England.

Summary

The Biometricians saw Mendelism, with its emphasis on observations of individuals as individuals, rather than as members of a long lineage, as a threat. Bateson replied to Weldon's questioning of the reality of Mendel's categories, with a polemic – *Mendel's Principles of Heredity: A Defence*. While many deplored its tone, the underlying logic was sound. It argued that when multiple elements determine a character, the distribution in the population is close to that predicted by Galton and Pearson, despite their faulty assumptions. At a conference in New York in 1902, Bateson was treated as a celebrity. Although the Biometricians scored debating successes, a critical turning point was Bateson's BA address in 1904. Furthermore, he showed that various inherited diseases could be understood in Mendelian terms and encouraged biochemical studies by Florence Durham and Muriel Wheldale. There were successive defections from the ranks of the Biometricians until, with the sudden death of Weldon in 1906, only Pearson was left. Around this Pearsonian nucleus, a new Biometrics was to emerge. Bateson saw the instant speciation that de Vries had observed in his evening primrose hybrids as an interesting aberration, not as an example of the fundamental discontinuity that must generally underlie the formation of a new species. Although the Mendelian battle was won, Bateson's personal position was insecure. Pressures from many quarters seem to have generated a restless urge to escape, which found expression in sudden departures to foreign parts. In 1906, the way ahead became clearer when the Darwin brothers gave research funds, and a lecture invitation arrived from America.

Chapter 10
Bateson's Bulldog

Bateson had many disciples, but was never himself of their number.

J. B. S. Haldane [1].

Of all the members of the rapidly growing Bateson School, the most stalwart and influential was the unflappable Charles Chamberlain Hurst (1870–1947) (Fig. 10.1), who inherited a flourishing nursery business at Burbage in Leicestershire. In the sense that there may be an enduring bond between comrades who have stood together under fire, so the friendship between these two leaders of Mendelism in England stood the test of time. Indeed, the largest single item in Bateson's surviving correspondence is that with Hurst. The fact that the correspondence began to tail off around 1910 signifies, not disenchantment, but victory in the Mendelian battle, increased usage of the telephone, and Hurst's war duties.

Hurst's intention to read Natural Sciences at Cambridge had been thwarted by ill health – possibly tuberculosis – from which he recovered. He became a member of the RHS (1891) and soon gained a reputation, both practical and academic, for his extensive knowledge of orchids. He became a strong Mendelian advocate and devoted a large amount of his wealth and time to the experimental breeding of plants and animals. Following his first meeting with Bateson in 1899, he was an important link with commercial interests and a steady ally against the Biometricians. While it is appropriate to refer to him as Bateson's bulldog, the canine quality displayed was not confrontational. The imperious Bateson was not easy to approach, and many tended to go to Hurst, rather than Bateson, when they had a genetic question or wanted a popular article written. As the years went by, and pressures on Bateson's time increased, this shielding role became of much importance. Gerald Leighton, a pathologist at Edinburgh University and Editor of *Field Naturalist's Quarterly,* wrote to Hurst in 1905 (September 4):

> You are probably accustomed by this time to the fact that whenever I write to you I am in search of information. It is your own fault. You should not be so jolly good natured; but

A. G. Cock, D. R. Forsdyke, *Treasure Your Exceptions*,
https://doi.org/10.1007/978-3-030-92099-9_10

Fig. 10.1 Charles Hurst

> since you are, will you kindly take up your pen, and write down on half a sheet of note paper exactly what is your idea of de Vries' theory of the origin of species by mutations, and wherein exactly it differs from Darwinism. … I am anxious to do de Vries justice, and above all things to be accurate.

Their work being labor intensive, the early geneticists saw every such enquirer as a potential pair of hands. Some four decades later, Leighton wrote to Hurst's wife Rona (May 10, 1949): "It must have been some time between 1903–1906 that I spent a very happy 10 days … at Burbage, back-aching days in a field of peas counting the 'smooth' and 'wrinkled'."

Spillman

It was to Hurst, not Bateson, that William Wilks, Secretary of the RHS and Editor of its journal, went for a review of a manuscript from William Spillman (1863–1931) who had studied wheat hybrids at Washington State Agricultural College. The work had been presented (November 1901) at the Annual Convention of Agricultural Colleges and Experimental Stations in Washington, DC [2]. Hurst saw that the data readily lent themselves to a Mendelian interpretation (January 21, 1903):

> The paper is biologically of the greatest importance, and in the large numbers with which it deals is, in my opinion, the most valuable confirmation of Mendel published since his day, indeed, in some respects it gives more facts than did he. Curiously, the author does not seem to have heard of Mendel, and hence does not make the most of his facts. … I cannot understand how the paper has been overlooked by the Mendelians but there seems to be no one in the U.S.A who has quite got hold of Mendel.

Wilks annotated Hurst's letter in red ink and returned it to him with the message: "Forgive this short way of answering – you would if you knew the overworked condition of your w.w.!" In this light, it is possible that the Spillman paper had been on Wilk's desk for some months before he got round to contacting Hurst.

Wilks agreed to publish the Spillman paper together with Hurst's extensive analysis [3]. In the meantime, perhaps influenced by a 1901 paper by Davenport – "Mendel's Law of Dichotomy in Hybrids" [4] – Spillman had himself seen the Mendelian light, a brief account of which was published in *Science* in November 1902 [5]. So, Mendel had won an important American advocate. Despite his failure to interpret his data fully, Spillman's presentation at the Washington conference in November 1901 is said to have impressed a future Assistant Secretary of State for Agriculture. In 1902, Spillman moved to a research position in the Department of Agriculture in Washington, DC [6].

Orchids

Hurst's first published articles in the late 1890s were notes on "curiosities of orchid breeding" resulting from a "few outlandish crosses" carried out "for the sake of experiment" along the lines of Charles Darwin [7]. There were descriptions of hybrids, their vigor, and fertility, which he hoped might "throw a light on the dark mysteries of evolution." In general, there was no dominance, first-generation hybrids being uniform in appearance and intermediate between their parents. Either parent could act as the pollen donor, leading Hurst to conclude in a lecture (October 12, 1897) that "the *determinants of* the characters of the father parent are packed up in the small compass of the pollen cell, and those of the mother parent in the egg cell." Since the pollen was "almost wholly nucleus," then "it follows that in the *handing down* of characters from one generation to another, the nucleus is the all-important matter" (Our italics). Citing Weismann's observations, Hurst further noted that:

> The most important elements in its constitution are certain rod-like fibres, looped threads, or round grains called 'chromosomes,' which change their appearances from time to time, but which nevertheless generally remain constant in the same species. These are present in the nucleus of every cell, but in the germ cells *at the time of fertilization*, not only is their position and shape altered, but *their number is apparently decreased by one half*, and ... the essence of fertilization consists of the removal of one half of the nuclear element from the egg cell of the mother and replacing them by an equal number from the pollen cell of the father.

This was the established view at that time and is noted here because, shortly thereafter, Hurst met Bateson, who some came to think did not appreciate the case for chromosomes (Chap. 13). Like the classical hybridizers, Hurst observed in 1897 that:

> Hybrids of the second and third generations are very variable indeed, having a wide range of variation, sometimes reverting wholly to the parent species or parent hybrid .... The more hybrids are crossed with one another the more related they become, and, consequently,

> reversions to hybrids of the first generation, and indeed to pure species, become more frequent.

Hurst added a historical footnote to this statement when preparing a volume of his collected papers in 1914: "The discovery of Mendelian segregation some three years later has made it possible to account for these facts in a more definite and precise form." The date 1914 is ominous. Due to the First World War, the volume did not appear until 1925 [7].

In 1898, Hurst observed that some of the horticulturalists' orchid hybrids were identical with orchids found in the wild, which were therefore "natural hybrids." He concluded that it was likely that "intercrossing between different species is carried on in a state of nature to a far greater extent than was formerly supposed." He also noted that when crosses failed, it was often due to a failure in one sex, which, in the cases he studied, was the male – the pollen parent.

Like Bateson (Chap. 8), Hurst read a paper at the 1899 RHS International Conference on Plant Breeding and Hybridization on a hot July day in a "stifling marquee" at Chiswick [8]. Hurst saluted the insights provided by microscopic examinations of germ cells but thought there was "a danger that … we may lose sight of the broad facts of inheritance as manifested to us in the outward characters of plants and animals." New varieties had great sales value, and as a businessman, he saw that "the inheritance or non-inheritance of varietal characters is most important. If a useful or ornamental variety is capable of transmitting its good qualities to its offspring, then its own natural value becomes greatly enhanced, and in the course of a few generations a more or less permanent race may be established." However, there were "many exceptions," to the breeders' rule "breed only from the best." His practical concern was that of getting a new line with attractive characters to "breed true."

Characters were recognized as "varietal" (relating to a variety within a species), "specific" (relating to a species), and "generic" (relating to a group of species, or genus). Here, the orchid grower was dependent on systematists, such as Rolfe at Kew Gardens, who classified plants, usually on the basis of their appearance. Best breeding results were obtained with characters deemed specific and generic, but still there were many exceptions. Hurst had obtained numerous fertile hybrids "from very distinct species," which according to physiological definitions of species should have been reproductively isolated from each other. He noted two species that had been classified as within "distinct *genera* by our best botanists; and I think that the most easy-going systematist would admit them to be [at least] distinct *species*, yet hybrids raised between these two genera are very fertile."

Thus, prior to meeting Bateson, he had learned from his own experience that there was no necessary correlation between species as classified anatomically ("morphological species"), and species as classified physiologically (with respect to fertility barriers; "physiological species"). He had sought guidance from Galton's Law of Ancestral Heredity, but noted that in some cases the law was "disturbed by *Partial Prepotency* [i.e. dominance]." In his 1897 lecture, he had considered that characters would blend [7]:

> I am not aware that the number of nuclear fibres in the cells of Orchids has yet been ascertained, but in the closely allied order of Liliaceae, in the giant plant Lilium Martagon, M. Guignard has recently [1891] observed that while the ordinary cell contained twenty-four nuclear fibres [chromosomes] the ripe egg cell had but twelve, apparently showing that it was ready for fertilization. … So that if we take Cypripedium x Leeanum as our typical hybrid, we find that its nuclear fibres would be made up one half from its parent C. Spicerianum and the other half from its other parent, C. insigne; both parents being pure species, their own nuclear fibres would of necessity be pure and true (their ancestors for many generations having been the same species as themselves). These nuclear fibres being, as we have seen, the bearers of the hereditary characters, and determining the building up of the future individual, it is manifest that the hybrid C. x Leeanum is of necessity an equal blend of its two parental species.

Unlike Galton, he did not go beyond this to consider what might have been lost when the chromosome number was halved during gametogenesis, and whether this loss might have been random or biased. Nevertheless, albeit from a somewhat different perspective, he was wrestling with the same problems as Bateson. Their encounter was timely. Bateson saw the horticulturalists as major allies and, in April 1900, agreed to be a RS delegate to the RHS Scientific Committee which was discussing ways of arousing interest in science among practical growers. The first surviving correspondence between the men is dated March 1902 when Bateson, having seen a note by Hurst in the *Gardeners' Chronicle*, wrote proposing to visit him at Burbridge (Chap. 8).

New York

For Bateson in 1901, life continued much as usual. There were trips to London to view paintings (June 4), and to Suttons' Nurseries, Reading (August 8). In August, Beatrice was away and Bateson was at home with the children. He wrote (August 25) telling her that Wilks had invited him to represent the RHS at the Second International Conference on Plant Breeding and Hybridization, to be held in New York in 1902, for which £75 were offered to defray expenses. Hurst was on the planning advisory committee but did not attend.

The three-day conference opened in New York (September 30) with an address by Bateson. He stuck very much to his title – "Practical Aspects of the New Discoveries in Heredity" – noting the difficulties of the "breeder or seedsman" who, when trying to fix "some strain of a new variety … finds a number of rogues which are not true to the character which he desires to put on the market – rogues which he is unable to eliminate. Formerly he said it was only a question of time; he must hoe out the rogues and go on, and he will gradually fix his type." Bateson gave various examples of the futility of this, including the well-known example of the Andalusian fowl:

> You may go to the poultry shows and buy the winning Andalusians, thinking that they will breed true. But they will not. Andalusians have been bred for at least forty or fifty years, and there is no good reason for thinking that they breed any truer now than formerly. … The

> Andalusian is almost unquestionably a heterozygote form made by the union of the *black gamete* with the *white-splashed gamete.*

Bateson did not, of course, imagine that the gametes themselves were "black" or "white-splashed." He employed his usual short-hand for gametes transmitting the elements that determined the black and white-splashed feather characters. While it certainly made for easy communication to a partially nonacademic audience, with hindsight it seems probable that in the long term Mendelism would have been better served by more pedantic definitions.

There was a warm discussion after Bateson's address. One speaker praised *Defence*: "I expect to use this book as a basis for all our work in plant breeding." Another declared: "Professor Bateson" has thrown light on "many … points which have puzzled many of us who are practical workers in this very interesting field." Professorial status was also accorded by Spillman, who wrote to Hurst (January 21, 1902):

> Professor Bateson charmed us all when he came over here and should he ever return he will receive an enthusiastic welcome. Not only is he a student with a very fertile mind, but he is a most charming man personally. I wish very much you could have been present at our New York conference. It was really a great meeting. It is true that of the fifty-one papers or more, no matter what the subject, everyone dealt with Mendel's law more or less; for as soon as Bateson's paper was read there was an excited state of mind until the end of the meeting.

Hurst submitted a paper which was read on his behalf immediately after Bateson's [9]. The title "Notes on Mendel's Methods of Cross-Breeding" indicated that he was now well into the Mendelian camp. At the outset, he expressed "the hope that the hybridists and breeders of the New World, with their progressive ideas, their many opportunities, their vast system of experimental stations, and their practical knowledge of breeding, will take up and test the matter on a much larger scale than we can hope to do in the Old World."

Referring to his studies on orchids, Hurst noted that "in any statistics of inheritance a definite result can only be determined by taking each *single character* separately as a distinct unit, completely ignoring for the time being the individual plant made up of many characters. Mendel apparently was the first to see this clearly." The key was to begin only with "constant and fixed races … so that … ancestry has been practically the same for many generations. … De Vries, Correns, Tschermak and Bateson have all for the most part followed or carried out Mendel's method by crossing *constant races*, and it is quite possible that some of their apparent exceptions to Mendel's results may have been due to their crossing peculiar races which were not really so fixed and constant as they believed them to be."

In the following discussion, the possibility was raised that a constant race might become inconstant due to the sporadic appearance of new "mutation forms." The questioner was H. R. Roberts who decades later was to produce a landmark text on the history of hybrid breeding [10]. Spillman commented: "This matter of hybridization is different from mutation. We may have mutations in our homozygotes and we may have mutations in a hybrid plant [heterozygote], so that we must not confuse them. When we are dealing with hybrids let us overlook the mutations that occur."

Hurst's paper could not have been better placed. It was followed by one that was read for de Vries. Another paper by William Cannon of Columbia University, describing microscopic observations of gamete formation in hybrids, will be considered in Chap. 13. The day after the conference, there was an excursion up the Hudson River to the Vanderbilt estate where Mrs. Vanderbilt showed the visitors around the gardens and farm.

Words

The close communication between Hurst and Bateson meant that they were able to present a common front with respect, not only to the Mendelian interpretation of their data, but also to the words employed. Hurst published much less and generally in lower-profile journals, but his writings were clearer and less nuanced partly because he omitted the rhetorical flourishes that Bateson was seldom able to resist. Once Bateson had understood something, he tended to think his readers would understand likewise, so, while concentrating ferociously on the problem at hand, he tended to employ short-hand terms like "zygote" when he really meant "adult," and to imply that the gametes bore characters themselves rather than the determinants of those characters.

Hurst's New York paper had dealt entirely with appearances without going into the basis of those appearances. A further paper entitled "Mendel's Principles Applied to Orchid Hybrids" was published in 1903 [11]. Here, he wrote of "determinants of," or "determinants for," a character. And rather than employing the Batesonian slang "purity of the characters," he wrote "purity of the determinants." The paper was of interest in that Hurst's initial attempt to apply Mendel's principles to orchid breeding had been unsuccessful because he had chosen the color of the dorsal sepal of a flower as a conspicuous single character. Subsequently, he had found this to be a "composite" (compound) of three other distinct, independently inherited, characters, which in various combinations produced the sepal color. When he took the three separately, he found, in each case, "accordance with Mendel's theory."

In a later address (June 1904) to the Leicester Literary and Philosophical Society, his terminology had changed a little [12]. Now, like Bateson, he tended to refer to germ cells as containing a "factor of" or "factor for" a character. But he sometimes seemed to slip back to Batesonese noting, with emphasis, that "*the true unit of heredity is not the individual* [organism], *but the* [factor for a] *single character*." Perhaps guided by our interpolations within square brackets, the modern reader may here pause. If by "the true unit of heredity" Hurst really meant, not the single character, but that which *determines* the single character, then he was declaring that what we now call a gene is a more fundamental unit of heredity than the individual containing that gene. A few more steps and Hurst would have crossed the threshold to the revolutionary concept if the 1960s now encompassed by Richard Dawkin's handy phrase "the selfish gene" [13]. Decades earlier Charles Darwin, Thomas Huxley, and Samuel Butler had prepared the ground (see Chaps. 5 and 19), but even if the concept had been pondered in the early 1900s, no one took the crucial step of documenting it.

Rabbits

In 1902, Hurst had begun crossing two breeds of rabbits, one known as "White Angora" (a long-haired race) and the other as "Belgian Hare" (although it was really a rabbit, hares being a different species). His observation that the long hair property was recessive to short hair was soon confirmed by Castle and shown also to apply to guinea pigs [14]. However, it was not until the experiments were well advanced that Hurst received external funding. In 1904, the BA awarded him £30 for rabbit experiments under the supervision of a committee of three – Bateson and J. Stanley Gardiner, with foreign input from Charles Minot of Harvard. There was also a grant from the US-based Elizabeth Thompson fund. It seems that Minot, acting as Secretary for this Fund, believed that Hurst was working under Bateson (September 30, 1904):

> Your application to the Elizabeth Thompson Science Fund for a grant in aid of 'Experimental studies on heredity in rabbbits' – these studies to be conducted under your direction by Mr. C. C. Hurst – has been carefully considered. I have the pleasure to add that the Trustees voted to assign you a grant of one hundred and fifty dollars ($150), which sum is found enclosed.

Hurst wrote to Bateson (October 25, 1904): "Many thanks for your cheque £30 – 17 – 3 from the E. T. Fund which is indeed very welcome. I note the conditions, etc. [a report sent to the Trustees], which I will comply with in due course. ... It was very good of Prof. Minot and you to get it for me."

Hurst gave a brief report of his studies at the fateful August 1904 BA meeting in Cambridge (Chap. 9). Later in the year, a more extensive presentation was made at the Linnean Society (December 15). Bateson, all set for his first "pilgrimage" to Brünn, regretted that he would miss the meeting. Hurst replied (December 9, 1904):

> I am sure you will not miss much, as most of it is the old story. I hear from the Secretary that there is likely to be a good discussion, and if the opposition turns up I shall miss you sadly. I rather think though, that after the smashing you gave them at Cambridge they have lost heart. W. appears to be silenced, and K. P. seems to be giving off a few dying kicks before adapting Mendel to his purpose! I am glad to see in *Science* (p. 698) that Davenport has come quite round, which is certainly a mutation for a biometrician.

Hurst's conclusions from the rabbit studies were "the same in principle" as those already drawn by Cuénot in his mouse experiments [15]. Hurst made his distinction between "characters" (or "unit characters") and "factors" explicit – the former being "somatic" (i.e., observed in the organism) and the latter being "gametic" (i.e., deduced to be transmitted from breeding studies):

> A study of the somatic characters and gametic factors of … individuals brings out the important fact that certain individuals, identical in appearance and with precisely the same ancestry, differ, in a regular and permanent manner in their breeding potentialities. … The true measure of heredity therefore is neither the somatic character of the individual nor of its ancestors, but its gametic constitution [the constitution of the gametes that formed it], and, in our present state of knowledge, this can only be determined by experimental breeding on Mendelian lines.

Shortly thereafter (January 23, 1905), further rabbit studies were reported to the Leicester Literary and Philosophical Society [16]. Here, Hurst referred to the "germinal constitution" of an organism and made a distinction between "somatic formulae" and "gametic formulae." Accompanying Hurst on many of his public appearances was a set of rabbit skins demonstrating the inheritance of various coat colors and markings. Over the years, Bateson and many others came to borrow these skins to illustrate their talks, and the set of skins traveled the land.

Royal Society Report

Although formally received on May 18th, 1904, the RS did not publish the second report (*RS-Report 2*) on "Experimental Studies in the Physiology of Heredity" until 1905. In the interim, various notes were added, the last being dated February 1905. The delay was partly due to caution on the part of the RS, since there had been many costly late corrections and additions to the first report, and partly due to obstruction by a reviewer. Regarding costs, the Secretary of the RS, Archibald Geikie, wrote (October 21, 1904): "I am desired to remind you that the corrections on the first report to the Evolution Committee amounted to as much as the whole cost of setting the manuscript in type." He requested that corrections be made before, not after, typesetting. It was originally determined that studies of Biffen and Hurst would be included, but by October, Biffen's patience had expired and he withdrew his contribution. The anonymous reviewer's criticisms were not so much over substance as style:

> The literary composition of these bulky memoirs seems to me below the standard that might reasonably be exacted by the Royal Society. An unnecessary burden is thrown on the reader by their cumbrous, confused, and imperfect presentation of objects, arguments and conclusions. The handwriting shows that some part of these faults is due to off-hand composition, without adequate revision.

Bateson replied to Geikie (October 23, 1904): "In so far as they relate to other considerations, the comments of your referee shall have such attention as they appear to merit." In private correspondence to Bateson, Hurst was less restrained (October 25):

> I was astounded to read the copy of the Sec's R.S. communication: I am almost sure that the phrasing of the referee is that of K. P. and after seeing him for the first time at the B.A., I can quite imagine his indulgence in such superior sneers. It is however somewhat strange

> that the R. S. has, through its self-appointed critic, only just discovered a lack of literary polish in certain biological memoirs! I suppose henceforth we must devote ourselves entirely to fine writing, and facts and arguments will be a secondary matter. On this basis the work of such men as Grant Allen and Henry Drummond will be infinitely preferable to that of Charles Darwin! I wonder if the literary composition of the 'Origin' would have passed the referee of the R. S.? … I am fully conscious of my lack of academic training in science, but in regard to literary composition I am not aware that excellence is wholly confined to the Schools!

RS-Report 2 was in two parts, the first by Bateson, Saunders, and Punnett was on plants, concluding with a section on poultry. There was then a summary with discussion of recent work from other laboratories (Coutagne with silkworms in France, Cuénot with mice in France, and Castle with guinea pigs and rabbits in the USA). In the second part, Hurst described his own experiments with poultry that had begun in 1901.

The report contributed much to the ongoing debate on hybrid sterility and the role of chromosomes. What later became known as linkage was again referred to as "coupling" of characters, and reference was made to "recombination of characters." There were still heavy demands on readers, such as "blue Andalusian gives off gametes, black and white splashed." Perhaps reflecting lingering thoughts on homotyposis (Chap. 6) and anticipating later work on somatic mosaicism, there was an inconclusive discussion of "the relation of *gametic* differentiation to the normal differentiation between *somatic* parts of the same body."

The wide-ranging nature of characters amenable to Mendelian analysis was evident: "In fowls … there is no reasonable doubt that such features as late or early feathering, shortness of wing-quills and tail, peculiar qualities of voice, forms of constitutional weakness, follow rules closely similar to those detected in features more amenable to critical study." There were also indications that "such features may be gametically coupled with others more easy to deal with, for example, certain types of pigmentation." Hurst in his part of the report described "5560 observations" on crosses between "four pure breeds" of poultry, resulting in "close approximation to the Mendelian expectation." No correlation (i.e., coupling) was found between characters indicating that each was "a unit character with an independent inheritance." These extensive and expensive experiments were self-funded, but Hurst's wealth had its limits and breeding was discontinued after the F_2 generation.

Ben Battle

In the pre-Mendelian era, Hurst had once gathered measurements on schoolchildren for Pearson. Referring to those days, Hurst later wryly remarked: "He trusted my judgement then and published the results." Still just "pre-Mendelian," in March 1900 Pearson submitted the eighth of his long series of papers on "Mathematical Contributions to the Theory of Evolution," which dealt with the inheritance of coat color in horses and of eye color in man, using the Galtonian approach [17]. In 1903, he was to conclude that "nothing corresponding to Mendel's principles" had emerged in these studies [18]. By virtue of his own work on horses and human eyes, Hurst became Mendel's, and hence Bateson's, major defender against Pearson.

Bateson, being a FRS, was Hurst's entrée to publishing under RS auspices. Bateson submitted the work on horse colors on Hurst's behalf on November 4th 1905. As Chairman of the Zoological Committee, Weldon arranged the printing. A postcard from Bateson to Hurst read: "Your paper comes at meeting Dec. 7, 4-30. Do come. K.P., W. and Co. are going to try conclusions once more, and will no doubt make a great effort." As the day approached, Bateson was in constant communication with Hurst. A notice in *The Times* stated that a general discussion on statistical methods was expected. The fateful day arrived, and the two met in London at an Austro-Hungarian Café near the British Museum. Rona Hurst described the scene [19]:

> Bateson was in a complete state of nerves and Hurst suggested a bottle of champagne as a celebration-plus-pick-me-up, but Bateson was horrified at such levity beforehand. Coming out they hailed a hansom and by a ridiculous coincidence ... the horse drawing it happened to be a liver chestnut (a very dark form). Bateson had never encountered one before and sitting in the hansom in close proximity to its dark chocolate tail, he became more and more upset, confusing it with the black tail of the bay. Hurst's explanations were in vain; when they got out Bateson insisted on the driver's giving him some of the hairs for Miss Durham to analyse, which she happily found to be truly chocolate and so recessive to black. ... The room was crowded. This had so come to be regarded as a great test between the rival biologists that everyone who could went to see the fun. Lord Rayleigh [Arthur Balfour's brother-in-law] was in the chair and as a physicist he was probably somewhat impatient with this biological quarrel.

Hurst was the last speaker. Mischievously, he gave his paper the same title as Pearson's paper of 1900 – "On the Inheritance of Coat Colour in Horses" [20]. He began by noting the impracticability of producing his own data by direct breeding experiments. So, he had adopted the approach taken by Pearson – the examination of pre-existing data in Weatherby's 20 volume *General Stud Book of Race Horses*. He noted that Pearson's "statistical methods did not disclose any intrinsic differences in the heredity of the several colour types." He then quoted the conclusions Pearson had drawn on the inheritance of eye and coat color in a 1903 paper in *Biometrika* [18]: "Nothing corresponding to Mendel's principles appears in these characters for horses, dogs, and men." Regarding horses, Pearson had added: "It is the same with every coat colour taken, its relative constancy depends largely on the extent to which it has appeared in the ancestry, and one by one, black, bay, chestnut, grey, must be dismissed by the Mendelians as neither 'recessive' nor 'dominant,' but as marking 'permanent and incorrigible mongrels'." To this, Hurst responded: "A careful examination of the *Stud Book* records so far fails to give any support to Professor Pearson's statement: on the contrary, the records show clearly, for instance, that bay and brown are Mendelian dominants to chestnut, which is recessive." Rona's account continued [19]:

> No sooner had he finished than the quibbling began which rapidly developed into a heated discussion. Weldon denied the whole thing, giving many examples from the stud books to back up his statements, which it was quite impossible for Hurst to refute without recourse

> to the volumes involved. Bateson, to illustrate a point, brought out a fowl-skin. One of the biometricians appealed to the Chairman that since the question was on horses they should stick, and not bring in poultry. Rayleigh called Bateson to order on this point, which provided the last straw. He completely lost his temper, slammed the skin into his bag, said he withdrew the paper and returned to his seat.
>
> Hurst, however, was not going to give in like this. It was his first appearance at the Royal Society and to him it was no more alarming than many experiences he had had on political platforms. He firmly stood his ground and gave the biometricians direct denial, defending the cause so stoutly that Weldon calmed down and accepted the offer that more facts should be produced on paper by both sides to ascertain the correct position.
>
> Hurst said afterwards that the only clear remembrance he had was a view of the shocked, strained faces of Miss Saunders and Punnett as Bateson walked off, which gave him [Hurst] an added stimulation to hang on. Going out to get his coat, he encountered Karl Pearson who was furious at this behaviour from his one-time colleague. 'One thing, you shall never be a Fellow here as long as I live,' he said, a vow that was faithfully kept, for although Hurst was put up after the war there was always this opposition to prevent it in spite of all the efforts of his many powerful friends.

Years later, Punnett recalled the tramp back to the Herringhams in Wimpole Street [21]:

> Together with Miss Durham we set out on foot for somewhere north of Oxford Street where we were due to dine with the Herringhams. It had happened that at this same meeting Bateson had read our communication on the 9:7 ratio in the sweet pea and had illustrated it with a large hanging diagram. So through the length of Bond Street he marched, grim and silent, shouldering a 6 feet role of paper with Miss Durham and myself trotting behind him in a state of mixed apprehension and amusement. Dinner that evening was a glum affair.

Bateson determined that a more intensive examination of the stud books was required. He expressed his high admiration for Hurst, writing from the Herringhams' (December 7): "You have pluck enough for anything! – and today that quality, thank goodness, came to our relief. We were simply filled with admiration of the tone of your answer. I felt utterly collapsed." Hurst replied (December 9).

> My only regret is that I have been the occasion of getting you all into the mud. I am still of the opinion however that we were in no sense beaten, and at the worst it was a drawn battle. Frankly I do not believe that they can produce the figures presented on Thursday! In the first place, I regard it as a physical impossibility to have got them in the time *critically*. The Stud Book is full of pitfalls for the unwary, and records got out in a hurry will not stand the test of time, only those who have worked at the S. B. for months can appreciate the difficulties. It was of course impossible to follow Weldon's figures and answer them on the spot and I felt that all that could be done was to hold them at bay for the time. I was afraid you would think me rough and blustering, but their constant interruptions and endeavours to silence me, set my back up! After all, what did they do? They simply produced a few more exceptions which appeared numerous because they left out the bulk. ... In view of Thursday, I have already got into the collar and have started to make a thorough analysis of the Stud Book. It means a grind but I do not mind, and if it takes me 20 years, I will now see it through! I will look up Ben Battle for you at once, but I want to do the thing systematically and not sporadically as they have done.

Weldon's argument had centered around a horse named Ben Battle and whether it was chestnut colored. If a true chestnut, then it should have been a recessive and hence bred-true, producing only chestnuts when crossed with chestnuts. Rona Hurst relates: "It was now a case of all hands to the plough. Mrs. Bateson, Miss Saunders and Punnett all took a share. Another set of studbooks was bought for them to use at Cambridge, while Hurst and his sister worked hard at Burbage." Bateson took a break (December 11) and accompanied by Florence Durham spent a week in Nancy visiting galleries, museums, and theaters; but their main purpose was to tap Cuénot's expertise on coat colors. Bateson wrote (December 14):

> Saw Cuénot and his mice today. He is very firm that pure yellow is dominant over agouti [grey]. We could not shake him about this. F. has not tried it. The uncomfortable thing about it is that if this is true, chestnut may dominate sometimes in horses? Cuénot is a rather dull man. He shows all his stuff, but did nothing for our entertainment. He talked of showing us the town, though he seemed rather relieved when we did not insist on his escort. It is a good thing we saw him and got clearer as to his statements and ideas. ... Your post cards {3} just come. B. Battle perhaps is one of these blackish chestnuts. It is rather provoking.

They turned homeward (December 16) by way of Brussels, where Florence advised him on a suitable present for the boys. There was then a Christmas break with Anna at Bashley, perhaps accompanied by the stud books, and then a further meeting with Hurst. On December 30th, Bateson sent a telegram to Hurst: "Ben Battle was bay or brown, am writing [later]." This was followed by a letter. A certain Mr. G. H. Verrall of Newmarket, handicapper to the Jockey Club, had written (December 29) pointing out that, although Ben Battle was recorded as a chestnut in the Stud Book, he appeared in the Racing Calendar as a bay or brown, and so far as Verrall ever knew never ran as a chestnut. More evidence followed in support of this. Bateson was exuberant adding as a post-script "P. S. Isn't Ben Battle ripping?" Rona commented [19]:

> Poor Bateson – he was certainly being initiated into all the complications of coat-colour with which Hurst was only too familiar having been born and bred in a county seething with horses of every type, and which had made him stick to the purely recessive character of the chestnut, the absence of black points, in the same way that in the eye colour he was only dealing with the presence or otherwise of pigment in front of the iris, refusing to be drawn into all the multiplicity of the actual colour, obviously due to a considerable number of factors.

At the next RS meeting (January 18, 1906), Weldon formally presented his criticisms in the form of a paper ("Note on the Offspring of Thoroughbred Chestnut Mares") but admitted that chestnuts bred true within a small percentage of error [22]. Bateson re-presented their horse paper which was approved [20]. Hurst later wrote to Bateson: "I am perfectly satisfied with Thursday's battle, tho' of course I longed to be in it (being such a bloodthirsty creature!) yet you did so well, there seemed no need for me to intervene. We probably stayed up too late on Wednesday, but it is a good thing we did!" Bateson later summarized the matter in his book *Principles*:

> The key to the phenomena was of course the fact that chestnuts mated with chestnuts breed true – with rare and dubious exceptions. It would seem at first sight impossible to devise a system of tabulation which could fail to disclose so prominent a feature. Nevertheless, Professor Pearson's correlation tables, which were compiled from the records of more than 6,000 horses, were made in such a way that the colours of the sire and dam could not be

taken into account *together*. ... Since however the colours of both sire and dam are recorded, and must indeed have been actually extracted from the Stud Book for the purposes of the tabulations, the investigators, by refraining from an inspection of these data till they had been separated, placed themselves at a gratuitous disadvantage. The true nature of the inheritance was therefore not discovered.

Hurst was able to add a note (January 31, 1906) to the paper he had submitted in November. Here, he concluded that "the distinct properties of chestnuts must be ascribed to segregation rather than ancestry, seeing that their behaviour in heredity is entirely different from that of bays and browns, though their ancestral composition may for several generations have been the same." He went on to cite evidence from a paper that Weldon had mentioned at the January 18th meeting, which "gives as I now find, extensive tables drawn from German sources, showing that within a small margin of error chestnut (*fuchs*) breeds true," as would be expected when recessives are interbred.

Weldon returned to the studbooks with increased vigor. They accompanied him to Rome where he went for a holiday, and to an inn in England at Easter. Letter after letter from Weldon to Pearson was "filled with Studbook detail." On April 13th, "overworked and overwrought," Weldon died following what appeared to be an attack of influenza (see Chap. 9). Hurst's horse coat color paper was published the same day [20]. He sent a reprint, and a short letter of condolence, to Pearson and got a sharp reply. Hurst wrote to Bateson (May 9):

I received an extraordinary reply in which he states that 'only a few days before his death he (Prof. Weldon) condemned in stronger language than I have ever heard him use of any individual the tone and contents of the Note added to your paper. It is a judgment in which I believe every man who has the interests of science at heart will concur.' What do you think of that? So far as I remember I wrote simply a few words of regret at the sad occurrence, just as I really felt, but it seems that it was foolish of me to write at all! However, I feel that I did my duty, however quixotic it may have been. Does it not suggest that they may have been rather hardly hit? Well, they asked for it on Dec. 7th, did they not? We did not complain then!

Bateson knew better than to write to Pearson. Three years later, he summarized the Biometricians' defeat in *Principles*:

Of the so-called investigations of heredity pursued by extensions of Galton's non-analytical method and promoted by Professor Pearson and the English Biometrical school it is now scarcely necessary to speak. That such a work may ultimately contribute to the development of statistical theory cannot be denied, but as applied to the problems of heredity the effort has resulted only in the concealment of that order which it was ostensibly undertaken to reveal. A preliminary acquaintance with the natural history of heredity and variation was sufficient to throw doubt on the foundations of these elaborate researches. To those who hereafter may study this episode in the history of biological science it will appear inexplicable that work so unsound in construction should have been respectfully received by the scientific world. With the discovery of segregation it became obvious that methods dispensing with individual analysis of the material are useless. The only alternatives open to the inventors of those methods were either to abandon their delusion or to deny the truth of Mendelian facts. In choosing the latter course they have certainly succeeded in delaying recognition of the value of Mendelism, but with the lapse of time the number of persons

> who have themselves witnessed the phenomena has increased so much that these denials have lost their dangerous character and may be regarded as merely formal.

Spillman, who had now shifted from wheat to animal studies, wrote (June 9, 1906) to congratulate Hurst on the horse paper: "These results of yours with horses will be of very great service to me in my lecture work this summer. Some of our biologists are a little incredulous concerning the data I have presented on this question, and your work makes it easier for me."

Third Genetics Conference

Hurst was married on June 6th 1906. At the end of the month, he read a paper at the Third Conference on Hybridization and Plant-Breeding [23]. An "outward" character was equated with a "zygotic" character. The latter character might be seen in the adult as the result of a single "determiner" having been transmitted by gametes when the zygote was formed: "In … simple cases the outward or zygotic character of a pure plant or animal is presumably represented in the germ-cells or gametes by a single factor or determiner. In other cases, however, the zygotic character, although apparently simple, is really compound, being represented in the gametes by more than one factor."

Having spoken so often of "unit characters," it was only a matter of time before the term "unit factors" (for Mendel's "elements … that determine") appeared, as it did repeatedly in his paper. As examples of "compound characters," he noted that "red flower-colour in sweet peas and stocks is due to the association of two gametic factors, purple colour to three factors, while hoariness in stocks has been shown … to be due to no less than four distinct gametic factors." The compound nature of these characters was usually not immediately obvious, requiring Mendelian analysis for its demonstration:

> The coloured coat of the rabbit is due to the meeting of two distinct gametic factors, one of which may determine the presence of the pigment, while the other determines the colour of the pigment. If, for instance, C be present, the animal will be coloured, if absent it will be white; if both C and G be present it will be coloured grey, while if C and B be present it will be coloured black. …

He then turned to presence-and-absence ideas:

> Mr. Bateson has suggested that in such cases the coat-colours may be due to at least three pairs of gametic factors, viz. (1) presence (C) and absence (c) of colour; (2) presence (G) and absence (g) if grey; (3) presence (B) and absence (b) of black; presence being dominant and absent recessive in each case. In that case the gametic formula of the 'Belgian Hare' would be (C + G + B), and that of the 'White Angora' would be (c + g + B), both being homozygous in B.

In all cases, "presence" and "absence" referred to presence/absence in/from *gametes*. Trying to extend this scheme to peas, he wondered:

> Whether many other of the apparently simple Mendelian characters are not also compound in their gametic constitution. ... In cotyledon colour in peas, might not the character pairs be really presence and absence of yellow on a basis of green, rather than the contrasting yellow and green? Is it not possible that many of the so-called contrasting pairs of Mendelian characters are really compound, and that the true unit-characters are simply presence and absence? ... The Mendelian contrasting pair, yellow and green, might be regarded as presence and absence of pigment on a basis of green. On this view, the characters yellow and green would belong to two distinct pairs instead of one as Mendel supposed, and these would be presence (Y) and absence (y) of yellow, and presence (G) and absence (g) of green, presence being dominant over absence. The gametic formula of the pure-breeding yellow pea *based on* green would, on this view, be (Y + G), and the zygote yellow owing to dominance. The gametic formula of the green pea would be (y + G) and the zygote green.

Here, the Mendelians, failing to think in chromosomal terms, seemed to be sowing confusion regarding the determining elements that were either allelic (one on the maternally-derived chromosome and one at the same position on the corresponding paternally-derived chromosome) or nonallelic (distributed along a chromosome in different positions, or on different chromosomes). Hurst, who had thought chromosomally from the start, should have known better.

Shull

It was perhaps Hurst's clarity that made it easier to attack his work than Bateson's. Commenting in 1909 on Hurst's account of the "presence and absence" hypothesis, George Shull at Davenport's Station for Experimental Evolution, who had studied under Tower in Chicago (Chap. 11), noted [24]:

> In a number of cases the presence and absence could be read quite as well backward as forward, and it will doubtless be impossible in many cases to decide which is the positive character and which its absence. Thus in the contrast between a yellow and a green pea, the yellow is described as present in the yellow pea and absent in the green pea. What is to hinder us from describing the green as present in the green pea and absent in the yellow one?

Furthermore, Hurst often implied that for a character to be designated the "presence-character," it should also be the dominant character, but Hurst's own data did not support this. "Both Bateson and Davenport appear to have tacitly agreed that the dominance of absence over presence is a difficulty for the 'presence and absence' hypothesis, but both have taken occasion to explain that what appears to be the absence of a character may really be the presence of a positive inhibiting factor."

In a lecture (May 4, 1909) to the RHS [25], Hurst's solidarity with Bateson was signified by increased usage of "we," but not in a royal sense:

> We regard tallness ... as due to a definite germinal factor *present* in the tall pea, but on the other hand we regard dwarfness as simply due to the *absence* of the tall [germinal] factor from the dwarf pea. Thus in the *presence* of the tall factor the pea is tall, while in its *absence* the pea is dwarf. Tallness appears to be dominant simply because it is present, and in its absence the seemingly recessive character is manifest. We prefer, therefore, to regard 'pres-

> ence' and 'absence' of tallness as the two contrasting characters, rather than tallness and dwarfness. At first sight this may appear to be a distinction without a difference. But in reality the difference is important and promises to lead to far-reaching consequences, for it means that each hereditable factor is a unit that may be distinct in its inheritance from all other factors.

The presence and absence concept was illustrated with a diagram (Fig. 10.2). While seeming to dig himself into a deeper hole, it is of interest that in his diagram Hurst represented individual organisms as rectangles and gametes as circles. Within both, the germinal factors were represented as rod-like bodies, which may have been his way of silently recording his adherence to chromosome theory. "Presence" was simply recorded by making the interior of the rod black. "Absence" was recorded, not by removing the rod and leaving a space, but by leaving its interior white.

Despite absence, there was still *something* there. This we may see as Hurst's pictorial representation of Bateson's "residue" or "base." Shull took up the "residue" idea in a 1910 paper [26]:

> Since we are discussing the Mendelian process as one of germinal analysis it is appropriate to discuss … the 'insoluble residue.' … Aside from … cases which show a distinctly non-Mendelian mode of inheritance, it must be remembered that Mendelian analysis can be made only in the presence of differential unit-characters possessed by individuals *capable of life and of sexual reproduction*, and that therefore, there can be no test, except under rare circumstances, of the Mendelian nature of the more fundamental vital characters. This leaves it an open question whether the whole of the germ-plasm is a complex of such genes as those which give rise to the phenomena of unit-characters, or whether, with its wonderful general powers of assimilation, growth and reproduction, it consists of a great nucleus of which the genes are relatively superficial structural characteristics.

A later listing of definitions by Shull stated [27]:

> Presence and absence hypothesis. The hypothesis that any simple Mendelian difference between two individuals, results solely from the presence of a factor in the genotype of the one individual, which is absent from that of the other. Presence and absence of unit-differences, as a convenient *method* of describing the results of genetic experiments, should be carefully distinguished from the presence and absence hypothesis. The *method* is purely objective and entirely free from hypothetical implications.

A "unit-character" was also defined: "A character or alternative difference of any kind, which is either present or absent, as a whole, in each individual, and which is capable of becoming associated in new combinations with other unit-characters." From his studies with the red campion (*Lychnis dioica*) Shull concluded [28]: "Mutation … depends upon reversible modifications of some permanent element or organ, rather than upon the origination of a new unit, and its disappearance. This interpretation bears both upon the nature of mutation and upon the real significance of the 'presence and absence' hypothesis."

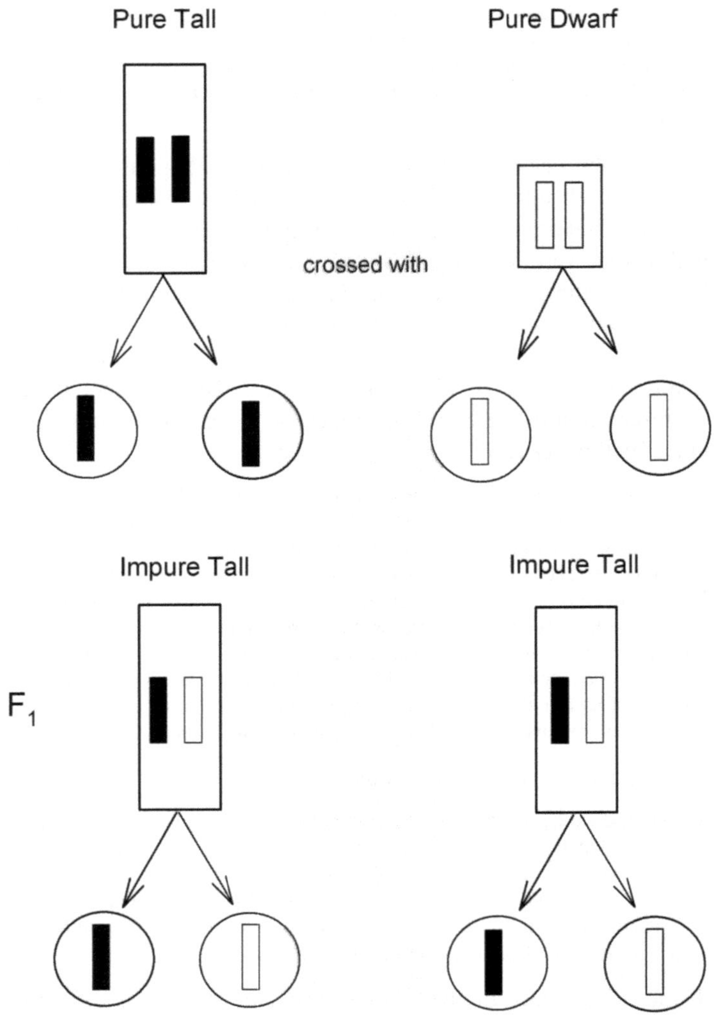

Fig. 10.2 Hurst's representation of presence and absence in peas. Adult parental plants are either tall (large vertical rectangle containing two small rods filled in black) or dwarf (small vertical rectangle containing two small rods filled in white). Designated "pure," each parent is homozygous for the elements determining the height character, which are indicated by the paired rods, one being derived from the grandparental father, and one being derived from the grandparental mother. When gametes (circles) are produced, the two rods segregate to different gametes, which thus correspond to either the grandfather's determining element or the grandmother's determining element. Since the parents are homozygous, the gametes of each parent have the same height-determining elements. The gametes from the tall and dwarf parental plants meet to produce child plants (F_1) that are all tall (large rectangles) – tall being dominant over dwarf. Since the children contain different determining elements (i.e., they are heterozygous for these elements) Hurst designated them as "impure." They, in turn, each produce gametes of two types in equal numbers that would be expected to generate an F_2 generation (not shown) with, on average, three tall children and one dwarf child

The Gene

One of Schull's papers, submitted in December 1908 and published in 1909 [24], displayed an early adoption in the English language of the word "gene" (Greek: "*gennao*" = to breed), which had been introduced in Danish by Wilhelm Johannsen as denoting "an internal something or condition upon whose presence an elementary morphological or physiological characteristic depends." In a review in 1911 [29], Johannsen, after acknowledging the initiation of "the genotype conception of the present day" by Galton and Weismann, began by commenting on language:

> It is a well-established fact that language is not only our servant, when we wish to express – or even to conceal – our thoughts, but that it may also be our master, overpowering us by means of the notions attached to current words. This fact is the reason why it is desirable to create a new terminology in all cases where new or revised conceptions are being developed. Old terms are merely compromised by their application in antiquated or erroneous theories and systems, from which they carry splinters of inadequate ideas, not always harmless to the developing insight. Therefore I have proposed the words 'gene' and 'genotype' and some further terms, as 'phenotype' and 'biotype' to be used in the science of genetics. The 'gene' is nothing but a very applicable little word, easily combined with others, and hence may be useful as an expression for the 'unit factors,' 'elements,' or 'allelomorphs' in the gametes, demonstrated by modern Mendelian researches. A 'genotype' is the sum total of all the 'genes' in a gamete or zygote.

Unwittingly, Johannsen created a future problem for "developing insight" in that he limited the genotype to genes. We now know that our DNA contains much more than conventional Mendelian genes, and often "genotype" and "genome" are used interchangeably as indicating the total store of information contained in DNA. The term "phenotype" has better withstood the test of time:

> All 'types' of organisms, distinguishable by direct inspection or only by finer methods of measuring or description, may be characterized as '*phenotypes*.' Certainly phenotypes are *real things*; the appearing … 'types' or 'sorts' of organisms are again and again the objects for scientific research. All typical phenomena in the organic world are *eo ipso* phenotypical, and the description of myriads of phenotypes as to forms, structures, sizes, colors and other characters of living organisms has been the chief aim of natural history.

While the word "genotype" was readily adopted, there was some confusion about its meaning. In 1912, Schull confessed [30]:

> As one of the chief offenders, I wish to publicly repent my misuse of the term and to heartily join the movement to limit the word 'genotype' as used in the literature of genetics, to the fundamental hereditary constitution of an individual. The use of this word both for the hereditary constitution and for the group of individuals possessing an identical hereditary constitution, will lead to much confusion if continued. The word which Dr. Jennings says is much needed 'for a concrete, visible group of organisms' that are 'all with the same hereditary characteristics,' has already been supplied … by Dr. Johannsen, and his word 'biotype'.

Shull was also concerned with the distinction between "pure line" and "clone":

> Genotype, the fundamental hereditary constitution or combination of genes of an organism. Biotype, a group of individuals possessing the same genotype. Pure line, a group of individuals traceable through solely self-fertilized lines to a single homozygous ancestor.

> Clone, a group of individuals of like genotypic constitution, traceable through asexual reproductions to a single ancestral zygote, or else perpetually asexual.

Shull's equation of "fundamental hereditary constitution" with "combination of genes" sufficed for much of the twentieth century. He subsequently amended his definition of clone by omission of "of like genotypic constitution" (although we now know this often applies), since he wanted it to be "a purely genealogical term and involving no implication whatever as to the genotypic equality of the individuals included in the single clone."

But new words, presumably on the advice of scientists who might not necessarily have been geneticists, were now beginning to enter dictionaries, such as Funk and Wagnall's *New Standard Dictionary of the English Language*. In 1914, Shull sent Bateson a draft paper where he amended the dictionary definitions to forms that he believed would "meet the approval of most geneticists." Bateson replied (March 7, 1914):

> 'Gene'? for instance, although claimed non-committal, seems to me *suggestive* of particles! I prefer 'factor,' though it has a nasty smack of mathematics about it. … *Genotype and Phenotype* are rather good and useful words. I have not used them, but I think I shall. I don't really like 'gene.' It is not well-formed. Much better stick to factor, a word of common speech, until we begin to know what sort of things they are physiologically. I often suspect they are matters of *arrangement* and not of the actual materials in the usual sense. *Genotype.* Is your definition right? I take the word to mean an organism (or group of organisms) considered in reference to its somatic characteristics.

The latter revealed continuing confusion in Bateson's mind. He then turned to one of his favorite words:

> *Segregate.* Ha ha! If genes are not corporeal particles how do they come to manifest 'independent movements'? I had not read this before. Doesn't it suggest that 'gene' gives a wrong picture? Perhaps it would be better to say 'The process by which the members of an allelomorphic pair are separated at or before the formation of gametes'.

Shull replied (March 14, 1914):

> I do not see any reason why 'gene' suggests to you a particle, since it has been so often defined in such a manner as to specifically deny this limitation. I also like 'factor' very much and use it frequently. The only difficulty I found with the word 'factor' is, not its 'smack of mathematics,' but the fact that the word 'factor' alone is often not sufficient, since there are many factors which are non-genetic. To be perfectly clear one must use the phrase 'genotypic factor' or some equivalent modifier, but the meaning of the phrase is rendered exactly by the single word 'gene' if the latter is kept to the non-committal definition with which it was launched.

He then added a comment on Bateson's idea of a gene as an "arrangement," a point Bateson was to repeat later in Australia (Chap. 13):

> I, too, have been inclined to think rather of *arrangements* than of independent material particles in connection with the genotypic factors or genes, the 'figures of speech' which have most frequently come to mind in connection with them being 'side chain,' 'polymerisation,' 'isomerisation,' but I have just as resolutely refused to adopt these as I have material particles and entelechies, feeling that the only safe attitude is the non-committal one.

Although neither of them knew it, they were both writing about DNA, which we now know to be a polymer formed by the sequential joining of four bases (as nucleotides), thus constituting an *arrangement* of bases (Part VI). Shull then turned to Bateson's seeming misunderstanding of the genotype–phenotype distinction:

> The words 'genotypic' and 'phenotypic' are not properly applied to groups of individuals, but only to hereditary constitutions of those individuals, the 'genotype' being the actual constitution, the 'phenotype' being that part of the genotype which can be inferred by the inspection of an individual. The genotype is not supposed to be completely analyzable [by methods then available], but partakes of the same non-committal qualities as the word 'gene.' 'Biotype' is the correct word to use for what you proposed to let the word 'genotype' represent, i.e. a 'group of organisms considered in reference to its genetic constitution.'

Finally, there was segregation:

> I have laughed with you concerning the definition of the word 'segregate,' but I do not agree with you at all that my expression 'independent movement,' implies that the genes are *corporeal particles*. If it does, some other figure of speech should be inserted in place of 'independent movement,' since I had no such idea in mind when using these words, and could have used 'factor' in the same connection, quite as well as 'gene.' Are not loci capable of independent movement in the language of mathematics? Vortices and waves are likewise capable of independent movement, and yet by such independent movement they do not prove themselves to be 'corporeal particles'.

In a letter to Morgan (May 31, 1914), Bateson continued to protest: "If we must have one of these derivatives, why on earth not gens simply? Why 'genes'? It merely looks barbarous and is impossible to pronounce. Already I hear people saying eugenic." In his 1914 Croonian Lecture [31], Morgan's colleague, Edmund Wilson, used the word "gen" without the "e" at the end.

Although originally meant for *Science*, Shull's definitions were eventually published in *The American Naturalist* in 1915 [27]. His definition of "factor" was: "An independent inheritable element of the genotype whose presence makes possible any specific reaction or the development of any particular unit-character of the organism which possesses that genotype; a gene or determiner." And "gene" was: "An element of the genotype; a genetic factor; a determiner." On the other hand, "genotype" remained as: "The fundamental hereditary constitution or sum of all the genes of an organism."

The Pack

Apart from Hurst and the female workers (Chap. 8), there were others in "the pack" who, while seldom baring their teeth, in their various ways played important roles. Of most immediate importance was Punnett, whose book *Mendelism* (1905) ran into several editions with translations in many languages [32]. Nevertheless, *Mendelism* received a hostile review in *Nature* by Darbishire [33]. Punnett introduced a way of expressing the interactions of parental gametes through a two-dimension matrix ("the Punnett square;" see Figs. 10.3 and 10.4). Since they met on a day-to-day basis, there was very little written correspondence between Punnett and Bateson. After giving an address on "Mendelism in Relation to Disease" at the RSM in 1908, Punnett was asked by Yule why, if brown eyes were dominant over blue, they were not more prevalent in the population. This led Punnett to ask his fellow cricketer, the mathematician Godfrey Hardy (1877–1947), the same question. In a twinkling, Hardy came up with what is now known as the Hardy–Weinberg equation. *Nature* declined to publish, perhaps because "Mendelism" was in the title, but the journal *Science* was more accommodating [34, 35]. Yule had implied that when one character was dominant there would be a 3:1 equilibrium in the population. Hardy said no.

Yule was impressed. He commented on this "lapse on Pearson's part" to Punnett (April 16, 1908):

> Absolutely correct and solves the whole difficulty! I am kicking myself for never having seen it. The fact is that in my New Phytologist articles of 1902 I took the population as starting from a single *D* x *R* cross, which gives the 1:2:1 ratio as stable among the descendents, and *it never occurred to me that any other ratio could be stable*. It's extraordinary how stupid one can be. Pearson also follows in the same track in his Phil. Trans. paper 'On a generalized theory of alternative inheritance,' and so *he* also gets the 1:2:1 ratio.

Next to Hurst, Bateson's most prolific correspondent was Robert Lock (1879–1915). In 1902 (April 11), Bateson wrote to Beatrice from Tours concerning: "My little

Fig. 10.3 Reginald Punnett and William Bateson, 1907

Fig. 10.4 Robert Heath Lock (1879–1915)

companion … [who] knows no French of any kind, and when I say *no* French, I mean it. 'What is French for 'to eat'?'. … German, he reads with fair ease. He has a prodigious memory for verse; knows most of Hamlet, Macbeth and Gilbert and Sullivan by heart, and can recite more ephemeral verses than I ever heard of."

In August, having passed first class in the Cambridge Natural Science Tripos, Lock obtained a Studentship to work for 2 years at the Royal Botanical Gardens at Peradeniya in Ceylon (now Sri Lanka). On arrival, he found that J. C. Willis, the Director, "after all his fine talk of acres of land and hordes of coolies, has greatly reduced his labour [force] … and gone off without leaving any instructions for my assistance." When Willis returned, it was Lock's task to update him on genetics (May 28, 1903): "Willis arrived here about a week ago and I am staying with him for the present and am starting to educate him up to the proper level of Mendel! He has begun on your book, and on Castle."

Like Bateson two decades earlier, Lock had his home support staff, notably his father, who kept him supplied with the latest books and journals, and collated materials for job applications when opportunities arose. Despite his academic exile in the colonies, Lock had the time, and a command of German, that allowed a close and careful reading of the relevant literatures. The correspondence with Bateson was by no means one-sided (but Bateson's letters have not survived). Lock was often apologizing for trespassing on Bateson's time and for spelling errors (which Bateson seems to have pointed out). To Bateson's consternation, Lock commented positively on de Vries' *Die Mutationstheorie* (April 16, 1904):

> De Vries gets spendider and splendider! And I look forward to the final part with bated breath. Anyone who reads the book with understanding must see that he has completely cut away the ground from beneath the feet of the Weldon-Pearson School. What can they bring forward against his host of cases where selection isolates 'new' forms already existing? Do you not begin to believe in the genuineness of mutation in Oenothera?

Lock was also carefully reading the works of Darwin and an earlier work of de Vries: "I suppose you have noticed Darwin's case of first-cousin marriages (like alkaptonuria) *Animals & Plants* volume I, page 460." And "Has anyone noticed Darwin's case of 'mutation' (?!) on p. 246 *Animals & Plants* volume II in geranium? It looks an exactly parallel case to Oenothera." And "I have been dipping into de V's 'Intracellular Pangenesis' lately; it is wonderful that he should have published in 1889 a hypothesis which explains all his subsequent facts, but I fancy you don't sympathise with pangenesis."

In principle, the correspondence was all that a free and open scientific correspondence should be. Lock gave fair warning concerning a paper he was writing (March 23, 1904): "I am going to support de V. in my paper unless you annihilate me first." Then, having received Bateson's reply, he responded (May 2, 1904): "I am squashed, but I can still wriggle, and I still think that de V's account explains the facts so far known for plants." There was also considerable appeal to underlying theory (June 12, 1904):

> What I want to hear is a case of resolution which it is quite certain that no twisting of dominance can fit into the scheme of a single gametic segregation of characters in the proportion 1:1. Because any other form of segregation seems to me to present enormous theoretical difficulties. Thus Weismann (don't think I have any faith in him) shows a reduction to ½ of 'ancestral germ plasms' to be an absolute necessity in formation of gametes. You have only to write 'Anlagen' for germ plasm and you have practically Mendel – assuming the Anlagen of one kind all move together (which is indeed assuming the essence of Mendel). ... I am not sure whether you think in terms of 'Anlagen.' Correns, de V., and Tschermak all seem to do so and I find it impossible to avoid such assistance [from theory]. I fancy this process of thought renders de V's account of Antirhinum more obvious. The combined effect of 0 and 7 Anlagen is to give red flowers. You see I am incorrigible.

Shortly thereafter (July 26, 1904), he confessed to "see my error in the case of the term Compound Character. ... A discussion at 6 week intervals is sadly unsatisfactory. ... We should have had the whole thing straight in half an hour if we had been able to talk it over." Yet his thinking was still very much influenced by de Vries: "As to practically infinite subdivision of Anlagen at [the] maturation of germ cells, Weismann discussed the point very logically (essays pp 355-358) and I think his view can be made to fit very nicely with Mendel's. De V's view allows for a very large number of latent Pangens (Anlagen). This is all very theoretical, but I do not see how to avoid thinking in some such terms."

Lock sailed home in September 1904 by way of Japan and America. Letters from Bateson were waiting in San Francisco where: "They are all very excited here over de Vries' recent visit." In October, Lock wrote on Mendelism in maize in *Nature*, and he returned to work as Curator at the Herbarium in the Cambridge Botany School supported by a College Fellowship. In his influential book *Variation, Heredity and Evolution* [36], he supported the view that the Mendelian factors were located on chromosomes and warmly acknowledged his mentor:

> Adequately to acknowledge Mr. Bateson's influence upon these pages is a more difficult matter, and not the less so because I have deliberately refrained as far as possible from consulting him whilst the book was in course of preparation, in order that it might retain if possible some traces of individuality. It is therefore clear that he is in no way responsible

> for its deficiencies. But apart from the fact that I am conscious of having quoted his ideas at more points than could be possible acknowledged *seriatim*, I owe to Mr. Bateson both my first introduction to the science of genetics, and a continual fund of encouragement in the prosecution of studies connected with it.

The book, which was updated in successive editions, was less technical and easier to read than Punnett's, partly because Lock had tended to present Bateson's viewpoints without his caveats. It was recommended by Wilson in his lectures at Columbia University. Herman Muller, one of the famous student triad who worked in the "fly room" at Columbia University (Chap. 13), described it as more influential than even Wilson's lectures [37]. But the book seemed not to impact Lock's career prospects. There was a second tour to Ceylon (1908–1910) where, during Willis' absence, he was Acting Director. In 1908 (November 22), Lock wrote from Ceylon:

> I doubt the view that Dominant and recessive are *really* the presence and absence of something, though it is convenient to speak of them in that way. If you think of these things as particles there must be something in the Recessive to occupy the space corresponding to that occupied by the dominant particle? ... All the F_1 tobaccos I have seen are obviously intermediate in every observable character. In these cases the D and R particles are both taking a share in producing the visible result – and the mechanism is surely the same whether there is dominance or not.

The first edition of his book (1906) tended to toe the Batesonian party line. He left as "hardly ... capable of a practical solution" the question of a "residuum" that might be left if all the Mendelian characters were subtracted. In 1909, there was a much updated second edition, and in 1911, a third edition with fewer updates. The differences between the first and later editions indicated both his mentor's new awareness of Butler, and his own awareness of the importance of reproductive isolation as had been set out by Romanes and Gulick. Thus, in the later editions, there were references to Butler's main books on evolution, to the three volumes of Romanes' *Darwin, and After Darwin*, and to Gulick's studies of snails. A photograph of Galton was moved from an early chapter on "Biometry" (first edition) to a new chapter on "Eugenics" (later editions). The book was later hailed as demonstrating an early understanding of the chromosomal basis for linkage of characters that had been hinted at by Bateson in 1902 (Chap. 13), and for kindling the statistical ideas of Fisher (Chap. 9) [38].

From his remote location, in 1909 Lock applied unsuccessfully for the Directorship of the John Innes Horticultural Institute (Chap. 13). Early in the 1914–1918 war, he was Chairman of the government's Vegetable Drying and Fruit Preservation Committee but died of complications of influenza in 1915 at the age of 36. His wife, Bella, arranged for Doncaster to edit later editions of the book (1916, 1920). Lock had first met Bella in 2008 on the boat to Ceylon when she was visiting her colonial administrator brother, Leonard Woolf [39], who was later to marry Virginia, a daughter of Leslie Stephen (see Chap. 19). Thus, Lock was Virginia Woolf's brother-in-law.

Summary

The Mendelian movement in Britain was led by the imperious Bateson from academia, and by the unflappable Hurst from his Burbage nursery. Initially known for his orchid studies, with relatively minor external support, Hurst expanded his operations to a wide range of plants and animals, putting his knowledge of horses to good effect when interpreting data from racing stud books. Having long been involved with the RHS, he provided for Bateson both an important link to commercial interests and a shield against those who requested popular articles and lectures. Through his work on the coat colors of horses and the eye pigments of humans, he was able successfully to combat Pearson. Although he published less often and in less visible journals than Bateson, Hurst's papers displayed a greater precision of thought and a clearer terminology. He had learned from his own early experience that there was no necessary correlation between species as classified anatomically, and species as classified physiologically (with respect to fertility barriers). With Shull, he strove to disentangle the ideas associated with the "presence and absence" theory that Bateson and Punnett were advocating to explain the coupling of characters, and to pin them down on definitions. This effort was greatly aided by Johannsen's introduction of the terms "phenotype," "genotype," and "gene." Schull agreed that Mendelian analysis was limited, and that genes might be only superficial aspects of the genotype beyond which there might be "an insoluble residue." Other members of "the pack" included Bateson's ex-students, Punnett and Lock, both of whom wrote influential textbooks. Lock gave the works of de Vries a detailed reading and corresponded about them with an enthusiasm that Bateson did not match. His text, *Variation, Heredity and Evolution*, indicates an early understanding of the chromosomal basis for linkage of characters and statistical aspects of heredity.

Chapter 11
On Course (1907–1908)

In many well regulated occupations there are persons known as 'knockers-up,' whose thankless task is to rouse others from their slumber. ... If I have knocked a trifle loud, it is because there is a need.

William Bateson, Defence

He had knocked hard, some thought too hard, but certainly hard enough to attract attention. Many liked what they heard and wanted more. As word of Bateson's impending Silliman Lectures spread, invitations flooded in and the scope of his impending visit expanded. Davenport knew what needed to be said (January 3, 1907): "I feel that very much of the future of our science depends upon your attendance." Thus, it came about that one July day in 1907 Bateson was at Liverpool docks boarding the Carmania for New York. There was no need to wait for a favorable wind. Sail had long given way to steam. Bateson was no longer in irons. He was back on course.

Lull

But the events of 1906 had taken their toll. Bateson's schedule for the first part of 1907 appears less stressful. There was time for closer attention to the work of his junior colleagues. Prominent among these was Muriel Wheldale, who had worked on antirrhinums prior to her graduation in 1904 when she obtained a first-class degree in the Natural Sciences Tripos. She had a Bathurst Studentship from 1904 to 1906, and then, until 1908, lectured and demonstrated for Saunders who had taken a leave fellowship to spend more time on her research. In early 1907 (February 2), Bateson suggested the use of the terms "epistatic" and "hypostatic" in her paper on flower color that was to be read at a RS Meeting (February 21), and later (March 31) advised on the proof corrections.

A. G. Cock, D. R. Forsdyke, *Treasure Your Exceptions*,
https://doi.org/10.1007/978-3-030-92099-9_11

In the interim, he took a spring holiday in France. In a letter from Boulogne (February 17), he referred to the stormy channel crossing that Olga Zacharias, daughter of a Hamburg botanist, would have experienced. Olga lived with them for several months as the boys' governess, and she was succeeded around August 1907 by another German, Miss Roth. Perhaps, it occurred to Bateson that they might also help his own German. Soon he was collecting art in Lille (February 22). In a note added later (1928), Beatrice recalled that on hearing that a painting by Antoine Watteau was to be auctioned, Bateson wired her to come at once with some money: "I had just enough cash to obey and hurried off from disconcerted household and nursery. He met me at Bologne .... I was left in the hotel whilst the auction took place, for fear that my presence should 'put the price up.' He bought it for £7 but had to pay cash which emptied our pockets. We got to Calais and had to sit there in a dingy hotel till the Bank sent relief."

There was a visit to the British Museum (April 9) and dinner at the Herringham's, but Wilmot "was called to a case and we had a rather dismal evening." Bateson's St. John's colleague, the physiological botanist F. Blackman, accompanied him to Happisburgh, on the Norfolk coast (April 21): "I have been 3 times in the sea and feel much refreshed. Blackman is a pleasant companion." It was perhaps on this occasion that Bateson discovered the Hill House Hotel that was to become a favorite writing retreat. Then the lull ended. Newton died. Academic musical chairs began. But Bateson was off to America!

America

The Carmania departed on July 30th and docked at New York on August 8th. Bateson was met by Alfred Mayer who conducted him to the Hotel Lafayette, and they went for a walk through the woods at Cold Spring Harbor. Later there was a visit to the Biology Laboratory of the Brooklyn Institute. "My first lecture went well" (August 9). He was impressed at the scale of Davenport's operations: "They are doing Duck, Fowls, Canaries, Grylluss, Drosophila, Clover, Sunflowers, Pigeons, Oenothera, Sheep, Poppy, several Lepidoptera, several beetles, etc, etc, etc." Then on to the Marine Biological Laboratory at Woods Hole (August 13): "I only retain faint control over my movements. I am everyone's prey, being torn to pieces by my admirers. No exaggeration, I assure you!" By the 19th, he was in Boston for the beginning of the Seventh International Zoology Congress. He reported (August 20): "Today we are going to see Sargent's place. He is, you know, Miss Willmott's friend. ... On Saturday I am to go for the weekend to Osborn's place up the Hudson. Probably the 'Willmottisation' will be considerable there."

On August 23rd, Bateson addressed the section on Cytology and Heredity at the Congress. Jacques Loeb and Edmund Wilson had spoken the previous day, and it is likely that they were in the audience, as may have been Michael Guyer who was to speak later. Bateson's title was "Facts limiting the Theory of Heredity:"

> Stripped of all that is superfluous ... genetics stands out as the study of the process of cell division. For if we had any real knowledge of the actual nature of the processes by which a cell divides, the rest would be largely application and extension. ... We may behold its minutest visible details .... Two centers form: the parts collect round each. The two halves withdraw ..., and there are then two cells instead of one. The likeness of these two cells we call heredity; their difference we call variation. If the two cells remain constituent parts of one body, we may speak of their likeness as symmetry or repetition; and their points of unlikeness we then call differentiation.

Acknowledging that he was "speaking in a country where ... a great school of cytologists has arisen," he doubted that microscopy would provide a solution to this "phenomenon unparalleled in the physical world, like consciousness, a distinctive property of living matter. By no confection of chemistry or mechanical contrivance can we yet fit together a system which will dichotomise and grow, dichotomise and grow, repeating the process again and again as long as certain materials are supplied to it." After warning that care must be taken in too readily regarding chromosomes as "the bearers of [factors corresponding to] hereditary characters," he turned to "profounder questions" where "the results of experimental breeding are beginning to limit the problem," thus revealing "that the bodily characters may result from the transmission of distinct unit-factors," which "exist in alternative or allelomorphic pairs, of such a nature that only one member of any one pair can be carried by a gamete." However, returning to his presence-and-absence ideas (Chap. 10), he held that "Mendelian phenomena" could be expressed in terms of a "simpler system," where "allelomorphism may be represented as consisting essentially, not in the presence of separate factors for the dominant and for the recessive characters, but in the *presence* of something constituting the dominant character which is *absent* from the recessive gametes."

In these opening remarks, there was less ambiguity than usual regarding the word "factor" as denoting something transmitted in gametes to form a new organism, which then either did or did not display a corresponding "character." Because a recessive character was not displayed, it certainly simplified matters to regard the corresponding gametic factor as absent. Much rested on what was then meant by the abstract notion of a gametic factor. We now know that the difference between a dominant and recessive allele may depend on a single base "letter" in a DNA sequence (see Part VI). The entire DNA sequence is present, but if one base letter has been substituted for another, then it would not be incorrect to regard the allele responsible for the dominant character as absent.

Tied in with this, Bateson had slipped into the habit of using the term dominance in a purely operational sense meaning that only one of two characters was manifest, whether or not they were members of an allelomorphic pair. He had disregarded the possibility that there might be different systems underlying the observation. But he now made clear that certain pairs of characters believed to be allelomorphic alternatives to each other might actually be members of different allelomorphic pairs. Nevertheless, there could be some interaction between the two characters such that either a new "compound character" emerged, or one excluded the other. He pressed for adoption of his newly coined words "epistatic" and "hypostatic":

> Till lately we spoke of the relations between grey colour of the mouse to the black colour in terms of dominance. Those terms, strictly speaking, should only be applied to members of the same allelomorphic pair. We can best express the relation between the grey and black by the use of the metaphor 'higher and lower,' and I therefore suggest the term *epistatic* as applicable to characters which have to be, as it were, lifted off in order to permit the lower or *hypostatic* character to appear.

Like many of the terms Bateson introduced, these have endured. However, having adopted a presence-and-absence viewpoint, he was confronted with the fact that from the heterozygous organism (DR), containing dominant (D) and recessive (R) alleles, equal quantities of gametes separately conveying D ("presence") and R ("absence") would arise. On the other hand, from a homozygote (DD), containing dominant (D) and dominant (D) alleles, equal quantities of gametes separately conveying D ("presence") and D ("presence") would arise:

> Allelomorphism, as we are becoming more and more disposed to believe, consists in the separation of a positive something from the absence of that something: More correctly, perhaps, we should say that the thing which conveys a certain power segregates, leaving in that cell-division no representative of that power behind. This allelomorphism is the one fact of which we have the clearest proof. It may govern, as we have seen, features of the utmost diversity. What then is that allelomorphism? An essential phenomenon of cell division it is not: for in homozygous organisms the products of division are alike. Any theory of heredity must include and recognise both these two kinds of division in its purview. We seek vainly as yet for a scheme by which these two sorts of division may be represented.

This seeming inconsistency in the presence-and-absence hypothesis had made necessary the further postulate of "two sorts of [meiotic] division," for which there was no evidence. Bateson was getting into a deeper mess because in 1907 he had little notion either of the biochemical basis of the characters he was observing (Chap. 8), or of genetics as an informational science (Chap. 19):

> Most astonishing is the fact that the same systems of transmission should be followed by characters which, by whatever test they be judged, must be supposed to be most diverse in physiological causation. Naturally when we are dealing with changes in colour, for instance, or in the reserve materials of a seed, we surmise that the critical factor is a certain ferment, or rather, the power to produce that certain ferment. … The diversity of these ferments must anyhow be very great, and it seems very strange that all these multifarious potentialities should exhibit gametic allelomorphism. … Farabee has shown that the peculiar condition of the human digits in which the fingers and toes have only two phalanges each, behaves as a simple dominant. … I cannot at all readily conceive how any ferment or other transmissible substance can be supposed to be responsible for such a variation as this. … If we are to bring the two phalanged digits into line with the rest of these observations we shall have to make an extreme demand upon the specific powers of chemical substances.

Bateson could understand how a chemical change of substance A into B might somehow be catalyzed by a ferment (enzyme), and hence be dependent on the factor (gene) that gave "the power to produce" that ferment, but he had no concept of regulatory genes – genes regulating development – mutation of which could result in one less finger joint.

The Congress agenda included a paper by Hurst on rabbits, which was to be read for him in his absence. At some point after it reading, the paper was lost. Hurst later related to Beatrice (February 12, 1928):

> Owing to an unexplained misunderstanding the results on the yellow rabbits were never published. The paper in which they were embodied was read at the Zoology International Congress in Boston 1907 (I believe your husband took my paper with him as I could not go) but owing to some misunderstanding it was not printed in the report and I have not been able to recover the M. S., and no copy of it was kept.

With the main business of the Congress over, Bateson returned to the Hotel Lafayette (August 24). From his perspective, the meeting had been "a stupendous success. … I never before felt what an exhilarating thing it is to speak to a really large and enthusiastic audience. It is dreadfully intoxicating!" He had occasion to seek the services of a New York dentist but was sufficiently recovered the following day to take the train to "Osborn's place" at Castle Rock, Garrison, overlooking the valley of the Hudson River. Regarding Osborn's obvious wealth, Bateson noted: "Pierpont Morgan is Osborn's uncle." A few days later (August 29), writing from New York, he reported that: "We all (foreigners) were presented to [President Theodore] Roosevelt at his private house yesterday."

There was time to fill before his Silliman Lectures. He moved to a farmhouse at Alstead, New Hampshire, near the residence of W. Field, a school master and entomologist, whom he visited for meals. Then, he moved (September 10) to Brookline, near Boston, where he stayed with the Sargent's, departing (September 19) to see an expert on violets, Ezra Brainerd, the President of Middlebury College, in Vermont. Bateson later wrote to Brainerd (October 11) noting that his results from crossing violets were "certainly the first satisfactory example of anything like straight Mendelism between distinct species in the strict sense and I have little doubt that the whole thing must become classical. If the significance of sterility can be run down we shall be a lot nearer to knowing what 'species' really means."

At the Addison Hotel, Middlebury (September 23), he was bored: "There is however some talk of a chess opponent being found for me. If that really comes to pass, I may stay on a few more days and get on with my lectures. Otherwise the wet and the solitude combined will probably drive me back to Boston." He stayed (September 26): "Lecture-preparing going very slowly …. Happily one of the Professors can put up a decent game of chess." A week later (September 30), he was in Boston, then on to the Graduates' Club at Yale (October 2). There was an excursion to New York (Belmont Hotel) for his first Brooklyn lecture (October 3), and then a return to Yale University for the Silliman Lectures.

His 11 lectures were to be given on 11 consecutive days "barring Sunday." For his first lecture (October 8), he reported an audience of 120 with many standing, so in future a bigger lecture hall would be used. Furthermore, a new invitation to lecture had arrived from Harvard, so he would be delaying his departure (until November 5). At the next Silliman lecture, there was an "enormous crowd, about 300," and at the next, he reported 300 including about 50 women. By October 13th, attendance was dwindling: "With introduction of gametic coupling I have thinned them down a bit. … I am inclined to change my mind concerning the Harvard lectures." But his attempt to cancel Harvard met with resistance and he relented.

Now the most exhausting part of the trip began. Writing to Beatrice from New York (October 20), he set out his itinerary in tabular form noting that most lectures would be worth $50. He first traveled to Cornell (October 21). Then there was a 15-hour train journey to Columbus, Ohio, where he gave one lecture (October 23). At Urbana, Illinois (October 24), he gave three lectures, and then (Sunday October 27) traveled to Chicago, where he stayed at The Quadrangle Club and met William Tower. By this time, his pile of US dollar checks was growing, but he translated into pounds when writing to Beatrice: "Tomorrow I shall try to invest about £300, securities being lower than they have been for ten years owing to the Wall Street 'flurry'. Yale paid £250, the other half to come when M. S. received."

After an informal talk and dinner at Chicago (October 28), he communicated some displeasure when writing to Beatrice en route back to New York (October 29): "Tower's story about the beetles" was dubious. "I fear I shall have trouble with him in the future. … Tower is the one blot on my expedition." Bateson's intuition seemed correct when in 1915 Tower resigned after the discovery that his beetles had been manually colored to support his case [1]. In a letter in *Nature* (1919), Bateson recalled the meeting with Tower when the Kammerer controversy emerged after the First World War (Chap. 21):

> The copious and astonishing observations said to have been witnessed by Professor Tower, of Chicago University, and by Dr. Kammerer of the Vienna Versuchsanstalt, naturally called for exceptionally careful examination. The results of both these authors had been widely accepted, and had begun to pass current in the text-books. In the case of Professor Tower's paper, as I demonstrated in my book [*Problems*], close textual criticism revealed features which suggested that implicit confidence should be postponed pending confirmation – a conclusion to which I had come when, on a visit to Chicago in 1907, I had seen illustrative specimens which Professor Tower was good enough to show me. Professor Tower's results are still quoted … but we have for some years awaited fresh light on the facts or any explanation of the difficulties to which I directed attention.

Between October 30th and November 1st, Bateson stayed at the Belmont Hotel and the Hamilton Club, gave more lectures at the Brooklyn Institute, and visited Davenport who had suddenly been taken ill with appendicitis. The last stop was Harvard (November 2): "Only three more lectures to give … am getting very stale." There was a Sunday motor-ride to North Shore, and he departed on schedule from New York arriving at Liverpool (November 16) having missed a RS meeting (November 14) where Hurst had presented his work on the inheritance of eye colour in humans.

Eyes

A paper on Hurst's eye work had been sent to the RS by Bateson (May 7, 1907). However, Sydney Hickson, Chairman of the Zoological Committee, wrote (June 4): "The referees are not altogether favourable, partly on the ground that the result is not in itself of great importance, and partly on the ground that the determination or classification of eye colour is to some extent a matter of opinion rather than a matter of fact." Bateson replied (June 6): "As to the importance of the result, when I undertook to communicate this paper it did not occur to me that any question could be raised as to the great importance, both scientific and practical, of the discovery which Hurst has made. I am astonished that anyone to whom this paper could have been referred should have expressed the opinion you quote" The letter continued with an account of the extensive literature of the subject and declared that "this paper provides the single definitive contribution to the physiology of normal as opposed to morbid heredity in man. Consequently, it must stand as a model for future researches on this subject." This was but the first round of a chain of correspondence, of which Hurst was kept informed:

> Any member of the Committee who desires to inspect Hurst's families is welcome to do so, and Hurst will so far as possible give him facilities. He has however been asked by Section K (Botany) to receive an excursion of the Section during the meeting of the British Association at Leicester. The immediate object of the visit will be to enable the members of the section to see his breeding experiments on plants and animals now in progress at Burbage (14 m. from Leicester). In responding to their request he has already invited the Section, and will now arrange that as many of his schoolchildren as possible – perhaps 100-150 – will be on view in the afternoon, for the inspection of those biologists who may care to attend. It is evident that these children cannot be subjected to frequent examination and he therefore suggests that the delegate of your Committee should if convenient come on that occasion.

Hurst wrote to Bateson (June 28):

> With regard to the eye paper I will defer discussion until I see you, but I cannot help saying that the correspondence betrays a state of things which seems to me disgraceful, one rarely sees such a combination of ignorance and insolence in the worst grades of society, and to find it in a presumably intellectual grade is somewhat of a shock! The only bright spot is your brilliant exposure of them, and I feel personally deeply grateful to you for presenting my case so clearly and well. I seem to be unfortunate in my RS papers and I am beginning to wonder if, for your own sake and that of Mendelism, you ought not to throw me overboard! I must be quite a Jonah to you! They do not seem to have objected to Lock's or Miss Wheldale's papers. Why are they so hard on me? I cannot but fancy that there is some knavery behind the scenes.

So, Bateson having departed for America, they came to Burbage on Friday August 2nd, 1907. With both parents and children in attendance, the scientists could see the eyes for themselves. The majority accepted Hurst's findings. His paper was published in 1908 [2]. In agreement with Mendelian principles, he had found that eye colors fell into two discontinuous groups: dominant brown (duplex) eyes where the iris had both anterior yellow or brown pigment and posterior black pigment; recessive clear blue (simplex) eyes where the iris had posterior black pigment only. In albino individuals, both anterior and posterior pigments were absent, and blood vessels seen through transparent tissues gave the eyes a pink appearance. To make these determinations, eyes had to be examined with a hand lens. In this way, a host of confounding variables, including age-related changes in pigmentation and structure, could be excluded.

Galton's data were unreliable. He had collected records on eye colors from various correspondents and had not examined the eyes himself: "It is highly probable that many of Mr. Galton's correspondents would record certain forms of self-coloured duplex eyes as 'hazel,' and certain forms of 'simplex' eyes as 'dark grey.' … In the nature of the circumstances … it cannot, of course, be expected that the observations were critical in regard to the presence or absence of anterior pigment in the iris." Harking back to the December 1905 confrontation with Pearson at the RS (Chap. 10), Hurst's paper further noted that: "It was on Mr. Galton's … data that Professor Karl Pearson based his memoir 'On the Inheritance of Eye Colour in Man,' and afterwards concluded that nothing corresponding to Mendel's principles appeared in the characters for eye-colour in man." Prior to publication, Hurst was able to add that "Professor C. B. Davenport independently arrives at similar conclusions, pointing out the Mendelian inheritance of eye colour in man." [3]. However, the matter was far from settled. It was to reemerge at a debate at the RSM (see below).

Musical Chairs

Back home, academic musical chairs was in progress. In July, Bateson had formally applied for Newton's Chair enclosing a copy of his 1904 BA Address. As expected, Adam Sedgwick was elected (July 23). Bateson's friend Lister, who had sat on the appointment committee, felt a word of explanation was called for:

> I think no one was inclined to estimate your claim lightly either from work accomplished or as the leader of an active subject in the future. The wish was in fact very generally expressed that a suitable University post may be found for you – and I hope it may prove not an empty one. On the other hand, I know you feel how strong Sedgwick's claims are. I confess for my own part, that I could not bring myself to vote for a result which would reduce him to a subordinate position in the laboratory of which he has been the head so long.

Bateson warmly congratulated Sedgwick (July 23): "You have received a distinction very fully earned; and I feel as I have done with regard to few appointments of late, that the right choice has been made." Sedgwick replied (July 24): "I thank you most heartily for your very kind and generous letter. I sincerely hope that I may be able to do a good … one or two for Zoology, and that it will not be long before the important branch of it which you follow with so much distinction receives adequate recognition in the university."

Of course, now Sedgwick's previous position was vacant. While in America, Bateson learned that he had been appointed Reader, noting (October 30): "Very satisfactory about the Readership." But on his return, he found that it had not been named as a Readership in Genetics, as he had assumed, and carried a salary of only £100 a year. Bateson saw red! A *Report of the General Board of Studies on the Readership in Animal Morphology* had noted (November 7): "Although the appointment of Mr. Bateson to the readership would not make all the provision for teaching in Zoology that may be required, the Board would welcome it in view of the distinguished character of his work and the importance of his teaching in University education, and because there seems to be at the present moment no other way in which his services can be secured to the University."

Word went round that Bateson was not going to accept the position. Seward of the Botany School wrote (November 15) approving of this but advising that he make quite clear that his refusal was not related to the salary, but to the discipline. "To allow you to lecture in Genetics as Reader in Zoology would not in reality be a recognition of Genetics by the University. I still think your opinion as to not taking the post is right." However, Francis Darwin was more conciliatory (November 15): "I really don't see that the Report is so grudging as you feel it to be. The Zoologists say they ought to have a Reader for pure Zoology, but that they willingly give it to you 'in view of the distinguished character of your work', etc., and because you must be kept here at any price. … I still think it would be a mistake, for the reasons I gave, to decline it. You have been teaching Genetics all this time as a Zoologist. Why not go on?"

Sedgwick knew his Bateson (November 17):

> As an old friend of yours, I venture to write on the subject of your proposed action with regard to the Readership of Zoology. You have to take into account the following facts. (1) You were willing to hold a Zoological appointment, as shown by your candidature for the Professorship of Zoology. From this it is fair to conclude that you then thought that Genetics form an important part of that subject. (2) You allowed me to understand, after definite inquiry, that you were willing to hold the Readership of the subject if it were changed to Zoology. (3) Acting on this, a Committee of the Special Board, … the Special Board itself, and finally the General Board, all spent a considerable amount of time and labour in drawing up a report recommending a plan which they had every reason to think would be satisfactory to you. … .

Bateson replied the next day (November 18). After some preliminary sparring, he concluded:

> In view of what you, Harmer and Lister have said, I think I did misinterpret the tenor of the Report. I took it to mean 'We don't much mind him having the Readership, for a time, but he should understand we would prefer a real Zoologist.' It scarcely seemed to me consistent with self respect to take an appointment thus given on sufferance. I am glad to know that this was not the meaning of the Zoologists. Understanding that and looking at the matter as a whole I have come to the conclusion that I must digest my disappointment and let the Report go to discussion without overt protest.

Sedgwick replied (November 19): "I rejoice at your decision. I am sure it is right. This I think is sufficient answer to your interpretation of the report, so far as I am concerned. … I sympathize with you thoroughly in your disappointment that an adequate position has not been provided for you. I sincerely hope that it will not be long before it is possible to remove that reproach from the University." Shortly thereafter (November 26), there was a formal discussion of the Report of the General Board of Studies. The *Cambridge University Reporter* related that Drs. Marr and Sedgwick had protested that the £100 salary was "undignified." It should have been at least £300. Sedgwick had given a long speech about the importance of genetics "not only for Zoology, not only for the University, but for the world at large." As it turned out, Bateson did not have to swallow his pride for long.

The musical chairs had other ramifications. Sedgwick wished to introduce changes in the Zoological Museum that Harmer, the Museum Superintendent, opposed. Bateson and Sedgwick's Demonstrator, Stanley Gardiner, were drawn into the matter. Sedgwick wrote to Bateson (February 12, 1908) asking for help. Bateson (February 13) urged compromise and patience, and to Stanley Gardiner wrote (February 23) noting: "To do this now will need great forbearance on both sides, but as sensible and high-minded people, I have confidence that Sedgwick and Harmer will be able to use such forbearance." Later (April 25), Harmer wrote to Bateson concerning the suppression of a report on his handling of the Museum, but eventually the matter blew over.

Meanwhile, there were exciting developments in both Bateson's science and his personal fortunes. Beatrice, who was at Brighton with sick children, received word (January 8) that: "De Vries has sent me his [new book] 'Plant–Breeding.' Acknowledging it will be a delicate task!" A day later, she received a telegram and shortly thereafter a letter with the time of writing, 6:45 a.m., underlined. This recorded what, with hindsight, can be seen as a major "eureka" moment in Bateson's life, as will be told in Chap. 13.

The second development was equally portentous. Things were going on much as usual. Beatrice was informed (January 31) that Mrs. de Vries has sent presents for the boys (perhaps late Christmas presents). He wrote a testimonial for Miss Wheldale (February 14), who was looking at a position at the University of Sheffield. Later (April 23), he supported her unsuccessful application for a Newnham College fellowship. She was successful the following year, and in the interim, she held a post in London. In February, Bateson gave two lectures in Newcastle (at 12 guineas a lecture), the first to a Sunday audience estimated at 3000 people. Then, it was reported (February 24) that the Council of Cambridge University Senate had received an offer to found a Chair of Biology, specifying the teaching of Genetics [4]. The Council had under consideration:

> A generous offer ... by a Member of the University who wishes to remain anonymous. It has come to his knowledge that there is a desire on the part of the biologists of Cambridge to celebrate in 1909 the centenary of Darwin's birth and the jubilee of the publication of the *Origin of Species*, by founding a Chair of Biology, the occupant of which shall devote himself to those subjects which were the chief concern of Darwin's life-work. Convinced of the great importance of the subjects with which such a Professorship would be concerned, and of the peremptory need that such subjects should obtain immediate recognition, the benefactor offers, under an arrangement approved by the Financial Board, to pay to the University £300 a year for five years, provided that the University establishes for that period and before June 30, 1908, a Professorship of Biology of the minimum annual value of £500. ... The further condition is made that it shall be the duty of the professor or professors elected during the period of five years above mentioned to teach and make researches in that branch of Biology now entitled Genetics (Heredity and Variation).

The specification "genetics" meant there could be only one candidate. Bateson's election to the position was formally announced on June 8th. With the new duties, it was not appropriate for him to continue his stewardship of the St. John's kitchens and farms. Furthermore, he had to relinquish entitlement to a pension associated with his college fellowship. So, in Beatrice's words "his position had more of dignity than security." Nevertheless, he was able to continue the fellowship itself (about £240/annum). The anonymous donor was "a Member of the University," a term which could include anyone who had graduated from the university, and the conclusion has been drawn that it was Arthur Balfour, who had been Prime Minister from 1902 to 1905 [5].

The appointment brought fresh attention to Genetics – so the time seemed right for seeking funds. Hoping for distribution under the auspices of the RS, Bateson drew up a circular "Experiments in Genetics." It specified that:

> A fund is urgently needed for the purposes of erecting and maintaining a building of moderate size in Cambridge, where breeding experiments with small animals, such as rabbits, canaries and insects could be carried on. It is likely that the University would be willing to provide a site for the purpose if the cost of the building could be assured. In the experiments on plants much assistance is given by the University Botanic Garden, but, for want of labour, the scale on which the work can be carried out is far too small. It is estimated that in order to provide the necessary accommodation and to carry on the experiments for a period of five years, the sum of £2000 is required.

Attached was a list of previous donors, including Mrs. Herringham (£300), Lord Peckover (£50), Beatrice (£105), and the five Darwin brothers (£200). A similar list of donations was added to another circular under the auspices of the Cambridge University Association: "Appeal for Funds to Build and Equip a School of Genetics." Both circulars mentioned that the Carnegie Station for Experimental Evolution at Cold Spring Harbor had an annual endowment of £4000. A letter to Bateson from Seward (July 4), indicated that there would almost certainly be objections to the RS Circular on the grounds that the funds would be going exclusively to Cambridge, so the circular may not have been sent out. The impermanence of Bateson's position became clear when he requested use of some central land at the Downing Ground. The Cambridge Financial Board declared (May 28) that it might "*let* to you personally the plot of ground at a small rent (probably minimal) under a written agreement providing for notice to quit on either side."

Bateson's elevation led to musical chairs for Punnett. Stanley Gardiner had been promoted to Lecturer in Zoology, so in 1908 Punnett returned to the Zoology Department as Sedgwick's Senior Demonstrator. Then in 1909, Harmer left the Museum of Zoology to take over Lankester's position as Director of the National History Museum in London, so Punnett became Superintendent of the Museum of Zoology. The music started up again later in the year when Sedgwick obtained a Chair at Imperial College. *The Athenaeum* announced (December 18, 1909) that Sedgwick had move to London, and that Stanley Gardiner had been appointed to the vacant Chair of Zoology.

Linnean Celebration

In 1858, a year before the publication of *The Origin of Species*, Darwin and Wallace had presented their ideas on natural selection to the Linnean Society. To mark this event, the Linnean Society held a Darwin-Wallace Celebration (July 1, 1908).

Medals were awarded to some of the survivors (Galton, Haeckel, Hooker, Lankester, Strasburger, Wallace, and Weismann). Bateson described the occasion in a letter to Beatrice (July 1):

> Proceedings just over. Though the remarks were at times obviously intended to irritate those who have dared to move beyond 1859, the event as a whole was not more galling than I expected. When I arrived I found I had been seated (seats numbered) *next Pearson*; which seems a gross breach of ordinary consideration and manners. He appeared disposed to make civil chatter – but I said no more than decency required. [Pearson described it differently, see below] Poor old Hooker got through the ordeal well. And of course he was cheered to the echo, but Wallace's spirit struck me as really surprising. His paper was full of point. Galton is dreadfully shaky – very lame – I don't know, but I suspect he has had a stroke, though one hopes not that. The way Galton and Wallace have gone down hill since I last saw them made me feel rather dismal. Their whole business of life is such a little thing? Of course their speeches treated recurringly of the theme of amazement that the orthodoxy of 1858 could see nothing in the new doctrine. No one seemed to hear '*De te fabula narratur*'! [The Roman poet Horace wrote: "*Quid rides? Mutato nomine de te fabula narrator.*" – "Why do you laugh? If you change the name, the story is about you."]
>
> Lankester naturally was the most blatant. He emerged from among the widows and orphans to tell us that the pure doctrine of Darwinism was still living in Edward Poulton, though works like his, pursued with minuteness, etc., could not take the public's mind so easily as theories which pretended to deal with wider issues. Three of these had come forward (like Theudas [a charlatan Messiah] in the days of their taxing, was what he wanted to say): Romanes; de Vries on a basis of somewhat unsatisfactory facts; Mendelism; but none had been able to add to the wisdom of 1858.
>
> Scott also must needs speak of 'Biometry and Mendelism both attempting to study Heredity by accurate methods.' It amused me to hear him introduce Lankester and Galton as the leaders of the 2 great developments following Darwin, *Morphology*, and *Genetics* (sic). Old Lord Avebury surpassed himself. It is a pleasure to hear a gentleman anyhow, and what he said was really for once not trite. Miss Willmott was there and perhaps I shall meet her tonight.

Pearson's account of the interaction with Bateson was somewhat different [6]:

> Some wag on the Linnean Executive had placed William Bateson in the chair adjacent to mine. I awaited his coming with expectations, determined that our greeting should disappoint the wag. But Bateson refused it, sat sideways on his chair, with his back to me, the whole of the medal distribution, and no doubt the wag was amused by what was simply pain to me – pain, that a distinguished biologist should refuse to join harmoniously with a biometrician, however despised, in a common service of reverence to one so immeasurably greater than either of us.

After the meeting, Wallace in *Contemporary Review* [7] objected to the reporting of "the Darwinian Jubilee of the Linnean Society" in the popular press, "the larger portion of which consists of 'Objections,' among which we find such statements as the following: … 'Where does natural selection come in then? Nobody knows exactly where, … but the whole question of the origin of species is as much a mystery today as when Sir John Herschel called it 'the mystery of mysteries'.'" Wallace then provided a critical examination of the three main alternatives to natural selection that had been advanced by "the Neo-Lamarckists, the Mutationists, and the Mendelians." The first were attacked in the person of the American paleontologist, Edward Cope, the second were attacked in the person of de Vries, and the last were attacked in the person of Mendel, with some brief allusions to Robert Lock's book *Variation, Heredity and Evolution*. The works of Tower in Chicago were employed to refute the first. The works of Thiselton-Dyer and Poulton were employed to refute the second. The works of J. A. Thomson and Charles Darwin were employed to refute the third.

Bateson was not mentioned by name, but his presence was pervasive. While not evident from the text, remarkably, all the views Wallace expressed were essentially *identical* to those held by Bateson on all three groups. Like Wallace, he opposed the Lamarckists and the Mutationists (de Vries). Like Wallace, he had articulated, and would continue to articulate, the point that Mendelian phenomena were not shedding light on the process of species formation. In Wallace's words, for "the great questions associated with the name of Darwin," or for the "essential part in the scheme of organic development," Mendelism was a "mere side-issue of biological research."

Wallace agreed that Mendelism had "a certain value for a comprehension of the mysterious phenomenon of inheritance," and for "the study of disease in all its strange forms," but deplored what he saw as "the complex diagrams and tabular statements which the Mendelians are forever putting before us with great flourish of trumpets and reiterated assertions of their importance." In short, as Darwin had shown so well, "hybridization or the intercrossing of very distinct forms had no place whatsoever in the natural process of species formation." The editor of *Contemporary Review* asked Bateson if he would reply, but Bateson declined.

In 1908, Poulton wrote an "introductory" chapter to a collection of his *Essays on Evolution, 1889–1907*. The essays were dedicated to Meldola who as President of the Chemical Society had promoted the 1906 jubilee celebrations of the founding of the dye industry and also had impeccable biological credentials, having translated Weismann's *Studies on the Theory of Descent* in 1882, and having long been associated with the Entomological Society [8]. However, instead of *introducing* the essays, Poulton's chapter was devoted to Bateson's writings on "Mutation and Mendelism," which were held to be "injurious to Biological Science, and a hindrance in the attempt to solve the problem of Evolution." The chapter was said to have been written under consultation with "a number of the leading zoologists and botanists in this country," who had "agreed with the general line of argument, and felt with me that the protest was called for." It attacked "Bateson and those who follow him," particularly Robert Lock. While sympathetic "with the efforts of the energetic and enthusiastic workers on Mendelian problems at Cambridge," Poulton thought the problems had been given "extraordinarily exaggerated importance," and deplored "the quite unnecessary depreciation of other subjects and other workers." Punnett replied to this with almost Batesonian fury in an article (October 1908) in *The New Quarterly* [9].

Elementary My Dear

As was her custom, old Annie Bateson had taken a place for the summer to which everyone was invited. This year (1908) she had taken a villa at St. Briac, near Dinard, where Beatrice joined her with the boys. Bateson remained behind to work on *Principles*. He wrote (July 31) enclosing John's school report, which was good, and declared (August 2): "I have been getting on at a fair pace, and the Press have a good bit of my stuff on hand. I am well into Chapter XI." He was increasingly disaffected with the suffragettes:

> An imbecile circular tells me that the "Banners" [Exhibition of Banners that had been used in the June 13th London Suffrage march] are to be shown here – with an inaugural meeting – Professor J. Ward in the Chair! I should like to hear Ward on Banners and wonder if he would be as illuminating as when he speaks on Monism – about the same I expect. From the w. p. b. [waste paper basket] I rescue these papers for your edification. The Parish magazine is the only literature which reaches a depth lower than the Suffrage.

But the lure of France was great. He wrote (August 5): "Miss. D[immer, who was his mother's companion] and I both come by the Saturday night boat, arriving Sunday morning. I don't think any time will be freer than this for me, and Punnett stops through August here. Probably I shall return after a fortnight."

The letters resumed (September 2) when he returned to Cambridge where Shull was visiting prior to accompanying him to the BA meeting in Dublin at which Francis Darwin was to give the Presidential Address (Chap. 19): "Shull, in the long run, is a bore. I shall be dead sick of him by the end of the time." Later from the BA meeting Bateson wrote: "It is a mistake to take his [Shull's] remarks about Pearson seriously. He is quite an uneducated man and is one of the many who got committed by publishing a lot of biometrical stuff early on. Now of course they are indisposed to admit they were fooled." Irish hospitality was much in evidence and there was a visit to the Guiness Brewery where the entertainment was described (September 3) as sumptuous. Adulation such as had been received in Boston was missing (September 5):

> The attitude of my 'fellow-workers' here is very different from that they assumed in Boston. I no longer play the part of Queen Bee, with my subjects arranged in radiating groups! However, there was some genuine interest shown yesterday, and I hope to get home on my audience in the sex-debate on Monday. Poulton was present in my audience, but I had not the good fortune to draw him. I quite think I shall leave on Tuesday, reaching London 10.30 p.m.

In fact, he arrived at 11.30 p.m. to find the hotel he had booked was full. He tramped the streets, trying hotel after hotel, none of which would even let him sit down. Eventually, he found a sofa at 3 a.m. in his old favorite, the Midland Hotel, from which he soon departed to catch the early train back to Cambridge [10].

In addition to *Principles,* he now had to write the inaugural lecture that was expected from the new Professor of Biology. For this he retreated to the Hill House Hotel. A letter (September 21) reported that the sea water was still at a swimmable temperature. Later (October 18) he told Beatrice that a fellow guest at the Hotel was a Mr. Thompson, whom they had met on a previous visit or heard about from their host, Mr. Gilbert Cubitt: "Mr. Thom[p]son is ... still here and has the drawing rooms. In all respects a layman. Yesterday he was very busy – writing, so far as I could judge, and I think his use of Hasbro' is much the same as my own. He is an elderly man with a lot of gray hair, and looks like someone who plays a part in the world."

Although the Bateson's may not have been aware of it, a frequent visitor to the Hotel was Arthur Conan Doyle of Sherlock Holmes fame, who used names from newspapers and acquaintances in his novels. One of the main characters in "The Dancing Men," published in 1903 was one "Gilbert Cubitt." The name "Thompson" occurs in one of his novels, and his work on spiritualism was aided by two Americans, the Thompsons. The skeptical reader may here huff in Batesonian fashion and declare "I will believe when I must, but not before." Stronger evidence might be Doyle's use of Bateson's name in a novel. What better novel than *The Adventure of the Naturalist's Stock Pin*? Here we find, not "Bateson," but "Romanes"!

Treasure Your Exceptions

Bateson gave his inaugural lecture (October 23, 1908) on "The Methods and Scope of Genetics," which was published "in a little red book." Here, he counseled beginners to "Treasure your exceptions!" and declared "to the study of Heredity, pre-eminently among the sciences, we are looking for light on human destiny." Referring to the relative roles of "nature" and "nurture," he predicted that "man's views of his own nature, his conceptions of justice, in short his whole outlook on the world, must be profoundly changed .... We may live to know that to the keen satirical vision of Sam Butler on the pleasant mountains of Erewhon there was revealed a dispensation, not kinder only, but wiser than the terrific code which Moses delivered from the flames of Sinai."

A year on from his Boston address, he was still wedded to the presence-and-absence formula and was still using the term "factor" very loosely. Citing Hurst, he noted that a blue eye was "due to the absence of a factor which forms pigment in the front of the iris." Here, "factor" was an enzyme or ferment through the action of which a character was established. He then spoke of "the presence or absence of elements which are treated as definite entities when the germ cells are formed." Furthermore, "such qualities as the formation of pigment in an eye" were "due to the transmitted elements or factors." Here, "elements" and "factors" were used synonymously as something transmitted by gametes. As to "the actual nature of those factors," it was recognized that "several of them behave much as if they were ferments." Here, he had slipped back to "factor" as active agent, not as "transmitting element." He hastened to correct this: "But we must not suppose for a moment that it is the ferment, or the objective substance [on which the ferment acts], which is transmitted. The thing transmitted can only be the power or faculty to produce the ferment or objective substance." Thus, a *gametic factor* was equated with a "power or faculty to produce" whereas a *character-forming factor* could be a ferment or its substrate (the substance on which an enzyme acts). Small wonder that in 1909 Thomas Morgan, in an article entitled "What are 'Factors' in Mendelian Explanations?" [11], joked:

> In the modern interpretation of Mendelism, facts are being transformed into factors at a rapid rate. If one factor will not explain the facts, then two are invoked; if two prove insufficient, three will sometimes work out. The superior jugglery sometimes necessary to account for the result, may blind us .... We work backwards from the facts to the factors, and then, presto! Explain the facts by the very factors that we invented to account for them. ... I realise how valuable it has been to us to be able to marshall our results under a few simple assumptions, yet I cannot but fear that we are rapidly developing a sort of Mendelian ritual by which to explain the extraordinary facts of alternative inheritance. So long as we do not lose sight of the purely arbitrary and formal nature of our formulae, little harm will be done.

Punnett was little help. In the third edition of *Mendelism* (1911), he was still pondering whether the "two bodies" – substrate and ferment – that gave a flower its color, "exist as such in the gametes or whether in some other form." Indeed, Bateson was digging himself progressively deeper into a hole. As facts multiplied, to factors that "coupled" he added factors that "repelled" so driving each other into separate gametes. All this was to be explained more simply by Morgan in 1911 (Chap. 13). Nevertheless, perhaps distracted by his loose terminology, Bateson's deeper messages were lost to Morgan. Bateson saw experimental breeding as a promising bridgehead that would help a move toward the more profound questions that lay beyond:

> We are obliged to examine the constitution of the germ cells by experimental breeding .... But cumbersome as this method must necessarily be, it enables us to put questions to Nature which never have been put before. She, it has been said, is an unwilling witness. Our questions must be shaped in such a way that the only possible answer is a direct 'Yes' or a direct 'No'. By putting such questions we have received some astonishing answers which go far below the surface. Amazing though they may be, they are nevertheless true; for though our witness may prevaricate, she cannot lie.

Continuing, Bateson likening the construction of a gamete to the taking of drops of fluid (an organic tincture in which an "ingredient" is dissolved) from a set of bottles in a medicine chest:

> There is one such chest from which the male gamete is to be made up, and a similar chest containing a corresponding set of bottles out of which the components of the female gamete are to be taken. But in either chest one or more of the bottles may be empty; then nothing goes in to represent that ingredient from that chest, and if corresponding bottles are empty in both chests, then the individual made on fertilisation by mixing the two collections of drops together does not contain the missing ingredient at all. It follows therefore that an individual may thus be 'pure-bred', namely alike on both sides of his composition as regards each ingredient in one of two ways, either by having received the ingredient from the male chest and from the female, or in having received it from neither. Conversely in respect of any ingredient he may be 'cross-bred', receiving the presence of it from one gamete and the absence of it from the other.

He further elaborated:

> In our model we may represent the phenomenon of segregation in a crude way by supposing that the bottles having no tincture in them, instead of being empty contained an inoperative fluid, say water, with which the tincture would not mix. When the new germ cells are formed, the two fluids instead of diluting each other simply separate again. It is this fact that entitles us to speak of the purity of the germ cells. They are pure in the possession of an ingredient, or in not possessing it; and the ingredients, or factors, as we generally call them, are units because they are so treated in the process of formation of new gametes, and because they come out of the process of segregation in the same condition as they went in at fertilization.

Of interest here is that "absence" is now not equated with nothing, but with a supporting medium – "an inoperative fluid" – in which no ingredients are dissolved. Of more interest is the postulated *existence*, whether "full" (containing tincture) or "empty" (containing water), of male and female "bottles" in the parental organism. Is this just a vague metaphor, or did the bottles perhaps signify Bateson's "base" or "residue" upon which the Mendelian unit-factors (genes) rested? Writing from Ceylon, Lock commented (December 21, 1908): "Very many thanks for the Inaugural address. I was especially pleased with the remarks on Erewhon and criminals. I recently had the pleasure of sitting on the jury here in 5 murder trials in 5 consecutive days."

The RSM Debate

In an August 1908 letter, Bateson had dourly noted: "R. S. Medicine is to have a full-drawn discussion of hered. of disease in November and bids me. No doubt this means a set-to with Pearson." However, the intense pressure of the year's events was beginning to tell. The backlog of unfinished work was mounting. In response to an enquiry (September 24) from Yale concerning his progress with the *Problems* manuscript, Bateson explained that *Principles* and a *RS-Report* to the Evolution Committee would have to be finished first: "Another difficulty has been that I found myself very tired and done up in the summer and had to take a month off, which I do not usually need."

The RSM discussion involved many speakers, mainly medical, and occupied four evenings at weekly intervals. It was reported in successive issues of *The Lancet*, and as a *Royal Society of Medicine General Report* [12]. As Bateson predicted there was a set-to, but his role was relatively minor, perhaps because he was weary, but also because the role of Mendelian advocate was taken by those who now recognized that Mendelian principles could be productively applied to the interpretation of patterns of inheritance in humans. Prominent among these was George P. Mudge of London Hospital Medical College. Bateson spoke on the first evening (November 11). While claiming to simplify, he equated a tall plant with a *tall germ-cell*, thus again spreading confusion regarding the distinction between phenotype and genotype:

> When we cross our tall plant with our short plant we imagine the meeting together of a germ-cell which is tall with a germ-cell which is short; and we may represent the result diagrammatically by putting them together – a tall line and a short line. … Now the discovery which Mendel made was, that in all cases to which his rules applied, when dissimilars meet in one individual there is, on formation of the germ-cells, a separation between the two characters which came in. That may be represented diagrammatically in a crude way by picturing the germ-cells, male and female, as a mixture of long lines and short lines, the long lines representing the germs carrying tallness, and the short lines as the germs destitute of that quality. … The importance of this representation of dominant as due to a factor present, and the recessive as the condition which results form the absence of that thing, will appear distinctly when I come to speak of the inheritance of disease. The consequence of the combinations of the cells produced by hybrids, females with males, is obviously that in some cases there will be the meeting of long with long, and in some cases the meeting of short with short, and in other cases of *long* female with short male, or long male with short female.

Perhaps indicative of Bateson's state of exhaustion, the confusion would have been compounded by an error ("short" not "long") in the latter sentence in the proofs of the published account (corrected above and italicized by us), which was not corrected in the final RSM version. Bateson continued: "The result will be that where those cells are distributed at random, three of the offspring appear tall and one appears short. The short plants thus reappear because they contain none of the long element. Of the tall plants thus produced, some will be pure to tallness, containing two 'doses' of the tall factor, others will again be cross-bred, containing only one 'dose' of it." There was also some confusion in his examples of the sex-linked inheritance of the relatively benign eye disorder, color-blindness, which was to recur in the first edition of *Principles* (Chap. 12).

Speaking on another evening (November 18), Pearson declared that "there is no definite proof of Mendelism applying to any living form at present." Regarding pedigrees in cases of albinism, he considered that "the principles required by Mendel fail utterly." Albinism being considered a recessive condition, he noted the high statistical improbability that albino children would appear among the two sets of offspring of a human female who had been married twice to apparently normal males who were not related either to her or to each other. Yet, such cases were recorded. Thus, "Mendelism has done an immense service in setting a large number of people experimenting and collecting, but I am perfectly sure it is too early in the day to assert that it holds for man – and I would go a step further: that it holds for any plants or animals."

At this point, Bateson interrupted saying that in his address at the earlier meeting "I showed the pedigrees of albinism as an example of a character which did not follow our rules." To this Pearson replied: "I am extremely glad to hear that. I am sorry I should have mistaken Professor Bateson's meaning at the last meeting. I thought it came as one of the things which were, somehow, supposed to follow the Mendelian rules." He turned to Bateson: "I understand, then, that it differs in man from the rules for animals?" Bateson replied to the affirmative. "You hold it to apply to animals?" Bateson replied "Yes, in many animals."

On the final evening (December 2), it was Mudge's turn. In a long address, Mudge set out the main features of the dispute between the Mendelians and Biometricians, applauding the former and criticizing the latter. Interrupted from time to time by Pearson, his history included the extensive studies of Hurst on eye pigments that had been corroborated by Pearson's former ally Davenport, and the controversy over whether Ben Battle was a chestnut recessive. As for the example of human albinism that Pearson had presented:

> The Mendelian explanation is that this woman and both the husbands must be carrying the albinism recessive. Professor Pearson said that if one knew the number of albinos in the country where this case occurred, they could calculate the probability of the woman meeting two husbands in succession in whom albinism was also recessive: he showed it, by the Theory of Probability, to be something very remote. The validity of that judgement wholly depended on where the case occurred, and where the individual in question came from.

When on questioning Pearson referred to Glasgow, Mudge referred to his own studies from that region which indicated that "recessive albinism on a relatively large scale must have been carried into Glasgow and Edinburgh for generations past. In fact, Professor Pearson's case is just what the Mendelian would have been led to predict, and it is very kind of him to thus confirm the Mendelian prediction."

In his reply, Pearson "frankly confessed he did not trust Captain Hurst's judgement. A priori he did not see why Captain Hurst's judgement as to what was a blue eye should be any better than Mr. Galton's. Everything depended on the way the eyes were put into categories." This unwillingness to acknowledge gradations in characters was a fundamental weakness of the Mendelians.

> Captain Hurst wrote a paper, published by the Royal Society, and said that a chestnut horse was a horse that had *no black* in it, but one of the hard questions in practice was to distinguish a *black* chestnut from a black horse; and every intensity of chestnut, from the lightest red to black, could be found, and the exceptions – which Mr. Mudge called errors of record – to the rule of chestnuts breeding chestnuts were a function of where the parents were taken in the scale of chestnut! With the existence of a *single* exception fell to the ground all the talk about dominants and recessives and 'pure' gametes.

On reading a report of his first address in *The Lancet*, Pearson considered he had been misquoted and wrote (November 28) a letter of complaint which was published (December 5, 1908). After giving some examples of the alleged inaccuracies, he concluded: "I feel compelled to repudiate entirely this account as representing what I said. It differs very materially from the report [page proofs] I have just received from the Royal Society of Medicine." The Editors of *The Lancet* added a note regretting that the "great condensation of technical speeches" might dissatisfy authors but declaring that Pearson's "suggestion ... that we modified the report in any way after receiving it from the reporter's hands should not have been made. We join in asking anyone who is at all interested in Professor Pearson's views to await the publication of the report to be published by the Royal Society of Medicine, when time and space will allow of a fuller rendering of the speaker's words and where, it seems, Professor Pearson will be able to revise the proofs."

The report of the final evening when Mudge attacked Pearson appeared in a later issue of *The Lancet* (December 12). To this, the Editors added a note entitled

"Professor Karl Pearson and The Lancet." Here, they observed that, having been advised by persons present at the debate that *The Lancet* report was "substantially accurate," they had compared it with the "official version of the Royal Society of Medicine as it was taken down," and had found that "our reporter's version tallies well with the version placed at our disposal by the Royal Society of Medicine, while both versions differ from what Professor Karl Pearson says that he said." The note concluded: "Professor Karl Pearson has impeached our accuracy, relying for the proof of his charge upon the version of his words to be published in the report of the Royal Society of Medicine. … This … seems to us a funny thing for a student of the exact truth to do."

Eager for blood, Mudge later wrote to Hurst, who had not attended the meeting, noting that in the presence of many witnesses Pearson had made remarks concerning his judgment that might be considered libelous. Hurst did not rise to the bait. Bateson wrote to Mudge (December 4): "As to Pearson's attack on Hurst, I advise leaving matters where they are, so far as controversial action is concerned. The aim of controversy is to interest, and that is amply done."

Summary

In 1907, prior to a three-month tour in America, Bateson attended to the needs of his research associates, discovered a writing retreat on the Norfolk coast, and worked to get Hurst's eye studies published by the RS. Adam Sedgwick ascended to the Chair made vacant by the death of the Professor of Zoology. The tour was a "stupendous" scientific, political, and financial success. In August, Bateson addressed the Seventh International Zoological Congress in Boston, and in October gave the Silliman Lectures at Yale. The following lecture tour included Chicago, where he crossed swords with William Tower. On his return, with some arm-twisting from Francis Darwin and Sedgwick, Bateson swallowed his pride and accepted the position of Reader in Zoology. In February 1908, he learned officially that an anonymous donor had endowed a Chair in Biology that was to be dedicated to Genetics. Obviously, there could only be one candidate. Wallace spoke at the fiftieth anniversary celebration of the famous papers he and Darwin had presented to the Linnean Society in 1858, and Bateson now found that he and Wallace were closer on evolutionary questions than he had supposed. Regarding the origin of species, Mendelism was a "mere side-issue." In his inaugural lecture, Bateson counseled beginners to "treasure your exceptions," and mentioned the "satirical vision" of Butler. However, he continued to expound his confusing "presence and absence" doctrine, and to employ the term "factor" very loosely. This may have distracted attention from his deeper messages. Pressed by Yale for a manuscript, Bateson confessed to exhaustion. Fortunately, the brunt of Pearson's attack in an RSM debate was borne by a new Mendelian enthusiast, Charles Mudge, and by the Editors of *The Lancet*.

Chapter 12
Darwin Centenary (1909)

I well remember receiving from one of the most earnest of my seniors the friendly warning that it was a waste of time to study variation, for 'Darwin had swept the field'.

Bateson, Principles

To mark the centenary of Darwin's birth, the Council of the Cambridge Philosophical Society invited essays for a celebratory volume, *Darwin and Modern Science*, to be published by Cambridge University Press. Bateson was heavily involved both in deciding who should be invited and as a contributor. By 1909, his manuscript of *Principles* was in the hands of the Press, which then prodded him for his essay. The writing did not go easily, but the result would encapsulate in its clearest form a view of evolution he had been developing from foundations laid by Galton and Romanes decades earlier, with an added historical depth, and perhaps some subconscious insights, from the new-found works of Samuel Butler.

Scraps

The ten representatives of the Philosophical Society and the Press Syndicate who met to plan the volume of essays included Bateson, Francis Darwin, Adam Sedgwick, David Sharp, Arthur Shipley, and Albert Seward, who was to be the Editor. Their work began in early 1908 when a list of possible contributors was prepared, a priority order agreed upon, and invitations sent out. Predictably, when Pearson's name came up, there was a scrap. Seward wrote to Bateson (March 19, 1908):

> You ask me if I think you spoke 'with too much heat': I do. As you know I am incompetent to express an opinion on Pearson's work. ... My confidence in your judgment leads me to accept your decision as to the value of Pearson's work, but he is unquestionably recognized

A. G. Cock, D. R. Forsdyke, *Treasure Your Exceptions*,
https://doi.org/10.1007/978-3-030-92099-9_12

> by many people … as a man whose views are worthy of attention. In the circumstances it would be unwise to ignore him. … If Pearson is all wrong, his work and influence will not last long; but it is hardly the duty of the Committee responsible for a book containing a collection of Essays purporting to represent different schools of thought – all of which cannot be right – to say 'We draw the line at Pearson as we are assured by one who has carefully examined his methods that his contributions are valueless'.

Pearson heard a grape-vine version of the meeting and relayed it to Galton (January 9, 1909):

> It was suggested that I should be asked to contribute a paper to the memorial volume. Bateson said that, if I were asked he would have nothing more to do with the Committee or the Volume. Sedgwick said: 'Are you the Pope?' and the incident ended in laughter, with the compromise that I should be asked to write on a definite topic – which did not permit of my breaking a lance for the threatened stronghold of Darwinism.

Seward wrote to Bateson (April 6, 1908): "Pearson declines and so, after some consideration, does Galton. … Lankester is still unasked, but I am rather inclined to ask him for an article on Darwin's Predecessors as the Committee would not agree to omitting him altogether. … De Vries accepts and will write on Plant-breeding." It is likely that Bateson's input resulted in Lankester's name being put low on the list, and it did not appear in the final volume.

After Christmas (1908), Beatrice took the children to Eastbourne. In a letter to her at an *en route* residence (December 30), Bateson reported that he was reading Samuel Butler. Perhaps he envied Butler's elegant prose, for in the next letter (January 2, 1909), he regretted that: "My paper is dreadfully stiff. The fact is I am so stale I can't get it lively." Forever confined by academic terms separated by short and long ("the long") vacations, his restlessness broke out the following day:

> I am heartily sick of this paper. It is crawling to an end at last and I think I must get a thorough change before next term is on me. Yesterday afternoon I had some chess with Deighton, but I missed the most obvious things and haven't played so badly for years. My thoughts turn towards Berlin *for one week only*! What do you think? It costs about £4-8-8 2nd class return. I dare say it would be cheaper than Paris *in the end*. Of course I have thought of Eastbourne too, but I seem to want something more distinct. I wish you poor old thing could come somewhere, but there it is. Berlin or London is my choice. I dare say London will have it. This afternoon I am going to sit with F. Darwin a bit. He is just up.

It appears that he completed, but did not submit, the essay. It remained on his desk and in his mind. By January 6th, he was alone in Berlin. There followed a stream of letters describing visits to theaters and art galleries, and meetings with dealers. He was ecstatic (January 9) about a performance of Macbeth, perhaps indicating an improved understanding of German. Then on to Osnabrück: "No. I feel sure that my Essay is not one of my best things. It is all scraps and I expect if you had been by, you would have insisted on neutralizing some of the more acid remarks. This may yet be done in proof." He visited an art dealer and then moved to Melle in Hanover, reporting (January 11) that he planned to be home on Wednesday "in time to prepare for my harangue on Thursday evening in London." By January 13th, he was home, and the "harangue" had mellowed to a "discourse": "The Press is poking me up for copy, so it is just as well I have got back. Tomorrow I am to give my discourse at the

South London Entomological Society." The scraps of his essay having been put in order, there was now time to attend to his research associates and to invite to Grantchester various foreign scientists who would be visiting Cambridge for the Darwin Centenary celebrations in June.

Perhaps because she was often in a remote laboratory rather than in the gardens and hen houses at Grantchester, in 1909, there was much correspondence between Wheldale and Bateson. He wrote (January 23) that her paper on antirrhinum colors was "down for Thursday next," at the RS. And a few days later (January 26): "Seward says he will be glad to give you a place in the Botanical Laboratory. Go and see him about it." Later in the year (May 25), Bateson was "very glad to hear of your election," presumably to a Newnham fellowship. Her work was gaining attention internationally, and in June, she corresponded with Erwin Baur (1876–1933) in Germany.

David Sharp, Curator at the Museum of Zoology, was retiring. So in December, Bateson distributed a circular seeking contributions for a parting gift, and in March 1909, the gift was formally presented. He wrote to Davenport (February 17) inviting him to stay at Grantchester, adding (as if an enticement were needed) that de Vries might also be their guest. Edmund Wilson accepted (April 2), and Bateson wrote to Davenport again (May 15) affirming that he was expected to stay with them. They all agreed to come, and de Vries would be accompanied by his wife and daughter.

Principles

Bateson's third book had the same title as *Defence* (1902) less the subtitle. *Mendel's Principles of Heredity*, here referred to as "*Principles*," was quite different from its earlier namesake and was officially published in March 1909. Wheldale contributed many of the colored figures. A reprinting was called for in August. It set out the ideas Bateson had presented in various papers and addresses, with the addition to his usual list of characters (e.g., height, seed color) those of "instinct and conduct," perhaps reflecting his new-found awareness of Butler. There was the usual look-back at the decline of breeding studies after 1859, and praise of Weismann for questioning the Lamarckian notion, shared by Darwin, "that 'acquired characters' – or, to speak more precisely, parental experience – can really be transmitted by offspring." Having apparently forgotten – if ever aware of – the studies of Galton and Romanes designed to test Darwin's "provisional hypothesis of pangenesis," he declared that "to Darwin and his contemporaries … no proof of the physiological reality of the [pangenesis] phenomenon was thought necessary." Nevertheless, high praise was reserved for Galton's other writings and for the "clear foreshadowing of that conception of *unit-characters*" advanced by de Vries in his *Intracellular Pangenesis*. When recounting the basic facts of Mendelism, there was the usual implication that gametes could bear the characters themselves, so this had to be followed by the usual qualification:

> Since the fertilized ovum or *zygote*, formed by the original cross, was made by the union of two germ-cells or *gametes* bearing respectively tallness and dwarfness, both these elements entered into the composition of the original F_1 zygote; but if the germ-cells which that zygote eventually forms are bearers of *either* tallness *or* dwarfness, there must at some stage in the process of germ-formation be a separation of the two characters, or rather of the ultimate factors which cause those characters to be developed in the plants.

The "fact of *segregation*" was again hailed as "the essential discovery," so that "segregation thus defines the units concerned in the constitution of organisms and provides a clue by which an analysis of the complex heterogeneity of living forms may be begun." The definition of the units, although a major advance, was a mere "clue" allowing no more than a beginning. It may be presumed that Bateson was writing here of the "ultimate factors," and he went on to imply that these units could vary in size, and might be so small as to be sometimes undetectable:

> There are doubtless limits beyond which such analysis cannot be pursued, but a vast field of research must be explored before they are reached or determined. It is likely also that in certain cases the units are so small that no sensible segregation can be proved to exist. As yet, however, no such example has been adequately investigated; nor, until the properties and laws of interaction of the segregable units have been much more thoroughly examined, can this class of negative observation be considered with great prospect of success.

This suggested that Bateson was seeking something deeper, and perhaps smaller, than the standard Mendelian elements that we now equate with genes. It may have prompted Haldane to remark in 1957 when giving the Bateson Lecture at the John Innes Horticultural Institute [1]:

> Bateson ... never accepted the word 'gene' with its rather wide connotations. ... Bateson used the neutral word 'factor,' .... Now [today] ... we could define a factor exactly. We could say that because in a particular chromosome an adenine residue has been substituted for a guanine, in, say, the 25473rd nucleotide counting from the free end of the longest chromosome, the plant makes a polyphenoloxidase with rather different properties. A factor, I suggest, can be anything from a difference of a few atoms in a single nucleotide, to an inversion or the presence of an extra chromosome; for these too are inherited in a Mendelian manner. If this is so, and a similar analysis of extranuclear factors is possible, all evolution is the accumulation or loss of factors.

While *Principles* mentioned "the late Professor Weldon," the biometric viewpoint was attacked indirectly as "Galton's system:"

> It may be useful to specify the distinctive features of Mendelian inheritance which differentiate the cases exhibiting it from those to which Galton's system of calculation – or any other systems based on ancestral composition – can apply. (1) In the Mendelian cases, in which the characters behave as units, the types of individuals considered with respect to any pair of allelomorphic characters are three only, two being homozygous and one heterozygous; while according to such a system as Galton's the number of possible types is regarded as indefinite. (2) The Mendelian system recognizes that purity of type may be absolute, and that it may arise in *individuals* of the F_2 or any later generation bred from heterozygotes. The views based on ancestry regard purity of type as relative, and arising by the continued selection of numbers of individuals. (3) In Galton's system no account is taken of dominance, a phenomenon which plays so large a part in the practical application of any true scheme of heredity.

The differences between the Galtonian and Mendelian schemes were held as "absolute and irreconcilable:"

> When Mendelian phenomena were first recognized it was naturally supposed that some classes of cases would be found to conform to the Mendelian scheme and others to the Law of Ancestral Heredity. With the progress of research, however, almost all the cases to which precise analytical methods have been applied have proved to be reducible to terms of Mendelian segregation. … Professor Pearson and others … have of late defended their position by arguing that there is no fundamental incompatibility between the Laws of Ancestral Heredity and the conclusions of Mendelian analysis. The matter would not be worth notice were it not that the same proposition is being freely repeated by several writers seeking some convenient shelter of neutrality. It is to be observed however that the supposition of an underlying harmony between Mendelian and biometrical results was not put forward by the biometricians until every possible means of discrediting the truth of Mendelian facts had been exhausted. These attacks having failed, we are asked to observe that the Law of Ancestral Heredity was meant as a statement of statistical consequence, and is not concerned with physiological processes.

This was also an occasion to display how easy it had been to be led off-track. He described the frequencies of the various types of comb in chickens (rose, pea, walnut, and single) as an example. It had been found that what they had thought were single characters, walnut (W) and single (S), were in fact compound characters resulting from the interaction of either dominant non-allelic factors R and P (giving walnut) or of recessive nonallelic factors r and p (giving single). R was allelic to r, and P was allelic to p:

> The interpretation of the facts was at first by no means easy, and I am sorry to have been responsible for the promulgation of a quite erroneous suggestion regarding them. … In the early years … knowing that R and P were [both] allelomorphic to S, I came to regard them as also allelomorphic to each other. This idea led to confusion, but we know now that no case justifies such an application of the principle of allelomorphism. A rose comb is not due to an elemental factor which can segregate from the pea comb factor. The two factors belong to distinct allelomorphic pairs and in the gametogenesis of the heterozygote [each] segregates from its own allelomorph, which is simply the *absence* of the factor in question. The single comb contains neither R nor P. … We may describe the rose as R no P, and the pea as P no R. It is convenient to use capital letters for dominants and small letters for recessives. … The walnut comb is RP, while rp gives the single.

Bateson here conceded that what he initially thought was an allele, S, was in fact the expression of the chance meeting of two different recessive factors r and p, in the absence of the corresponding dominants R and P. Thus, there were 16 possible individual genotypes each with distinctive allele pairs (as in Fig. 9.4 of Chap. 9).

There were other sources of confusion in plant breeding. First, there was the phenomenon of female self-reproduction resulting in what was essentially an identical copy – in other words, self-cloning (parthenogenesis or "virgin birth"). In the case of hawkweeds, this had exasperated Mendel [2]. Second, and even more confusing, were some of Hurst's orchids, where fertilization with a foreign pollen stimulated only self-reproduction – there was no genetic input by the pollen:

> In all cases where irregular results [of breeding] are observed, we may have to reckon with two possibilities: (i) actual parthenogenesis, or the development of unfertilized ova without fertilization; (ii) a phenomenon tantamount – as regards heredity – to parthenogenesis,

occurring after fertilization. In both cases the offspring are purely maternal. The latter exemplifies the conception of Strasburger and Boveri, that fertilization may consist of two distinct processes, the stimulus to development, and the union of characters in the zygote.

What were the ultimate factors that determined character units? While Bateson still slipped glibly from "factor" to "character" and then back to "factor" – with "unit" used interchangeably for each – when these words were understood in context, there was little ambiguity:

What the physical nature of the units may be we cannot yet tell, but the consequences of their presence is in so many instances comparable with the effects produced by ferments [enzymes], that with some confidence we suspect that the operations of some units are in an essential way carried out by the formation of definite substances acting as ferments. ... The heredity of characters consists in the transmission of the power to produce something with properties resembling those of ferments. It is scarcely necessary to emphasize the fact that the ferment itself must not be declared to be the factor or thing transmitted, but rather, [the thing transmitted is] the power to produce that ferment, or ferment-like body. ... We have to recognize that this antecedent power must be of such a nature that in the cell divisions of gametogenesis it can be treated as a unit, being included in one daughter-cell and excluded from the other at some definite cell-division. As we have no knowledge as to the actual nature of the factor – and only a conjecture as to whether it is a material substance, or a phenomenon of arrangement – we are not in a position to hazard so much as a guess respecting the physical process of segregation.

Bateson here suggested that "a material substance" might be somehow different from a "phenomenon of arrangement." He did not say what it is that might be arranged. Clearly material substances, such as the bases in a DNA sequence, can be arranged. So, DNA can be construed as both "a phenomenon of arrangement" and a material substance. Even if Bateson was thinking of the arrangement of waves in a pattern, like ripples in the surface of a sandy beach, the medium conveying the waves is a uniform, pliable, *but still material*, substance (Chap. 4). The distinction between the ultimate factor as an "antecedent power" that had the property of transmitting "the power to produce" a ferment, and the ferment itself, was explicit. But we search in vain for the word "information" (Chap. 19).

Another distinction was that "segregation is *not* a process of chemical separation. Its features point rather to mechanical analogies." Bateson then gave, as an example of the latter, the precipitation of a substance which would then be decanted off from the solution that contained it. At other times, he spoke of an organic fluid that would not mix with water (Chap. 11). There is the implication that a process which is "chemical" cannot at the same time be "mechanical." Thus, a key could be mechanically (physically) separated from its lock, but neither lock nor key would be chemically changed in the process. In similar fashion, a sword could be mechanically separated from its scabbard (see Chap. 17), and one DNA strand could be mechanically separated from its partner (see Part VI).

Despite his preoccupation with the Mendelian characters that had been shown transferable through breeding studies, Bateson had not lost sight of the "residue" idea and its possible connection with the distinction between species. Under the heading "Possible Limits to Recombination," he wrote of the need to detect the "limits to these possibilities of transference" as "one of the more important tasks still awaiting us." He continued:

> Then again we must surely expect that these transferable characters are attached to or implanted upon some basal organization, and the attributes or powers which collectively form that residue may perhaps be distinguishable from the transferable qualities. The detection of the limits thus set upon the interchangeability of characters would be a discovery of high importance and would have a most direct bearing on the problem of the ultimate nature of Species.

He also urged that the term "factor" be used in the context of latency, only when it was certain that the factor existed in the organism at the level of what came to be called the genotype: "For the present … the expression 'factor,' qualified as necessary as unseen, seems sufficiently precise" for alluding to "factors which may … exist without making their presence visible." Furthermore, "the term 'latent' is only admissible in application to the *factors* … not to the *characters*, except loosely to those which are actually present but hidden owing to the operation of some epistatic factor."

In the Bateson archives, there are bundles of press clippings corresponding to each of his books. *Principles* was reviewed generously and widely in journals ranging from *The Atheneum* to the *Live Stock Review*. The review in *Nature* even described Bateson as one "on whom Darwin's mantle has fallen." The reviews were didactic rather than critical. That Mendelism was valid was no longer in question.

At this time, Olga Zacharias' mother came to visit and, while Beatrice was away at Robin Hood's Bay, Bateson took her to town. He wrote from the Midland Grand Hotel (April 26, 1909): "Just back from 'Englishman's Home' [a play] with Frau Zacharias. … Tomorrow I am taking her to South Kensington Museums which she has never seen. Olga is alright, now reported out of risk." The following day, he told of art purchases and meeting up with Frau Zacharias to see *School for Scandal*. Then he learned that *Principles* had been published (April 29): "My book is out. … I took Frau Z. to the Wallace [Gallery] yesterday … then to the sale … the first auction she had ever seen!" He summed up his art purchases (April 30): "The result of 4 days is that I have bought 5 lots aggregating £58-5. This may make you jump, but I believe it is not unwise." However, he had combined pleasure with academic matters:

> In the afternoon I got involved in a disagreeable affair. I opposed the move of the Zoological Society house to Regent's Park. My speech was on the whole well received, and I thought I should get about an even vote, when the maniac Cunningham started a tirade against the Council [presumably agreeing with Bateson]. After a few minutes the house was in uproar. I never saw such a scene. He declined to sit down. Said nothing should prevent him from finishing his speech, and it was the meanest thing to bull baiting by a crowd you can imagine. The consequence was the meeting went right round, and about 50 people voted with me to perhaps 400 against. Some one rubbed it in by saying 'Professor Bateson's supporters say, … etc.'. I feel very sore. Mitchell of course triumphing.

Late in May Bateson retreated to the Hill House Hotel (May 23): "I have been in the sea several times and begin to freshen up, but very little writing done." There was a trip (June 2) to Oxford University at the behest of the Junior Scientific Club which had invited him to give the Boyle Lecture. Earlier lecturers in the series had been Ernest Rutherford (on radioactivity) and Pearson (on eugenics). At that time, Bateson declined to follow the usual rule and have his lecture published, but in a later letter from the Hotel (March 15, 1910), he declared that he was "getting my Boyle Lecture into shape," and soon (March 26) that the "Oxford lecture [was] done after a fashion" [3].

Centenary

On January 1st, 1909, the centenary of the birth of Charles Darwin was celebrated at Johns Hopkins University. The star visitor from England was Poulton with leading American scientists – Davenport, Castle, Wilson, Osborn – in attendance. Later, there were meetings at Oxford (February 12) and Cambridge (June 22). Events at Cambridge began with a reception in the Fitzwilliam Museum [4], followed next morning by formal addresses at Senate House by Oscar Hertwig (Germany), Elie Metchnikoff (France and Russia), Osborn (United States), and Lankester (a last minute stand-in for Lord Avebury). The scene at Cambridge was described by William Rothenstein [5]:

> Francis Darwin invited my wife and myself to attend the many functions. What impressed me most was the scene outside Senate House, where representatives from universities throughout the world, come together to receive honorary degrees, walked in procession, wearing their robes. … Mrs. Huxley, Thomas Huxley's widow, and Sir Joseph Hooker, both well over eighty, were staying with the Darwins. Were old world manners and charm, I wondered, more common in the past, or do they come with mature years? Mrs. Huxley certainly had them, with a surprisingly alert mind. It was a touching site to see old Sir Joseph Hooker with Francis Darwin's little grand-child in his arms. I thought what a wide period would be covered if the infant lived to the scientist's great age. A great commemoration dinner, to which I was bidden, was given in the Hall at Trinity College, at which Mr. Balfour presided. … My wife looked on, with the Darwin ladies, from the balcony above, and heard the speeches. How proud Francis Darwin, and the other sons, must have been, at the homage paid to their father.

The celebratory volume, *Darwin and Modern Science*, contained essays from many of those encountered on these pages. The volume was described by Meldola in *Nature* [6] as "monumental" because "it stereotypes the collective thought of our age," and brings "together the views of the different and often antagonistic writers." While saluting the "twenty-eight essays by English and foreign experts – every name being that of a recognised authority," Meldola was "struck by the omission of certain names which we should have liked to see on the list of contributors. The names of Alfred Russel Wallace and Francis Galton are conspicuous by their absence. Biologists would no doubt have been glad also to have read essays by Henry F. Osborn of Columbia University, by Sir Ray Lankester, and by Karl Pearson.

'Biometricians' are not represented. Presumably there are valid reasons for this omission, but the loss is ours nevertheless." As for the question "How stands the species question after passing the ordeal of half a century?" Meldola answered:

> The only theory of species formation which still holds the field is that of natural selection which suggested itself to Darwin after reading Malthus in 1838, and independently to Wallace in 1858. ... It is perfectly clear that the historian who in the distant future ... consults the present book in order to ascertain what platform had been reached in the year 1909 cannot but arrive at the conclusion that at the time of the publication of this memorial volume no alternative theory of the origin of species had survived the test of scientific criticism. No more effective mechanism of organic evolution than that offered to us by Darwin fifty years ago has up to the present time been suggested. We may degrade natural selection from the position assigned to it by Darwin, or ... we may attach quite a small value to it as a factor, or we may eliminate it altogether. If so, the other factors remain to be discovered, and we must declare that we are still without a theory of the origin of species.

Meldola would appear to have read Bateson's essay, but for him as for all others, it was far from an eureka moment. Like the Mendel paper decades earlier, the essay (see below) created not one ripple on the intellectual landscape of its time.

Bateson and Beatrice both being home at Grantchester during the celebration, there is little record as to how they and their guests interacted. Beatrice later annotated a letter from Bateson to Hurst (June 25, 1909): "Madame de Vries & daughter stayed with us, but the Professor stayed in a hotel in Cambridge and turned his back on Will at the Conversazione in the Fitzwilliam." Bateson was in Paris for a week at the beginning of August. Davenport had noted the philatelic interests of the Bateson boys, since they wrote to him from Robin Hood's Bay (August 16) thanking him for stamps he had sent.

In November, after some months of negotiation, Bateson was offered the directorship of the John Innes Horticultural Institute (Chap. 14). Later in the year (December 1), he gave the Huxley Memorial Lecture at Birmingham on "Mendelian Heredity." Here, he restated Huxley's concern that the sterility of interspecies hybrids was a major difficulty for Darwin. The academic term completed, he departed for Berlin, writing (December 18): "Great luck today. I called on Erwin Baur – find him a delightful and most capable man – the best Mendelist *ausserhalb* England, I think, to judge from short visit. There is a Congress on, with some project of Teubner's in hand, and Johannsen and several others are here! We are to lunch together shortly. Isn't that luck? My German will now go up by leaps and bounds."

Modern Lights

Bateson's Centenary essay "Hereditary and Variation in Modern Lights" displayed him at the height of his power shortly before the emergence of the Morgan school in the USA. It also displayed his increasing awareness, but not full understanding, of the ideas of "the astute" Samuel Butler. Bateson pointed to the long history of those who had considered "the question of the mutability of species" and "the nature of specific difference," citing Butler's *Evolution, Old and New* (1879). He challenged Charles Darwin's assertion that prior to 1859 few "seemed to doubt about the permanence of species," noting that "the literature of the period abounds with indications of 'critical expectancy'." As for the post-*Origin* decades:

> In contrast to ... immense activity elsewhere, the neglect which befell the special physiology of Descent, or Genetics as we now call it, is astonishing. ... The class of fact on which Darwin built his conceptions of Heredity and Variation was not seen on the highways of biology. It formed no part of the official curriculum of biological students, and found no place among the subjects which their teachers were investigating.

He hailed Weismann's "thorough demolition of the old loose and distracting notions of inherited experience" (Lamarckism), about which there had been "a certain vacillation ... in Darwin's utterances." As for natural selection, examples of "the doctrine '*que tout est au mieux*'" had been "discovered with a facility that Pangloss himself might have envied." It was time to "cease to expect to find purposefulness wherever we meet with definite structures or patterns. Such things are, as often as not, I suspect rather of the nature of tool-marks, mere incidents of manufacture, benefiting their possessor not more than the wire-marks in a sheet of paper, or the ribbing on the bottom of an oriental plate renders these objects more attractive to our eyes." Romanes' ally John Gulick had made just this point in 1872 [7]. In the late twentieth century, the analogy was extended by Gould and Lewontin to the spandrels in the Cathedral of San Marco [8]. Bateson continued:

> The characters of living things are dependent on the presence of definite elements or factors, which are treated as units in the processes of Heredity. These factors can thus be recombined in various ways. They act sometimes separately, and sometimes they interact in conjunction with each other, producing their various effects. All this indicates a definiteness and specific order in heredity, and therefore in variation. This order cannot by the nature of the case be dependent on Natural Selection for its existence, but must be a consequence of the fundamental chemical and physical nature of living things. ... Genetic Variation is then primarily the consequence of additions to, or omissions from, the stock of elements which the species contains.

As for the nature of variation, so much was unknown that it was no surprise that "medleys of dissimilar occurrences are all confused under the term Variation. One of the first objects of genetic analysis is to disentangle this mass of confusion." If "*fluctuational* variations, due to environmental and other accidents," were excluded, then "real, genetic, variations" would be left. These "must be expressed in terms of the factors to which they are due." In particular, there was the variation accompanying the formation of a new species: "In the light of present knowledge it is evident that before we can attack the species problem ... there are vast arrears to be made up. ... We have been taught to regard the difference between species and variety as one of degree. I think it unlikely that this conclusion will bear the test of further research."

Then where were we to start? "What one bit of knowledge would more than any other illuminate the problem?" We must understand cell division, but "this may be looking too far ahead." Bateson then named "the secret of inter-racial sterility," the problem Huxley had reiterated decades earlier, as the "one piece of more proximate knowledge which we would more especially like to acquire." Allied *species* might be able to crossbreed to produce viable offspring, but that offspring was always sterile. Allied *varieties* were *always* able to crossbreed to produce viable offspring, and that offspring was always fertile. This property defined species as species and varieties as varieties. But how could we explain presumed varieties (races) that crossed to produce sterile offspring? What was the cause of such sterility? And how was sterility to be investigated by breeding if sterility, by definition, prevented future breeding?

> *Failure to divide* [to produce gametes] is, we may feel fairly sure, the immediate 'cause' of the sterility. Now, though we know very little about the heredity of meristic differences, all that we do know points to the conclusion that the less-divided is dominant to the more-divided, and we thus are justified in supposing that there are factors which can arrest or prevent cell division. My conjecture therefore is that in the case of sterility of cross-breds we see the effect produced by a complementary pair of such factors.

In what sense was Bateson here using the term "factor," and what did he mean by "a complementary pair"? The following paragraphs are probably Bateson's most inspired, perhaps because he argued, in Galtonian fashion, from first principles. He first considered *two* species that were sufficiently allied to allow crossing and the production of viable offspring:

> It should be observed that we are not discussing incompatibility of two species to produce offspring (a totally distinct phenomenon), but the sterility of the offspring which many of them do produce. When two species, both perfectly fertile severally, produce on crossing a sterile progeny, there is a presumption that the sterility is due to the development in the hybrid of some substance which can only be formed by the meeting of two complementary factors. That some such account is correct in essence may be inferred from the well-known observation that if the hybrid is not totally sterile but only partially so, and thus is able to form some good germ cells which develop into new individuals, the sterility of these daughter-individuals is sensibly reduced or may be entirely absent. The fertility once reestablished, the sterility does not return in the later progeny, a fact strongly suggestive of segregation.

In other words, should the two species be so closely allied that the sterility among progeny (offspring) was incomplete, then factors which *failed* to complement – so allowing an individual to be at least partially fertile – would most likely continue to fail to complement when each found itself in companionship with a new factor in a future progeny, which would thus also show some degree of fertility. He next considered the case of two members of *one* species the normal members of which, when crossed with each other, would produce fertile progeny:

> Now if the sterility of the cross-bred be really the consequence of the meeting of two complementary factors, we see that the phenomenon could only be produced among the divergent offspring of one species by the acquisition of at least *two* new factors; for if the acquisition of a single factor caused sterility the line would then end. Moreover each factor must be separately acquired by distinct individuals, for if both were present together, the possessors would by hypothesis be sterile. And in order to imitate the case of species each of these factors must be acquired by distinct breeds. The factors need not, and probably would not, produce any other perceptible effects; they might, like the colour-factors present

in white flowers, make no difference to the form of other characters. Not till the cross was actually made between the two complementary individuals would either factor come into play, and the effects even then might be unobserved until an attempt was made to breed from the cross-bred.

Thus, the factors which distinguished these members of the species from their fellows would be latent. They would become patent, generating a sterility-of-offspring phenotype, only if by chance there were a cross between *certain* individuals, at least one of which would be rare (since the normal abundant types were generally fertile when intercrossed). How often could this happen?

Next, if the factors responsible for sterility were acquired, they would in all probability be peculiar to certain individuals and would not readily be distributed to the whole breed. Any member of the breed also into which *both* the factors were introduced would drop out of the pedigree by virtue of its sterility. Hence the evidence that various domesticated breeds, say of dogs or fowls, can when mated together produce fertile offspring, is beside the mark. The real question is, Do they ever produce sterile offspring? I think the evidence is clearly that sometimes they do, oftener perhaps than is commonly supposed.

And how was this to be investigated?

These suggestions are quite amenable to experimental tests. The most obvious way to begin is to get a pair of parents which are known to have had any sterile offspring, and to find the proportions in which these steriles were produced. If, as I anticipate, these proportions are found to be definite, the rest is simple.

Sterility-of-offspring would be considered as a parental phenotype, so the fact that such sterile individuals could not be subject to further examination by breeding would not matter. The parents of those individuals complemented sufficiently to produce the phenotype, but the complementary factor of one parent might be more prone to produce the phenotype than the complementary factor of the other. Through segregation (further crossbreeding of each parent), these complementary factors might be distinguishable. Finally, Bateson returned to crosses occurring between members of *two* species, again emphasizing the independence of the sterility-of-offspring phenotype from other phenotypic characters:

In passing, certain other considerations may be referred to. First, that there are observations favouring the view that the production of totally sterile cross-breds is seldom a universal property of two [allied] species, and that it may be a matter of individuals, which is just what on the view here proposed would be expected. Moreover, as we all know now, though incompatibility may be dependent to some extent on the degree to which the species are [have been classified as] dissimilar, no such principle can be demonstrated to determine sterility or fertility in general. For example, ... the hybrids between several *genera* [so deemed] of orchids are perfectly fertile on the female side, and some on the male side also, but the hybrids produced between the turnip (*Brassica napus*) and the swede (*Brassica campestris*), which, according to our estimates of affinity, should be nearly allied forms, are totally sterile.

From the foregoing, it seems likely that Bateson was primarily thinking of gametic factors that complemented in a *negative* sense in the hybrid to prevent it from forming gametes. Thus, the hybrid was sterile, so demonstrating that the parents were reproductively isolated from each other, a property that would define them as members of distinct species. This line of argument is reminiscent of that of Romanes, whose hypothesis of "physiological selection" Bateson had applauded as a young man (Chap. 5).

Bateson stressed that the sterility-of-offspring phenotype would not necessarily correlate with any anatomical phenotype. Organisms would not be anatomically differentiated *prior* to factor acquisition. Likewise, Romanes had held that further differentiation, be it anatomical or physiological, would tend to *follow*, not precede, the development of reproductive isolation [9]. The sterility phenotype could emerge whether or not the parents were differentiated from each other in other ways. They could then be "distinct breeds" only in the physiological sense that the offspring of one breed when crossed with offspring of the other would produce offspring that could not be crossed, because they were sterile.

One modern commentator [10] has interpreted Bateson's paragraphs as implying his belief that the gametic factors determining the sterility-of-offspring phenotype were the Mendelian elements determining character-units (i.e., genes; see Chap. 24). However, Bateson had clearly distinguished the "sporadic" sterility attributable to defects in genes from the "regular" sterility that occurred when members from two allied species were crossed (Chaps. 13 and 28). Indeed, as indicated above, it was pointed out decades later by one who knew Bateson's work well, that his use of the general term "factors" (not "genes") allowed him to encompass any form of information that was contained in the chromosomes, for all such information would segregate [1]. In his essay, Bateson did not consider whether the sterility factors might relate to the "residue" discussed in *Principles*. He also did not consider the converse argument. Instead of two factors working together (complementing) to *impede* fertility, there could have been two factors that failed to work together to *promote* fertility (e.g., promoting the correct pairing of two homologous chromosomes). This alternative interpretation (closer to our modern understanding; Part VI) was later made explicit in *Problems* (Chap. 14) (Fig. 12.1).

Having consulted Punnett, Shipley echoed aspects of Bateson at a BA meeting in Winnipeg, Canada, in August 1909 [11]: "Mutations are variations arising in the germcells and due to causes of which we are wholly ignorant; fluctuations are variations arising in the body or "soma" owing to the action of external conditions. The former are undoubtedly inherited, the latter are very probably not. But since mutations … may be small and may appear similar in character to fluctuations, it is not always possible to separate the two things by inspection alone."

Final RS Reports

Bateson was no longer battling the establishment. He was the establishment. There was no longer difficulty finding a publishing outlet. There was no longer a dependence on the *RS-Reports*. These now became vehicles for the deposition of data

Fig. 12.1 William Bateson

supporting and extending previous studies rather than for the presentation of new analyses. *RS-Report 4* was submitted on May 18, 1908, with revisions added up to July 20th. Most data could now be interpreted in terms of dominance and recessive, "presence and absence," and epistasis and hypostasis. In addition to the usual trio of Bateson, Saunders and Punnett, the report included studies by Killby on stock, by Florence Durham on coat color in mice and sex-inheritance in canaries, and by Doncaster and the Reverend Raynor on sex-inheritance in moths. Doncaster conceded that their original assumption that males and females each had two types of gamete that determined sex (i.e., they were each heterozygous with respect to sex) was more complicated than the alternative suggested by Bateson and Punnett, which had received support from new studies. Durham reported that her results on sex-inheritance in canaries were "closely parallel to those discovered by Doncaster." Today we know that in moths, butterflies and birds the female is the heterozygous sex (Chap. 13).

The last report, *RS-Report 5*, was received on July 6, 1909, with revisions added up to September and contained separate papers by Wheldale, Marryat, and Sollas. Bateson had long championed the need for a better understanding of the "physiological chemistry" of floral pigments, and Wheldale's results were a major step in this direction. She acknowledged that the plants studied "were kindly provided by Prof. Bateson and Miss Saunders, to whom I am much indebted not only for material to work upon, but also for advice and information in connection with the genetics of flower colour." From extracts of homogenized plant tissues, various pigments and their precursors ("chromogens") could be isolated. The chromogens would be acted upon by enzymes ("ferments" such as "oxidase") to produce the pigments.

The chromogens and ferments each corresponded, either directly or indirectly, to a Mendelian character unit. For example, wild-type antirrhinums have magenta flowers and "the magenta anthocyanic pigment may be regarded as an oxidation product of a chromogen, in nature allied to the flavone series of coloring matters; the oxidation is brought about probably through the agency of an oxidase. The loss of power to produce the oxidase (representable by a Mendelian factor) from the zygote gives rise to a variety bearing *ivory-white* flowers." It was concluded "that the character, flower colour, in *Antirhinum* must be represented by at least six Mendelian factors, the presence, absence, and combinations of which, give rise to the numerous horticultural varieties."

Journals

Mudge had his "Mrs. Herringham." Her name was Mrs. Rose Haig Thomas. But she was more than a provider of funds – she reared pheasants for Mudge's Mendelian studies. Mudge wrote to Hurst (October 31, 1907) inviting him to come to what in fact turned out to be a preliminary organizational meeting for a new society and a new journal. Mrs. Haig Thomas planned to entertain at the Ritz a "small gathering of people interested in hereditary problems" (Cunningham, Bernard, Darbishire, Archdall Reid, Mudge). Hurst declined the invitation. Mudge reported to Hurst (November 3) that the meeting had been a success despite Archdall Reid (Chap. 15) not turning up. The group had plans for a Biological Society for discussions of matters relating to heredity. They wanted Hurst to agree to be a "vice President." The next meeting would be in January 1908 when Cunningham would give a paper on "Crossed Races in Man."

At this point Hurst agreed to get involved. The "Mendel Society" began with informal "At Homes" sponsored by Mrs. Haig Thomas. Hurst contributed on "Mendelism and Sex." Later he agreed to publish in the new *Mendel Journal*. Among his many activities, Hurst was just completing *The Orchid Stud Book*, the product of a collaboration with Rolfe of Kew Gardens. Hurst sent a copy to Bateson with his apologies (February 9, 1909): "I was obliged to omit all references to Mendelism from the O. S. B. because unfortunately Rolfe is a bitter opponent in this respect, otherwise he is an excellent colleague." Hurst did not mention the journal on that occasion; it seems that he informed Bateson of it later (November 3, 1909):

> I knew that the 'Mendel Journal' was about to be issued at the expense of Mrs. Haig Thomas and under the editorship of Mudge and saw it announced in 'Nature' and 'The Times,' but I understood that it was merely a publication of the papers read at the Mendel Society's at Homes. From Mudge's [recent] letter however it appears that a much more ambitious role is intended, in fact a replica of 'Biometrika' on Mendelian lines! That of course raises a big question. Mudge writes 'Now that the first number is out … I am hoping that Bateson, Punnett, and yourself and others may feel inclined to form an editorial staff and run the Journal to the full extent of its usefulness.' … We shall see how the Journal goes!

Having learned of Bateson's new appointment to the John Innes Institute (Chap. 13), Hurst wrote (November 29, 1909).

> I have only just read of your new appointment. The Trustees are most fortunate to secure your services and I sincerely trust that you will be given a free hand. So far I can hardly realize the situation, but it seems to me that it will be a fine thing for horticulture to have you at the head of such an institution. ... In view of the step you have taken, I am wondering whether I am doing the best thing in practically giving up plants for horses? What do you think of the 'Mendel Journal'? I am sorry Mudge gives so much prominence to the Biometricians. I think it is unwise just now. I rather feel inclined to accept Mrs. Haig Thomas' invitation to join the editorial staff if only to put on the brake!

It was not the first genetics journal. That was the *Zeitschrift für induktive Abstammungs und Vererbungslehre* founded in 1908 by Baur. Hurst was still optimistic about the prospects of the new journal and naïvely unaware that Bateson might see it as an attempt to preempt his own plans (February 22, 1910): "I am doing my best to pull the 'Mendel Journal' straight and I hope that you will be more satisfied with the 2nd and following numbers. I will not ask you to help with it until I get it into decent form. I am however asking Punnett, Biffen, Wood and a few others to help me pull it round and a word from you would help me materially."

But, unknown to Hurst, Bateson and Punnett were attempting to gain support for their own journal to be called the *Journal of Genetics*. In February 1910, the way ahead became clear, and Bateson wrote informing Hurst of the new development. Hurst replied (February 28): "I need hardly say that I shall be proud to support the new Journal, and if necessary send you some of my stuff. ... I quite understand that in the circumstances none of you will be able to join in the *Mendel Journal*. The question now arises, is it worth while to go on with the latter venture?" He withdrew his support. Bateson wrote to Hurst (March 5): "I am rather glad to hear that you are segregating from the M. J. It is a stupid undertaking." Why did he deem it stupid? Perhaps he did not like the name of an individual – Mendel – attached to that of a journal?

The *Journal of Genetics* appeared in November 1910, but for the first decade was largely inward looking – a continuation of the *RS-Reports* in another form. The first issue was not unrepresentative with five articles, four or which were from the Bateson stable – Pellew, Pellew, Saunders and Doncaster. The fifth was by Redcliffe Salaman who had staked out a position in potato genetics. Mrs. Haig Thomas continued her pheasant studies and published them in the journal, either coauthored with Geoffrey Smith, a Fellow of New College, Oxford (1913), or as sole author (1914, 1916), or with Bateson (1917).

Another contributor to the journal was J. B. S. Haldane. In 1908, while still in his teens, Haldane and his sister Naomi began breeding guinea pigs in their Oxford garden and soon discovered linkage groups. As a student at Oxford (1911–1914), he switched from mathematics to classics and never obtained a formal degree in science. In 1914, he enlisted in the Black Watch regiment. Recognizing the possibility that he might not return from the battlefront, he wrote (March 18, 1915) asking

Bateson to assist Naomi, if necessary, in completing a paper on their work for the *Journal of Genetics*. The paper entitled "Reduplication in Mice," which appeared in the December 1915 issue [12], was an extension of their childhood studies. The final draft, prepared, literally, in the trenches in war-time France, appeared under the authorship of Haldane, (New College, Oxford), A. D. Sprunt (deceased in the war), and N. M. Haldane (later the author Naomi Mitchison).

Although wounded both on the western front and in the Mesopotamian campaign, Haldane survived. Displaying a penchant for elliptical citation, Haldane referred to his 1915 paper in 1926 [13]: "If Bateson had merely demonstrated the truth and importance of Mendelian heredity the world would be his debtor. For in its essential manifestations it is so simple that I have known a child of fourteen apply it with complete success to practical breeding." Despite a shaky period in the early twenties when printing costs rose steeply, the *Journal of Genetics* thrived. Haldane had a long association with it, eventually becoming its editor and transporting the editorial office to India, where it remains to this day.

Summary

Bateson was involved, both as organizer and contributor, in the preparation of a volume of essays – *Darwin and Modern Science* – to celebrate the Centenary of Darwin's birth. Thus, in 1909, several publications emerged – his Centenary essay, his advanced treatise *Mendel's Principles of Heredity*, and the last of the *RS-Reports*. In *Principles,* Bateson affirmed the need to determine the limits of Mendelian analysis. The factors determining the observed character units were "transferable qualities" distinct from some basal level of organization, or "residue," that might limit their transferability and so, in some way, might relate to the ultimate nature of species. Bateson's Centenary essay encapsulated in its clearest form the view of evolution he had developed from foundations laid by Galton and Romanes. The difference between species and variety was real. Unlike members of allied varieties (within a species), crosses between members of allied species (recently derived from a common ancestral species), produced sterile hybrids. Thus, a few individuals within a species would have developed complementary factors that, on meeting in their hybrid children, would inhibit the cell divisions that produced gametes. Although able to grow into healthy adults, they would be sterile; hence, their parents would be reproductively isolated from each other, but not necessarily from other members of their species. A new species could emerge when two such parents, reproductively isolated from members of the main species but not from each other, produced fertile hybrids. Mudge's attempt to found a new journal was undermined by Bateson and Punnetts' *Journal of Genetics*.

Chapter 13
Chromosomes

We have reason to believe that the chromosomes of the father plant and mother plant, side by side, represent blocks of parental characters.

William Bateson (1902)

The Mendelian factors, today known as genes, were abstract entities, inferred entirely from the results of breeding experiments. Initially, there was nothing to tie them to any structure within cells. Bateson was among the first to see a link with chromosomes, and he and Saunders cited Guyer in this context (Chap. 8). However, as evidence for a chromosomal location became stronger, Bateson's doubts increased. It was not a question of Bateson not seeing the wood from the trees. His idea of what constituted the wood differed from that of others. They saw a collection of genes. He saw genes plus something else that was somehow related to the question of the origin of species. What was the something else and where was it located? Was it on the chromosomes like the Mendelian factors? Or was it elsewhere, perhaps outside the nucleus in the cytoplasm? In vain he looked to Galton's latent "residue" and to Guyer's cytoplasmic "substratum" for solutions.

Pre-Mendelian Observations

The importance of the cell nucleus and the chromosomes it contained became evident in the 1880s. Theodor Boveri (1862–1915) showed in 1888 [1] that female ova could be induced to develop without fertilization into offspring that resembled their single parent ("virgin birth" or "parthenogenesis"). Thus, maternal materials alone (nuclear and/or cytoplasmic) sufficed for development. On the other hand, when he used a male sperm – essentially a mobile nucleus – to fertilize an ovum from which the nucleus had been removed, development was again achieved, with the offspring

A. G. Cock, D. R. Forsdyke, *Treasure Your Exceptions*,
https://doi.org/10.1007/978-3-030-92099-9_13

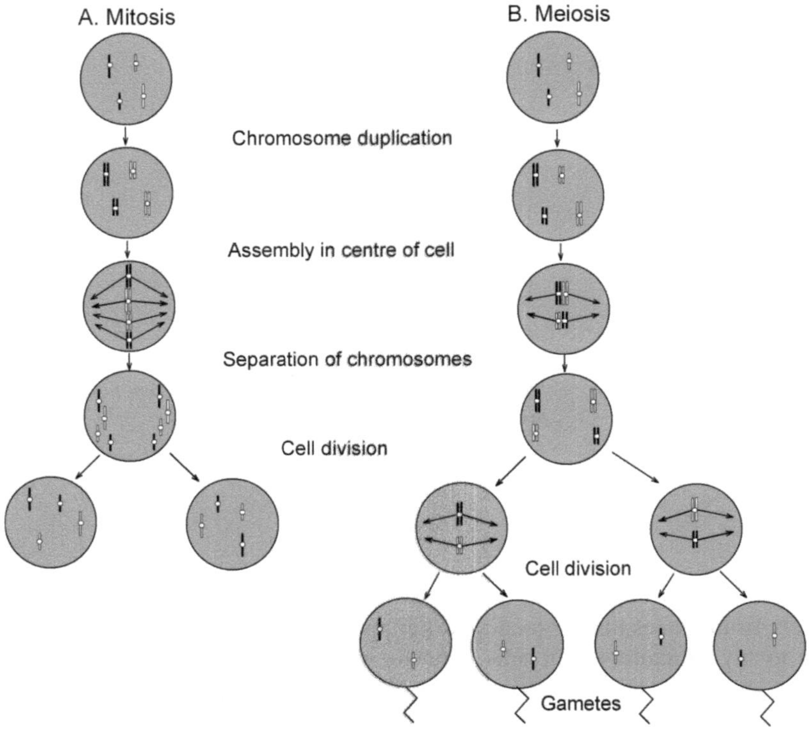

Fig. 13.1 Comparison of mitosis and meiosis. Diploid cells contain two chromosome pairs (rods), one of each pair being derived from the organism's father (filled in black) and the other from its mother (filled in white). In mitosis (**a**), each individual chromosome, having duplicated itself (thus making the cell transiently tetraploid), assembles in the center of the cell. In meiosis (**b**), the assembly involves pairing of the corresponding paternally derived and maternally derived chromosomes. In mitosis, following longitudinal splitting of each duplicate, the cell then divides to restore diploid status. In meiosis, the paired chromosomes part ("disjunction"), and the cell divides to restore diploid status (reduction division). Since orientation during assembly in the center of the cell of origin is random, sometimes the parental types get separated (as shown), and sometimes they remain together (i.e., black with black, and white with white). Apart from some local exchanges of segments during their pairing (Fig. 13.2), the duplicated parental chromosomes remain intact. A second "equational" division assigns half of each duplicate to gametes (which are hence haploid). Since the second division is more like mitosis, the early workers referred to it as the "homotypic" division, whereas the first division was the "heterotypic" division

in this case resembling the father. Egg cytoplasm seemed to play just a supporting role [2].

At that time, microscopists (cytologists) in continental Europe saw that when a cell divided into two cells ("mitosis") the nucleus did not divide *en masse* but first resolved itself into a definite number of thread-like (Greek: *mitos* = thread) deeply staining "colored bodies," each of which split longitudinally. The two parts so generated then separated, each going to a different child nucleus (Fig. 13.1a).

In 1883, Wilhelm Roux suggested that mitosis was for the precise division of cell contents and described the threads or colored bodies as "centers of a series of qualities," which were arranged in linear series along them, so that by longitudinal splitting the qualities would be evenly distributed to the two child cells [3]. The potential individuality of these colored ("chromatic") segments was enunciated in 1887 by Boveri [4]: "I regard the so-called chromatic segments or elements as individuals, I might say as the most elementary organisms, which lead their independent existence in the cell." The bodies were called chromosomes in 1888 [5].

Mitosis was not the only type of cell division. When the gametes (sperm, egg) were formed (gametogenesis), there was a different type of division. The word "mitosis" was used for both types of division until 1905 when Farmer and Moore named the latter "meiosis" [6]. At a BA meeting in 1887, Weismann described the halving of the number of chromosomes (to the haploid number) during gametogenesis (Fig. 13.1b). On fertilization to produce the zygote, the parental number (diploid number) was restored [7].

By the 1890s, it was appreciated that the numbers and shapes of chromosomes were constant per species, and it was deducible that since only half those of each parent was passed on to a child, then half would not be passed on. The chromosomes could be considered to be *directly* "passed on" in that they could be seen in parental cells, and in the gametes, and in the child cells. But there was no visible evidence of characters such as fingers or tallness being directly "passed on." *Something else was involved.* In 1891, Henking described the chromosomes as "conjugating" two-by-two at an early stage of meiosis ("synapsis"); this was followed by their separation ("disjunction") at what was called the "reduction" division (because the chromosomes in a cell were reduced from the diploid number to the haploid number). The evolutionary implications of this were discussed in the German literature, notably by de Vries in his *Intracellular Pangenesis* (Chap. 3), but there was also much discussion in the literature to which Bateson had more ready access [8–10].

Joining the Dots

Although there was no mention of chromosomes in his address to the Second International Conference on Plant Breeding and Hybridization in 1902, it would have been in character for Bateson to have wandered from his text and to have interacted informally with numerous people, many of whom would have read *Defence*. We may be sure that the topic of chromosomes arose. Indeed, other papers included one by William Cannon (1870–1958) of the Department of Botany at Columbia University [11]. In his paper – "Some Cytological Aspects of Hybrids" – Cannon summarized work on the cytology of gametogenesis and asked: "What are the relations of the cytological to the experimental researches of hybrids?" He began by noting that Mendel's peas

> behave as if the bundles of inheritance which were derived from the parents of the original cross were kept separate, and were delivered as such to the succeeding generation, and, as if in fertilization, these were united in all possible proportions. … In other words, according to the laws of Mendel, we might expect that the chromatin derived from the primitive parents maintained its individuality, and was disposed in such a manner at the time of the maturation mitosis [meiosis] that the resulting sex cells [gametes] were not *hybrid*, but *pure*.

The paper then considered studies on sterile hybrids by Jeul (Chinese lilacs), and on both fertile and sterile hybrids by Guyer (pigeons) and by Cannon himself (cotton). It was concluded that "normal divisions lead to fertility, the abnormal to sterility," with the implication of a cause-and-effect relationship. If there were abnormal divisions, then either an insufficient number of gametes would be formed and/or they would be malformed. The individual would to some degree be sterile. This meant that *its parents* were reproductively isolated from each other in that the continuation of the line was impaired.

In the discussion of Cannon's paper, Bateson gave one of his earliest pronouncements on the role of chromosomes, which could contain "particles *representing*" or "fragments *representing*" the observed characters of an organism:

> I am afraid there is a little difficulty … in the way of ever hoping to analyze ultimately by the microscope the characters in the way that Mendel's law teaches us to believe they might be analyzed. Because it is very true that in Ascaris and in a number of other forms referred to we have reason to believe that the chromosomes of the father plant and mother plant, side by side, represent blocks of parental characters, that is not enough to help us trace out ultimately the different parental forms of gametes. To do that you would have to have particles representing each parental character, not merely the whole block of chromosomes representing the father plant and the mother plant; you would have to have fragments representing each of the constituents of the father and each of the constituents of the mother, and they would again combine in the various combinations that we must expect.

Bateson's "fragments" or "particles" were the "gemmules" of Darwin, the "ids" of Weismann, the "pangens" of de Vries, and the "genes" of today. The modern reader might here pause, expecting Bateson to take the next step. Was he going to relate the collecting together of certain "fragments representing each of the constituents" in a "block," to the possibility of their "coupling" (later known as linkage)? Bateson did not take that step.

Although not listed as such, also in attendance was Walter Sutton (1877–1916), a student of McClung (see below). Prior to joining Edmund Wilson at Columbia University in the fall of 1901, Sutton had studied the stages of spermatogenesis in grasshopper testes (from spermatogonia, to spermatocytes, to spermatids, to mature spermatozoa). He had been impressed with the individuality of each chromosome, noting "a separate membrane around … each of the chromosomes" as observed in normal testes, so that "if … an attraction be postulated between ids [genes] of different chromosomes, it would be difficult to see how … an id from … one" would "reach its mutually attracting id … of another" [12]. Thus, the notion was in the air that the pairing of two chromosomes might reflect a more fundamental level of pairing *between the genes they contained*.

A few weeks after the conference (October 17), Sutton submitted a paper "On the morphology of the chromosome group in *Brachystola magna*" (grasshopper) to

Biological Bulletin, where it was published in December [13]. Referring to the pairing of homologous maternal and paternal chromosomes as their "copulation," and citing earlier work of Montgomery (see below), he noted:

> When the two [chromosomes] copulate … in synapsis, the entire chromatin basis of a certain set of qualities inherited from the two parents is localized for the first and only time in a single continuous chromatin mass; and when in the second spermatocyte division, the two parts are again separated, one goes entire to each pole, contributing to the daughter-cells the corresponding group of qualities from the paternal or the maternal stock as the case may be.

On this occasion, there was no allusion to the intermixing (recombination) of chromosomal materials representing paternal and maternal characters. Using Bateson's playing card analogy (Chap. 9), from first principles it might have been deduced that if one wanted to evenly split the red suites from the black suites, a discrimination based on color alone would be needed. All the red cards would be placed on one pile and all the black on another. On the other hand, if one wanted to mix red with black, but maintain equality (e.g., two kings in one pile, two kings in the other), then pairing followed by random allocation to one of the two piles would be needed. The fact of pairing *prior* to allocation to a pile was highly significant. However, attention was focused on a connection between pairing and Mendel's observations. Sutton's paper concluded:

> The eleven ordinary chromosomes which enter the nucleus of each spermatid are selected one from each of the eleven pairs which make up the double series of the spermatogonia. … I may finally call attention to the probability that the association of paternal and maternal chromosomes in pairs and their subsequent separation during the reducing division … may constitute the physical basis of the Mendelian law of heredity.

A few weeks later, Wilson elaborated on the work of his Columbia colleagues in *Science* (December 19). Cannon reiterated his viewpoint in the next issue (December 30) of the *Bulletin of the Torrey Botanical Club* [14]. On this occasion, his interest was the origin of variation. He suggested that any variations from parental types seen in hybrids were not the result of "irregular maturation mitoses" such as were seen in sterile individuals but were in some way "caused and brought about by" the *normal* mitoses (meiosis) seen in fertile individuals. So, "variation in the hybrid offspring must come about either because the maturation mitoses were such as would induce them, or quite independently of these nuclear divisions, since, in fertile hybrids [variants still emerge but] the mitoses are normal."

Continuing our playing card analogy, by January 25, 1903, Sutton had realized that all red cards would not go to one pile and all black cards to the other. He submitted a second paper entitled "The Chromosomes in Heredity," which was published in April [15]. Since different chromosome pairs would orientate randomly prior to segregation, his "results gave no evidence in favor of the parental purity of the gametic chromatin *as a whole*" (our italics):

> The position of the bivalent chromosomes in the equatorial plate of the reducing division is purely a matter of chance – that is, that any chromosome pair may lie with maternal or paternal chromatid indifferently toward either pole irrespective of the position of other pairs – and hence a large number of different combinations [i.e. recombinations] of maternal and paternal chromosomes are possible in the mature germ products of an individual. …

> It is this possibility of so great a number of combinations of maternal and paternal chromosomes in the gametes which serves to bring the chromosome-theory into final relation with the known facts of heredity; for Mendel himself followed out the actual combinations of two and three distinctive characters and found them to be inherited independently of one another and to present a great variety of combinations in the second generation.

Sutton also referred to the results of Saunders on stocks (*Matthiola*) that had been reported in *RS-Report 1*. Here, he noted, there were "two cases of correlated qualities which may be explained by the association of their physical basis on the same chromosome." In short, he saw a possible basis for coupling (linkage). He had taken the step that Bateson had omitted to take, and he now spelled it out quite unambiguously: "Some chromosomes at least are related to a number of different allelomorphs [genes]. If then, the chromosomes permanently retain their individuality, it follows that all the allelomorphs represented by any one chromosome must be inherited together, … The same chromosome may contain allelomorphs that may be dominant or recessive independently."

Sutton also suggested a chromosomal basis for the "mosaic" fruit of *Datura* (thorn apple) that displayed patches of the dominant character (thorny) on otherwise clear skin (as had been reported by Bateson and Saunders in *RS-Reports 1*): "If each [somatic] cell contains maternal and paternal potentialities in regard to each character, and if dominance is not a common function of each of these, there is nothing to show why, as a result of some disturbing factor, one body of chromatin may not be called into activity in one group of cells and its homologue in another." Bateson and Saunders had interpreted this either as "indicating that the germ cells may also have been mosaic," or as "an original sport [mutation] on the part of the individual" (i.e., a somatic mutation to thorny of certain cells consistent with Sutton's "disturbing factor" – later found to be a virus – having somehow locally unmasked a character). Bateson in 1909 (*Principles*) considered somatic mutation as a likely explanation for a phenomenon that had intrigued Darwin – the appearance of a smooth-skinned "glabrous" variety of peach (a nectarine) as a "bud-sport" on a dominant hairy-skinned ("hoary") peach tree:

> Bud sports … in which nectarines appear on peaches must be interpreted as meaning that in the formation of that bud or cell from which the branch, or fruit, or part of a fruit, derived its separate existence, the element or factor for the peach-character was omitted. Therefore at some cell-division, evidently a *somatic* division, segregation of the allelomorph for hoariness must have taken place, and we are thus obliged to admit that it is not solely the reduction-divisions [of meiosis] which have the power of effecting segregation.

In his papers, Sutton frequently cited Bateson, and years later he described Bateson's talk at the New York conference as the catalyst for his ideas on the relationship between Mendelian character units and chromosomes [16]. Bateson may have catalyzed others. Spillman wrote to Hurst (1903):

> I have just read an important paper by Mr. Sutton of Columbia University, New York, on the chromosomes in heredity. … It is the most suggestive of anything I have read recently, and illustrates another case of coincidences. His theory is almost identical with the one I had formulated, but had not published for lack of time to write out. I hope you will read it, for I regard it of fundamental importance.

In Holland, de Vries was more aware of chromosome dynamics than Bateson. In an essay that was presented at Haarlem in 1903, and translated in 1910 into "an enlarged form" for inclusion as an appendix to the English translation of *Intracellular Pangenesis* [17], de Vries noted that when chromosomes conjugated "the individual parts of the nuclear [chromatin] threads would be mated one by one." As in 1889, he assumed the existence of "special units, special granules in the nuclear threads for the visible characters of the organisms," which would imply "as many units in the nucleus, as a plant or animal possesses individual characters." In this case, if "nuclear threads of the paternal and the maternal pronuclei lie together in pairs, each granule can enter into communion with its corresponding unit in the other pronucleus." How would this come about?

> We assume ... that the cooperation comes about in such a way that the individual units in the stretched threads lie in the same numerical order. Then ... we can imagine that all the like units of the two pronuclei lie opposite each other ... [so] we may assume a simple exchange of them. Not of all (for that would only make the paternal pronucleus into a maternal one) but of a larger, or even only a smaller part. How many and which, may then simply be left to chance. In this way all kinds of new combinations of paternal and maternal units may occur in the two pronuclei, and when these separate at the formation of the sexual cells, each of them will harbor in part paternal, in part maternal units.

de Vries here envisaged recombination between units representing characters (genes) held together on a chromosome. Furthermore, he envisaged what we now recognize as a *homology requirement* between the genes themselves:

> Every unit can only be exchanged for a like one, ...which ... represents the same hereditary character. ...The children must inherit all specific characters from their parents, and they must also transmit all of them to their own progeny. This exchange must hence be accomplished in such a way that every pronucleus retains the entire series of units of all the specific characters, and this result can evidently be obtained only when the interchange is limited to like units.

de Vries then turned to crosses between "allied elementary species," where this homology requirement might not be met and, although hybrids might be formed, they would be sterile:

> Every character and every unit corresponding to it, which in a crossing is present in one species and lacking in an older one, forms a special point of difference. ... The threads no longer fit ... when the units are exchanged before the formation of the sexual cells. ... The greater the number of points of difference, the more numerous are the gaps, and the more will the cooperation of the two nuclei be interfered with.

Boveri declared in 1904 [18] that he had long been considering the relationship between chromosomes and Mendelian characters. Indeed, despite his emphasis on the individuality of the chromosomes, Boveri [4] had envisaged the transfer between homologous chromosomes (recombination) of characters that were normally linked *within* a chromosome (Fig. 13.2):

> [If] in successive breeding two traits should always appear together or disappear together, then it would be permitted to draw the conclusion ... that the *anlagen* for these two traits are localized in the same chromosome. And furthermore: if a hybridization experiment included numerous traits, and if it should be found in successive breeding that the number

of combinations in which the separate traits can occur, is greater than it would correspond to the possibilities of recombination of the chromosomes present, then it would have to be concluded that the traits localized in a chromosome can go independently of each other into one or the other daughter cell, which would point to an exchange of parts between the homologous chromosomes.

In short, when the facts came out, the relationship between chromosomes and Mendelian characters was so obvious that everyone working in the field recognized it. Some scurried to touch all theoretical bases and secure documentation, others did not. To many of Bateson's associates, the evidence appeared overwhelming. Why did Bateson, who was among the first to "join the dots," become increasingly reluctant to think in terms of genes sequentially arranged along linear chromosomes?

Sex Chromosomes

Within the cell nucleus, there is a small circular body, the nucleolus, which we now know is concerned with the synthesis of ribosomes – the cytoplasmic structures responsible for protein synthesis. Henking in 1891 had observed an element in male insect nuclei that he considered to be an accessory nucleolus. Since it stained like a chromosome, Montgomery in 1898 named it the "chromatin nucleolus" [19]. Paulmier in 1899 called it the "small chromosome" and thought it might serve as a collecting center for discarded "ids" [genes] that represented characters being lost by the species [20]. Clarence McClung (1870–1946), noting that half the spermatozoa had it and half did not, named it "the accessory chromosome," and suggesting it was the inherited "determinant" of sex. In diagrams of grasshopper chromosomes, he labeled it with an "X" [21]. Later, it was formally named the X-chromosome by Wilson. It turned out that grasshopper males had only one, whereas females had two (expressed as ♂ XO, ♀ XX). Thus, rather than being an *extra* chromosome in males, the solitary X reflected a *lack* of one member of the XX pair normally found in females. In some insect species (beetles), a very small companion to the X was observed in males [22]. This was named the Y chromosome (i.e., ♂ XY, ♀ XX).

In 1900, in the second edition of his textbook *The Cell in Development and Inheritance* [10], Wilson summarized the evidence that chromosomes differed from each other qualitatively. Boveri had a similar view [23], and in 1902 he correlated chromosome deficiencies with abnormalities in larval development, but did not relate this to Mendel's work [24]. McClung's view that the X-chromosome was a determinant of an organism's sex was touched upon by Bateson and Saunders in *RS-Report 1* and was further elaborated by Bateson in 1904 in his application to the Balfour Fund (Chap. 9):

Whenever sexual organisms breeding together produce a mixture of forms, there is … *prima facie* reason to suspect that the mixture is due to differentiation of the germs [gametes]. The most familiar case is sex itself. A population consisting of males and females has so many features in common with the differentiated offspring resulting from the segregation of characters among the germ-cells of cross-bred organisms that it is impossible to avoid the suspicion that the two phenomena are similar in causation. A categorical proof of this conclusion would make a remarkable advance in biology.

In *RS-Report 2,* Bateson went further, arguing that if an individual of *one* sex produced gametes of *both* sexes (e.g., a male spermatozoon and a female spermatozoon), then an individual of the opposite sex would need to produce only sexually undifferentiated gametes (e.g., an ovum that was not specifically male or female):

In Report I we indicated the possibility that sex may be ultimately a phenomenon of gametic segregation comparable with that seen in ordinary Mendelian cases. In an interesting essay, Castle (Bull. Harv. Mus. 40, 1903) has greatly amplified this suggestion, and has shown reasons for believing that the segregation of male spermatozoon from female spermatozoon and of male ovum from female ovum may occur at the reduction division in gametogenesis. While admitting the likelihood of this suggestion, we feel that for the present it should be received with caution. In particular, we doubt the conclusion that *both* ova and spermatozoa (after a reduction division) are always bearers of either the male or the female character. It seems more likely that special cases will present special phenomena in this respect. As yet the evidence most applicable to the decision of the question is thus derived from those reciprocal crosses which give dissimilar results, and in view of the idiosyncracies of these cases, we incline to expect that sometimes the male element, sometimes the female, will be found to be responsible for sexual differentiation, and that the similar differentiation of both elements is not likely to be universal.

He considered it probable that in canaries the cinnamon color of feathers was sex-linked and noted that similar cases of sex-linkage had been suggested by Castle. The report also referred to "new evidence as to the individuality of the chromosomes and the discovery that a true transverse division occurs in the chromosomes of plants at [the] reduction [division], are clearly favourable to the hypothesis that the reduction-division is the critical moment [of segregation of characters]." From this, it is clear that (despite use of "transverse" rather than "longitudinal") Bateson was convinced that chromosomes were differentiated from each other, in keeping with roles as transmitters of the determinants of Mendelian unit characters.

Sex-Limited Inheritance

Although the word "chromosome" was not used, the relationship between segregation and the preferential affliction of one sex in certain diseases arose in an address Bateson gave to the Neurological Society in February 1906 (Chap. 9). Physicians were well aware of various "sex-limited inheritances," where "the peculiarity is manifested generally, if not always, by members of one sex, say the male; but the females, though unaffected themselves, may transmit it to their male offspring." Thus, "unaffected males do not transmit the condition, though the unaffected females may do so." For color-blindness, the pattern of inheritance had been established by a Swiss ophthalmologist in 1876, and Bateson noted the work of Edward Nettleship (1845–1913) in England. He also mentioned his friend Herringham's work on peroneal muscle atrophy [25].

Despite much correspondence with these physicians, Bateson did not always get his clinical facts straight. In his Neurological Society address, he noted that "the popular belief that such a [sex-limited] condition is transmitted *only* by females is, of course, a mistake. In hemophilia, for example, there are many cases of transmission by affected males." This incorrect statement was repeated in the review he submitted to *Progressus rei Botannica* in August 1906 (Chap. 9). Given this belief, it was easy to infer that hemophilia was simply dominant in males and recessive in females. We now know that the trait is linked to the X-chromosome. A male can transfer this chromosome to his offspring, but, by virtue of it being an X-chromosome, that offspring is always a female, who will have acquired another, *normal*, X-chromosome from her mother. Accordingly, the *recessive* hemophilia trait is not expressed, and she becomes a healthy carrier of the trait to, on average, half her sons (Fig. 13.2).

In his *Progressus* review, Bateson referred to the possibility, which he and Saunders had raised in *RS-Report 1*, that sex might be "a consequence of gametic segregation," and noted that Doncaster [26] had "put forward an ingenious hypothesis on similar lines" based on studies of the inheritance of color variation in moths (*Abraxas grossulariata*). Given the "observed facts," it seemed possible to Bateson "that haemophilia and colour-blindness were dominant in males and recessive in females." However, he added that "this hypothesis does not agree well with such numerical results as exist, for I find that many more males are affected (with haemophilia) in these families than would be expected on a simple Mendelian scheme, and too many females transmit." In short, Bateson was stumped! He did not begin to get at the truth until Nettleship, having seen the first edition of *Principles*, corrected his misinformation.

In *Principles,* Bateson had listed as the three main features of several sex-limited conditions: "1. They affect males more commonly than females. 2. They may be transmitted by affected males but are not … transmitted by unaffected males. 3. They are nevertheless transmitted by the *unaffected females*. Apparently normal women, daughters, or sisters of the affected males, may thus transmit the condition to some of their *sons*." Nettleship informed Bateson that statement 2 was incorrect: *all* of the 23 sons of color-blind fathers he had studied were normal [27]. Thus, transmission was exclusive to females, who were usually themselves unaffected, and usually only *some* of their sons were affected (Fig. 13.2). Nettleship and others had a total of seven cases of color-blind women who were thus presumably homozygotes. These women had among them 17 sons, *all* of whom were color-blind (Fig. 13.3). Having learned this, Bateson saw that there was "a clear indication of dimorphism among the sperms, such that those destined to take part in the production of females bear the colour-blindness factor, while those destined to fertilize the male ova [those destined to make the zygote male] are free from this factor." The correction was added to the August 1909 reprinting of *Principles,* but, despite an eureka moment (see below), his continuing confusion is suggested by the reference to "male ova."

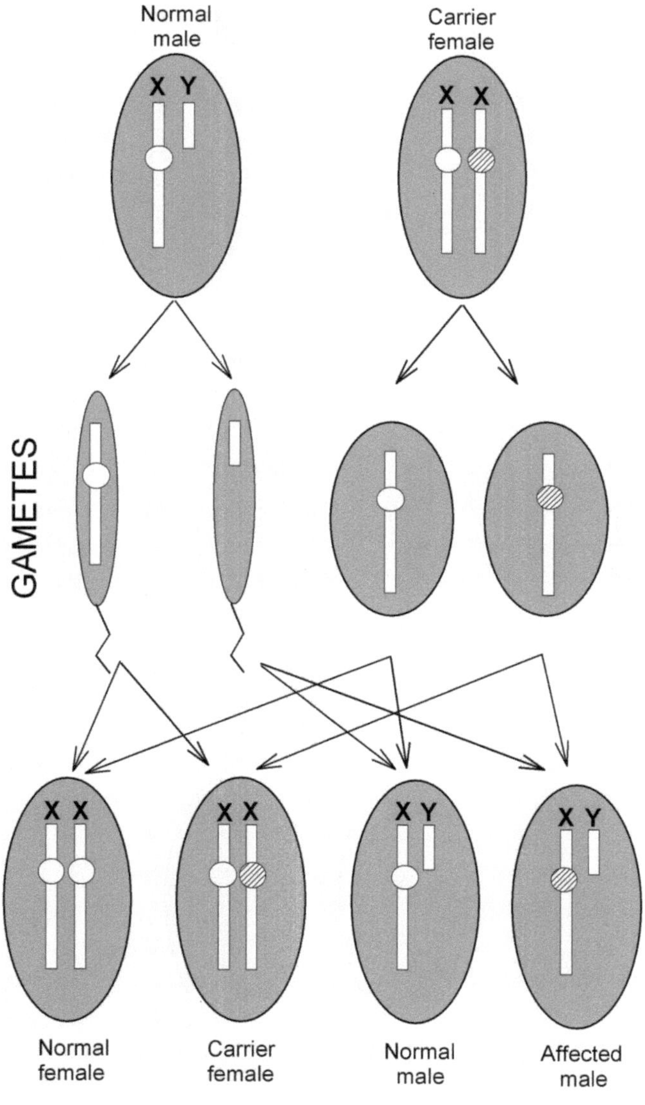

Fig. 13.2 Sex-linked characters. When the mother is heterozygous (only one of her two sex chromosomes is affected), half the sons are affected. Small open circle: normal copy of a gene. Small, slashed circle: abnormal copy of the gene. Since the gene is recessive, the normal copy of the gene in heterozygous "carrier" females suffices to keep them healthy

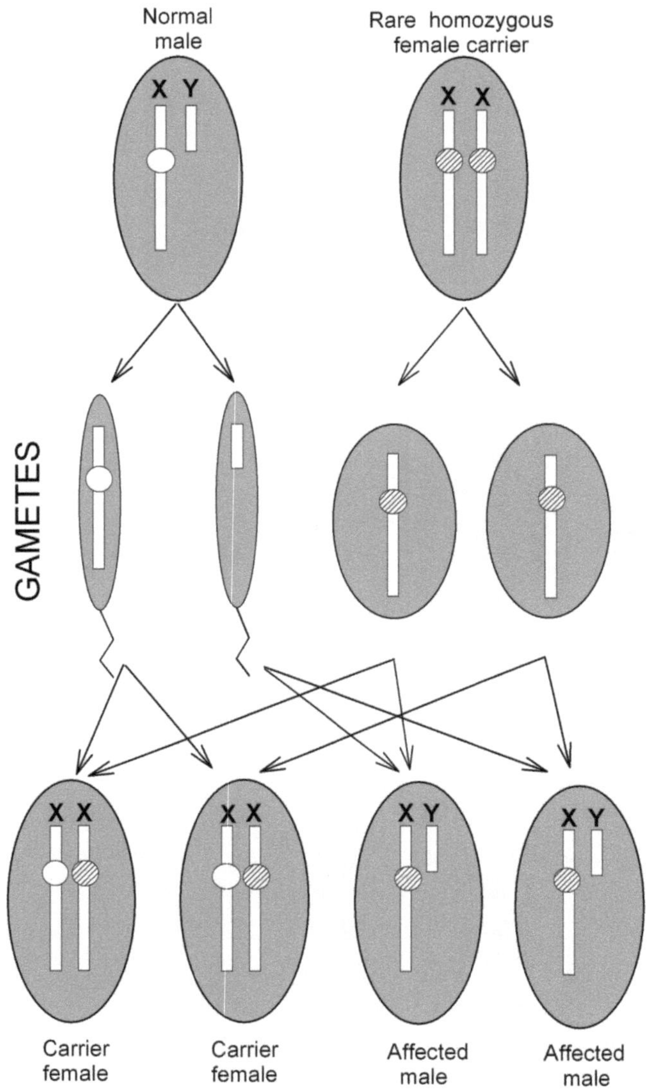

Fig. 13.3 Sex-linked characters. When the female carrier is homozygous, all her sons are affected. For details, see Fig. 13.2. Since the female parent has two copies of the abnormal gene (i.e., no copy of the normal gene), she is affected, but her daughters are not

Differentiation of Chromosomes

Concerning "the facts of cell division," in his *Progressus* review Bateson had declared that "all that has been observed by cytologists is consistent with the results of experimental genetics." An "important advance" was "the recognition of a definite differentiation among the chromosomes," and he had noted with approval

Wilson's "evidence of a visible dimorphism among the sperm cells of certain insects" that "corresponds with differences existing between the somatic cells of the males and females of these species."

Much of this was repeated in Bateson's Boston address in August 1907 (Chap. 11). As was his custom, at the outset he mentioned as a problem "the phenomena of specific difference" and cautioned that "formidable difficulties" confronted the conventional chromosome wisdom:

> In the visible rearrangement of the chromosomes … occurrences so tangible and striking are witnessed that the observer can hardly avoid exclaiming, "This is the essential process of heredity," or "Those chromosomes which I can watch and count must be the physical basis of hereditary likeness." Attractive and stimulating as these wonders are to behold, the essential is still beyond. Heredity began in the explosion which impelled the chromosomes on their courses. If it were possible to identify the chromosomes ever so clearly as the physical bearers of hereditary characters, the problem of division [of cells] would remain, and I am strongly led to expect that it must be in some new light on the causation of the division that the way to attack the essential problem will be found.

As usual, "bearers of hereditary characters" was Batesonian shorthand for "bearers of determiners of hereditary characters." Whether chromosomes were the bearers was, for Bateson, "the more immediate question" about which he did not presume "to a definite opinion," but it was not "the essential problem." Regarding the immediate question, except for the sex-determining "accessory chromosome," no one had "been able to connect a cytological difference with a character difference in any instance." Regarding the accessory chromosome:

> The definiteness of the facts is evident beyond all question, and whether the accessory body is in these types the "cause" of femaleness or only associated with that cause, we have at last the long expected proof that sex is determined by the germ cells, so far as these specific cases are concerned. In these cases we may even go farther and declare that the female is homozygous in femaleness [e.g. XX], while the male is heterozygous in sex [e.g. XY].

Bateson's chromosomal agnosticism rested largely on differences between *species*, for "no correspondence with chromosome numbers and complexity of structure has ever been asserted to exist. Low forms may have many [chromosomes]; highly complex types may have few [chromosomes]. Then, on the contrary, very closely allied types may show great differences in these respects." He was thinking about chromosome differences between species in the *same frame-of-reference* as anatomical differences between species. Closely related species were recognized as such because of their overt anatomical similarities. Similarly, Bateson thought that they should resemble each other at the microscopic level (which they often do), and this microscopic similarity should extend to chromosome number and morphology. He viewed chromosomes in the same context as other intracellular structures, not as unique entities:

> In *Aphis* Miss Stevens … has shown how wide a diversity may be presented by the chromosomes of forms so alike as to have passed for one species. These differences prove both too little and too much. I cannot but believe that all this evidence points to the conclusion that we are about to find among the chromosomes one more illustration of the paradoxical incidence of specific difference, not the fundamental phenomena on which that difference depends.

To back this up, Bateson gave examples where *anatomical* features critical for the classification of one group of species, paradoxically, were of no use in classifying another group of species. Thus, "the specificity of the chromosomes may conform in general to these other phenomena of specificity."

Anatomical or physiological differences may be "varietal," "specific," or "generic" (Chap. 9). Having agreed that the *within-species* differences between the two sexes could be correlated with the "accessory chromosome," the most Bateson would agree to was that other *within-species* "varietal" differences might have chromosomal correlates: "All that has been witnessed regarding the behaviour of the chromosomes is in fair harmony with the expectations which our Mendelian experience would lead us to form respecting the hypothetical 'bearers' of varietal differences." That he specified *varietal* differences, not just "differences," or "character differences," denotes an unwillingness to extend this generalization to the anatomical or physiological differences that distinguished species and genera.

The Heterozygous Sex

In January 1908, the boys had colds and Beatrice took them to the seaside for a break. On the 9th, she received a telegram: "Silky problem almost certain solved. Solution very exciting. Bateson." Shortly thereafter, she received a letter he had begun writing at 6:45 a.m.:

> I have almost certainly solved the silky problem, and a great part of sex with it! Last night I began to try to think over sex for my book. I started by reading what I had written for *Progressus*. Coming on Doncaster's moth story I felt it most unsatisfactory, and then I tried working it with the hypothesis that ♀s are all heterozygous in sex, being ♀♂ [having two types of gamete], ♂ being homozygous ♂♂[having one type of gamete]. To my surprise I found it went quite smoothly and saved a lot of assumption. So then I tried the Silkies with this clue – the exact contrary to what I had assumed before Christmas. I stuck to it all evening and by dodging it about in various ways, before midnight I had got a scheme which *approximately fits all the facts*. It is so simple and on the whole fits the facts so well I feel sure it can't be far out. I wrote it out for Punnett and went to bed at 1 a.m. but only slept a little! I got up at 5.45 as I could not stay in bed and have been touching up my letter to P. When it gets light I shall run in a wire for you. I feel rather like I did on the morning of January 11, 1889 [when they were engaged] – Very pleased with myself – only perhaps a little more *certain* I am on the right track. Also the risks incurred are not so great, because hypotheses can be amended: wives less easily.

Excited, he tramped through the early morning snow, sent off the telegram to Beatrice at 7:30, and knocked at Florence Durham's house at 8:00. He was answered by a maid in nightshirt, who informed him that her mistress did not breakfast until 9:30. Nevertheless, he "communicated the joyful intelligence in her chamber," and then discovered he had lost his ring in the snow.

Shortly thereafter, a paper – "The Heredity of Sex" – with Punnett as coauthor, was submitted to, and quickly published by, *Science* (1908). In it, they stated simply: "On general grounds it seems to us probable that one and not both sexes of the same organism will be shown to be heterozygous for sex and that the approximately equal output of the two sexes in ordinary cases is a consequence of this." It is evident (Fig. 13.2) that if one sex produces equal numbers of two kinds of gamete, one conferring maleness and one conferring femaleness, and if both kinds are equally able to unite with the gamete of the other sex, then there should be approximately 50% males and 50% females in the population. This would occur irrespective of which sex was the donor of the two kinds of gamete. Thus, they continued: "There is, we think, no reason *a priori* why in nature generally dominance should be a special property of one sex alone. We rather anticipate that dissimilarity will be found between the great groups in this respect."

The paper was perhaps accepted on the advice of Wilson. That year, he was Chairman of the Zoology Section at the meeting of the American Association for the Advancement of Science in Baltimore. His address on "Recent Researches on the Determination and Heredity of Sex" was published in January 1909 [28]. He opened with the point that it was customary to refer to gametes arising from males as "male gametes" and gametes arising from females as "female gametes." However, new work on "sexual predestination" indicated that the gametes from *one* individual, if that individual were "heterozygous" in respect to sex, were *themselves* of two sexes, and this sex determined the sex of the organism to which they contributed. One individual could produce both male gametes that would transfer maleness and female gametes that would transfer femaleness.

Often it was not possible to tell by microscopy whether a gamete was male or female, although in some species ova showed this dimorphism. Wilson attributed to Castle the idea that both sexes were heterogametic, to Correns the idea that the male sex was heterogametic and the female homogametic, and to Bateson the idea that the female sex was heterogametic and the male homogametic. He noted that Bateson "further suggests that different species or groups may differ in respect to the sex that is homozygous." As Bateson later acknowledged, the suggestion had been made by Geoffrey Smith in 1906 [29], but it had also been hinted at in *RS-Report 2* (see above). Under the Castle scheme, only gametes conferring opposite tendencies could unite, so "selective fertilization must be assumed." This to Wilson seemed improbable. Bateson set out the case in his 1909 Centenary essay (Chap. 12):

> If the accessory body is really to be regarded as bearing the factor for femaleness, then in Mendelian terms female is *DD* and male *DR*. The eggs are indifferent and the spermatozoa are each male *or* female. But according to the evidence derived from a study of the sex-limited descent of certain features in other animals the conclusion seems equally clear that in them female must be regarded as *DR* and male as *RR*. The eggs are thus each either male or female and the spermatozoa are indifferent. How this contradictory evidence is to be reconciled we do not know.

This was restated in *Principles* (1909): "It would *a priori* seem most probable that one sex is heterozygous in sex and the other homozygous." It is today still not known why species differ in their "choice" of which sex will be heterozygous. The question might be approached by asking why certain countries chose to drive on the left side of the road, and others on the right? In *Principles,* Bateson repeated his belief that naturally the common expectation we all share is that the reduction division [of meiosis] is the critical moment [of segregation of the character-units]. At that division the number of chromosome-elements, of which the nucleus is formed, can be seen to be halved. Up to this point the nucleus of each daughter-cell seems to be simply a repetition of the nuclei of the body and may be supposed to contain all the elements which they contain. But when the number of these elements is halved, the germ-cell begins to acquire its own specific features, and we may without much difficulty imagine that if two daughter-cells are to be differentiated allelomorphically from each other, the differentiation will come about at this reduction, or meiotic division, as it is called.

With his usual clarity, Hurst relayed Bateson's viewpoint on the mechanism of inheritance of sex in a talk given (March 29, 1909) to the ephemeral Mendel Society [30]:

The individual distinctness of the sexes ... indicates the complete segregation of maleness and femaleness. This segregation of the sexes and their occurrence in equal numbers at once suggested the well-known Mendelian ratio of 1:1 , ... the result of mating a Mendelian hybrid with an individual carrying the [pure] recessive character, and it indicates that one of the two sexes is a dominant to the other. ... If we tentatively regard the female as a dominant hybrid and the male as a pure recessive (or *vice versa*), then they will be respectively symbolized as *D R* and *R R*. ... The expected offspring from a *D R* parent mated with an *R R* one, will consist of equal numbers of *D R*'s, which in this case will be females – if we regard the female character as dominant – and of *R R*'s, which will be males.

Hurst drew diagrams to illustrate his talk, with both gametes and individuals represented as circles. These contained elongated rods, as in Fig. 13.2, but in this case the rods were unambiguously referred to as chromosomes. His belief in the chromosome theory was explicit: "There are reasons for believing that the different factors which determine the hereditary characters of organisms are carried by certain nuclear bodies called the chromosomes."

Unfortunately, as Bateson and Punnett deliberated on what they saw as the "coupling" or "repulsion" of characters, they became enchanted by a new "reduplication hypothesis" (see below). The relatively simple view of only one sex being heterozygous then became more complicated. In 1911, in a paper in the *Journal of Genetics* on fowl, where females are the heterozygous sex, Bateson and Punnett presented the results of breeding studies and there was no mention of chromosomes:

It must be expressly stated that the suggestion that females are heterozygous for *femaleness* is offered without prejudice as to the possibility that males may also be heterozygous in *maleness*. The systems followed by the descent of colour-blindness in Man and by that of the white eye recorded by Morgan in *Drosophila* clearly point to the existence in those cases of a repulsion between a factor for maleness (*M*) and factors respectively for colour-blindness and for the red eye. The operation of the system of sex-limitation is similar in all these examples, the only difference being that in the one group the repulsion is from the factor *F*, in the other from the factor *M*. Recognition of the existence of factors both for femaleness and for maleness of course involves the assumption that ova bearing *F* can only be effectively fertilized by sperms not bearing *M*, and *vice versa*. For that supposition no independent evidence yet exists, and we note that Morgan has made observations on *Cumingia* (Mollusca) distinctly unfavourable to it. At present however we think it is the most acceptable account of the facts ascertained both as to the heredity and the variability of sexual characters.

Bateson did not abandon the idea of two types of ova in humans. In 1914, in Australia (Chap. 14), while acknowledging that spermatozoa were of two types, he proclaimed that "there is evidence that the ova also are similarly predestined to form one or other of the sexes, but to discuss the whole question of sex-determination is beyond my present scope." Lingering doubts about the heterozygous sex were communicated to a biochemist (H. Onslow) many years later (March 6, 1921):

> We have proof that in man ♂ is [the] heterozygous [sex] and in Lepidoptera ♀ is [the] heterozygous [sex], but I am not *thoroughly* satisfied that both sexes cannot be heterozygous … I don't think they are, but I just keep a look out that way. We must not be bluffed by X chromosomes. I have no doubt Doncaster went wrong in his interpretation of the *lacticolor* experiments by thinking that he *had* to make the ♂ heterozygous on account of the X chromosome then rising in the West.

Sporadic Sterility

At the time of *RS-Report 1,* Bateson and Saunders had known of "no Mendelian case in which fertility is impaired" (i.e., sterility that we would now attribute to a genic defect). By 1903, they were studying the *regularly* appearing sterility of hybrids that were derived by crossing the succulent species *Kalanchoe bentii* and *Kalanchoe flammea.* In a letter to Thiselton-Dyer (May 17, 1903), Bateson observed: "It is provoking that we seem unable to make progress with the 'causes' of these sterilities. All that seems visible is that the pollen grains can't divide properly."

However, in *RS-Report 2,* Bateson, Saunders, and Punnett reported studies, mainly carried out by Punnett, on crosses between certain races of the sweet pea (*Lathyrus odoratus*). Among these *race-hybrids,* they observed that the F_2 generation contained "a large proportion of plants in which the male organs, at least, were sterile. These anthers were in various states of imperfection, some small, others fair size, but withered, or 'contabescent,' to use Darwin's term." Among 227 F_2 plants, "there were 173 fertile, 54 sterile, or 3.25: 1, a ratio which makes it probable that this sterility is a recessive character transmitted in a Mendelian fashion." They now had a Mendelian case where fertility was impaired. Further evidence that a Mendelian factor was involved came from the observation that the sterility character was coupled to another character, leaf axil color (see below).

Today, we would interpret this type of sterility as a malfunction due to a mutation in a gene determining one of the many cell components required to bring about successful development of the anthers. In this case, the component seemed to be required in the reduction division of meiosis (see below). Whatever the ultimate basis of the sterility, Bateson and coworkers pointed out that this *sporadic* event appearing in a *within-species* cross (unpredictable in that it had not been anticipated when the interracial cross was first performed) contrasted with the *regular* (predictable) sterility appearing when allied species in general were crossed. *RS-Report 2* continued:

Contabescent anthers were seen from time to time in many families .... This sporadic sterility has not been particularly studied. It is of interest to compare this example of the definite appearance of sterility, at least on the male side, with the familiar [regular] occurrence of sterility in cross-breds. Such a relationship has often been supposed to indicate remoteness of kinship, yet here a closely comparable effect occurs in F_2 as the result of a cross between two types which must be very nearly related. Mr. Gregory in a careful examination of the pollen-genesis, found that the divisions were normal up to the reduction-division, when the chromosomes form shapeless knots and entanglements, failing to divide.

In 1904, Gregory had reported studies of meiosis in ferns "in connection with other cytological work bearing upon the Mendelian hypothesis," which led him to "a consideration of the significance of the reduction division in connection with Mendelian segregation." He noted [31]:

The occurrence of a qualitative reduction [division] in plants as well as in animals is extremely important as affording a possible provision for that purity of the gametes, in respect of allelomorphic characters, which is demanded by Mendel's hypothesis. The work of Boveri upon the qualitative differentiation of chromosomes, supported by that of Sutton, McClung, and others, affords strong evidence in favour of the theory that the development of certain characters in the zygote corresponds with the presence of certain chromosomes or groups of chromosomes in the nuclei.

Gregory drew attention to Sutton's above-quoted remark: "that a large number of different combinations of maternal and paternal chromosomes are possible in the mature germ-products of an individual." This, together with earlier observations of Hacker, led Gregory to suppose "a more comprehensive symmetry, which probably underlies the production of the different types of gametes in Mendelian hybrids." He then cited *Defence,* where Bateson had written:

It is impossible to be presented with the fact that in Mendelian cases the cross-bred [hybrid] produces on an average *equal* numbers of gametes of each kind, that is to say, a symmetrical result, without suspecting that this fact must correspond with some symmetrical figure of distribution of those gametes in the cell-divisions by which they are produced [i.e. during meiosis].

Gregory continued: "On the hypothesis that the segregation of characters occurs at the reduction division, we shall expect that the mitoses [meiosis] in a Mendelian hybrid will be perfectly regular, and in our present condition of inability to recognize qualitative differences between chromosomes alike in form, we should further expect that the mitoses [meioses] will differ in no *visible* way from those of the pure paternal and maternal races." He then referred to coupling (correlation of sterility with axil color) in the above "race-hybrids" of the sweat pea, samples of which had been provided for his study by Bateson:

The sterility which characterizes many hybrids follows upon the abortive development of the sex cells, and the suggestion has been made that this may be due to the inability of the hybrid to separate [segregate], in the formation of the gametes, the characters which were united in the hybrid zygote. It is well known that sterile plant-hybrids are particularly characterized by abortive development of the pollen .... Among the offspring of a race-hybrid of *Lathyrus odoratus* fertilized with its own pollen, Mr. Bateson obtained a number of individuals which failed to form good pollen. In the plants with coloured flowers the sterility was, with few exceptions, correlated with the development of a somatic character – the

> sterile plants generally possessing a green leaf axil, while the fertile coloured plants with rare exceptions had red axils. ... The irregularity makes its appearance only in the heterotype [reduction] division. ... Since the equation divisions are quite normal, this would seem to indicate that the union of the chromosomes in synapsis is such as to prevent any subsequent separation, the result being that no sex-cells can be organized, since the essential condition of a qualitative separation of the chromatin is not fulfilled.

Guyer

Concerning hybrid sterility among sweet peas, Gregory cited *RS-Report 1* which contained a long footnote on similar studies in sterile and fertile hybrid pigeons reported in two abstracts in successive issues of *Science* (February 1900) by Michael Guyer (1874–1959), who had been "in ignorance of Mendel's work" [32]. Guyer had noted that "peculiarities in chromosome formation may point perhaps to a tendency in the chromatin of each parent species to retain its individuality." In this case, seeming to make the same error as Cannon (that maternal and paternal chromatin would remain entirely separate), Guyer saw that some of the new gametes would have chromatin from one parent and some from the other, and observed that "reversion" of the offspring of hybrids (the F_2 generation) to parent species "may be due to the persistence of the chromatin of only one species in one or both germ cells." However, Guyer added that "the other variations in the offspring of hybrids may be due, perhaps, to the varying proportions of the chromatin of each species in the mature germ cells." Thus, Guyer entertained the notion of some intermixture of chromosomes (i.e., recombination), but did not specify how this might come about.

From this, we see that Bateson was closely familiar with the work of Guyer. Furthermore, Bateson recognized that the sterility observed could sometimes be of the type we now call "genic," and that there could be co-inheritance (linkage) of a sterility character and a color character. Such rare "sporadic" genic sterility between "types very nearly related" was *unpredictable* in occurrence, and this distinguished it from the *predictable* occurrence of sterility in the majority of if not *all* offspring when there was "remoteness of kinship" between the parents.

In March 1899, Guyer had submitted to *Zoological Bulletin* an early report of his PhD studies at the University of Chicago under Charles Whitman (1842–1910). In crosses between white and brown pigeon varieties, the first generation (F_1) were all brown and fertile [33]. When these were bred among themselves, a mixture of white and brown emerged in the second (F_2) generation, the browns tending to be male and the whites tending to be female. One of the white females was sterile, of strange appearance and voice, and with degenerate ovaries that were devoid of mitotic figures. Guyer speculated that her "disreputable" appearance was secondary to the ovarian abnormality, which was primary, and "might in some way be connected with hybridization," an idea rendered "by no means implausible" by his ongoing studies that included hybrids derived from crosses between members of different pigeon species.

The latter hybrid studies were reported in 1900 in his *Science* abstracts (see above) and in his thesis (dated May 10, 1900, but not formally published until November 1902). The thesis described studies of chromosomes during meiosis, where in fertile hybrids the parental chromosomes came together in pairs, but in non-fertile hybrids, due to "incompatibility between the chromosomes from the two different species," instead of pairing there was "irregularity in division" and the chromosomes assorted in independent groups ("separate spindles"). Thus, Guyer noted [34]:

> In the spermatocytes of normal pigeons there is no satisfactory evidence other than the remarkable decrease in the number of chromosomes to indicate the exact nature of the division, but in hybrids, owing to certain exaggerations which frequently occur, especially in hybrids from very distinct species, some very significant facts come to light.

Thus, abnormal "exaggerations" (Bateson would have said "exceptions") could reveal facts that were not so easily discernible in normal types. Guyer considered there to be a "union" or "conjugation" of maternally derived and paternally derived homologous chromosomes, but in non-fertile hybrids the "pairing" necessary for formation of the bivalent chromosomes characteristic of the synapsis stage "comes about with more or less difficulty or not at all." Studies of such hybrids could provide a valuable "index" of what was occurring in normal hybrids:

> The irregularity in division of the primary spermatocytes which appears in hybrids between very different species, is but an index of what occurs in normal crosses. In the latter, instead of separate spindles and non-fusion of chromosomes, a true union occurs, but the bivalent chromosomes ultimately divide in such a way that the respective plasmas occupy different cells. There is a separation of the paternal and the maternal chromosomes which had fused [paired] during synapsis.

The gametes could be considered "pure" because they contained "qualities of only one [of the parental] species." Guyer equated the chromosomes with the "chromatin" observed in the nuclei of interphase cells and was fully aware of the need to relate chromatin to the characters observed in offspring:

> In chromatin we have a substance which, for all we know of its nature and actions, seems to be intimately bound up in the phenomena of inheritance. It is reasonable to suppose, therefore, that it constitutes at least a part of the material basis for the variation of the germ. The question then arises as to whether there is any correlation between the distribution of chromatin as seen in the germ-cells of hybrids, and the marked variability which characterizes the offspring of fertile hybrids.

Guyer concluded with a statement of what we now refer to as Mendel's law. He had derived this independently, albeit possibly with some statistical input from his colleague Davenport, and albeit assuming that "predominantly" characters of one parent would be inherited in one germ-cell package:

> There seems to be no other interpretation … than that in the many normal mitoses [meiosis] of the bivalent chromosomes which occur, the chromatin of the father and mother are set apart so that the ultimate germ-cells are what might be termed "pure" cells; that is, a given egg or sperm-cell contains exclusively, or at least predominantly, qualities from one parent. The offspring from fertile hybrids of the same parentage might then be similar to the mixed type of the original hybrid, or revert to one of the grandparent types, dependent upon the chances of the various cells for union at fertilization. If a spermatozoon and an egg containing characteristics of the same species [elements determining characters of the same type]

> unite, then the reversion will be to that species [type]; if a sperm cell containing the characteristics of one species happens to unite with an ovum containing characteristics of the other species, then the offspring will be of mixed type again. By the law of probability the latter will be the more prevalent occurrence, because there are four combinations possible, and two of the four would result in the production of mixed offspring, while only one combination could result in a return to one of the ancestral species.

This situation would correspond to the case of de Vries's "monohybrid" crosses, involving a single character. Following the 1902 New York conference, where Guyer's work was mentioned but not formally presented, it was often cited [35]. Guyer confirmed that in 1900 he had been unaware of Mendel's work, in a paper entitled "The germ cell and the results of Mendel," which he presented at a conference in April 1903 and published in May in a journal published by the University of Cincinnati where he was then based [36]. Here, he recognized the possibility of the random assortment of each member of a homologous pair of chromosomes, thus providing a mechanism for the intermixing of parental chromosomes, and aligning his viewpoint with Sutton's:

> Fertile hybrids from closely related forms ... display spindles normal in appearance. We may suppose, however, that there is the same tendency for the respective parental characters to separate, though the incompatibilities of the plasmas are not sufficient to prevent the formation of bivalent chromosomes and normal spindles. It does not necessarily follow that in the ensuing division all of the characters of one parent will be set apart in a separate cell [gamete], as was the case in the abnormal mitoses. If such a separation were actually to occur, then the offspring which returns to the grandparental type must revert in not only one given character, but in all characters; that is the reversion would be complete. The Mendelian law, however, confines itself to a given character, and if any other character is chosen, although it will follow the same law, it does so without any reference to the first character, so that offspring may be pure with respect to a particular character, yet also possess other characters of a mixed nature, or even pure characters of the other parent. In the case of these milder fertile crosses, then, where reversions follow the Mendelian law, the germinal incompatibilities must be narrowed down to the qualities themselves rather than confined to the respective germ plasms as a whole.

While not considering the possibility of linkage, Guyer then went further proposing that, whereas randomness would be characteristic of hybrids between closely related types, when less related the mixing would be impaired. Hence, in cases where sterility was incomplete, to some degree the "qualities or characters" or each grandparent might be passed intact to the offspring:

> These qualities must separate and each take up its abode in a different germ cell irrespective of whether the other qualities of that particular germ cell are of a different parentage or not. The cases in which the entire plasmas are segregated are then probably but magnified images of what occurs among the specific qualities of the milder crosses. The interesting possibility arises that if fertile hybrids can be secured from widely different species the plasmas of which must be more incompatible than those of nearly related forms, such hybrids will give rise to offspring in which there is reversion, not only of one character, but of many or all characters in the same individual, due to a more thorough segregation of the parental germ as a whole. In other words, the farther apart the parent species are, the more complete will be the return in any given offspring which shows reversion.

A Substratum

In his 1900 thesis, Guyer wrote of a "general substratum" (which we may tentatively equate with Bateson's "residue") that contributed species specific characters upon which would be superimposed individual-specific characters. Only the latter were held to be conveyed by the chromosomes. The species-specific characters he supposed would be conveyed in some way by the cytoplasm.

In his *Progressus* review of August 1906, Bateson drew attention to *RS-Report 1,* where, with Saunders, he had considered the possible existence of something quite fundamental, which was concerned with differentiating two species, but was somehow different from allelomorphic Mendelian character units (i.e., it was non-genic). They thought this fundamental something might be "a property of the residue or basis upon which the allelomorphic characters are implanted" (Chap. 8). In the interim, de Vries had addressed the issue in his two volume *Die Mutationstheorie* (1901) and his *Species and Varieties, Their Origin by Mutation* (1905). He had "defined somewhat strictly the differences between specific and varietal distinctions, asserting that it is the latter alone which exhibit Mendelian heredity." Thus at least three biologists, Bateson, de Vries, and Guyer, were in agreement in holding that whatever formed the basis for differentiating two groups as species *might not* be subject to conventional Mendelian analysis. Bateson wrote (August 1906):

> Mendelian segregation proves the unity of characters. Specific differences we must suppose are built up of characters. Is it a sound deduction that specific differences come into existence by the addition or elimination of such characters? Now it is scarcely necessary to insist that plenty of the characters which are now known to segregate would be far more than sufficient to constitute specific differences in the eyes of most systematists .... We may even be certain that numbers of excellent species universally recognized by entomologists or ornithologists ... would, if subjected to breeding tests, be immediately proved to be analytical varieties [e.g. F1 hybrids]. But this is not enough. We must eventually go further; and, supposing such tests to be applicable on a comprehensive scale to great numbers of natural forms, we must ask whether the results of such an investigation will show first that certain kinds of differences segregate and that certain other kinds do not segregate; and secondly whether we shall then recognize that it is to the non-segregating that the conception of species attaches with the greater propriety.

On the claim of de Vries to have found various "non-splitting" (i.e., non-segregating) characters, which might underlie species formation, Bateson commented:

> It is not easy to suppose that the features, breadth of leaves, and length of flowering stem – named by de Vries as non-splitting characters in *Lychnis* – are of this fundamental nature. Feeling the impossibility of now defining the segregating from the non-segregating, I am unable to follow de Vries in the further step which he has taken in assigning a definite physiological reason for the differences between these classes.

In March 1907, Guyer submitted a paper to *Science* entitled "Do offspring inherit equally from each parent?" [37]. Here, he argued that the "so-called 'unit characters'" of "the hybridist" were only "superficial individual and specific qualities" since "the crossing of any but comparatively closely related forms" was prohibited by species barriers. The fundamental characters, which make an organism, for example, "an animal and a vertebrate of a given genus and species," were "common to both parents." Or, as he later stated [38]: "Bi-parental inheritance, if extended to all the details it has been assumed to embrace, would be largely a matter of duplicating identical protoplasmic constituents." He argued for a greater role of the cytoplasm, and hence of the maternal contribution by way of the cytoplasm-laden ovum, in inheritance. Furthermore, "we must also recognize, as pointed out by Davenport and others, that this [chromosomal] theory is not in harmony with certain facts of … non-Mendelian … inheritance." Thus, "both cytoplasm and nucleus are involved specifically in inheritance, the cytoplasm of the germ cell representing the more stable and constant form of the animal, and the chromatin the more individual and variable characteristics." Here, he cited Davenport as sharing Bateson's skepticism regarding the chromosome hypothesis.

In 1907, Guyer spoke on "Deficiencies of the chromosome theory of heredity" [39] at the Seventh International Zoological Congress (Chap. 11). For him, the chromosome theory was the "theory which regards the various morphological parts of an adult as specifically predetermined by corresponding anticipatory units which reside in the chromosomes of germ-cells." At that time, there was a prominent decoy in the literature. He cited Emil Godlewski's claim that "non-nucleated pieces of sea-urchin eggs, fertilized by sperm … produced larvae *exclusively of the maternal type*," thus emphasizing the role of the cytoplasm. Likewise, Bateson in *Principles* mentioned: "The recent work of Godlewski gives … strong reason to believe that heredity in Echini may be governed by the cytoplasm of the egg." [40]. Furthermore in his Centenary essay, admitting to a view "no doubt contrary to the received opinion," Bateson wrote:

> All attempts to investigate Heredity by cytological means lie under the disadvantage that it is the nuclear changes which can alone be effectively observed. Important as they must surely be, I have never been persuaded that the rest of the cell counts for nothing. What we know of the behaviour and variability of chromosomes seems in my opinion quite incompatible with the belief that they alone govern form, and are the sole agents responsible in heredity.

Guyer went far beyond Bateson in disparaging the beliefs that "a multitude of particles bearing incipient hereditary qualities must exist in the germ cells," and that "the chromatin masses which make up the individual chromosomes were assigned the role of being the actual bearers of the so-called heredity qualities." To the contrary, Guyer argued that there was "no evidence which will warrant us in assuming that the nucleus bears or makes, as it were, self-sufficient morphological units which at just the right time pass out and take up their proper position in the cytoplasm and with the more of less passive cooperation of the latter expand into the structures required." For "heredity is a problem of the handing on of metabolic energies already established, rather than of the transmission of a series of determinative units which create a wholly new organism." Furthermore: "I see no more necessity for postulating specific anticipatory characters [i.e. genes] … than I do of regarding yeast, or flour, or milk as in itself a specific determinant of a loaf of bread." Indeed, "it would seem that we might account for the so-called Mendelian phenomena by attributing to the chromosomes simply chemical and physical differences without endowing them with morphological entities."

Compared with Guyer, Bateson was a moderate! Bateson himself in *Materials* had made the point that the new organism was made afresh, "just as the wax model has gone back into the melting pot before the new model was begun" (Chap. 4). Not thinking in informational terms, Guyer seemed unable to admit that the preparation of a loaf of bread would require more than just the ingredients. Where was the cook? Guyer's thinking was more in line with that of Spencer (Chap. 19), but nevertheless, he was warmly applauded by Spillman [41]. Guyer's doctoral supervisor, Whitman, who was still at the University of Chicago, was as skeptical as Guyer. He doubted in 1909 that both "the species" and "its characters" could be interpreted in terms of the unit characters referred to by de Vries [42]:

> The idea of unit characters ... as distinctive elements that can be removed or introduced bodily into the germ does not appeal to me as removing difficulties, but rather as hiding them .... I am strongly persuaded that his [de Vries'] hypothesis of unit characters fails as a guide to the interpretation of the species and its characters. It is true that a great amount of work on Mendelian heredity seems strongly to support the unit-character hypothesis, and that cytology offers some further support.

It is of interest that, of all the cities visited by Bateson during his 1907 tour, he came away from Chicago the most distressed (Chap. 11). But the case for a role of cytoplasmic factors in non-Mendelian inheritance was a real one [43], albeit more of a decoy than a contributor to an understanding of evolutionary processes. And Bateson was in good company in considering it. Jacques Loeb in America [44] and Johannsen [45] continued to think along these lines.

Rejuvenation

Guyer's derivation of Mendel's law was purely descriptive, based on cytological appearances. Like Guyer, Montgomery in 1901 postulated a "union" or "conjugation" of maternally derived and paternally derived homologous chromosomes. However, he provided an explanation for conjugation – "rejuvenation." This hypothesis for the adaptive basis of sexual reproduction resembled those of Galton and Butler [46], and was supported by studies on unicellular organisms ("infusoria"). Montgomery wrote of the "maintenance of chromosomal individuality ... through all generations of the germinal cycle," but held that

> this conclusion by no means implies that a chromosome of one generation is actually the same as the chromosome of another. ... [For] new substance must continually be elaborated by chromosomes during the rest stages. ... But nevertheless, it seems very probable that a chromosome of one generation is a derivative of a particular chromosome of the preceding generation, and that the chromosomes may thus be said to maintain themselves as entities through successive generations.

Concerning the repeated cycling of organisms – from gamete to zygote to embryo to adult to gamete – through the generations (Fig. 13.2), Montgomery noted:

> Each such cycle is succeeded by a similar one, and so on indefinitely for an indefinite number of cycles. Now it is unthinkable that a cycle should be without a beginning; it must have been gradually evolved, and some particular stage in it must have been the starting point. What was this first stage? An answer is necessary before we can enter into the discussion of the meaning of the synapsis stage. It appears to me most probable that the stage of conjugation of the germ cells must be considered the starting point. For from the studies of R. Hertwig and Maupas on Infusoria, it appears probable that conjugation or fertilization is essentially a process of rejuvenation: cells may divide and reproduce for a number of generations asexually, but there comes a period when the cellular vitality diminishes, so that no further reproduction is possible except after rejuvenation by conjugation with another cell. When thus rejuvenated by admixture of substances from the other conjoint, the cell starts upon a new period of generation – the period of conjugation thus being the commencement of the cycle.

Montgomery next reviewed evidence relating to the significance of the pairing of chromosomes – the synapsis stage – and concluded:

> These considerations render it very probable that in the synapsis stage is effected a union of paternal and maternal chromosomes, so that each bivalent chromosome would consist of one univalent paternal chromosome and one univalent maternal chromosome. … The synapsis stage … may be considered the stage of the *conjugation of the chromosomes*. When the spermatozoon conjugates with the ovum there is a mixture of cytoplasm with cytoplasm, of karyolymph with karyolymph, possibly also an intermixture of other substances; but there is no intermixture of chromatin, for the chromosomes then, as we have seen, remain more separated from one another than at any other stage. … But after this beginning stage of the germinal cycle [in the gonad], the repulsion of the paternal for the maternal chromosomes gradually diminishes, is generally no longer recognizable in the last of the spermatogonic and ovogonic divisions, and in the synapsis stage instead of repulsion we find a positive attraction between the paternal and maternal chromosomes.

Why did this occur?

> The reason … is obvious; it is evidently to produce a rejuvenation of the chromosomes. From this standpoint the conjugation of the chromosomes in the synapsis stage may be considered the final step in the process of conjugation of the [parental] germ cells [that in humans would have occurred decades earlier]. It is a process that effects the rejuvenation of the chromosomes; such rejuvenation could not be produced unless chromosomes of different parentage joined together, and there would be no apparent reason for chromosomes of like parentage to unite. At the same time the so-called "reduction in number" of the chromosomes is effected, but this is probably not primal, but rather a necessary result of the conjugation of the chromosomes.

The rejuvenation would involve chemical exchanges:

> Now R. Hertwig and Maupas have shown for *Infusoria* that the two conjoints remain for only a certain period in apposition, and that when the interchange of nuclei necessary for rejuvenation has been accomplished the conjoints separate. Of course, it is not a true analogy to compare conjugating *Infusoria* (i.e. whole cells) with conjugating chromosomes (i.e. portions of cells). But still it is very probable that the two chromosomes unite temporarily for the same reason that two *Infusoria* do, that is, for an interchange of substances.

In the course of his long paper, replete with 238 figures, Montgomery challenged the powerful McClung at several points. Scientific controversy was not unique to England! Bateson never cited Montgomery, and as we shall see, he never understood the key step Montgomery had taken in relating the behavior of chromosomes to "rejuvenation." A half century later, Gregory Bateson came to see this in terms of error correction [46]: "By return to the unlearned and mass-produced egg, the on-going species again and again clears its memory banks to be ready for the new" (see Part VI).

Dosage

When a gene is present in a varying number of copies, its dosage can be said to vary. When the dosage is zero, the gene is absent, and the corresponding character is absent. When a given dosage satisfies the needs of the organism, then the character will be fully developed. In this circumstance, increasing gene dosage will not change the character. Such dosage considerations were considered by various workers when considering possible explanations for genetic dominance. In his essay on "Fertilization and hybridization," de Vries wrote [17]: "An active and a latent [inactive] unit are not essentially different in their cooperation from two active ones; a fact which may probably be best explained by the assumption that two cannot accomplish more than one already does." Bateson made a similar point in *Principles* (1909) referring to the homozygous dominant as DD and the corresponding heterozygote as DR:

> In cases where pure dominants are recognizably distinct from the heterozygous dominants, it must naturally be supposed that two "doses" of the active factor are required, one from the paternal, and another from the maternal side, in order to produce the effect. … Dominance indeed is not often so pronounced that a practiced observer cannot distinguish the DD types from the DR's with fair certainty by a thorough and minute examination. As examples in which heterozygotes are indistinguishable from pure dominants, may be mentioned tall and dwarf in Peas and Sweet Peas, coloured flowers and white flowers in Sweet Peas and Stocks, hoary and glabrous in Stocks. In all these the one "dose" of a dominant factor is sufficient to produce the full effect.

George Shull visited Bateson in December 1908, and the same month he submitted a paper on gene dosage to *American Naturalist* [47]:

> Having arrived at the conclusion that all Mendelian characters are dependent upon chemical relations, we may return to the question of dominance. … A fundamental principle in this connection is the law that the extent of a reaction between two chemicals is determined by the amount of that reagent which is present in less relative quantity, and not by the one which is present in excess. When the positive homozygote [DD] …, and the heterozygote [DR] … are alike, i.e. when there is complete dominance of presence over absence, it may mean that already the presence of the one unit [D] … of the heterozygote is sufficient to result in the maximum reaction, in which case the doubled factor [DD] … of the positive homozygote can do no more. When, on the other hand, one unit [D] … is not sufficient to produce a maximum reaction with the other factors present, the [DD] … of the homozygote produces the corresponding character in greater intensity, and the heterozygote will be intermediate between the two homozygous parents.

Similar ideas on dosage were independently advanced in 1910 by A. R. Moore in California [48] and were resurrected later by Haldane, leading to a fierce dispute between Fisher and Wright in the 1930s [49].

Fruit Fly

Galton recognized that breeding studies were best performed with organisms that bred rapidly generating large numbers of offspring in each reproductive cycle. Insects seemed ideal for this purpose. He tried his hand with moths (*Natural Inheritance*, 1889):

> The moths *Selenia Illustraria* and *Illunaria* are chosen for the purpose, partly on account of their being … double brooded; that is to say, they pass normally through two generations in a single year, which is a great saving in time to the experimenter. They are hardy, prolific and variable, and are found to stand chloroform well, previously to being measured and then paired. … Among other reasons for choosing moths for the purpose, is that they are born adults, not changing in stature after they have emerged from the chrysalis and shaken their wings. … An intelligent and experienced person can carry on a large breeding establishment in a small room, supplemented by a small garden.

However, Galton's "somewhat extensive series of experiments with moths" failed because of difficulties of maintaining the stock. While looking for "small, fast-breeding, mammals," he came across extensive records (studbooks) for dwarf bloodhounds, known as Bassett hounds, and interpreted the data as supporting his Law of Ancestral Heredity [50].

Shortly after his appointment to the Evolution Committee in 1897, Bateson was studying three "species" of butterflies (*Pieris napi*, *Pieris egeria*, and the Alpine variety *Pieris bryoniae*). At the end of the year, he was able to report "a number of crosses between *napi* and var. *bryoniae* were made and a large stock of the pure forms have been reared for next season's work." In correspondence with Galton, he hammered home the advantages of insects (undated, circa 1897): "Among wild animals, insects are especially tolerant of captivity and in many other ways are exceptionally suitable for breeding experiments. They can be easily preserved in vast quantities without expense, and their hard parts … which show many specific characters, last indefinitely without any special preparation at all." In *RS-Report 1,* Bateson reported failed experiments with lepidoptera, which were "exterminated by disease." Nevertheless, Bateson continued to suggest that these offered "unusually good opportunities for investigating problems of heredity."

Weldon began rearing moths in London, and continued this at Oxford, but there had been high death rates [51]. It had proved difficult to feed large numbers of insects and to count the number of scales in wing color patches. The work came to nothing. In correspondence (August 14, 1899) Weldon remarked: "For the sake of these caterpillars I have, at the risk of personal liberty and reputation, stolen from the roadside one hundred square feet of clover turf, the property of the Lords of various Manors in the neighbourhood. The little ruffians have now eaten all this clover."

Neither Bateson nor Weldon had realized that formidable technical problems would have to be overcome before breeding studies became feasible. One solution was to find and collaborate with a lepidopterist who had already overcome them. This was the route taken by Doncaster in his collaboration with Raynor (Chap. 12). Another was to work patiently, step-by-step, to overcome the technical problems, before tackling the exciting questions that more and more people were wanting to ask. In the modern era of "mission-orientated" research, this route tends to receive accolades retrospectively rather than prospectively (i.e., it is not a good topic for a grant application).

Castle at Harvard University took the latter route with fruit fly in the early 1900s. With a host of student helpers, technical obstacles were overcome, and in 1906, he reported that inbreeding caused a decline in fertility, which could be more than counterbalanced by the selection of the most productive among closely inbred pairs [52]. Using the Castle group's "banana technique" in milk bottles, the work was continued at Davenport's Experimental Evolution Station [53], from which Thomas Morgan obtained stock of the normal, red-eyed fly. It was the Morgan school at Columbia University that fully exploited what can now be seen as a major breakthrough.

In 1907, a student bred the flies in the dark for numerous generations to see if there was an inherited loss of vision – perhaps Lamarckism in action. Sometime in 1910, Morgan saw through his hand lens a single male fly with white eyes. As Pasteur had remarked, chance favors the prepared mind. Morgan's was prepared. White eyes were seen very rarely in females. Crossing studies indicated that the corresponding gene was associated with the unpaired X-chromosome in males. He sent Bateson (July 7, 1910) a manuscript entitled "Sex-limited inheritance in *Drosophila*," which he had just submitted to *Science*, and offering to send him fruit fly samples.

Morgan

The authorship of the following can hardly be doubted?

> Since the number of chromosomes is relatively small and the characters of the individual are very numerous, it follows on the theory that many characters must be contained in the same chromosome. Consequently many characters must Mendelize together. Do the facts conform to this requisite of the hypothesis? It seems to me they do not. … The absence of groupings of characters in Mendelian inheritance seems a fatal objection to the chromosome theory, so long as that theory attempts to locate each character to a special chromosome. … If Mendelian characters are due to the presence and absence of a specific chromosome … how can we account for the fact that the tissues and organs of an animal differ from each other when they all contain the same chromosomal complex?

Yes, here in true Batesonian fashion, chromosomes are held to "contain" characters themselves (rather than "elements representing" them) and the validity of the chromosome hypothesis is questioned. But no, this is not another Bateson quotation.

It is from an August 1910 paper by Morgan [54]. Morgan had received an early training in morphology, and for many years his thinking was close to Bateson's. In his 1903 book *Evolution and Adaptation,* Morgan took an anti-Darwinian stance, and in a 1909 address [55] he agreed that Darwin had unwittingly slowed progress: "So extensive were the facts of variation accumulated by Darwin, so penetrating was his analysis of the facts, so keen was his insight, and so wise his judgement as to their meaning, that for thirty years afterwards little of importance in this direction was added. In their amazement at Darwin's accomplishment, zoologists forgot that he had opened the door leading to an unexplored territory."

In a paper "On the Inter-Relations of Genetic Factors," read at the RS in March 1911 and published in July, Bateson and Punnett discussed "coupling" and "repulsion." They noted that in certain cases "particular combinations occurred in the gametes with greater frequency than others." Thus, for certain pairs of alternative characters (allelomorphs) – T and t, and Y and g – if one parent had T and Y together, and the other parent had t and g together, then T and Y would tend to remain together in the offspring, as would t and g. This they described as "coupling" between T and Y, and between t and g (Fig. 13.1b). On the other hand, if one parent had T and g together, and the other parent had t and Y together, T and Y would tend to remain separate in the offspring, as would t and g. This they described as a "repulsion" between T and Y, and between t and g. Bateson and Punnett declared: "We have as yet no probable surmise to offer as to the essential nature of this distinction, and all that can yet be said is that in these special cases the distribution of the characters in the heterozygote is affected by the distribution in the original pure parents."

Shortly thereafter, the first issue of their new *Journal of Genetics* contained a new hypothesis of genetic linkage. Why was it "that the heterozygote produces a comparatively large number of gametes representing the parental combinations of factors and comparatively few representing the other combinations"? Bateson had long pondered "the moment of segregation." Indeed, it seems to have dominated his thinking. While agreeing that segregation most likely occurred in the reduction division of meiosis, in these strange linkage cases it was proposed that the segregation had occurred somatically within the parents prior to gametogenesis, so that the gonadal cells (e.g., spermatogonia), which through their divisions would produce the gametes, were *already* biased toward certain combinations of characters.

This can be regarded as a carryover from Bateson's *Materials* thinking of the 1890s, when he was concerned with differentiation of the fertilized ovum into tissues and organs, and with the reduplication of parts that Pearson was later to refer to as "homotyposis" (Chap. 6). Underlying Bateson's chromosome doubts was his attempt to relate the apparent "segregation" of characters among tissues during development (e.g., the character blue might "segregate" with eyes, but not with fingers) with the segregation of elements determining characters during gametogenesis. This would lead to future studies of somatic mosaics and bud sports. Thinking along these lines, Morgan wrote to Bateson (April 2, 1910):

> Of late I have been wondering (to myself) whether the two classes of Mendelian germ cells may not owe their origin to the same kind of change that takes place in somatic cells, when they produce regional differences, rather than be due to a quantitative separation of materials already different. In other words, whether the fundamental differences in cells may not arise at certain critical divisions, which divisions do not involve the quantitative or qualitative separation of substances in the ordinary sense.

Bateson and Punnett were unable to explain why their postulated early somatic (rather than late germline) segregation of paired characters (e.g., T and Y) should be specific for some pairs of characters and not others. As stated in their July paper, why were some factors "able to take precedence of the others in such a way as to annul the present repulsion with subsequent coupling as a consequence"? Nevertheless, they concluded with a bold flourish:

> In view of what we now know, it is obvious that the terms "coupling" and "repulsion" are misnomers. … Now that both phenomena are seen to be caused not by an association or dissociation, but by the development of certain cells in excess, these expressions must lapse. It is likely that terms indicative of differential multiplication or proliferation will be most appropriate. … The various systems may conveniently be referred to as examples of *reduplication*, by whatever means the numerical composition of the gametic series may be produced.

From this, we may discern that in 1911 Bateson was struggling. Punnett, based back in Cambridge, could not provide the criticism he needed. For intellectual challenge, short visits to foreign experts had to suffice. Bateson wrote to Baur (April 13, 1911) enthused with the idea of early somatic segregation as the explanation for linkage.

> I am coming more and more to think that the subepidermal layer [in plants] must really be the germinal tissue, already separated from the soma, like the germ-tissue of so many animals – according to the Weismannian view. … Then I am wondering whether we have not been on the wrong track in supposing that segregation happens in the maturation divisions [of gametogenesis]. May it not be all got over in the stages when the sub-epidermal layer is formed? … It was *coupling* which makes me think segregation must come early. We now have the series 63 : 1 : 1 : 63 and even 127 : 1 : 1 : 127, namely in series of 256 members [c. f. Fig. 9.2]. Surely such a series can only be formed when the cells are all held together in a close system of symmetry – and this is only true of the early stages of development. There is of course the great difficulty, that I cannot see how the cells, thus differentiated, get distributed into their places all over the plant. Still I feel it is not fatal. … I do so wish you would give us a week here. … You are the only man who would understand the questions involved.

Enter again Morgan. Sometime in late 1910 or 1911, he would appear to have had an eureka moment: "I venture to contrast Bateson's hypothesis with the one I have to offer." These words mark the end of the Batesonian era of dominance in the biosciences. Bateson would remain a major force, but in September 1911, with a few crisp sentences in a one-page paper in *Science*, Morgan seized the crown [56].

The paper was entitled "Random segregation versus coupling in Mendelian inheritance" and cited only Bateson and Punnetts' July paper in the *Proceedings of the Royal Society*. Moving briskly from the "agnostic" position he had held a year earlier, Morgan declared, quite simply, that Mendelian factors were associated together on a chromosome. There was no need to invoke repulsion of factors. Coupling could be explained by physical linkage. From a parent derived from a Tg (tall, green) and tY (small, yellow) cross, a chromosome containing T and g would go into one gamete, and a chromosome containing t and Y would go into another gamete. Thus, T would remain with g and be "repelled" from Y, and t would remain with Y and be "repelled" from g:

> In place of attractions, repulsions and orders of precedence, and the elaborate systems of coupling, I venture to suggest a comparatively simple explanation. ... The results are a simple mechanical result of the location of the materials in the chromosomes, and of the method of union of homologous chromosomes, and the proportions that result are not so much the expression of a numerical system as of the relative location of the factors on the chromosome. *Instead of random segregation in Mendel's sense we find "association of factors" that are located near together in the chromosomes. Cytology furnishes the mechanism that the experimental evidence demands.*

Do not imagine from this that the Batesonian walls came tumbling down overnight. The reduplication hypothesis was still in contention. But opinions initially swayed by the simplicity of Morgan's chromosome location hypothesis began to shift more decisively as supporting evidence accumulated. While in his 1915 paper (Chap. 12) Haldane did not question the reduplication hypothesis, after the war he was more critical [57].

Morgan's hypothesis made the important prediction that linkage would be disturbed if there were a break in a chromosome between two linked genes, an event deemed more likely the further the genes were from each other. So tractable was the fruit fly as an experimental system that, within a few years, the young Morgan team – Calvin Bridges (1889–1938), Alfred Henry Sturtevant (1891–1970), and Herman Muller (1890–1967) – had confirmed the prediction.

Indeed, by 1914 Bridges had even entered the enemy camp. He took data that Punnett and Gregory had continued to interpret in terms of reduplication and calculated the relative locations of the genes on the chromosomes of sweet pea and primula [58]. Conversely, Sturtevant applied the reduplication hypothesis to his own extensive fruit fly data – which were much more satisfactory from a statistical viewpoint – and found "no sound reason" to support it [59]. Finally, in 1914, Muller discovered the missing link – genes on the small fourth chromosome. The number of linkage groups in fruit fly was now equal to the number of their chromosomes [60]. Shortly thereafter, Sturtevant established the first linkage map [61]. In an amazing *tour de force,* each of the three large chromosomes was covered, with 2–6 genes being mapped on each of them. All the great mass of later work on linkage in *Drosophila* served to add, in more and more detail, to the magnificent framework Sturtevant had provided in 1915, without materially affecting any of his conclusions. In 1923, Punnett was able to report for sweet pea that the number of linkage groups was equal to the number of chromosomes [62].

The publication of the Sturtevant paper would seem to have been delayed because the journal of first choice for publication had been Bateson's own *Journal of*

Genetics. In 1912, the journal had published an earlier paper by Sturtevant, where he was critical of recent studies on color inheritance in horses. The new submission was an act of confidence in Bateson's open-mindedness and dedication to truth, however unpalatable it might appear. Sadly, Bateson did not see it this way. In a letter to Morgan (May 31, 1914), Bateson declined the paper, but left the door open for resubmission:

> If there were any difficulty in finding a journal willing to publish Sturtevant's paper we would gladly take it. The purpose, however, of sending it to us, I understand from your letter to be the hope of originating a discussion of the subject here. None of us is the least likely to take the matter up and we think that publication of the paper in *Journ. Gen.* would merely look odd. On the other hand I myself consider the evidence against chromosome representations is gradually getting stronger but I am in no hurry to get into controversy on the subject. If it will give – on mature consideration – you or Sturtevant any satisfaction to "trail your coat" in the *Jour. Gen.* send it along again. It seems more dignified to keep each to our own ground. Send it to Baur (editor of *Z.i.A.V*) or one of your own journals. Why not? We all see these things.

Disappointed, Morgan approached Punnett, who wrote to Bateson following his return from Australia (November 25, 1914): "I heard a little while back from Morgan regretting we would not publish Bridges and Sturtevant. I'm afraid he thinks we are afraid of publishing criticisms scored against us! ... What do you say to asking Morgan to get Sturvevant or someone to write an account of the Drosophila work properly pulled together and *translated into our jargon*?" Punnett was implying that he did not understand the Morgan group's work. This may also have applied to Bateson (Chap. 17). Even so, in 1916, Bateson was able to criticize the paper in some detail in a review of the group's first book (Chap. 16). Possibly it was a need for economy that lay behind the exclusion of Sturtevant's 54-page paper. When it appeared in the *Zeitschrift für induktiv Abstammungs- und Vererbungslehre* in 1915, the journal stated that it had been received on March 15, 1914. This does not match the date of Bateson's formal letter rejecting it for the *Journal of Genetics*; possibly the actual rejection was earlier, and the latter letter was just a response to a request from Morgan for further clarification as to the basis of the rejection. Nevertheless, the episode may have brought home to Morgan that America needed a journal equivalent to the *Journal of Genetics*. A rival, *Genetics*, was established in 1916 with Shull as editor, the first article being "Nondisjunction as proof of the chromosome theory of heredity" by Bridges.

While the Morgan group's discoveries were proceeding apace, Bateson was in Australia (Chap. 14) declaring that "all that can be made visible by existing methods has been seen, and we come little if at all nearer to the central mystery. ... Not only does embryology give no direct aid, but the failure of cytology is, so far as I can judge, equally complete." Furthermore, alluding to the reduplication hypothesis of gametic coupling, he spoke of "characteristics, essentially independent" that "may be associated in special combinations which are largely retained in the next generation [i.e. coupling], so that among the grandchildren there is numerical preponderance of those combinations which existed in the grandparents." With some handwaving, he now declared that "segregation can happen at earlier stages of differentiation" of the embryo, and that he was "entirely sceptical as to the occurrence of segregation solely in the maturation of the germ cells." Coupling was "a new phenomenon of polarity in the organism. We are accustomed to the fact that the fertilized egg has a polarity, a front and hind end for example; but we have now to recognise that it, or the primitive germinal cells formed from it, may have another polarity shown in the groupings of the parental elements."

Summary

Guyer derived Mendel's laws from observations of developing germ cells. He related the chromatin seen in the nuclei of nondividing cells to the chromosomes seen in dividing cells and ascribed the hybrid sterility resulting from crosses between members of allied species to an incompatibility between parental chromosomes that prevented their pairing in meiosis. Montgomery thought this pairing facilitated a "rejuvenation" of the chromosomes. Sutton thought pairing reflected a fundamental attraction between the genes they contained. de Vries thought similarity between paternal and maternal copies of a gene would promote chromosome pairing (i.e., the chromosomes would be compatible), and segments could then be exchanged ("recombination"). On the other hand, dissimilarity ("the threads no longer fit") would impede pairing (i.e., chromosomes would be incompatible). Bateson saw *between-species* hybrid sterility as a *predictable* regularity. In contrast, sporadic *within-species* sterility was attributable to *chance* genic defects (mutations). Furthermore, one of the sexes might have two types of gamete, one determining maleness and one determining femaleness; but which sex was the two-gamete sex varied with the species. Aware of Guyer's work, Bateson saw chromosome movements during meiosis as explaining the distribution of the Mendelian factors (genes) among offspring; the chromosomes contained "particles" or "fragments" *representing* the observable unit characters. However, as others came to agree with this, Bateson grew skeptical. Sutton, Boveri, and, eventually, Morgan argued that co-inheritance of characters in groups ("coupling") could be explained by linkage of the corresponding genes on one chromosome. Bateson held tenaciously to an alternative "reduplication hypothesis." Behind his chromosome doubts was his attempt to relate the distribution of characters among tissues during development (e.g., the blue character to eyes not to fingers) with the segregation of elements determining characters during gametogenesis. There was also his quest for something beyond genes (perhaps cytoplasmic) that would explain the origin of species.

Part III
The Innes Years

Chapter 14
Passages (1910–1914)

I am not one of those who have traveled along a set road towards an end that I have foreseen and desired to reach. … Nevertheless, I have strayed into no field in which I have not found a flower that was worth the finding.

Samuel Butler [1]

After 49 years in Cambridge, it was time to move on – a physical move and a move into a more public arena. Bateson was now not just on-stage, he was center-stage – an elder statesman of science. The multitude that had flocked to Grantchester now flocked to the John Innes Institute at Merton in Surrey – his court, his Camelot! It was also time to declare enough regarding the cataloging of more and more examples of species demonstrating Mendelian inheritance. The Mendelian battle was won. The species question remained. Perhaps a look at the many "exceptions" that he had always advised should be "treasured" might provide the vital clue? Perhaps this was the time to examine more thoroughly the many strange rogue, chimeric, and variegated forms that sporadically appeared among his stock? And would he now find time to reconsider his readings of the newfound Samuel Butler? However, while in the first decade of the century Bateson may have overtaken events, now events were surely to overtake him. The scientific world having digested the Mendelian lesson was moving on. And war clouds were gathering.

Goodbye Cambridge

The will of John Innes, who died in 1904 having made a fortune in property transactions, provided for the establishment at Merton in Surrey of a "School of Horticulture." As not unusual where a large sum is involved (in this case at least £300,000), the will was contested [2]. Eventually, in April 1909, there was a

A. G. Cock, D. R. Forsdyke, *Treasure Your Exceptions*,
https://doi.org/10.1007/978-3-030-92099-9_14

meeting of an Advisory Council, which included Biffen, Farmer, David Prain (1857–1944; who had taken over Kew Gardens from Thiselton-Dyer in 1906), and Sedgwick (now at Imperial College, London). Apart from a director, the Council envisaged salaries of the order of £300/annum for a "plant breeder," a "mycologist," an "entomologist," and a "biochemist." So the director would be supported by a mature team of experienced peers. The Council appointed two subcommittees that would report back to it – a Management Committee and a Selection Committee (Biffen, Farmer, and Prain), which would recommend a director. Bateson was not among the 30 applicants, 3 of whom were short-listed – B. T. Barker, Gregory, and Lock, who was then acting director of the Royal Botanical Gardens at Peradeniya. The Council deemed none satisfactory.

Lock's hopes had been high, perhaps thinking that even if he did not get the directorship there might be a supporting role. He may have appreciated neither that Bateson had held aloof but was unspokenly in the running, nor that Bateson's fair-mindedness had been sorely tried by his support of de Vries and even, on occasions, of Pearson. Recently (January 19, 1909), Lock had written to Bateson declaring that a result with indent peas "justifies Pearson's statement as far as peas are concerned." Lock happily reported (March 8) that his textbook had sold so well that the publisher was proposing a second edition, and added: "What are they going to give the [future] Chief of the Innes Institute? And will they give it to me? The two last are very serious questions indeed. That's a job I would stick to if they'd make it worth my while." A month later, he wrote asking for a testimonial.

After informal negotiations between Farmer and Bateson that included a visit to Merton in October, it was resolved (November 3, 1909) that Bateson be offered the dual post of "director" and "plant breeder" at what was to be called the John Innes Horticultural Institute. The position carried a salary of £1000/annum and a residence, the Manor House, the former home of John Innes surrounded by two acres of land. Bateson accepted, noting that the retirement age would be 65 and trusting that the Council would make appropriate pension arrangements. He wished to continue experiments on small animals, but Farmer persuaded him not to bring the matter up "till you will have educated the Council generally." The Council later (August 1910) approved his formal "Application for Permission to Conduct Experiments in Breeding Animals," which specified that three quarters of an acre of land would be needed for poultry, canaries, and bees, and noted that "nothing whatever of the nature of vivisection or experiments giving pain is contemplated." The appointment was applauded in *The Gardeners' Chronicle* (November 27, 1909):

> With Kew as the leading school of scientific botany, with Rothamsted engaged in the solution of soil and allied problems, and with the Innes Institute undertaking investigations into the causes and modes of prevention of plant diseases, into the laws of inheritance and their applications to plant breeding, and into the many obscure problems of what may be called applied plant physiology, we may well hope to develop in this country the finest system of scientific horticulture in the world.

Among the letters of congratulations was one from the "anonymous donor" of the Cambridge Chair (December 9, 1909):

> Dear Sir, I am grateful to you for so courteously informing me of your possible resignation of the Chair of Biology – a resignation which I have since learned has become an accomplished fact. Though I cannot but regret the loss which Cambridge will sustain by your departure, I rejoice that an opportunity of such value for the advancement of the study of Genetics has been afforded. I am dear Sir, Very Truly Yours, The Founder of the Professorship of Biology.

Bateson submitted a "Report to Council" (November 25, 1909) proposing to build up staff gradually, beginning with a Garden Superintendent – Mr. Allard, of the Cambridge Botanic Gardens, for whom a residence would be provided. Next, he needed "a Punnett," but was not yet ready for a nomination. Lock had written (December 20, 1909) to congratulate on the appointment, but Bateson would not consider him. Indeed, the nomination never came about. Bateson was keen to get started with three or four "high and coveted" studentships, valued at £200–250 per annum, and some "minor studentships" of £50–100/annum, to be held mainly "by women students who wish to devote themselves to research in various fields but can scarcely afford to earn nothing." For the high studentships, he named in the first instance Muriel Wheldale, who had worked with him since 1902 and currently held a Newnham College Fellowship, and William Backhouse, who was Biffen's Demonstrator in the School of Agriculture.

Believing that in the long term Bateson intended to build up senior staff, the Council, having already allowed Bateson to upgrade the position of director to that of both director and plant breeder, now allowed him to downgrade the positions of mycologist, entomologist, and biochemist. In Beatrice's words (*Memoir*): "The Council and Trustees showed him generous confidence, giving him a free hand to design and plan the gardens and laboratories, and to manage and arrange the whole work of the institute." However, "for months – until he had established contact of a sort with the London schools – his exile from laboratory and library depressed him. He feared at first the isolation for his students." It would seem to have been a time to reapproach Doncaster, who had just published *Heredity in the Light of Recent Researches*. But it was not to be. In 1910, Doncaster moved from Birmingham to superintend the Museum of Zoology in Cambridge.

Passage to India

In early 1909 (February 10), there was a strange letter from Mrs. Herringham, who was piqued that Bateson had refused her something:

> You hit very hard by your refusal. I thought – well – that you would have been more – what seems to me – fair and far-sighted. Will you look round the question again? Would you even send some arguments or let me with some – well, no, that latter is impossible. I can't write to you for your private benefit a special essay. It is true this is not a propaganda piece of work – but you have given me one of the nastiest turns I have ever had.

Whatever this means, it may not have concerned Wilbur, who sent his hearty congratulations on the John Innes appointment (November 20). The letter, however, was addressed to Beatrice, not to Bateson:

> This is really magnificent. I suppose the idea of Heaven is different for each: but I should think that for a philosopher the natural Heaven would be a place where he could have *free scope* and ample means and equipment for experiment. If so I imagine your husband is somewhat in the position of Enoch who is popularly supposed to have been translated [to Heaven] while yet alive. The exact position of Heaven has always been disputed. Hell of course is well known. It is in the centre of the earth and the restricted life and general pressure is no doubt one of its chief torments. I own I did not expect to find Heaven at Merton and I don't think that situation has ever before been suggested. "You never can tell." I think when you come to try it, you will find that London is near enough to make amends for Cambridge. Of course it means the loss of many friends, but you are both young enough and active enough to make swarms of new, and you will have a charming place for Summer Saturdays.

At that time, Christiana Herringham was on her second passage to India with an assistant from the Hornsey Art School, Miss Dorothy Larcher. Her first passage had been with Wilmot in November 1906 prompted by their son's regiment being posted there. On learning of the intended visit, Laurence Binyon of the Department of Prints and Drawings at the British Museum had suggested that they try to see the ancient Buddhist wall paintings in caves at Ajanta, not far from Hyderabad. In the damp, bat-ridden caves, they had found the paintings in a sad state of decay. Thus, Christiana had acquired a new project – to copy the paintings and foster recognition of their importance, both in India and England. She arrived at Hyderabad on December 15, 1909, and over a 6-week period made a start on the copying project.

Her third passage was in October 1910, and she was accompanied by another art enthusiast, William Rothenstein [3], and by Miss Larcher (who took an earlier boat). This time they meant business. After 3 months of copying in the gloom and stench of the caves, they returned to Bombay. Here they met up again with Rothenstein, who noted her bedraggled appearance and became alarmed. Strange thoughts of being pursued by vengeful Indians angry at her tampering with holy relics began to obsess her. Shortly after her return, she was diagnosed with "systematized delusional insanity with marked depression." She spent the rest of her life in private mental institutions. There is a cryptic comment from Bateson to Beatrice in 1912 (September 7): "Much surprised with news of Mrs. H. and wonder how much is to be trusted."

It is of interest that the novelist Edward Forster (1879–1970) was a dinner guest of Wilmot Herringham in 1912. Some aspects of Mrs. Herringham and Miss Larcher may be found in his novel *A Passage to India* (1924). Although she lived until 1929, Christiana knew none of this. In 1914, she knew not that she had become Lady Herringham as the result of the conferral of a knighthood on Wilmot who was Vice-Chancellor of the University of London. And, fortunately, she knew not of the death of her son, who saw action in France in the early days of the war [4].

Passage to Ceylon

Punnett succeeded Bateson as Professor of Biology at Cambridge, serving from 1910 to 1912 when the 5-year term of the professorship expired. A university newspaper [5] hailed Punnett as "a worthy successor, both in mind and pen, to the founder of the Cambridge School of Genetics." It was noted that

> as a general rule, the presence of a dominant character effectually masks a character of an opposite kind. The amateur of Genetics might reasonably come to the conclusion that Culture and Athleticism, Book-lore and Biology, should be Mendelian allelomorphs, and therefore incapable of manifesting themselves in the same individual. And yet a critical analysis has revealed the astounding fact that all these qualities are simultaneously manifested in the subject of this notice.

Punnett was left the poultry and sweet pea stocks, which, with the move from the Grantchester house, had to be relocated. He found space on the University Farm for the poultry pens and incubators, and the sweet pea work continued on a rented allotment in the Botanic Garden. So he carried on, but largely at his own expense. For a brief period, with funds from the Board of Agriculture, Punnett was able to have his own "Punnett," P. C. Bailey, but the latter was lost in the war.

Punnett's disagreements with Poulton had become focused on mimicry (Chap. 6). He tended not to antagonize his opponents. Indeed, it would seem he could enjoy their company for many weeks during a long ocean voyage. Between them they decided that the issue warranted a journey to a tropical country where the phenomenon could best be observed. In the summer of 1909, they sailed to Ceylon with the parasitologist Clifford Dobell. They were entertained at Colombo by Arthur Willey, Director of the Colombo Museum, and at Peradeniya by Lock. In 1910, Punnett published a paper on "Mimicry in Ceylon Butterflies with a Suggestion as to the Nature of Polymorphism." There were two further papers on Mendelism and mimicry, and in 1915, a major text, *Mimicry in Butterflies*.

By the time of expiry of Punnett's professorship, Balfour and his friend Lord Esher had persuaded a wealthy donor (Mr. William Watson) to subscribe £20,000 to establish the Arthur Balfour Chair in Genetics. Esher and Balfour themselves gave a house (Whittingehame Lodge) and land for the new department. Bateson was invited to return (letter from Balfour; June 25, 1912). He declined (June 26, 1912): "If there were difficulty in finding a man thoroughly qualified, then the case would be different; but I am sure Mr. Punnett, who worked in partnership with me for several years and succeeded me in the Professorship in Biology, is in every way worthy to be appointed to the new post."

So, Punnett continued. In 1912, he was elected a FRS and held the first Genetics chair in Britain for 28 years. Whittingehame Lodge became his residence. As Jane Austen had noted, a man with such good fortune "must be in need of a wife." Indeed, in 1913 at the age of 38, he married Eveline Froude, who was 3 years his senior and had different interests from his own. She encouraged him to live a life outside the university [6].

Moving In

Jane Austen and Benjamin Disraeli lent color to Bateson's final months at Grantchester. As had been the custom of his Aiken grandfather, he went swimming (at Happisburgh) despite the cold and reported (March 25, 1910):

> I am revelling in *Coningsby* [by Disraeli]. It is packed with bits that either are or ought to be quotations, a genuine classic. Concerned as it is with the struggle against the House of Lords over the Reform Bill, it is just the book for these times. There is a vast difference, however; for Manchester and the big towns were real powers and the people that had made them were real and powerful men, whereas now it is the mere multitude which is trying to destroy what is above it. The humour is Voltairian but amazingly human. I had thought Disraeli was nothing but upholstery and glitter, but all through one feels there is a man with real feelings if he would only let them out.

And as for Austen (April 17, 1910): "*Sense and Sensibility* shows J. A. at her worst as a human being. The comic bits are delicious but she must have had a really hard heart. The men are mere sticks, except Sir John." And to Beatrice at Robin Hood's Bay (August 31): "*Life of C. Bronte* an extraordinarily interesting book, which I had never read before."

In February, Bateson gave a lecture in Dublin and another at the Royal Institution, London, on "Heredity in Sex." He resigned his St. John's College Fellowship but was promptly elected an Honorary Fellow. He worked to ensure that Lynch, who had given so much help at the Botanic Gardens and now advised on the equipment that would be needed for the new institute, was made M. A. *honoris causa*. There were also visits to and by various scientific colleagues. In January, he visited Baur in Berlin, and there was some correspondence concerning the inheritance of the chlorophyll-forming organelles (chloroplasts) through the maternal line due to their cytoplasmic location. Raymond Pearl wrote (February 3) from Maine Agricultural Experimental Station saying he was planning to visit Europe in the spring: "We are getting a lot of interesting Mendelian results in our poultry work here and I am particularly anxious to talk these things over with you. … I hope to see something of Mr. Hurst and something of his good work. … I hope you will not regard it as too great an imposition … for me to descend upon you for a few days." The visit was in April.

The Bateson archives contain Beatrice's note that in the 1990s: "Ray Lankester and Co. blackballed WB at the Savile Club." There are two 1914 letters on this. One was from Farmer (February 17) asking if he could propose Bateson for membership. The other was Bateson's reply (February 18): "I became long ago reconciled to the fact that the club had rejected me, and much as I should like to be among my Savilian friends more often than is possible, it seems to me that questions of this kind, once determined, cannot be reopened with complete satisfaction to either party." However, albeit at a late hour, Bateson had been elected in 1910 to the inner sanctum of British science, the Athenaeum, where he could read the newspapers, play chess, and encounter some of the prime movers in politics and science. He took the occasion of his first visit to use the club notepaper (April 12, 1910):

> You will be glad, I am sure, to see that I have had the courage to penetrate hither. … Balfour was very amiable and professed great interest and sympathy. He did not however seem to know where the endowments were to come from. … I mentioned casually to F.D. yesterday my suggestion that regeneration may be something like the re-formation of a wave, or a raw ripple and he seemed to think it rather pretty. I have nearly done that much ruminated chapter [for *Problems*] and hope to read it to [a] typist before the end of the week.

This is of interest since Beatrice's annotations record that a professional secretary was not appointed until 1924, implying her continuing help as set out in her *Memoir*:

> Intimate letters he wrote freely and fluently; in all other forms of composition he wrote slowly and with difficulty, never content until he had found "*le mot juste*." If he was critical of others' contributions, he was doubly critical of his own. His first manuscript would be one word superscribed over another, all inserts and deletions, fair copy succeeded fair copy in like case; often until the small hours of the morning we sat up, he writing, I copying, until at last he was satisfied that he had found the one way to say exactly and indubitably what he meant to say.

The writing of *Problems* continued at the Hill House Hotel with letters to Beatrice detailing his continuing obsession with waves and ripples (March 27): "Struggling with my chapter, but I fear it is dull work. However in some way or other it is dragging to an end. If I could get quit of this part, I would take up some easier parts." And (March 28): "My wave idea is not difficult, and so far it is an idea for an idea. It has no body, or anything to take hold of, but as an idea for an idea I think it is a good one." From time to time, there were grim forebodings: "The coast guards had revolver practice on the shore. Their shooting didn't strike me as very good, but I daresay the German's is no better."

From Grantchester, he wrote (April 10) to Beatrice who was with the children at Robin Hood's Bay: "Stupid game of chess with Lady Jebb yesterday, but the poor old lady does enjoy it. Book moving a little." At that time, Cambridge was getting a state-of-the-art telephone system with an appropriate directory. Bateson discerned that the "Trumpington exchange list of numbers is arranged in strict order of social dignity and importance" – low numbers counting high in this respect. His letters were typically high-spirited and full of humor. Beatrice was seldom chided. His next letter (April 14) was an exception: "Your remarks on ripples don't show your usual intelligence. There is no difficulty on score of *irregularity*. Ripple marks may be as regular as the engine-turning on a watch case. You really must think me a wiseacre! All the same what you say shows me that the lay mind may not know the characteristic appearance of ripple-mark, and perhaps I ought to begin by describing it."

Despite Beatrice's input, he was dissatisfied (April 16): "My chapter, such as it is, [has] at last gone to [the] typist. Don't offer congratulations please, for it is not of much value I fear. I think there is an idea floating around it, but I can't pretend that the idea has been caught tight." The recalcitrant chapter behind him, there was a return with Beatrice to Happisburgh as described in her *Memoir*:

> Except for bathing and eating the fine local shrimps, we had no pastime in view. The attraction of the coast was, he said, its absolute nullity, and without lectures to prepare or even proofs to correct, we did not rightly know how we should bear this absoluteness. *En route* came the inspiration – we would paint. We had to change trains at Norwich; we rushed into the town, found an artists' colour-man and fitted ourselves out with paint board and brushes. From that day nearly all holidays and many happy leisure hours at home were devoted to painting.

There was devotion to painting at another level. With his newly won affluence, it was less easy to resist the temptations of a London art auction (June 30, 1910): "Let me clear my breast: I spent £14-4, and did *really well.* ... I should think £1500 worth of stuff went for £100 and it may have been worse than that."

In July as packing began for the move, the boys were sent to holiday with Aunt Edith Bateson at Robin Hood's Bay. The elder two were then sent to Saint Faith's boarding school in London, so Gregory was the only one to actually move to Merton. The break with Cambridge was not easy. Gregory later reminisced [7]: "Living at Merton Park, we had a sort of family myth which lent a crystal halo to Cambridge and especially Grantchester. I grew up feeling in a gentle way that we had been turned out of, or perhaps had deserted, the Garden of Eden."

The attractions of the university setting were also likely to influence future candidates for positions. Despite Bateson's eminence and the proximity of Merton to London, to some the prospect of working in the grounds of a remote country house could not compare with Cambridge. Furthermore, Punnett and Saunders were still colleagues, but now they were also competitors. By virtue of their location, they would have the first pick from among "the brightest and the best" that graduated from Cambridge. Bateson wrote from Grantchester to Wheldale (July 9): "I am at last empowered to offer you a studentship." But she first wished for further training in biochemistry. Bateson replied (September 7): "You are right to take a period with Nierenstein. ... To what extent we will be able to organize chemical work on an extensive scale I hardly know, but I see no reason why we should not."

Various Merton citizens began dropping by to welcome the new arrivals. Bateson escaped (September 3) to the BA meeting at the University of Sheffield, where he received an honorary degree. He stayed at Brinkburn Grange with Major Strange and his wife Maggie who was Bateson's cousin. One evening they dined out at the "rich, very rich" Hoyle's, the guests of whom included Marie Stopes, who was to become well known for her books advocating contraception and sexual license. Although Bateson's "intimate letters he wrote freely and fluently," we can be sure he picked his words carefully in his letter home to Beatrice:

> Amongst them, quartered on the house was the egregious Dr. Miss Marie Stopes – really, I believe an able person, but full of minx-like ways, preserved 10 years too long. She tried to get up an understanding with me across the table – for she was sharp enough to take in the situation – but I didn't much respond. Bridge after, in which M. S. disgraced herself with calumnies that I was glad I had put her down.

Late in September 1910, Bateson departed on his second "pilgrimage" to Brünn (Chap. 20). On his return, there was a letter (October 15) from Yale concerning *Problems*. He replied: "Looking at the matter without any self-deception I recognize that the real difficulty is that I am not thoroughly ready with a new book in my head. When I came over to Yale I thought I had such a book in outline or I should not have accepted the invitation." He proposed various alternatives of which he preferred "prolonging the understanding that you are to have my book when it can be prepared." Yale accepted this (November 29), but asked for a time commitment. Bateson replied (December 13) that he would do his best to send the manuscript by January 1, 1912.

Managing

Apart from the writing of papers and annual reports, correspondence with other scientists and commercial breeders, and entertaining visitors, Bateson had taken on considerable managerial duties. The scale of operations was vast compared with Grantchester. Without any "Punnetts," it was bottom-heavy rather than top-heavy. Bottom-heavy meant there was no one, save Beatrice and Allard, to share the load. Bottom-heavy meant many inexpensive, rather than few expensive, personnel. And each of that many had strengths and weaknesses that had to be understood and catered to. There was no shortage of applications. Indeed, on hearing of his appointment, Constance Garnett wrote (December 4, 1909) concerning her young brother-in-law Arthur Garnett who was on the staff at Kew, but had incurred the wrath of the powers-that-be by agitating for higher wages for gardeners. Was there likely to be an opening for him at the John Innes?

The first Annual Report for 1910 documented that Backhouse had been appointed, but Wheldale had deferred until June 1911 in order to obtain biochemical training at Bristol. Miss Caroline Pellew from Reading had been appointed to a minor studentship, and Miss Dorothy Cayley, a mycologist from Reading, was to receive a similar appointment in 1911. Those who had taken part in the year's activities were Backhouse (fruit trees), Cayley (fungoid pests), Durham (canaries), Gray (sugar beets), Gregory (*Primula sinensis*), Marryat (*Mirabilis*), Mudge (*Petunias*), Pellew (peas), and Richardson (strawberries).

By the time of the Annual Report for 1911, two further studentships had been accepted (Bailey and C. B. Williams) and the list of those using the facilities had expanded: Backhouse (plums), Bailey (plant diseases), Mrs. Nora Barlow (née Darwin; trimorphic flowers), Cayley (fungoid rots), Durham (chemistry of pigmentation, canaries, bees), Gates (*Oenothera*), Gray (sugar beets, cyclamen), Gregory (*Primula sinensis*), Mrs. Dorothea Lister (née Marryat; *Mirabilis*), Mudge (*Petunia*), Pellow (peas), Richardson (strawberries), Saunders (doubleness in *Petunias*), Thursby-Pelham (*Primula* hybrids), Williams (insect pests), Willmott (interrelations between species), and Wheldale (chemistry of plant pigments).

Wheldale pursued an independent path. This meant going to wherever the intellectual climate, facilities, and funding best met her needs. In early 1911, Bateson wrote (January 10): "I quite think you were right to improve your mind at Bristol. My only fear is that you may soar out of our range.... Don't however get spoilt for plain doings." Keeping her options open, much of her time was spent in the laboratories at University College, London. Bateson enquired (March 10) about her plans and was (April 10) "very glad that Hopkins approves." Later in the year, there was a letter of condolence (September 8) about an illness, but in 1912 (June 1) he caused some alarm: "You will not, I am sure, suspect me of want of sympathy with your work, still it is clear that we must give a preference to those who actually require the opportunities we provide." He hastened to soothe her anxieties (June 6): "Your suggestions sound to me rather panicky! I see no reason for any extreme action of that kind at all. The best course I think would be for you to keep something going at

University College concurrently. … Meanwhile please rest assured that you are 'quite the kind of worker' we want at the Institution." Anxieties needed soothing again in 1915 (November 2) as her treatise on pigments neared completion: "Things are not so bad as all that! The introductory pages do need a thorough reformation. The rest is not amiss. The last section (chemistry and Mendelism) is quite readable and good." In 1916, Cambridge University Press published *The Anthocyanin Pigments of Plants*.

The emerging importance of Genetics, and Bateson's fame, were ready passports to posts overseas for his male coworkers who foresaw few prospects for advancement at home. In 1912, Backhouse accepted a position in Argentina to study corn breeding. To Beatrice, Bateson wrote (August 6): "Looking for a successor to Backhouse is discouraging work." Returning exiles, such as William Balls (1882–1960), flushed with the publication of *The Cotton Plant in Egypt*, felt it appropriate to drop by to see the director [8]. A letter (August 1912) revealed that sometimes this was not to the director's liking: "Nettleship came and was both soothing and delightful. Then, unannounced, came our brother Balls. I also speak of him as brother, for one might forget he is human. He stepped straight out of Wells. Every speech was *bardé* with 'another nail in the coffin', or 'not for yours truly', or other conversational spices. He made me feel ill in my inside and spoilt Nettleship's visit for all of us." Despite this, Bateson's review (1913) of Balls' book was kind, although he thought there was a lack of coherence and it read more like a collection of journal articles than a book.

With hindsight, the failure formally to appoint Wheldale as biochemist, and to appoint other senior staff who might more vigorously have challenged his views, appears as a major mistake. There was an attempt to get the Council to agree to the appointment of a pathologist (Kraus) from Austria, but Prain noted (October 17, 1910) that "*the engagement of foreigners* will … lead to a searching of hearts." Even several senior foreign workers who came for short-term visits could not compensate for the institution's bottom-heaviness. These international visitors included N. I. Vavilov (1913–1914) from Moscow (cereals), T. Kusano (1914–1915) from Tokyo (self-sterility in fruit trees), and K. Matsui (1919–1920) from Tokyo (cytology). Bateson initially had no direct access to his governing Council. The matter was resolved by his attending all Council meetings. That Bateson soon established a warm relationship with the chairman, David Prain, is revealed in the copious surviving correspondence.

Another difficulty became known within the Institute as "the rumpus." Minor discontents among the inhabitants of the nearby village of Merton came to a head in 1917, stirred up by a local schoolmaster. There were too few local boys on the gardening staff and the work of the Institute was insufficiently practical. In general, the Institute was said to be departing from the terms of its trust deed, especially in neglecting the interests of the local people. A public meeting was held, a report was sent to, and discussed at length by, the local District Council – all duly reported in the local newspaper. The issue rumbled on until April 1918 when the John Innes Council issued a statement that the Institute followed the procedures of Kew

Gardens regarding the appointment of "young gardeners" who must first "have worked for at least four years in good gardens or nurseries."

Visits and Visitors

The Manor House at Merton welcomed friends from Cambridge days and the world of art. Among them was Geoffrey Keynes (Chap. 1), whose brother Maynard had recognized the economic importance of biology and had been reading Pearson. Maynard wrote to Bateson (November 9, 1910): "Geoffrey tells me you would like to see my articles on K. Pearson in the Statistical Journal. ... The further I press my researches into K. P.'s productions, the more convinced do I become that the man's a *liar* – and modern methods of controversy have no adequate means of dealing with such."

Binyon became a frequent visitor and went with Bateson to Paris (February 1911) continuing a friendship in art that, in Beatrice's word, "ripened" in later years. However, a letter from Bateson (December 13, 1910) to his sister Edith about an exhibition of post-impressionist art, as championed by Roger Fry (Chap. 8), indicates that he could be as skeptical about art as about science:

> You certainly must manage to see the post-impressionists. I feel sure you would be interested uncommonly. A little of the work is really strong and is evidently going to have an important influence on the younger generation. The rest is not easy to distinguish from bluff, and I suspect that this is about what it is. I have seen some foreign "secessions," but never anything so brazen as this show. If you painted with your foot, standing back to the canvas, you could do as well. The gallery is generally full, and seldom does one see picture-gazers so intent, all laughing and chatting and enjoying themselves, instead of staring dully as if in Church. You really must come. We can put you up when you like.

A glimpse of their affairs is provided by a series of letters to Beatrice on holiday at Robin Hood's Bay in August 1911, probably accompanied for part of the time by Miss Pellew, who was quite becoming one of the family. Edward Bateson had settled back in England. He married Margaret Corbett in 1910. Bateson wrote cryptically (August 5): "Saw Ned and his wife yesterday at my mother's. She [Margie] has been ill and shows it, but they seem much pleased with each other's society still. No signs of her becoming the fruitful vine." By this time the Grantchester house, which had taken a long time to sell, had been sold to J. J. Lister, who, with his new wife (Marryat), was among the visitors. And (August 10): "Eva de Vries came in and sang her little songs last night – very pretty. Listers come tonight."

The labor movement was on the rise, and a strike of railway workers had drawn out others in sympathy. Bateson was concerned about an impeding shortage of supplies for feeding his house guests, adding (August 10): "As Miss Pellew is on strike she may as well fill her idleness with reading Tschermak (enclosed). I should like to know if there is anything in it." There followed (August 20) comments on the strike and the final settlement between the companies and the unions: "The article on Syndicalism interests me. It shows that Labour at last has found out that democracy is incompatible with its ideals. Syndicalism is bound to be the next programme. The Welsh, as usual, stand out as the worst of the population."

Olga Zacharias' brother paid a visit (August 21): "Gotthard Zacharias and GCMS [G. C. Moore Smith] called yesterday. G. Z. is very warlike. He hopes his government will not give way. He thinks they can easily dispose of France and that England will not be able to do them much harm. If many Germans are talking in this strain, there is real danger." The description of the arrival of Saunders the next day was laden with sarcasm (August 22): "ERS arrived at the Castle last night and had the honour of dining with the Director. It is understood she is prolonging her stay." And later (August 23): "ERS still sitting on my shoulder."

Utility of Research

At the end of August (1911), there was the BA meeting at Portsmouth. Bateson presided over the Agricultural Subsection and Edward Russell (1872–1965) was Recorder. In his address, Bateson praised the new national development grant system, but regretted the associated waste of labor and expense in the issuing of frequent "agricultural bulletins purporting to give the results of practical trials and researches." This was done "to satisfy a public opinion which is supposed to demand rapid returns for outlay, and to prefer immediate apparent results, however trivial, to the long delay which is the almost inevitable accompaniment of any serious production." Pressure for quick returns was inclined "to compel premature publication" and to "put the investigator into a wrong attitude towards his work." Admittedly, "there come moments when a series of obvious improvements in practice can at once be introduced, but this happens only when the penetrative genius of a Pasteur or a Mendel has worked out the way into a new region of knowledge, and returns with a treasure that all can use."

To try to direct research, to tell a researcher "that he must not pursue that inquiry further because he cannot foresee a direct and immediate application of the knowledge he would acquire" was "a course detrimental to the real interests of the applied science." Indeed, "if Mendel's eight years' work had been done in an agricultural school supported by public money, I can imagine much shaking of heads on the County Council governing the institution, and yet it is no longer in dispute that he provided the one bit of solid discovery upon which all breeding practice will henceforth be based." He concluded by praising "the Cambridge School of Agriculture" as now "a force for progress in the agricultural world." Yet 15 years ago his own urgings "for some study of the physiology of breeding" had "found little favour."

The practical benefits that might derive from discoveries in research were becoming generally recognized. It was likely these benefits would accrue first to the nation where the discoveries were made. In Germany in 1911, the Kaiser Wilhelm Gesellschaft established several research institutes with Boveri directing the new Institut für Biologie. In Britain, Lloyd George's "Road Improvement and Development Bill" (1909) contained provisions for expanding scientific research.

With a budget of nearly £2,000,000, in early 1910, a Development Commission was set up as a permanent body of eight Commissioners under Lord Cavendish's chairmanship [9].

In 1912, Daniel Hall, one of the Commissioners, ceded his Directorship of Rothamsted Experimental Station to Edward Russell. The Board of Agriculture Development Commission led by Hall had established the Plant Breeding Institute at Cambridge under Biffen, whose Mendelian studies had led to the marketing of a rust-resistant line of wheat in 1910. A committee, chaired by Bateson, was asked (January 13, 1912) "to consider in what way scientific research with a view to economic results can best be conducted into the subject of Animal Breeding." In particular, there was the possibility of the establishment of an Animal Breeding Institute. The committee's recommendations were against such centralization of research: "Scientific research in animal breeding with a view to economic results can best be promoted by the encouragement of individual investigators. ... It is as yet premature to establish one particular institution at which such investigations should be pursued." This may not have pleased Hurst and Ewart, who had proposed expanding breeding studies on horses. However, the threat of war was a far greater stimulus for their line of research.

Horses for War

In the spring of 1909, Burbage was visited by two authorities, Nettleship (eye specialist) and Redcliffe Salaman (physician-virologist), who were both convinced of the soundness of Hurst's eye work. The Burbage Nursery (now fully in Hurst's possession following the death of his father) was renamed "Burbage Experimental Station for Genetics" to regularize (presumably for tax purposes) the receipt of grants and the diversion of nursery profits to purely scientific work. However, Hurst did not inform Bateson of this name change until 1911, perhaps because it did not come fully into effect until that date. As a research station, it was required to produce Annual Reports, and the first was not due until 1912. Hurst wrote to Bateson (March 3, 1911):

> I am seriously thinking of turning my place here into an experimental station. My private experiments have now encroached so much on my time, land and labour that the outcome seems inevitable. The only difficulty I see is that the experiments will to a certain extent have to be self-supporting, but that I think can be got over by one thing helping another. My idea is to take over the 120 acres of nurseries and seed-grounds here and work up all the experiments I can get out of them with the aid of my present staff. The principle aim would be of course genetic. I propose to organise the station in three departments, Agriculture, Horticulture and Forestry, which I think will include everything available (except Eugenics in Man). ... I would be glad to know what you think of it, and whether you would kindly allow your name to be associated with the station as a Consultant, together with Punnett, Biffen, Wood, Hall, or anyone else you would advise I should ask. I am not mentioning the matter to anyone until I hear from you.

Hurst's scheme seems to have surprised Bateson. After a meeting at the John Innes, Hurst wrote (March 16, 1911): "I have quite decided to take your advice for the present at any rate to run the Experimental Station entirely on my own. The thing is to be launched on the 25th inst. as quietly as possible. I hope in time it may achieve that for which it has been designed." One can surmise that, for the first time, Bateson began to see Hurst more as a competitor than a colleague. Although the John Innes was independently endowed, an experimental station devoted to genetics at the scale Hurst proposed could draw on other potential funding sources. It is unlikely that Bateson underestimated Hurst's entrepreneurial skills. Indeed, Hurst was already in the running for large grants for the breeding of horses.

Horses were still of major economic importance in transport and agriculture, and, of course, there was an abiding interest in racing. Despite the costs, long gestation time, and low brood size, these were good practical reasons for genetic studies with the aim of improving the various breeds. Yet, it required the threat of an impending war for allocation of substantial government funds. In 1911, on invitation, Hurst wrote for *The Times* [10]:

> The military authorities have lately experienced some difficulty in securing enough saddle-horses suitable for their purpose. Further, the experience of the South African war showed the vital necessity of retaining in this country a large reserve of horses suitable for military purposes, over and above the 3000 or so required annually for the peace establishment. It is evident, therefore, that, apart altogether from economic considerations, the decay of hunter breeding in this country has become a matter of urgent national importance.

Hurst welcomed a recent £40,000/annum development grant for horse breeding and the establishment of an advisory council with two veterinary experts. A scheme drawn up by the Board of Agriculture had proposed ways of identifying the best horses for breeding purposes, when necessary, purchasing those that might otherwise be sold abroad, and arranging suitable breeding. However, Hurst warned:

> The outward appearance, or even the good performance, of an animal is not necessarily any guide to its breeding potentialities. The only safe test of a stud animal is the stock which it produces; however good-looking it may be or whatever good it may have done, it is no good for high-class breeding purposes unless it is known to have thrown good stock. All this goes to show the vital necessity or applying Mendelian methods to the breeding of stock.

He hoped that a "small portion" of the grant would be invested "in some definite scientific experiments, which … should … ultimately make the whole grant unnecessary" since, given the knowledge, the breeders would do the rest.

> If the Board decided to carry out some scientific experiments, the question of the establishment of an experimental breeding station or a Government stud-farm might have to be considered. … Fortunately Mendelian experiments deal with small numbers of individuals rather than with large masses, so that there would be no need to conduct the experiments on a large and expensive scale.

Bateson wrote (June 6, 1911) asking Hurst to contribute a paper on horse breeding at the Portsmouth BA meeting. Hurst replied (June 14):

> I have written to Russell that I will read a paper on the "Application of Genetics to Horse-breeding" which I hope will cover the ground you wish me to take. I find that this horse

> business is taking a great deal of my time. We had a rare do at the last meeting! The Board of Agriculture actually had the impertinence to put up a man named Bruce to run down Mendelism. He quoted Pearson by the yard, and ended by quoting yourself, Punnett and Wood in support of his contention that horses were unsuitable for experiments! I was so annoyed at the move that I simply went for him and told him that he did not know what he was talking about, and I don't think he will trouble us again. … Ewart thinks that in the end we made a deep impression; at the same time I do not care for the tactics which were evidently designed to catch us napping and make us look foolish. I fancy that they thought that Ewart and I would not exactly pull together, but we do.

Hurst and Ewart [11] were appointed as advisers on horse breeding for the armed forces (Horse Breeding Committee of the Development Commission), and in August 1911 Hurst set out their proposals for establishing a line of homozygous hunter stock at the BA meeting [12]. He also spread the Mendelian message more directly to breeders through forums such as the newly established *Bloodstock Breeders' Review* [13].

Genetics Congress

Under the aegis of the Societé Nationale d'Horticulture de France, the Fourth International Congress of Genetics was held in Paris in 1911. Johannsen wrote (September 8) saying he would be staying with the organizing secretary Philippe de Vilmorin (1872–1917), and that there was much he would like to discuss with Bateson before a trip to America. Bateson informed Wheldale that they would be leaving for Paris on September 18 and would return, by way of Majorca, around October 12. In the course of the conference, a Standing Committee was appointed under Bateson's chairmanship to organize future conferences (Baur, Johannsen, A. Lang, Lotsy, Herman Nilsson-Ehle, W. F. Swingle, Tschermak, Vilmorin).

In November, Bateson received the Victoria Medal of the RHS. How close he came to his deadline (January) for submission of the *Problems* manuscript to Yale is not clear, but the 1912 correspondence contains no hint of anguish in this respect. In February, he began the Fullerian Lectures at the Royal Institute, a 3-year appointment involving eight lectures a year at £83/annum plus expenses. The lectures were reported serially in the *Gardeners' Chronicle*. There was also the Herbert Spencer Lecture at Oxford, which was given (February 28) for a stipend of £25 to "a thin house in a very large room" (Chap. 15). He reported to Beatrice (February 29) that he had visited an art gallery where he met Salaman, who was in Oxford to give a lecture on heredity to a Jewish club: "They had heckled him a good deal, and he says that the young Oxford Group are still disposed to question the whole Mendelian case." Beatrice was away in March, and Bateson's letters to her were mainly concerned with the suffragette movement and trips to the London theatres (March 2):

> I am sending *Times* with a fuller account of the new female idiocy than you will get in Yorkshire. … It is perhaps as well they should all be on guard before the general upheaval begins. … I am being quite gay. On Thursday I took Miss Pellew to *Mind the Paint Girl* –

amusing, but not very. Tonight after an educative visit to some Museum with Gregory, I am to go with Hermia [Durham] to *Women and Wine*!

There was an encounter with a policeman (March 3):

Hermia showed me Lilla's name in a list of arrested, but I am glad to hear she does not propose to do anything. Mrs. Pankhurst got 2 months, but as a policeman with whom I traveled remarked "Without hard labour it's no use giving her that, Sir." The fact is they are very nearly within the scope of the Mental Defect Bill. My policeman said he had thought there was some right on their side, but he don't see "what they want to smash my face for." They are believed to have another abomination planned for tomorrow.

As usual, the Bateson matriarch took a house by the sea for the summer – this time at Sheringham – where she was visited by Beatrice who retrospectively annotated the correspondence: "She was distinctly *difficult* sometimes, though always generous and kind and devoted to her grandchildren." There was some talk by Bateson of "calling John to order" in light of an unfavorable school report (August 4): "He is much too easy going." The routine work at Merton continued (August 7): "Miss Pellew came back and we put in a day's work yesterday. In the evening we went to 'Hindle Wakes' partly on Florence's recommendation and partly on O. S's [Owen Seaman in *Punch*]." Bateson and Pellew set out to join the family party at Sheringham, but rail traffic was suspended due to a damaged embankment (August 29): "We found that Sheringham was impossible but that Dersham was probably accessible. So she [Pellew] suggested I should join the party there." At some point, they made their way to Dundee (September 7), where the BA meeting was beginning. Shortly thereafter he visited Baur (see below).

In November 1912, Bateson gave a lecture in Groningen in Holland, probably at the invitation of Lotsy. Meanwhile, an address entitled "Unsound Mendelism developments, especially as regards the presence and absence theory," was given at the Royal Dublin Society by James Wilson of the Royal College of Science for Ireland [14]. Wilson had communicated his concerns to Bateson and Punnett. In reply, Bateson wrote (June 17):

This difficulty would be removed if we gave up writing any symbol at all for recessives. For some time past I have contemplated doing this, and have discussed such an amendment of our symbolism with others, but there are some conveniences in the present notation which make it perhaps worth maintaining. When once it is realised that when we write *a* for the absence of *A*, we do not mean that the creature "contains" or "is bearing" anything corresponding to *A*, but merely that there is no *A* in it, I think the difficulty disappears. … To use your comparison: the animal or the plant is the bookcase, right enough, and the factors are the books – so far as we have yet gone – of undetermined number. Sometimes it is useful, in comparing two libraries, to use a symbol to show that a particular book is absent from one of them, but that is all that is meant when the little letters are written in.

Wilson replied (June 19):

I fancied I had found a fallacy in a theory which has been used a great deal, and as you and Punnett are chiefly concerned, and are not like Pearson, who told me in private correspondence when I found him wrong that my brains were not good enough for the work, I thought I should like to put the matter before you both before putting it in type. … The word *absence* is ambiguous and is sometimes used in one sense, sometimes in another. If *a* were

> used only for the non-presence of *A* and as having no connection with *A*, there would be no trouble.

In further correspondence (August 27), Wilson regretted that Bateson and Punnett would not give him "a hearing in the Journal of Genetics."

Problems of Genetics

Since his Silliman Lecture in 1907, Bateson had been working on his final book, *Problems of Genetics*, which emerged in August 1913: "To construct a true synthetic theory of Evolution it was necessary that variation and heredity instead of being merely postulated axioms should be minutely examined as phenomena." Genetics had ceased "to be merely a method of investigating theories of evolution or of the origin of species," and had become an "instrument by which the nature of the living organism may be explored."

From this, it might appear that the question that had so occupied him in the past – the origin of species – was no longer a central concern. But within a few lines, he was back on topic: "Nowhere does our new knowledge of heredity and variation apply more directly than to the problem what is a species and what is a variety?" Bateson was aware "that some very eminent systematists regard the whole problem as solved. They hold as Darwin did that specific diversity has no physiological foundation or causation apart from fitness." This was not so. While admitting that "comparison between forms from dissimilar situations contributes something," he postulated that it was "by a close examination of the … genetic behaviour of familiar species when living in the presence of their nearest allies that the most direct light on the problem is to be obtained." His old theme – the insufficiency of natural selection – was not far behind. The differences between various species from the genus *Veronica* that "grow side by side in my garden" had not arisen because of their survival value, not because of their utility, but because

> none of their diversities was so damaging as to lead to the extermination of its possessor. When we see these various Veronicas each rigidly reproducing its parental type, all comfortably surviving in competition with each other, are we not forced to the conclusion that *tolerance* has as much to do with the diversity of species as the stringency of Selection? … The control of Selection is loose, while the conformity to specific distinction is often very strict and precise, and no less so even when several closely related species co-exist in the same area and in the same circumstances. The theory of Selection fails at exactly the point where it was devised to help: *Specific* distinction.

Sticking to floral examples, he went on to compare two *Lychnis* species that were presumed to share a common parental species:

> But whatever the common progenitor may have been, if we are to believe that these two species have been evolved from it by a gradual process of Natural Selection based on adaptation, enormous assumptions must be made regarding the special fitness of these two forms and the special unfitness of the common parent, and these assumptions must be specially invoked and repeated for each feature of structure or habits distinguishing the three forms.

> Why, if the common parent was strong enough to live to give rise to these two species, is it altogether lost now, or at least absent from the whole of Europe? Its two putative descendents, though so distinct from each other, are, as we have seen, able often to occupy the same ground.

Then, returning to, but not citing, the physiological selection of Romanes, Bateson gave examples of the importance of "internal constitution" where, of two allied species, one showed great variability and the other did not:

> When we compare the polymorphism of one species with the fixity of another, and attempt to determine the causes which have led to these extraordinary contrasts, two distinct lines of argument are open to us. We may ascribe the difference either to causes external to the organisms, primarily, that is to say, to a difference in the exigencies of Adaptation under Natural Selection; or on the other hand we may conceive the difference as due to innate distinctions in the chemical and physiological constitutions of the fixed and variable respectively. ... We cannot declare that Natural Selection has no part in the determination of fixity or variability; nevertheless looking at the whole mass of fact which a study of the incidence of variation provides, I incline to the view that the variability of polymorphic forms should be regarded [more] as a thing tolerated than as an element contributing directly to their chances of life; and on the other hand that the fixity of monomorphic forms should be looked upon not so much as a proof that Natural Selection controls them with a greater stringency, but rather as evidence of a natural and intrinsic stability of chemical constitution. ... As soon as it is realized how largely the phenomena of variation and stability must be an index of the internal constitution of organisms, and not mere consequences of their relations to the outer world, such phenomena acquire a new and more profound significance.

Discarding for the time being the homeotic subdivision (Chap. 4), he pointed to two broad classes of variation:

> The only classification that we can yet institute with any confidence among the phenomena of Variation is that which distinguishes ... variations in the processes of division from variations in the nature of the substances divided. Variations in the processes of division are most often made apparent by a change in the number of parts, and are therefore called *Meristic* Variations, while the changes in actual composition of material are spoken of as *Substantive* Variations. ... That there may be a real independence between the Meristic and the Substantive phenomena is evident from the fact both that Meristic changes may occur without Substantive Variation, and that the substances composing an organism may change without any perceptible alteration in its meristic structure. ...[Thus] the study of genetics has on the one hand a physical, or perhaps more strictly a mechanical aspect, which relates to the manner in which material is divided and distributed; and also a chemical aspect, which relates to the constitution of the materials themselves.

Next, a distinction was made between division and differentiation:

> There is the event by which the cell *divides*, and the event by which the two halves or their descendants are or may be *differentiated*. It is common knowledge that in some cell-divisions two similar halves, indistinguishable in appearance, properties, and subsequent fate, may be produced, while in other divisions daughter-cells with distinct properties and powers are formed. We cannot imagine but that in the first case, when the resulting cells are identified, the division is a mechanical process by which the mother-cell is simply cut in two; while in order that two differentiated halves may be produced, some event must have taken place by which a chemical distinction between the two halves is effected.

In some respects, the book was a regression to Bateson's pre-Mendelian days, representing the unwritten second volume of *Materials*, for which abundant notes may be found among Bateson's archived papers. While Mendelian analysis had much to offer, save for the identification of segregatable factors, it had so far contributed relatively little to progress with the problems that had concerned *Materials* – the basis of the repetition and differentiation of parts during development. Bateson was still much in awe of the problem of establishing and maintaining symmetry and pattern – today referred to as "developmental genetics" or "evo-devo." Perhaps with the new ideas of Einstein in mind, he considered that "it is … a problem rather for the physicist than for the biologist," for "it is in the geometrical phenomena of life that the most hopeful field for the introduction of mathematics will be found."

As for the classification of substantive variations – sometimes the "chemical sports" of Garrod – there was here a need for "a knowledge of the chemistry of life far higher than that to which science has yet attained." Nevertheless, Bateson anticipated "that future analysis will recognise among the contributing elements, some which are intrinsic and inalienable, and others which are extrinsic and superadded." He then foresaw – it now appears correctly – a twentieth century with biochemistry dominant: "It may well be that before any solution is attained, our knowledge of the nature of unorganized matter must first be increased. For a long time yet we may have to halt." Yet, what could be the underlying nature of variation? Here, he envisaged something close to retroviruses that we now know can become part of a host cell's DNA:

> If we could conceive of an organism like one of those to which disease may be due becoming actually incorporated with the system of its host, so as to form a constituent of the germ-cells and to take part in the symmetry of their divisions, we should have something analogous to the case of a species which acquires a new factor and emits a dominant variety. … The appearance of recessive varieties is comparatively easy to understand. All that is implied is the omission of a constituent. How precisely the omission is effected we cannot suggest, but it is not very difficult to suppose that by some mechanical fault of cell division a power may be lost. Such variation by unpacking, or analysis [opening up] of a previously existing complex, though unaccountable, is not inconceivable. But whence came the new dominants? Whether we imagine they are created by some rearrangement or other change internal to the organism, or whether we try to conceive them as due to the assumption of something from without, we are confronted by equally hopeless difficulty. … If they were proved to enter from without, like pathogenic organisms, we should have to account for the extraordinary fact that they are distributed with fair constancy to half the gametes of the heterozygote.

At the proof stage shortly before the book was to go to press, perhaps anticipating some criticism from Morgan, Bateson added:

> The degree of dominance becomes … the deciding criterion by which we distinguish the existence of factors. But it should be clearly realized that in any given case the argument can with perfect logic be inverted. We already recognize cases in which, by the presence of an inhibiting factor, a character may be suppressed and, purely as a matter of symbolic expression, we might apply the same conception of inhibition to any example of factorial influence whatever. … We may extend this mode of reasoning to all cases of genetic variation, and thus conceive all alike as due to loss of factors present in the original complex. Until we can recognize factors by means more direct than are provided by a perception of their effects, this doubt cannot be positively removed.

Bateson next turned to the role of genic differences in species formation, with special reference to "the mutation theory" of de Vries who had been deeply impressed by the phenomenon now known as polyploid speciation (to be discussed in Chap. 16). Bateson still thought, correctly, that de Vries' evening primrose mutants "provide illustrations of physiological phenomena of the highest importance in the study of genetics at large," but he considered the evidence as "still ambiguous" and knew "of no considerable body of facts favourable to that special view of Mutation which de Vries has promulgated."

Nevertheless, Bateson raised the general question, to which we will return: "Can we suppose, that in general, closely allied species and varieties represent the various consequences of the presence or absence of allelomorphic factors [genes] in their several combinations?" He noted that the reproductive barrier between species meant that "in most of the examples in which it has been possible to institute breeding experiments with a view to testing the question, a greater or lesser sterility is encountered." In short, if lines cannot be crossed then Mendelian analysis is no longer possible. Mendelian analysis is an approach to within-species variation, not to between-species variation. Under the heading "variation and locality," Bateson next turned to the question of reproductive isolation:

> One outstanding feature is hardly in dispute, namely that prolonged [geographical] isolation is generally followed by greater or less change in the population isolated. Groups of individuals which from various causes are debarred from free intermixture with other groups almost always exhibit peculiarities, but on the other hand, cosmopolitan types which range over wide areas are on the whole uniform. ... The barriers to intercourse may be seas, deserts, prairies, mountain-chains, or circumstances of a much less obvious character which isolate quite as effectually.

As examples of the latter, he mentioned seasonal isolation as between early and late broods of moths, or early and late flowering species of flowers. Under this heading, Gulick in *Evolution Racial and Habitudinal* had described two races of *Cicada*, one with a 13-year life cycle and one with a 17-year life cycle. Many had followed up on Gulick's pioneering studies on the distribution of island snails (1872); however, the descriptions tended to be anatomical rather than physiological (i.e., whether different lines would cross with each other). Yet hybrids were sometimes seen. Bateson concluded: "I incline ... to agree with Gulick who, after years of study of the local variations of the Achatinellidae, came to the conclusion that it was useless to expect that such local differentiation can be referred to adaptation in any sense." In short, "the creature persists not merely by virtue of its characteristics but in spite of them, and the fact of its persistence proves no more than that on the whole the balance of its properties leaves something in its favour."

The naturalist in Bateson then took off with detailed descriptions of various species and varieties as recorded by the systematist: "When two forms are found coexisting in the same area they are usually recorded as one species if intergrades are observed, and as two species if the intergrades are absent. On the other hand when two forms are found occupying separate areas, ... then forthwith each is named separately either as species or subspecies." Still in pre-Mendelian mode, he described his own studies of the distribution of moths throughout Europe carried out in the 1890s, ending up with the thought:

> As to *why* a variety should increase in numbers we have nothing but mere speculation to offer .... That such survival and replacement may reasonably be taken as an indication that the replacing race has some superior power of holding its own I am quite prepared to admit. Nevertheless it seems in the highest degree unlikely that the outward and perceptible character or characters which we recognize as differentiating the race should be the actual features which contribute effectively to the result.

As for Lamarckism, the possibility of the inheritance of acquired characters was still entertained on theoretical grounds "of late revived by Semon and ... by F. Darwin" (Chap. 19). What was needed was definitive evidence. The *cause célèbre* here was to be the Kammerer affair (Chap. 21). The latter was not an isolated incident. In *Problems*, under the heading "Causes of genetic variation," Bateson severely criticized various studies purported to support Lamarck, noting work on Colorado potato beetles by Tower of the University of Chicago, that had been subjected to "destructive criticism" by at least one other. There were also preliminary indications, described by W. T. MacDougal, that radiation might be mutagenic (an observation that was later to be placed on a firmer footing by Muller). Bateson concluded: "I do not doubt that evidence of this type will be greatly extended. As a contribution to genetic physiology these facts are very important and interesting, but I cannot think that anyone, on reflection, will feel encouraged by such indications to revive old beliefs in the direct origin of adaptations."

Problems concluded with a repetition of Bateson's views on hybrid sterility. Variation *within* a species resulted in varieties (races) that differed in the Mendelian character units that we now equate with genes, but this was quite different from the variation that separated allied species such that, if crossing were possible, the offspring were sterile. He regarded "the power to produce a sterile or partially sterile hybrid as *a distinction in kind*, of a nature other than those we perceive among our varieties. ... The distinction ... is so striking and so continually before the eyes of the practical breeder that he can scarcely avoid the inference that when he meets a considerable degree of sterility in a cross-bred he is dealing with something belonging to *a distinct category*, and not merely a varietal feature of an exceptional kind." Furthermore, "though we cannot strictly define species, they yet have properties which varieties have not, and ... the distinction is *not merely a matter of degree*." The various types of sterility should not be lumped together:

> No doubt all breeders know that sterile animals and plants occasionally appear in their cultures, but it is more in accordance with the probability that the sterility in these sporadic instances should be regarded as due to [genic] defect, than that it should be thought comparable with that of the sterile hybrids. For their sterility must, by all analogy with results elsewhere seen, be attributed not to the absence of something, but to the presence and operation of complementary factors leading to ... inhibition of division [inhibition of gametogenesis]. ... The distinction between these several kinds of sterility was of course not understood in Darwin's time. The comparison, for example, which he instituted between the sterility of "contabescent" anthers and that of hybrids no longer holds, for at least in those cases in which the nature of contabescent anthers have been genetically investigated ... they proved to be a simple recessive character [i.e. genic].

It was a mistake to regard varieties that differed from the main line in various character units as steps in a process that might lead to a new species:

> All constructive theories of evolution have been built on the understanding that what we know of the relation of varieties to species justifies the assumption that the one phenomenon [i.e. variety] is a phase of the other [i.e. species], and that each species arises, or has arisen, from another species either by one or several genetic steps. In the varieties we have accustomed ourselves to think that we see these steps. … Among the plants and animals genetically investigated are many illustrations of striking and distinct varieties. Many of these might readily enough be accepted as species by even the most exacting systematists, and not a few have been so treated in classification; but when we have examined their relationship to each other we feel, not merely that they are not species in any strict sense, but that the distinctions they present cannot be regarded as stages in the direction of specific difference. Complete fertility of the results of intercrossing is, and I think must rightly be regarded as, inconsistent with actual specific difference [the difference between species]; and of variations leading to that consequence [formation of a new species] no clear indication has yet been found. … In spite of all we know of variability, nothing readily comparable with the power to produce a sterile hybrid on crossing with a near ally, has yet been observed spontaneously arising, though that characteristic of specificity is one of the most widely distributed in nature.

Thus, "the first step is to discover the nature of the factors which by their complementary action inhibit the critical divisions and so cause the sterility of the hybrid." These factors were "ingredients which cause greater or lesser disturbance in the processes of cell-division, and especially the processes of gametic maturation, when they are united by fertilization with complementary ingredients."

As when he had reviewed *Materials* two decades earlier, when reviewing *Problems* Cunningham again pointed to the derivative nature of Bateson's ideas [15]:

> He shows in a very convincing manner how impossible it is to maintain that the slight differences by which species are distinguished have any utility or relation to habits or mode of life. The subject has never been discussed more forcibly or with greater literary ability, but at the same time it is surprising that no mention is made of the fact that the subject has been discussed before. Readers unacquainted with the history of theories of evolution might suppose that Professor Bateson had discovered the uselessness of specific characters. As a matter of fact, this objection to the theory of natural selection is discussed by Darwin himself in the Sixth Edition of the "Origin of Species," and he attempted to remove it by attributing such characters to correlation [with hidden adaptive characters], or the argument that they might have uses of which we were ignorant. Romanes developed very thoroughly and in detail the thesis that natural selection was a theory of the origin of adaptations, and that specific characters were in most cases not adaptations, and the present writer has published many papers to the same effect with special reference to the flat fishes.

Whether Bateson saw or in any way responded to Cunningham's point is not clear. But we will be noting later (see Chap. 25) that the tracing and recording of the true origins of a discovery was more difficult a century ago than today.

Baur

In September 1912, Bateson went on holiday with Germany's first Professor of Genetics, based at Berlin Agricultural College – Erwin Baur [16]. It was an opportunity to sound him out, not only on science, but also on war. After a brief solo visit to the Berlin art galleries (September 27), Bateson was taken by Baur to Potsdam and they ended up at Tambach, where they walked in the woods (October 1):

> I get on very comfortably with my companion. He is easier to travel with than I expected, but he is also duller. The end of his ideas and experiences is very soon reached. The whole range, not only of the arts (including music) is sealed to him, but all literature too. He knows not even Goethe or Schiller and has in fact read literally nothing in any language not in science. Hatred and contempt for the French … is his leading idea outside Mendelism. The feeling against us is not quite what I had supposed. It seems they don't want *our* possessions, but they are determined to have the foreign possessions of France and Spain, and why we should prevent them they cannot understand. It seems to them mere cussidness on our part. There is evidently great excitement here and he says that in the past year the warlike spirit has grown universally. He thinks it quite certain that they will fight whenever the moment of strain comes like last year. He says it is quite impossible that such a crisis should be outlived again without an outbreak. His desire is that the Germans and the English should combine and simply crush out the other nations, whom he regards as mere trash and doomed in any case to extinction in about 200 years. The notion that France has made untold contributions to civilization is quite new to him, and he doesn't understand how any one can support such a proposition. A simple soul, you see. He will last me well for a week or so, but I would rather have someone more complicated if … walking for a fortnight.

It is likely that Bateson and Baur considered the venue of the next Genetics Congress, thinking that it might be in Copenhagen in 1915. In early 1913, Bateson was in Paris, both hunting down art treasures and attempting to sell parts of his own collection. On his return, he wrote to Baur (January 31) noting Johannsen's ill-health and doubting that they could rely on him to organize the Congress. Berlin or San Francisco might be better? Baur responded (February 16) pointing out the difficulties of Berlin. Bateson reiterated (February 25) that "poor Johannsen" should not undertake a congress "before we have assurance that he is really fit for it, which as yet, he evidently is not." Two weeks later (March 7), Bateson was informed that he had been nominated to preside at the 1914 BA meeting in Australia. He promptly wrote to Baur (March 9) pointing out that, coming so soon after the Australia meeting, a conference in May 1915 in San Francisco seemed less appealing. He suggested Berlin 1916.

Beatrice noted (*Memoir*) that Bateson's own health now came into question. He had a minor heart attack and thenceforth suffered from intermittent anginal pain: "We were never quite free from anxiety; he felt pain, lurking round the corner waiting to pounce on him. The gay sense of physical well-being was lost." No longer was Beatrice trying to keep up with Bateson as they walked. Bateson wrote to Baur (May 9): "I have, I am sorry to say, been ill – some gastritis of unknown origin – or I would have written before. … I have lost much time, and though well enough while making no exertion, I am still not really up to the mark." Again, he suggested Berlin 1916. The correspondence continued, and after consultation with other members of the Standing Congress Committee, a proposal from Baur to hold the 5th International Congress in Berlin in 1916 was accepted.

Bateson was empowered to invite speakers to the meeting in Australia, and he invited Baur and Morgan, who had received a complimentary copy of *Problems*. While declining the invitation, Morgan thanked him for *Problems* and enclosed a copy of "my little book on heredity and sex." In September, there was the 1913 BA meeting at Birmingham. Here, a major attraction was an excursion to view Hurst's operations at Burbage.

After Christmas, Bateson was off to Berlin again. Miss Pellew had preceded him to visit friends, and they met from time to time (galleries, concerts, and theaters). Of course, they visited Baur and with his help a German translation of *Principles* was negotiated. Shull and his wife had recently visited Baur. Bateson wrote (December 31): "Frau Baur much livelier than before. She had been very shy I think. Poor Miss Pellew has been a sad failure. They say they *can't* understand anything she says. Shull and Frau Shull speak lovely English which they follow perfectly, but Miss. P. is unintelligible. … So she has got rather shelved."

On the first day of fateful 1914, Bateson wrote: "I don't think I can stick Parsifal. Miss. P and a friend are going, but my time would be better spent over [a] chess-board." He found a British paper and commented on the latest conferral of knighthoods, etc. (January 2): "Saw list of honours in *Daily Mail* while waiting for my chess this afternoon. Better lot than usual." Finally, there was a visit to the Agricultural College (January 4): "Went over to Baur's laboratory. … We fetched Miss. P. out of library and had a good business lunch. … He has a pleasant fresh mind, but I don't find I get actual new ideas from him. Still, feel that the visit has 'jogged my scientific liver' very usefully." From correspondence with Wheldale, we learn that he gave (March 14) a "tea paper" at Imperial College, and later (March 23): "I am going to rusticate at Burbrig Gap Hotel, near Eastbourne this week to write 'Address' [for Australia]." Beatrice added: "The 'rustication' was after his angina-like attack – his first real knock-down. He was very unwell at this time" (Fig. 14.1).

Passage to Australia

Bateson, Beatrice, and Saunders boarded the *Ascanius* of the Blue Funnel Line (June 22). Beatrice wrote to old Annie Bateson (July 12) that there was no "Marconi News" available "so we have no papers and don't know whether civil war or peace rules Ireland." The BA party was 68 in number. There was no band, and Beatrice sometimes entertained on a badly tuned piano. On Sunday, Mrs. Poulton played hymns. The ship stopped at Las Palmas in the Canary Islands, where there was sightseeing. During a short stay at Cape Town, they first learned "of the coming European disaster" heralded by "news of the Serajevo murder." Beatrice related: "Will was frightfully anxious at once, saying, 'there will be war,' and when we laughed at him: 'You will see, this is no laughing matter.' The satisfaction at being

Fig. 14.1 Deck games

cut off the vulgarities of the suffragettes, and the unruliness of Irish patriots, was lost in restless longing for fuller, up to date news" (Fig. 14.2).

Prior to his Sydney address (Chap. 15), Bateson spoke in Melbourne. He began by proclaiming that heredity "in its scientific sense is not older than Herbert Spencer." For organisms "are formed as pieces of living material split from the body of the parent organisms. Their powers and faculties are fixed in their physiological origin." As for the nature of the fundamental "elements, or *factors* as we call them," it was obvious that they were "in some way directly transmitted by the material of the ovum and of the spermatozoon." But it was "unlikely that they are in any simple or literal sense material particles. I suspect rather that their properties depend on some phenomenon of arrangement." The idea of arrangement was deemed fundamental to variation, and he appeared to have modified his previous view that variation would be understood in terms of additions and subtractions:

> That which is conferred in variation must rather itself be a change, not of material, but of arrangement, or of motion. The invocation of additions extrinsic to the organism does not seriously help us to imagine how the power to change can be conferred, and if it proves that hope in that direction must be abandoned, I think we lose very little. By the rearrangement of a very moderate number of things we soon reach a number of possibilities practically infinite.

The modern reader will tend to see here DNA and the rearrangements or substitutions of its four bases in ways that change the information it conveys (see Part VI).

Fig. 14.2 Bateson, Beatrice, and Becky Saunders

But there is little in Bateson's writings, save for these few words, the *Daily Mail* allusion (Chap. 6), and the above discussion of books in bookcases with Wilson, to suggest that his ideas were approaching those of Butler and Hering (Chap. 19).

In these remarks, Bateson was not referring to rearrangements of the Mendelian genetic factors (genes) themselves, which was a separate issue: "Analytical breeding proves that it is according to the distribution of these genetic factors … that the characters of the offspring are decided. … An organism cannot pass on to an offspring a factor that it did not itself receive in fertilization." Variation was now seen "as a distinct physiological event" so that varieties within a species "are simply terms in a series of combinations of factors separately transmitted, of which each may be present or absent." Bateson considered as "done with," Darwin's "notion that large differences can arise by the accumulation of small differences." Such small phenotypic differences were often ephemeral, nontransmissible, effects of environment ("conditions of life") – *nurture* not nature. Furthermore, the "intergrade" forms found in areas of overlap between two species, which "used to be taken to be transitional steps," were now seen as "merely mongrels between the two species."

Bateson thought Lotsy "on unsafe ground" in proclaiming crossing to be "the sole source of variation." Furthermore, the rare and sporadic emergence from a long-established variety of a new spontaneously arising form of plant, breeding true from the start, was unlikely to be a long-concealed recessive. The rare and "most interesting 'mutations' recorded by Professor T. H. Morgan and his colleagues in the fly, *Drosophila*" were also of this type. To Bateson, such examples of *original* variation seemed "practically conclusive." Thus, the genetic factors (genes), seemingly "permanent and indestructible," which by their recombinations could generate "a profusion of forms," could also, albeit rarely, "undergo a quantitative disintegration, with the consequences that varieties are produced intermediate between the integral varieties from which they were derived." In short, Bateson was anticipating what we now generally ascribe to an amino acid-changing base substitution in a DNA sequence (see Part VI).

He seemed to wander somewhat from his usual strict distinction between *differences between varieties* that, because extended breeding studies were possible, could be subjected to Mendelian analysis, and *differences between species* that, because extended breeding studies were not possible, could not:

> An organism is pure-bred when it has been formed by the union in fertilization of two germ cells which are alike in the factors they bear; and since the factors for the several characteristics are independent of each other, this question of purity must be separately considered for each of them. A man, for example, may be pure-bred in respect of his musical ability and cross-bred in respect of the colour of his eyes or the shape of his mouth. Though we know nothing of the essential nature of these factors … we feel justified in the expectation that with continued analysis they will be proved to be responsible for most, if not all, of the differences by which the varying individuals of any species are distinguished from each other. I will not assert that the greater differences which characterize distinct species are due generally to such independent factors, but that is the conclusion to which the available evidence points.

However, it should be noted that he was writing here on factors responsible for the observed differences that *characterize* species, which had themselves been identified by systematists as distinct *by virtue of those characters*. As Gates was to point out (see below), this did not necessarily relate to the *process* by which those species initially diverged from a common ancestor – the origin of species. Bateson still pondered the possibility of the "operation of systems of descent quite other than those contemplated by the Mendelian rules," noting that "I myself have expected such discoveries, but hitherto none have been plainly demonstrated." Thus, "speaking generally, we see nothing to indicate that qualitative characters descend … according to systems which are incapable of [Mendelian] factorial representation."

Nevertheless, while proclaiming "this is no time for devising a theory of Evolution," he was not beyond toying with the idea of representing evolution as "an unpacking of an original complex which contains within itself the whole range of diversity which living things present." In this way, he challenged the received view "that evolutionary progress is from the simple to the complex," and asked, "whether after all it is conceivable that the process was the other way about." While it is difficult to see this in modern terms, perhaps it explains his above reference to "the rearrangement of a moderate number of [pre-existing] things"? This aside, he summed up more simply:

> The isolated events to which variation is due are evidently changes in the germinal tissues, probably in the manner in which they divide. It is likely that the occurrence of these variations is wholly irregular, and as to their causation we are absolutely without surmise or even plausible speculation. Distinct types once arisen, no doubt a profusion of forms called species have been derived from them by simple crossing and subsequent recombination. New species may be now in course of creation by this means, but the limits of the process are obviously narrow.

While the meeting was in progress, war was formally declared. Bateson worked to get German scientists safely repatriated. He failed in the case of two (Graebner and Pringsheim), who were interned in Australia. For Bateson, the voyage home was uneventful but was not by way of India as he had planned. Not so for Baur. He returned by way of Java and was apprehended by the British in the Mediterranean. However, he escaped and returned to Berlin. Some were not so fortunate. Michael Pease had gone from Cambridge to Germany in 1913 for postgraduate research and was interned for the duration of the war. Richard Goldschmidt (1878–1958), who had just joined Boveri's Institut für Biologie in Berlin (he headed the Genetics department from 1914 until 1935), was traveling back from a collecting trip in Japan by way of the United States. The British sea blockade kept him there, and for a brief period when America entered the war in 1918 he was interned in a civilian prison camp.

Gates

Although often agreeing with Bateson, several British colleagues had reservations. Chromosome studies were proceeding in the laboratories of Doncaster (moths), Farmer (lilies, liverworts), Gates (*Oenothera*), and Gregory (*Primula sinensis*). Doncaster wrote (January 5, 1912) that he could no longer "help feeling that the cytological results of recent years are so closely in accord with what one must assume in Mendelian segregation that the two things must be closely connected." The most forceful in his arguments was the Canadian cytologist at Bedford College, Reginald Ruggles Gates (1882–1962). In 1907, he had reported on the chromosomes of de Vries's evening primrose variants (Chap. 9). For a while, Gates was married to Marie Stopes, and later became involved in the eugenics movement. In his text *The Mutation Factor in Evolution* [17], Gates agreed with Bateson that Lotsy's disregard for mutations per se was unwise:

> The origin of a true mutation must be regarded as a process entirely distinct from its subsequent inheritance. … The fact that mutations and hybrid segregation may bear a superficial resemblance to each other has led several writers to the false conclusion that any mutations that occur in a hybrid race are necessarily a result of the previous cross. … Now that mutations are known to occur in pure species, it can never again be assumed that because

> mutations appear there has been previous crossing [between different types]. … Although [such] crossing may in some cases increase the frequency of mutations or even initiate a condition of germinal instability, yet there is no necessary relation between crossing and mutations. For the latter may occur in the absence of crossing, which shows not only that mutation is an independent process but that it is in many cases … due to other causes than hybridisation.

By "hybridization," Gates here meant the outcrossing between members of two races or allied species that would produce hybrid F_1 individuals. Like Bateson, Gates also cautioned against lumping all forms of sterility, often manifest as "bad pollen grains," together:

> The only feature which all mutations have in common is that they result from germinal disturbance in the organism, and it is obvious that such disturbances may be brought about by a variety of agencies. One peculiarity which mutants not infrequently share with hybrids is sterility. A condition of partial or complete sterility is, therefore, not in itself a proof of hybridization, for sterility may arise suddenly in connection with the origin of a mutation, as in the pollen of *Oenothera lata* and the ovules of *Oenothera brevistylis*. The presence of bad pollen grains is therefore not necessarily an indication of crossing. … We are inclined rather to regard the high frequency of bad pollen grains as a result of the peculiar cytological condition of Oenothera, in which the chromosomes in meiosis are very loosely paired and hence form irregular combinations … which may be incompatible with development [e. g. autopolyploids; see Fig. 16.1 of Chap. 16]. The weak attraction between homologous chromosomes, which results in this loose pairing, may be merely an indication of some fundamental peculiarity in the condition of the germ plasm.

While not attributing any "fixed purpose to the organism," Gates appealed from time to time to "the principle of orthogenesis, whatever its explanation may be." By this, it seems he meant what Bateson attributed to the "internal constitution" of an organism, not something mystically directing the path of evolution. Gates drew further examples of mutation from studies of bacteria on culture plates, which have a distinctly modern ring. Thus, in 1907, Massini had observed the sporadic appearance of lactose-fermenting colonies on his plates. The new property was inherited by child colonies. And in 1914 Henri had demonstrated that ultraviolet light was mutagenic in bacterial cultures:

> If a bacterium can undergo a constant change of function, the same may reasonably be expected to happen in a [eukaryotic] chromosome. … It is probable that in ultimate nature it is an alteration in the chemical constitution of the chromosome or a portion of it. This may be thought of as a stereochemical rearrangement in the complex molecule of the *nucleic acid* or as some other type of chemical change involving the formation of slightly modified protein substances (our italics).

Gates saw no contradiction between this and what we now regard as a genic mutation:

> The hypothesis that each [chromosomal] change is connected with the alteration of a particular pangen [gene], gives the process a "local habitation and a name" [quote from Shakespeare's *Midsummer Night's Dream*] but does not add to our knowledge of it. But if we assume that the change is concerned with a particular chromosome or portions of one, we make the matter still more concrete, and the hypothesis can … be verified by observation and experiment. In this way the chromosome hypothesis (which is already proven in

certain cases) should, we think, be used as supplementary, and not contradictory, to the pangen hypothesis of de Vries.

Bateson's ideas on mutation were becoming increasingly challenged. In 1912, James Wilson had spoken of "Unsound Mendelism developments" (see above). Gates agreed. Indeed, "the neo-Mendelian philosophy of evolution, founding everything on the presence-absence hypothesis, has led to conceptions which sometimes border upon the grotesque." He was particularly concerned about recessive mutations and "reversions" (now sometimes seen as due to "back mutations"; e.g., DNA base A changing to base T for which the corresponding "forward mutation" would be base T to base A; see Part VI):

With regard to the origin of recessive mutants which Bateson believes are easily accounted for by "some slip in the accurate working of the mechanical process of division" by which "a factor gets left out," it seems more probable that the change occurs by the loss or alteration of an activity on the part of a chromosome or other cell constituent. This being the case, the character or activity may not always be lost irrevocably, but may occasionally reappear, causing a "reversion."

In his Melbourne address, Bateson had speculated that all changes might be the result of progressive losses from some greater preexisting whole. Gates compared this "unpacking theory" with Bonnet's ancient preformation theory (the Russian doll "emboîtement" model): "It is scarcely thinkable that biologists today could be induced to return to a conception of evolution as crude and elementary in its way as was this eighteenth century theory of Bonnet in embryology." Gates' concern was reiterated in 1955 by Richard Goldschmidt, who declared [18] that "the early Mendelians … assumed a skeptical attitude towards evolutionary speculations, which found its extremest expression in Bateson's Australian address (1914) with the embarrassing idea of evolution by loss of inhibitors." Actually, Bateson's remarks can now be construed as examples of his open-mindedness (see Chap. 18).

Gates did not like Bateson's viewing variations in chromosomes in the same light as variations in other characters:

Loose statements regarding variation are frequently made, implying that variations in chromosome number are no more significant than fluctuation in any external feature, such as the number of petals in a flower. But … chromosomes come in a unique category. They are almost the only primary morphological features transmitted as such directly from the previous generation. The constitution of the nuclei is determined at the time of fertilization, while in higher organisms all other features of the adult … are secondary in origin, developing as the result of interaction between the nucleus and cytoplasm of the cell.

Winge (Chap. 16) made the same point in 1917 [19]:

Zoologists have endeavoured to make … comparison, for instance, between the constancy of the number of chromosomes and that of the fingers on a human hand. … Such parallels are in my opinion unjustifiable. In the first place, no other element within the walls of the cell can be compared with the chromosomes. … and even if other elements in the cytoplasm have the power of dividing, and even of reaching the embryo cell with both sexual cells, there is nevertheless in none of these anything approaching the regular system of division which distinguishes the chromosomes. … It must naturally be altogether unwarrantable to compare the constancy of chromosomes with the occurrence of a fixed number of limbs,

> etc. in a species. Members, segments, stamens, feathers, etc. are as matter of fact not transmitted at all as such, but merely as morphologically non-differentiated factors inherent in the sexual cells, whereas the chromosomes are literally inherited, by direct transmission from cell to cell, from generation to generation. Nothing in living nature is more directly hereditary than these very chromosomes, and I can only regard as incomprehensible, that anyone should attempt to pass off the numerical constancy of such elements with a reference to the fact that other elements of a totally different nature also are constant.

That "Bateson and Punnett neglect the cytological facts entirely" was Gates' sad conclusion:

> The truth is that Mendelism is a theory of inheritance, and as such is not adapted to deal with the question of origins at all. It is false logic to assume that the inheritance of a character *necessarily* throws any light at all upon its origin. Characters of a race which have been acquired gradually may be suddenly lost or altered and thus give a Mendelian pair [alleles]; or characters which have suddenly appeared may be gradually modified, by crossing with different species or by other means. It is curious how many have been misled by the logical fallacy above mentioned, and assume that if they can prove that the *inheritance* of a new type is Mendelian, they have at the same time shown its *origin* to be a Mendelian phenomenon. … It would seem that the failure of the modern Mendelians to recognise the limitations of Mendelism, both as a method and as a doctrine, is the chief source of weakness in Mendelism at the present time. Mutation deals with origins in so far as they can be considered discontinuous; Mendelism, on the other hand, concerns itself with discontinuity in inheritance.

Finally, Gates turned to the question of the origin of species arising through genic differences in the chromosomes of two races that had "become isolated" in some way:

> As one part of the origin of species we have to consider the origin of Mendelian characters. The writer's conception is that every such character, whether dominant or recessive, arises through an alteration in the chromosome, or a change which affects, and thus becomes incorporated in, a chromosome. If, in the course of time, a number of such changes take place in the different chromosomes of two races which have become isolated, we may in this way obtain two distinct species which Mendelise in a number of characters when crossed.

Bateson corresponded with Gates in 1920 (April 24): "I shall not be very greatly surprised if some connexion is proved to exist between the number of linkages [linkage groups] and the number of chromosomes. That was a very good suggestion. I shall believe it when it can be established by reasoning which is not circular, and free from alternative and mutually interdependent hypotheses." Gates replied 2 days later:

> You say the chromosomes appear to show the same kind of variation as any other structure. I am certain this attitude could only be taken up by one who is not a cytologist. I know of no cytologist of any repute at all who does not admit the remarkable constancy of structure in the chromatin material, even though they try to explain that constancy without any theory of "individuality."… As regards the reasoning which connects the chromosomes in detail with heredity, the arguments are far more many-sided and convincing, nay conclusive, than biologists frequently accept. The Drosophila paper of Bridges is chiefly remarkable because it presents the largest single body of data linking together the cytological and breeding facts point by point. The reasoning is so conclusive that I see no escape from it except by denying the well-established facts of cytology.

Bateson repeated his points in other correspondence, a favorite theme being that chromosome theory has "too many bunkholes" allowing retreat and a shifting of position. The points he raised were all, in their time, valid and pertinent. His criticisms can be viewed as the purely scientific objections of a person examining a large body of theory and results, who carefully distinguishes between those parts he is prepared to accept and those that he rejects, pending further evidence. They are not the objections of a backwoodsman determined to reject the theory in toto, and at all costs, on philosophical or similar grounds. Indeed, in some ways chromosome theory was to Bateson what quantum mechanics was to Einstein – an incomplete approximation to something very fundamental that lay beyond [20].

Summary

The John Innes years began with passage of the Bateson family from Cambridge to the Manor House at Merton in Surrey. There were also longer passages, to India by Mrs. Herringham, to Ceylon by Punnett, and to Australia by Bateson. The Mendelian lesson had been digested worldwide, and now events took control, tending to distract from his immediate goal – the understanding of phenomena associated with somatic segregation. Challenges came not only from trans-Atlantic sources, but also from home-grown cytologists. Having made no senior appointments at the Institute, Bateson hoped to jog his "scientific liver" by combining continental cultural tours with visits to Baur in Berlin. But rather than new ideas, he gained a dour apprehension of German militarism. Bateson had major input into the restructuring of British science partly prompted by ambitious developments in the United States and Germany. Hurst and Ewart took advantage of expanded military budgets to apply Mendelian principles to horse breeding. In 1913, Bateson's long-awaited *Problems of Genetics* emerged. For solutions, we must look to the "internal constitution" of an organism, but we might be stalled for many years pending more fundamental advances in biochemistry. Again, the distinction between hybrid sterility resulting from a species cross, and the sporadic sterility arising from fresh mutations, was stressed. Although Bateson mentioned Gulick, a reviewer chided him for failing to mention Romanes. In his 1914 BA address, Bateson noted that the properties of the Mendelian factors [genes] might "depend on some phenomenon of arrangement."

Chapter 15
Eugenics

Democracy is the combination of the mediocre and inferior to restrain the more able.

William Bateson (1919)

The title of Galton's Huxley Lecture to the Royal Anthropological Society (October 1901) was "The Possible Improvement of the Human Breed under the Existing Conditions of Law and Sentiment." His solution was "eugenics," a word he had coined. The times were auspicious. Delay in achieving victory in the Boer War had been attributed to the lack of fitness of the soldiers, mostly recruited from the "working class." In 1865, Galton had contributed an essay on "Hereditary Talent and Character" to *Macmillan's Magazine* [1]. Here, he held that eminent men begat offspring who were significantly more eminent than the general population. Such talent was deemed to depend more on an individual's inherited characteristics than the more stimulating environment that inherited wealth could provide. He supported this with a display of pedigrees showing that able fathers had able sons. However, his views were challenged, notably by the Swiss botanist Alphonse de Candolle [2]. Much of the rest of Galton's life was spent trying to understand heredity and the relative roles of "nature and *nurture*," a term he had introduced [3]. Ironically, his vigorous eugenic advocacy may have distracted attention from the conceptual foundations he had provided for Bateson's work (Chap. 3). Around 1909, Bateson probed his own genetic roots, discovering more than one "skeleton in the closet." There were several cases of insanity and alcoholism among his collateral relatives. This suggests a reason for his eugenic caution.

A. G. Cock, D. R. Forsdyke, *Treasure Your Exceptions*,
https://doi.org/10.1007/978-3-030-92099-9_15

Nature and Nurture

From time to time, major transforming influences – the industrial revolution, war, computers – realign social forces and prompt debate. At the beginning of the twentieth century, advances in biology began to transform public attitudes toward the nature–nurture question. Eugenic ideas came to be taken seriously. The Balfour Conservative Government established a "Committee on Physical Deterioration." Would Galton's "actuarial" method or the new genetic methods provide the most enlightenment?

In 1903, Pearson gave the Huxley Lecture on "The Inheritance of Mental and Moral Characters in Man and Comparison with the Inheritance of Physical Characters." He was "worried by the fact that the mentally better stock is not reproducing as it used to do" [4]. In 1904, he chaired a meeting of the newly formed Sociological Society (May 16). Here, Galton addressed an audience that included Weldon, the surgeon George Archdall Reid (1860–1929), the philosopher Victoria Welby (1837–1912), the playwright George Bernard Shaw (1856–1950), and the science fiction writer Herbert G. Wells (1866–1946). Bateson was unable to attend (his third son Gregory was born on May 9). Like some others, he had received an advanced copy of the address, to which he responded with a written communication that was to be read in his absence.

Speaking on "Eugenics: Its Definition, Scope and Aims," Galton envisaged a future where the "average quality" of the nation was improved [5]: "The tone of domestic, social and political life would be raised. The race as a whole would be less foolish, less frivolous, less excitable and politically more provident than now. Its demagogues who 'played to the gallery' would play to a more sensible gallery than at present." To this end, he suggested (i) "wide dissemination of a knowledge of the laws of heredity," such knowledge being "the *actuarial* side of heredity," and (ii) "collecting records on a large scale of thriving families," such thriving families being those "in which the children have won for themselves distinctly superior positions to those who were their class-mates in earlier life." Regarding the choice of a marriage partner, "the passion of love seems so overpowering that it looks like folly to try to direct its course," but "plain facts do not confirm this view." As for negative eugenics, "the community might be trusted to protect itself against accepting … those whom it rates as belonging to the criminal classes." Various communications were then read. Bateson returned to his debate with the biometricians:

> Though these "actuarial" methods were appropriate to an incipient stage of the enquiry, means of attacking the problem directly and with greater effect are now well developed. In nearly every case to which the method of experimental breeding has been applied, it has been possible to show that the phenomena of heredity follow precise laws of remarkable simplicity, which the grosser statistical methods had necessarily failed to reveal.

In the ensuing discussion, Wells agreed that the new Mendelian investigations might better shed light on the issue, but he was concerned that the role of nurture had not been adequately excluded, noting the prevalence of what we would now call "white-collar crime" among the upper classes. When a member of the audience pointed out that Shakespeare was of humble origin, Weldon rose to concede this but extolled the virtues of the biometric approach. (Since then, the case has been made that Shakespeare was a "front man" for the real author, the Earl of Oxford.) Through correspondence with Bateson after the meeting (June 12), Galton came to "quite understand ... your point, and to a great extent agree with it." But he lamented: "What are we humans to do if any 'eugenic' progress is attempted? We can't mate men and women as we please, like cocks and hens, but we could I think gradually evolve some plan by which there would be steady, though slow, amelioration of the human breed."

In 1904, Galton established the "Eugenics Record Office" at University College, London, and Pearson became director of the Biometric Laboratory, funded by the Draper's Company. Indeed, Bateson received an enquiry (February 23, 1905) about his own pedigree from the "Francis Galton Research Fellow in National Eugenics," Edgar Schuster (Chap. 9), who had now relocated to London. There is no record of Bateson's reply. Correspondence with Galton continued. Despite his advancing years, Galton had understood the Mendelian message, and a letter from him (October 1, 1905) even suggested a variation of the square method that Punnett had introduced to sort out the distribution of characters (Figs. 9.3 and 9.4 of Chap. 9). Galton died in 1911 and bequeathed £48,000 to establish the Galton Professorship of Eugenics, with Pearson as the first Chair.

Pearson continued his Editorship of *Biometrika*, whose editorial board included Davenport and later Pearl. When in 1910 Davenport proposed to publish a pro-Mendel paper, Pearson abolished the board. In a letter to Bateson (April 2), Morgan commented: "I suppose you have heard that Pearson has turned off Pearl and Davy from his Journal. I am informed that they are properly incensed. It seems not only high-handed, but a fatal blunder as to policy, for these are the only two men in this country who could have given any credit to his school." Based in the transformed Eugenics Record Office – renamed in 1907 as the "Francis Galton Laboratory for the Study of National Eugenics" – and with the help of statistician Julia Bell (1879–1979), Pearson initiated the *Treasury of Human Inheritance*, the first volume of which was published in 1912 [6]. Later to *Biometrika* he added a new journal, the *Annals of Eugenics* – subsequently renamed *Annals of Human Genetics*.

Mimicry

In the course of his career, Galton had demonstrated formidable powers of intellect, yet the possibility that "nurture" had overwhelmed "nature" is raised by his poor academic record at Cambridge. The same can be said of the early Cambridge years of George and Francis Darwin. George failed entrance scholarships at both St. John's and Trinity Colleges. Francis failed the Trinity examination in 1869. Both went on to excel in their final examinations. Ironically, the power of nurture became manifest to the Bateson's when Gregory, who, since the age of six had attended a local school near Merton, showed signs of picking up the local accent. In 1913, the

9-year-old boy was packed away to a private boarding school, but not before being coached by his brothers on the art of mimicry [7]:

> My mother saw there was a crisis ahead. There was very little use to ask my father about it. So she handed me over to my brothers for religious education. They took me for a walk, and told me that at school I would probably sleep in a dormitory, and that I'd better watch what the other boys did, that when they knelt down to pray, that I should do the same. And that if I said the alphabet over eight times, that that would be long enough and then I could get up. So I went to school with complete authoritarian approval for conforming to that which I did not believe in.

Mimicry was not in the repertoire of Darwin's grand-daughter Gwen Raverat, who in *Period Piece* related how she faced the inquisition of the senior girls when she was sent away to school:

> On the first Sunday there, all the big girls ... got round me in a ring and began asking me questions. "Don't your really believe in Adam and Eve, Gwen? How do you think you were made then?" ... When we went off in a crocodile to church next Sunday, I determined to testify to my faith. And when they all knelt down to pray on first going in, I sat up straight and didn't kneel; and when they bowed, I didn't bow; and when they mumbled the Lord's Prayer, I didn't mumble; and I felt a real martyr.

National Physiology

Eugenics was of much interest to medical people. Following up on his earlier *The Present Evolution of Man* (1896), Archdall Reid published *The Principles of Heredity*, which Bateson reviewed for *The Speaker* (October 14, 1905) as "a valuable addition to the popular literature of evolution." Despite its portentous title, the book drew attention to "the plain facts," as first proclaimed "to a sceptical world" by Galton. "Questions of grave national anxiety" – "the alarming increase in the relative numbers of the insane, the utility of teaching the minds of starving children, the relation of the State to the unemployed" – were all "problems of national physiology," namely eugenics. Indeed, the book had been promoted by a newspaper, the *Morning Post*, which was concerned with "the bearing of heredity on deterioration and thence indirectly on universal [rather than selective] conscription [into the armed forces]." Bateson doubted that Reid had properly discriminated among the various evils faced by societies since the remedies appropriate for some might not be appropriate for others. Nevertheless, Bateson foresaw, correctly as students of twentieth-century history will agree, that there might be dire consequences:

> What ... will happen when ... enlightenment actually comes to pass and the facts of heredity are ... commonly known? One thing is certain: mankind will begin to interfere; perhaps not in England, but in some country more ready to break with the past and eager for "national efficiency." Mr. Galton has suggested a selection at the top, with State encouragement of families of superlative quality. More probably, and we suspect more effectively, selection will begin by elimination at the bottom. ... [While] contemporary socialism strives for the elevation of the unfit ... that of the future will probably aim at their extinction. Ignorance of the remoter consequences of interference has never long postponed such

> experiments. When power is discovered man always turns to it. The science of heredity will soon provide power on a stupendous scale; and in some country, at some time, not, perhaps, far distant, that power will be applied to control the composition of a nation. Whether the institution of such control will ultimately be good or bad for that nation or for humanity at large is a separate question.

Preparing for the second edition of his book, Reid corresponded with Hurst, who gave him "a very clear idea of Mendelism" (Reid to Hurst. August 15, 1905). Reid later noted (November 11, 1905): "I should be only too pleased to see a hostile criticism of it before it is published. If I sent it to a man like Bateson I should not get anything reasonable. He would denounce it and me as things unholy."

In 1906, Shaw, an admirer of Samuel Butler, considered eugenics favorably in *Man and Superman*, and was quoted in Lock's book. In his childhood, Shaw had been told that "you cannot make a silk purse out of a sow's ear," and he now solemnly declared: "The bubble of heredity has been pricked: the certainty that acquirements are negligible as elements in practical heredity has demolished the hopes of educationalists .... Being cowards, we defeat natural selection under cover of philanthropy: being sluggards, we neglect artificial selection under cover of delicacy and morality." However, later the opposite thesis began to intrigue him. In *Pygmalion* (written 1912), Shaw argued that, through education, it was indeed possible to make silk purses out of sows' ears. He revealed the "my fair lady" latent in the flower-seller Eliza Doolittle, and the high oratorical skills latent in her dustman father.

In a series of letters (*Nature* 1907), Lock sprang to the defense when Mendelism was criticized by Reid, who did not appreciate that characters contributed to by multiple genes could appear to blend in hybrids, as was the case with human mulattos [8]. Lock was soon joined by Mudge and Cunningham, who was less restrained than Bateson in pointing out errors in Reid's 1905 book.

Caution

In *Principles,* Bateson cited an 1823 text of the surgeon William Lawrence [9]:

> The hereditary transmission of physical and moral qualities, so well understood and familiarly acted on in the domestic animals, is equally true of man. A superior breed of human being could only be produced by selections and exclusions similar to those so successfully employed in rearing our more valuable animals. Yet in the human species, where the object is of such consequence, the principle is almost entirely overlooked. Hence, all the native deformities of mind and body, which spring up so plentifully in our artificial mode of life, are handed down to posterity, and tend, by their multiplication and extension, to degrade the race.

However, Bateson was cautious:

> Whatever course civilizations like those of Western Europe may be disposed to pursue, there can be little doubt that before long we shall find that communities more fully emancipated from tradition will make a practical application of genetic principles to their own populations. The power is in their hands and they will use that power like any other with

> which science can endow them. The consequence of such action will be immediate and decisive.
>
> For this revolution we do well to prepare. Interference may take one or both of two courses. Measures may be taken to eliminate strains regarded as unfit and undesirable elements in the population, or to encourage the persistence of elements regarded as desirable. … To the naturalist it is evident that while the elimination of the hopelessly unfit is a reasonable and prudent policy for society to adopt, any attempt to distinguish certain strains as superior, and to give special encouragement to them would probably fail to accomplish the object proposed, and must certainly be unsafe. … Whereas our experience of what constitutes the extremes of unfitness is fairly reliable and definite, we have little to guide us in estimating the qualities for which society has or may have a use, or the numerical proportions in which they may be required. … More extensive schemes are already being advocated by writers who are neither utopians nor visionaries. Their proposals are directed in the belief that society is more likely to accept a positive plan for the encouragement of the fit than the negative interference for the restraint of the unfit.
>
> Genetic science … gives no clear sanction to these proposals. It may also be doubted whether the guiding estimate of popular sentiment is well-founded. Society has never shown itself averse to adopt measures of the most stringent and even brutal kind for the control of those whom it regards as its enemies.

The eugenic pronouncements of those knowledgeable in the area became much sought after. Davenport was a major spokesman in America. Hurst, in an address (February 10, 1908) on "Mendel's law of heredity and its application to man," echoed Bateson [10]: "With a knowledge of Mendel's law, it is possible to foretell, to a certainty in many cases, what Mendelian characters will appear in the offspring of two parents, and in what proportions these will be likely to appear. The practical value of this knowledge depends entirely on its application. That delicate question, I must leave to your imagination." One of the Darwin sons, Major Leonard Darwin, invited Hurst to address the Eugenics Education Society (founded in 1907). Hurst, perhaps more than anyone in Britain (with the possible exception of Mudge in London), was working on the human characteristics – skin, eye, and hair color – that were often used to discriminate human races. To these he had added handedness, susceptibility to tuberculosis, and musical temperament. In his address (November 16, 1911), Hurst described studies of parents and their children among a population of 3000 in and around the village of Burbage, stressing that "the advantages of personal observation seem far to outweigh the disadvantages of small numbers in a limited area." His conclusions were cautious [11]:

> Before we can venture to apply the scientific principles of Genetics to human life we must first make our foundations sure. For this reason, I am convinced that a good deal of spade work in human Genetics will have to be done before any considerable amount of practical good can be accomplished in Eugenics. Eugenics is simply applied Genetics, and *sound Eugenics can only be founded on sound Genetics.*

Meanwhile at a meeting of the undergraduate members of the Cambridge Eugenic Society in Trinity College (November 10, 1911), Ronald Fisher, then an unknown student, but later to become a leading Biometric critic of Bateson and eventually to assume Punnett's Balfour Chair in Genetics, took "Heredity" as his topic [12]:

> Suppose we knew, for instance, twenty pairs of mental characters. These would combine with over a million pure mental types, each of these would naturally occur rather less frequently than once in a million, or in a country like England may occur in 20,000 generations; it will give some idea of the excellence of the best of these types when we consider that the Englishmen from Shakespeare to Darwin (or choose whom you will) have occurred within ten generations; the thought of a race of men combining the illustrious qualities of these giants, and breeding true to them, is almost too overwhelming, but such a race will inevitably arise in whatever country first sees the inheritance of mental characters elucidated.

Spencer Lecture

In 1912 (February 28), Bateson gave the Herbert Spencer Lecture at Oxford on "Biological fact and the structure of society." He spoke of a widespread "uneasiness" and feeling that "changes of exceptional magnitude are impending." Mankind was "fast nearing one of those great secular changes through which history occasionally passes." He hailed Spencer as the philosopher whose teachings helped "men to see themselves as they really are, stripped of the sanctity with which superstition and ignorance have through all ages invested the human species." However, the lecture was concerned more with politics than religion – namely, the balance between socialism and individualism, and the extent to which politicians should consider (i) "that man is an animal, subject to the same physical laws of development as other animals" and (ii) "the control which civilized communities have acquired over the forces of nature." The "community at large" was naturally asking "how far the outcome of these interferences with what have usually been regarded as natural forces will bring good or evil to the societies which attempt them."

Bateson noted that Marshall, the Cambridge economist, was full of Malthusian gloom, having calculated that by the year 2400 "the population will then be 1000 for every mile of fairly fertile land." Yet, Bateson noted that "we are in fact passing through a phase which is quite exceptional in the history of a species – exceptionally favourable if you will – and it is in a decline in the birth-rate that the most promising omen exists for the happiness of future generations." Thus, Bateson was most concerned with differential fertility among social groups, citing Spencer's 1873 remark "that social arrangements which retard the multiplications of the mentally-best, and facilitate the multiplication of the mentally-worst, must be extremely injurious." Similarly, Bateson noted that "every legislative encouragement given to one class and every repression given to another has an effect on the future of the race."

True to his Galtonian roots, Bateson was a believer in progress and the critical role of the elite in that progress. He referred to the inventor of the steam engine, James Watt, as "that remarkable mutation," for "it is upon mutational novelties, definite favourable variations, that all progress in civilization and in the control of natural forces, must depend." Yet, his recent readings of Butler may have had some impact. He quoted *Erewhon* (1872), where a young prisoner convicted "of the great crime of labouring under pulmonary consumption" was sentenced to "imprisonment with hard labour for the rest of your miserable existence." Bateson lamented the "sickening cruelty of the courts" and doubted the wisdom of the legislators. He praised "the polymorphism of man." Society was rich, full of "natural, genetic distinctions which differentiate us into types and strains – acrobats, actors, artists, clergy, farmers, labourers, lawyers, mechanics, musicians, poets, sailors, men of science, servants, soldiers and tradesmen. Think of the diversity of their experience of life." The problem was "to find a system by which these differentiated elements may combine together to form a coordinated community, while each element remains substantially contented with its lot. To discuss this mighty problem in its full scope I have neither qualification nor desire." He paused to clarify his position:

> You will think, perhaps, that I am about to advocate interference by the State, or by public opinion, with the ordinary practices and habits of our society. There may be some who think that the English would be happier if their marriages were arranged at Westminster instead of, as hitherto, in Heaven. I am not of that opinion, nor can I suppose that the constructive proposals even of the less advanced Eugenicists would be seriously supported by anyone who realised how slender is our present knowledge of the details of the genetic processes in their application to man.

The State could intervene in two ways, positive and negative:

> If we picture to ourselves the kind of persons who would be infallibly chosen as examples of "civic worth" – the term lately used to denote their virtues – the prospect is not very attractive. We need not for the present fear any scarcity of that class, and I think we may be content to postpone schemes for their multiplication.

As for negative eugenics, while "democracy regards class distinction as evil; we perceive it to be essential." He held that the "grades" of society had to "find their right places," for "if everyone were determined to play first fiddle no orchestra could be got together." Yet for the "hopelessly unfit" there was "one perfectly clear line of action which we may agree to take. ... When it is realized that two parents, both of gravely defective or feeble mind, in the usual acceptance of that term, *do not have any normal children at all*, save perhaps in some very rare cases, ... no one can doubt that the right and most humane policy is to restrain them from breeding." Under this heading Bateson included "the practice of sterilizing criminals of special classes," but held this to be "the very utmost length to which it is safe to extend legislative interference of this kind, until social physiology has been much more fully explored. Beyond that, if there is authority to go, it is not drawn from genetic science." He included here various genetic diseases, such as those with hereditary cataract:

> Though ... many remain grievous burdens to their families, or to the public funds, and though they could probably be eliminated after a few generations without difficulty by legislative interference, that would be a very dangerous course. They are not necessarily useless persons, nor are their own lives necessarily miserable. There are many healthy and active types which are a far greater nuisance to their neighbours and reproduce themselves with equal exactitude. ... We all have grave defects, not least those who contribute much to the happiness of the world. ... We are made of fragments of diverse races, all in their degree contributing their special aptitudes, their special deficiencies, their particular virtues and vices, and their multifarious notions of right and wrong.

With this, Bateson returned to the area of religion, adopting the Panglossian attitude that he so often deplored, yet lamenting the laissez-faire democracy had bestowed on the ill-informed:

> It is much better that we should be of many sorts, saints, nondescripts, and sinners. Posterity is likely to discover that to eliminate sinners there is only one way – that which St. Paul pointed to us when he wrote that "where no law is, there is no transgression." Science knows nothing of sin save by its evil consequence.
>
> In all reverence she inverts the ancient saying and proclaims that the sting of Sin is Death. It is not the tyrannical and capricious interference of a half-informed majority which can safely mould or purify a population, but rather that simplification of instinct for which we ever hope, which fuller knowledge alone can make possible.
>
> As science strengthens our hold on nature, more and more will man be able to annul the evil consequence of sin. Little by little the law will lapse into oblivion, and sins which it created will be sins no more. The great and noble work which genetic science can do for humanity at the present time is to bring men to take more true, more simple, and if so inexact a word can be used intelligibly, more natural views of themselves and of each other.

Concerning the notion of "survival of the fittest" in the context of human societies, Bateson was also much opposed to the "social Darwinism" associated with the teachings of Herbert Spencer:

> I lay stress on this aspect of the social problem because I have seen several times of late the claim put forward that the teaching of biological science sanctions a system of freest competition for the means of subsistence between individuals, under which the fittest will survive and the less fit tend to extinction.
>
> That may conceivably be a true inference applicable to forms which, like thrushes, live independent lives; but so soon as social organization begins, the competition is between societies and not between individuals. Just as the body needs its humbler organs, so a community needs its lower grades, and just as the body decays if even the humblest organs starve, so it is necessary for society adequately to ensure the maintenance of all its constituent members so long as they are contributing to its support.

Congress 1912

Bateson's Spencer Lecture was issued as a pamphlet and reviewed in various journals in April. A few months later, 750 delegates met for 6 days in London (July 24–30). It was the occasion of the First International Eugenics Congress, organized by the Eugenics Education Society, with Leonard Darwin presiding. Among the "vice-presidents" who lent their names to the cause were the Right Honorable Winston Churchill (the First Lord of the Admiralty), Lord Alverstone (the Chief Justice of Britain), and August Weismann. At the inaugural banquet, Arthur Balfour spoke of Weismann's separation of germ plasm from the rest of the body and argued against Spencer's Lamarckian notions of improvement within a few generations: "The idea that you can get a society of the most perfect kind by merely considering certain questions about strain and ancestry and the health and the physical vigour of various components of that society – that I believe is a most shallow view of a most difficult question." Leonard Darwin in his presidential address spoke of the nature–nurture dichotomy and of the need to use social means to augment natural selection and avoid breeding from "inferior stocks" [13]. There followed addresses by international scientists of high repute, including the new Professor of Biology at Cambridge, Reginald Punnett [14].

An Infection of Dullness

In 1913, Bateson preached eugenic caution at the Seventeenth International Congress of Medicine, in London. He was now able to extend the list of pedigrees that displayed inheritance of disease. In some cases, a trait was dominant, in others recessive. "Sons of colour-blind males do not inherit the peculiarity, and therefore cannot transmit it. The daughters of colour-blind fathers inherit it, and though it does not appear in them, probably all of them have the power of transmitting it to their sons." The caveat "probably" indicates that he was still not thinking in chromosomal terms. The prediction of sex chromosome linkage is that carrier daughters should *always* have the power, and *do* transmit, on average, to half their sons (Fig. 13.2 of Chap. 13):

> Now in colour-blindness, … repulsion … may be represented as acting in the germ cells of the male between the factor for maleness and the factor for colour-blindness; such that the sperms destined to become males carry factor *M* [for maleness], but do not carry the factor for colour-blindness, which passes entirely into those sperms which are destined to produce femaleness. It is evident that females heterozygous for colour-blindness are not colour-blind, and that colour-blind females can only be produced by the meeting of two germ cells both bearing the affection. It follows from the hypothesis that the sons of colour-blind women will all be colour-blind, and all the records of such families with which I am acquainted, with a single doubtful exception, are in agreement.

Having gained the attention of his medical audience, Bateson then turned to some basic science. "Whereas some factors are continually transmitted in their entirety, others are liable to be broken up by what I regard as a process of quantitative fractionation occurring in the mechanical dissociation of the elements at certain critical cell divisions." When and where were those divisions? Two years after the Damascus-like revelation by Morgan that "coupling" would be simply explained if the Mendelian factors were linearly arranged on chromosomes (Chap. 13), Bateson could still declare that some "segregation may occur in somatic divisions," in keeping with the reduplication hypothesis. This would explain "the fact that in certain cases the parental combination influences the distribution of factors among the gametes so that the distribution among the grandchildren is different according to the way in which the characters were combined in their grandparents."

As for eugenics, "whatever influences may be brought to bear by hygiene or by education, the ultimate decision rests with the germ cells." Bateson welcomed the recent Mental Deficiency Bill as "in principle, a wise beginning of reform," but viewed with "disquietude" the "violent measures" that were being adopted in certain parts of the United States:

> It is one thing to check the reproduction of hopeless defectives, but another to organize a wholesale tampering with the structure of the population, such as will follow if any marriage not regarded by officials as eugenic is liable to prohibition. … Nothing yet ascertained by genetic science justifies such a course, and we may well wonder how genius and the arts will fare in a community constructed according to the ideals of State Legislatures. Philologists tell us that, by an irony of development, the word "dull" comes from the same original which in Dutch has become "dol" – mad, and is better known to most of us in the German equivalent "toll." But I anticipate that, connected as the ideas may be, we might, by ridding our community of mania, leave it gravely infected with dullness.

They Will Try

All this sounds moderate, but a year later from the Presidential podium at the Sydney BA meeting, while professing to ask his audience "merely to observe the facts," Bateson's remarks were less ambiguous, so providing ammunition for extreme elements:

> Inveighing against hereditary political institutions [e.g., the House of Lords in the United Kingdom], Tom Paine remarks [that] the idea is as absurd as that of an "hereditary wise man," or an "hereditary mathematician," and to this day I suppose many people are not aware that he is saying anything more than commonly foolish. We, on the contrary, would feel it something of a puzzle if two parents, both mathematically gifted, had any children *not* mathematicians.

How were we to deal with "inferior strains?"

> Modern genetic discovery … cannot fail to influence our conceptions of life and ethics. … From the population of any ordinary English town as many distinct human breeds could in a few generations be isolated as there are now breeds of dogs. … Even as at present constituted, owing to the isolating effects of instinct, fashion, occupation, and social class, many incipient strains already exist. … The powers of science to preserve the defective are now enormous. Every year these powers increase. This course of action must reach a limit. To the deliberate intervention of civilization for the preservation of inferior strains there must sooner or later come an end, and before long nations will realize the responsibility they have assumed in multiplying these "cankers of a calm world and a long peace." The inferior freely multiply, and the defective, if their defects be not so grave as to lead to their detention in prisons or asylums, multiply also without restraint. … The union of such social vermin we should no more permit than we should allow parasites to breed in our own bodies.

The medical profession bore its share of responsibility:

> Something too may be done by a reform of medical ethics. Medical students are taught that it is their duty to prolong life at whatever cost of suffering. This may have been right when

> diagnosis was uncertain and interference usually of small effect; but deliberately to interfere now for the preservation of an infant so gravely diseased that it can never be happy or come to any good is very like wanton cruelty. In private few men defend such interference. Most who have seen these cases lingering on agree that the system is deplorable, but ask where can any line be drawn. The biologist would reply that in all ages such decisions have been made by civilised communities with fair success both in regard to crime and in the closely analogous case of lunacy.

The cleric and the lawmaker could not escape responsibility:

> The real reason why these things are done is because the world collectively cherishes occult views of the nature of life, because the facts are realized by few, and because between the legal mind – to which society has become accustomed to defer – and the seeing eye, there is such that can hardly be combined in the same body. So soon as scientific knowledge becomes common property, views more reasonable and, I may add, more humane, are likely to prevail.

Liberated from the orthodoxies of the past, a new era of human endeavor was rapidly approaching:

> Hitherto superstition and mythical ideas of sin have predominantly controlled ... [our] powers. Mysticism will not die out: for those strange fancies knowledge is no cure; but their forms may change, and mysticism as a force for the suppression of joy is happily losing its hold on the modern world. As in the decay of earlier religions [when] Ushabti dolls were substituted for human victims, so telepathy, necromancy, and other harmless toys take the place of eschatology and the inculculation of a ferocious moral code. ...
>
> We are witnessing an emancipation from traditional control in thought, in art, and in conduct, which is likely to have prolonged and wonderful influences. Returning to freer or, if you will, simpler conceptions of life and death, the coming generations are determined to get more out of this world than their forefathers did.

But he cautioned:

> Everyone must have a preliminary sympathy with the aims of the eugenicists both abroad and at home. ... The spirit of such organizations, however, almost of necessity suffers from a bias towards the accepted and the ordinary, and if they had power it would go hard with many ingredients of Society that could be ill-spared. ... It is not the eugenicists who will give us what Plato has called divine releases from the common ways. If some fancier with the catholicity of Shakespeare would take us in hand, well and good; but I would not trust even Shakespeares, [when] meeting as a committee.
>
> Let us remember that Beethoven's father was a habitual drunkard and that his mother died of consumption. From the genealogy of the patriarchs also we learn ... that ... the founders ... of the arts and sciences ... came in direct descent from Cain, and not in the posterity of the irreproachable Seth. ... Genetic research will make it possible for a nation to elect by what sort of beings it will be represented not very many generations hence, much as a farmer can decide whether his byres shall be full of Shorthorns or Herefords. It will be very surprising indeed if some nation does not make trial of this new power. They may make awful mistakes, but I think they will try.

And try they did. Even in the land where he spoke there was legislation to remove "half-caste" children from their aboriginal mothers and isolate them in a native settlement in Western Australia. This story was told years later by Doris Pilkington in *Follow the Rabbit-Proof Fence* [15]. In Germany in 1921, Baur coauthored *Outline of Human Genetics and Racial Hygiene*, which came to dominate genetics teaching in Germany, and was translated into English in 1931 [16]. The book contained a seemingly authoritative hierarchy of races leading up from "blacks," "Mongols," "Alpines," and "Mediterraneans," to the white-skinned "Nordic" pinnacle. That the biological differences were fundamental had gained support from papers that described the first attempts, now known to be in error, to show chromosomal differences between black and white people.

Slavery

The distinction between races and species was fundamental to the eugenics issue, sometimes serving to justify slavery, albeit not in ancient Greece [17]. In antebellum America, black slaves were required for menial labor and were traded like horses, sometimes even with support of the prestigious Royal Society [18]. They were actively "dehumanized" [19]. Thus, it was all too easy to slip from seeing slaves as racially different, to seeing them as a different biological species, more akin to apes than to humans [20]. To such "scientific facts," American naturalists such as Louis Agassiz and Samuel Morton gave their authority [21]. In this respect, the studies of Darwin and Mendel, and their later elaboration by Bateson – all of whom were *not* primarily concerned with the slavery issue – were to become of major importance. Morton had set the stage in 1847 [22]:

> It was taught by Buffon, John Hunter, and other naturalists of the past century, as is yet assumed by some learned men of the present day, that the hybrid offspring of two distinct species of animals, are incapable of reproducing their kind; thus making hybridity the test of specific character [the test of belonging to a distinct species]. It follows, according to this supposed law of nature, that if mankind embraced [consisted of] several species, the intermixture of these would go no further than to produce a sterile variety. But since all races are capable of producing, with each other, a progeny more or less fertile, it is inferred that they must all belong to one and the same species. This is the question at issue.

After extensively reviewing the evidence from various animals and plants – such as the sterility of the hybrid offspring (mule) between a horse (one species) and an ass (another species) – Morton concluded that "the mere fact that the several races of mankind produce with each other, a more or less fertile progeny, constitutes, in itself, no proof of the unity of the human species." While Charles Darwin was critical of Morton's analysis [23], he did not contradict it. For much of his life, Darwin was hesitant in drawing a distinction between races and species, tending to see a continuum with one leading to the other [24].

Salt of the Earth

In keeping with his liberal roots, Bateson concluded his Sydney Presidential address by sniping again at the social Darwinism of Spencer:

> To the naturalist the broad lines of solution to the problems of social discontent are evident. They neither lie in vain dreams of a mystical and disintegrating equality, nor in the promotion of that malignant individualism which in older civilisations has threatened mortification of the humbler organs, but rather in a physiological coordination of the constituent parts of the social organism.
>
> The rewards of commerce are grossly out of proportion to those attainable by intellect or industry. Even regarded as compensation for a dull life, they far exceed the value of the services rendered to the community. Such disparity is an incident of the abnormally rapid growth of population and is quite indefensible as a permanent social condition.
>
> Nevertheless capital, distinguished as the provision of offspring, is an eugenic institution; and unless human instinct undergoes some profound and improbable variation, abolition of capital means the abolition of effort; but as in the body the power of independent growth of the parts is limited and subordinated to the whole, similarly in the community we may limit the powers of capital, preserving so much inequality of privilege as corresponds with physiological fact.

A year later, he returned to "physiological fact" in a short piece entitled "Education and evolution" for Pitman's *New Educator's Library* (Chap. 13). Could education (nurture) result in "racial changes"? When "applied for even a few generations to composite populations, universal education can effect remarkable changes," but these would merely provide opportunity "for the more intellectual individuals in the various classes of the community to improve their position." There would be "rearrangement of the constituent members." However, he doubted "whether various non-European races submitted to our education system will be found capable of assimilating themselves to our mental standards." Perhaps harking back to the mimicry issue of the 1890s (Chap. 6), he disparaged people who might appear to have risen by "the exercise of the faculty of *imitation,*" which could "present a semblance of uniformity," while disguising a "congenital want of aptitude."

Did the war experience (Chap. 16), or perhaps Shaw's plays (Chap. 19), temper Bateson's attitudes? In 1920, he accepted the Eugenics Education Society's invitation to give its "Galton Lecture" (*Eugenics Review*; April 1921). He chose as his title "Common sense in racial problems." At the outset, he declared that "I have never seen my way to take a definite part" in the Society's activities or "even to become a member," but he was not "out of sympathy with" its objects. "Nevertheless, the pursuit of truth is one thing and its application another. Few have combined these objects with success." For "genetics are not primarily concerned with the betterment of the human race or other applications, but with a problem of pure physiology," so "the eugenicist and the geneticist ... should not be brigaded together." Indeed, "alliances between pure and applied science are as dangerous as those of spiders, in which the fertilizing partner is apt to be absorbed."

While still maintaining "the propriety and … the humanity of exercising control over" the propagation of "the feeble-minded members of the population," he was now opposed to such steps regarding the insane since "the forms of mental disease are manifold and still most imperfectly distinguished," and this included environmentally induced states "so obvious as the syphilitic origin of general paralysis." Indeed, he "would specially emphasize a doubt whether from the point of view of society, which is that in which we are here concerned, families which have suffered from definite stigmata may not contribute at least their proper share to the success and delight of mankind." He also was opposed to "the sterilization of habitual criminals" on grounds that we knew little of "the genetics and aetiology of criminality." Furthermore, "pending the institution of a proper classification system" as to what is criminal, he wondered whether such a classification might not include

> army contractors and their accomplices the newspaper patriots. The crimes of the prison population are petty offences by comparison, and the significance we attach to them is a survival of other days. Felonies may be great events locally, but they do not induce catastrophes. The proclivities of the war-makers are infinitely more dangerous than those of the aberrant beings whom from time to time the law may dub as criminals. Consistent and portentous selfishness, combined with dullness of imagination are probably just as transmissible as want of self-control, though destitute of the amiable qualities not rarely associated with the genetic composition of persons of unstable mind.

To Bateson, eugenics could emerge as a "perversion," in the form of "a cold and ascetic faith" where "priests in black gowns will be walking their rounds, and binding with briars our joys and desires." He confessed he could not "read without a shudder" Galton's disparagement of "the Bohemian element in our own race" that was "destined to perish" and, as far as Galton was concerned, "the sooner it goes the happier for mankind." Bateson had heard similar remarks in Germany before the war when it was "the Latin races" that were destined to perish. While Galton was discerning regarding literature, Bateson saw his remarks as indicative of "a lurking contempt for the other arts" that he could not share:

> In the eugenic paradise I hope and believe that there will be room for the man who works by fits and starts, though Galton does say he is a futile person who can no longer earn his living and ought to be abolished. The pressure of the world on the families of unbusiness-like Bohemians, artists, musicians, authors, discoverers and inventors, is severe enough for all conscience. In well-ordered communities their support should be a first charge of the State. They are literally the salt of the earth, without whom the savour of life would be flat and wearisome indeed.

From "the wholesome teaching of biology," there should emerge "a wise and pagan sense of the facts, teaching us to see things as they are." Such "knowledge must surely make for width and generosity, not for narrowness and constraint." In this context, he cautioned moves to prohibit the sale of alcoholic drinks: "To abolish wine because men get drunk is like abolishing steel because men fight with it." Prohibition is "an act of tyranny," and those who perpetrated it will "not stop there. Neither, tobacco, nor art, nor literature is safe." Censorship had failed in the past, but, hinting darkly at the growing power of the press – in years ahead to be transformed into the "mass media" – he noted that in those days "the art of raising waves of emotion for political purposes was imperfectly understood."

Nevertheless, Galton had shown "that nations, unconsciously by their own acts of policy, favouring one class or discouraging another, change the genetic composition of succeeding generations." Galton had given as "simple illustrations of this theme" the example of how "the celibacy of the clergy kept down the numbers of the intellectual strains," and how "strict catholic families" threatened "to replace other components of the population for the reason that they obey the Church's ban on the practice of restricting the number of children by methods until lately almost universally adopted by the prudent remainder."

Conditions "favourable to other breeds or 'genotypes' as Johannsen calls them" had supervened so that "the landed gentry" of the last century had literally "gone to the dogs," their place being taken "by financiers and tradespeople." Meanwhile, the "institution of abundant scholarships and other machinery for detecting and encouraging the abler children, … must gradually remove, or as some would say raise the more intelligent elements out of the industrial classes, thereby sensibly lowering the mental capacity of these classes." As a result of this stratification, the various classes in a society would become more clearly defined. The "intellectual class" would become more clearly distinguishable from others.

The Intellectual Class

In his Galton Lecture, Bateson noted that the "development of mankind" has been "arrested for about a millennium by the domination of the Church. It may be suspended indefinitely by the edict of the proletariat. We may have made the world safe for democracy, but we have made it unsafe for anything else." All this heralded the doom of "the intellectual class" with which he identified:

> To those who have witnessed the rapid transition from a period when learning, the arts, and even pure science stood high in general reverence, to the present time when science is tolerated as a source of material advantage, when chaos is acclaimed as art, and learning supplanted by schools of commerce, the rarity not merely of intellectual producers but of intellectual consumers will need no further demonstration. … Their existence is precarious indeed.

Faced with the pending disappearance of the intellectual class "merged in the unsegregated mass of the dominant population," we should be "recognizing this fact and allowing it prominence in political philosophy." He suggested reform of the system of collecting population data so that "returns to the Registrar-General" will reveal "how the birth-rate is distributed among the various grades."

> We are a heterogenous group of dissimilar beings, and it is time that the greatness of this dissimilarity were brought home to all civilized communities. No one perhaps at this time would venture to assert that men are born equal, but few realize *how* unequal they are. By an ingenious calculation Galton found that in many types of competition the difference

> between the performance of the winner and that of the second man is commonly three times that of the distance separating the second man from the next best, ... Has anyone considered the implications of this natural heterogeneity for political economy? Who dare tell voters that? ... Equality of political power has been bestowed on the lowest elements of our population. This is nearing the final stage of democratic decay, in which the lowest not only have the power but exercise it, a sequel which the next generation may witness. ... Our immediate posterity will learn something of the consequences of un-applied biology. The force of the intellectual and professional class is assuredly prodigious and there is a bare chance that they may exert it in some coordinated form, against which the rest could of course offer no effective resistance. Recent history does not encourage us to expect that any such thing will come to pass. The truth has been recognized too late.

Many of the comments cited above, made at the beginning of the twentieth century, would be deemed as "outspoken" and "politically incorrect" at the beginning of the twenty-first. Thus, readers can judge for themselves whether any eugenic "progress" of the form Bateson hoped for has been made, or is likely to be made, in the decades ahead. We may note the experience of China that for decades opted for a "level-playing field" strategy – couples, *whatever their status*, were restricted to one child. This was deemed an error because the system could be gamed to produce more boys.

Genetics and Eugenics

In the early 1920s, planning began in the United States for continuing the International Genetics Congresses. Bateson was asked to advise. Should Genetics and Eugenics share a common forum? Morgan wrote to Bateson (December 3, 1920):

> Thank you for the letter ... concerning the program for the next meeting of the International Genetic Congress. On its receipt I immediately copied those parts ... which seemed to me ought to be known by the geneticists in this country, and sent out a letter to a dozen or fifteen men asking ... whether it was desirable to invite the Genetic Congress to meet with the Eugenic Congress. With one exception, which didn't count for much, they all agreed that it was undesirable to take this step, making the geneticist appear as the tail to the eugenic kite. ... The geneticists here have been very little active in the Eugenic Congress, although some of them show an interest and none of us has any objection to a genetic section meeting with the Congress. In fact, several, or all, perhaps, of the geneticists might take part in such a section, although I do not find very much enthusiasm even for this. I need not tell you what sort of a show the eugenic crowd is apt to put up, and my impression is that they are grasping at a shred of respectability by drawing in the genetic group. However, even a shred is better than nothing at all, so that I see no reason why we shouldn't, to some extent at least, take part in the proceedings.

In 1925, Bateson turned down an invitation to lecture on eugenics from Alan Cock's mentor, Michael Pease (January 28, 1925):

> I never feel Eugenics is my job. On and off I have definitely tried to keep clear of it. To real Genetics it is a serious – increasingly serious – nuisance diverting attention to subordinate and ephemeral issues, and giving a doubtful flavour to good materials. Three times I have

> come out as a sort of Eugenicist, yielding to a cheap temptation, and on each occasion I have wholly missed even that humble mark; I don't mean to try again. My Galton lecture, which I thought would be famous clap-trap, had the unique distinction of being the only Galton lecture to which no single newspaper would make allusion, much less report. I infer it got home on to somebody's nerves all right. I have tried to republish these papers with others more or less cognate, but publishers know their public, and refuse with contumely. My Eugenic career I regard as closed, and serve me right for dabbling in taboo waters. The kind of thing I say on such occasions is what no reformer wants to hear, and the Eugenic ravens are croaking for Reform.

To a request from Julian Huxley (February 1, 1925) to coauthor a letter on mental defectives to *The Times*, Bateson replied (February 2): "I like to keep out of newspaper debates of a semi-political type. … Guard yourself as you like, you cannot in newspaper controversy avoid assuming responsibility for the policy advocated by the side on which you come in, and I am by no means easy about compulsory sterilization."

Summary

Galton foresaw that a better "average quality" would mean a better-informed populace less likely to be swayed by biased electioneering. Democracy would be strengthened. However, he stressed the "actuarial" aspects of the underlying science, rather than genetics. Positive eugenics would favor those deemed more likely to contribute to "national efficiency," while negative eugenics (e.g., sterilization of criminals) would disfavor those deemed less likely to contribute. H. G. Wells welcomed the new genetic approach but advised that those who had attained "superior positions" in society, and the so-called "criminal classes," might not necessarily be distinguishable. While stressing the power of nature over nurture in *Man and Superman*, Shaw later explored the power of nurture over nature in *Pygmalion*. Bateson doubted the wisdom of those who would determine "civic worth." Furthermore, self-marketing devices (mimicry) could conceal what was really a "congenital want of aptitude." While hailing the inventor of the steam engine as a "remarkable mutation," he delighted at the range of human types and saw the challenge as that of harmonizing diversity. Through knowledge of genetics, men would take "more natural views of themselves." Nevertheless, Bateson predicted that in the not-too-distant future, even though genetics "gives no clear sanction," in some countries, eugenic measures would be tried. Lines would more likely be drawn correctly in a world that no longer cherished "occult views of the nature of life" or was imbued with "superstition and mythical ideas of sin." But in democratic societies to draw lines appropriately was often politically incorrect. Thus, "we may have made the world safe for democracy, but we have made it unsafe for anything else."

Chapter 16
War (1915–1919)

The merest spark may set all Europe in a blaze, but though all Europe be set in a blaze twenty times over, the world will wag itself right again.

Samuel Butler [1]

Eras are marked not only chronologically but also by attitudes and practices. Formally, the twentieth century ended in the year 2000. Some would argue that the twentieth century actually ended in 1989 with the change in mindset that found expressions in the dismantling of the Berlin Wall, the revolt in Tiananmen Square, and the fatwa sentencing of Salman Rushdie. Formally, the nineteenth century ended in 1900. In practice, nineteenth-century attitudes lingered on until 1914. Then Europe was set ablaze. The carnage continued until 1918. Then in 1919 came an influenza pandemic. The twentieth century was up and running.

Mobilization

The great powers mobilized for war. On his arrival back from Australia (November 1, 1914), Bateson found that many of the men at the John Innes Institute (both scientific and gardening staff) had joined the war effort. So had some of the women. Durham was into laboratory work at a local military hospital, and Cayley became a tool setter at Vicker's airplane works. She later worked on tetanus for the Royal Army Medical Corps. At Burbage, Hurst had had a long association with the Leicester Territorial Regiment. When in 1912 the National Reserve for Home Duties was established, he took on its organization in his area and attended lectures on wireless telegraphy at Birmingham University. In 1914, he wound down his Burbage operations and became a Signals Officer in the Royal Engineers involved in training and East Coast defenses. Back in Cambridge Punnett, while continuing the duties of his Chair, served in the Food Production Department of the Board of Agriculture.

A. G. Cock, D. R. Forsdyke, *Treasure Your Exceptions*,
https://doi.org/10.1007/978-3-030-92099-9_16

Bateson did not attempt to dissuade his sons from enlisting. The eldest, John, joined the army in 1916 after graduation from Charterhouse School. Martin left Rugby in 1918 and was training in photographic reconnaissance for the Royal Air Force when the war ended. Bateson's position, as set out in an address on "Science and Nationality" to the Yorkshire Natural Science Association in 1918, was unambiguous: "I am no advocate of the pacifist creed. The duty of self-defence is one which no Government can decline. I have never doubted that such a duty fell upon our Government in 1914. For the climax then reached, the world had been long in preparation. To have averted that catastrophe the policy of nations must have changed its course long years ago."

However, Bateson saw much patriotism as the misplaced result of a narrow education system, and decided, even in wartime, to press for education reforms. As for the causes of the war, although fueled by many deeply seated antagonisms, Bateson thought it could be regarded "as almost purely a commercial matter," reflecting a fundamental conflict over markets, and the power of what Eisenhower was later to call "the military-industrial complex" (letter to C. H. Ostenfeld in Denmark; January 1915). Whereas the skills of some science establishments could be diverted to the war effort (e.g., the Galton Laboratory turned to ballistics and the analysis of economic data), Bateson determined to press on with the scientific work of the Institute as best he could. Of immediate concern was access to German scientific journals. Some were acquired by way of Ostenfeld (July 15, 1915):

> It is pleasant ... that the sciences and the arts are going on all the time, but who is to follow us middle-aged people? The pitiful thing is the destruction of young men coming on. All the best are in it – the pick of the best breeds in the country. Inevitably the succession – in learning, I mean – must be broken throughout Europe, and I greatly fear that a more or less permanent state of war, for a generation at least, is the most probable outcome of it all.

So far was Bateson from being a pacifist that in 1915 he twice (on the same day) wrote to the chemist Henry Armstrong, urging that he take up with the War Office the weapon potential of an arsenical gas that had been used to eradicate the prickly pear cactus in Australia: "Its deadliness ... does not seem to me a drawback" [2]. This theme was taken up in the postwar years by Haldane in his *Callinicus: a Defence of Chemical Warfare* [3]. When interviewed decades later, Mrs. H. B. Pease, who as Miss Helen Wedgwood worked at the John Innes in 1917, recalled that she had been an adherent of the "No Conscription Fellowship," but Bateson had scathingly disparaged conscientious objectors [2].

Natural Knowledge

At a time when the number of war dead was already in the hundreds of thousands, Bateson was invited to address the Salt Schools, at Saltaire, Shipley (December 7, 1915). While some extracts were published in *The Evening News* (December 10), perhaps considering its political content it was not published fully until Beatrice made it part of her *Memoir*. Bateson must have felt very secure. If the John Innes Trustees had got wind of it, there might have been repercussions, but none fatal to his cause.

To put it bluntly, the British education system, by which Bateson primarily meant male education and the private school system, was a tragedy. Education was a "treatment" to which we subjected our children and we would do well to follow the "suggestion of Sir Joseph Thomson and send the children to school for the holidays and keep them at home to learn." For:

> Of all mad institutions in England, this, to our friends abroad, seems perhaps the maddest. Those unfamiliar with England can scarcely believe that for several days at three seasons of the year the London railway stations are blocked with streams of cabs distributing thousands of unhappy children at random to all parts of the country. But their homes have to be run just the same in their absence. After beginning school, boys get almost nothing from the home-life. Without the incalculable waste that this displacement involves, the children of Germany or Holland for example, live in their own homes, get a better education at a mere fraction of the cost.

Bateson was here praising the education system of a nation with which England was at war. He went on to praise "the German mind" for "that terrible habit of looking at things frankly" and hinted that it was in Germany that the printing press had been invented. Meanwhile, Cambridge was transformed. "The students are gone. The colleges, even the classrooms, are full of soldiers." Europe was burning. Passions were high.

> The tragedy that is overwhelming Europe … [resulted from] … the uncontrolled growth of national sentiment. After a long quiescence these feelings, stirred by the writers and speakers of many countries, have once more burst into flame, threatening the destruction of civilization. This spirit of nationality, whether it masquerade under its more glorious name of patriotism or comes forth without disguise in its true shape as selfish pride, is but a poor thing in the light of natural knowledge.

He considered that natural knowledge "should be the basis of education." By this, he meant "something wider and more inclusive than is commonly denoted by the word science," which had "come to bear a narrow sense" implying "work done in laboratories" and "technical knowledge." For it was "in the spirit rather than in the subject-matter of education" that the fault lay. Indeed, there were "books of Homer and not a few passages from the Bible, which, illustrated and expounded by a competent teacher," could serve "to convey essential parts of science as fruitfully as many an hour's work in a laboratory."

Education was more than just training. Through education the student should acquire "the truths that enable man to see himself in his true position in time and space, [and] the knowledge that gives him a sense of balance and proportion in his progress through life." And by knowledge he meant natural knowledge – chemistry, physics, and biology – and history, not just dates, but "the true broad history of man, the history of civilization, the rise and disappearance of races, … knowledge that will give an understanding of the composition of a mixed population like our own, its capabilities and its limitations." For "natural knowledge is the basis of all power, the one source of rational conduct; it is the light which shows man in his true natural perspective, that makes him at home on this planet, and steadies him among events."

He criticized "stupid school games," and the overwhelming emphasis on classics, where Latin and Greek "were not taught as literature, nor even as languages, … [but] as lessons, mere pedantic lessons, something laboriously contrived in the classroom, for the classroom, without value or meaning in, or relation to, the real life of the world." The "profound neglect of science" was "something specifically British, and it would be difficult to find any other civilised country in which a comparable state of things exists. Such a charge is a grave one, but anyone who reads the newspapers or who observes the utterances of our public men knows that it is true." All this endured "because the people like it so." Indeed, "it has become the fashion of our race to demand an ingredient of humbug;" for "we live in a world of make-believe and pretence."

From natural knowledge would be derived an appreciation that a human population "is a medley of many kinds of dissimilar individuals" and "when the multiplicity of mental types is understood it will be realized that education should be as varied as it is possible to make it." Patriotism "was glorious," but it could be misdirected through ignorance of others. It was an "emotion" that "moves most deeply those who know least of other nations and often very little of their own. The chained dog who has never left his yard is the readiest to fly at strangers. They are strangers and that is enough. So it is with nations. Knowledge, more and more knowledge of natural fact, is the only cure for these ephemeral emotions."

What, indeed, was "the meaning of nationality" in an "age of rapid travel, of continual immigration and intermarriage"? For "we are almost all the mixed product of many breeds. … We are all men, born into a splendid and terrible world in which for a while our lot is to enjoy and to suffer. The one reasonable aim of man is that life shall be as happy as it can be made with as much as possible of joy and as little as possible of pain. There is only one way of attaining that aim: the pursuit of natural knowledge." The address concluded with a moving appeal for recognition of our common humanity:

> We are all citizens of one little planet. We are as it were a ship's company marooned on an unknown and mysterious island. There is no time to quarrel about our origins. We have food to find and shelter to prepare. Of what that island can provide for our comfort we know still very little. Let us in peace explore the place. It is full of wonderful things and for aught we know we may yet find even the elixir of life.

Also in 1915, Bateson wrote on "Evolution and education" for *The New Educator's Library*. This was not published, apparently for logistic reasons, until 1922. Here he set out to dispel Lamarckian conceptions and show that "the germ cells of which the offspring are composed possess from the beginning ingredients determining their powers and attributes." Thus, it was "not in the power of the parent, by use, disuse, or otherwise, to increase or diminish this total." His intent was more to attack the "brilliantly expounded" views of Butler and Hering on "unconscious memory" (Chap. 19) than to comment on the education system, but the overall impression would have been that Bateson was not hopeful regarding the long-term effects of education (i.e., of nurture):

> Applied for only a few generations to composite populations, universal education can effect remarkable changes by re-arrangement of the constituent members. Opportunity is given for the more intellectual individuals in the various classes of the community to improve their position. ... As a result of this sorting process, a considerable reconstitution of the layers of the classes may be effected, ... but the alternation accrues by changes in the distribution of opportunity and the selective process, not by any physiological transmission of the cumulative effect of education.

A brief window on Bateson's political activities is provided by correspondence with Beatrice who was away in April 1916. At that time, Coats and Jones were candidates for parliamentary election, and he wrote (April 19): "With hesitation I voted for Jones. At least it is a vote against Asquith whom I have seen. Whether Jones, whom I have not seen, has greater merits I don't know. ... I fear I rather voted for Lloyd George yesterday and don't feel proud of it." A few days later the results of the election were in, there were interesting war developments, and he was off with Miss Pellew to hear a choir (April 21):

> Asquith seems to have averted the "National Disaster". ... Relief to me that Jones did not actually get in. He had enough support to give them a shake. A most friendly letter from Morgan. It seems some of their stuff was lost in *Prinzessin Caecilie*. Interesting to see what notice Germany will take of Wilson now. Not impossibly they may say that, as the whole force of the U.S.A is against them, they must retire. I look on the intervention of the U. S. as meaning that they (the U.S.) see their interests lie in preservation of European Balance of Powers, just as ours do. They cannot afford to have either England or Germany crushed, and I expect they foresee that in the new submarine campaign we (England) can make no effective answer. W. B. Going with C. P. to Messiah today.

Two Cultures

Bateson was not alone in pressing for educational reforms. In his Presidential Address at the York BA meeting in 1906 (Chap. 19), Lankester had lamented the "neglect of science" in public life. He blamed "the defective education, both at school and university, of our governing class" that reflected an institutionalized contempt for science at the highest levels. In 1916 (May 3), Lankester organized a conference on "The Neglect of Science," which was held in the rooms of the Linnean Society in Burlington House. The Report of the Conference Proceedings was highly critical of the Schools and Colleges, where Greek and Latin too often dominated the curriculum. It was proposed that knowledge of natural sciences should be an examination requirement for entrance both to universities and to the civil service. The Report was signed by 36 "men of science," which included Armstrong, Biffen, Ewart, Chalmers-Mitchell, William Osler (Oxford Professor of Medicine), Pearson, Poulton, Shipley (now Master of Christ's College), and J. A. Thomson – but not Bateson.

A reply, engineered by F. G. Kenyon of the British Museum, entitled "Limitations of Science: a Plea for Tradition," appeared in an issue of *The Times* (May 4) that also contained a report on the Burlington House conference. The reply was signed by 23 "men of note, mostly statesmen, ecclesiastics, or professors," and reiterated many of the arguments Bateson had made in the course of the "for classics" debate (Chap. 23). Following the signatures there was a postscript: "Professor W. Bateson desires to add his signature as in general agreement, provided that natural science be first recognized as an indispensable part of secondary education." Bateson seemed to want to have his cake and eat it. A letter in *The Times Educational Supplement* two days later commented on this [4].

In 1917, Bateson repeated his arguments in *Cambridge Essays on Education* to which he contributed a piece on "The Place of Science in Education." It was in many ways an early example of the "two cultures" theme that would be explored by Charles Snow in his 1959 Rede Lecture at Cambridge. Bateson kept the mimicry issue alive: "The attitude or pose of the average Englishman towards education, knowledge and learning, is largely a phenomenon of infectious imitation." He wrote of "fatuous sports" and declared "those who govern the Empire" to be scientifically ignorant:

> We may be reluctant to confess the fact, but though most scientific men have some recreation, often even artistic in nature, we have with rare exceptions withdrawn from the world in which letters, history and the arts have immediate value, and simple allusions to these topics find us wanting. Of the two kinds of disability, which is the more grave? Truly gross ignorance of science darkens more of a man's mental horizon, and … is far more dangerous than even total blindness to the course of human history and endeavour; and yet it is difficult to question the popular verdict that to know nothing of gravitation though ridiculous is venial, while to know nothing of Ananias is an offence which can never be forgiven.

In 1918, Lankester and H. G. Wells published an influential essay *The Aim of Education*. Shortly after the war, Bateson produced further essays on the role of classics in education (Chap. 22).

The Morgan School

Meanwhile, while Europe bled, it was business-as-usual for American geneticists. Morgan and his group at Columbia University were opening a new frontier that would establish the major role of chromosomes in heredity. He wrote to Bateson (December 22, 1914):

> I was glad to hear that you all got safely back to England. … Your address published in this country has aroused a great deal of interest and, as you no doubt intended it should do, not a small amount of opposition. We ourselves are going to get after you soon in a small book that we are writing on "The Mechanism of Mendelian Heredity" for arguing from the nature of characters to the nature of factors. Meanwhile, I send you my heartiest best wishes for a Merry Christmas.

Morgan and his three colleagues threw down the gauntlet in their Preface [5]: "Since the chromosomes furnish exactly the kind of mechanism that the Mendelian laws call for; and since there is an ever-increasing body of information that points clearly to the chromosomes as the bearers of the Mendelian factors, it would be folly to close one's eyes to so patent a relation." Furthermore, while not including Lock's book (see our Chap. 10) in their bibliography, they observed that he had in 1906 "called attention to the possible relation between the chromosome hypothesis and linkage," but had not mentioned Janssens (see below).

Of course, the editors of *Science* asked Bateson for a review, which was published in October 1916. This gave Bateson an opportunity to switch to the new nomenclature – linkage, not coupling; mutations, not factorial varieties; gens (almost "genes"), not factors or elements; sex-linked, not sex-limited – and Bateson did so at some points. There was also the opportunity to praise, and Bateson's was lavish: "The advances made are on any estimate many and of quite exceptional significance." – "There can be no doubt as to the extraordinary value of the *Drosophila* work as a whole." – "Not even the most sceptical of readers can go through the *Drosophila* work unmoved by a sense of admiration for the zeal and penetration with which it has been conducted, and for the great extension of genetic knowledge to which it has led – greater far than has been made in any one line of work since Mendel's own experiments."

From Bateson's lips, there could be no greater praise than comparison with Mendel. Yet the review was sober, not ecstatic. He began by drawing attention to the startling parallel between the sex-linkage displayed by color-blindness in man and that displayed by eye color in *Drosophila*. Furthermore, "half a million flies have been bred, with the result that the data respecting the genetics of *Drosophila* now surpass those from any other animal or plant. ... If we go further, and accept the whole scheme of interpretation without reserve we are provided with a complete theory of heredity, so far as proximate phenomena are concerned." This was not an occasion for Bateson to elaborate on *non-proximate* phenomena. Regarding sex-linkage:

> Morgan concludes that such limitation is in reality only a special case of that complete or partial association of factors in their parental combinations which was first recognized as coupling and repulsion. These phenomena may in fact be all one. They are examples of linkage between factors, the second factor involved in the case of sex-limitation being that for sex. The fundamental identity of these linkage-phenomena had naturally been suspected. Difficulty, however, lay in the peculiarity of sex-limitation, that in it the linkage has never been observed to be other than complete. The new theory ... represents this distinction in a simple and readily conceivable way, so that we are at once attracted.

Seeming to have forgotten that he had himself edged toward it in 1902 (Chap. 13), Bateson regarded the speculation as "legitimate" and "extraordinarily promising" although "far-reaching," that "all the factors are linked together in groups, and that the number of independent groups is that of the haploid chromosomes. This number in *Drosophila* is four, and it is claimed that, on genetic analysis, the various factors of *Drosophila* can be proved to be so interrelated as to constitute four linkage groups and no more." However, a similar match between chromosome number and linkage groups had not yet been observed in other organisms. Furthermore, in cases other than sex-linkage, the linkage was often incomplete, for which some interchange or "crossing-over" between homologous chromosomes had to be postulated. To explain this, Morgan had to appeal to the "twisting and interlacing of chromosomes in synapsis" that had been observed by Janssens in amphibia, and to the "very fragmentary observations" of Stevens and Metz in fruit fly. Remarkably, the finding that sex-linked characters were always completely linked in fruit fly fitted in with another "curious and significant" fact:

> The formation, then, inside a linked group, of factorial combinations other than those which entered with the parents, is ascribed to crossing-over from one chromosome to its fellow or mate. At an early stage in the work, the curious and significant fact was observed that in the male no such crossing over took place in regard to the various factors which had been proved to be *sex-linked*. The cytological interpretation of this discovery was ready at hand.

In early work, the Morgan group had found only one X-chromosome in *Drosophila*, so there was no "real mate" to crossover with and "therefore in the case of sex-limited characters linkage is complete." However, in the book the group now reported two sex chromosomes in the male (X and Y), so, in principle, crossing over could occur. The fruit fly solution was to prohibit crossing over for *all* chromosomes in males:

> We meet, however, a fact which is much more difficult to harmonize with the theory, though constituting one of the most novel and remarkable discoveries made in the *Drosophila* work. Not only do the sex-linked factors show no crossing over in the male, but experimental breeding shows that in the male there is no crossing over even of the factors composing the other groups. *Crossing over, in fact, in Drosophila, turns out to be exclusively a phenomenon of the germ cells of females*. This is a genetic discovery of the first magnitude, whatever its ultimate significance, but the cytological interpretation of crossing over must now bear a very considerable strain. … It is with some surprise that we find neither in the book nor in the material previously published any coherent discussion of the difficulties so created. If further cytological work shows that the chromosomes of the female twist and anastomose, but that those of the male do not, then chromosomal theories of heredity will receive a very remarkable support.

Nevertheless, "recombination … within the limits of a linkage group" had led on "to a further and very remarkable speculation," for if the heredity factors were "arranged in a row, like a string of beads, along the length of the chromosome," then "the amount of crossing over can thus be interpreted as an indication of the relative positions of each factor in such a series. Upon this follows the great thesis of the book: that this series is in fact a row of points along each of the four chromosomes." Bateson noted that when crossover percentages approached 50% it was not possible to distinguish between factors on separate chromosomes, or at two ends of one chromosome (Figs. 9.1 and 9.2 of Chap. 9). Furthermore, there was the theoretical possibility of multiple crossovers. Finally, while there were data consistent with linkage, control data showing "no linkage between members of distinct [chromosomal] systems" was "admittedly meager."

The highest praise was reserved for the work of Bridges on the abnormal non-separation of members of a chromosome pair at meiosis ("non-disjunction"), resulting in gametes that either carried both or no chromosomes of a particular type [6]. Concerning sex chromosomes, this meant that a fly, instead of being a normal male of constitution XY, could be XXY. This "very fine achievement" had produced a fascinating set of phenotypes, so that it was "difficult to see how we can deny that the sex-linked characters have some very special relation to the sex-chromosomes." Nevertheless, Bateson would not relent: "In our present ignorance of the nature of life we cannot distinguish cause and effect in these phenomena and it is not possible to attach any satisfactory meaning to the expression that the sex-linked factors are 'carried' by a chromosome."

Finally, Bateson turned to the apparent homogeneity of chromosome structure: "It is inconceivable that particles of chromatin or of any other substance, however complex, can possess those powers which must be assigned to our factors or gens. The supposition that particles of chromatin, indistinguishable from each other and indeed almost homogenous under any known test, can by their material nature confer all the properties of life surpasses the range of even the most convinced materialism."

Bateson sent his review to Doncaster, who replied (November 2, 1916) recommending a paper by Muller on crossing over in *American Naturalist*, adding "doesn't this look as if the linear arrangement with constant distances were a fact?" The evidence on the role of chromosomes was set out later in Doncaster's *Introduction to the Study of Cytology* (1920).

Gamete and Zygote

Since he was reviewing Morgan's book, Bateson had to mention Morgan by name. But Bateson preferred the elliptical approach even when Morgan's results backed his own view. In their fly studies, the Morgan group had produced many gradations of eye color from red to white, and Herbert Jennings in the United States had deduced from the existence of such intergrades that the process of variation must be continuous as Darwin had proposed. In 1917 in his Sidgwick Memorial Lecture entitled "Gamete and zygote," Bateson addressed this issue, although not acknowledging Morgan:

> The fly, *Drosophila*, normally red-eyed, produced a white-eyed variety. In generations of intercrossing between the red-eyed and the white-eyed, pink-eyed flies occasionally appeared. By more intercrossing cherry-coloured, eosin-coloured, etc., etc., were produced, until there is now a series which may reasonably be described as continuous. The ingredient on which the red eye depends is *commonly* treated in division as an integral unit, but sometimes its integrity is impaired. Offered such a series the evolutionist of 25 years ago would of course have inverted the truth, seeing clearly that the white was derived from the red by the occurrence of a gradual change towards white, whereas the truth was that the white came directly from the red and the transitional forms afterwards. Examples of such disintegration abound. It is the proof of *integration* that is doubtful and elusive.

As for the Morgan group's chromosome experiments, in 1917 Bateson could still declare:

> Others ... declare that each property of the organism is determined by a specific particle of nuclear material, and believe that as the result of certain very remarkable experiments they are even able to decide the order in which these particles are grouped. I mention this inter-

esting line of enquiry to illustrate the scope of modern genetic analysis, though for the present I am unconvinced of the cogency of the arguments employed.

At this time, there was a brief nod toward the ideas of Montgomery (Chap. 13), although he was not mentioned by name: "Whatever be the rationale of fertilization, it must take into account the conjugation of ciliated infusoria and the unicellular algae." And he was forced to acknowledge the growing importance of laboratory work. New studies, confirmed by Julian Huxley, were suggesting how cells might become organized into tissues:

H. V. Wilson's regenerated sponges are wonderful enough. The live tissues of a sponge were pulped through the meshes of a fine cloth, falling as a layer of cells to the bottom of the vessel. These cells then joined up again into small masses, and those derived from the different layers migrated into their appropriate positions, thus forming new sponges from the *débris*. This coherence may perhaps be regarded as realizing the suggestion … that the shapes of young cells are such as to imply the existence of forces of attraction acting on each other …. But of course in this case the cells were already differentiated before the dissolution, whereas when groups of embryonic cells separate to form a bud they carry with them also the power of differentiation.

His proclaimed "lay discourse" may have created some confusion regarding his materialistic views. At the outset, he stated that "of life unattached to matter we have no evidence at all, and, however minute they may be, it is by the matter which the two germ cells contribute, and by nothing else, that the powers of the animals and plant are fixed." Yet when it came to the manner of segregation of characters among offspring and the thoughts of "our American colleagues" in this respect, he was much vaguer:

In critical cell divisions, especially those immediately antecedent to the formation of gametes, segregation happens, and we must express that process as a breaking up, a tearing apart, a sorting out, or by the use of some similar figure. Nevertheless, we should keep our minds open to the possibility that the thing broken up is not actual material, whether of the nucleus or the cell plasm. Many of our American colleagues interpret the visible features of cell division as literally conforming to this material scheme, and it is not to be denied that following that guiding principle they have been led to far-reaching positive discoveries. Complex, however, as the living cell must be I am unable to see in it a material heterogeneity so vast; and I incline to the expectation that the heterogeneity of the determining elements as factors lies rather in forces, of which the cell materials are the vehicle, than in the nature of the material itself.

Here, Bateson implied that the determining elements (genes) could not be homogeneous (uniform, like a flat sandy surface). So in what way could they be heterogenous (nonuniform, like the ripples in a sandy beach)? Perhaps he was thinking of vibrations as when sound waves were recorded on the uniform medium of a gramophone record (e.g., analog information), rather than the "phenomenon of arrangement" mentioned in his Melbourne address (e.g., digital information).

Lay persons in the audience would probably have benefited from his exposition of the differences between plants and animals: "The germ cells of animals – as opposed to plants – are formed from material definitely reserved from a very early stage in individual development [e.g. in the gonads]. This fragment of material undergoes unnumbered divisions, of which the free eggs and sperm are the final products." In plants, while the capacity to form gametes is localized to anthers and ovules, it was also more generally distributed: "The animal is a closed system growing by intercalation, the plant is an open system growing by division at the apices. … In some ways an animal zygote is like a hollow ball with gametes inside it, able to become similar balls when liberated, while a plant is like a stocking, knitting itself forwards at both ends by means of growing points. The growing point has perpetually the power of making gametes."

Bateson considered this fundamental difference explained differences between some "mosaic" animals and plants. For example, mutations to "piebald" in animals could immediately be fixed by breeding, as was the case with picotee carnations. But from time-to-time bizarre lines of carnations arose with irregular segments and stripes, which could never be made to breed true. This was elaborated in a later lecture with the same title – "Gamete and zygote" – given (February 15, 1918) at one of the popular weekly evening series held at the Royal Institution in London, with Lord Rayleigh in the chair:

> Plants which are mosaics or patchworks, presenting as mixtures of allelomorphic characters *not subordinated to geometrical control*, are usually incapable of being bred true. The geometrical disorder is an indication that the distribution is a mere fortuitous collocation of dissimilar elements, and not a genetically transmissible pattern. Some of the gametes in such a plant will carry one or other of the components; others arising themselves as mosaic cells may repeat the mixture (c.f. mosaic Azaleas, Carnations, the Bizarria Orange, etc.). Animals, however, having mosaic patterns may often be readily established as pure breeds (Sheeted Cows, Dalmation Dogs, White Bantams having one or more small grey ticks, etc.). The geometrical relations of gametes to zygotes are therefore quite distinct in animals and plants.

In an abstract of the lecture, Morgan and Bridges were named and doubts repeated:

> From certain elaborate and valuable experiments Morgan and his associates have convinced themselves that the properties of organisms are determined by particles of nuclear material arranged in a predictable linear order. Nevertheless, alternative hypotheses have not hitherto been adequately investigated. The evidence most favourable to the chromosome scheme of heredity is that furnished by the phenomenon called "non-disjunction" by Bridges. Here usual zygotic types were consistently produced in association with the visible presence of definite aberrations in the chromosome numbers. Striking as this evidence is, it must still be possible to doubt whether the relations are certainly those of cause and effect.

Bateson also took the opportunity to again contradict Lotsy's claim (Chap. 12), which had been further elaborated in a book in 1916 [7], that variation could be produced by the shuffling of that which preexisted, so that "original variation" was an unnecessary abstraction.

Rogues and Chimeras

Among Bateson's treasured exceptions were some far-from-ordinary "rogues," a term applied by horticulturalists to unexpected types of plant that appeared from time to time. When investigated, such plants often turned out to be rare recessives or the result of the transport of foreign pollens by insects (accidental cross-fertilization). However, the wild-looking rogues with pointed leaves that Pellew discovered among her culinary peas (*Pisum sativum*) were very special. Her studies of the phenomenon, albeit unexplained, did much to convince Bateson of the importance of continuing to explore the moment-of-segregation theme that he had long enunciated as a problem.

In 1915, Bateson and Pellew reported in the *Journal of Genetics* that backcrosses of rogues with the parental type *always* yielded rogues (indeed in appearance these rogues could be more roguish than the original rogues). The offspring of a rogue were *always* rogues. Whatever the phenotype, rogues bred true. In Bateson's words, "the types [i.e. normal plants] can throw rogues and the rogues cannot throw types." So, it seemed that "the types contain something which the rogues do not contain. This something, however, is different from the ordinary Mendelian factor both in the effects of its presence and in the manner of its distribution among gametes."

The rogue phenomenon affected certain characters and not others, which would remain true to type: "Clearly … there is no general exclusion of the contribution of the type parent, and it is only the features special to the type which are excluded." When rogues first appeared, the change in appearance was often best appreciated at the adult stage, the juvenile plants looking more like the parents. However, when a rogue was allowed to self-fertilize, or was crossed with other rogues, the changed appearance was evident from the start: "From the genetic evidence it is clear that in order to influence the somatic structure beyond the juvenile stage, or to appear in the germ cells, the character, whatever it is, must be introduced from both sides of the parentage."

Bateson and Pellew concluded that "the rogues arising as the offspring of types are … frequently heterozygotes formed by the union of type and rogue gametes, and since they always breed true, in them also the type-elements must be lost in some somatic stage." Thus, the old idea of somatic segregation that had given rise to the then rapidly fading Bateson–Punnett reduplication hypothesis of linkage (Chap. 13) might in this case be applicable: "We incline to think these indications point to some process of somatic segregation which prevents the type-elements from reaching the germ cells of the cross-bred plant." From further experiments reported in the *Proceedings of the Royal Society* (1915), it was concluded:

> The genetic constitution of the F_1 plants raised by crossing types with rogues was especially remarkable. These plants, as young seedlings, are intermediate between types and rogues, but … as they mature they become normal rogues and behave genetically exactly like pure-bred rogues, producing only rogues as offspring. We conjecture … that a segregation of factors takes place in the soma, such that the type elements are left behind in the base of the F_1 plant and are thus excluded from the germ lineage.

In 1920, they were able to further affirm that "rogues crossed with types … give F_1 plants which as seedlings show evident indications of the type-characters, having parts much larger than those of rogues at the same age. But these plants at an early stage, usually at some node below that at which the first flower is borne, *change to rogues*, producing stipules, leaves and eventually pods, like those of rogues." They concluded that "the persistent recurrence of rogues among the offspring of types must indicate some liability to an error in cell-division. Once the abnormality has occurred, of which pointed leaflets are the ostensible indication, there is a progressive change in successive generations such that, assuming equal fecundity in all classes, the progeny would in a few generations consist of rogues in overwhelming proportions."

Many of the phenomena Bateson studied, such as gametic coupling, found explanations in his lifetime. However, a century was to pass before these rogues were to approach a solution. In 1958, Brink ascribed the change to rogue phenotype in maize to "paramutation" [8]. This is now recognized as an example of transgenerational epigenetic inheritance (a form of Lamarckian inheritance favored by Butler but deplored by Bateson), which often involves differential methylation of DNA bases without an immediate change in base sequence (as when a grave accent over the letter "e" in "gène" gives it a meaning in French different from "gêne"; [9, 10]).

Another intriguing phenomenon was that of chimeras. It had long been known that parts of plants if cultivated could give rise to new plants (vegetative propagation). This revealed that the information for the entire plant was latent in each part of it, the presumption being that a plant produced from a stem-cutting would be the same as a plant produced from a root-cutting. At a meeting of the RHS, Bateson was said to have sat up with great excitement on learning that this was not always so. In 1916, he reported in the *Journal of Genetics* that from the roots of pinkish white-flowered *Bouvardia* red-flowered plants could be derived. This meant that a plant could be chimeric with respect to its ability to give rise to offspring displaying a particular character. Bateson concluded:

> Having in view the fact … that these various chimaeras arose as seedlings, their peculiar constitution must be recognized as having been produced by somatic segregation. How the [potentials to produce the] characters are distributed among the embryonic layers of the plants is however as yet uncertain. … Pending further evidence it is natural to interpret these cases as examples of heterozygous plants in which there has been somatic segregation of a factor at an early stage. On a general survey of the phenomena of chimaera-production and bud sports, the indications suggest that such a segregation may occur at many and perhaps at *any* cell divisions by which the parts of the embryo are constituted, or the organs of the plant are differentiated. It may well be that segregation is most commonly relegated to the division of the germ cell cycle, but I am unwilling to regard segregation postponed to the reduction division [of meiosis] as a process distinct in kind from those somatic segregations of which bud sports are the visible manifestation.

Gregory and Winge

Polyploid forms with giant cells containing multiples of the normal (diploid) number of chromosomes were noted in the tall plants that appeared occasionally among clumps of evening primrose (see de Vries in Chap. 9 and Gates in Chap. 14). Bateson was drawn again to the phenomenon when his colleague Gregory studied a "giant" form that had been noticed one day at Sutton's nurseries [11]. This was a

Chinese primrose (*Primula sinensis*) that had twice the standard number of chromosomes in both somatic cells and gametes. The results of his crossings led Gregory to believe that his plants "possess a double set of factors," suggestive of a "definite connection between chromosomes and factors, yet, on the other hand, the tetraploid number of chromosomes may be nothing more than an index of the quadruple nature of the cell as a whole." Edmund Wilson in 1914 interpreted Gregory's finding as implying that all the Mendelian character units, or "gens," must be "correspondingly doubled" in polyploid species [12].

Another student of polyploidy was Öjvind Winge (1886–1964) in Denmark. In June 1916, he completed a review entitled "The chromosomes. Their numbers and general importance," which was published in 1917 [13]. Bateson wrote to Ostenfeld (November 19, 1917): "I today received also a paper by Winge of your Carlsberg Laboratory on Chromosomes. It looks interesting. Please thank him for it." The 275-page paper was in English except for a few quotations and, while referring to "certain authorities" who would not admit chromosomes "as the seat of genotypic disposition," did not mention Bateson by name. Winge held that normally parental types were "in harmony" so that within their offspring their chromosomes would pair correctly during meiosis to generate fertile gametes. However, "less marked harmony" was sometimes present:

> Genotypic differences might be present within one and the same species, causing the extreme forms, on intercrossing, to act very much as do entirely different species – i.e. produce offspring in which the parent chromosomes are not normally paired, which again would involve the further consequence that the formation of sexual cells [gametes] must proceed abnormally or even not take place at all.

Winge pointed out that disharmony (hybrid sterility) would be cured if, prior to seeking a pairing partner, the chromosomes doubled, so generating a polyploid organism where there would be "a quantitative, but not a qualitative difference, just as 1 molecule of NaCl + 1 molecule of NaCl continues to be NaCl" (see Figs. 9.5 and 9.6 of Chap. 9).

> We must then suppose that this [disharmony] will be visibly expressed by the fact that the chromosomes derived from the two [parental] gametes will not unite in pairs at all, but distribute themselves throughout the primary [gamete-forming] cell of the zygote [adult], as if no dualistic relation of any kind existed. If the chromosomes are to find a partner, then each of the chromosomes in the zygote [adult] must divide, for thus indirectly to produce a union of the chromosomes [because there is then a harmonious partner for pairing], and we must assume that this is realized in the hybrid zygotes which have any possibility at all of propagating – in accordance with what we know from experience as to the behaviour of pairs of chromosomes.

> The hybrid … thus produced will then have 4x chromosomes, taking the number for each of the parent gametes as x. After this, either of the chromosome pairs will have the power of further separating by reduction division, transmitting one set of chromosomes from either parent to each of the gametes – in which case we have a new hybrid organism with the qualities of a pure species and a "double" chromosome number; i.e. containing the sum of the chromosome numbers in the parent species. … We may doubtless assume that most of the species exhibiting "double" chromosome numbers are hybrids formed in this manner. … [Thus a] mutual affinity between the parent chromosomes in pairs is … present in normal

> plants. Where such affinity is lacking, as frequently ... in the case of hybrid forms, the situation described [above] ... arises, and may lead to ... the formation of apogamous [sterile] species. ... The hereditary differences are not themselves visible, it is not the morphological differences in the chromosomes, but variations of a far more delicate nature ... [that] must rest on some chemical conditions ... as lie far beyond the range of our present knowledge.

Why was the pairing ("parallel conjugation") so important? Winge came to similar conclusions as Montgomery (Chap. 13), namely that pairing chromosomes could complement each other, a defect at one point in one chromosome being correctable by the other: "The position might be illustrated ... by pointing out that two ships at sea without losing their 'individuality' might well exchange certain wares in order to complete their stores; the chromosomes can doubtless carry on a similar traffic during their conjugation in the synapsis stage." To illustrate this, Winge characterized a chromosome of maternal origin as "a b c d e f g h i," which was "genotypically dissimilar" from "a x c d e f g y i," the corresponding defective paternal chromosome:

> The importance of this pairing must, I imagine, be that the chromosomes two by two supply each other with missing chromatic parts, possibly ides or pangenosomes [genes] ... in case such should have been lost during the ontogenesis [development] ... and which are necessary if the organism (species) is fully to retain its disposition, i.e. remain genotypically unaltered. During the synapsis and subsequent stages ... the chromosomes ... have to a greater or lesser degree effected an interchange of substances, so that while each in the main still has its original composition, all are none the less affected by the temporary pairing in the points where they have individually suffered loss during the ontogenesis of the double organism.

Winge also entertained the possibility that "the two chromosomes have ... lost the same qualities, and are thus unable mutually to replace these." This would occur when there was a "close ... relationship between gametes" that would arise if they had a "common derivation." Thus, "the fact that self-fertilization or inbreeding in many cases gives bad offspring or fails altogether must ... be due to lack of mutual complementation, as a result of the over close relationship." But there was an "argument against" this. "Many plants ... continue to develop and flourish even when self-pollination constantly takes place." In this circumstance (asexual reproduction), Winge held that natural selection would have freer rein so that "the less adapted, or the abnormal, here ... perish in the struggle for life."

The other extreme would be when there was a very distant relationship between the parental gametes. This might sometime occur within a species (see above) but was seen most clearly in crosses between allied species. In this circumstance, there might be a developmental failure ("first test") so that no adult organism would arise (hybrid inviability), or if an adult could arise it would fail to produce viable gametes (second test): "Once the hybrid has passed the first test, which decides whether it is capable of independent vital action at all, it develops with often surprising luxuriance [hybrid vigor], until the inadequacy of the sexual products puts an end to its further propagation [hybrid sterility], and the biotype produced dies with the individual." This "inadequacy of the sexual products" (i.e., a failure of gamete formation) could be related to the extent of pairing seen when meiosis was examined microscopically. Thus, "the sterility of the offspring is in inverse proportion to the power of conjugation of the chromosomes."

In this context, Winge mentioned Guyer's pigeon studies (Chap. 13) where the allied species had the same number of chromosomes, and later studies of Rosenberg [14], who had crossed two allied species one of which had 20 chromosomes (*Drosera rotundifolia*) and the other, a polyploid, had 40 chromosomes (*Drosera longifolia*). So, the corresponding gametes had 10 and 20 chromosomes, respectively:

> The hybrid plant contained 30 chromosomes in the somatic cells, this being the sum of the haploid number of both parents. ... In the hybrid ... the chromosomes conjugated in such a manner that the 10 *D. rotundifolia* chromosomes united with 10 of those from *D. longifolia*, the remaining 10 of the latter distributing themselves irregularly throughout the daughter cells. ... On reduction division therefore, it was possible to discern the presence both of paired and unpaired chromosomes. ... [and] the pollen was sterile.

In his 1918 address on "Gamete and zygote" (see above), Bateson considered polyploidy as an intracellular twinning:

> Remarkable phenomena of polarity are exhibited by the "giant forms" best known in certain plants. Just as a zygote [shortly after fertilization] can divide to form a twin-pair [which can grow to become adult individuals], so can the material which normally is distributed as two plants be compounded as a single "individual." The case most studied is that of *Primula sinensis*, described by R. P. Gregory. On at least two occasions "giant" plants containing double the usual number of chromosomes have arisen. Both of these sets of giants exhibit the unique property of being totally unable to breed with the plants from which they were derived, though fertile with other giants. Similar cases are known in Banana (Tischler). Species of Chrysanthemum are described as having respectively 9, 18, 27, 36, 45 chromosomes. Geoffrey Smith and others have observed giant sperms in animals which presumably contain extra chromosome materials.

In the modern view, due to chromosome entanglement at meiosis, "intracellular twins" arising within an individual organism (autotetraploidy) would usually not continue the line. This would be less likely when chromosomes from allied species were twinned (allotetraploidy; see Fig. 16.1).

Ear of Dionysius

The war led to a reemergence of spiritualism. This was unsettling for many parents whose sons had been killed. Bateson could hardly overlook the matter since the Balfour family was involved. The "Ear of Dionysius" dispute involved Winifred Coombe-Tennant, later a representative of the British Government at the League of Nations, who at that time worked as a medium under the name of "Mrs. Willett." In 1910, in the course of a séance Willett wrote "Dionysius's Ear. The Lobe." Arthur Verrall, a Cambridge classicist, saw in this an allusion to a text by Thucydides that Mrs. Willett, having no training in classics, could not have known of. In subsequent séances after Verrall's death, Willett wrote similar phrases and implied that they came from the deceased Verrall. The March 1917 issue of the *Proceedings of the*

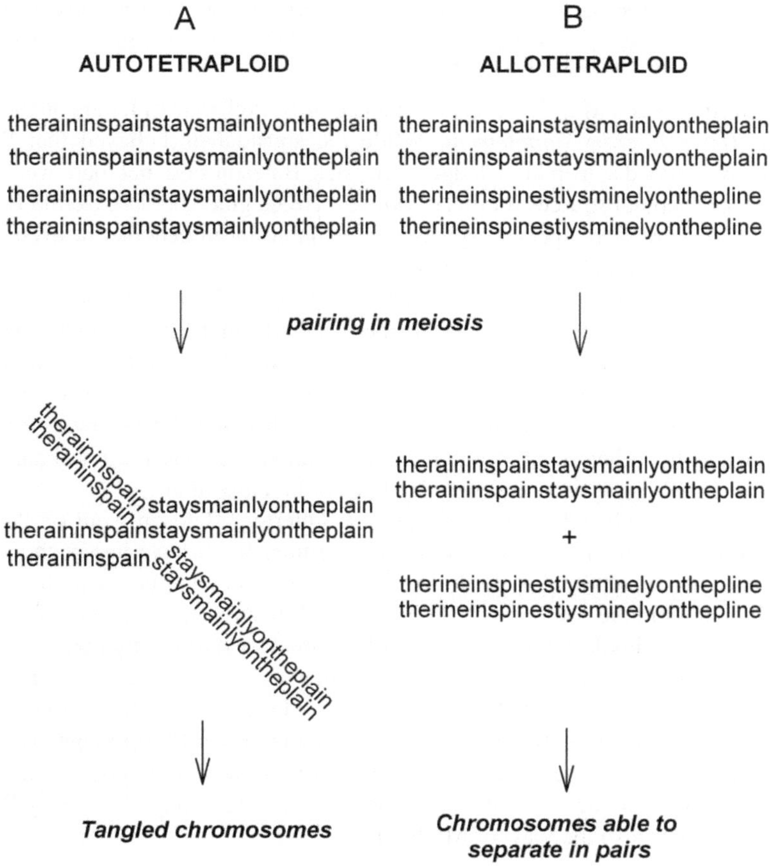

Fig. 16.1 Intracellular "twinning" works best when chromosomes come from allied species. Within a species, the four chromosomes of an "autotetraploid" (**a**) have the same "accent" so that in meiosis segments of one chromosome can pair with corresponding segments of *any* of the other three chromosomes. Thus, the four chromosomes tend to become entangled and cannot properly separate in pairs (carry out "disjunction"). However, chromosomes of allied species differ sufficiently in their "accents" to prevent mispairing. When two such diploid species form an "allotetraploid" (**b**), each chromosome has a unique pairing partner and gametes can be formed (see Figs. 9.5 and 9.6 in Chap. 9)

Society for Psychical Research contained an article entitled "The ear of Dionysius: further scripts affording evidence of personal survival" by Arthur Balfour's brother Gerald:

> The Ear of Dionysius is a kind of grotto hewn in the solid rock at Syracuse and opening on to one of the stone quarries which served as a place of captivity for the Athenian prisoners of war who fell into the hands of the victorious Syracusans after the failure of the famous siege so graphically described by Thucydides. A few years later these quarries were again used as prisons by the elder Dionysius, Tyrant of Syracuse. The grotto of which I have

> spoken has the peculiar acoustic properties of a whispering gallery, and is traditionally believed to have been constructed or utilized by the Tyrant in order to overhear, himself unseen, the conversations of the prisoners.

Between April 5 and May 31, Bateson engaged in an exchange of letters on the topic in *The Times Literary Supplement*. While the spiritualists believed that Willett would have needed a deep knowledge of classics, Bateson said that there were readily available popular sources for this information (e.g., guide books). Gerald Balfour replied that there were special aspects not available in guidebooks or even in the *Encyclopaedia Britannica*. With this the issue faded.

Life went on. There was correspondence (June 4, 1917) with Wheldale who had befriended a highly talented fellow biochemist (unfortunately a cripple), and she would soon need to be addressed as the Honorable Mrs. Onslow. Bateson remarked on the kind review of her plant pigments book in *Nature*. There was also more correspondence with Ostenfeld (June 15): "I get further and further away from the chromosome hypothesis and Morgan and his friends. Last year I did hesitate somewhat, but I can't apply it to the everyday facts of heredity at all."

In September while Beatrice was away at Otterton, Bateson went on "a short Surrey walk." On his return, he reported (September 6): "I have just been through revision of my Education article. It does not run as easily as I could wish. Bits are all right, but there are some jolts." He began an obituary notice for Lock, for which he reread Lock's book (September 11): "I have made a start on my notice of Lock. I don't like the book better on closer acquaintance." Even the issue of Palestine briefly arose (September 16): "About once a week I receive papers on Zionism. Not sure why. Perhaps your views of my origin are shared by Sir Philip Magnus or some other coreligionist. If they are right, my position is becoming serious .... I have always said the real question which the war raises is the custody of the Holy Places. This evidently is also the view of Mr. Schiff."

Meanwhile, Bateson, with some guidance from Dobell, had been looking down a microscope at unicellular flagellate organisms that rotate as they swim. Did they rotate in a particular direction? In her *Memoir,* Beatrice related that on one occasion he had envisaged relating the direction of coiling displayed by some climbing plants with the handedness of the chemicals they contained (i.e., Pasteur had discovered that some chemicals rotate light to the left, others to the right). Sometime in late 1917 and early 1918, he believed he had observed that when traveling toward light flagellates rotated one way ("right handed"), and that when traveling away from light they rotated the other way ("left handed"). In great excitement, he submitted a paper entitled "An observation on the influence of light on the direction of rotation in flagellates and fusoria" to the Linnean Society. Shortly thereafter (March 15), he concluded that the probability was in favor of an optical illusion and quickly withdrew the paper.

At the Front

John Bateson enlisted in 1916 and became an artillery officer. Later in the year, his father sailed to Rouen, where, under the auspices of the YMCA, he gave lectures in biology, and even manned a counter selling tobacco, candles, soap, and buttons to the troops. A letter (December 22) revealed that his first lecture

> was *not* a great success, I fear. Conditions dreadful. Long narrow room with no illumination but the lantern, so that I could never light up and see my audience .... Many crept away under cover of the darkness – a few no doubt interested. The accepted cue is to talk down to them. They live chiefly on comic songs, sentiment, and religion in many forms. ... The R. E. men asked me after the lecture whether man could be descended from monkeys. If so, how had he come by an immortal soul? ... These were men leaving for the front next day and so – well – I temporized.

John Bateson was back in England for Christmas, and Beatrice later noted: "John ... had gone into training at Exeter RAF, and I and Martin and Gregory spent the Xmas holidays in lodgings ... down there. There were a lot of Canadians in training and the drunkenness, etc. going on in the camps was appalling." In January, Bateson's lecturing ceased when he was placed in a military hospital with a throat and chest infection; here he was, nevertheless, able to read *Madame Bovary* and correct page proofs for *Nature*. His condition improved to an extent sufficient for him to return to England, where he spent a further three weeks in bed. In November 1917, John returned home with a minor injury and was awarded the Military Cross for bravery.

Bateson returned to France in January 1918 and encountered Wilmot Herringham, who ran the Royal Army Medical Corps. To his knighthood he had now added the title of Major General. In an account of his war experiences, Herringham praised the practice of inoculation against typhoid that had led to fewer infections among British troops than among the French [15]. In some ways he was very different from Bateson – deeply religious, and an accomplished team-player and committee-man. He thought that the war might have been prevented and regretted that warnings had not been heeded. After the armistice, he was not inclined to forgive the German people: "They pose now as gentle and trusting people misled by their rulers."

It seems that, having given some support to the John Innes Institute in the form of short-term training fellowships, Hall's Development Commission was now keen to expand its assistance. However, Bateson was wary: "The drag in the economic direction is likely to prove too strong to be successfully resisted. It becomes, therefore, the more desirable that at least one institution should retain its financial independence complete, though prepared to co-operate with state-aided institutions in all possible ways." Thus, an opportunity was lost to recruit a senior researcher (and possible successor).

Bateson was also wary in the spring of 1918 when approached by the Ministry of Information for a propaganda article that would convey the idea that the dynamism of British Science had been sustained during the war years. He declined (April 19) noting that by closing the Museums for the period of the war the Government had "gravely injured the cause of science and learning, and advertised to the world the contempt in which such Institutions are held in the country. Whatever propaganda may be disseminated by the Ministry of Information, that action is likely to be remembered abroad and its consequences persist." In a follow-up letter (April 27), he noted that during the war years the "Prussian Government has … started a new Institute at Potsdam for purely scientific research in genetics, which was opened on May 1916."

In the summer of 1918, Bateson's mother died at age 89. A string of tragedies followed. First, John Bateson, who had returned in July, was killed in action (October 14). Bateson had just written to him from Ingleborough, asking him to write to Ned (September 25): "We are a large and merry party here – the Barlows, Raverats, ourselves, and until this morning, Martin. … I hope you sent a note to Edward. E. Bateson Esq., Maadi, Cairo, Egypt." At age 42, the head-gardener at Merton, Mr. Allard, whom Bateson had first hired as a boy, died in the influenza epidemic. The same epidemic brought Reginald Gregory's life to an end. Bateson wrote to Ostenfeld (November 27, 1918):

> Our eldest boy was killed at the very end – some 14 days before the Armistice was signed. He was a brave, good, boy. He won the Military Cross for a piece of bravery. Like us he also had no illusions as to the war. The second [boy] has not been out [at the front]. … You will have heard that [Daniel] Hall's eldest boy was killed, and the second boy terribly crippled. He has an arm almost entirely paralyzed and one lung nearly useless. We have had other sorrows. My Superintendent Gardener, a most gifted and exceptional man died of pneumonia, and now we have lost R. P. Gregory, also of pneumonia. He used to work here a great deal. It is difficult to keep up much spirit.

In her *Memoir*, Beatrice commented: "Characteristically putting aside his own unhappiness, Will set himself at once to help the widows of these two friends." Martin Bateson took the death of his brother badly (Chap. 19). In 1920, Doncaster died of cancer at age 42 only a year after assuming the Chair of Zoology at the University of Liverpool.

So, the war was over. The survivors returned to their peacetime duties. Miss Cayley was reappointed in 1919. Major Hurst returned to Burbage, where his wife had died during the war years. In his absence, much of the land had been turned to food production. Michael Pease returned to work with Punnett in Cambridge. In Austria, Paul Kammerer (Chap. 21) returned from his role as military censor to find his collections of breeding animals, designed to demonstrate Lamarckism, had decayed. Many of the male scientific staff of the John Innes, either before or during the war, had left for other lands: Backhouse to Argentina, C. B. Williams to Trinidad, Arthur Bailey to Egypt (Fig. 16.2).

The world was in disarray. Bateson was increasingly drawn into the international arena, both scientifically and politically (Chap. 22). He became a member of the influential University Grants Committee. Yet, amidst all this, the issue of the origin of species remained. He wrote (January 29) to the Swedish botanist N. Heribert-Nilsson:

> With more and more experience I have steadily settled down to the conviction that species and varieties are physiologically distinct in their natures. … It is not possible to define a category of distinctions which apply rather to varietal than [to] specific differences. The distinction between the two kinds of relationship, if ever they can be defined, must be defined by means of genetic tests. We are a long way from such a definition yet.

Fig. 16.2 Bateson wearing a fez with Miss Cayley (left), Beatrice, and Miss Pellew (1919)

Extra-National Future

In his 1918 address on "Science and nationality" (see above), Bateson reflected on the war and the future of mankind:

> A great cry has gone up in all the land; and not in our land alone, but through all the earth, for is there a house where there is not one dead? Caught in the wheels of a hideous destiny the young men of the nations and the innocent boys have been torn to pieces. … They went in the high call of Duty. The altar upon which they bled bears the glorious names of Patriotism and Duty. From the enemy cities and from their quiet villages has poured another stream of youth to perish at the self-same shrine, calling alike on Patriotism and Duty, with, we must believe, an equal devotion. So it has been from the beginning; must it continue so to the end?

The war reflected the failure of centuries of influence by "priests and lawgivers." It was time to ask whether "the makers of natural knowledge" might help? Nationalism was seen as the main problem:

> Nationality … is a sentiment, picturesque and within limits laudable, yet in essence accidental and ephemeral, capable of being turned to effects, often good so long as they last, but impermanent, adding nothing to *universal* good, and commonly a pretext for the grossest forms of selfishness and cruelty. Transitory as the fruits of patriotism must be, it is pathetic to observe that this force is especially invoked by the nations in their pitiful striving for terrestrial permanence and immortality.

Bateson's answer was that we must move beyond nationalism, even beyond internationalism in the form of the proposed League of Nations, to something that had always been with us as "science, art and letters," namely *extra-nationalism*:

> Men of science, whose calling familiarizes them with epochs, whose measure of achievement transcends the reckonings of statesmen, judge greatness by another scale. We have our units, and the commonality have theirs. Seldom even are the two estimates commutable. For us a man is great not according as he has succeeded in influencing the ephemeral destinies of some artificial group on whom the lawyer has conferred the title of a State, but rather as he has extended thought or penetrated new provinces of knowledge. We speak sometimes of science, art and letters as an international domain. More truly we should think of them as *extra-national*.

And how was extra-nationalism to be attained? "It is for us men of science and our brethren in the arts and letters to lead." And how were they to lead? By using all measures available to sway public opinion:

> The last fifty years have been a period of deterioration and of lowering in public ideals. The ascent may be less easy. Yet it is comforting to reflect that in private manners we have in the same period advanced. It *was* public opinion which abolished slavery in the West Indies overcoming vast financial interest. It might be impossible for Wilberforce and Clarkson to carry such a reform at the present day, but it *is* public opinion that has suppressed public cruelty to animals and children during our own lifetimes, and when a few years ago men suddenly gave up the habit of spitting in railway carriages, they changed their manner not for fear of a forty shilling fine but at the bidding of fashion. The force of imitation, once developed, binds all but the rare exceptional men, and normal man, being almost incapable of independent thought, once bound by fashion is as powerless to escape as the hypnotized subject to resist the commands of the mesmerist. If the powers of the school, the church and the press were exerted to interest the world in extra-national things as they have been exerted to inflame national ambitions and to glorify war, the public opinion we long for may yet be created.

There was no place in his scheme for a "World-State" achieved either by consensus or by the hegemony of one "resolute, unscrupulous and dominant State." Within a democratic system, people could be educated so that "the public mind" would "be turned to other thoughts and future elected governments would then be less inclined to wage wars." The media could exert an influence for the betterment of humankind, but an alternative outcome was not overlooked:

> Those who control the press and the cinema can now in a few years inoculate the mass with any requisite opinion, whether poisonous or prophylactic. Democratic governments, intending a course of action, do not in modern conditions offer that measure directly. They prepare the way for a while by propaganda, judiciously exhibiting selected materials, arranged and timed to appear so as to produce a desired effect upon the passive minds of their populations, who presently find themselves thinking what they were meant to think, as they imagine, of their own mere motion. If the national leaders are sincere in their professed desire to abolish war they must proceed by this propagandist method, and they might then perhaps be successful.

However, Bateson had mixed feelings on democracy because, to him, the path to extra-nationalism required a deep reverence for the contributions of the exceptional:

> The progress of civilization has resulted solely from the work of the exceptional men. The rest merely copy and labour. By civilization I mean, here as always, not necessarily a social

> ideal, but progress in man's control over Nature. ... Wars may decide the destinies of nations; short of extermination they do not decide the destiny of *man*. ... And just as science is an extra-national possession, so is great art. Shakespeare, Beethoven, Rembrandt, Raphael, ... the poets and the artists who have seen deepest into the heart of man, the makers of beauty, the creators of delight, the pioneers of emotion; in them shall all nations of the earth be blest. Their calm and mighty works soar eternally beyond the noises of temporal ambition, high above the plane on which the nations grapple. In their presence the voices of the partisans are hushed.

Yet could such gifted individuals survive in a democratic society? For democracy was "a system which confers equal political power on individuals, in defiance of genetic inequality." Bateson had no political alternative to offer. He proposed neither monarchy nor oligarchy, whether benign or tyrannical. In one breath, he criticized democracy, yet in another he proclaimed:

> Does someone say that the minds of common people cannot be made to see ...? Low as may be our estimate of the common man, we can fairly rate him higher than that. We need the help of the writers and poets [to make this clear to him]. What epic theme of Titans or of blood can stand in grandeur beside the story of Promethean man, tearing their secrets from the elements, building, bit by bit, by his genius and toil the dazzling fabric of knowledge by which he shall surely scale the heavens? That is our *Paradise Regained*.

Besides all this, how petty were ideas of nationhood. In one breath, Bateson was declaring:

> As between individuals, so between nations, there is similar inequality. When even in Europe we observe there are teeming populations which have scarcely made any significant contribution to art, learning or science throughout their history. ... There are proletariat races as there are proletariat families and their redemption does not lie in statesmanship. The unequal distribution of illustrious men among the nations is a biological fact.

Yet, in another breath:

> If we inquire what does constitute a nation, we soon find that neither common racial origin, nor identity of language, or of religion, or of manners is essential. The most acceptable definition is probably that which declares that people compose a nation when they feel themselves to be a nation. But if we inquire how they come to share this feeling which has no necessary dependence on genetic relationship, on collocation in space, on common language, or common customs or beliefs, the answer is by no means obvious. National sentiment ... is the power of those whom we call patriots, to given them their more noble title, thus to polarize the peoples. In their wake follow the journalists, the contractor and the manufacturer of armaments. The scene is set. Catch-words are chosen, insults bandied and the play begins.

But biologically, however it may have been in the past, today there was no such thing as a homogeneous community. And since outbreeding was associated with increased vigor (hybrid vigor), this could be a good thing:

> In various degrees the chief nations are mongrelized, and we British are probably the most mongrel of all. I hasten to add that to this physiological fact I believe some of our national efficiency is largely due, partly because much differentiation of type means great variety of aptitudes, an essential to a large industrial community, but also because I am fairly sure ... that in special cases, cross-breeding even in mankind does contribute to vigour.

> The fact however, which I wish more to emphasize, is that by the workings of the phenomenon of genetic segregation a man's children may possess few of the transferable ingredients which characterized him, his grandchildren may possess none at all, and of his collaterals it is practically certain that few will contain so much of his that he need feel any personal satisfaction or humiliation in their performances. …
>
> Looked at coldly in the light of physiological knowledge, what is called the tie of blood is therefore in modern times exceedingly slender, and in all likelihood many of us contain no more of the elements that went to the making of Shakespeare and our heroes than the modern Greek contains of Zeus or Phoebus, despite the frequent alliances which those deities contracted with the daughters of man.

There is here an allusion to the fact that genetic recombination always destroys individuals while usually preserving genes, so all the antics of living and reproducing, subsumed under the name "life," can be seen as aimed at preserving what we now refer to as "selfish genes." While the reaction of Bateson's audience is not known, in the light of the events of the century that followed, readers may feel some wisdom in his closing remarks:

> Let extra-national progress be recognized publicly as the highest and the one indisputable good, in which all may share, and let it be known that in comparison, national pride is small and trivial, and so palpable a truth may not impossibly spread among the leaders of men. Fashion will do the rest.
>
> Those who contribute to extra-national advancement are certainly in all ages few, but though separated in time and space, *they* truly have ingredients in common. The bond which unites them is a thousand times more real than that which unifies a modern nation. If their collective consciousness could be awakened, as that of each separate nation has been, it might constitute a definite force for the direction of public opinion.
>
> Truth and beauty, science and art, wisdom and loveliness, these are extra-national possessions. They are the only aims which in the long run are worth pursuing. They are the treasure whose glory cannot pass away.

Nature Jubilee

Fifty years after its inception, in 1919, a jubilee issue of *Nature* appeared (November 6) with articles by Lankester, de Vries, Ewart, Bateson, and many others. The actors were old ones, and they knew their lines well. Lankester recalled Huxley beginning lectures and laboratories in elementary biology in South Kensington, with three assistants – Foster, Lankester, and Rutherford – and with input from Thiselton-Dyer on plants. He related how in 1869 he had gone to Vienna to learn from Stricker methods of sectioning and embedding biological materials for microscopic study. Without naming Caldwell, he praised "English workers" for "the development of the microtome and the methods of producing long ribbons of consecutive sections." His concluding remarks suggested that old antagonisms were still alive:

> Various serious attempts have been made to improve upon or to add to Darwinian theory, perhaps to its detriment. One example of this is Romanes's notion of physiological selection. Another is the attention given to the experiments and conclusions of the Abbé Mendel. Mendel's conclusions differ but little from those contained in Darwin's own work, …. No doubt the breeding experiments which are now carried out in the name of Mendel might equally well be performed in the name of Darwin. The importance of this work was little assisted by those interested in Mendelism, when in the early days they called it a "new science."

The title of de Vries' contribution was "The Present Position of the Mutation Theory." He still held that his evening primrose results were "quite analogous to the species-producing steps of Nature." Bateson's paper on "The Progress of Mendelism" declared the Morgan group's observations "the most attractive of all," but being limited to the fruit fly the extent to which they applied to plants was still unclear. Various studies, including the rogue phenomenon, showed that the "moment or moments of segregation" were not necessarily always related to meiosis. Finally, he returned to the "weak spot" Huxley had so often referred to. Although sterile offspring could be consistently produced when allied species were crossed:

> We are still without any incontrovertible example of co-derivatives from a single ancestral origin [i.e. organisms from the same species] producing sterile offspring when intercrossed. Thus, one of the most serious obstacles to all evolutionary theories remains. The late R. P. Gregory's evidence that tetraploid primulas, derived from ordinary diploid parents, cannot breed with them, though fertile with each other, is the nearest approach to the phenomenon, but the case, though exceptionally interesting, does not, of course, touch this outstanding difficulty in any way.

Summary

Bereft of staff, the wartime John Innes Institute analyzed rogue and chimeric forms for evidence that segregation of genes occurred prior to the reduction division of meiosis. The rogue phenomenon was an example of what is now known as epigenetic inheritance. Bateson visited France to lecture to the troops. Otherwise, he continued editing the *Journal of Genetics* and gave public addresses of increasingly political content. While not a pacifist, he ascribed the war to misplaced patriotism and a competition for markets. He challenged the democratic principle but suggested no alternative. With little enthusiasm, he supported Lloyd George rather than Asquith. He proposed an expansion in science education and recognition of our common humanity ("extra-nationalism") centered round the pursuit of "natural knowledge." But the growing understanding of genetic principles indicated that there were limits to what education ("nurture") could achieve. The Morgan school's *The Mechanism of Mendelian Heredity* was praised, but the book was mainly based on studies in a species (fruit fly) where males did not undergo genetic recombination. Moreover, the homogeneity of chromosomes seen under the microscope made it difficult to imagine they could have material heterogeneity. Nevertheless, studies of Gregory and Winge suggested that there were differences between parental chromosomes that, in extreme form, could impede their meiotic pairing in offspring, resulting in sterility. Thus, the line would come to an end. The parents would be reproductively isolated from each other. If the offspring were polyploid, each chromosome would have a pairing partner and the line could continue. In the last days of the war, John Bateson was killed.

Chapter 17
My Respectful Homage (1920–1922)

I am to be in Canada and U.S.A. for a few weeks at Christmas. Great hopes of my conversion are entertained ... in certain quarters.

Bateson to Julian Huxley (October 20, 1921).

Honorary degrees from universities and honorary memberships of foreign learned societies flowed in. The RS conferred the Royal Medal in 1920. The new generation that had learned Mendelism as part of the accepted wisdom of its time, looked to Bateson for guidance. He kept them on their toes. Julian Huxley's inductions were not permitted to stray beyond the facts merely to make the tale simpler. Haldane had to iron out the ambiguities in his famous "rule" paper. Sewall Wright's breakfast musings caused an explosion. But the new generation that had arisen around Morgan in the West continued to keep Bateson on his toes. To it a respectful, but qualified, homage was due.

Genetical Society

A major forum for the meeting of the generations was the Genetical Society. Largely at the instigation of Saunders, the Society got underway in 1919 (June 25) in rooms of the Linnean Society. Many of the people encountered in these pages (26 in all) gathered in Burlington House to propose that Bateson take the chair, and that Arthur Balfour be invited to become the first president. Elected officers were Bateson, Saunders, and Arthur Sutton (all vice-presidents), and Pellew and Punnett (as secretaries). The minutes of the meetings in the early 1920s were clearly recorded in (presumably) Pellow's handwriting and were countersigned by Bateson. They relate that the Society met in various places, including commercial nurseries (e.g., Allington Nursuries near Maidstone on the invitation of Edward Bunyard), and museums and research stations (e.g., Rothamsted Experimental Station on the

A. G. Cock, D. R. Forsdyke, *Treasure Your Exceptions*,
https://doi.org/10.1007/978-3-030-92099-9_17

invitation of John Russell). The Society had no formal connection with the *Journal of Genetics*.

The first meeting was at Cambridge (July 12, 1919). There were demonstrations, and talks by Punnett and Haldane, now a Fellow of New College Oxford and author of a new paper in the *Journal of Genetics* that opposed the reduplication hypothesis [1]. The proceedings concluded with dinner at St. John's College. The third meeting (February 14, 1920) was at the John Innes Institute. There were 44 present with 11 "visitors" that included Beatrice. The group then traveled to London for dinner at the Pall Mall Restaurant. At the fourth meeting (April 21, 1920), there was a discussion on "the heredity of sex in plants and animals." Gates gave "a communication on the relation of chromosomes to sex determination emphasizing the significance of the cytological features observed by Morgan, Bridges and others."

Also at that meeting: "Mr. J. B. S. Haldane made a communication on sex ratio and unisexual sterility in hybrids. He stated that in all the cases which he had been able to collect the following rule holds: if one sex is absent or infertile in a species cross, it is the one which from other evidence may be inferred to be heterozygous in sex." This was a first airing of the famous "Haldane's rule," which turned out to be of high relevance to the problem of the origin of species [2]. Bateson "congratulated Mr. Haldane on the detection of this new and remarkable principle which, on first hearing, appeared consonant with the evidence." This was affirmed by Mr. Lewis Jones and Miss Florence Durham. Julian Huxley, a grandson of Thomas Huxley, "put forward a possible explanation of Mr. Haldane's rule, with some observations on the influence of metabolism in modifying the sex ratios."

The biochemistry of breeding was touched upon by Cunningham, who "spoke on the origin of somatic sexual characters, explaining the hormone theory, especially in its bearing on the development of secondary sexual characters." From this, it can be discerned that the membership included those who bred, observed, and counted (i.e., the "Mendelians"), and another group more interested in the nature of underlying character differences. These two categories roughly approximate to what became known as transmission genetics and developmental genetics. This was to lead to organizational problems (Chap. 22). Later, there emerged a third category – population genetics – the study of the spread of Mendelian factors through populations [3]. In this respect, it should be noted that the "visitors" at the fourth meeting included Ronald Fisher from Rothamsted (at the next meeting, he was elected a member).

The 11th meeting was the most exciting. It was held in 1922 on the occasion of the visit of Morgan and Sturtevant to the John Innes (see Chap. 18). The 13th meeting (February 17, 1923) was a visit to Lord Rothschild's museum at Tring:

> The Society had the privilege of inspecting a magnificent series of Lepidoptera illustrating polymorphism, geographical variation, and multiformity of the sexes, and demonstrations of special points were given by Lord Rothschild and Dr. K. Jordan. In the afternoon, Dr. R. Goldschmidt gave an account of his experiments on the production of intersexes in *Lymantria*.

On this occasion the "visitors" list included "Mrs. Bateson," the new "Mrs. Hurst," and the John Innes Institute's new cytologist "Mr. Newton" (W. C. Frank Newton was elected to the membership at the next meeting). Bateson invited the visitor from Germany, Richard Goldschmidt, to stay overnight with them at the John Innes, adding, "Do not trouble about clothes [i.e. formal dress]." He wrote (June 19) congratulating Goldschmidt on his new book on *Ascaris*: "The work, as a whole, is certainly remarkable and important – especially when one reflects upon the difficult conditions in which it was carried out, and much congratulation is due to you on it. In case you may write again, I enclose a coupon."

The 15th meeting was held at Oxford in 1923 (July 15): "In the Hope Department Professor Poulton kindly arranged a very fine exhibit of Lepidiptera illustrating sexual polymorphism, mimicry, and seasonal dimorphism." Poulton was himself listed as a "visitor," together with Beatrice and others.

Haldane's Rule

Having already noted the tendency of hybrid sterility to affect one sex more than the other (Chap. 8), Bateson had a considerable hand in getting the rule paper into shape for the *Journal of Genetics*. Haldane submitted a draft in late 1920, and Bateson replied (December 1):

> After reading your MS [manuscript] the impression remains that you have come on a principle of value – I mean in the distinction between the homozygous and heterozygous sexes in their liability to sterility or suppression. The paper nevertheless as it stands strikes me as somewhat post-impressionist. I should have preferred *either* a note making your main suggestion, *or* a properly developed and documented paper. Your materials are mostly of a topical description. The literature on hybridism is considerable. Do you mean to collect and analyse it? If you set out Harrison's Bistora in detail, surely more should be made of Standfuss. It is disconcerting to see such a familiar book quoted as "from Colleague" .... Then you refer to Goldschmidt's lecture Amer. Nat. vol. L (1917 should be 1916), but not to his main work .... This is not a mere point of pedantry. I don't see in your paper any adequate treatment of the *intersexual males*, which have steadily assumed great importance in G's later publications. They don't fit quite easily into your scheme, nor I suspect into Goldschmidt's. ... The sex-linkage part of your paper is scarcely so solid as the other. There are of course hints of something coming there, but I don't see what.

There followed numerous detailed suggestions. Haldane replied (undated):

> Many thanks for your notes and apologies for my delay in answering, due to laziness, pressure of work and a small operation on my nose. I will get the Kuhn-Archiv and get Standfuss's Works (not to be had in Edinburgh) out of Poulton. Daphnia is interesting. I also have to read Goldschmidt's book before I send you that paper again. I am just off for a fortnight in the south of France. Remember me to your wife and son and Miss Pellew.

The paper steadily approached its final form, but there were several last-minute additions (undated letter to Bateson):

> I am so sorry, but Punnett has just sent me the reference to another species cross which fortunately obeys the rule. ... There seems to be no doubt of the sterility of the one bull, so

> it seems worth including, especially as the number of cases in mammals and Diptera is not as large as I should like. … I should be very glad if you could tell me of any other cases for or against my views that you know of. I have, I admit, left out Durham and Marryat's result(s) for Pink-eyed ♂ x Black-eyed ♀ (namely 7 ♂, 19 ♀), which are against me, as the numbers are not very large and sex-ratios seem pretty wild in canaries. But if you think I ought to put it in I will. I am sorry to trouble you, but the questions raised are so important that it is worth collecting all the evidence. I only refrained from giving the bad evidence in full because it would fill a good many pages, and is almost all either in Guyer's or Whitman's paper, but if you think it would look impressive I don't mind putting it in.

The paper, entitled "Sex Ratio and Unisexual Sterility in Hybrid Animals," appeared in the October 1922 issue of the *Journal of Genetics*. In another study, Haldane assumed that progressive melanism in moths was adaptive (Chap. 6). From the available data, he calculated a high "selection coefficient." Whether he attempted to publish this in the *Journal of Genetics* is not known. More likely, he anticipated opposition from Bateson and prudently submitted it to the *Transactions of the Cambridge Philosophical Society* [4].

An article by Julian Huxley on "The Regulation of Sex" in *The Athenaeum* (January 14, 1921) provoked an immediate, but private, response from Bateson (January 17):

> In the Athenaeum this week I read an article over your initials which gave me rather a shock. It contains not one, but several, expressions in the most positive way, for which, as far as I know, no real evidence exists. One passage, on which the rest largely depends, strikes me as absolutely misleading.
>
> In the third paragraph you say "Extended microscopical work co-operated with the results of genetic experiment to show that a special modification of the Mendelian mechanism – in modern terms, of the chromosome apparatus – existed in practically every type of higher animal and plant to ensure that the sexes should arise in roughly equal numbers in each successive generation"! How can you justify this? We know practically nothing of the inheritance of sex in higher plants. In the birds and the mammals (man) we may perhaps claim to know from genetic observations that [there is] inheritance of sex, but I am not aware of anything beyond the unconfirmed work of Guyer (which in so far as birds are concerned happens to point the wrong way) at all supporting your statement as to the chromosome apparatus, which is almost exclusively based on a few insects.
>
> The two paragraphs about insects give no hint of a caution. You give Goldschmidt's speculations as statements of fact without a word to indicate how flimsy the whole case is, nor any mention of the prodigious difficulties besetting his interpretations – as for instance the appearance of *fertile* females as sisters to the otherwise all-male families, or of the paradoxical transition from *sterile intersexual females* to *fertile males*, which happens when we pass from "Höchstgradise distersexualität" to "völlige Geschlechten examlehr." If you were here I would like to take the article sentence by sentence and cross-examine you on it.

Huxley mounted a spirited defense but had to admit that the article was meant for the general reader, and he supposed this gave him license to go beyond the facts.

Meanwhile, the usual round of activities continued. In 1920, Bateson gave the Croonian lecture (see below), and in June he gave evidence to a Prime Minister's Committee on the teaching of the Classics (Chap. 23). In the archives, there is the typescript of an address labeled "Kew" (February 19, 1921) suggesting a trip to Kew Gardens. In the same month, he gave the Galton Lecture (Chap. 15). In March 1921, there were lectures at Croydon (March 1) and at the National Physics

Laboratory in London (March 21). Later (May 13), there was an address at the Royal Institution on "Sex." In September 1921, he joined Beatrice and Gregory who were visiting the Raverats in the south of France. At that time, Martin entered the Royal Academy of Dramatic Arts.

Beginning on November 3, 1921, there were lectures on "Genetics" at London University. The cryptic notes for the lectures read:

> (i): Began as study of Variation. Mendelism made much of that insipid, and so became study of heredity. Were relieved when we got *mutations*, a term much abused from the first. Breeding work has revealed state of things utterly unexpected.
> (ii): Selection theory fails at spec. diff.
> (iii): Theory of linkage ... *Drosophila* features. 4 chromosomes haploid. About 150 pairs all in 4 linkage series. ... Maps ... allowance for double crossing over and interference.
> (iv): Sex chromosomes.
> (v): Non-disjunction.
> (vi): Haldane's rule ... somatic segregation.

Segregation

Segregation continued as a major concern. Bateson set out his aims in his Croonian Lecture, which was published both in the *Proceedings of the Royal Society* (1920) and in *American Naturalist* (1921):

> The later developments of Mendelian analysis have been ... an attempt to elucidate the scope and nature of segregation. Mendel proved the existence of characters determined by integral or unit factors. Their integrity is maintained by segregation, the capacity, namely, to separate unimpaired after combination [in a diploid organism] with their opposites. Our first aim has been to discover specifically what characters behave in this way, whether there is any limit to the scope of segregation, or [whether there are] any characters or class of characters which are determined by elements unable to segregate simply. The second object has been to decide the time and place in the various life-cycles at which segregation occurs.

He began by noting that substantive (e.g., color) differences could often be analyzed in Mendelian terms, but repetitions of parts (meristic differences such as in finger number), and continuous quantitative differences, were sometimes less amenable. Was there any difference between the segregation of unit factors (genes) corresponding to characters that systematists had found of value in distinguished species, and the segregation of unit factors corresponding to other characters?

> The supposition that segregation is concerned solely with characters of a superficial or trivial nature has been long ago disproved. Baur's antirrhinums, the study of which was continued by Lotsy, were an excellent demonstration to the contrary, for they provided many illustrations of segregation in features, the "specific value" of which [utility for classification] no systematist would question. ... Many geneticists are inclined to the view that segregable characters should be pictured as implanted on an irreducible base which is outside the scope of segregation, but no means have yet been devised for testing the reality of the conception.

As for "the time and place in the various life-cycles," the "moment," when segregation occurred, Bateson now began to backtrack a little: "Though unconvinced, I cannot deny that linkage and crossing-over may well be represented provisionally as effected during synapsis. The scheme previously offered by Punnett and myself as a diagrammatic plan capable of representing these phenomena is certainly far less attractive. … It must be granted that no indication that gametic linkage results from somatic differentiation has yet been obtained."

In other words, the "reduplication hypothesis" of coupling was wrong. Nevertheless, he remained adamant that from the evidence in plants "it is clear that in a wide view of living things segregation cannot be exclusively a property of the reduction division, and for the present, it should be regarded as a possibility which may occur at any division in the life-cycle." He also remained adamant that "there is no body of evidence that the number of linkage systems agrees with that of the chromosomes, a primary postulate of Morgan's theory. *Drosophila* is the only example which has been adequately investigated." Moreover, Bateson noted that Wilson, Morgan's colleague at Columbia, was also recommending caution, noting that "the genetic development of the chromosome theory has far outrun the cytological" [5].

Segregation was also the topic of Bateson's Leidy Lecture in 1922 (see below). The lecture, however, was not published (in his own *Journal of Genetics*) until 1926 and contained many post-1922 references, including one to a 1923 paper [6], where Johannsen had referred to the "Presence-and-Absence hypothesis as 'now abandoned'." But in the foreword to the lecture, Bateson noted that since 1922 he had "learned nothing which appears to dispose of the view to which I am inclined." Referring to the pre-Mendelian era, he added: "We do well … to remember that that long spell of dullness from which we were so lately emancipated, ensued as the direct consequence of a too facile acquiescence to impermanent doctrines. Curiosity was too easily allayed. We are in no such danger yet, but … even as regards the outline of genetic principles, finality has not been attained."

Again, there were two aims: "First, what in the act of segregation separates from what? Secondly, when, at what moment or moments in the life-history of an animal or plant, does this separation occur"? For the homozygous state, segregation resulted in "the formation of two similar products." For the heterozygous state, two dissimilar products resulted. We were still "wholly ignorant … of the proximate elements upon which these powers depend." He still referred to the elements as allelomorphs, or alternative forms of an element. In many cases, there were just two alternatives in a species, so both could occur in a diploid individual if it were a heterozygote, but only one if it were a homozygote (i.e., two identical copies). In some cases, there could be several allelomorphs, distributed among various members of a species, with never more than two per diploid individual. Sometimes the corresponding characters were manifest as a quantitative series. The phenomenon of "multiple allelomorphs" is now referred to as *polymorphism*.

Bateson referred to the Presence-and-Absence hypothesis as the "quantitative conception," or a "quantitative system of representation," or "the quantitative mode of expression," and continued to defend it as "in such complete harmony with the tenor of the evidence, that I have no serious doubts of its correctness." For, even when criticizing Presence-and-Absence, Morgan had stated that "it is a characteristic of 'multiple allelomorphs' that the same character is affected." Bateson agreed:

> This is true, and the fact is of great significance. The statement should be amplified, and may be put in the form that … factors composing a multiple allelomorphic series produce an effect in degrees quantitatively different. The factors composing any such series may thus be arranged in a descending scale. This is exactly what is expected. The factor put in on one side is absent, wholly or in part, from the other, and, on segregation, the two qualities which respectively were combined in fertilisation, reappear. Why Morgan should declare that "only one kind of absence is thinkable," I do not understand.

Arguing along the same lines as Hurst, who had represented absence as an empty rectangle (see Fig. 10.1 in Chap. 10), Bateson continued:

> We should not assert that because a sovereign is absent from a purse, that [the] purse must contain nothing. … We have been asked [by Morgan] to conceive of each allelomorph in any pair as equally a positive something. In a multiple series, are we then to suppose that, as the allelomorph *of each grade* in the series, a corresponding positive something exists? Such an interpretation verges on absurdity, and the multiple series are a clear demonstration that allelomorphism is not a haphazard relationship between two independent features.

The reader of today may see Hurst's empty rectangle, Bateson's empty purse, and Morgan's positive something, as indicating a DNA molecule with a series of base "letters" (Part VI). The presence of a particular base at a particular position may correlate with one character (the purse is full). The absence of that base (i.e., often replaced by an alternative base) may correlate with the disappearance of that character (the purse is empty). But the DNA molecule with its series of bases (the purse) *remains*. Bateson did, however, concede: "In one of their contentions my critics are right. The course of the evidence may convince us that of each allelomorphic pair one is positive and the other negative, but we have not yet the means of distinguishing with complete certainty which is the positive and which the negative."

Goldschmidt

Among those who offered to provide German translations of *Principles* was Richard Goldschmidt, an assistant to Richard Hertwig at the Munich Zoological Institute. He began corresponding with Bateson around 1910. Unfortunately, a compliant publisher could not be found. At a conference on Evolution in Munich in 1911, Goldschmidt, speaking on "Evolution in light of recent genetics," expressed enthusiasm for the "entirely new spirit" kindled by the genetic approach and fully subscribed to the role of chromosomes. Bateson wrote (January 25, 1913):

> As you may know, I am always inclined to scepticism as to the chromosomes being the *only* bearers of hereditary factors and I will believe when I *must*, but I should like to wait till then. … I am ashamed to say that it is only in the last few weeks that I have come to a proper acquaintance with your own text book, of which I lately heard much in Holland. I wish I had read it before. … My own book is sadly out of date now. Soon it must be rewritten in great

> measure. I like most of your book. Nevertheless, I incline to be much more critical towards Tower, MacDougal, and above all, Kammerer. I shall not believe these stories till they are independently confirmed. My greetings to Hertwig.

Goldschmidt replied (January 27): "I am familiar with your scepticism toward the chromosome theory, but I believe that the day is near when you will have to endorse the theory." Despite his enforced stay in the United States during the war, Goldschmidt had been busy, and another book appeared in 1920. This treatise on sex determination, *Mechanismus und Physiologie der Geschlechtsbestimmung*, was reviewed by Bateson in *Nature* (1921). Considering Goldschmidt's studies of "intersex" types among gipsy moths, Bateson cautioned the need to disentangle the "influence which primarily causes … [sexual] differentiation to proceed in one direction rather than in the other," from "proximate mechanisms by which the effects of sexual differentiation are produced."

Shortly after his meeting with Goldschmidt at the Tring Museum (see above), Bateson received a letter from Morgan (March 23, 1923) saying that he was "interested in getting your opinion of Goldschmidt. We have known him long and under a good many different conditions. I have, as you know, a rather high opinion of his work, as work, but believe, as apparently you do also, that he runs off the track on theoretical matters."

To the Fly Room

The first edition of Wheldale's *Practical Plant Biochemistry* appeared in 1920 and soon went into further editions. A letter of thanks from Bateson (May 28, 1920) indicated that chromosomes and vibratory ideas were seldom far from his thoughts:

> Your handsome book is very pleasant as a gift, though as you well know, it is mostly on a plane which I can never reach. We have got a copy for the laboratory where I don't doubt it will be really used. All the same I have just been trying to get something out of it. Morgan's "genes" must, I think, be approaching molecular size, and if that is so, would they not be under-going movement sufficient to alter their positions sensibly? I don't know what sort of matter chromosomes are when living. Do you? I should think they must be gels – probably with more fibrous material running through them. May not very sensible movement along their long axes take place? Don't trouble to write if my question is unanswerable, or merely foolishness, which is more likely.

But by the end of 1921, the time for letter writing was over. The issue had to be confronted. Armed with advice from Binyon as to where to seek art treasures, Bateson sailed (December 10) on the Cunard Line's *Scythia* for the land of the automobile, prohibition, disdain for public smoking (especially by women), and the chromosome. It was his fourth, and last, visit to America. Morgan met him at the dock in New York (December 19) and took him to his house, which included teenage children. Bateson's notebooks record that during the visit: "T. H. M. bets

W. B. £1 that England prohibits alcohol (viz. spirits, wine and beer) before the U.S. prohibits tobacco." A letter home (December 20) reported:

> Morgan has a rough good nature that attracts, but I have just the same impression I got 14 years ago, that he is of no considerable account. His range is so dreadfully small. Off the edge of a very narrow track he is not merely puzzled but lost utterly. Yet there is no denying the fact that by intensive methods they have got a long way. He is totally free from pretence – is almost without shame in his ignorance – I mean of things scientific, but curiously enough I note a sensitiveness when he is convicted of less professional lapses.

He then proceeded to criticize the food, Morgan's attire, and the dirty hands of the son who nevertheless redeemed himself as "one of nature's engineers" by working "wonders with the light fittings."

On the first morning, Morgan took him to the famous "fly room" where he met Sturtevant and Bridges, and in the afternoon he "stole an hour" at the Metropolitan Museum and went to the New York Aquarium (December 20): "Coming out of the Aquarium we were run into by a man who shook my hand warmly. ... I had to ask him who he was. H. G. Wells (Doing the Peace Conference for the N. Y. Sun). We talked for some time. I had, as on former occasions, to remind him that I was *not* BATHER! He knows me right enough, except the name, so it isn't as bad as it sounds." Wells' recollection of the meeting is given in Chap. 25.

After more peering down microscopes, he wrote to Beatrice (December 24): "Most of each morning devoted to chromosomes. I can see no escape from capitulation on the main front. The chromosomes must be in some way connected with our transferable characters. About linkage, and the great extensions, I see little further than I did." Likewise, he wrote to Miss Pellew: "Cytology here is such a commonplace that everyone is familiar with it. I wish it were so with us. Bridges inspires me with complete confidence. Sturtevant I have only seen casually as yet. ... I wish I liked Morgan better. I think he has made a great discovery, but I can't see in him any quality of greatness." High regard for Morgan's young colleagues was reinforced (December 26):

> Saw something of Sturtevant yesterday, and thoroughly liked him. He and Bridges are quite different from the type I expected. Both are quiet, self-respecting young men. Sturtevant has more width of knowledge. Bridges scarcely leaves his microscope. I wonder whether they are not the real power in the place. Morgan supplies the excitement. He is in continual whorl and very active and inclined to be noisy. Nothing that he says is really interesting or original. I like him and I don't. ... I am heartily glad I came. I was drifting into [an] untenable position, which would soon have become ridiculous. The details of the linkage theory strike me still as improbable. Cytology however is a real thing – far more important and interesting than I had supposed. We must try to get a cytologist.

It might be thought that, in suggesting Sturtevant and Bridges were the "real power," Bateson was influenced by antipathy to Morgan. However, the view accords with that of another Morgan associate, Muller, who later portrayed him as being dragged struggling into the twentieth century by his three innovative research students [7]. The same point was made by Wilson years later: "Morgan's three greatest discoveries ... were ... Bridges, Sturtevant and Muller" [8].

Breakfast at Ithaca

After Christmas with the Morgans, Bateson headed for Toronto (December 27) and Morgan remained in New York. The meeting of the American Association for the Advancement of Science opened the next day (see below). Bateson wrote (December 29) from the York Club, where he stayed for the next two weeks save for an excursion to Ithaca:

> Last night's address went well. I had to announce my conversion on the main point: "that chromosomes are definitely associated with the transferable characters," is how I express it. Much enthusiasm over this of course – but as a candid man I don't see how any other view can now be maintained. … In absolute whirl here. … Evening after speech, etc. shook 60-70 hands, some horny, 2 sweaty. … Enthusiasm intoxicating and supplies place of drink very fully for a while.

After the meeting, Bateson and Wright took the train east to continue discussions with biologists at Cornell University. In view of his quest for "minds of first-rate analytical power" (*Problems*, 1913), it is likely that Bateson had high hopes for young numerate biologists such as Wright. As later related by Wright [9], there was a stormy breakfast at an Ithaca hotel, where Bateson "was by now fed up with Wright's quantitative analysis and bitterly exploded, condemning the use of biometrician-like statistics in genetics." Writing to Beatrice on the way back (January 1, 1922) Bateson related that he had met the "corn men" at Ithaca, who desired to rival the "fly men" of New York.

> Spent today at Ithaca with them and others who came on with me there. I liked Emerson very much. They are pathetic in their simplicity, knowing nothing whatsoever outside Genetics. They are some 150 machines for grinding out genetics. It can't go on like this. Genetics is the very top of the fashion. I have had more speaking to do than in Australia, I think. Dined with the Sigma Xi, with the Am. Soc. Naturalists, with the Am. Soc. Zoologists, and each time was turned on – this last a more formal occasion in which I did less well, having to produce my own positive contributions which were a bit meager. In the orthogenesis discussion I also wore a bit thin, I fear. Osborn irritated me with repeating the old stories, and I set him down. He was very cross and I did perhaps show a touch more temper than was courteous. … Bridges traveled with me to Toronto and on to Ithaca – two long days' train. He and the others talked Genetics of the utmost technicality nearly the whole day yesterday in the train for 9 hours. I detached three to play weakish bridge to the great surprise of the whole party (one of our players was an archeologist and one an entomologist – only one geneticist joined out of many).

There was an overnight stay in Buffalo, and he wrote the next morning (January 2): "What really is beautiful is the simplicity of at least those types who are my colleagues here. They literally are children. They throw themselves into what they are doing exactly like children making mud pies. One can't help admiring it – though *à la longue* I should be terribly bored." Later in the day he was back at the York Club, where he was to remain while giving lectures with his "Mendelism without tears" slides. He reported (January 4) on his first lecture: "Room held about 350 naturally – quite 400 in it, at least 50 standing all the time and they say 500 turned away! Today we move to Convocation Hall which holds some 1300. As I must soon speak of prophases, mitochondria, etc, we shall probably soon return to the 350 room en route for the coal hole which may suffice for the 5th discourse." The second lecture had an audience estimated at 1100 (January 5): "It is settled. I am a great *popular* lecturer. That's me, take it or leave it. … You were quite right to make me bring rudimentary slides. I wish I had more. People dull and boring for the most part. Religion pulls the strings."

The next day there came a dinner at the Toronto Faculty Club, where he gave an address on "Twins." Some surviving rough notes contain a list of examples of twins and also the words "ripple mark," "general implications. Vortex", "idea of node and internode and of wavelength." The following day (January 7): "Taken on a drive through environs yesterday by a widow – deeply wealthy – who called me 'Sir' all the time. If intended to keep me at a distance, think superfluous. … Canada, judging by Toronto, *very* distant from US. Like a tame England. … Not at all like Australia. Much quieter. I fancy myself joining the US, but not, oh not, Canada." There was another popular lecture on January 9: "Today the Luncheon Club and my shouting lecture in the big hall. … They still hope to hear me black Darwin's eye, I believe, but spirits evidently fell last time when I told them how to map a chromosome (an art which I still imperfectly understand by the way)."

As on his previous visit, he now headed west, but only as far as the University of Michigan (January 11). Then there was a return to New York. After an interlude with friends (Chappells, Allyns) at New London, Connecticut, he went on to Boston with the Wheelers (January 17): "Mrs. Wheeler is a great force here. She is an Emerson of purest blue – but a large human soul." He saw Castle at the Bussey Institute and gave a popular evening lecture to the Boston Natural History Society. Mrs. Wheeler took him to Wellesley College for lunch. He treated the ensuing address as an after-dinner speech and smoked a cigar during his discourse. "My ancient friend Miss Wilcox came over to meet me – I never liked her 38 years ago. She hadn't moved an inch except towards the grave, and even that progress scarcely perceptible."

There was a lecture at Yale (January 19), where he stayed at the Graduates' Club, but the following day in New York the strain began to tell: "Davenport met my train … but I am voiceless and shall take day in bed. …Voice very creaky at Yale yesterday and lecture moderately dull. Same lecture night before in Boston had gone well." He remained at the Pennsylvania Hotel in New York (until January 23), canceling a proposed lecture at Columbia University, but got to Cold Spring Harbor. Finally, invited by McClung, there was a trip to Philadelphia (January 24). He visited the Wistar Institute and the University of Pennsylvania, gave the Leidy lecture (see below), and then sailed from New York on the *Scythia* (January 26). Even so, certain important bases had not been touched. He wrote to Shull (March 3): "To miss Princeton was a great disappointment to me." And in a later letter (August 2) he noted: "Muller and his friend [probably Altenburg] by an unfortunate chance I missed." However, Muller recalled that they had met [10].

Evolutionary Faith

On the evening of December 28, 1921, in Convocation Hall at the University of Toronto, Bateson gave the plenary address – "Evolutionary faith and modern doubts." With reluctance, he was persuaded to read a 500-word précis to the press in advance. Quite possibly some reporters filed their copy based on the précis alone. In any case, the coverage was extensive and in his files there are 20 clippings from US and Canadian newspapers. He began in classic style:

> The unacceptable doctrine of the secular transformation of masses by the accumulation of impalpable changes became not only unlikely but gratuitous. … The evolutionist of the eighties was perfectly certain that species were a figment of the systematist's mind, not worthy of enlightened attention. Then came the Mendelian clue. We saw the varieties arising [in our breeding studies]. Segregation maintained their identity. The discontinuity of variation was recognized in abundance. Plenty of Mendelian combinations would in nature pass the scrutiny of even an exacting systematist and be given "specific rank."
>
> In the light of such facts the origin of species was no doubt a similar problem. All was clear ahead. But soon … less and less was heard about evolution in genetic circles and now the topic is dropped. When students of other sciences ask us what is now currently believed about the origin of species we have no clear answer to give. Faith has given way to agnosticism for reasons which on occasions such as this we may profitably consider.

Mendelism had thrown light upon the origin of varieties, but not of species. Bateson considered this had come about because "as we have come to know more of living things and their properties, we have become more … impressed with the inapplicability of the evidence to these questions of origin." But for many in Bateson's audience, this, his major point, would have been passed as just opening chatter, leading up to his much-awaited pronouncement on chromosomes:

> We have turned still another bend in the track and behind the gametes we see the chromosomes. For the doubts – which I trust may be pardoned in one who had never seen the marvels of cytology, save through a glass darkly – cannot, as regards the main thesis of the *Drosophila* workers, be any longer maintained. The arguments of Morgan and his colleagues, and especially the demonstrations of Bridges, must allay all scepticism as to the direct association of particular chromosomes with particular features of the zygote.
>
> The transferable characters borne by the gametes have been successfully referred to the visible details of nuclear configuration. The traces of order in variation and heredity which so lately seemed paradoxical curiosities have led step by step to this beautiful discovery. I have come at this Christmas season to lay my respectful homage before the stars that have arisen in the west.

He was to elaborate on chromosomes at an after-dinner address to the American Society of Zoologists the following evening [11], so continuing his plenary address he returned to the species problem, which remained unresolved. Most present would have thought he had left the chromosome issue. Actually, the species problem was at the very heart of his chromosomal skepticism:

> Variation of many kinds, often considerable, we daily witness, but no origin of species. … That particular and essential bit of the theory of evolution which is concerned with the origin and nature of *species* remains utterly mysterious. We no longer feel as we used to do, that the process of variation now contemporaneously occurring is the beginning of a work

> which needs merely the element of time for its completion; for even time cannot complete that which has not yet begun.
>
> The conclusion in which we were brought up, that species are a product of a summation of variations, ignored the chief attribute of species that the product of their crosses is frequently sterile in greater or lesser degree. Huxley, very early in the debate, pointed out this grave defect in the evidence, but before breeding researches had been made on a large scale no one felt the objection to be serious.

Then, harking back to his Centenary essay of 1909, Bateson considered the type of change that might generate sterile hybrid offspring when one organism mated with another of an allied species. Whereas for most variations, including those in the large collection of the *Drosophila* workers, it appeared that "the elements have been lost," the much rarer "variations by addition," were probably at the root of hybrid sterility:

> If species have a common origin, where did they pick up the ingredients which produce this sexual incompatibility? ... The variations to which inter-specific sterility is due are obviously variations in which something is apparently added to the stock of ingredients. It is one of the common experiences of the breeder that when a hybrid is partially sterile, and from it any fertile offspring can be obtained, the sterility, once lost, disappears [i.e. the offspring on further crossing produce fertile offspring].

This was a narrowing, perhaps to simplify, of the scope of the "ingredients" compared with that given in *Problems*. In his 1913 text, the complementary ingredients causing incompatibility had been held capable of working either positively or negatively (by inhibiting something that exerted a positive effect). Nevertheless, Bateson stuck to his theme that whatever the ingredients were, they were nothing to do with what had come to be known as genes:

> Analysis has revealed hosts of transferable characters [i.e. characters that can be transferred in breeding studies]. Their combinations suffice to supply in abundance series of types which might pass for new species, and certainly would be so classed if they were met with in nature. Yet, critically tested, we find that they are not distinct species and we have no reason to suppose that any accumulation of characters of the same order would culminate in the production of distinct species.

So, what were the ingredients? "Specific difference therefore must be regarded as probably attached to the base [residue] upon which these transferables are implanted, of which we know absolutely nothing at all. Nothing that we have witnessed in the contemporary world can colourably be interpreted as providing the sort of evidence required." And "nothing that we have witnessed" included, of course, *chromosomes*.

Another old theme was the parallel to be drawn between the evolutionary process that divided two groups within a species such that they become *two species*, and the evolutionary process that divided two groups within a species such that they become *two sexes*:

> We now have to admit the further conception that between the male and female sides of the same plant these ingredients may be quite differently apportioned, and that the genetic composition of each may be so distinct that the systematist might without extravagance recognise them as distinct specifically [as if male and female forms were distinct species]. If then our plant may by appropriate treatment be made to give off two distinct forms, why is not that phenomenon a true instance of Darwin's origin of species?

Thus, understanding the process of differentiation that led to two sexes might help us understand the process of differentiation that led to two species. Finally, Bateson returned to the agnosticism theme, but it is likely that the reporters were already rushing to the phones:

> I have put before you very frankly the considerations which have made us agnostic as to the actual mode and processes of evolution. When such confessions are made the enemies of science see their chance. If we cannot declare here and now how species arose, they will obligingly offer us the solutions with which obscurantism is satisfied. Let us then proclaim in precise and unmistaken language that our faith in evolution is unshaken. … Our doubts are not as to the reality of evolution, but as to the origin of *species*, a technical, almost domestic, problem. Any day that mystery may be solved.

The Toronto address – soon printed both in *Science* and *Nature* – generated a flurry of correspondence in both the general and the scientific literature. Writing in 1989, Alan Cock refrained from summarizing the address "because I do not fully understand some of Bateson's arguments" [12]. Can it be wondered that decades earlier Bateson had been misunderstood?

First to record his objections in print was Osborn (*Science*, February 24). The address of "Professor William Bateson the distinguished representative of the University of Cambridge and British biology" has made "a very regrettable impression" [13]. This charge was supported by lengthy quotations from newspapers. The morning after the address, headlines in the Toronto *Globe* had declared: "Distinguished Biologist from Britain Delivers Outstanding Address on Failure of Science to Support Theory that Man Arrived on Earth through Process of Natural Selection and Evolution of Species." Correspondence in the same paper the following day was headlined "The Collapse of Darwinism." Osborn held that such reports "could not be dismissed as mere newspaper talk of no import. They are called forth by the fact that many of the statements in Bateson's address … are inaccurate and misleading, especially those relating to the origin of species, natural selection, and the infertility between species." Osborn was reminded of an admonishment of T. H. Huxley "that before delivering any of his popular addresses he very carefully wrote out every word he intended to say, lest in the heat of enthusiasm … he might say something which would give a wrong impression of the truth."

In a brief reply – "Genetical analysis and the theory of natural selection" – Bateson stood his ground: "The divergence between the conceptions to which genetic analysis introduces us and the doctrines of which Professor Osborn has been so long a distinguished champion is indeed wide." He disclaimed the excuse that Osborn had proffered: that speaking extempore, he had failed to say exactly what he meant. He had followed virtually verbatim the written text "without serious modification."

Two months later, another critical voice was raised – this time from Bateson's side of the Atlantic – in an article by "Our scientific correspondent" in the *Times* [14]. The article, likely by Chalmers-Mitchell, had two themes: (i) recent attempts to suppress the teaching of evolution in the United States and (ii) popular confusion between Darwinism (i.e., natural selection) and evolution, and the way this had played into the hands of the US fundamentalists. He lamented that "those keen about the pursuit of knowledge are tempted to be indifferent as to its diffusion. … This attitude is unwise, and is really a survival of the time when knowledge was regarded as the property of a [higher] caste, [and was] not to be shared with, or … [understood by], the people. It is most foolish when, as in the case of evolution, a scientific doctrine is in conflict with popular prejudices or endowed opinions of which there are always champions ready to confuse the issues in ignorance or in optimism."

Mitchell mentioned "Professor Bateson" in only one paragraph, relating how William Jennings Bryan had exploited Bateson's Toronto address in his propaganda. "No one except Bateson and a few of his group had been so foolish as to expect" that Mendelism would "explain the origin of species." People were "most concerned about the descent of man from lower forms of life. In that sense of Darwinism, the fact of evolution is accepted by every biologist of scientific repute in every country of the world. … But because we know that a thing has happened, it does not follow that we know what brought it about."

Bateson responded two days later with a letter on "Darwin and evolution: limits of variation," in which he defended his skepticism, his crucial point being the failure so far to obtain by experimental breeding variants that yielded sterile hybrids when intercrossed. Because his Toronto audience had consisted largely of qualified people, he had scarcely thought it necessary to guard against the "comic misinterpretations" that followed. This invoked a tart rejoinder from Mitchell [15]. Bateson, he said, must indeed be "modest" if he thought his "pontifical statement at Toronto as to the failure of 'all theories of evolution based on descent with modification,' could have been confined to a specialist audience." He brandished a quotation from Osborn – as "one of the highest living authorities" – to show that "even zoologists did not accept Bateson's pronouncements as a calm statement of the esoteric difficulties in their subject."

Chromosome Complementarity

The matter did not stop there. In June 1922, there were two thoughtful letters in *Nature*. The first was from Cunningham, who since the 1890s had been alone in criticizing Bateson for failing to acknowledge Romanes (Chaps. 5 and 14). Now, he was almost alone in attempting to deal substantively with his arguments [16]. Finding Bateson's reasoning "very difficult to understand," Cunningham thought that

> the idea of a specific base distinct from specific characters seems merely false metaphysics. How can we conceive of an organism without characters, or characters without an organism? Perhaps Mr. Bateson means that unit characters such as those which can be transferred [as genes] in Mendelian crosses might all be taken away, and still an organism would be left with non-Mendelian characters. What are these characters? He does not tell us.

As for the origin of new characters: "Mr. Bateson's difficulty seems to be merely that we do not know how they came into existence. We can, however, scientifically form the conclusion that they originate by some change or development in the chromosomes, not directly dependent on any corresponding external stimulus." Cunningham also challenged "the sterility of species hybrids" as "the 'chief attribute' of species. It is neither a universal nor necessary characteristic, and all we can say is that we do not know how it arises in certain cases." Cunningham's main concern was that Bateson's address had

> implied disparagement of those who have [in fact] not ceased to discuss evolution. There is more in evolution than the origin of species. Mr. Bateson himself has contributed largely to the proof that the distinctions between species have nothing or little to do with adaptation, but at the same time he has failed to realise the true nature and importance of adaptation itself. … The origin of species is a very important problem, but it is not the whole, or the most important part of evolution. The origin of adaptations is not the same problem as the origin of species, and the methods of modern genetics have very little bearing upon it. To this subject biochemistry and physiology and, yes, "early Victorian" notions of recapitulation, had much to contribute.

The second *Nature* correspondent, C. R. Crowther, a Plymouth-based physician, had come much closer to grasping Bateson's message. He was concerned with the mechanism of chromosome pairing that would require some form of lock-and-key complementarity, which he saw in terms of a sword fitting into its scabbard [17]. After noting, like Cunningham, that Bateson's reasoning was "very difficult to follow," Crowther pointed out that

> homologous chromosomes … have to co-operate to produce the somatic cell of the hybrid, and their co-operation [for development] might be expected to require a certain resemblance; but for the production of sexual cells [for gametogenesis] they must do more, they must conjugate; and for conjugation it is surely reasonable to suppose that a much more intimate resemblance would be needed. We might, therefore, expect, on purely theoretical grounds, that as species and genera gradually diverged, it would be increasingly difficult to breed a hybrid between them; but that, even while a hybrid could still be produced, a fertile hybrid would be difficult or impossible, since the cells of the germ-track would fail to surmount the meiotic reduction stage when the homologous chromosomes conjugate. This is exactly what happens: the cells go to pieces in the meiotic phase.

Crowther cited Bateson's remark that "the conclusion that species are a product of a summation of variations, ignored the chief attribute of species, that the product of their crosses is frequently sterile in greater or less degree." This left Crowther "frankly puzzled," for "the proposition is certainly not self-evident." Surely, if the sterility of an offspring were due to a failure within that offspring of homologous chromosomes to conjugate, it mattered little whether the lack of complementarity responsible for that failure was produced by one large variation, or by the summation of many smaller variations. That Crowther was here thinking of primary variations occurring at the chromosome level, rather than visible anatomical variations of the sterile individual, was made explicit: "If a sword and a scabbard are bent in different directions, it will happen sooner or later that the sword cannot be inserted, and the result will be the same whether the bending be effected by a single blow, or whether it be, in Dr. Bateson's words, 'a product of a summation of variations.' Is this illustration apt? The sword and the scabbard are the homologous chromosomes."

Next, perhaps believing that Bateson had not moved on from the monstrosities arising in one generation as set out in *Materials* in the 1890s, Crowther opted for the lack of complementarity between pairing chromosomes as being most likely due to a summation of their variations:

> It would even seem that the argument is exactly contrary to Dr. Bateson's statement of it: it seems easier to imagine sterility arising from a gradual modification, spread over a length of time, and involving many chromosomes, than from the half-monstrous variations chiefly studied by Dr. Bateson and his school, variations which appear to affect only a few chromomeres, and those by loss alone.

Modern readers may see here a hint of the idea of a chromosomal "accent" that must match between homologues for pairing to succeed (see Chap. 5 and Part VI). Bateson replied to Cunningham and Crowther in one letter, entitled "Interspecific sterility." He did not elaborate on what he had meant by "the base upon which these transferables are implanted," but his argument that "we still await the production of an indubitably sterile hybrid from completely fertile parents" was repeated:

> Dr. Cunningham has taken exception to my speaking of this interspecific sterility as the chief attribute of species, but he will not dispute that it is *a* chief attribute of species. The races of fowl might, as he holds, on account of their enormous divergences, be without impropriety compared to natural species, ... but inasmuch as they do not show interspecific sterility they do not help us to understand how that peculiar property of species arose in evolution. In contemporary variation we witness the origin of many classes of differences, but not this; yet by hypothesis it must, again and again, have arisen in the course of evolution of species from a common ancestry. The difficulty is no new one; but I emphasized it because naturalists should take it more seriously than they have done hitherto.

With Crowther, he had much to agree:

> It is, as he says, not difficult to "imagine" interspecific sterility produced by a gradual (or sudden) modification. That sterility might quite reasonably be supposed to be due to the inability of certain chromosomes to conjugate, and Mr. Crowther's simile of the sword and the scabbard may serve to depict the sort of thing we might expect to happen. But the difficulty is that we have never seen it happen to swords and scabbards which we know to have belonged originally to each other. On the contrary, they seem always to fit each other, whatever diversities they may have acquired.

So, "now that a great deal of experimental breeding is in progress, watch should be kept for such an occurrence. I by no means declare that the event cannot happen, but, so far as I know, it has not been witnessed yet." Cunningham had trotted out the usual instance of the instant speciation observed by de Vries in his evening primroses, but Bateson still thought that "the [general] applicability of that example is exceedingly doubtful."

Qualified Conversion

Bateson's Toronto "conversion" was not so dramatic as it sounds [18]. He sought "a comprehensive theory of heredity." He sought not only a theory explaining why the characteristics of two child cells either were, or were not, the same as those of the parent cell that had divided to produce them, but also a theory explaining why the characteristics of offspring either were, or were not, the same as those of the parent or parents from whom they were derived. Such a theory would encompass both the anatomical differentiations (development of the embryo. i.e., "ontology") that had been dealt with in *Materials* and the species differentiations that might result in new species and higher taxa ("phylogeny"). In essence, there was very little change from his previous position, and this was summarized in his Leidy lecture in Philadelphia:

> Having in view the various facts and considerations here enumerated I think we shall do genetical science no disservice if we postpone acceptance of the chromosome theory and its many extensions and implications. Let us distinguish fact from hypothesis. It has been proved that, especially in animals, certain transferable characters have a direct association with particular chromosomes. Though made in a restricted field this is a very extraordinary and most encouraging advance. Nevertheless, the hope that it may be safely extended into a comprehensive theory of heredity seems to me ill-founded, and I can scarcely suppose that on a wide survey of genetical facts, especially those so commonly witnessed among plants, such an expectation would be entertained. For phenomena to which the simple chromosome theory is inapplicable, save by the invocation of a train of subordinate hypotheses, have been met with continually, as even our brief experience of some fifteen years has abundantly demonstrated. Through all this work, with ever increasing certainty the conviction has grown that the problem of heredity and variation is intimately connected with that of somatic segregation, and that in an analysis of the inter-relations of these two manifestations of cellular diversity lies the best prospect of success. Pending that analysis, the chromosome theory, though providing much that is certainly true and of immense value, has fallen short of [the] essential discovery.

Nevertheless, he did take steps, albeit at a late hour, to remedy the John Innes Institute's deficiency in cytology. In 1922, he appointed Frank Newton, a student of Helen Gwynne-Vaughan, a leading cytologist. Newton's contributions were cut short by his death in 1927. On the recommendation of E. S. Salmon of the South Eastern Agricultural College at Wye, a second person, Cyril Darlington (1903–1981), was appointed in 1923 and encouraged by Bateson to join Newton in the cytological work. Darlington became highly eminent in the field and eventually director of the John Innes [19]. He, among others, carried the banner of the chromosomal (as opposed to genic) theory of the origin of species through the twentieth century [20]. When interviewed by Alan Cock in 1973, Darlington wryly recalled that Bateson would be quite willing to spend a few hundred pounds on rare botanical books for the Institute library but was reluctant to spend a lesser sum on a first-rate microscope.

Nothing displays so clearly Bateson's reservations about chromosome theory than his correspondence with the protozoologist Clifford Dobell, who wrote (May 20, 1924):

> I don't know the value – on the purely genetical side – of much of the evidence advanced by the Drosophilologists: indeed, I can't read a lot of their stuff. But on general grounds,

> and on the cytological side, I regard Morgan and Co. with contempt. The whole thing is a colossal stunt – one of the biggest bluffs they have ever put on even in America. I can't help thinking that if somebody, who understands their breeding experiments with *Drosophila* – I certainly don't – could get their real figures and analyse them properly, the whole of the evidence would evaporate away.

Bateson replied (May 22):

> We here, who have to face the thing every day, are with you that in the crude form propounded by Morgan, it won't do. But we are equally clear that a great part of it – probably the main part – is substantially true. We are not at all inclined to believe in the "rosary" theory of crossing over. Nevertheless the transferable characters, almost beyond question, are somehow specially attached to the chromosomes. I don't like it, but I see no way of escape. Like you, I can't read Drosophila papers in more than a superficial way.

As for bluffing:

> I used to think as you do; though with the long experience I had from 1890 onwards of imposters of several types I got to know the class pretty well and always felt that the Columbians were not of that sort. They are conceited and cliquish beyond belief, but they are not out to deceive themselves or us. Their weakest point is their profound ignorance of anything but the topical and trite in genetics outside Drosophila, and their complete satisfaction with their ignorance. ... There are endless reasons why the chromosomes should not do all the things that they are supposed to do, but these reasons have somehow to be reconciled with the fact that, by representing the chromosomes as doing these things, we get a consistent representation not merely of linkage, etc., etc., but of a great many quite unforeseen facts, e.g. those associated with the idea of non-disjunction.

Dobell was only moderately chastened (May 25):

> I wrote in my usual flippant and disrespectful fashion .... I certainly did not mean to say – or imply – that Morgan and Co. are imposters. I have never thought that. When I called their work a "stunt," I meant that it was a fine show – and precious little else. ... Anyhow, I take comfort from what you say in your letter: for it is clear that you have not been properly converted. ... The most extraordinary thing in your letter is your remark that "like me, you can't read *Drosophila* papers in more than a superficial way." This absolutely stumped me! ... (If you really thought this was the true gospel you would have it all by heart by now.)

With his views perhaps moderated, in 1925 Dobell published a paper entitled "The chromosome cycle of the Sporozoa considered in relation to the chromosome theory of heredity." Unable to imagine what is now referred to as the differential expression of genes in different tissues, he considered the sex-linked red eye of the fruit fly [21]:

> Now in the course of development the X chromosome is divided and distributed to every cell in the body of the insect. It is finally present not only in the cells forming the red eye, but also in those forming the legs, the wings, the gut, and all the other organs. The X chromosome therefore has no specific relation to the form or function of any particular cell which possesses it. It is impossible to grant that it has any particular relation to the red eye. For this to be so, it would be necessary to show that the X chromosome is present in the eye and in no other organ.

However, as Johannsen had articulated in 1923 (echoing de Vries' *Intracellular Pangenesis*), the "genotypical factors" are "represented throughout the individual" and it is "local conditions" that determine whether one gene will be expressed while another is not expressed [6]:

> The properties or rather the possibility of their realisation (I should say [of] the genotypical factors in question) may be represented throughout the individual; the local conditions in the different regions of an organism may prevent their appearance – and in many cases a special property can only be obviously manifest in specialized organs, for instance the colour of the iris in the eyes, the special negro-pigments in the surface of the body and so on.

Johannsen, like some in Germany [22], was one of those who continued to resist the chromosome theory (Chap. 18). He postulated a "*Grundstock*," a core of the genotype ("*das Zentrale*") possibly located in the cytoplasm [23]. This was something more profound than the inheritance of cytoplasmic organelles (mitochondria, chloroplasts). In 1930 in his *Theories of Development and Heredity*, the marine biologist and biohistorian Edward Russell (1887–1954) dealt at length with the reservations concerning chromosomes of Johannsen and Dobell [24] – but Bateson was not mentioned.

Honors and Sadness

Punnett was going from strength to strength, and his text *Heredity in Poultry* was to appear in 1923. In 1922, the issue of nominating him (but not Saunders) for the Darwin medal arose. In 1917, the United States had entered the war, and Osborn received the medal in 1918. Biffen had it in 1920. Bateson wrote to the RS Secretary, W. B. Hardy (February 12, 1922):

> I feel great embarrassment about the Darwin medal question. Already on two occasions – I am not sure there have not been three – I have drawn up a "statement of claim" on Punnett's account, at the request of members of Council. On one occasion a grotesque award was made to Osborn, and on the other, one to Biffen, which though satisfactory to the laity is unjustifiable on the facts. Useful as Biffen's work has been, it is trifling as a contribution to Science. While Punnett and Morgan have not received this award, to give it to lesser men is not creditable to the Society, and no great compliment to the recipients. ... Our partnership was a successful one, but if this kind of thing happens, it becomes too dangerous for a junior to enter into a partnership with a senior. Neither of us contributed much more than the other. ... I put Punnett before Morgan because it was Punnett's work which made Morgan's possible.

This time Punnett got the medal.

In April 1922, only a few months before an important meeting of the Genetical Society at the John Innes (Chap. 18), the Bateson's were devastated when Martin Bateson committed suicide (Chap. 19). Beatrice noted in her *Memoir*: "Months of despondency and dejection followed. Even the garden and laboratories almost failed to rouse him." In May, Bateson resigned from the University Grants Committee on being elected a Trustee of the British Museum. Beatrice noted: "This honour gave him extraordinary pleasure; it was an expression of confidence that he felt he must stir himself to justify. No foreign travel or medical *régime* could have helped him as did this unexpected appointment. Gradually, from fortnight to fortnight, his interest in his new responsibilities grew, and with his interest grew his pleasure." At that time (May 22), he was offered a knighthood, which he declined.

There were other unhappy events. A letter from Bateson to Wheldale (July 4) expressed regret at the death of her husband Onslow (1890–1922). The obituary by Punnett in *Nature* [25] described how in his teens Onslow had become paralyzed below the waist following a diving accident. Although confined to a wheelchair, he had turned a room in his Cambridge house into a laboratory and had worked with Wheldale, under Hopkins' guidance, on the genetics of plant pigments. Another death was that of David Sharp. Bateson wrote to Beatrice (September 1): "He was a very good and useful friend. Without him I don't suppose I should ever have made any start with entomology. … Johannsen comes on Sunday afternoon till Monday lunch." Apart from Johannsen in September, and Morgan and Sturtevant in June (Chap. 18), there was also a visit by Davenport in October 1922. In August, the Batesons joined Gregory on holiday in Switzerland shortly before his entry into St. John's College as an undergraduate.

Summary

Despite his Victorian-Edwardian past, Bateson largely understood the bizarre postwar world. Unfortunately, it did not understand him. In many quarters, his salutation of the work of the Morgan school and his recognition that the chromosomes were carriers of Mendel's units were seen as capitulations. However, his speeches were skillfully framed to reveal his abiding dissatisfaction. Still unresolved was the problem of how the limits of a species were maintained yet could be transgressed to produce a new species. Perhaps there was an "irreducible base" or "residue" upon which characters subject to Mendelian analysis (i.e., segregatable or "transferable" characters) were implanted. Understanding the differentiation that led to two sexes might aid the understanding of the differentiation that led to two species. His caveats being disregarded, remarks at a Toronto conference were seized by the press and provided ammunition for those who opposed the teaching of evolution in schools. Cunningham thought Bateson's irreducible base was "false metaphysics." Bateson conceded to Crowder that he had been wrong to imply that small variations in chromosome structure, by their cumulation, could not bring about the incompatibility that would become manifest as hybrid sterility and hence would have the potential to originate a species. Haldane presented his famous "rule" observation at a meeting of the newly founded Genetical Society, which had no formal connection with the *Journal of Genetics*. Bateson was greatly depressed by the suicide of Martin Bateson in April 1922.

Chapter 18
Limits Undetermined (1923–1926)

I never had an argument with him – and I had many – without the absolute conviction that he would no more hesitate to admit himself in the wrong if I could convince him, than to tell me that I was talking nonsense if, as was more usual, I failed to do so.

J. B. S. Haldane [1]

Knowledge of limits can help a decision between alternative scenarios. If limits are undetermined, few scenarios can be dismissed. When the age of the earth was construed in thousands of years, divine creation appeared plausible, but when construed in billions, Darwinian evolution came into its own. In his dotage, Bateson continued to insist that there might be an irreducible "base" or "residue," of unknown extent, that might be beyond the limits of conventional Mendelian analysis.

Major Hurst and Rose Ploidy

The war had increased support for research on homegrown foods, which included poultry products. Since newborn chicks were half male and half female, an easy method of sexing chicks appeared advantageous. Enter genetics. Punnett showed how, through judicious crosses, males would emerge red and females white. This method did not come into use until long after the war, eventually ceding to the method, long practiced in the Far East, of noting the size of the copulatory organ in the cloaca. It seems the Cambridge geneticists had not discovered this.

While much of Hurst's stock was lost during the war years, poultry experiments were continued at Burbage by J. B. Perkins with funding from the Ministry of Agriculture's Development Commission. In 1919, Hurst returned, and in 1921 his final reports on the genetics of egg production and fecundity in fowl [2, 3] were presented to the National Poultry Society (July 20), to the First World Poultry Congress at The Hague (September 5), to the BA in Edinburgh (September 13), and

A. G. Cock, D. R. Forsdyke, *Treasure Your Exceptions*,
https://doi.org/10.1007/978-3-030-92099-9_18

to the Second International Congress of Eugenics in New York (September 22–28). Hurst then abandoned chicken work, which continued to be a major interest of Punnett. With Michael Pease and the support of the National Poultry Institute (founded at the end of the war), Punnett expanded the operation at the Cambridge Genetical Research Institute. In 1930, a separate poultry unit was established under Pease. So, Punnett lost his "Punnett."

In 1922, Hurst was married again, to his cousin Rona (1897–1980), and began the study of rose cytogenetics as part of the Doctor of Philosophy program just instituted at Cambridge. For this, he received an honorary Bachelor of Arts degree. Bateson had encouraged Hurst to get a cytologist, not thinking that he might want to do the cytology himself: "The real difficulty in cytology begins in the interpretation of the sections, not in the cutting of them. It is very easy to go wrong without years of experience. However, I know you have plenty of courage." So, at the age of 52, Hurst became a "fellow commoner" at Trinity College, which gave him rights to dine at high table and other privileges. Rona related in her unpublished text, *The Evolution of Genetics* [4], that on first attending dinner "he was jovially entertained by a delightful farmer-like person," who turned out to be the physicist Ernest Rutherford. While it would seem appropriate that Hurst work with Punnett, the latter retained some hostility toward cytogenetics. Hurst joined Seward at the Botanical School and grew plants in the Botanical Gardens allotments that had been used by Bateson and Saunders.

In 1923 at meetings of the BA at Liverpool (September 14) and of the Genetical Society in London (December 8), Hurst presented his results: "Chromosomes and characters in Rosa and their significance in the origin of species" [5]. Various polyploid rose species appeared as multiples of a basic set of seven chromosomes (haploid septet). He distinguished five "duplicated forms" (diploid species) each displaying distinctive character sets:

> The tetraploid species showed the combined characters of two distinct diploid species, while the hexaploid species showed the combined characters of three distinct diploid species and the octoploid species showed the combined characters of four distinct diploid species. ... These observations suggested that the four double septets in the octoploid species were not simple reduplications of the double septet of a single diploid species, but, on the contrary represented the four differential septets of four distinct diploid species in combination.

The five distinct haploid septets of chromosomes were designated A, B, C, D, E, which when duplicated as diploids (14 chromosomes) were designated AA, BB, CC, DD, EE. Each of these could be correlated with a certain character set. Various combinations produced the polyploid types (e.g., octoploid: AA + BB + CC + DD, or AA + BB + CC + EE). A decaploid species with five double septets (70 chromosomes) was not observed. Hurst concluded:

> The five differential diploid species of *Rosa* are real discontinuous species, with no intergrading or transitional forms, since each has its own set of at least 50 taxonomic characters which are distinct from the other four sets of characters. Each has a septet of chromosomes which is double in the somatic cells [e.g. AA] and single in the gametes [e.g. A] and

> presumably the five differential sets of taxonomic characters are represented in the five corresponding differential septets of chromosomes.

Thus, it appeared that, just as in a regular diploid (e.g., AA) where the paternally derived septet (A) and the maternally derived septet (A) existed in a common nucleus (e.g., A + A), yet they retained their individualities, when sets of diploids came together in a common nucleus (e.g., AA + BB), they also retained their individualities. But whereas during gametogenesis the paternal and maternal contributions to a diploid set could partially *lose their individualities* due to meiotic recombination (e.g., recombination between paternal A and maternal A, and between paternal B and maternal B), each diploid set would *retain* its overall individuality, recombining *within itself* but *not* with other diploid sets in the *same* nucleus (e.g., no recombination between A and B). Indeed, "as a rule, each septet works equally with, but independently of, the other septets with which it may be associated. So far no cases of blending nor of the general dominance of one septet over another have been found."

In other words, although existing within the same organism and sharing a common nucleus, failure to recombine meant that the different septets in a polyploid plant were *still as reproductively isolated from each other as when they had existed alone in independent diploid rose plants* (i.e., when they no longer shared a common nucleus). We now see this as showing that reproductive isolation does *not* necessarily require physical isolation. Despite being within a common nucleus, two sets of chromosomes can still be isolated from each other to an extent sufficient to prevent inter-set recombination. *Thus, isolation can be a property of the chromosomes themselves* (see Part VI).

Also of interest to Bateson was Hurst's claim (unpublished) that the seven chromosomes could line up in strings, indicating a level of organization above the chromosomes themselves. Bateson queried this since his cytologist, Newton, thought that if it were a general phenomenon others would have noticed it. Bateson considered the implications of the string idea for the distinction between species and varietal characters (August 9, 1923):

> I can't follow when you speak of *group* characters. Supposing that one can distinguish group characters from varietal characters. I gather that you mean that *qua* string the chromosome material carries the group characters and *qua* chromosomes it carries the varietal characters. Is that right? I can see this may be true. But the distinction of group from varietal characters always seems to be a very uncertain matter.

Here, Bateson perhaps pondered whether the concept of a residue (beyond genes) was a function of some higher-ordered structure. Again (August 17):

> I can't see how, granting that the chromosomes have the attributes which we must recognise as belonging to them, fresh attributes are to be assigned to them by reason of their being in strings. ... As I understand, the various differences in composition, *inside* AA etc., should give the Mendelian varieties, not species. Is that right? I am lost somewhere. ... Of course I have a strong disposition to believe that species, etc., are predestined from the beginning, and that some such account as yours *should* apply – but I can't see how it does apply.

In January 1924, Stanley Gardiner, on discovering how cramped Hurst was at the Botany School, and having seen a preliminary draft of his thesis, offered him a room in the Zoology laboratory.

Morgan at Merton

The 11th and "most memorable" meeting of the Genetical Society was at the John Innes Institute on June 2, 1922. The minute book records that the visitors included Morgan and Dr. and Mrs. Sturtevant. There were various demonstrations and talks, and Lancelot Hogben of Edinburgh University was elected a member:

> At 4-30 the Society reassembled in the Library, the Chair being taken by the Earl of Balfour, President of the Society. Professor Morgan then gave an address on "Mutants in *Drosophila*," and was followed by Dr. Sturtevant, who instituted a comparison between *D. melanogaster* and *D. simulans*. With his accustomed felicity the President offered the thanks of the Society to its distinguished visitors, and delivered a short address on the trend of evolutionary thought. In thanking the President for his address, W. Bateson raised the problem of the nature of species, and in the discussion which followed diverse speakers offered diverse contributions. Eventually came to an end the most memorable meeting in the annals of the Society.

Years later, Rona, who was then Hurst's fiancée, described her "somewhat awesome experience" of "initiation into the genetical world":

> We were introduced to the famous visitor, and seizing Hurst's hand he exclaimed "Why Hurst! I thought you were killed in the war – I haven't heard from you for ages!" ... He was an attractive man of good appearance and pleasant manners – at the time he struck me as being all brown, brown-suited, brown-haired and brown-bearded. He was naturally very pleased by his great reception in England where he visited round to the various research places and the Royal Society put on a big show for him.

Then they wandered into the greenhouses to see exhibits of work in progress that Saunders was demonstrating:

> Here we soon encountered Miss Saunders, very imposing in her dark tailored suit and shirt blouse with stiff linen collar and tie, surmounted by a severe black hat, in fact the typical Blue Stocking as envisaged in pre-war days, and indeed the fashionable dress of all women who claimed equality with men .... To my horror, Hurst was taken away by a colleague and I was left alone with the Gorgon. She nobly started explaining an intricate experiment to me and I did my best to look intelligent, but it was Greek to me at that time. It was a blazing hot day and the temperature in the greenhouses was tropical. I began to wonder uneasily how long I could endure it in spite of my thin silk frock as against her costume. Suddenly she turned and flashed upon me that wonderful smile which could irradiate her whole face and was so much part of herself. "Come on!" she said, "Let's get out of here – I shall be a grease-spot in a minute!" Her use of colloquial slang astonished me – she too was human after all; I was her devoted slave from that moment and through the future years when our roses shared with her stocks the famous genetic allotments used by Bateson at the Botanical Garden .... She was shy and reserved behind all her apparent composure, but when one got to know her she was a woman of great charm and culture, allied to the magnificent brain which had backed Bateson up quite selflessly and unassumingly from the beginning. Her

> careful and well-designed experiments had always formed a steady and solid background to the general genetic story. Hurst had a great respect and admiration for her, regarding her as a wonderful stabilizing influence in the earliest and most difficult years of the research in heredity.

Soon, it was time for tea and more talk:

> Presently, the "inner circle" was invited in to tea in the very attractive Manor House, and I perforce with them, but by this time I had lost my fears and enjoyed watching the clever hostessing of Mrs. Bateson, tall, handsome and golden-haired, keeping all her guests – even me – in the picture. Morgan gave an interesting discourse on the Drosophila work, much above my head at this time, although I was deeply intrigued and inspired by it, having read his books I could at least follow the main points. He had brought microscopical slides of the Drosophila chromosomes and for the first time I peered down microscopes at these minute bodies which were to become the main focus of my life for some years to come.

Then there came the parting:

> Bateson had been very helpful in getting Hurst in at Cambridge but there was no chance of private conversation till the end when he made the excuse of coming out with us to show us the right bus and so seized a few minutes. ... He waved us goodbye and stood for a moment on the pavement, a rugged figure. One hears of leonine men; Bateson certainly was one both in appearance and behaviour, disguising his sharp observation under an almost sleepy appearance and rather drooping head, to be awakened to a sudden alertness at once by anything of importance. "Isn't he marvellous?" said Hurst, as the bus carried us away.

Scopes Trial and Intellectual Freedom

Bateson's Toronto address heightened the smoldering antagonism to the teaching of evolution in the United States that would soon influence various State legislatures. The speeches of William Jennings Bryan made it a campaign issue. Six thousand people flocked to the New York Hippodrome (April 2, 1922) to hear him denounce Bateson, among others, in an address on "God and Evolution." The *New York Times* quoted Bryan the next day: "It is more important to know the Rock of Ages [a hymn] than the age of rocks. ... Geology is good in its place, but it does not come first. ... Recently an evolutionist refused to eat lunch with me in Kentucky. If I can afford to eat with a man who says he is the son of an ape, what possible objection should he have to me?" The matter came to a head in August 1922 when F. E. Dean, a superintendent of schools at Fort Sumner in New Mexico, was dismissed, and later there was to be the famous trial of John Scopes.

In 1923, *Nature* asked Bateson to comment on "The revolt against the teaching of evolution in the United States." He reminded readers of the caveats with which he had closed his Toronto address, pondering whether it really mattered "that the people of Kentucky ... should be rightly instructed on evolutionary philosophy." He restated his concerns that democratic rights might be exercised capriciously by those who did not have access to, or were unable to evaluate, the relevant information:

> The chief interest of these proceedings lies in the indications they give of what is to be expected from a genuine democracy which has thrown off authority and has begun to judge for itself on questions beyond its mental range. Those who have the capacity, let alone the knowledge and leisure, to form independent judgements on such subjects have never been more than a mere fraction of any population. We have been passing through a period in which, for reasons not altogether clear, this numerically insignificant fraction has been able to impose its authority on the primitive crowds by whom it is surrounded. There are signs that we may be soon about to see the consequences of the recognition of "equal rights," in a public recrudescence of earlier views.

By 1925, 15 states had introduced legislation to ban the teaching of evolution. In Tennessee in February, a bill was enacted making it unlawful "to teach any theory that denies the story of divine creation as taught by the Bible and to teach instead that man was descended from a lower order of animals." In July, the Scopes trial began and *Nature* in a supplement entitled "Evolution and intellectual freedom" gave various theologians and scientists, including Bateson, an opportunity to air their views. Bateson considered that "the Tennessee trial is something more than a curiosity in the history of civilisation," but rather it was "the symptom … of a strain in the social fabric which sooner or later may end in catastrophe," for "if the true convictions of our own people could be ascertained, I do not suppose they would be found to be very different from those of Tennessee."

As to the "still larger considerations which lie behind," such as "whether a State stands to gain or lose by the encouragement of intellectual freedom in comparison with others which control or suppress truth," this was "a problem on which political philosophers have exhausted the arts both of eloquence and sophistry." Certainly, "scientific men are not required to pronounce," for "no universal solution, independent of time and place, can be expected." He concluded: "Our liberty is vital; and to suppose that movements of this magnitude in the United States have no significance for ourselves is to cherish a very dangerous illusion."

Hormone Jargon

Bateson wrote to Beatrice while on an art expedition to Amsterdam (March 25, 1923). But, perhaps indicating that the shock of Martin's death was still evident, that year there were fewer separations and hence fewer letters between them. Gregory was now in his first year at Cambridge, dutifully studying biology as his parents had hoped, and correspondence with him increased. In early 1923 (March 2), there was an exchange concerning Gregory's purchase of a secondhand microscope. Then, the Cambridge Natural History Society invited Bateson to Kammerer's address concerning the nuptial pads (*Brunstschweilen*) of the midwife toad (Chap. 21). Bateson wrote to Gregory (April 22):

> I have begged off, saying that my criticisms are well known, and that since K. did not appear at the Vienna Congress, he evidently has nothing further to say. I should think the invitation came as an amazing surprise to K. who is I understand discredited in Germany as much as here. … As regards Alytes, it is obvious that the right way to investigate the

> question is to make a careful case study of the natural wild species right through the year, and then to show when and how the "Brunstschweilen" are formed in detail, and exactly on what parts, etc, etc. We were first told they come on the thumbs – just as a little ornament to the story. Now we are told they come anywhere on the arms. I wonder if any come on the legs too? (He has had 12 years in which to do all this since he was challenged.) Even if the story were true, to regard such an occurrence as any proof of the transmission of adaptation – in the sense in which the conception is invoked to explain the evolution of specific forms – seems to me absurd.

Bateson thanked Muriel Onslow (May 20) "for [an] interesting account of Kammerer's meeting." There was continuing intellectual engagement with Gregory (May 28):

> Your questions and comments have set me reading K's *Proteus* paper. ... I am taking up the *Alytes* story again. To my eye the dark mark on the *Alytes* hand was not in the least like a *Brunstschweilen*, and I don't believe it was one at all. Moreover, it was most certainly in a place where no *Brunstschweilen* could be. I understand my letter will be in *Nature* this week.

Bateson's commentary on Gregory's undergraduate thesis (July 16, 1923) is indicative of the growing division between Mendelism (transmission genetics) and the underlying physiology:

> I have looked through a good deal of your dissertation and consider it remarkably well done. ... I want to keep clear of the "hormone" jargon. The idea can be expressed without it. To us the word has pretentious associations. It is typical of the physiologists' way of thinking – attending to the *means* by which an effect is brought about, and diverting attention from the fundamental factors which put the system into operation and contribute the means.

The usual demands were attended to. There was a lecture in Birmingham (April 26, 1923), a visit to the Chelsea Flower Show (May 29–31), a visit from Nilsson-Ehle in June, and the BA meeting in Liverpool. There was a conference in Amsterdam, a Genetical Society Meeting in London (December 8), and a few days later, a lecture at St. Thomas' Hospital.

Willis

Tablets of genetic wisdom had been brought down from Cambridge to Ceylon by Lock in 1903 (Chap. 10). A major beneficiary was J. C. Willis, author of the *Dictionary of Flowering Plants*, and director of the Royal Botanical Gardens at Peradeniya. Willis later moved to a similar position in Rio de Janeiro (1912–1915) and then retired to Cambridge. He was editor of the *Empire Cotton Growers' Review* for which Hurst wrote articles. In 1922, Willis informed Bateson (June 22) that a new multiauthor book, of which he was a major contributor, would shortly be out, and that he and Yule would be presenting papers at a BA meeting in Hull (September 11) as part of a session on "The present position of Darwinism." Bateson declined to attend (June 27):

> Darwin did a fine thing, but the veneration of a great name may become a serious obstacle, and it is absurd to attribute to him a penetration which he did not possess. … I would rather keep out. Deductive reasoning attracts me very little and I mistrust conclusions based on it, though I daresay your ideas are logically sound. I am satisfied that species did not arise as Darwin supposed, but I do not see how they did arise. If you would show me one real species contemporaneously arising, that would interest and please me enormously. Till someone can, I remain an agnostic.

Willis replied (June 29):

> In my paper … I am … pointing out that Darwinism rests upon at least five bold assumptions, for none of which is any proof forthcoming, and then I go on to say that the theory was primarily devised to explain Geographical Distribution, Morphology and Evolution, and if it can explain none it had better be dropped. It has become, as I describe in my book, a "limiting factor" in progress.

Bateson's new Cambridge spy, Gregory, attended a lecture by Yule on "Age and area" at St. John's (October 28, 1922) and sent his father an account the next day.

> I asked him … whether he did not think that it would have been more reasonable to suppose that the speed of growth of a genus depended on the number of *individuals*, rather than on the number of existing *species*. He replied that throughout this work, which was only in its infancy, it had been assumed that the number of species was the important factor.

Willis' book, *Age and Area: a Study of Geographic Distribution and Origin of Species*, was reviewed in *Nature* by Bateson in 1923 under the heading "Area of distribution as a measure of evolutionary age." This "new venture" by Willis advocated "the theory of mutation in its crudest form" and contained an article by de Vries, "which Willis clearly had at the back of his mind" as justifying:

> the bold assumptions lightly made in the doctrine of Age and Area …. [For] from the first the meaning of the Oenothera work was ambiguous. The researches of Renner and of Heribert-Nilsson have now shown that those early suspicions were justified, and that the "mutations" of Oenothera are not genuine illustrations of the origin of species by variation in descent from a pure form. Had de Vries grasped the implications of Mendelian analysis, he could never have so interpreted them. … The few words in which he conveys his benediction on this new venture should be read with caution and reserve by persons unfamiliar with the history they purport to relate.

Present in Bateson draft version, but *absent* from the final review (perhaps because he realized he was mistaken), was a comment on pangenesis (to be discussed later in this chapter):

> The conception of pangenesis was unfortunately the least penetrative of Darwin's ideas, now only a historical curiosity; moreover to suggest that Johannsen's "gene," a brief name for the Mendelian unit, derives from the pangene by trifling abbreviation, implies a lapse of memory which may puzzle the ill-informed. The two words denote wholly distinct conceptions of the hereditary nature of living things, the one obscure [de Vries' pangene], the other certainly in great measure true [gene].

Willis wrote to Morgan, who informed Bateson (March 23, 1923) that Willis had said "you gave a 'fair but generally hostile review of his book'."

The idea of a species spreading in ever-widening circles over the surface of the earth, much as a colony of bacteria spreads over the surface of a culture plate, was something likely to appeal to the mathematical mind. Predictably, the RS contacted Bateson (June 25) when Yule submitted a paper containing "a mathematical theory of evolution based on the conclusions of Dr. J. C. Willis F. R. S." Godfrey Hardy was to review the maths and Bateson (with the assistance of his new recruit Newton) the biology. Bateson thought the assumptions were "so far detached from natural fact that the work ought not to be treated as a serious contribution to Evolutionary Theory." Hardy thought the maths sound, and saw models, however imperfect, as a useful way of approaching a complex topic; however, the assumption that the rate of spontaneous mutation would not depend on the number of mutable units (organisms, species, genera) was questionable. The paper went on to be read at the RS (February 4, 1924).

Johannsen and "Memory"

Johannsen's understanding of evolutionary biology was close to Bateson's [6]. In 1923, Johannsen's article, "Some remarks about units of heredity" (Chap. 17), argued Bateson's most fundamental point that, apart from genes, there was some underlying "base" or "residue," which had some bearing on the question of the origin of species (Fig. 18.1):

Fig. 18.1 Johanssen and Bateson (1923)

> By far the most comprehensive and most decisive part of the whole genotype does not seem to be able to segregate in units; and as yet we are mostly operating with "characters," which are rather superficial in comparison with the fundamental Specific or Generic nature of the organism. This holds good even in those frequent cases where the characters in question may have the greatest importance for the welfare or economic value of the individuals. We are very far from the ideal of enthusiastic Mendelians, viz. the possibility of dissolving genotypes into relatively small units, be they called genes, allelomorphs, factors, or something else. Personally, I believe in a great central "something" as yet not divisible into separate factors. … The problem of Species … does not seem to be approached seriously through Mendelism nor through the related modern experiences of mutations.

And where was that "great central something" if it was distinct from the genes of the Mendelians? Johanssen noted that "*Chromosomes* are doubtless vehicles for 'Mendelian inheritance,' but *Cytoplasm* has its importance too. … Cytoplasm is perhaps more prone to 'memory'." Thus, he shared Bateson's reservations on the role of chromosomes in inheritance, agreeing that they were the location of genes, which were indeed inherited, but that they were not necessarily the location of the "central something" (Johannsen) or the "residue" (Bateson), that seemed to lie beyond genes. While neither he nor Bateson used the term "information" in this context, with his reference to "memory" Johannsen went beyond Bateson in recognizing some form of *stored information* (Chap. 19).

As for "the possibility of dissolving genotypes" Johannsen did not shrink from supposing that "in a remote future we might be able to do some Homunculus-work, viz. to construct organisms through the addition or artificial combination of discreet factors, stored perhaps in bottles or small tubes!!" This precisely describes the operations of the genetic engineers who emerged from biochemistry in the 1970s. Anticipating the writings of George Williams on the disruption by recombination of the constellations of "selfish" genes that make each organism an individual, Johannsen continued:

> Gametogenesis with chromosome-reductions, accompanied by reformations and, as it were, partial rejuvenescence of cell-structures, must in some way act as if especially organized for *obliterating* the individual's personally "acquired characters," which as a rule totally disappear in sexual reproduction. … *Continuity in inheritance*, the cardinal idea of Aristotle, is – as applied to Mendelian heredity – represented by the continuity of chromosomes in the forthcoming generations – but [is] greatly complicated by disjunctions and recombinations of chromosome pairs. This hereditary continuity is … dissolved into … regular periodic discontinuities: Mendelian heredity always operating with discreet genotypical elements [genes]. Hence differences are here always discontinuous as chemical constitutional differences. Phenotypes however may show discontinuous as well as all degrees of continuous variation!

Referring to the ancient Greeks, Bateson had opined similarly in 1918 (Chap. 16) and Williams expressed likewise in 1966 [7]:

> Socrates' genes may be with us yet, but not his genotype, because meiosis and recombination destroy genotypes as surely as death. It is only the meiotically dissociated fragments of the genotype that are transmitted in sexual reproduction, and these fragments are further fragmented by meiosis in the next generation. If there is an ultimate indivisible fragment it is, by definition, "the gene" that is treated in the abstract discussions of population genetics. … I use the term *gene* to mean "that which segregates and recombines with appreciable

> frequency." ... A gene is one of a multitude of meiotically dissociable units that make up the genotypic message.

Johannsen had discarded the "presence and absence" hypothesis (a rare disagreement with Bateson; see Chap. 17):

> The nature of the genotypical units hitherto observed is highly problematic. When we regard Mendelian [allelic] "pairs," ... it is in most cases a *normal* reaction (character) that is the "allel" to an *abnormal*. Yellow in ripe peas is normal, the green is an expression for imperfect ripeness as can easily be proven experimentally .... "No starch" in maize is evidently an abnormality and so in the many cases upon which Bateson – as it seemed with full reason – founded his for a time highly useful and suggestive but now abandoned, hypothesis of "Presence and absence": the "normal" almost always positive and dominant, the "abnormal" being (in a morphological spirit) expressed as a "loss."

Between one allele and its partner "we might discover that there is one single genotypical point of difference between them, [and] this difference may probably consist in an alteration of the 'chemism' [chemistry] at a special point of a chromosome. Now such alterations may be more or less different, and where several such differences exist in a certain locus of a chromosome we have the so-called '*multiple allelomorphs*' [polymorphs]." The modern reader can see this in terms of various base substitutions at various points in the DNA sequence of a gene (see Part VI).

Having referred to the "morphological spirit" that had influenced early thinking, Johannsen pressed – unsuccessfully as it turned out – for a revision of some of Bateson's terminology. Thus "multiple allelomorphs" was cumbersome and "ought to be replaced – the 'morph' eliminated. 'Allelogene' seems a more neutral word. Perhaps the best expressions are 'multiple allelos' and 'multiple allelism,' or – to be purely Greek – 'polyallelism'. At any rate multiple alleles are (for the chromosome theory) different states (chemisms) in the same locus of a chromosome."

Progress in Biology

In February 1924, Bateson gave three lectures at Leeds at weekly intervals. His notes show that he first covered genetic linkage and mapping of genes in primroses. The second covered fruit fly chromosomes, and "Difficulties ... Goldschmidt's *Lymantria* intersexes." The third covered "Problem of species. By selection? By crossing? By direct adaptation? How should we know? ... Specificity as contrasted with characters. Plenty of variation in *characters*. None yet in specificity. Sterility. Tetraploids, etc. ... Somatic segregation."

In March, the "vitalist" Hans Driesch visited Merton, and Bateson gave a lecture on "Somatic segregation" to the Imperial College Natural History Society during which he mentioned the ancient Greek botanist, Theophrastus. There was a celebration to mark the Centenary of Birkbeck College, London (March 12), and Bateson gave an address on "Progress in biology" (see below). In May, the Genetical Society met at Daniel Hall's "farm." In June, at the Linnean Society, Bateson exhibited bud-sports among pelargoniums to illustrate somatic segregation, and the next day he gave a lecture at St. Mary's Hospital on "Sex," where he considered "freemartins" and "intersexes." The Imperial Botanical Conference was held at Imperial College (July 12) with Prain as president. A session on the economic possibilities of plant breeding was chaired by Bateson, as vice-president. One afternoon there was a visit to the John Innes Institute, and the accompanying brochure offered displays of work in progress, noting that "tea will be served at the Manor House from 4 to 5 p.m."

The Scandinavian Mendel Society invited Bateson to tour various plant breeding institutes. He took a boat to Copenhagen (August 3), where he met Johannsen, Clausen, and Winge, and visited art galleries. He wrote (August 6): "Very tired in evening. *Festessen* by Johannsen – much the same company as the previous day. He toasted me with much affection and reminiscence – and when I tried to answer I couldn't and had to stop. However, that passed off and we spent a pleasant evening in Bot. Garden, a wonderful large place, right in the centre of the city." There was a lecture at Åkarp, Sweden (August 7), and then travel by boat to Stockholm, where Bateson met Nilsson-Ehle and discovered some lingering doubts about chromosomes (August 10): "The chromosome cult here is after all only a half-hearted affair. Evidently misgiving has intervened and I hammered a few wedges into the cracks." Beatrice wrote in her *Memoir*: "The friendly welcome from his colleagues gladdened him. … He came back in better health and spirits than for some time past."

Bateson's "Progress in biology" speech at Birkbeck College was featured in two successive issues of *Nature*. Its broad scope was indicated by his mentioning both an original Samuel Butler (1612–1680; Chap. 25), and a later, and unrelated, Samuel Butler (1835–1902; Chap. 19). He began by asserting that "biology," a "new word" coined by Treviranus in Bremen in 1802, was not just the "old things" – zoology and botany – in a "new form," but had "connoted a new thought" – a new way of thinking. Soon, advances in microscopy had led to "the recognition of the cell." Nevertheless, disparaging dependence on instruments, he considered that "surely the essential feature which … differentiates living creatures from all other systems whatever, might have been distinguished by ordinary observation, at any time. … The conception of the cell as a unit was necessary to give anything like accuracy to this knowledge, but it was not essential." Furthermore, even before Darwin's revelation came "like some meteor of the heavens," people like William Lawrence, who in 1818 introduced the word "biology" to the English-speaking world, were not limited by "theological obsessions." Lawrence "flouted authority with great enjoyment," but nevertheless was "genuinely convinced … that the mutability of species was contrary to observation."

In France, Godron and others had noted that "the variability they observed did not result in the production of new species, and that in particular … the new forms derived from a single common origin, when interbred, did not produce offspring of impaired fertility as so many genuine species do." Darwin had clouded this issue. Although "the progress of the last twenty years" now allowed fruitful discussion, "no general principles governing the incidence of interspecific sterility have been ascertained."

There was "utter darkness before the Mendelian dawn," but "Mendelian analysis … has not given us the origin of species. … We now have to recognise that the transferable characters do not culminate in specific distinctions." The further discovery of "Morgan and his colleagues … that some, probably all, of this group of characters are determined by elements transmitted in or attached to the chromosomes" was "of the greatest … brilliancy," but "as to what the rest of the cell is doing, apart from chromosomes, we know little. We think that in plants the presence or absence of chloroplasts may be a matter of extra-nuclear transmission. Perhaps the true specific characters belong to the cytoplasm, but these are only idle speculations."

At that time, Bateson considered "the future of biology" to lie "not in generalization, but in closer and closer analysis." Since he called for more of the approach that we now term "reductionist," we should not be surprised that Bateson's more general message – his non-genic message – was not perceived or, if perceived, did not form a basis for experimentation in the decades ahead. The genic juggernaut was gaining a biochemical momentum that would carry it through the twentieth century and into the twenty-first [8].

The "Weak Point"

Bateson expanded these thoughts a year later when *Nature* invited contributions to celebrate the centenary of the birth of Thomas Huxley. Among the contributors were Lankester, Thiselton-Dyer, Lloyd Morgan, MacBride, Osborn, and Poulton, who also gave the "Centenary Lecture" at the Royal College of Science (May 4, 1925). Bateson's topic was "Huxley and evolution." The "great architect of academic morphology" and "champion of evolutionary doctrine" had been "an incomparable master" of that "most precarious art, demanding imagination and a large knowledge of human nature," namely, "the direction of public opinion." But he had never let Darwin forget "the weak point" – the need to explain hybrid sterility. "Nothing that has happened since at all mitigates the seriousness of this criticism." Huxley had made the crucial distinction between speciation as a morphological phenomenon and as a physiological phenomenon. He had declared that Darwin "*has* shown that selective breeding is a *vera causa* for morphological species; but he has not yet shown it a *vera causa* for physiological species" (see Chap. 28). Nevertheless, Huxley had hoped that if we persisted we would succeed in creating physiological species by appropriate selection and crossing. Bateson doubted:

> Whether many of those best acquainted with modern genetics are so sanguine as Huxley was, that by the most carefully devised system of experimentation are we in the least likely to produce a physiological species by selection [i.e. differential survival on the basis of character differences]. Rather have we come to suspect that no amount of selection or accumulation of such variations as we commonly see contemporaneously occurring can ever culminate in the production of that "complete physiological divergence" to which the term species is critically applicable. With entire candour Huxley reiterated that if this were the necessary and inevitable result of all experiments, the Darwinian hypothesis would be "shattered."

Yet, Bateson saluted Huxley's "admirable scientific judgment" in criticizing (what the modern reader, with hindsight, can view as Darwin's most important scientific contribution) his chapter on pangenesis in *Animals and Plants under Domestication*. Darwin had submitted the chapter to Huxley three years prior to its publication in 1868, and Huxley had advised him to omit it. It will be recalled (Chap. 3) that in 1889 de Vries, in his *Intracellular Pangenesis*, had referred to Darwin's "gemmules" as "pangens" that, as the material determiners of characters, were linearly arranged on the chromatin threads within the nucleus. Pangens being constrained to *intracellular* existence, de Vries had cleansed pangenesis of Darwin's "use and disuse" (i.e., the Lamarckian component). From "pangens" came Johanssen's abbreviation, "genes" (Chap. 10). It was perhaps Bateson's annoyance with de Vries's unwillingness fully to embrace Mendelism that led him to an insufficient appreciation of the profundity of his earlier work, which, as de Vries had not disguised, was an elaboration of Darwin's pangenesis.

The Huxley Centenary in May 1925 was one aspect of a busy year. Visitors to the John Innes included Balfour (February) and Davenport (July). The Institute also hosted the Fruiterers' Company (June 18) and the Genetical Society (June 20). Gregory sent photographs of his Galapagos voyage, and new innovations in radio communication permitted monitoring of his progress. Bateson wrote triumphantly (April 27) to Beatrice who was staying with the Arnold Forsters at St. Ives: "Got Marconi's to accept a message for Gregory." The titles of the presentations at a scientific meeting in the United States caused some amusement (April 29): "I have a program of Nat. Ac. Sci. gathering at Washington … including Henry F. Osborn: Rejoinder to William A. Bateson. … 'Rejoinder' indeed. Two errors in one short title." There was a lecture to the Wimbledon Medical Society (May 15) and presentations at the Linnean Society in May and June. While he was away in Russia (see below), the death of Francis Darwin (September 19) was announced. There were lectures in October (Gardeners' Lecture) and November (University College Medical Society). Early in 1926 (January 30), he began preparing a lecture for the "Cage Bird Show." It was never delivered.

Russia Again

In 1923, Bateson, Johannsen, and Morgan were elected foreign members of the Russian Academy of Sciences. This did not expedite negotiations when Bateson applied for a visa in 1924. However, together with Henry Miers, D'Arcy Thompson, and Maynard Keynes, Bateson was able to attend the bicentennial meeting of the Russian Academy of Sciences in Moscow in 1925. Keynes was accompanied by his ballerina wife: "I have made acquaintance of Mrs. J. M. Keynes and like her. She is about the smallest lady I ever saw." Bateson officially represented the British Museum. Among other foreign scientists were Richard Goldschmidt and Marie Curie. But many, having noted the expulsions of thousands of "bourgeois" foreign students during the revolution, did not attend.

Departure (August 31) was from Hull for Copenhagen, and then on to Stockholm (September 2). A steamer took them to St. Petersburg (September 3), which had been renamed Leningrad. Here, there was privileged access to the Hermitage art collection, although officially closed: "Today is 'Science Holiday' throughout Russia! Vavilov's place on enormous scale. Says some 400 people, of whom about half are trained. Has seeds of 13,000 varieties of wheat." Then on to Moscow (September 10). "I am glad I came. I have learned some new things and above all I am glad to have seen Vavilov's work and installation which is really good and must develop into a fine institution if he gets his chance." Bateson had brushed up his Russian prior to the visit, and his fluency had returned: "I do all simple things in Russian now and am not infrequently put on to interpret!" The research establishments were impressive: "They have a lot of work going on, some of it quite promising, and a vast number of students. ... I get more flattery here than is good for me, but after a good few kicks in the past, I can lap up some treacle."

Bateson gave an account of the visit in *Nature* [9]: "The gathering had been organized largely with an eye to its propaganda value," and they had been entertained lavishly. Although "readers of *Nature* will not expect a report on social conditions," he could not refrain from noting "groups of artisans going home from work in rags," and the shabbiness of the clothing of professional "men of refinement and learning." From various official speeches (Fig. 18.2):

> One conclusion very plainly emerged, that the revolutionary government is perfectly sincere in its determination to promote and foster science on a very large scale. Signs were not wanting that science, especially perhaps in its applications, is regarded by the present governors of Russia as the best of all propaganda. It was interesting to hear the faith that the advancement of science is a first duty of the State proclaimed by professional politicians. We ought perhaps not to inquire too closely whether they and we mean the same thing by the term science. ... Next in importance after communism, the tenets of Leninism assert the doctrine that science is the basis of happiness. Religion is to be eradicated as a vice; science and the arts are to be promoted in its place.

Fig. 18.2 N. I. Vavilov and Bateson

Many palaces and stately homes had been transformed into research institutions, for example, the Institute of Zoological and Botanical Research under Professors Philiptschenko and Dogiel had been set up "in the house and parks of the Leuchtenberg family at Peterhof." The most extensive of the new organizations was Professor Vavilov's Institute of Applied Botany and Plant Breeding, which aimed to "provide breeds of cereals and other agricultural plants for the various parts of Russia." Vavilov had written to Bateson in 1924: "Much of my time is taken up with the organization of the new experimental station in the environs of Petrograd. You will be rather astonished to hear that the country house we live in was … presented by the late Queen Victoria to her godson the ex-Granduke Boris Vladimirovitch. The country seat is very charming, the house itself quite English in style. During the past four years, unfortunately, the main buildings were occupied by comrades."

Bateson related that the final banquet in Moscow had seated 1200 people and "in a city teeming with beggars, we tried in vain to conjecture any system by which admission to the table, or exclusion from it, may have been determined. Of liberty we saw no sign." As for the likelihood of the experiment with communism being successful "judging from the portion of the population lost in the first experiments, those who survive to benefit from the ultimate deliverance will be few." Bateson's indictment was severe: "We here are accustomed to think of science and learning as flourishing best in quiet places, where they may come to slow perfection, under systems providing a reasonable measure of personal independence and security. Present conditions in Russia have brought about the very contrary, and among the grave indications of disharmony, which every visitor observes, the want of freedom is by far the most serious."

There were some sad footnotes to this episode. A letter from Philiptschenko from Leningrad University (April 12, 1925) requested some reprints from the *Journal of Genetics*: "It is of great importance to be in possession of some of them on behalf of procuring an allowance (pension) for the widow of the late Issayev." And among the Bateson papers, there is a philosophical document entitled "The English-speaking world and Siarus" ostensibly written by "a cooperator," but perhaps written by Vavilov's wife, "Katya" (Yekaterina N. Sakharova), or his physicist brother S. I. Vavilov. "Siarus" is an anagram for "Russia." There is also a letter from Katya (January 15, 1926), probably seen by Bateson shortly before his death:

> I am rather interested in the fate which befell "The English speaking world and Siarus," and if you could write a few words about it to Prof. S. I. Vavilov who is going now to Germany where he intends to stay for some months I should be greatly indebted to you. Prof. S. I. Vavilov is a physicist and he will work in Berlin where he will point out his address for you.

In fact, there was a Berlin address at the end of the letter, which went on to describe her plans to emigrate with her son to Canada, and asked his advice: "I have no acquaintances there but I have written previously to some agricultural stations in N. Dakota and elsewhere and I got answers that knowing 4 languages and having the degree of an agronomist – it might be possible to get a living there." Vavilov's brother may have been involved both in the plan and in circulating the Siarus document. The marriage ended in 1927, and N. I. Vavilov remarried. His brother became president of the Soviet Academy of Sciences and died in 1951.

In the short term, Vavilov's institute thrived. At Vavilov's invitation, one of Morgan's gifted triad, Muller, emigrated to Russia (Chap. 24). But one of Vavilov's students, Trofim Lysenko, succeeded in exploiting the disharmony among geneticists in the West to convince the politicians that the Lamarckist route was the one to follow. Muller saw what was coming and retreated by way of the Spanish Civil War to the University of Edinburgh, and thence back to the United States. At the Lenin All-Union Academy of Agricultural Sciences in 1939, Lysenko interrogated Vavilov:

Lysenko	I understand from what you wrote that you came to agree with your teacher Bateson, that evolution must be viewed as a process of simplification. Yet in Chapter Four of the history of the party it says evolution is increase in complexity.
Vavilov	When I studied with Bateson, ...
Lysenko	An anti-Darwinist.
Vavilov	No. Some day I'll tell you about Bateson, a most fascinating, most interesting man. ...

In 1940, Vavilov was arrested, to be replaced by Lysenko as director of the Institute of Genetics. Two years later, Vavilov died in a Stalin gulag [10].

The Russian interest in Lamarckism was evident in 1926 when the Communist Academy (Moscow) offered Kammerer a position for which a laboratory would be built. Before this could come about, Kammerer had shot himself on a high Austrian mountain path. His suicide note asked that his body be given to the university for dissection: "Perhaps my worthy academic colleagues will discover in my brain a trace of the qualities they found absent from the manifestations of my mental activities when I was alive" (see Chap. 21). He requested no religious ceremonies and that his wife abstain from wearing mourning. A film by Lunacharsky, on the inheritance of acquired characters (*Salamandra*), was based on the alleged persecution of Kammerer. The plant-breeder Ivan Michurin (1860–1935) became Lamarckism's highest authority. He studied fruit trees, which are often hybrids and have, we now know, very complex genetics.

Limits

Bateson contributed articles to the 12th edition of the *Encyclopedia Britannica* (1922). The closest we have to Bateson's last word on evolution is in the revisions and additions he was making for the 13th edition shortly before his death. There were two major articles: (i) *Mendelism*, with sections headed analysis, phenotype and genotype, variability, mutation, inter-specific sterility, evolution, and adaptation; and (ii) *Genetics*, with sections headed linkage, cytological interpretations of genetic phenomena, and bearing on evolution theory. Apart from reasserting that the *limits* of Mendelian analysis were *undetermined*, and that the elements "detachable" by Mendelian analysis were implanted on an irreducible "basis" (residue), there was a strange return, even though as a "symbolism," to the old "unpacking" idea that Gates had found so objectionable (Chap. 14). However, it must be conceded that, in the absence of knowledge of the size of what we now refer to as the genome (i.e., the *limits* of the genome were *undetermined*), whether a mutation arises afresh, or is the release of something that up to that time had been inhibited, is irrelevant to most of Bateson's arguments. Indeed, a kaleidoscopic reshuffling of a limited number of elements can also be considered to release something previously inhibited. Thus, the unpacking argument can be seen as an example of Bateson keeping his options open until the biochemists could complete the picture.

Mendelism

- *Analysis*: "Organisms ... are now recognized as largely possessing attributes behaving as units and as such capable of being detached and transferred to any other type with which cross-breeding can be effected. The *limits* governing this principle of segregation and recombination are still *undetermined*." (Our italics).
- *Phenotype and Genotype*: "We have ... to distinguish the organism as it outwardly appears to be, from that which it actually is by genetic composition, a distinction which Johannsen conveniently expressed by the use of the terms *Phenotype* for the former and *Genotype* for the latter. ... This fundamental distinction ... must constantly be remembered."
- *Variability*: "As regards the de novo appearance of dominant characters the evidence is less abundant. Morgan and the American geneticists have made prominent several instances of this kind in *Drosophila* (fruit fly), of which the spontaneous origin of 'eosin' ... eyes in a white-eyed strain may be cited. Admitting provisionally these examples are free from objection, they are nevertheless extraordinary events. ... But ... nothing absolutely forbids us from inverting the representation of positive and negative factors by extension of the concept of inhibitors of which many are familiarly known: so we may express the apparent addition of a new element as a loss of one which when present had repressed the new attribute. This symbolism, though admittedly objectionable when dominance is complete, does without strain apply to all cases in which the heterozygote is intermediate, and a large range of alleged new dominants can be covered. In so far as this conception applies, evolution is conceived as a process of unpacking, a progress consisting in the loss of component elements."
- *Mutation*: "The term 'mutation' ... is now generally accepted to denote definite genetical variations which are sensibly discontinuous."
- *Inter-Specific Sterility*: "The new forms whose productions we witness are never new species. In *Primula sinensis* about 20 pairs of factorial differences have been determined, which in their several combinations present an amazing polymorphism. A systematist, if he met these forms in nature might ... take many of them for distinct species. But interbred, they and their products are perfectly fertile. Polymorphism like this ... avails us little as material out of which true specific differences can be supposed to develop. The conspicuous defect in the evidence for the origin of species by common descent remains. ... No one has yet raised

types from a common origin which when interbred produce sterility of the kind and degree which is one of the commonest attributes of crosses between natural species. By whatever concatenation of arguments theories of evolution have been constructed, that most essential link has never been supplied. The lapse of time is occasionally invoked in the hope of rectifying this and similar evidential defects; a strain which has been maintained distinct for a long period being thought more likely to show interracial sterility when crossed with its progenitor than one newly separated. Reasoning of this kind, plausible enough in scholastic days, is not acceptable in an age of chemistry, nor may we suppose that that which is never begun will be attained by mere effluxion of time. The more genetical experience extends, the more serious does this hiatus in the evidence become."

- *Evolution*: "The lines of argument converging to support the theory of common origin are so forcible and so many that no alternative can be entertained. … Common descent, though rarely if ever a proposition demonstrable in any detail, ranks as an axiom. … Parts of the apparatus by which the validity of this claim was enhanced have fallen into desuetude. In particular, the modern geneticist assigns to Natural Selection a subordinate and inconsiderable role. … We no longer look for utility in the details of a peacock's feather than in the iridescence of a Roman bottle, or in the regularity of basaltic prisms."
- *Adaptation*: "It is not merely in regard to the mode by which species have arisen that agnosticism has prevailed. While unwilling to accept adaptation, with Darwin, as a summation of happy accidents, we have no alternative to offer, nor is there in the recent attempts of various experimenters to find that organisms transmit to their posterity structural emendations in response to parental experience anything which sensibly alleviates the difficulty. Most of these claims are obviously faulty and few require serious notice."

Genetics

"Modern theories of evolution are based on the assumption that species have arisen by descent with modification, and that the constancy and diversity which living things manifest in their reproduction provide a sufficient basis for these conceptions. It is significant that as a result of the preliminary work done under the new inspiration, attention has been largely diverted from these more philosophical aims. … The scope and character of these discoveries are referred to below. Their immediate consequence has been that the development of evolutionary theory has been tacitly suspended or postponed, and activity is concentrated on the exploration of genetical physiology, the theoretical evaluation of the knowledge thus gained being relegated to the future. … Organisms may now be represented as aggregates of units [genes] which confer upon them their various attributes. The degree to which an organism may be thus resolved is as yet undetermined, but there is presumably a limit to the process, as it is natural to suppose that the detachable elements are implanted on a *basis* which for a given type is irreducible."

- *Linkage*: "The terminology followed … is that introduced by T. H. Morgan, to whom progress has been largely due. It is … convenient to distinguish the case in which the two dominants (A, B) are introduced together by the parent as '*coupling*', and the converse as '*repulsion*,' but the physiological process is now recognized as being … the same in both cases. … The factors thus linked have plainly no connection with each other as regards the effects which they produce in the zygote, but may concern the most dissimilar characters. … From the existence of such cases and from certain other considerations, it has been urged, especially by the American geneticists, that the method of representation by presence and absence is incorrect, and that a negative allelomorph should be treated as a real entity. There is no valid means of deciding this question as yet. The probability is perhaps that the absence should always be regarded as relative only. As a mode of symbolic expression, the representation of the two allelomorphs as differing quantitatively is often convenient, though certainly not universally applicable."
- *Cytological Interpretations of Genetic Phenomena*: "Soon after the discovery of Mendelian analysis the plausible suggestion was made that the behaviour of the chromosomes in the course of the maturation divisions [of meiosis] was consistent with what might be expected if they were actually the bearers of the segregable factors. … The inference is drawn that the factors composing each linkage group are borne on one chromosome. Developing this conception, Morgan suggests that the factors are arranged in the chromosomes as beads on a string, each having a position normally fixed in relation to the rest. … The relative 'loci' of numerous factors have been determined …, and the fact that this can be done forms a strong argument for the belief that … the factors must be disposed in linear systems. … The defect of the theory … is that it rests on many subordinate hypotheses which are not all capable of independent verification."
- *Bearing on Evolutionary Theory*: "On studying a variable species critically it is found that the various forms cannot all produce each other as was formerly assumed, but … [are] terms in a series of combinations of definite factors. Such series are no evidence of contemporary variability. Many of the terms can be separated in the homozygous condition, and thereafter may breed perfectly true. Even such an appearance of variability as that seen in polymorphic species is frequently not above suspicion of being the consequence of a cross, more or less remote. Contemporary variation certainly may occur, but of the contemporary origin of new species, or of the occurrence of genetic changes which can be colourably interpreted as likely to lead to the production of incipient species in a strict sense, no indication has been found. That the forms of life have evolved from dissimilar precedent forms, we know from the geological record, but as to the process by which this evolution has come to pass we are still in ignorance. All that can be declared with any confidence is that variation most commonly arises as an error of cell division, and that conceivably new species have so arisen."

Summary

Bateson overcame the depression that followed Martin's suicide. Hurst moved from "trade" and crossbreeding in the field, to academia and cytology in the laboratory. His future wife Rona's introduction to the world of genetics was at a 1922 meeting of the Genetical Society, which was attended by Morgan and Sturtevant. Gregory Bateson reported back on Cambridge events – Kammerer's visit, and a lecture by Yule. Bateson warned that excessive attention to gene expression ("the *means* by which an effect is brought about") could divert attention from the genes themselves. Concerning the teaching of evolution in the United States, repercussions from Bateson's Toronto address were evident in the political speeches of Bryan, the dismissal of a school superintendent in New Mexico, and the Scopes trial. Bateson cautioned that Britain was not immune to the challenges this implied. In various articles, Bateson reiterated the Huxleyan "weak point" in Darwin's theory – that the offspring of crosses among members of a species are not sterile. The transferable characters corresponding to genes shed no light on the origin of species. Undetermined were the limits, both of the individual variability that was subject to Mendelian analysis, and of the material basis of that variability. Given this, various scenarios remained in contention. In 1925, Bateson returned to a Russia transformed by a communism that, despite its lavish support for science, he could not abide.

Part IV
Politics

Chapter 19
Butler

Buffon planted, Erasmus Darwin and Lamarck watered, but it was Mr. Darwin who said, 'That fruit is ripe,' and shook it into his lap.

Samuel Butler, 1887 [1].

Consideration of Samuel Butler comes late, not because his contributions to evolution were late, but because they were late in impacting Bateson's life. Even then, like most of his contemporaries and those who came after, Bateson never really understood what he was driving at. Butler's significance is twofold. First, together with Ewald Hering in the 1870s, he declared heredity to be the transfer of stored *information*, a concept that included a distinction between genotype and phenotype although these words were not then in use. Second, he developed what was, for his time, a scientifically coherent basis for the inheritance of acquired characters, otherwise known as Lamarckism. Although the latter stamps him as a Victorian – a creature of the nineteenth century through and through – the former places his thinking at the forefront of evolutionary bioinformatics (EB), a discipline that may be to the twenty-first century what genetics and biochemistry were to the twentieth [2–5].

The Darwins, Butlers, and Batesons

Charles Darwin was born in Shrewsbury, England, in 1809. For several years (1818–1825), he was at Shrewsbury School where he crossed swords with the headmaster, the classical scholar Samuel Butler – a dire foreshadowing of future confrontations with the latter's grandson of the same name. Recalling his schooldays, Darwin wrote [6]:

> I am one of the root and branch men, and would leave classics to be learnt by those alone who have sufficient zeal and the high taste requisite for their appreciation. … I was at

A. G. Cock, D. R. Forsdyke, *Treasure Your Exceptions*,
https://doi.org/10.1007/978-3-030-92099-9_19

school at Shrewsbury under a great scholar, Dr. Butler; I learned absolutely nothing, except by amusing myself by reading and experimenting in chemistry. Dr. Butler somehow found this out, and publicly sneered at me before the whole school for such gross waste of time; I remember he called me Pococurante [a person concerned only with trifles], which, not understanding, I thought was a dreadful name.

Darwin left the school to join his brother at the University of Edinburgh and went on to Cambridge (Christ's College) in 1827.

William Henry Bateson also attended Shrewsbury School, and he also crossed swords with Butler [7]:

The 'beef row' as it is commonly called, was the second and last case of insubordination which Dr. Butler had to deal with in the course of his long head-mastership. … The outbreak occurred on the appearance of a round of beef on the dinner-table of the Doctor's hall, whereon the boys one and all – by arrangement preconcerted among themselves before they had seen the beef – left the hall, declaring that the meat was not fit to eat. Dr. Butler required an apology from the preposters, among whom were several of his most brilliant pupils – Robert Scott, James Hildyard, Brancker, J. W. Warter, and Bateson, to mention no others. As they all refused to apologise he dismissed them, with the saving clause that on apology being made he would receive them back. Of course the parents of the boys made them apologise.

From Shrewsbury there issued a stream of prize-winning scholars that soon established the school's (and Dr. Butler's) reputation. In those days, prizes were few and candidates had to write several examinations over a period of days. It took a Bateson to bring to light the possibility that Cambridge was not above some hanky-panky in this respect [7]. A letter to Butler from the Reverend Thorp of Corpus Christi College in June 1830 extolled Bateson's virtues and regretted that, by only a fine margin, he had been bettered by a certain Wilson. In February 1831, Thorp wrote again having got wind of Dr. Butler's dissatisfaction with the outcome. Butler responded: "As you fairly enter upon the subject of the last examination of one of my pupils, I will be quite candid in my reply." He gave an account of the examinations as told to him by Bateson. It seemed that Thorp had questioned Bateson prior to the examinations about the areas in which he was especially prepared, and the examination papers were biased toward those in which he was least prepared, namely "English themes." Thorp declared a connection to be impossible since:

The plan of the whole written examination is arranged long before the candidates present themselves, and … all of it is [pre] printed which can be so …. After the examination was over I spoke as kindly as I could to Bateson, and commended his examination. He made a pettish reply (the only time he did so), and appears to have impressed you with very incorrect notions of the treatment he received here.

Butler replied that he had written down Bateson's words on his return and that his integrity was not in doubt:

He is a boy whose manners are blunt, but whose word may as safely be relied on as that of any person I know. … He enquired of several of the other candidates whether you had asked them similar questions, and … they uniformly answered in the negative. … Further, … so far as the printed papers were concerned he doubts not that the whole was previously arranged, but that the subjects for the English themes were brought in with the ink wet, and they thus formed an exception to the rest of the examination. … Under the circumstances I

> have stated I cannot but think myself excusable for the view I took of them. I rejoice to find myself mistaken, and beg in my turn to apologise to you for my error.

However, he did not let the matter rest. A letter in April from the Reverend Symons of Wadham College stated:

> Your account of the effect which the result of the election at Corpus has had on the school I read with deep concern. It is too painful to dwell upon, and hopeless thoroughly to explain. All I will say upon it is that I am inclined to attribute very much of it to a faulty but long-established mode of examination, and to an erroneous judgement, rather than to any intentional want of principles. I am, however, much gratified that you allowed my wishes to have any weight in determining you to forbear making the case public.

Butler resigned in 1836 on being appointed Bishop of Lichfield. He was succeeded by one of his star pupils, the Reverend Benjamin Kennedy, who adopted many of his teaching methods and eventually became Professor of Greek at Cambridge. Butler had a great influence on the development of the public schools and universities but published very little.

William Henry Bateson was accepted by St. John's College, Cambridge, where he was a contemporary of Butler's son Thomas, with whom Darwin was acquainted. All three were intended for the church. Darwin turned to extracurricular biology, eventually graduating in classics and mathematics in 1931 before setting off on his voyage on The Beagle. Thomas Butler became Canon at Langar Rectory in Nottinghamshire, where his son Samuel was born in 1835. In his fictional *The Way of All Flesh* [8], the younger Samuel later described the trials of a Victorian childhood and his years (1848–1854) at Shrewsbury under Kennedy. William Heitland, William Bateson's brother-in-law, was also at Shrewsbury School (1862–1867). He considered that in Butler's *The Way of All Flesh* "Dr. Skinner is Kennedy, drawn by a somewhat cruel hand" [9].

William Henry Bateson was third in the Classics Tripos (First Class) in 1936. He was at St. John's College most of his life, beginning with a fellowship that required that he not marry (a rule not revoked until 1882). In 1846, he became Senior Bursar, and in 1857, Master, when he became free to marry. Thus, compared with his contemporaries Charles Darwin and Thomas Butler, he had children relatively late in life (see Chap. 1). William Henry was acquainted with the younger Samuel Butler who began religious studies at St. Johns in 1854.

Having studied mathematics and classics, Butler graduated in 1858. Much to the consternation of his father, he declined to be ordained [10]. On Saturday October 1, 1859, Charles Darwin noted in his diary that he had finished correcting the proofs of his great work, *The Origin of Species*, which appeared in November [11]. That same Saturday, Butler sailed for New Zealand with Justus von Liebig's *Agricultural Chemistry* and the aim of becoming a sheep farmer. Shortly after his arrival, he obtained a copy of Darwin's book. Gregor Mendel is also likely to have obtained a copy in the early 1860s [12, 13]. Thus, while Mendel was breeding peas and studying Darwin in Moravia, Butler was breeding sheep and studying Darwin in New Zealand. This inspired his first articles on evolution, which appeared in *The Press* of Christchurch and soon received Darwin's commendation [14]. The sheep breeding was financially successful and in 1864 Butler returned to England.

Erewhon

Butler lived a short walk from the greatest library in the world – that associated with the Reading Room of the British Museum. With a small, assured income and, other than an annual visit to Italy, relatively inexpensive tastes, Butler had, what was denied the professional "men of science" whose writings he devoured – time. However, unlike the men of science whom he came to challenge, his interests were not focused. He was attracted to painting and literature through which he hoped he might further his financial independence.

In 1870, Butler collected together some earlier publications with titles such as "Darwin Among the Machines," "The World of the Unborn," "The Musical Banks" and "An Erewhonian Trial," and wrote his most "successful" literary work, a satire on the materialism, pseudo-idealism and academic pretentiousness of his times – *Erewhon* (the title is close to "nowhere" backward). Exploring a distant mountain range in a country with an uncanny resemblance to New Zealand, the hero discovered a lost civilisation where conventional views of right and wrong were inverted and the inhabitants learned strange lessons in the "Colleges of Unreason." Sprinkled with evolutionary anecdotes, it was completed in 1871 and went on sale in 1872 with the support of a loan to cover the publisher's expenses [15]. There was some correspondence with Darwin and two visits to his house at Downe in Kent. The following seemingly convoluted paragraph on seed germination indicates the already advanced state of Butler's thinking:

> The rose-seed did what it now does in the persons of its ancestors – to whom it has been so linked as to be able to remember what those ancestors did when they were placed as the rose-seed now is. Each stage of development brings back the recollection of the course taken in the preceding stage, and the development has been so often repeated, that all doubt – and with all doubt, all consciousness of action – is suspended. … The action which each generation takes – [is] an action which repeats all the phenomena that we commonly associate with memory – which is explicable on the supposition that it has been guided by memory – and which has neither been explained, nor seems ever likely to be explained on any other theory than the supposition that there is an abiding memory between successive generations.

Then as now, the word "memory" referred to the physiological function responsible for the recall of stored information. Today in the computer-age we use "memory" to denote stored information itself. Whatever the sense, although he did not use the word "information" at that time, Butler was here describing heredity as the transfer of stored information. This was made explicit in 1880 in his book *Unconscious Memory* [16]:

> Does the offspring act as if it remembered? The answer to this question is not only that it does so act, but that it is not possible to account for either its development or its early instinctive actions upon any other hypothesis than that of its remembering, and remembering exceedingly well. The only alternative is to declare … that a living being may display a vast and varied *information* concerning all manner of details, and be able to perform most intricate operations, independently of experience and practice. (my italics)

Shaking his head in bewilderment "at the vanity which has induced so incapable and ill-informed a man gravely to pose before the world as a philosopher," Romanes gave *Unconscious Memory* a hostile review in *Nature* in 1881 [17]:

> We can understand, in some measure, how an alteration in brain structure when once made should be permanent, … but we cannot understand how this alteration is transmitted to progeny through structures so unlike the brain as are the products of the generative glands [gametes]. And we merely stultify ourselves if we suppose that the problem is brought any nearer to solution by asserting that a future individual while still in the germ has already participated, say, in the cerebral alterations of its parents.

Butler's attempt to reply to this was initially stonewalled by the editor, Norman Lockyer, a friend of Romanes. It required a letter to the publisher, Macmillan, threatening legal action, to secure publication (*Nature* Feb. 3, 1881). Butler's work has been dealt with in my text *Evolutionary Bioinformatics* [5]. The reasons why the Victorians were so unsympathetic are only touched upon here, but they partly explain the delay before Bateson began to take him seriously.

Unconscious Memory was dedicated to Richard Garnett (1835–1906), whose father had been acquainted with Dr. Samuel Butler. Garnett, the father of David Garnett (Chap. 2), had worked at the British Museum Library since the age of 16. He was Superintendent of the Reading Room from 1874 to 1884, later becoming "Keeper of Printed Books." Butler regarded him as "the best informed man I have ever met." Garnett and another user of the Reading Room, Miss Eliza Savage, both commented on draft versions of *The Way of All Flesh* and of some of his other works. In 1886, Garnett wrote a testimonial for Butler when he applied (unsuccessfully) for the Slade Professorship in Fine Arts at Cambridge. In 1892, Garnett and Miss Jane Harrison (Chap. 4) attended a lecture by Butler at the Working Men's College on the "Humour of Homer" [18].

Spencer's "Physiological Units"

In 1863 in his *Principles of Biology*, Herbert Spencer postulated that cells contained "physiological units," which today we can best equate with proteins. These had a property, termed "polarity," that conferred an ability to aggregate specifically, just as molecules aggregate into crystals with distinctive structures. Thus, Spencer wrote:

> If then, this organic polarity can be possessed neither by the chemical units [small molecules in the cell protoplasm], nor the morphological units [cells themselves], we must conceive it as possessed by certain intermediate units, which we may term *physiological*. There seems no alternative but to suppose that the chemical units combine into units immensely more complex than themselves, complex as they are; and that in each organism the physiological units [polymeric macromolecules] produced by this further compounding of highly compound atoms, have a more or less distinctive character. We must conclude in each case, some slight difference of composition of these units, leading to some slight difference in their mutual play of forces, produces a difference in the form which the aggregate of them assumes.

This, in a nutshell, describes how three of the four main classes of macromolecules within cells (proteins, lipids, and carbohydrates) interact with each other to generate the phenomena of life. By virtue of their structures and/or catalytic activities, these macromolecules confer the distinguishing features of cells, tissues, and organisms (the phenotype). Molecules (chemical units) present in all cells (e.g., amino acids), combine to produce macromolecules (e.g., proteins). Spencer admitted that his physiological units would not, in themselves, explain the phenomena of heredity: "A positive explanation of heredity is not to be expected in the present state of biology." Something was missing.

Heredity as Information Transfer

In 1868 in his "provisional hypothesis of pangenesis," Darwin went further (Chap. 3). Inspired by the new work showing the great proliferative powers of microorganisms, most of which were not visible with the microscopes then available, he postulated "gemmules" as minute heredity units with the ability to self-replicate. Gemmules were not themselves physiological units but carried from generation to generation what we would now refer to as the "information" for the establishment of new physiological units (e.g., new proteins), and hence, new organisms. Thus, Darwin (1868) wrote of "formative matter" and "formative elements," noting that "the child, strictly speaking, does not grow into the man, but includes germs which slowly and successively become developed and form the man." He equated "germs" with the "formative matter," which "consists of minute particles or gemmules." These were fundamental units, each associated with a particular inherited character and capable of independent multiplication by "self-division." We can, of course, speak of a baby *developing and forming* a man or of a seedling *developing and forming* a mature plant, but neither baby nor man, neither seedling nor plant, has the property of "self-division" (i.e., duplicating as a whole to make an identical copy of itself). Today we can best equate gemmules with genes [19]. In 1878, in his first major work on evolution – *Life and Habit* – Butler supposed "a memory to 'run' each gemmule" [20].

Unknown to Butler, a step closer to the idea of heredity as information transfer was taken in 1870 by the Prague physiologist Ewald Hering (1834–1918). Initially heredity had been seen as the process by which phenotypic characters themselves, or the related "physiological units," were passed *forward* from parent to child. However, Hering, like Butler, turned this round. He put the onus on the embryo, which *remembers* its parents. The newly formed embryo is a passive recipient of parental information (genotype), and this information is used (recalled) by the embryo to construct itself (i.e., to generate phenotype from genotype). Heredity and memory are, in principle, the same. *Heredity is the transfer of stored information.* This powerful conceptual leap led to new territory. Evolutionary processes could thenceforth be thought of in the same way as mental processes. Hering [21] considered that: "We must ascribe both to the brain and body of the new-born infant a far-reaching power of remembering or reproducing things which have already come to their development thousands of times in the persons of their ancestors." More simply, Butler wrote [22]:

> There is the reproduction of an idea which has been produced once already, and there is the reproduction of a living form which has been produced once already. The first reproduction is certainly an effort of memory. It should not therefore surprise us if the second reproduction should turn out to be an effort of memory also.

Indeed, Bateson's mentor, William Brooks (Chap. 2), had considered there to be "a profound truth' in the view of Hering that 'Heredity is Memory'" [23]:

> It is plain that this power [to adjust to the environment] depends upon experience, but experience depends upon "memory." So we may state, with truth, that in a certain sense, life is memory. – We know memory, however, only in connection with organization, and if it is true that heredity, the power of an organism to reproduce its like, is simply the memory, by the ovum, of the experience of its ancestors, we must believe that there exists in the ovum an organization of some kind to correspond to each of these past experiences.

There is a glimmer of this in *Materials* with the analogy of the wax model and the melting pot (Chap. 4). Given that heredity would require the recalling of stored information, what physical and/or chemical form would that information take, and how would it be transmitted? At that time, spectral studies had associated each element with a particular wavelength of light, and sound was also understood in terms of frequencies and wavelengths. Hering in 1870 suggested that memory was based on the transmission of molecular vibrations, and in 1871, St. George Jackson Mivart in London suggested that variation between organisms might reflect an "upset of the previous rhythm of the physiological units of the living organism" [24]. This is not so far from the mark. Macromolecules, by virtue of their distinctive structures, have distinctive resonance frequencies. Although today we think of DNA macromolecules replicating based on "key-and-lock" chemical complementarity, a slight alteration in structure of a macromolecule would also slightly alter its vibration, or "rhythm." Butler, however, was cautious [16]:

> Professor Hering … goes into the question of what memory is, and this I do not venture to do. I confined myself to saying that whatever memory was, heredity was also. … I am not committed to the vibration theory of memory, though inclined to accept it on a *prima facie* view. All I am committed to is, that if memory is due to the persistence of vibrations, so is heredity; and if memory is not so due, then no more is heredity.

Bateson in his "vibratory theory" (Chap. 4) gave some support to this viewpoint [25]. However, Galton in 1875 [26] had considered that the "mutual affinities" of the "organic units" responsible for the transmission of inherited characters had "simplicity and sufficiency" enough to explain the specificity of their interactions. He did not invoke vibrations or rhythms. He implied some degree of chemical (key-and-lock) specificity, giving as example Charles Darwin's observations on the exquisite species-specificity of fertilization, so that "with unerring certainty" the stigma of a flower of one species would accept only pollen from another member of its species. Perhaps he was also influenced by the ability of particles, if they have similar structures, to aggregate in "lines like those which attach the blood corpuscles face to face in long rouleaux [like piles of coins] when coagulation begins." The specificity of this – corpuscles of a species preferentially aggregating with their own kind – has since been demonstrated [27].

Hering was cited by the German evolutionist Ernst Haeckel in 1876 in his Lamarckian essay *Die Perigenesis der Plastidule oder die Wellenzeugung der Lebenstheilchen* (literally "The Generation of Waves in the Small Vital Particles," the latter being cells – called "plastidules" by Haeckel). He dismissed Darwin's pangenesis as an "airy nothing" – a reference to Shakespeare's *A Midsummmer Night's Dream*: "And as imagination bodies forth, the forms of things unknown, the poet's pen turns them to shapes, and gives to airy nothing a local habitation and a name." Haeckel sent Darwin his essay [28], which Darwin forwarded to Romanes a few days later, noting [29]:

> As you are interested in pangenesis, and will some day, I hope, convert an "airy nothing" into a substantial theory, I send by this post an essay by Haeckel attacking Pan[genesis] and substituting a molecular hypothesis. If I understand his views rightly, he would say that with a bird which strengthened its wings by use, the formative protoplasm of the strengthened parts became changed, and its molecular vibrations consequently changed, and that these vibrations are transmitted throughout the whole frame of the bird, and affect the sexual elements in such a manner that the wings of the offspring are developed in a like strengthened manner. … How he explains reversion to a remote ancestor, I know not. Perhaps I have misunderstood him, though I have skimmed the whole with some care. He lays much stress on inheritance being a form of unconscious memory, but how far this is part of his molecular vibration, I do not understand. His views make nothing clearer to me; but this may be my fault. No one, I presume, would doubt about molecular movements of some kind. His essay is clever and striking. If you read it (but you must not on my account), I should much like to hear your judgment, and you can return it at any time.

Haeckel's attempt to substitute his perigenesis for Darwin's pangenesis was brought to the attention of the English-speaking world by Lankester in a *Nature* article [30]. Heredity was "a form of unconscious memory," as had been proclaimed by Hering. At that time, Butler, after a distracting period in Montreal trying to salvage unwise investments, was busy on *Life and Habit* [20]. Two people, Edwin Clodd and Francis Darwin, advised him on the relevant literature. Clodd was a friend of Grant Allen, who on various occasions reviewed Butler's books [31]. Clodd drew Butler's attention to Mivart's anti-Darwinian *The Genesis of Species* [24]. Francis Darwin, shattered after the death of his first wife in childbirth in 1876, was then based near London, and he and Butler would often meet. Francis pointed out the similarity between Butler's and Hering's ideas in September 1877, only three months before the publisher's release of *Life and Habit*. While not mentioned in the book, Hering was fully acknowledged by Butler in a later letter to *The Athenaeum*.

Lamarckism

The parallels he had drawn between memory and heredity suggested to Butler a set of alternatives that had some plausibility at the time. He opted for an agency *internal* to an organism that would, in small steps, bring about variations that would accumulate to the organism's advantage. Darwin's argument that natural selection would cruelly send the weaker to the wall and select the fittest to survive sufficed "to arouse instinctive loathing; ... such a nightmare of waste and death is as baseless as it is repulsive" [18]. To buttress this feeling, Butler argued along the lines of Ernst Brücke who had viewed unicellular organisms as elementary versions of organisms considered higher in the evolutionary scale [32]. Microscopic studies of unicellular organisms, such as amoebae, showed them to possess organelles analogous to the organs of multicellular organisms. Thus, an amoeba, far from being an amorphous jelly-like mass of protoplasm, was seen to extrude "arms" (pseudopodia), fashion a "mouth," and digest its prey in a prototypic stomach (digestive vacuole). If it could achieve this degree of sophistication, then perhaps it could, in an elementary way, also think? Butler was quite open with his premises [33]: "*Given* a small speck of jelly with some power of slightly varying its actions in accordance with slightly varying circumstances and desires – *given* such a jelly speck with a power of assimilating other matter, and thus, of reproducing itself, *given* also that it should be possessed of a memory and a reproductive system ..." (my italics). He was also inclined to appeal to "people in ordinary life," or to "plain people" [18]:

> The difference between Professor Weismann and, we will say, Heringians consists in the fact that the first maintains the new germ-plasm, when on the point of repeating its developmental process, to take practically no cognisance of anything that has happened to it since the last occasion on which it developed itself; while the latter maintain that offspring takes much the same kind of account of what has happened to it in the persons of its parents since the last occasion on which it developed itself, as people in ordinary life take things that happen to them. In daily life people let fairly normal circumstances come and go without much heed as matters of course. If they have been lucky they make a note of it and try to repeat their success. If they have been unfortunate but have recovered rapidly they soon forget it; if they have suffered long and deeply they grizzle over it and are scared and scarred by it for a long time. The question is one of cognisance or non-cognisance on the part of the new germs, of the more profound impressions made on them while they were one with their parents, between the occasion of their last preceding development and the new course on which they are about to enter.

Thus, Butler downplayed chance (luck) and championed an *intrinsic* capacity for bias (cunning) as the way advantageous characters acquired by parents would be transmitted to their children. In short, he appealed to the doctrine of the inheritance of acquired characters as advocated by Lamarck [34]. Butler even went so far as to assert, in Romanes' words [35], "that a future individual while still in the germ has already participated ... in the cerebral alterations of its parents." However, like Hering, Butler maintained a strictly materialist position. He used the words "intelligent" and "design," often separately, and sometimes together [33], but never in a way as to suggest the involvement of an agency *external* to the organism [18]:

> The two facts, evolution and design, are equally patent to plain people. There is no escaping from either. According to Messrs. Darwin and Wallace, we may have evolution, but are in no account to have it mainly due to intelligent effort, guided by ever higher and higher range of sensations, perceptions and ideas. We are to set it down to the shuffling of cards, or the throwing of dice without the play, and this will never stand. According to the older men [Lamarck], the cards did count for much, but play was much more. They denied the teleology of the time – that is to say, the teleology that saw all adaptation to surroundings as part of a plan devised long ages since by a quasi-anthropomorphic being who schemed everything out much as a man would do, but on an infinitely vaster scale. This conception they found repugnant alike to intelligence and conscience, but, though they do not seem to have perceived it, they left the door open for a design more true and more demonstrable than that which they excluded.

Butler soared beyond the comprehension of his contemporaries on the wings of his conceptual insight that heredity was the transfer of stored information. Although Butler's dalliance with Lamarckism won little support, we recognize today that the robustness of the underlying idea led him closer to solutions to fundamental biological problems such as the origin of sexual reproduction, the sterility of hybrids, and aging [20].

For example, since parental gametes can transfer genetic information to offspring only while the parents are reproductively active (i.e., young), Butler deduced that offspring would receive no information from parental gametes to help them cope with aging. Today, we consider this to be true, but for entirely different reasons – namely, that it is natural selection acting on the young, not on the old, that increases genetic fitness (as measured by number of descendants produced) [36]. Over evolutionary time continued reproduction would have been prevented by early death from natural causes (predation, disease), not by aging itself. The germline normally does not contain information for extending lifespan long after the mean age of reproduction. Exceptions may occur in the case of species where grandparental care of offspring is important (e.g., man) and in species that for various adaptive reasons have incidentally developed additional ways of protecting DNA against damages that can accelerate aging (e.g., a tortoise shell protects against predators, but should also be an effective radiation shield [37]).

For his time, Butler's *correct* equation of heredity with memory (i.e., the transfer of stored information) was beyond the pale, but his *incorrect* Lamarckism seemed plausible. Spencer was a Lamarckist and, at Darwin's behest, Romanes spent several years fruitlessly seeking to show that gemmules containing information for acquired characters could be transferred from parental tissues (soma) to the gonads [38]. And as late as the Darwin Centenary in 1909, Haeckel was proclaiming that "natural selection does not of itself give the solution of all our evolutionary problems. It has to be taken in conjunction with the transformism of Lamarck, with which it is in complete harmony" [39].

Discrete Inheritance

Butler regretted that the men of science had not done their homework regarding early sources (see below), but he did not extrapolate this to the current literature. He assumed that they were conversant with this literature and, where deemed relevant, made it known in their publications. He identified – perhaps because Charles Darwin by way of Francis had suggested it – "one apparently very important problem, which I do not at this moment see how to connect with memory, namely, the tendency on the part of offspring to revert to an earlier impregnation" (e.g., to appear more like the grandparents than the parents; [20]). However, it is possible that he did not know of the private correspondence between Darwin, Huxley, and Hooker where they discussed Naudin's studies of plant hybrids in France [19]. These indicated that parental characters might be inherited discretely (discontinuously) rather than blending (continuously) in their offspring. Thus, after a period of latency, the characters might emerge unchanged in future generations, even though appearing to be absent in immediate offspring. Perhaps privy to this correspondence, Romanes had been thinking along similar lines (Chap. 8). Butler may have got some wind of it, since he wrote in *The Way of All Flesh*:

> It often happens that the grandson of a successful man will be more successful than the son – the spirit that actuated the grandfather having lain fallow in the son and being refreshed by repose so as to be ready for fresh exertion in the grandson. A very successful man, moreover, has something of the hybrid in him; he is a new animal, arising from the coming together of many unfamiliar elements.

However, the prevailing view was that parental characters were blended in their offspring. Jenkin (Chap. 3) saw members of a biological species as enclosed within a sphere and noted that there would be a tendency for individuals to vary toward, rather than away from, the centre of the sphere. Given blending inheritance, it was difficult to see how a *rare* new variant organism, even if personally advantaged by virtue of the variation, would find a mate with a similar variation. Rather, the variant (e.g., white) would be swamped by crossing with the abundant non-variant type (e.g., black) to produce a blend (e.g., grey). With further crossing with the non-variant type (black), the descendants of the variant would soon be indistinguishable from the original non-variant type: "Any favourable deviation must … give its fortunate possessor a better chance of life; but this conclusion differs widely from the supposed consequence that a *whole* species may or will gradually acquire some one new quality, or *wholly* change in *one* direction and in the *same* manner."

Easy access to a mate with the same mutation had been a strength of the Lamarckist doctrine of the inheritance of acquired characters. According to Lamarck (and Butler), organisms subjected to the same environmental provocation would *simultaneously* adapt and so be in possession of the same variant character. Being co-localized in the same environment there would then be little difficulty in their finding a mate with the same adaptation [18, 33]. But, unknown to most of the men of science (and hence to Butler), by 1865, Mendel had obtained evidence for discontinuous, non-blending, inheritance. This fact emerged only two years before Butler's death in 1902. Thus, although Butler was probably not aware of it, his Lamarckian ideas rested on unstable foundations.

There was also an anti-Lamarckian argument that Butler had acknowledged but never satisfactorily dealt with [20]. Among colonial insects, there is usually a greater separation of the germ line from the soma than in other organisms. For example, the germ line is personalized in the form of the queen bee. The soma is personalized in the neuter worker bees. Through "use or disuse" the workers might come to strengthen or deplete characters in a Lamarckian fashion. But how could the corresponding Darwinian "gemmules" transfer back to their queen. Apart from work showing the systemic spread of RNA within a plant, which is probably an antiviral defence [40], there is no evidence that gemmules (i.e., genes) can transfer from soma to germ line *within* a body, let alone *between* bodies (that do not interact sexually). Many of the phenomena associated with colonial insects have found explanations in terms of modern selfish gene theory [41].

Debate with Romanes

Butler wrote to Francis Darwin (Nov. 1877) about *Life and Habit*:

> Nothing would surprise me less than to see something sprung upon me in reviews ... which cuts the ground completely from under me; and, of course, I neither expect nor give quarter in philosophical argument. We want to get on to the right side; and neither your father, I take it, nor I care two straws *how we get on to the right side* so long as we get there. Neither do we want half refutations nor beatings about the bush. We want to come to an understanding as to what is true and what is false as soon as possible; and we know well that we score more by retracting after we have been deeply committed, than by keeping on to our original course when a new light has been presented to us.

The book, usually dated 1878, was formally published in early December 1877, and Francis Darwin responded (Dec. 28): "I have read [it] with great pleasure. ... I should have agreed with you more if you had said that memory and reproduction (or growth) are both consequences of the same property of matter – the property which makes a series of molecular states follow each other in a certain order because they have done so before." The book was to be warmly reviewed by Alfred Wallace in *Nature* (1879) as a remarkable work, both original and logically complete, that was "in great part complementary" to Darwinian ideas [42].

Unlike Hering, a scientist who had made a one-time sortie into the professionally hazardous area of evolutionary speculation, for the next two decades the polymathic Butler, while claiming disingenuously to "know nothing of science" [20], continued to develop his theme, frequently employing the technique of subject-object inversion of which he was master. For example, *Life and Habit* contained the memorable phrase: "A hen is only an egg's way of making another egg". In the twentieth century, Dawkins was to argue similarly that a body is only a gene's way of making another gene [41]. Jenkin had noted in 1867 that for "an impartial looker-on" the best way to question the Darwinians was to "admit the facts, and examine the reasoning." This was the approach taken by Butler: "The contention against me is that I have made no original researches, but have, as a general rule, taken my facts at second hand. ... But what are the Darwin, Huxley and Tyndall people good for if we cannot rely upon their facts and proceed to make deductions from them."

Late in his life Butler declared that "I have never written on any subject unless I believed that the authorities on it were hopelessly wrong. If I thought them sound, why write?" This was evident in *Life and Habit*, and even more so in his second major work, *Evolution, Old and New* (1879). This was a major indictment of the Darwinians for failing to acknowledge their intellectual antecedents – Buffon, Doctor Erasmus Darwin, and Lamarck [33]. Writing anonymously in *The Athenaeum*, Cambridge philosopher James Ward declared [43] that:

> No one previous to Mr. Butler has shown in detail how far the earlier biologists had advanced towards a denial of the older views of the origin of species by special creation. ...The three great masters of Darwinism have treated of the more recent speculation on the origin of species; but neither the historical chapters of Haeckel's 'History of Creation,' nor the historical sketch prefixed to the 'Origin of Species,' nor Prof. Huxley's article on 'Evolution' in the *Encyclopaedia Britannica*, deals quite fairly with the immediate predecessors of Mr. Darwin. Mr. Butler, coming to his subject with a polemical interest in favour of the older inquirers, has more sympathy with their views and expounds them more correctly.

The book, to put it mildly, disturbed the Darwinians, particularly Romanes who, by virtue of his deep interest in both evolution and mental function, was best prepared to understand Butler. While in her account, Romanes' biographer discretely omitted Butler's name [38], it was Romanes who wrote to Francis Darwin (Dec. 14, 1889) referring to Butler as "a lunatic beneath all contempt – an object of pity were it not for his vein of malice."

In his 1884 book *Mental Evolution in Animals*, Romanes – today acknowledged as one of the founders of evolutionary psychology – used the term "hereditary memory" in the context of instinctual behavior where there had been no opportunity for learning. This, or similar terms, had been used by three previous authors: Spencer in his 1855 *Principles of Psychology*; the Regius Professor of Modern History at Cambridge, Charles Kingsley, in an article concerning bird migration (1867); George Lewes in his *Problems of Life and Mind* (1874). However, Hering in 1870 and Butler in *Unconscious Memory* (1880) had gone much deeper. Butler later pointed out that "none of these writers (indeed no writer that I know of except Professor Hering of Prague) ... has shown a comprehension of the fact that these expressions [i.e., "hereditary memory"] are unexplained so long as 'heredity,' whereby they explain them, is unexplained; and none of them sees the importance of emphasizing Memory, and making it as it were the keystone of the system."

Certain activities manifest early in life, before there has been an opportunity to learn them, are referred to as instinctual. Human babies smile at about five weeks of age, and this occurs even in babies that are born blind. In his *Mental Evolution in Animals,* Romanes was happy to consider instinctual activities, such as bird migration and nest-making, as dependent on inherited ("ready-formed") information [44]:

> Many animals come into the world with their powers of perception already largely developed. This is shown … by all the host of instincts displayed by newly-born or newly-hatched animals. … The wealth of ready-formed *information*, and therefore of ready-made powers of perception, with which many newly-born or newly-hatched animals are provided, is so great and so precise, that it scarcely requires to be supplemented by the subsequent experience of the individual. (My italics)

But much to Butler's frustration, he would not move beyond this [35]:

> Mr. Romanes … speaks of 'heredity as playing an important part *in forming memory* of ancestral experiences;' so that whereas I want him to say that the phenomena of heredity are due to memory, he will have it that the memory is due to heredity …. Over and over again Mr. Romanes insists that it is heredity which does this or that. Thus, it is '*heredity with natural selection which adapt* the anatomical plan of the ganglia;' but he nowhere tells us what heredity is any more than Messrs. Herbert Spencer, Darwin and Lewes have done. This, however, is exactly what Professor Hering, whom I have unwittingly followed, does. He resolves all phenomena of heredity, whether in respect of body or mind, into phenomena of memory. He says in effect, 'A man grows his body as he does, and a bird makes her nest as she does, because both man and bird remember having grown body and made nest as they now do … on innumerable past occasions. He thus reduces life from an equation of say 100 unknown quantities to one of 99 only, by showing that heredity and memory, two of the original 100 unknown quantities, are in reality part of one and the same thing.

In March 1884, Romanes was taken to task in *The Athenaeum* (again by the anonymous James Ward) for not having the "literary courtesy" of citing Butler's work in *Mental Evolution in Animals* [45]. There followed a heated debate, enjoined by Butler, Lankester, Romanes, and Spencer.

Butler held that the cause of variation was not *external* (e.g., a supernatural creator bringing about intelligent design) but was *internal* to the organism (i.e., it was an explainable feature of an organism's biology). Whether the 22-year-old Bateson, then very much into acorn worms (Chap. 2), took any interest in this we do not know. However, we can note that James Ward anonymously praised Butler for tackling the problem of the origin of variation, a problem which the Darwinists had dismissed but which Bateson was to turn to in the 1890s [46]:

> Mr. Butler's service to students of descent consists in the prominent way in which he has called attention to the causes, as distinct from the fact, of evolution. Mr. Darwin once for all established the fact that species are evolved by descent with modification. … But when we come to the further question as to what are the causes which led to the variations out of which species were evolved, Mr. Darwin, with the true caution of science, declined to give any general answer. … It is true that he elaborated the theory of the survival of the fittest to explain how variations developed when once originated. … At this stage of evolution doctrine, after it had crushed and pulverized the theory of special [i.e. divine] creation and was on the look out for causes of variation, Mr. Butler's speculations come in. In seeking for a cause of variation, he at last found it in the "sense of need" on the part of the animal which would lead to the greater use of the organ and … to change in the organ itself [i.e. Lamarckism] …. Thus, Mr. Darwin would hold that the animal from which the giraffe has been evolved *chanced* to have a long neck, and was thus enabled to get at food inaccessible to his brethren, and then transmitted this quality to his descendents. Mr. Butler would rather say the long neck of the *Ur*-giraffe was due to no chance, but to the intelligence of the animal, which caused it to stretch its neck, and thus made it grow longer by use, just as the muscle of a blacksmith grows larger.

Butler had applied this principle to instincts, conceived as originally intelligent actions that became unconscious and were then, somehow, transmitted to the germ line. As far as we know, a pianist, having practiced a piece so well that it can be played virtually unconsciously, does not pass knowledge of that piece through the germ line to children. Yet, Hering and Butler pointed out that some psychomotor activities of comparable complexity are passed on [21]:

> Not only is there reproduction of form, outward and inner conformation of body, organs, and cells, but the habitual actions of the parent are also reproduced. The chicken on emerging from the eggshell runs off as its mother ran off before it; yet what an extraordinary complication of emotions and sensations is necessary in order to preserve equilibrium in running. Surely the supposition of an inborn capacity for the reproduction of these intricate actions can alone explain the facts. As habitual practice becomes a second nature to the individual during his single lifetime, so the often-repeated action of each generation becomes a second nature to the race.

Although Butler supposed "a memory to 'run' each gemmule," of course he knew the material basis of the storage of inherited information no more than he knew the material basis of the storage of somatically acquired mental information. We now know that DNA is the former, but still have little understanding of the latter. Butler implied that both forms of storage were the same. Current evidence does not support this. Just as segments of DNA (genes) contain the information for the construction, by some chain of events, of, say, a hand, so segments of DNA should contain the information for a particular pattern of cerebral "wiring" (an "anatomical plan of the ganglia") that would, by some chain of events, be manifest as a particular instinct. If and how such germ-line memories might relate to memories acquired during individual lifetimes ("cerebral alterations") is still unknown. Butler's genius was in seeing that inheritance and mental function, like today's computers (which he anticipated in his article "Darwin Among the Machines" [14]), are all "modes of memory," a property that distinguishes the living from the non-living [20]:

> Life is that property of matter whereby it can remember. Matter which can remember is living; matter which cannot remember is dead. *Life, then, is memory*. The life of a creature is the memory of a creature. We are all the same stuff to start with, but we remember different things, and if we did not remember different things we should be absolutely like each other. As for the stuff itself, we know nothing save only that it is 'such as dreams are made of'.

Butler Ignored

Butler took the opportunity of the publication of *Selections from Previous Works* [35] to document the dispute with Romanes. Although Butler had emerged the clear winner, it was pointed out by the anonymous James Ward [46] that: "Mr Butler chose to adopt a tone towards Mr. Darwin which could only alienate all who reverence the name of the great naturalist." Given the prominence of Darwin's supporters, who had petitioned successfully for his burial in Westminster Abbey near to Isaac Newton, it is not surprising that Butler's views did not thrive. Butler's own initial diffidence, and later bitter disdain for his contemporaries, may partly be to blame. His first evolution book, *Life and Habit*, began modestly claiming to be ignorant of science and hoping just to entertain. His last evolution book, *Luck or Cunning*, while chiding the Darwinians for being scientists, but not being literary (since they had chided him for being literary and not a scientist), proclaimed that he was writing for the future not the present.

The conflict with Darwin took a nasty turn in 1879. In February, an article on Erasmus Darwin appeared in the German journal *Kosmos*. In May, Butler's *Evolution, Old and New*, also dealing with Erasmus Darwin, was published. Charles Darwin sent a copy to the German author, Ernst Krause, who was revising the Erasmus Darwin article for translation into English. Darwin advised that he "not expend much powder and shot on Mr. Butler, for he really is not worthy of it. His book is merely ephemeral" [10]. In November the translation was published as *Erasmus Darwin, by Ernst Kraus, translated from the German by W. S. Dallas, with a preliminary notice by Charles Darwin*; it contained additions and deletions that, although he was not named, constituted an obvious attack on Butler: "Erasmus Darwin's system was in itself a most significant first step in the path of knowledge his grandson has opened up for us, but to wish to revive it at the present day, as has actually been seriously attempted, shows a weakness of thought and a mental anachronism which no one can envy."

Should we regard these words as an independent initiative by Kraus or the result of egging on by Charles Darwin? In any case, predictably, Butler took exception regarding both content and, more importantly, timing. Darwin's "preliminary notice" had mentioned the May appearance of Butler's book and implied that the negative remarks had been in the February German article *before* Butler's book appeared. This gave the false impression that Kraus had not had Butler's book in mind, and hence was entirely unbiased with respect to Butler.

As explained by Henry Festing Jones in his biography of Butler [10], Butler wrote to Darwin who replied admitting that Kraus had altered the article. It later emerged, though Darwin had forgotten it, that an early draft of his "preliminary notice" had stated: "Dr. Krause has taken great pains and has largely added to his essay since it appeared in *Kosmos*." However, these words were later deleted in error. This was not made known to Butler. Furthermore, Butler did not know, but he later heard from two sources, that Krause's written attack had initially been "of a very severe character," but had been toned down at Darwin's request. Indeed, Krause himself had later noted in *Nature* (Jan. 27, 1881):

> Since … I thought it desirable to point out that Dr. Erasmus Darwin's views concerning the evolution of animated Nature still satisfy certain thinkers, … I have made some remarks upon the subject … without however naming Mr. Butler. And I here emphatically assert, that although Mr. Darwin recommended me to omit one or two passages from my work, he neither made nor suggested additions of any kind.

With delightful irony, Jones continued [10]:

> Butler was not satisfied. He was ready to strain points, but no amount of straining would turn this into an explanation he could accept. It was admitted that, in the evolution of Dr. Krause's article from the German magazine to the English book, variations had arisen, but the question as to their cause and origin was left untouched. Butler took the teleological view that they had been put there on purpose, and thought it ought to have occurred to Mr. Darwin to mention them; especially as it did occur to him to mention that *Evolution Old and New* had appeared since the original article, of which he was giving a translation guaranteed to be accurate.

Having informed Darwin that he was so doing (Jan. 21), Butler brought the matter to public attention in a letter in *The Athenaeum* (Jan. 31, 1880). Perhaps having received an early copy from editorial sources, the Darwinians were already in a flap. Drafts of possible replies were circulated. What was to be done about this "clever and unscrupulous man?" Eventually, seeking the advice of Huxley, and against the advice of his son Francis, Darwin decided not to reply. Butler related the affair in *Unconscious Memory*, which was shortly to go to press, and in 1882, after Darwin's death, added a preface to the second edition of *Evolution, Old and New*:

> I have always admitted myself to be under the deepest obligation to Mr. Darwin's works; and it was with the greatest reluctance, not to say repugnance, that I became one of his opponents. I have partaken of his hospitality, and have had too much experience of the charming simplicity of his manner not to be among the readiest to at once admire and envy it. It is unfortunately true that I believe Mr. Darwin to have behaved badly to me; this is too notorious to be denied; but at the same time I cannot be judge of my own case, and that, after all, Mr. Darwin may have been right and I wrong.

In December 1884, Romanes had an opportunity to set the record straight in an article in *Nineteenth Century*. The title – "The Darwinian Theory of Instinct" – foretold that an apology would not be forthcoming [47]. Neither Hering nor Butler was named:

> Mr Darwin's theory does not, as many suppose, ascribe the origin and development of all instincts to natural selection. This theory does, indeed, suppose that natural selection is an important factor in the process; but it neither supposes that it is the only factor, nor even that, in the case of numberless instincts, it has had anything at all to do with their formation. … Just as, in the lifetime of the individual, adjustive actions which were originally intelligent may by frequent repetition become automatic, so, in the lifetime of the species, actions originally intelligent may, by frequent repetition and heredity, so unite their efforts on the nervous system that the latter is prepared, even before individual experience, to perform adjustive actions automatically which, in previous generations, were performed intelligently. This mode of origin of instincts has been appropriately called the 'lapsing of intelligence,' and it was fully recognized by Mr. Darwin as a factor in the formation of instinct.

This was pure Hering–Butlerism. The fact of prior "need" was also acknowledged:

> Instincts are not rigidly fixed, but are plastic, and their plasticity renders them capable of improvement or of alteration, according as intelligent observation requires. The assistance which is thus rendered by intelligence to natural selection must obviously be very great, for, under any change in the surrounding conditions of life which calls for a corresponding change in the ancestral instincts of the animal, natural selection is not left to wait, as it were, for the required variations to arise fortuitously; but it is from the first furnished by the intelligence of the animal with the particular variations which are needed.

As for the extent to which Charles Darwin had supported this, the best quotation Romanes could muster was from a personal letter. Concerning the anatomical precision with which a bee stings to paralyse its prey, Darwin wrote:

> Bees show so much *intelligence* in their acts, that it does not seem improbable to me that [they] … observed by their intelligence that if they stung them in one particular place, … their prey was at once paralysed. It does not seem to me at all incredible that this action should then become instinctive, i.e. memory transmitted from one generation to another.

Thus, Romanes' article tended to give the impression that Butler had been covering ground that had been fully explored by Darwin. By the 1870s, Darwin had moved well beyond classical Darwinism (i.e., the view of the over-arching power of natural selection) to the view that characters acquired or lost through "use or disuse," could be passed on to offspring (i.e., Larmarckism). As we saw in Chap. 5, Romanes embraced Darwin's new-found Lamarckism both ideologically and practically (i.e., years of fruitless experiments). Romanes became Zoological Secretary to the Linnean Society in 1881. Butler and Romanes met – or, at least, were in the same room – at a meeting of the Linnean Society in early December 1884. A paper of his friend, Alfred Tylor (in memory of whom Butler was to dedicate *Luck or Cunning*), was to be read by Jackson, the botanical secretary. Besides Romanes, others present included Thiselton-Dyer and ornithologist Henry Seebohm. After the paper, Butler was allowed to speak for six minutes, an opportunity he took to describe part of the paper that Jackson had omitted as too speculative for the Linnean Society. Seebohm invited him to dinner.

In 1887, Francis Darwin, who had edited the *Life and Letters of Charles Darwin*, added to Butler's sense of injury by including the statement: "The affair gave my father much pain, but the warm sympathy of those whose opinion he respected soon helped him to let it pass into a well-merited oblivion." Expressing regret that Francis Darwin had chosen to reopen the affair, Butler responded (Nov. 22) in *The Athenaeum* (published Nov. 26) pointing out that there remained no public correction of his father's misstatements. Romanes drafted a letter for Francis to use in reply, but Francis (Nov. 26) declined to do so [38]. Instead, Francis added what he thought might be an ameliorative footnote to a new edition of the *Erasmus Darwin* translation, which only added fuel to Butler's fire – again publicly displayed (*The Academy,* Dec. 17). In a letter to his sister Charlotte (May 18, 1888) Romanes wrote of commotion among "the saints" upon earth [38]: "One of these same saints has been behaving outrageously in print, and everybody is full either of jubilation or indignation at what he has been writing about Darwin and Darwinism. F. Darwin asked me to do the replying, and today I am returning proof of an article for the 'Contemporary Review'."

Butler's last work on evolution was entitled "Thought and Language." It took the form of an address first given in 1890 and, in revised form, in 1894 [18]. Here, without mentioning Romanes, he examined various aspects of mental evolution, a subject that had preoccupied Romanes in the 1870s. Butler approved of the Marquis of Salisbury's presidential address to the British Association at Oxford in 1894

(Chap. 5). While lamenting the recent death of Romanes, Salisbury's remarks, when translated in Butlerian terms, implied that "luck" would not do and that "cunning" was needed to explain evolution [48]: "Darwinian theory has not effected the conquest of scientific opinion; and still less is there any unanimity in the acceptance of natural selection as the sole or even the main agent of whatever modifications may have led up to the existing forms of life. Two of the strongest objections … appear still to retain all their force." First, Lord Kelvin had shown that calculations of the age of the earth based on its rate of cooling did not allow long enough for evolution to have occurred. Second Weismann himself had stated that "we accept natural selection … because all other apparent principles of explanation fail us." While conceding that the breeder, through artificial selection, could produce new types, Salisbury asked:

> But in natural selection who is to take the breeder's place? Unless the crossing is properly arranged, the new breed will never come into being. What is to secure that the two individuals of opposite sexes in the primeval forest, who have been both accidentally blessed with the same advantageous variation, shall meet, and transmit by inheritance that variation to their successors? Unless that step is made good, the modification will never get a start; yet there is nothing to insure that step, except pure chance. … It seems strange that a philosopher of Professor Weismann's penetration should accept as established a hypothetical process the truth of which he admits that he cannot demonstrate in detail, and the operation of which he cannot even imagine.

Huxley, who had introduced Salisbury, was obliged to sit silently through it all. Butler later commented (Aug. 20): "What is wanted is not to reconcile science and religion – let them fight it out – but to reconcile science and common modesty, accuracy, and straightforwardness which, so far as I can see at present, have a very righteous quarrel with her for innumerable insults she has heaped upon them."

In 1904, Francis Darwin counselled his sister Henrietta not to mention the Butler matter in her biography of their mother *Emma Darwin*. Under pressure from the Darwin family, he had edited out his father's comments on Butler when making the Charles Darwin *Autobiography* public in 1887 as part of *Life and Letters of Charles Darwin*. Grand-daughter, Nora Barlow, reintroduced the deleted section in a new version of the *Autobiography* in 1958 [49]: "Owing to my having accidentally omitted to mention that Dr. Krause had enlarged and corrected his article in German before it was translated, Mr. Samuel Butler abused me with almost insane virulence. How I offended him so bitterly, I have never been able to understand." Not content to let the matter rest, and with no fear of recrimination, Nora Barlow described Butler as a "bitter" person of "intense emotional virulence" with "persecution mania," who, egged on by his friend Miss Savage, liked Dr. Erasmus Darwin and "disliked the younger upstart [Charles] Darwin" to such an extent that "resentment had … warped his saner judgement." Yet, surprisingly, she concluded: "So in the end Francis Darwin … thought Butler had a real cause of complaint."

To some extent the Darwin's made amends to Butler through the husband of another grand-daughter, Margaret (a daughter of George Darwin and a sister of Gwen Raverat). She married Geoffrey Keynes in 1917. Through A. T. Bartholomew of the Cambridge University Library they became friendly with Festing Jones from whom Bartholomew inherited the office of Literary Executer to Samuel Butler. When Bartholomew died, this was passed on to Keynes who, with Brian Hill, became responsible for publishing *Letters between Samuel Butler and Miss E. M. A. Savage* and selections from *Samuel Butler's Notebooks* [50].

Butler's Science

In a lecture at the Royal Institution in March 1878 on "The Analogies of Animal and Plant Life" [51], Frances Darwin discussed movement in plants and instinctive behaviour:

> It seems to me that the presence of what Mr. Lewes calls 'thought consciousness' is not the crucial point, and that if it is allowed that the sensitive plant is subject to habit (and this cannot be denied), it must, in fact, possess the germ of what, as it occurs in man, forms the groundwork of all mental physiology.

He cited Romanes but not Butler. However, divorcing personal animosity toward Butler from the latter's science, in subsequent years Francis Darwin increasingly recognized his work. In a paper in 1892 "On the Artificial Production of Rhythm in Plants," he and Pertz cited Butler's *Life and Habit* [52] and noted that: "This repeating power may be that fundamental property of living matter which stretches from inheritance on one side to memory on the other."

Furthermore, in 1901, in a lecture "On the Movements of Plants," he noted that "if we take the wide view of memory which has been set forth by Mr. S. Butler (*Life and Habit*, 1878) and by Professor Hering, we shall be forced to believe that plants, like all other living things, have a kind of memory." This was made explicit in 1908 at his inaugural address as President of the BA in Dublin (Chap. 11). Here he noted his indebtedness to Semon's *Die Mneme* (Chap. 5) in a phrase reminiscent of Butler:

> In the case of memory the introduction of a link between one mental rhythm into another can only occur when the two series are closely similar, and this may remind us of the difficulty of making a cross between distantly related forms. Enough has been said to show that there is a resemblance between the two rhythms of development and memory; and that there is at least a *prima facie* case for believing them to be essentially similar. It will be seen that my view is the same as that of Hering, which is generally described as the identification of memory and inheritance.

Francis Darwin accorded Butler two brief footnotes, one referring to "his entertaining book *Life and Habit*," and the other to *Unconscious Memory* – the latter only as a source for a translation of Hering. The address ended in a distinctly Lamarckian note, again, highly reminiscent of Butler (and the later works of Charles Darwin): "Nor shall I say anything more as to the possible means of communication between soma and germ cells. To me it seems conceivable that some such telegraphy is possible." Bateson was in the audience and was not impressed (Sept. 3, 1908):

> F. Darwin's address grieved me a good deal. Very nicely put, full of pretty turns but oh! So thin! Such an address in [the] name of Science weakens the common front. If *he* can allow himself such vapourings, everyone will feel entitled to the same privilege. He must have seen me looking rather black, for he interjected that 'he felt apprehension of what Mr. Bateson would say to him when he descended from the platform and became again accessible'!

It seems that Bateson did not keep his views to himself. Two days later he wrote: "My rebuke to the President yesterday seems to have been well received, by my friends at least."

Qualified Recognition

Butler was aware of Bateson's *Materials*, but by 1894, his attention had shifted and it is doubtful that he read it. To what extent was Bateson aware of Butler and his four books on evolution? Despite being in almost day-to-day contact with Francis Darwin, there is little indication that in the 1880s and 1890s Bateson was aware. In the context of Galton, the word "*Erewhon*" appeared once in Bateson's lecture notes on "Heredity" dated November 1901 (see Chap. 5):

> First real modern progress due to Galton – *Hereditary Genius*, *Natural Inheritance* – by statistical methods. [He] Conceived *idea of a Law of Heredity* – a numerical expression of probability. Idea of correlation and of regression. When Galton wrote [there was] strong scepticism of the whole thing. *Erewhon.* Bach family. Comte's table. Nature and nurture.

This implies awareness that Butler had made an evolutionary contribution at about the same time as Galton. But, from correspondence with Francis Darwin around 1908 (see below), it is apparent that Bateson knew little of Butler's dispute with Charles Darwin.

Although completed in 1885 with the encouragement of Miss Savage (the nearest he had to a female soul mate), *The Way of All Flesh* was published posthumously. Hailing "the late Samuel Butler, in his own department the greatest English writer of the latter half of the XIX century," its merits were extolled by a major literary figure – George Bernard Shaw. Perhaps indirectly, this did as much as anything else to draw Butler to Bateson's attention. In June 1906, Shaw wrote in the preface to his play *Major Barbara*:

> It drives one almost to despair of English literature when one sees so extraordinary a study of English life as Butler's posthumous *Way of All Flesh* making so little impression that when, some years later, I produce plays in which Butler's extraordinary fresh, free and future-piercing suggestions have an obvious share, I am met with nothing but vague cacklings about Ibsen and Nietzsche, and am only thankful that they are not about Alfred de Musset and Georges Sand.
>
> Really, the English do not deserve to have great men. They allowed Butler to die practically unknown, whilst I, a comparatively insignificant Irish journalist, was leading them by the nose into an advertisement of me which has made my own life a burden. In Sicily there is a Via Samuele Butler. When an English tourist sees it, he either asks 'Who the devil was

> Samuele Butler?' or wonders why the Sicilians should perpetuate the memory of the author of Hudibras [an earlier, unrelated, Samuel Butler].
>
> Well, it cannot be denied that the English are only too anxious to recognize a man of genius if somebody will kindly point him out to them. Having pointed myself out in this manner with some success, I now point out Samuel Butler, and trust that in consequence I shall hear a little less in future of the novelty and foreign origin of the ideas which are now making their way into the English theatre through plays written by socialists. There are living men whose originality and power are as obvious as Butler's and when they die that fact will be discovered. Meanwhile I recommend them to insist on their own merits as an important part of their business.

In October 1908, in his inaugural address as Professor of Biology at Cambridge, Bateson spoke of "the keen satirical vision of Sam Butler on the pleasant mountains of Erewhon." After Christmas, he was reading Butler's *Evolution, Old and New*, writing to Beatrice who was en route with the children to Eastbourne (Dec. 30. 1908):

> Had high tea at 6 and sat till 11.45 reading Butler's other books. In edition 2 of Evol. Old and New he put a Preface which considerably softens the attack [on Darwin] in some ways. It was written as Darwin died after proofs were passed. The language is admirably chosen, and though it takes off some of the edge, the dignity and significance of the controversy gains. F. D. writes as you see. I fancy he is against putting anything on paper. Return it, please.

An undated letter to Francis Darwin was probably written at about this time:

> Two days ago I got a copy of S. Butler's 'Unconscious Memory.' I had never seen it before. Though it will be stirring up an old vexation, I cannot refrain from writing to you about it. What I want to say is that I feel very strongly that in some way or other the story ought to be cleaned up. Butler's writing was of course meant to give the greatest possible offence, and I can understand how your father may have decided that such an attack could be disregarded. So for his time it might be, but I do think that the material for an explanation should be put together and preserved, even if not published now. Butler's work is beginning to attract great attention, which your Address has of course increased. I do not like to think of such a book falling into really hostile hands hereafter, when no more defence can be produced. It is likely enough that you have prepared such an answer. If so my point is met. I can't tell you what a shock I had in reading that book. I felt as once or twice I have done when, by some accident, I have overheard something not meant for me to hear. I venture to send this letter because I remember that some time ago I asked you what the row with Butler had been, and you told me that one of your brothers had thought as I do. The obvious explanation that occurs to me is that Krause and Dallas were responsible. I do not know if either of them is still living.

In his 1909 centenary essay, Bateson repeatedly mentioned Butler's writings, albeit mainly as footnotes. Having lauded Darwin's demonstration that "the problems of Heredity and Variation are soluble by observation," a footnote read:

> Whatever be our estimate of the importance of Natural Selection, in this [Darwin's demonstration of it] we all agree. Samuel Butler, the most brilliant, and by far the most interesting of Darwin's opponents – whose works are at length emerging from oblivion – in his Preface (1882) to the 2nd edition of *Evolution Old and New*, repeats his earlier expression of homage to one whom he had come to regard as an enemy.

Butler's words of homage were then quoted. However, Bateson gave the considered opinion that: "Butler's claims on behalf of Buffon have met with some acceptance;

but after reading what Butler has said, and a considerable part of Buffon's own works, the word 'hinted' seems to me a sufficiently correct description of the part he played." Concerning Weismann's doctrine of separation of germ line from soma, Bateson noted: "It is interesting to see how nearly Butler was led by natural penetration, and from absolutely opposite conclusions, back to this underlying truth." He then gave an extensive quotation from Butler:

> So that each ovum when impregnate should be considered not as descended from its ancestors, but as being a continuation of the personality of every ovum in the chain of its ancestry, which every ovum *it actually is* quite as truly as the octogenarian *is* the same identity with the ovum from which he has been developed. This process cannot stop short of the primordial cell, which again will probably turn out to be but a brief resting place. We therefore prove each one of us to *be actually* the primordial cell which never died nor dies, but has differentiated itself into the life of the world, all living beings whatever, being one with it and members one of another.

It appears that, perhaps because of repeated disparagement by people whose opinions he valued, it took some time for Bateson to appreciate Butler. Bateson's library, now held in the John Innes Institute, contains most of Butler's works, but those which are dated, are dated around 1908–1909. Bateson's handwriting at the front of *Unconscious Memory* reads "W. Bateson. 22 Dec 1908." Pasted into the book is a copy of a letter from Butler's literary executor, R. A. Streatfeild (Oct. 21, 1909), which indicates that Bateson had written at some earlier date perhaps cautioning that the Darwin family might be upset if the book were reissued:

> Dear Professor Bateson, Many thanks for your letter. It was very kind of you to write and tell me of the passage in 'Luck and Cunning', which is perfectly delightful. I am afraid I had quite forgotten it. With regard to the republication of 'Unconscious Memory', I am afraid that I am now committed beyond recall. Fifield told me that it was often asked for, and I remember that Butler himself said something to me, not long before his death, about reissuing it, so I felt that perhaps it was my duty to republish it. I shall be very sorry if by doing so I give offence to any of the Darwin family, but I fear that the reissue of any of Butler's books is likely to annoy somebody or other. Yours truly, R. A. Streatfeild.

Bateson discerned Butler's influence on Shaw when he went to see *Fanny's First Play*, which opened in London in 1911. He wrote to Beatrice (Aug. 25):

> Last night with GCMS [Moore Smith] to *Fanny's 1st Play*. Never again will I say we have no actress! 'Mrs. Knox' – Cicely Hamilton is as good as can possibly be. D. Minto is good too, but it is a 'finished study,' not the life. You must go. The whole *propos* is stupid and a mere thread for Shaw's remarks, and I found the critics a bore, simply. The humour is almost pure Butler and one feels the 'Way of All Flesh' atmosphere through it all. I found myself wondering whether the parents or the children were the greater monstrosities, and on the whole I think perhaps the children were. The parents after all had a scheme which would work for a good while in a dull fashion, but there is not much lasting fun in bashing the police. I don't know when I have laughed so [much]. … Nora Barlow came yesterday. She will grow into a very formal little lady I expect.

In 1909, Bateson visited Verrières, the home of France's "Hurst," the horticulturalist Philippe de Vilmorin, a man "of the world who had shot big game and raced yachts in the best company, known everybody and seen everything" (Bateson 1917). A subsequent letter from Vilmorin (Oct. 12, 1909) indicated that Bateson's newfound enthusiasm had been shared: "I am very [obliged] … to you for your indication of Butler's books. Of course 'Erewhon' and 'The Way of All Flesh' are the best, but 'Life and Habit' is full of humour and deep ideas, although not quite 'digestible' it opens one's mind." Bateson went on to read some of Butler's later non-evolutionary works (April 14, 1916): "Have been enjoying Butler's 'Authoress of the Odyssey.' There is a really plausible case any how. As to the geographical thesis I am less impressed."

Butler's Partial Rehabilitation

Unconscious Memory was reissued in 1910 with a new forward by Marcus Hartog of Queen's College, Cork. Hartog had corresponded with Butler in 1891 and the following year in an article on "Problems of Reproduction" [53] had cited Butler's *Life and Habit* as anticipating the studies of Maupas and others, which had shown that sexual conjugation in protozoa could avert a bodily degeneration that might relate to aging in multicellular organisms (Chap. 13). And in 1897, in an article on "Fundamental Principles of Heredity," Hartog had declared that "the most satisfactory explanation, perhaps, is that put forward by Hering and Butler, the latter of whom has written with singular freshness and an ingenuity which compensates for the author's avowed lack of biological knowledge" [54].

However, in his 1910 forward to *Unconscious Memory*, Hartog declared that "I do not in the very least share Butler's views. … Butler everywhere undervalues the important work of elimination played by Natural Selection." Nevertheless, in 1914 in a paper entitled "Samuel Butler and Recent Mnemic Biological Theories," Hartog pointed out that Richard Semon's single reference to Hering and two references to Butler in *Die Mneme* barely reflected his debts to these authors. Comparing *Life and Habit* with *Die Mneme*, paragraph by paragraph, Hartog noted that "the confluence of his [Semon's] thought with Butler's is at this point absolute, and the same holds good for a great part of *Die Mneme*" [55, 56]. In short, Francis Darwin's indebtedness to Semon (see above) was misplaced [57].

Meanwhile, many were working to re-establish Butler's reputation, including his friend Jones, who in 1911 donated a Butler self-portrait (Fig. 19.1) to St. John's College. Jones was also a friend of William Rothenstein who painted Francis Darwin's portrait and frequently stayed with him at Cambridge. Here Rothenstein encountered Bateson and Beatrice; indeed, Bateson later supported Rothenstein for membership of the Athenaeum club. It may have been Rothenstein who awakened them to Butler since in his *Men and Memories* he later noted: "Festing Jones gave me an account of the quarrel between Butler and Darwin, which was new to Francis; and the misunderstanding was finally explained in a pamphlet written before his life of Butler appeared, a pamphlet wherein Francis Darwin as well retracted some of the hard opinions of Butler he had formerly published."

At Hartog's instigation, from 1908 until the First World War, there were annual "Erewhon dinners." The attendance, initially all male, grew from 32 in 1908 to 53 in

Fig. 19.1 Samuel Butler (self-portrait 1878)

1909 when George Bernard Shaw joined. In 1910, there were 58. Bateson commented (Aug. 15):

> I did not enjoy my Butler dinner so much as last year's. I was in seat of great honour, on left of Chairman and next to Forbes Robertson [Shakespearian actor]. Jones was delightful, but though Robertson and I had a great deal of talk he clearly felt that I did not come within his range and it was no use troubling about anyone so remote from reality. He had Fuller Maitland [music critic] the other side, who evidently suited him much better. Shaw was rather amusing and also Birrell, though … too long.

Later in the year, Jones was invited to speak on Butler at St. John's College with Bateson proposing a vote of thanks. The following year at the Erewhon dinner (July 14), with 75 in attendance, Bateson noted how appropriate it was that Butler's portrait was now at St. John's. Concerned with posthumously reconciling the differences between Butler and Charles Darwin, in 1910 Francis Darwin wrote to the Director of the Fitzwilliam Museum, Cambridge, apparently unaware that Jones was writing a biography of Butler. Darwin stated that he had letters from Huxley and Leslie Stephen that would be pertinent to a biography. This led in 1911 – with input from Francis Darwin (who shared the expenses) – to the publication by Jones of *Charles Darwin and Samuel Butler. A Step towards Reconciliation* [58]. This aimed to portray the dispute as a misunderstanding and was a prelude to Jones' *Life of Samuel Butler* that was to appear in 1919 [10]. Jones visited Bateson at the John Innes Institute and the Annual Reports document donations from him in 1913 and 1915. Meanwhile, attendance at the Erewhon dinners continued to increase, and at the seventh in 1914 ladies, including Mrs. Shaw, were present.

On occasions, Bateson would write of a "plan," implying some sense of information. Thus, in *Materials* he wrote: "In the case of Sex in the higher animals we are familiar with the existence of a race whose members are at least dimorphic, being formed either upon one plan or upon the other, the two plans being … alternative and mutually exclusive." But he never really appreciated Butler's equation of heredity with stored information (memory) and tended to throw this baby out with Butler's Lamarckist bathwater. In *Problems* (1913), Bateson wrote of the strong case made for the "transmission of acquired adaptations," which he dismissed as no better than creationism:

> Those who desire to see how strong it [the case] is, should turn to Samuel Butler's *Life and Habit*, and even if in reading they reiterate to themselves that no experimental evidence exists in support of the propositions advanced, the misgiving that none the less they may be true is likely to remain. … The suggestions that organisms had had from the beginning innate in them a power of modifying themselves, their organs and their instincts so as to meet [the] multifarious requirements [of the environment] does not materially differ from the more overt appeals to supernatural intervention.

As for memory, while not dismissing it out of hand, he cautioned against drawing parallels between "physiological and structural change" and "psychical analogies":

> The conception, originally introduced by Hering and independently by S. Butler, that adaptation is a consequence or product of accumulated *memory* was of late revived by Semon and has been received with some approval, especially by F. Darwin. I see nothing fantastic in the notion that memory may be unconsciously preserved with the same continuity that the protoplasmic basis of life possesses. That idea, though purely speculative and, as yet, incapable of proof or disproof contains nothing which our experience of matter or of life at all refutes. On the contrary, we probably do well to retain the suggestion as a clue that may some day be of service.
>
> But if adaptation is to be the product of these accumulated experiences, *they must in some way be translated into terms of physiological and structural change*, a process frankly inconceivable. To attempt any representation of heredity as a product of memory is, moreover, to substitute the more obscure for the less. Both are now inscrutable; but while we may not unreasonably aspire to analyse heredity into simpler components by ordinary methods of research, the case of memory is altogether different.
>
> Memory is a mystery as deep as any that ever psychology can propound. Philosophers might perhaps encourage themselves to attack the problem of the nature of memory by reflecting that, after all, the process may in some of its aspects be comparable to that of inheritance, but the student of genetics, as long as he can keep in close touch with a profitable basis of material fact, will scarcely be tempted to look for inspiration in psychical analogies. … We believe such things when we must, but not before.

Bateson followed this with discussion of some alleged cases collected by Semon of acquired experiences being transmitted to offspring, most of which were actually treatments (such as subjection to extremes of temperature) that produced no effects on the parents but did on their offspring. Of some significance for later developments was "the famous case of Schübeler's wheat." Plants raised from seed in Central Europe mature more slowly than similar plants raised in Norway, where the summer days are longer. Schübeler found that Central European seed grown in Norway acquired the property of maturing earlier which was retained on transferral back to Central Europe. Bateson considered that "without careful simultaneous control experiments this evidence is almost worthless," and was "surprised that Semon should claim these experiments as one of the chief supports for his views." It is not unlikely that Bateson discussed experiments of this nature with Vavilov when he visited Merton in 1913. Years later, a rival of Vavilov gained power in Russia through claims of the successful "vernalization" of wheat (Chaps. 18 and 21).

In 1915, Bateson wrote concerning "Evolution and Education" (Chap. 16):

> There is now scarcely any doubt that the germ-cells of which the offspring are composed possess from the beginning ingredients determining their powers and attributes; and that, with rare and doubtful exceptions, it is not in the power of the parent, by use, disuse, or otherwise, to increase or diminish this total. ... It is true that in the last decade some have again revived the view brilliantly expounded by Samuel Butler (*Life and Habit*, 1878), and also by Hering, that living things may, through their generations, have a continuous accumulation of 'unconscious memory'.

He then gave a series of examples of instinctual actions that appeared to have been inherited in that there had been no opportunity for somatic learning and concluded that:

> An apparatus is present ready to act if the appropriate stimulus is given at the right time; but for want of that stimulus it remains inoperative. It is tempting to suppose that the apparatus, the readiness to make the right response to various stimuli, is a manifestation of 'unconscious memory'; but since, as we have said, there is no good reason to suppose that even the simplest experiences of the parent are at all transmitted to a succeeding generation, the suggestion of continuous memory as applicable to education can only be defended on grounds which to the biologist are mystical and unconvincing.

It was a disciple of Butler, not Butler himself, who was named in an address Bateson gave in 1917 on "Gamete and Zygote," but Butler's message was again attacked:

> The evidence by which heredity is represented as unconscious memory is purely circumstantial. ... I understand, mainly from psychological considerations that Dr. James Ward has been led to believe in unconscious memory, as Mr. Edmond Holmes has been converted to metempsychosis. Offered these views as an alternative, I frankly prefer Genesis, which at least is simple, and makes no pretence to appeal to observation. The doctrine of transmitted experience as the origin of modification asks us to turn away from the whole course of genetic experiment.

Butler, however, would not go away. Laurence Binyon wrote (April 30, 1920): "I have to lecture at Oxford, Glasgow and Edinburgh: already written, thank goodness. Money must be made somehow .... I want to know if Butler's theory of unconscious memory is accepted, or how regarded now." How Bateson replied we do not know, but by 1924 he had brushed up his history of biology and in his Birkbeck College address (Chap. 18) he quoted from *Luck or Cunning* to illustrate Darwin's shilly-shallying over Lamarckism:

> His [Darwin's] most urgent task was to make evolution an acceptable principle, and one argument failing he would invoke the other, until in the edition of 1876 certain passages read uncommonly like Lamarck obscured. Seizing upon one which is, to say the least of it, ambiguous, the irreverent Samuel Butler makes the flippant comment: 'This comes of tinkering. We do not know whether we are on our head or our heels. We catch ourselves repeating 'important', 'unimportant', 'unimportant', 'important', like the King when addressing the jury in *Alice in Wonderland*'.

Perhaps the greatest tribute to Butler was made silently by Galton [59]. He wrote a novel *Kantsaywhere* in the style of *Erewhon*, where a wealthy Mr. Neverwas bequeathed funds to a governing council to promote eugenics. The inhabitants of Kantsaywhere were calibrated genetically following a written examination and accorded reproductive rights accordingly. Full rights went to those who gained diplomas from the Eugenics College. Galton completed the novel around 1910. It was not merely a mechanism to allow him to work out his ideas; a publisher (Methuen) declined it a few weeks before his death in January 1911.

So, yes, Butler was rehabilitated. Some Victorians heard him, the Edwardians more so. Yet modern historians have been largely silent. It seems that the Butlerian dog does not, indeed cannot, "bark in the night" – the pall cast by the Darwinians over the evolutionary biology of the nineteenth and twentieth centuries, remains. A purpose of this book, as in a previous book, *Evolutionary Bioinformatics*, has been to enquire why Butler did not get a hearing. His ghost haunts this book as it did Lipset's. However, in the tenor of his times, Lipset in 1980 dismissed Butler's science as "monochromatic but clever casuistry."

The Bateson Boys

Bateson's recognition of Butler came when his boys were at impressionable ages. He took their education seriously, and from their earliest days, they were being introduced to the natural world, learning the Latin names for plants and animals, and eventually being introduced to Darwin's books. Beatrice wrote (*Memoir*):

> Working as he did almost entirely at home, the children saw much of their father, and he soon established a delightful *camaraderie* with them. He was keenly interested in watching their development, and enjoyed helping them. He was determined that their education should be based on literature. Every morning for many years he read to them after breakfast, generally from the Old Testament, but sometimes from Bunyan or other fine prose, or even from Shakespeare.

Bateson did not want his children to grow up "empty-headed atheists." His remarks on the virtue of biblical knowledge are paralleled by those of Romanes, who described himself in his 1885 Rede Lecture as "utterly agnostic," but thought his children should be familiar with the bible "as a mere matter of literary education" [38].

In the Bateson library, there are two editions of Darwin's *Fertilization of Orchids*, one issued in 1862 with "W. Bateson" written inside the cover in Bateson's handwriting and one issued in 1904 with "John Bateson" written inside the cover. There are also three editions of Darwin's *The Variation of Animals and Plants under Domestication*: a first edition (1868), a second edition (containing numerous annotations and "W. Bateson, 1885" in the front pages in Bateson's handwriting), and a later issue (being a prize given to Martin Bateson at Rugby school).

The boys, especially John and Martin, were a close-knit group, and the fact that they were somewhat different from other children of their age did not appear a problem, least of all to the fleet of adoring maiden aunts who visited and were visited. Bateson did little by half-measures. When he was enthusiastic, he was very enthusiastic. Thus, it is likely that by 1909 the boys had come strongly under the influence of Butler. Martin was much affected. Indeed, he may later have seen his father as the Butlerian father in *The Way of All Flesh.* When Martin sent some poems home from school, his parents advised that making a living through writing poetry would not be easy. In 1919, he joined the air force, and Beatrice sent him *The Times Literary Supplement* and *Nature* but cautioned about openly flaunting *The Cambridge Magazine* before his working-class comrades who might resent his privileged background.

The death in action of John in October 1918 came as a great blow. Bateson tried to console Martin and sent him a copy of "Science and Nationality" (Chap. 16). To his son's bitter questioning, Bateson argued for the inevitability of the turmoil into which Europe had been thrown. Concluding a letter to Martin, who was then stationed at Yarmouth, Bateson wrote (Dec. 13, 1918): "It is always a prominent thought in my mind that strictly speaking such people as we are do not belong and are only here on sufferance. To think for oneself in most societies is a crime. But there it is! One has just to make the best of the situation and be thankful that we are allowed our niche."

In her *Memoir*, Beatrice records quite simply for 1922: "In April of that year we lost our second boy." This followed three years of turmoil. Following his father's advice, Martin had enrolled at St. John's College in 1919 to study zoology. Soon he was rereading the works of Samuel Butler so meticulously that he found a section in the original manuscript of *The Way of All Flesh* that had been omitted in the posthumously published version. It implied that the hero, Pontifex (i.e., Butler), had, unknown to the parents, secretly married and produced a child. In his letters to Bateson, Martin compared Butler to Einstein, Euclid, and Newton. With hindsight, this does not appear too far-fetched; but Lipset, writing in 1980, considered it "lavish." By now, Bateson had become quite disenchanted with Butler, writing to Martin (Nov. 16, 1919): "I had thought he [Butler] was my sort and he isn't. He comes out as anti-science not merely anti-scientist."

Martin travelled widely throughout Europe, took to gambling, and purchased a motorcycle with a sidecar. He obtained first class honours in the Natural Science Tripos in 1921. Concerning his membership of the Savile Club, Bateson wrote to Beatrice (Dec. 26, 1921): "I don't think you should worry about the Savile. If Martin gets in there, it will tend to keep him out of mischief, and he will meet a number of decent people who will perhaps raise his standard. Probably well worth the money." In the autumn of 1921, Martin, hoping to become a playwright, entered the Royal Academy of Dramatic Arts (RADA) in London. He fell in love with a fellow student, an 18-year-old actress with a working-class background who was engaged to another. To her, Martin must have seemed like someone from another planet. A play he began writing featured a son and father arguing about money, with the actress as a central character. He shot himself in Piccadilly Circus on John's birthday (April 22, 1922). The next day Bateson wrote to Gregory:

> At times like this, when our hearts come nearer to the surface than when all is going well, it is possible to speak more easily than one can in ordinary daily life. Looking back I can see that Martin had a certain instability – not much, but a little. If John had lived, possibly enough the two together would have got to work, and the interest which I feel sure they both genuinely felt in science would have steadied Martin through trouble. What exactly has been happening to him we don't know, but it seems that he had looked for affection where it could not be returned, and in sudden melancholy, he did this. He had lost faith in Science, and also his confidence in us – as so often happens when a young man grows up, and he has nothing to steady him through.
>
> What I want to say now is that to people like us, work, meaning the devotion to some purpose, the nobility and worth of which we cannot question, is the one and only thing that helps in time of trouble. I think that like John too, you have plenty of steadiness and I do not fear for your force of character; but if ever there comes a moment when the horrible tempter which inhabits most human hearts, raises the question whether anything is worth doing, then remember that no such doubts arise in happier times, when judgement is at its best – and crush those foolish doubts out of your mind.
>
> The faith in great work is the nearest to religion that I have ever got, and it supplies what religious people get from superstition. There is also this difference, that the man of science very rarely hears the tempting voices and very seldom needs a stimulant at all, whereas the common man craves it all the time. Of course there is great work that is not science – great art, for instance, is perhaps greater still, but that is for the rarest and is scarcely in the reach of people like ourselves. Science, I am certain, comes next and that is well within our reach – at least I am sure that it is well within yours. … To set oneself to find out something, even a little bit, of the structure and order of the natural world is, and will be for you I dare foresee, a splendid and purifying purpose, into which you can always withdraw in the periods of suffering that everyman must pass through. If you keep your eyes on that, the other things in life look so poor and small and temporary that the pain they give can be forgotten in the greater emotion.

Gregory left Charterhouse School for St. John's College in 1922 to read zoology. He and his father did a project together on partridge feathers which was published in the *Journal of Genetics* (1925). In the course of this, Gregory visited a museum in Geneva, Switzerland. During his residence there with one of his father's professional acquaintances, Fernand Chodat, he fell in love with the daughter and they became engaged. Gregory later recalled that his mother saw his fiancée as "a sturdy peasant." Abiding by his parents' wishes, the engagement discontinued. With it, went his interest in zoology. A chance meeting with a field anthropologist led to a decision to change subjects, and he went on to have an eventful career, which included being married for 14 years to fellow anthropologist Margaret Mead. As related by their daughter [60], Gregory's relationship with his parents, particularly Beatrice, had not been easy, and it was not until 1953 that he was ready to receive the many parental items stored by the Barlows "as if deciding finally that he was ready to deal with his childhood and youth."

His interests spread widely to psychiatry and linguistics, and to information and control theory as they emerged in the 1940s. He contemplated a book to be called *The Evolutionary Idea*, which would approach evolutionary biology from an informational perspective, so "finally completing his dialogue with his own father." It was never completed, but the underlying ideas – now seen as founding Evolutionary Bioinformatics (EB) – may be found in his other writings where the influences of his father and Butler are acknowledged [61–64]. When asked by Alan Cock in January 1975 about his father's view of the scope of genetics, Gregory replied [65]:

> He meant it to cover the entire field of determination by communication – it is what it comes to in the end you know. It's so *near*, all the time – the Bateson material, the Butler material, and various other things – are so *near* to saying the sort of things which we *now* know how to say, but they did not. All they knew was that the world that was not saying those things, was somehow wrong. But they really hadn't any idea how to make the next step. ... They were on the wrong step. They did not know what the right step was. A very curious feeling.

Gregory seems neither to have appreciated Butler's long New Zealand apprenticeship in biology nor that observation of nature was one of the goals of long weekend rambles around London and annual visits to Italy. In a lecture at the New York Harvard Club in 1970 Gregory noted [66]: "There were still some naughty boys, like Samuel Butler, who said that mind could not be ignored ..., but they were weak voices, and incidentally, they never *looked* at organisms. I don't think Butler ever looked at anything except his own cat, but he still knew more about evolution than some of the more conventional thinkers" (My italics). The author of *The Way of All Flesh* sought escape from parental authority. He had to make not only a living but a small fortune – certainly more than enough to live on – in order to indulge his tastes without further parental support. This concentrated his mind quite powerfully to the task of keeping his New Zealand flocks healthy and reproductive.

Summary

William Henry Bateson, Charles Darwin and, by default, Thomas Butler, all studied under Dr. Samuel Butler, the headmaster of Shrewsbury School. Among their sons were William Bateson, Francis Darwin, and Samuel Butler. While Gregor Mendel was breeding peas and reading *The Origin of Species* in Moravia, Samuel Butler was breeding sheep and reading *The Origin of Species* in New Zealand. He returned to a private life of art and science, visited Charles Darwin, and befriended Richard Garnett of the British Museum Reading Room and Miss Eliza Savage. While avowedly a non-specialist, Butler's approach, like that of Jenkin, was to "admit the facts, and examine the reasoning." Unaware of the work of Mendel and Naudin, in the 1870s Butler in London, and Hering in Prague, inferred that heredity was a form of memory (involving stored information), distinguished what are now known as genotype and phenotype, and embraced Lamarckism. Despite the latter, their views on memory can now be seen as at the root of a new science – Evolutionary Bioinformatics. Butler's questioning of the power of natural selection, and of the extent to which the Darwinians had attended to the history of their subject, led to his estrangement from Francis Darwin. Egged on by Charles Darwin, an attack on Butler by a biographer of Erasmus Darwin, led to public confrontation. While Charles Darwin stood aside, Romanes' attempts to defend him were successfully countered by Butler and his allies. Despite his close friendship with Francis, Bateson knew little of the dispute until after Butler's death. Bateson communicated his new interest in Butler to his sons, for one of whom they became an obsession. Bateson urged Francis to publish what was, in effect, a posthumous apology. But the Darwin dynasty did not forget. In the 1950s, granddaughter Nora renewed the attack.

Chapter 20
Pilgrimages

> *True greatness wears an invisible cloak, under cover of which it goes in and out among men without being suspected. … What then … is the good of being great? The answer is that you may understand greatness better in others, whether alive or dead, and choose better company from these, and enjoy and understand that company better when you have chosen it – also that you may give pleasure to the best people and live in the lives of those who are yet unborn*
>
> *Samuel Butler, The Way of All Flesh*

Mendel's sister gave up part of her dowry to assist his early education. Mendel was later able to help with the education of her sons. Ferdinand and Alois Schindler were delighted when their uncle's work was recognized. Ferdinand wrote to Bateson (Aug. 30, 1902):

> You may be surprised to hear that your excellent work 'Mendel's Principles of Heredity' came to the hands of a simple country physician who read … it with the greatest interest and pleasure. … He died Abbot Mendel, was a man of *liberal* principles, and it is not too much to say, he hated the Ultramontane propaganda and dissimulation. … He read … with the greatest interest Darwin's works … and admired his genius, though he did not agree to all principles of this immortal natural philosopher. … My uncle in the latter part of his life retired from development of scientific and evolutionary questions because he had many clerical enemies. He said often to us nephews that we shall find as his heritage, papers for publication, which he could not publish during his life. But we did not receive anything from the cloister, not even a thing for remembrance.

Pilgrimage That Was Not

Mendel presented his work at two meetings in Brünn early in 1865, and it was published in 1866. His discovery of the integral transmission from generation to generation of the elements (*merkmale*) corresponding to individual characters should have

A. G. Cock, D. R. Forsdyke, *Treasure Your Exceptions*,
https://doi.org/10.1007/978-3-030-92099-9_20

been seized upon by biologists worldwide. Had they done their homework and read the, albeit obscure, journal the *Verhandlungen des naturforschended Vereines in Brunn*, then a few years later students, such as Romanes and Bateson, might have been reading their Mendel along with their Darwin. The world might have "pilgrimaged" to the Brünn monastery where Mendel grew his peas. Mendel might have been encouraged, the Moravian powers that be might have been more inclined to support his operations, and students might have flocked to sit at his feet. In short, the history of late nineteenth-century and twentieth-century science could have been entirely different ("counterfactual" history) [1].

But the pilgrimage was not to be. It is perhaps not unfair to compare Mendel's "epiphany" in 1865 with Morgan's in 1911. Mendel discovered the character units (see Chap. 8), and Morgan, perhaps influenced by de Vries and Lock (Chap. 10), saw that linkage phenomena could be explained by the linear arrangement of character units along chromosomes (Chap. 13). Had Morgan been a monk working alone with his bottles of fruit fly in some poor up-state monastery, and had he published in some obscure parish journal, then it is quite conceivable that the Bateson–Punnett reduplication hypothesis would have reigned without challenge until Morgan was "rediscovered" decades later by someone like Fisher, Haldane, or Wright.

As it turned out, Morgan was already centerstage at the time of his epiphany, and it was published in the widely read journal *Science*. In a twinkling, he "seized the crown." Had it not been for the 1914–1918 war, the scientific world – and especially the European scientific world – would soon have flocked to his Columbia "fly room." As it was, the Columbia pilgrimages began in the post-war years, with Bateson the most prominent pilgrim of all.

It is conceivable that, had they known, Darwin and his bulldogs would have pilgrimaged to Brünn in the 1870s. But their challenge was a different one. Darwin's importance was not natural selection, which, as Butler pointed out, had been proposed decades earlier. Darwin's importance was that his wealth, connections, enthusiasm, and ideas combined to garner the support necessary for the long-awaited confrontation with religious orthodoxy. The "time was ripe." Lost in the heady optimism that the process of evolution might explain man's presence on earth, the details were not a major concern. And, as Bateson has so frequently reiterated on these pages, the detail that most concerned the early evolutionists was the determination, mainly through studies of comparative morphology and embryology, of plausible lines of descent. So, Mendel remained, and died, in his monastery (see Chap. 27). The pilgrimages began much later [2].

Pilgrimage 1904

There was a sudden decision (around Dec. 11) to take a holiday with Beatrice and visit Brünn. The minute book of the local Natural History Society in January 1905 records that "Professor William Bateson of St. John's College, Cambridge, paid us a visit in Brünn, undertaking, as he puts it, a pilgrimage to the city in which Abbot Gregor Mendel carried out the experiments and preliminary studies for his famous work." It was further noted that "he also visited the Königskloster in Altbrünn," but "unfortunately his efforts to discover Mendel's handwritten records proved in vain." If there were further Mendel manuscripts, they had been destroyed decades earlier. Bateson's signature may be found in the visitor's book of the Königskloster (Dec. 29, 1904), and there is a letter (Dec. 13) from Ferdinand Schindler arranging to meet Bateson in Brno (as Brünn had been renamed) on "Thursday at midday," and proposing a visit to Mendel's birthplace at Heinzendorf.

As part of the holiday, Bateson also visited Vienna, where he met Tschermak and Hans Przibram who shared his interest in art to the extent that his own works were exhibited at the Secession, a center for *Jugendstil* art. Przibram had purchased a former zoological exhibit hall and turned it into a research institute, the "Vivarium." Kammerer joined in 1902–1903 and his first paper (on Salamanders) was published in 1904.

Pilgrimage 1910

Bateson's second pilgrimage was to a formal celebration of Mendel organized through the efforts of a local high-school teacher, Hugo Iltis. A statue had been erected in the Klosterplatz and was to be unveiled on October 2nd. Contributors to a Festschrift would include Bateson, Baur, Cuénot, Hagedoorn, Hurst, Lotsy, Nilsson-Ehle, and Tschermak [3]. Bateson's attendance was partly motivated by the thought that he could again visit Przibram in Vienna, where Kammerer and his specimens purporting to demonstrate the transfer of acquired characters were lodged. Having received a letter from Bateson, Przibram replied (Aug. 28, 1910): "Would it not be possible for you to come to Vienna either going or coming from Brünn? I would be glad to offer you and Mrs. Bateson all hospitality during your stay at our town. … Then you would have the opportunity of seeing all our experiments and the new installations which I am fitting up to make the experiments more exact. For me it would be the greatest pleasure to be able to talk things with you and to hear your criticisms." Bateson recollected (*Nature* 1919): "Late in the summer of 1910 I unexpectedly was able to attend the *Mendelfeier* at Brünn, and was for some time in Vienna, having the privilege of being the guest of my old friend Dr. Przibram."

He planned to travel alone by way of Berlin and Dresden. Thus, before the dust had settled from the move from Grantchester to Merton in the summer of 1910,

Bateson was off for the usual round of art galleries, operas, and games of chess in Berlin and was sorting out "various fragments in my head for my Mendel oration." Of course, there was a visit with Baur (Sept. 25): "Baur takes my fancy more and more. He is so fresh and bright. Perfectly unaffected – doesn't mind going first through a door – and altogether a natural and humorous man. He has several good new things – says self-sterility in *Antirhinum* behaves as a recessive [i.e., a genic defect] – amazing if true." However, for Bateson, there seemed no reason why a scientist should not also be well-rounded. In Baur, as in others he came across, the lack of general knowledge and culture disappointed him. He is "so utterly immersed in his notebooks that I think his work may suffer from want of breadth." Baur took him to a "flying ground," but the wind was too strong for the primitive airplanes to take off. Bateson pondered the significance of the new flying machines: "I remain with the feeling that the chief secret is still undiscovered. They are amazing toys, but toys still."

In Dresden, he "saw the San Sisto first, and afterwards the Carela See," where he and Beatrice had skated together in the old days (Chap. 7). He wrote to her from Vienna (Sept. 28): "I had a long spell with Kammerer, and there is no denying the extraordinary interest of what he is doing. The 'Brünstschweilen' cannot be produced. Somehow or other I have hit on a weak spot there. … But he … comes uncommonly near showing that an acquired adaptation is transmitted. I don't like it, and shall not give in till no doubt remains." Przibram took him to see Oscar Wilde's play *Ideal Husband*, which Bateson considered "a sordid story – brilliant, of course, but not quite his best." He wrote (Sept. 29): "What a deadly dull evening we did have. I don't find Przibram a sympathetic companion. He is a very cold-blooded beast – a Jewish Doncaster. … Of humour he has hardly a trace. He touches water only, not even a cigarette. … I suspect him of having something of the Pococurante disposition." The following day he met Tschermak and traveled with him to Brno.

Brno was then a provincial town under a German (Austrian) monarchy, and the Germans tended to treat the Czechs as an inferior subject race. There was in the region a large indigenous German population, and the Mendel celebration was laden with politico-religious undercurrents – in particular, the relegation to a very minor role both of Mendel's own Augustinian monastery and of Czechs generally. There was an exhibition of documents from the monastery, set up, not in the monastery but in the "German House" where it formed part of a pre-celebration meeting for the participants. The inscription on the monument was in German alone.

The Abbot at this time was a Czech, Franciscus Salesius Bařina, who was the sole survivor among those friars who had been accepted into the monastery by Mendel himself. Only a mere handful of those present had ever met Mendel, even in a casual way. This made the exclusion of Bařina from any part of the official proceedings the more pointed. After the meeting, Bateson and Tschermak were made honorary members of the Natural History Society of Brno. On his return to Vienna, where he again lodged with Przibram, Bateson wrote:

The messe [German for mass] is over, the choirs have sung. The monument is unveiled in all its gross absurdity; we have eaten the Mediaeval Bankett – birds in their feathers and other Snyderesque dishes – a veritable feast à la Jordaens [Flemish painter specializing in feast scenes]; the speeches are spoken. Mélanie and Philippe [Vilmorin] have ejaculated many doubtful remarks which beguiled our weary homeward journey. ... I believe I spoke 4 minutes at the very top of my voice, but I did not reach to the other end of the hall Przibram says. ... The gist of my remarks was that though we grieved for Mendel's obscurity, it was no case for sorrow. To have looked, as M. did, once intimately on the face of a new Truth was as keen a pleasure as the world can give. ... Tschermak spoke fairly well. Iltis, the little local secretary, very well. The President, a Baron somebody [von Haupt-Buchenrode], referred to Mendel as a liberal [the German word *Freisinniger* carries overtones of a free-thinking, even heretical] priest – and somebody else – one of the Government officials said that if such a thing occurred again in the proceedings, he should leave at once. In Brünn it is just touch and go. The state of strain is extreme. It was arranged that no clerical or Czech speech was to be made, so the Kloster never once came onto the horizon. The Prälat was there, but no one attended to him, or his office. Every other sort of institution was toasted, and every other sort of bigwig had his health drunk, but this mitred Abbot sat in silence. I was only within a little of getting up and proposing the Kloster and rather wish I had done so. ... Starting homeward tomorrow night. Lotsy wants me to stop a night in Holland, and I may do so. ... The monument is banal and shocking. The peas climb up a *rock*. They have leaves on their peduncles! Underneath are two kneeling humans, typifying marriage and its consequences. These are better, but they are so small that they don't redeem the rest.

Hurst contributed a paper on "Mendelian characters in plants, animals and man" [4]. Systematic as usual, Hurst used the occasion to present a relatively exhaustive list of the characters of plants, animals, and man, which had been found "subject to Mendel's law of segregation." He concluded:

The ... list will serve, better than any words of mine, to illustrate the great and growing importance of Mendel's discovery. ... Thanks largely to the labours of Bateson and his coworkers, a new science – the science of Genetics – has been built up on a Mendelian basis. This science, with its modern methods of breeding, promises to provide solutions to many problems of heredity and variation hitherto obscure.

All that remains of Bateson's speech is the quotation on the brotherhood of man from Schiller with which he concluded: "Alle Menschen werden Brüdern." There were several letters to Beatrice from Vienna, but none from Brno, where he seems to have stayed only one night. In Vienna, he stayed with Przibram, met Kammerer, and with others (Baur, Lotsy, Nilsson-Ehle, Hagedoorn, de Vilmorin) was shown round their institute. Many of these "shared the same feeling of doubt" concerning Kammerer's work (*Nature* 1919). Przibram wrote (Oct. 12, 1910):

I would like to tell you how much I enjoyed your visit. On thinking over your criticism or scepticism as to Kammerer's experiments, I cannot find any ground for joining in the latter, although I certainly cannot agree to all of K's conclusions as to the mode by which characters are transferred to the young. I am not losing any time however to have K's results done over again by other of my students with a little variation of the method of investigation being better able to reveal some of the ways by which the transfer may occur – as I think.

Centenary 1922

As a result of the reconfiguring of Europe as part of the Treaty of Versailles (1919), what had been Moravia became part of the new state of Czechoslovakia. The Ministry of Education in the new government encouraged Iltis to complete a book on Mendel, now seen as a national hero, and in 1922 the centenary of his birth was duly celebrated in Brno. Here was another opportunity to go to Vienna and examine Kammerer's specimens. In a letter to Baur (Aug. 7, 1922), Bateson implied he would be going, stopping by in Berlin around September 16th to pick him up, but in the end, he did not attend. Davenport from the United States, and Michael Pease from Cambridge attended, and Pease also visiting Vienna. A *Memorial Volume* was published in 1925 [5].

Mendeliana

From time to time, Bateson toyed with the idea of a biography of Mendel. It was not to be. In 1905, he learned that Iltis had such a project in hand, and in 1909, Iltis told him, optimistically, that it would be completed the following year. In 1925, Bateson reviewed two historical books for *Nature* under the heading "Mendeliana." One was a collection of Correns's papers produced by the German Society for the Study of Heredity [6]. The other was Iltis's book, the first major biography of Mendel [7]. Both were in German, but in the years ahead, English translations, albeit abbreviated in the case of the biography, would follow:

> Correns was, of course, one of the rediscoverers of Mendel, in a sense, perhaps, *the* rediscoverer. The earliest papers in this volume recall that curious and diverting episode, and the cryptic nature of the first announcements. In view of all that has happened since, he may, in any case, find satisfaction in remembering that in 1902, sometime before linkage had been observed as an actual fact, he made a suggestion … as to the linear arrangement of elements on chromosomes and as to the exchanges between them, now spoken of as crossing-over, which in all essentials is that now adopted by the orthodoxy of the day.

Correns had discovered, as had Saunders in Stock, that there can be "genetical inequality between the eggs and pollen-grains of the same plant." While Bateson ('many of us in England') still interpreted this in terms of somatic segregation, Correns, being "unwilling to admit anything which conflicts with the strict chromosome theory," did not take this viewpoint.

As for the book on Mendel, Bateson ignored Iltis's call for a unification of Darwinism and Mendelism, and considered Mendel himself: "The few new facts and anecdotes now first made known, help us in some measure to reconstruct his personality, but the generation that knew him during the years of his scientific work had almost passed away before his fame began." Mendel had not fared well in the examination system: "His fame as a very successful teacher still survives in Brünn, and these incidents provide ironical commentary on the public utility of a highly regulated educational system." So, Mendel's work was published but: "He remained alone. After his immense labour he found not a single creature who understood, not one who believed him." And as to Nägeli's lack of support: "Perhaps we may draw the moral that a discoverer of something really new, wishing to find sympathy and encouragement, does not act wisely in appealing to the highest established authority on that particular subject." Bateson noted Iltis's evidence that, besides the work on peas and hawkweed, Mendel was breeding Fuchsia, birds, mice, and bees.

The extent to which Mendel was truly religious was of interest to Iltis who was "convinced that Mendel was virtually a freethinker, and only officially Catholic." To this, Bateson responded:

> Without much stronger evidence I should hesitate to accept Dr. Iltis's judgement, which is tantamount to a charge of active insincerity. Rather I should suppose that Mendel's position was that of numberless honest men in all ages the world over, who can take things as they find them. Nothing at all suggests considerations of faith or doubt had much interest for him, or that he was ever in the position of having to take a side on such questions. Probably they never troubled him one way or the other.

Iltis related that Mendel was considered by "a superior" as having "little aptitude for the care of souls." Furthermore, "ministration to the sick and dying distressed him so much as to induce serious illness, presumed to have been a sort of hypochondria." Bateson doubted that Mendel was of such tender disposition: "I imagine Mendel as a man full of practical good sense, with an exceedingly clear head, thinking in well-divided compartments, rarely disturbed by the eccentricities of genius." This viewpoint has gained support [8–10].

Centenary 1965

After the Second World War, Czechoslovakia was in the domain of the USSR. The Augustinian Monastery where Mendel had worked was closed in 1949 and fell into disrepair. When Stalin died in 1953, Lysenko lost absolute power but continued to be highly influential under Kruschev. The latter fell from power in 1964, and in 1965, Lysenko lost his position as director of the Institute of Genetics [11]. In the early 1960s, renovations began for the 1965 Mendel Centenary. The meeting of geneticists and historians in Brno for the Mendel Memorial Symposium marked the reunion of the genetics communities of the East and the West after the sad Lysenko years (Chap. 21). The occasion was marked by the official opening in the Monastery of the "Mendelianum" (a sub-department of the Moravian Museum) for the gathering of materials on the history of genetics [12]. The 1910 statue was transferred from the city square to the proximity of Mendel's garden, where it can be seen today, with the German inscription removed (Fig. 20.1).

(a) (b)

Fig. 20.1 Statue of Gregor Mendel as it was (**a**) and as it is now (**b**)

Summary

Mendel studied many species apart from peas and hawkweeds. Mendel's nephews wrote to Bateson delighted that their uncle's work was being recognized, but no trace of the further manuscripts they alluded to could be found in 1904 when Bateson visited Brno, then part of the Austro-Hungarian Empire. Largely through the activities of a local teacher, Hugo Iltis, there was a formal celebration in 1910 with the unveiling of a Mendel statue. Bateson proclaimed that all men were brothers, but the current Czech Abbot was marginalized, and the inscription on the statue was in German. Bateson stayed in Vienna with Przibram and was among those who were not impressed by Kammerer's examples of Lamarckian inheritance. In 1919, Brno became part of Czechoslovakia, and Mendel was recognized as a national hero, but Bateson did not attend the 1922 centenary celebrations. After the Second World War, Czechoslovakia came within the domain of the USSR, and the monastery fell into disrepair. By 1965, Lysenko's influence had decreased, and the world flocked to Brno to celebrate Mendel and his famous paper. Had they known of Mendel's work, the Darwinians might have pilgrimaged to Brünn in the 1870s, but attention was then primarily directed at combating religious orthodoxy and establishing plausible lines of descent.

Chapter 21
The Kammerer Affair

All the acquisitions or losses wrought by nature on individuals, through the influence of the environment in which their race has long been placed, and hence through the influence of the predominant use or permanent disuse of any organ; all these are preserved by reproduction to the new individuals which arise.

Jean-Baptiste de Lamarck [1]

Paul Kammerer (1880–1926) was an Austrian zoologist who conducted experiments purporting to demonstrate the inheritance of acquired characters – Lamarckism. This was an issue which, in the early years of the twentieth century, excited even more heated, and more evenly divided, debate than it does today. Kammerer's researches were seriously interfered with by the First World War when he lost most of his experimental animals. In 1926, the American zoologist G. K. Noble reported a detailed microscopic examination of a specimen that was crucial to Kammerer's claims [2]. Noble's conclusion was that the specimen was a deliberate fake, although he did not say, or even suggest, that Kammerer himself had done the faking. Six weeks after the publication of Noble's paper, Kammerer committed suicide.

Koestler's Account

Bateson had been one of the most persistent of Kammerer's critics, yet it would hardly be necessary to devote a whole chapter to Bateson's dealings with him, were it not for the misleading impression created by Arthur Koestler's account of Kammerer's work in *The Case of the Midwife Toad* [3]. Koestler's book is misleading in several respects, not least in the picture it gives of Bateson. Kammerer is portrayed as a martyr, hounded to his death by the biologically orthodox with

A. G. Cock, D. R. Forsdyke, *Treasure Your Exceptions*,
https://doi.org/10.1007/978-3-030-92099-9_21

Bateson at their head, baying for Kammerer's blood as loudly as any of his followers. I write "misleading" advisedly, for it is not so much a matter of outright errors and inaccuracies, though these play their part, as that whatever Bateson said or did – or failed to do – is presented in the worst possible light, usually without violence to the letter of the truth. Actions not susceptible of such treatment, on the other hand, are ignored or glossed over. Nor should it be assumed that the more knowledgeable of Koestler's readers are immune to this misrepresentation. Thus Stephen Jay Gould, who on other important points is highly critical of Koestler's work, writes [4]: "It [the Kammerer affair] merely shows that William Bateson was a very nasty man."

The controversy therefore concerns us in two ways. The first is its strictly scientific content. How far were Kammerer's claims (and equally Koestler's own additions to the Kammeronian canon) justified? How carefully controlled were the experiments? What precautions did he take to guard against false-positive results? What criticisms did Bateson advance against them, and were these reasonable scientific criticisms such as any revolutionary claim must expect to face, or were they merely the expression of blind anti-Lamarckian prejudice? The second has to do with Bateson's own character and personality. The picture of him given in Koestler's book – sometimes by direct assertion, at other points by thinly veiled hints – is certainly heavily biased. But does any of this have validity – are there grains of truth within the dross?

Inheritance of Acquired Characters

The doctrine of the inheritance of acquired characters was expounded by Jean Baptiste de Lamarck in his *Philosophie Zoologique* [1]. Lamarck did not, however, believe that *all* acquired characters are inherited: only those which the animal itself produces as an adaptive response to the environment, by its own efforts. Unsophisticated attempts to refute Lamarckism, e.g. by showing that amputation of the tails of successive generations of mice does not lead to mice born without tails, are therefore beside the point. For Lamarck, the inheritance of acquired characters played an important though not the prime role in evolution. The prime force was a "self-perfecting" tendency which, he held, drove organisms to higher and higher levels of organization.

Throughout the nineteenth century, the Lamarckian doctrine had a chequered career. Darwin admitted it as playing a subsidiary role to natural selection, but opposition to it crystallized around the writings of Weismann, who objected to it on a priori grounds. Since in animals (though not in plants) the germ-line (the precursor cells of the gametes) separates from the rest of the body (soma) at a very early stage of development, there was, Weismann argued, no physical means whereby adaptations in the soma could be transferred to the germs. By 1900, Weismann's view had come to predominate. In choosing therefore, at the outset of his career, to look for evidence of the inheritance of acquired characteristics, Kammerer was making a courageous decision: the subject was of the greatest importance, but it was also unorthodox.

Green Fingers

After an early "false start" when he studied the piano, Kammerer graduated in zoology at Vienna University. He immediately joined the staff of the Institute of Experimental Biology in Vienna under its director and co-founder Professor Hans Przibram, where he remained for the rest of his career. The Institute, or Vivarium as it was nicknamed, occupied the buildings of a former public aquarium, and was thus particularly well equipped with aquaria and chambers for keeping the lower vertebrates. This was quite possibly a factor influencing Kammerer's choice of experimental animals: three of his four main series of experiments were carried out with amphibia, the fourth with the sea squirt *Ciona.* Indeed, Kammerer seems to have been appointed initially not so much as an experimenter but as a person in charge of the aquaria and terraria. All who knew him seem to have agreed that he had an uncanny skill in keeping – and more important, causing to reproduce – amphibia and reptiles: the zoological counterpart of green fingers. It may have been his skills with amphibia and reptiles that got him the job with Przibram, so his opportunities may have shaped his interests. But in a paper published in 1906, Kammerer stated that his experiments with the midwife toad *(Alytes obstetricans)* were begun in 1894 – i.e. when he was a schoolboy. Whether these were experiments is the usual sense, or just attempts to keep and breed *Alytes* in captivity, is uncertain. Nevertheless, to some extent, his interests may have shaped his opportunities.

Alpine and Lowland Salamanders

Kammerer's first published experiments were with salamanders. There are two European species: the spotted salamander, *Salamander maculata,* which inhabits lowland pools, and an alpine form, *Salamander alpina.* In both, the eggs are retained in the mother's body until the embryos are *partially* developed, but in the alpine form, they are retained until a much later stage than in the lowland form. When released, *alpina* is virtually a miniature adult, whereas *maculata* is a tadpole larva with external gills. *Alpina* produces only two large young per brood, whereas *maculata* produces up to 50 much smaller larvae. These differences between the species are of obvious adaptive significance in relation to their natural habitats: the small *maculata* larvae would rapidly be swept away in a fast-running alpine stream.

What Kammerer did was to force *maculata* to reproduce in the manner of *alpina* by keeping it in a cold dry (i.e. simulated alpine) environment. It took four or five broods produced under these conditions to effect a change, but eventually *alpina*-like broods, consisting of two large fully developed individuals only, were produced. Moreover, in these modified broods, the remaining developing ova in the mother were aborted in their development so as to provide additional yolk material for the two embryos, just as happens in *alpina*.

More striking, when individuals of these modified broods came to maturity, they themselves reproduced in the modified manner – i.e. the environmentally induced modification was "inherited." Kammerer also carried out the converse experiment, i.e. he persuaded *alpina,* maintained in an environment simulating that natural to *maculata*, to adopt the reproductive habits of *maculata*. It is noteworthy, however, that Kammerer did not carry (or did not report) these experiments beyond the first generation (F_1), or even try to obtain further broods, beyond the first, from his F_1 individuals. If an environmentally induced change can be incorporated into the hereditary makeup in a single generation, then one might expect that restoration to the normal environment would lead to an equally rapid reversal of the change. Even the modified F_1 *maculata* females, after three or four broods, might well revert to normal *maculata* reproduction: an outcome which would diminish the Lamarckian import of the experiments, and which Kammerer failed to test.

Kammerer also carried out experiments on *maculata* involving the pigmentation of the skin. The normal pattern of specimens caught near Vienna is yellow spots on a black background. Kammerer found that specimens reared on a background of yellow soil gradually became, over a period of several years (the species needs four years to become sexually mature), more and more yellow. This was accompanied by a change in pattern: the spots coalesced into longitudinal stripes. Conversely, in individuals reared on black soil, the yellow spots slowly became smaller. In the offspring (reared on the same color of soil as their parents), these effects were enhanced, i.e., they were yellower (at the same stage of development) or blacker, as the case might be, than their parents. In subsequent generations, up to the third, the effect was increased until the black-reared strain became almost solid black, while the dorsal surface of the yellow-reared stock was almost fully yellow. He later reported that by rearing the tadpoles – which did not yet possess the yellow and black pattern – in tanks suffused with yellow light, it was possible to induce (in one generation!) a degree of yellowing as great as had needed three generations to achieve in the original experiments. If young, of the yellow stock say, were reared on black, their color reverted somewhat toward black, but they were still much more yellow than the foundation stock, thus confirming the claim that the induced changes were inherited.

Midwife Toad

While these experiments with *Salamandra* were in progress, Kammerer began parallel experiments with the midwife toad, *Alytes obstetricans*, which gains both its popular and its Linnean specific name from the part played by the male in rearing the eggs. Most frogs and toads copulate in the water and lay their eggs there, which then develop without further care from either parent. *Alytes* both copulates and lays its eggs upon land. As the eggs emerge from the female as strings of gelatinous masses, they are taken up by the male, who winds them round his own legs and back, and carries them around with him until the eggs hatch.

Kammerer kept *Alytes* at a high (25–30 °C) ambient temperature, but with cool water available: under these conditions, the toads spend progressively more of their time in the water (allegedly in order to keep themselves cool), and eventually (i.e., after two or more broods) copulate and lay their eggs there. Even in the first brood, laid upon land, the males do not attempt to wind the eggs around themselves, and the subsequent broods of eggs are probably too slippery for this to be done. In parallel with these changes in reproductive behavior, the eggs in successive broods become progressively smaller, less yolky, and more numerous – all changes which bring them into closer conformity with those of other species of toad.

Alytes eggs laid in the water are very difficult to rear, as they are very susceptible to fungal infections, but by dint of careful aseptic precautions, Kammerer managed to rear some to maturity, although mortality was heavy. When the offspring came to maturity, Kammerer found that those from eggs of the first brood (laid on land) were unmodified, whereas those from subsequent broods displayed the changed reproductive habits which their parents had exhibited. In later generations, moreover, these aquatic habits of reproduction were retained even in individuals reared in a normal (low temperature) environment.

These effects of high temperature were produced, like the changes in reproductive habit in *Salamandra*, in a single generation, which sufficed (contrast the case of coloration in *Salamandra*) to bring them forth in maximal degree. Kammerer reported one more change, which he did not notice until the third generation (F_3), yet which was to excite more controversy than his other claims. The males of most species of frogs and toads develop *Brunstschweilen* (nuptial pads) during the breeding season. These are swellings on specific parts of the hands – the exact location varies somewhat from one species to another – which are pigmented black and covered with tiny horn spines. The function of these nuptial pads is presumably to facilitate the grasping of the female round the neck during copulation. Between breading seasons, the swellings subside and the spines disappear. *Alytes*, exceptionally, does not have nuptial pads: it has been plausibly argued that in copulation on land, where the skin is less slippery, they would be superfluous. What Kammerer reported was that in the F_3 generation he found *Brunstschweilen* on the males (all of them!) during the breeding season. Retrospectively, he also found fainter indications of Brunstschweilen in males of the F_2 generation, but these were unpigmented and much less clear than those in F_3.

Sea Squirts

Kammerer's other main experiment was very different, both in its nature and in its experimental subject. He used not an amphibian, but the sea-squirt (Ascidian) *Ciona intestinalis*, and his procedure was repeated amputation of the siphons. *Ciona* is a filter-feeder: it subsists by extracting microorganisms and food particles from a stream of seawater passed continuously through its body cavity. The body consists of a central sac, with two stolons, which serve to anchor it to the substratum below, and two siphons above. The longer syphon is inhalant, the shorter exhalant.

What Kammerer found, in common with two earlier experimenters, was that if the syphons are cut off (it seems to be essential to cut off both syphons), not only are new syphons regenerated from the stumps but they grow to a greater length than the original ones. Even more remarkably, this effect is enhanced if a series of successive amputations is performed: this results in extremely long siphons with a "segmented" appearance due to ridges marking the lines of successive planes of amputation. The reader will by now hardly need to be told that Kammerer found that the elongation of the siphons (though not the segmented appearance) was hereditary – in the first generation.

The *Ciona* experiment does not constitute a valid test of Lamarckism. For when *Ciona* overcompensates for the removal of its siphon by regenerating one longer than the original, this can hardly be called an adaptation to its environment. (Regeneration per se may well be, but not the regeneration of a *longer* siphon.) It was to *adaptive* responses to the environment that Lamarck specifically restricted his claims. The *Ciona* experiment is much more akin to mutilation experiments, such as those in which Weismann cut off the tails of mice for 22 successive generations.

Blind Newts

Kammerer did other experiments: on lizards, where, in parallel with the *Salamandra* experiments, he claimed to change both the reproductive habits and the color, and also on the cave-dwelling newt *Proteus*. This last was perhaps the most spectacular of all his claims. *Proteus* normally lives in the dark and is both colorless and blind, the eyes, indeed, being mere vestiges. By rearing *Proteus* in alternating red light and daylight, he succeeded in obtaining specimens with quite large well-developed and apparently functional eyes. However, he did not, in this case, follow up the work by breeding from his modified specimens, being, as he said, sure that if he did he would only be greeted by the "ever and again reiterated, stereotyped objections against the inheritance of acquired characteristics." Nevertheless, the experiments described here in outline were the most widely publicized and give a good idea of the nature and scope of Kammerer's work.

Uniformity, Rapidity, and Inclusiveness

There are three remarkable things about all this. The first is that Kammerer's experiments were uniformly successful in a field where reports of negative results were frequent. The second is the rapidity with which his effects became hereditary:

usually in a single generation, and never needing more than five. Even where, as with color in salamanders, a cumulative effect over three generations was reported, Kammerer later found that the same effect could be achieved in one generation by using a different environmental manipulation.

For a modern comparison, we may take the work of Bateman, who did experiments on the "genetic assimilation" of five different characters involving the wings of *Drosophila* [5]. In the experiment which gave the most rapid result, it took five generations to reach a point where 46% of the individuals in the assimilated stock exhibited the character. At the other extreme, Bateman's slowest result was 8% in 30 generations. Paradoxically, Koestler regards work on "genetic assimilation" as in some way providing support for his own belief in the validity of Kammerer's claims. Its effect is rather the reverse, for it provides a means whereby apparently Lamarckian results can be explained by a purely Mendelian and selective mechanism.

The third remarkable thing is another aspect of the suddenness with which Kammerer obtained changes: it is that the change appeared from the start in *all* individuals of that generation (or in all individuals of the appropriate sex). Since natural populations almost invariably turn out to be genetically variable in respect of whatever character is investigated, one could expect that there would be genetic variation in the readiness with which induced changes of this kind become hereditary. This would lead one to expect – on Kammerer's own premises – that in the first affected generation only a small proportion of individuals would exhibit the hereditary induced change, the proportion increasing progressively in later generations.

Bateson's Request

Bateson's first contact with Kammerer was in 1910. He wrote (July 17) explaining that he had been reading his papers with interest and that he was devoting a chapter to this and cognate matters in a forthcoming book. Bateson himself and several colleagues were interested in seeing specimens from Kammerer's experiments. While most of the changes produced were matters of degree and age, to an extent that made them not easy to demonstrate, the *Brunstschweilen* of *Alytes* stood out as easily demonstrable. Would Kammerer lend him such a specimen? Bateson promised to treat it with every care and to return it promptly.

Kammerer replied (July 22) from Steinbach where he was on holiday. He promised to send, as soon as he returned to work, whatever specimens Bateson might need, although he was not sure whether he had any preserved males with *Brunstschweilen* or had any living males. However, he was sure that other available material was well fitted to Bateson's purpose – especially the color changes in *Salamandra*. He concluded by referring to a promise to Doncaster that Przibram had made on his behalf, to supply a series of tadpoles for the museum at Cambridge, saying that he intended to fulfill both promises early that autumn. Kammerer seems to have sent Bateson's letter back to Vienna without making it clear that he had himself already replied to Bateson. For a few days later, Przibram wrote to Bateson reiterating all the main points of Kammerer's letter.

Both Kammerer and Koestler seem, in different ways, to have misunderstood Bateson's motives in asking for the loan of a specimen – and specifically, for one with *Brunstschweilen*. Kammerer thought that it was for a photograph or drawing to illustrate Bateson's book. Naturally, from this point of view, spotted and striped salamanders would make a more striking illustration than a not very pronounced swelling on the thumb of a toad. Koestler, on the other hand, correctly perceived that Bateson's main point was that the nuptial pads were in some way more critical evidence, by virtue of their being a structure not normally found in that species, but he imagined that the rationale was simply that this made it impossible to fake matters by the simple expedient of substituting a specimen from outside the experiment. This was possibly – even probably – one factor in Bateson's choice: who, indeed, can deny that a structure outside the normal repertoire of variation of the species concerned is a better safeguard? But there is a more cogent reason, one less dependent on suspicions as to Kammerer's good faith: if the structure is one foreign to the species, the chance is thereby greatly reduced that it could be the result of manipulation – whether accidental or intentional – of the genetical variation present within the experimental stock.

Bateson paid a visit to Vienna in late September of 1910 (Chap. 20). Whether this was something he already had in mind to do in July or was something decided upon at the last minute is not clear; probably the latter, since if he had known that he would soon be in Vienna, he would hardly have needed to ask for the loan of a specimen. Perhaps, too, he timed his visit in the hope of seeing live *Alytes* with nuptial pads: *Alytes* normally has two breeding seasons each year, one in April and one in September. If so, he was to be disappointed: the autumn breeding season must have been unusually late that year, and his visit was too early by one or two weeks. Nor did he see any preserved *Brunstschweilen*, although he was shown material from all the other experiments. He seems also to have had some difficulty in convincing Kammerer and Przibram that the nuptial pads were indeed critical. What constitutes critical evidence depends on the particular hypothesis to which it is held to attest, and neither the hypotheses nor the general viewpoints were the same in the two cases of Bateson, on the one hand, and of Kammerer and Przibram, on the other.

At this point in his narration, Koestler concludes that "after that [the visit to Vienna] there was no more question of sending specimens." Why not? The visit seems to have passed off quite amicably, and Bateson subsequently maintained a cordial if sporadic correspondence with Przibram, although he seems never to have written directly to Kammerer again. In fact, in *Problems of Genetics,* he specifically states that during his visit he repeated his request for a specimen. No specimen ever arrived. Kammerer's other promise to provide specimens – to Doncaster at Cambridge – surely cannot be regarded as cancelled, or in any way affected, by Bateson's visit. Yet again, no specimens arrived.

Post-war Controversy

During the war, Kammerer seems to have lost most or all of his stocks although his stock of water-reproducing *Alytes* had already terminated in 1913 with the failure of the F_6 generation to reproduce. The controversy was reopened in 1919 with a long paper by Kammerer devoted specifically to the *Brunstschweilen,* which included replies to his three chief critics: George Boulenger, Bateson, and Baur. Here he published, for the first time, photographs of males showing *Brunstschweilen* with microscopic sections through the skin. Ernest MacBride, probably the leading and certainly the most ebullient British Lamarckian of the time, publicized Kammerer's paper in a letter to *Nature.* Kammerer, he declared, had "fairly taken up the gauntlet thrown down to him by Professor Bateson:" McBride promptly threw it back to Bateson.

In reply, Bateson quoted in full Kammerer's letter of July 1910, with its promise to send him a specimen. He also criticized the photograph of *Alytes* for its indistinctness and for the fact that it had been extensively touched up, and he raised an important query as to the location of the pads. The earlier papers had described them as being on the inner surface of the thumb whereas the one now shown was on the outer side of the little finger, a position where it could not come into contact with the female during copulation. He raised, too, a very pertinent question: why was there no photomicrograph of a pad in situ? This would have been technically quite easy and would have demonstrated the most critical feature – the horny spines – far more convincingly than any of the published photographs. In June 1920, Przibram wrote to Bateson: "You will have seen by our reprints … that Kammerer has succeeded in getting again the *Brunstschweilen* in *Alytes:* we have now such specimens and histological series at our museum, but as they are only very few I would not like to trust them to the post."

The years between 1920 and 1923 seem to have been relatively quiescent. Kammerer sent Bateson some microtome sections through the skin of a *Brunstschweile*: Bateson's comment on these was that "they may have been taken through real incipient rugosities, but the development is slight and ambiguous." He nevertheless showed them to MacBride (who found them convincing) and to various other persons (who, for the most part, did not). A number of British biologists visiting Vienna during these years were shown a specimen of *Alytes* with an alleged *Brunstschweile* on the right hand (that on the left had been removed for sectioning). As Bateson put it, they "reported verbally and variously on what they had seen." One party of visitors, however, was not vouchsafed a sight of this famous specimen: they were geneticists calling at Vienna after the International Genetics Congress at Brünn in 1922. Bateson himself did not attend this congress, but at least two friends (Pease and Tjebbes) reported to him by letter on what they had seen – or not seen – in Vienna. Kammerer himself was not there, and there was no material either of *Alytes* or of *Salamandra* to be seen.

In building up his argument, Koestler attaches considerable weight to a letter by Bateson (July 20, 1920), said to be addressed to "a zoologist, Mr. Martin," who was about to visit Vienna. In it he recommends him to see Kammerer's famous *Alytes:* "Make them take it out of the bottle and examine it quietly and leisurely with a dissecting lens. Note very carefully the position of the alleged *Brunstschweilen* on each side, and make sure that it is an *Alytes*." Further, "bear in mind that a good deal of grafting is nowadays possible." To send to a comparative stranger a letter containing such imputations, particularly the one about grafting, would indeed be a serious matter, but the letter was sent, not to the unidentified "Martin," but to Martin Bateson, his own son.

This puts a quite different complexion on the letter. (Koestler was perhaps led astray by the relatively formal signature: "Yours, W. Bateson." Bateson invariably signed himself thus in letters to his parents, sisters, and sons. Only in writing to his wife did he permit himself the slightly greater intimacy implied by "Yours, W.B.".) One expects a letter written within the family to be less inhibited: suggestions can be made which one would not dream of making to a stranger. Moreover, the letter was written, not to a fully trained scientist, but to a young man of 20, still an undergraduate. It was quite natural, therefore, for Bateson to spell out in detail the ways in which his son should examine the specimen and encourage him to bear in mind all possibilities, however remote.

English Tour

Kammerer's visit in 1923 was sponsored by the Cambridge Natural History Society – primarily an undergraduate society – and students predominated in the audience at Kammerer's lecture on April 30th, but a dozen or so senior people were also present. Gregory Bateson told me that Koestler strongly overplayed Gregory's role in organizing the visit. His main function was to try (without success) to persuade his father to attend (Chap. 18). Koestler makes a lot of Bateson's non-appearance, but there seem to be natural and innocent explanations of this. Bateson would certainly have known that Kammerer was to give a repeat performance in London soon afterward in the more sober atmosphere of the Linnean Society. Why, then, should he go to Cambridge, just to cooperate in the undergraduate pastime of egging on their seniors into a knockabout confrontation?

Koestler quotes most of Muriel Onslow's long account of the meeting in a letter to Bateson, though he wrongly gives her maiden name as Wheldon. More seriously, he manages to represent her as saying that Hans Gadow "interrupted" Kammerer's lecture at intervals. What she actually wrote was "Gadow interpreted at intervals" – Kammerer lectured in German. As Onslow's letter is typed, there seems hardly any excuse for the error, other than wishful thinking by Koestler. Onslow was sharply critical of the speeches made by Gadow, Stanley Gardiner, and MacBride. Koestler smugly concludes "Kammerer came off best of the lot" – ignoring the fact that Gardiner and MacBride were ardent Lamarckians and admirers of Kammerer.

The reminiscences from various biologists (W. H. Thorpe, G. E. Hutchinson, J. H. Quastel, L. Harrison Matthews) who had attended the meeting as undergraduates add little of substance to Onslow's contemporary account, except that all of them had found Kammerer a genuine and sincere person. What is also clear is that the character and atmosphere of the Cambridge meeting were just what Bateson had feared. It was an occasion at which the pro- and anti-Kammerer camps set forth their claims dogmatically, and certainly not an opportunity for a dispassionate discussion of the strong and weak points of his case.

Indeed, by this time, Kammerer had become so incensed by the way the scientific establishment (with a few prominent exceptions) simply ignored his claims, that he regarded any refusal (or even reluctance) to accept his results in toto, together with his own interpretation, as casting a slur on his personal integrity. One can see in Kammerer's writings a decade or more earlier the beginnings of his "outsider syndrome." People ignore or dispute my results partly because they run counter to current orthodoxy, but also because of the kind of person I am: an Austrian, a Jew, something of a dilettante (he had studied music), not a "fertige Akademiker." (Koestler makes no mention of Kammerer completing his degree course in zoology. He does say that Kammerer became "thoroughly fed up with the old-fashioned type of zoology taught at the University:" it seems possible that he abandoned the course prematurely and left without a degree.) Koestler lovingly returns, again and again, to embellish this theme of poor Kammerer, the despised and persecuted outsider.

It may be doubted whether Kammerer ever had the critical-analytical cast of mind to appreciate that revolutionary claims such as his demanded especially stringent examination of every aspect to exclude various categories of error or misinterpretation (quite apart from outright fraud). If he ever had the necessary critical acumen, by 1923 it had become deeply buried in a protective layer, whose main constituents were *amour propre* and a total emotional commitment to the truth of his own theories, which enabled him to see all criticism as stemming from nothing more or less than pig-headed prejudice.

Bateson, as we have seen in other contexts, was not a man to shrink from vigorous public controversy, but debate with so evasive a target as Kammerer – always ready with facile replies (which usually managed just to miss the questioner's crucial point) – he probably regarded as a sterile and undignified undertaking, especially in the atmosphere which would obtain at the Cambridge meeting. Muriel Onslow ended her letter recounting the events at Cambridge by telling Bateson: "You were well out of it." In his brief reply, Bateson emphatically endorsed that sentiment.

Kammerer repeated his Cambridge lecture at a meeting of the Linnean Society in London (May 10), and the now famous specimen of *Alytes* was again available for inspection. Only a brief abstract of the lecture and discussion was published, and Koestler gives most of his attention to charging Bateson with having given the specimen only a cursory examination. ("Could it be that he looked away for fear of being convinced?") If Bateson's examination was indeed brief and hardly searching, there is a simple explanation: he had already seen it.

H. Graham Cannon wrote in 1959 that there had been a private meeting in MacBride's room at the Royal College of Science, with just four people present: MacBride, Kammerer, Cannon, and Bateson [6]. Kammerer produced his *Alytes* specimen in a glass tube, and MacBride examined it "literally for a couple of seconds" (those who are already converts are easily convinced!) before passing it on to Cannon, and in turn to Bateson, who examined it with a hand-lens. Cannon alleges that Bateson then said to him, *sotto voce* "It looks to me like a spot of black ink." If this is correct, then Bateson's comment showed a quite remarkable prescience, for three years later the nuptial pad on this same specimen was shown to have been injected with Indian ink. However, it seems likely that this part of Cannon's account owes more to his own creative hindsight than to Bateson's prophetic powers. Cannon does not specify exactly when this meeting took place: immediately before the Linnean Society meeting seems the most likely.

Later in 1923, there was a fresh spate of letters in *Nature* (Bateson, Kammerer, Przibram, MacBride) – mainly going over old ground, with recriminations. Bateson wished to see the specimen again and to take it out of the jar so as to examine the dorsal side of the hand. (Why didn't he do so at the Linnean Society? He said he hadn't been told there was anything special to see on the dorsal side.) He offered £25 (later doubled) to pay the expenses of a courier to bring the specimen to London again. Przibram refused and counter-offered that Bateson should come to Vienna to examine it. Actually, Bateson had no special expertise to judge how similar the nuptial pad was to those of other amphibia. He had never worked with amphibia, and people like Boulenger and Gadow (who had) were much better equipped as critical examiners.

American Tour

Kammerer's lectures at Cambridge and London had been given at private meetings, to which the press was not admitted: brief accounts were eventually published in *Nature* and in the *Proceedings of the Linnean Society*. Nevertheless, a story about the Cambridge meeting appeared on the front page of the *Daily Express* the next day. This was the work of a staff reporter, a friend of Michael Perkins (a member of Council of the Cambridge Natural History Society). He had been debarred from the lecture itself but had managed to talk to Kammerer and others after the meeting. The headlines were indeed sensational: "RACE OF SUPERMEN – Scientist's Great Discovery Which May Change Us All – Hereditary Genius – Eyes Grow In Sightless Animals" – and the text was in similar expansive tone. A few days later, accounts clearly based on the *Daily Express* story (though apparently supplemented from other sources in at least one case) appeared in several American newspapers, with similar dramatic headlines.

These reports must have come as welcome advance publicity for the lecture tour in the United States which Kammerer planned to make in the autumn of 1923. This was to be a very different affair from the sober academic English visit. For

Kammerer, it was a matter of sheer bread-and-butter necessity: the tour was intended to raise funds just as much as to propagate his ideas. Kammerer resigned from his post in Vienna in 1925, with the intention of earning a living by lecturing and journalism. (It is not clear whether this was a voluntary resignation or whether he was "pushed." Przibram's whole Institute was by this time in severe financial difficulties because of galloping inflation.) Accordingly, the organization and publicity for the tour were put into the hands of commercial agents – "Europart," with an address in Broadway, New York. Kammerer was thus putting himself onto the well-known "American lecture-tour circuit," a circuit which many literary, scientific, or otherwise marketable Europeans have found to be a rewarding, if exhausting, source of income. Europart appears to have done its best for Kammerer, and from a financial point of view, the tour was a success, sufficiently so for Kammerer to return for a second tour in 1924. According to Koestler, he also gave popular lectures in continental Europe, again on a money-making basis.

As Koestler acknowledges, the distinctly brash publicity material, which surrounded the American lecture tours in particular, did further damage to Kammerer's reputation in the scientific world. It might be acceptable for a scientist to commercialize himself, to allow grandiose claims to be made on his behalf, and to reap the monetary rewards of this propaganda – but he must first convince his own scientific colleagues of the validity of his claims. Even Kammerer's sense of self-importance could not have concealed the fact that most biologists regarded his claims with strong suspicion. The natural result of the popular lecture tours was thus to confirm and strengthen the widespread tendency among biologists to regard Kammerer as a publicity-seeking charlatan.

Morgan sent Bateson samples of some of the publicity handouts prepared for Kammerer's tour, together with his own, decidedly tart, reply to an invitation to join a committee of sponsors. Morgan refused, mainly on the ground that the advertising material was so blatant: a good many other American biologists must have reacted in similar fashion. Indeed, it is not at all clear whether Europart ever managed to assemble a sponsoring committee that would impress the layman. No doubt the odd convinced Lamarckian or two could easily be roped in, but for the rest, they may well have had to rely on people who, whatever their eminence, worked in areas which were at best peripheral to biology. For example, there was the famous behaviorist psychologist J. B. Watson (the only American supporter of Kammerer named by Koestler, though he implies that there were others). Watson's remarks about Kammerer (quoted by Koestler) are indeed quite warmly sympathetic, but they are also carefully guarded ("We all want to believe his facts if they are true."). To an insider, Watson would hardly rank as an authoritative expert on Kammerer's work, but to the man-in-the-street, he was a famous scientist and therefore impressive.

One must have a good deal of sympathy with Kammerer in his dilemma. He had somehow to earn enough money to keep himself alive (presumably, too, he had contractual obligations to help support his divorced wife and their daughter). His lecture tours could resolve that difficulty, but at the same time, they would inflict further damage on his already shaky reputation as a scientist. I do not for a moment suppose that he was so egotistical as to be concerned solely, or even primarily, with his personal reputation. The future of Lamarckian ideas mattered to him greatly, and his commitment to them was both genuine and intense – too intense, perhaps! His character contained strands both of the crusading hero and of the martyr. Had he believed that by sacrificing his own personal reputation he would somehow advance the cause of Lamarckism in the longer run, he might well have followed such a course with wry satisfaction. But he was enough of a realist to know that the discrediting of Paul Kammerer would also be a severe blow to the progress of Lamarckian ideas in general.

A great many other Central European scientists (and not scientists alone!) at that time faced financial difficulties just as severe as Kammerer's. Nor can one escape the conclusion that the particular way out of his dilemma that Kammerer chose entailed his playing a role that he found distinctly congenial. A kind of hero's role, that is: a man much in the public eye, performing before admiring, even adulatory, audiences, extolling his own claims as a great scientist whose revolutionary discoveries could bring incalculable benefits to future humankind but who (here comes the martyr's role) suffered only neglect or abuse from the scientific establishment of the day, too far sunk in lethargy and dogmatism to appreciate his merits. Kammerer's complex character included prominent elements of the actor and showman, which he was able to exploit freely in his lecture tours.

Noble's Exposure

After the dust from the flurry of letters in *Nature* in 1923 had settled, the Kammerer affair remained relatively quiescent. A steady trickle of visitors to Vienna came to examine the one remaining specimen of *Alytes* but, by and large, skeptics remained skeptical, and converts held fast to the true faith. There were no further publications from Kammerer: he was too busy earning his keep by other means. Then, early in 1926, Noble, a 32-year-old American zoologist, came to examine the famous *Alytes* [2].

Even his first examination, done externally, revealed two features so disturbing that Noble obtained Przibram's agreement to a more thorough (and inevitably in some degree destructive) examination, including the removal of tissues for histological study. There were two disturbing features: (1) the skin covering the alleged nuptial pads was smooth and shiny, with no trace of the hard horny spines which would have made it rough to the touch, and were the chief diagnostic feature of any amphibian nuptial pad; (2) examination under a binocular microscope at moderate magnification showed clearly that the blackish coloration of the area was due not (as it should have been) to pigment in the epidermal spines (which in any case were lacking) but to pigment in the deeper, dermal, layer of the skin. Noble concluded, and his further investigations confirmed, that the pigment was almost certainly Indian ink.

Nobody – not Kammerer himself, and not even Koestler – has ever challenged Noble's main conclusion: that the specimen had been tampered with or "doctored up" by injection of Indian ink, and was, in crude terms, a forgery. As to whether this specimen had ever, at any stage, had genuine nuptial pads, Noble was unable to find any direct evidence one way or the other. There was the possibility, stressed by Przibram in a report published alongside Noble's article in *Nature*, that, through a combination of poor initial fixation, long preservation (for 12 years, apparently) and – one might add – overmuch handling, the epidermis had sloughed off in some areas (and what more natural than that the region most subjected to scrutiny would be thus affected?). This could account for the negative features: the smooth skin, lack of detectable spines or of natural pigmentation. Nevertheless, it seems to me that less direct evidence (discussed below) points rather strongly, despite the protestations of Przibram, Koestler, and others toward the conclusions that this particular specimen had never had true nuptial pads.

Noble was content to assert that forgery had been committed. Naturally, many people saw the finger of suspicion as pointing straight at Kammerer, and when, six weeks after the publication of Noble's article, Kammerer committed suicide, many saw this as tantamount to a confession of guilt. The point seems not at all so clearly determinable. There is one strong argument against the forgery having been done by Kammerer (or with his consent) – and here, for once, Koestler misses a trick in defense of his hero. Noble found that the thumb-base on *both* arms had been injected with Indian ink. Yet the universal testimony is that the specimen seen in Cambridge and London in 1923 (which is supposed to have been the same specimen that Noble examined) had a nuptial pad only on the left arm – the pad on the right side had been removed for sectioning. It is barely conceivable that Kammerer himself, dealing with (or attempting to reproduce) his own so-familiar specimen, would have made so gross an error, though any other forger might well have done so.

Koestler's own arguments against Kammerer as the perpetrator seem a good deal less secure. Principally, he argues that the forgery was a crude and clumsy attempt: Kammerer, as a highly skilled experimenter, would have done a much better and more convincing job. Perhaps, yet Noble's specimen appears to have been injected *into* the substance of the dermis. To achieve this would require a very fine needle and great skill (or, at least, practice). (By contrast, in the experiments by Hayden, done at Koestler's suggestion in an attempt to reproduce the state of affairs found by Noble, there is every indication from Koestler's account that the injections went right through the dermis, depositing the ink in the sub-dermal space. This probably explains why Hayden's artifacts were so unsatisfactory!)

If not Kammerer, the most likely candidate seems to be somebody on the staff of the institute – perhaps a technician – who, having accidentally lost, destroyed or seriously damaged Kammerer's precious specimen, attempted to cover up the loss by making a forged substitute. Or somebody was trying to be helpful in a less dramatic way: finding that the specimen had become less clearly convincing through fading with age, they tried to "restore" it by a little creative touching up. Either version seems not inherently implausible, and nothing stands against it.

Koestler's other suggestion was that an enemy of Kammerer did the forgery with the intention of discrediting him. Though not impossible, this seems very unlikely. It would be difficult for anyone not working in the institute to gain access to the specimen, and there is no evidence that any of Kammerer's colleagues hated him with sufficient intensity. My personal view is that there is only one fair answer to the question whether Paul Kammerer was guilty of forgery, and that is the evasive verdict peculiar to Scottish law: not proven. On the one hand, from all we know of Kammerer, I find it impossible to envisage him as a strict "sea-green incorruptible," a man who would never in any circumstances, or for motives however high-minded, do anything which bordered on forgery. (We have to remember that he could have seen the act, not as outright forgery, but merely as "reconstructing" valid evidence which he once had but had lost. He was utterly convinced of the soundness of his scientific case, but unfortunately a specimen was needed to convince those bigoted skeptics). On the other hand, the evidence against him is merely circumstantial and does leave room for "reasonable doubt." Besides, how many of us could safely swear that we would never, however strong the temptation, do anything comparable to what Kammerer is alleged to have done? Well, not perhaps exactly comparable, not so palpably dishonest as outright forgery, but something uncomfortably close to that rather fuzzy line separating honesty from dishonesty.

Summary

Bateson did not dismiss out-of-hand claims purporting to have demonstrated Lamarckian inheritance. Of these, the *cause célèbre* was that of Paul Kammerer in Vienna to have examples in several species. The most notorious was the water-induced acquisition by land-copulating male midwife toads of the pigmented hand pads (*Brunstschweilen*) that allowed water-copulating toad species to grasp females. This acquired character was inherited by offspring that had not been exposed to water. Following a visit to Vienna in 1910, Bateson grew increasingly skeptical concerning Kammerer's competence and/or good faith. In the early 1920s, Kammerer's visits to England and the United States received much public attention. A few months after Bateson's death, Kammerer's sole remaining midwife toad specimen was exposed as a fake by an American zoologist. Kammerer shot himself. Arthur Koestler's tendentious account of these events (1971) misrepresented Bateson and portrayed Kammerer as a martyr [7, 8].

Chapter 22
Science and Chauvinism

Heaven is the work of the best and kindest men and women.
Hell is the work of prigs, pedants and professional truth tellers.
The world is an attempt to make the best of both.

Samuel Butler, The Way of All Flesh

One's views can change considerably between one's twenties and one's sixties. Nevertheless, it is clear from his responses to taunting letters from "Mamma" during his Steppe travels that, at that time at least, Bateson was snobbish, racist, and intensely patriotic (July 1, 1887):

> You allude to the dyeing industry. I happen to know that what you say is true as far as it applies to the Aniline dyes. But I happen also to know that the industry was in the first instance invented by [the] English and has only comparatively lately migrated to Germany. I also know that on the authority of R. Meldola … nothing but technical training can bring it back again; which technical training means the establishment of a series of Technical Colleges, which will result in the rearing of vast hordes of persons like unto that ruddy party from Birmingham who paid us a visit about a year and a half ago. I therefore see the Aniline dye migrate to Germany without a pang.

As Beatrice relates in her *Memoir*, in other letters he waxed eloquently on the superiority, not just of the British but of the English race in particular (June 1887):

> A Russian is no more the equal of an Englishman, and a Negro is no more the equal of a white man than a Kirghiz pony is the equal of an English racer, or the Phylloxera the equal of the vine. … Of course I know that there is no test of universal application by which a man's worth can be estimated: this cannot be helped. But there are a set of qualities which we, following our instinct for want of a better guide, regard as denoting superiority. … For all these things it seems to me that we are, as peoples go, well to the fore. It is no light thing that nearly all the great inventions are of English origin.

However, as his speeches on eugenics later revealed (Chap. 15), these views were moderated and, in Panglossian fashion, he came to embrace diversity as the best of all possible worlds with each contributing, as he is able, to a harmonious whole. The task of reestablishing scientific communications at the end of the First World War

A. G. Cock, D. R. Forsdyke, *Treasure Your Exceptions*,
https://doi.org/10.1007/978-3-030-92099-9_22

was a stern test of this. We here consider Bateson's involvement with scientific organizations at both national and international levels.

Royal Society Calandruccio Enquiry

While always much involved with the RS, Bateson was never its President. But he was a member of its Council for a few years (1901–1903). Here he became involved in a priority dispute over the discovery of the life cycle of malarial parasites – a classic example of the power of the macro-observational skills of the naturalist combined with the micro-observational skills of the microscopist. A letter (Jan. 31, 1902) established Bateson's appointment to the "Calandruccio Enquiry Committee charged with investigating a claim advanced by Dr. S. Calandruccio of Catania to a share in the award of the Darwin Medal, allotted in 1896 to his senior colleague Prof. G. B. Grassi." The matter wound on for some decades and neither the outcome nor the scientific details need concern us here.

The main protagonists were Salvatore Calandruccio, Giovanni Battista Grassi, and Ronald Ross (1857–1832) of the Liverpool School of Tropical Medicine, who had established the link between mosquitos and malaria. A publication of Ross in 1900 (*Il Policlinico*. Nov. 1, 1900), which implied that Grassi had overstated his contribution with respect to discoveries Ross had made, had prompted Calandruccio, after consultation with Ross, to bring forward his own claim. Ross met with Bateson, Harmer, and Sharp in Cambridge (Feb. 9, 1902), after which Bateson noted: "He is giving evidence ... to support Calandruccio's charges against Grassi by showing that Grassi's behaviour towards his (Major Ross's) work has been similar to that now alleged against him by Calandruccio." The Bateson papers contain a brief from Ross (Feb. 5, 1902):

> In the first place we should consider how easily mistakes as regards priority may be made in connection with highly specialized subjects which possess a large literature. As a rule there are only a few persons who really have an intimate knowledge of such subjects; and these persons seldom trouble to publish minute discussions on priority. Hence errors and misstatements often remain at first uncorrected, and then pass from text-book to text-book until they are finally accepted as 'well-known facts' which it is no longer possible to contradict. Unscrupulous persons easily take advantage of this circumstance to distort the whole history of such a subject. Thus as regards malaria, there are really very few persons who possess at once a practical acquaintance with the disease itself, and a thorough knowledge of its literature; and in fact – as is well known to the few – this subject has given occasion to attempt after attempt of this nature.

Ross went on to give examples of such attempts, including the case of the mysterious "*Bacillus malariae*" that had been discovered in 1879 by Klebs and Tommasi Crudeli:

> These facts were confirmed and extended by some of the most distinguished Italian students of malaria .... Of course, no such thing as the *Bacillus malariae* ever existed. ... In 1880, however, Laveran discovered the true parasite of malaria; but for five years he was opposed by all the advocates of the *Bacillus malariae* – mostly Roman writers. Suddenly however,

> forced to abandon this error, they accepted Laveran's views, and then immediately endeavored to claim a share in his discovery. Even in their last works Marchiafava and Celli continued to pretend that they were the first to prove the existence of the parasite of malaria.

The questionable activities of Professor Grassi were then listed with the conclusion:

> It is impossible to touch upon the numerous other falsifications of Grassi .... The case which he has so skillfully and unscrupulously built up is shattered, partly by several slips which he himself had made in his own writings, and partly by the letters written to me from Rome by Dr. Edmoston Charles. Dr. Charles was intimate with Grassi and his colleagues when they were attempting to follow my work at the end of 1898; and he wrote to me a minute record of their progress. His letters are published by me with his consent. Grassi, though he evidently tried to obtain information from me through Dr. Charles, appears to have forgotten the probability that Dr. Charles might keep me informed regarding his own doings.
>
> To compare these letters with the priority claims of Grassi ... suffices, I think, to establish the complete mendacity of that person. Nuttall has examined several of these questions in the May numbers of the *Quarterly Journal of Microscopical Science* 1901; and I think it will be found that his decision generally accords with mine. ... Whether the numerous misstatements made by G. B. Grassi in connection with the history of malaria are innocent errors or are willful subversions of the truth must be decided by those who are quite familiar with the facts of the case.

Thus, Bateson, at a time when he was hard-pressed with his own controversies (Chap. 9), became embroiled in the malaria controversy, some of the documentation of which was in Italian and French and dated back to the early 1890s. In 1901, his Cambridge colleague, the immunologist George Nuttall (1862–1937), summarized some of the issues in a paper entitled "On the Question of Priority with Regard to Certain Discoveries upon the Aetiology of Malarial Diseases" [1]. In 1902, Ross gave a further account in his Nobel Lecture (Dec. 12). The experience perhaps served to remind Bateson that senior workers must be sensitive to the need to give their junior colleagues due credit, and the extreme caution that must be employed when examining scientific claims.

Society for Experimental Biology

The Genetical Society got underway in 1919 and seemed to grow in strength. But from the start, a division could be discerned between the Mendelians and the rest (Chap. 17). Furthermore, like a biological species, within a society there can be splinter groups that may interface with other constituencies, and from time to time fledgling societies emerge to cater to their needs. At the seventh meeting of the Genetical Society (June 1, 1921), it was resolved to hold a joint meeting with the Association of Economic Biologists on March 31 1922. Perhaps Maynard Keynes had a hand in this, but there is no record of the meeting. Perhaps the time was not ripe for economic biology. It thrives today under a variety of names.

At the eleventh meeting of the Genetical Society (June 1922) Lancelot Hogben (1895–1975) was elected a member. In the same year, he was recruited to Edinburgh University by Francis Crew (1886–1973) who had become Director of the Animal Breeding Research Department. Prior to the move, Hogben had been a lecturer in Thomas Huxley's old department, which had become part of the Imperial College of Science. Here he had met the son of H. G. Wells, and through him his famous (and wealthy) father, who expressed a willingness to provide start-up funds for a new journal. On the occasion of a visit to Edinburgh of Julian Huxley, then a Senior Lecturer in the Department of Zoology at Oxford, and Haldane, then a Lecturer at Cambridge, the topic of a new journal was raised. Crew agreed to help with the financing. The *British Journal of Experimental Biology* was off, the first issue appearing in October 1923.

To make abundantly clear that the new journal was for both plant and animal workers, Ruggles Gates, by then Professor of Botany at King's College, London, was invited to join the Editorial Board. As a source of contributions, both scientific and financial, it seemed wise that the journal should be backed by the members of a society. So a meeting was held at Birkbeck College, London (Dec. 21–22, 1923), under the Chairmanship of Crew, the Managing Editor of the journal. It was not just a business meeting. There were formal presentations on topics such as "Vasomotor Activity of Pituitary Extracts" (Hogben), "Amphibian Growth and Metamorphosis" (Huxley), and "Chromosome Form and Number" (W. C. F. Newton). Gates and Hogben were elected Honorary Secretaries of what would be called "The Society for Experimental Biology" (SEB). Among members of the first Council of the Society were Fisher and Huxley, and Bateson was co-opted in March 1924 [2]. Not surprisingly, there was much overlap of the membership list with that of the Genetical Society [3].

Bateson's initial reaction had been ambivalent. Active concern seems to have been aroused by the proposal (originating with Hogben) to hold a symposium on a frankly genetical topic – inbreeding – as part of the next meeting of the SEB. Punnett wrote to Bateson (Feb. 22, 1924) saying that he had no knowledge of the proposed inbreeding symposium (which it was then hoped would be held in Cambridge – in fact it was never held anywhere). Punnett thought the matter serious: "Something must be done." He had suspected something like this would happen ever since Huxley had proposed that the Genetical Society should take over and run the *Journal of Genetics*: "As I see it there is a 'B. M. G' ['Bateson Must Go'] party of ambitious young men who want to be in control of something, and to figure more prominently in the public eye than they do at present." If this were the case, Punnett thought that "there cannot be anything but antagonism between the Genetical Society and the Association of Experimental Biologists." Rather than risk this, he would much prefer to see the Genetical Society disbanded.

Hostility was absent from Bateson's letters on the subject. He was careful to agree that there was a need for the new society and displayed merely a pained surprise that others should not understand the broad scope which he attached to the term "Genetics." He conceived Genetics to be something wider than Mendelism – practically synonymous, in fact, with "Experimental Biology". So, he agreed that the best solution would be for the Genetical Society to disband itself. It was announced that a "secessionist meeting" of the Genetical Society would be held (Feb. 22, 1924) for consideration of "policy in view of the establishment of a new organization, the Society for Experimental Biology, with functions largely identical" to those of the Genetical Society. Bateson wrote to Huxley (Feb. 10) giving his view on the term "Genetics:"

> I always understood it to be synonymous with what we used to call 'Variation and Heredity,' and as such to include the consideration of the influence of conditions in the causation and control of those phenomena. ... I am by no means convinced that the Genetical Society ought now to be maintained. As a small body, in number not too great to visit private establishments, it had a distinct use. But when our activities were extended to the holding of meetings and reading of papers, we undertook a function which would be much better discharged by the new Society, with a larger membership and a greater variety of composition. There is evidently a real demand for such a Society as that for Experimental Biology and I am very glad to hear that this should be so. It is a natural development to supply a real need, and this should not give rise to any irritation or resentment.

This brought a prompt response from Huxley (Feb. 16, 1924), who was in the delicate position of being on the governing bodies of both societies. He protested that the overlap between the societies was not as great as Bateson had implied, and thought accommodations could be made, such as holding joint meetings. Hogben wrote (Feb. 16, 1924) hoping Bateson would accept co-option to the Council of the Society. Since Bateson's books, *Materials* (which he had first read as a schoolboy) and *Problems*, had played a seminal role in the development of his own thoughts, he would be hurt to think that Bateson considered the new Society as a challenge to the Genetical Society: "And it would never have occurred to me that you would fail to recognize that when you emphasized to an earlier generation of biologists the futility of seeking the living among the dead [i.e. the morphological approach], you were yourself paving the way for a bigger revolution in biological thought than the Mendelian renaissance." He hoped that Bateson would "recognise that Experimental Biology is wider in scope than Genetics (even if the latter is extended to include the entire physiology of reproduction)."

Bateson replied (Feb. 19, 1924), repeating the points he had made to Huxley on the scope of genetics:

> No indeed! I have never thought of 'the Mendelian renaissance' as an end in itself. To me Mendelism was a very welcome incident, coming at a time when it was greatly needed. The development which supervened on that beginning has been so dazzling that when one now says "Genetics", everything else looks in the shade. Genetics cannot be 'extended' to include the physiology of reproduction. It *is* the physiology of reproduction. ... It was encouraging to be asked to join your Council. Obviously I could not accept until the Genetical Society had decided its course. I suppose they will continue for a while and see what happens. If the situation is regularised there should be no objection to my accepting your invitation. It was Miss Saunders who forced the Genetical Society upon us. Our meetings have not been without their use, but I dare say they may before long be suspended without much loss.

At the fateful February 2nd secessionist meeting, it was indeed decided to "continue for a while and see what happens." Bateson informed Hogben of the outcome and accepted "with great pleasure" co-option to the Council. But a proposal of Huxley and Hogben for a joint meeting of the two societies in June would hardly be possible as the Genetical Society Annual General Meeting was already arranged for that month. Bateson suggested November 1924. In fact, the first joint meeting of the two societies was not held until 1938.

The SEB went ahead and held its meeting in June 1924. Of the 23 presentations or displays, five topics stand out as particularly genetical: "The Comparative Genetics of Rodents, with remarks on Homology" (Haldane); "Some Peculiar Linkages in Brassica" (Pease); "Intersexuality in the Pig" (J. Baker); "So-called Mutations in Pathogenic Bacteria" (A. D. Gardner); "Inheritance in Polymorphic Butterflies" (Poulson). But, at least as judged by Huxley (June 25), the Genetical Society was still thriving in 1925: "I enjoyed myself so very much at Merton on Saturday – I think it was the best meeting the Genetical Society has ever had – and certainly I don't think you need now worry about the Society dying the death as the result of the starting of the Society for Experimental Biology."

Internationalism in Science

The literature of science extends across national boundaries. Science is a transnational, even an "extra-national," activity (Chap. 16). At the personal level, scientific internationalism takes the form of the exchange of letters and papers between scientists of different countries. At a more formal level, there are scientific meetings, often initially called into being on an ad hoc basis. This may sometimes lead to the establishment of some permanent structure to facilitate continuity. At the Fourth International Congress of Genetics (Paris, 1911), such a continuing body was appointed (Chap. 14). But continuity was lost due to the war, and it took much of the next decade to reestablish it.

Issues such as scientific nomenclature and the standardization of measurements demand some continuous central coordination that periodic international conferences cannot easily provide. Thus, scientific organizations, such as the Genetical Society, which emerge initially as national organizations, may give rise to corresponding discipline-specific international organizations each with a central office and a permanent secretariat. These international organizations, in turn, make apparent the need for some form of overarching organization that may allow not only science to speak to science but also nation to speak to nation. International organizations, irrespective of their specific mission, may promote international peace and understanding (or the converse).

Such thoughts were among those that drove the movement in 1919 for establishment of what was to be called the International Research Council (IRC). Under IRC auspices, there would be various discipline-specific International Unions, such as the International Union of Biological Societies (IUBS). Bateson could hardly help being drawn into this [4].

Boycott

However, the times were not auspicious. The war years had heightened antagonisms both within countries and between countries. A noted example was the *cause célèbre* of the removal of the philosopher Bertrand Russell from his fellowship at Trinity College because of his public advocacy of pacifism, which had landed him in prison. Other less vocal pacifists among college fellows – such as the mathematician Godfrey Hardy – escaped unscathed. Bateson deplored the activities of men, such as Baur, his friend of pre-war days, who went about representing Germany as an injured innocent. Indeed, in October 1914, a large group of leading German scientists and intellectuals had signed a widely disseminated manifesto – "To the Civilized World" – that downplayed Germany's role in the initiation of hostilities. A counter-manifesto from the University of Berlin calling for an immediate end to hostilities had only three signatures, one of which was Einstein's.

Early in the war, the hostilities had affected people of German ancestry, such as the physicist Arthur Schuster (1851–1934), whose Jewish father had emigrated to England in 1869 from Frankfurt. In 1915, there were pressures (led by the chemist Henry Armstrong) to topple him as President-designate of the BA and as Secretary of the RS. Lankester considered that Schuster "ought out of consideration for his colleagues in the Society, to remove the ill feeling, which his presence evokes … by making a voluntary retirement" [5]. It says a great deal for the good sense of the RS Council that it maintained throughout the war a secretary of German birth and, despite pressures, did not follow the precedent set by the corresponding French Academy early in the war, by striking from its list of foreign members all German nationals.

In 1918 (Oct. 9–11), the RS convened the "London Conference" with delegates from Belgium, Brazil, Britain, France, Italy, Japan, Serbia, and the United States. Godfrey Hardy (not to be confused with the biochemist W. B. Hardy) was a delegate. At that time, the Allied victory was in reach but not yet quite achieved. Feelings of narrow nationalism and revenge were high. A headline in *The Times* revealed the mood of the conference, particularly as expressed by the French and Belgian contingents: "Boycott of German Scientists." Against this overwhelming sentiment, Hardy could not prevail. In the same month, *Nature* published an article on "International Relations in Science" by Lankester, who opposed personal contacts with German scientists. Lord Walsingham wrote in *Nature* (1918) deploring the pre-war attempts of German naturalists to usurp the system of scientific nomenclature in favor of the German language, adding: "Let us trust that for the next twenty years at least, all Germans will be relegated to the category of persons with whom honest men will decline to have any dealings." One British scientist prefaced a scientific paper with the remark: "No quotations from German authors published since August 1st 1914 are inserted. *Hostes humani generis*" [6]. Another proposed that no publications in the German language should be cited. The Belgian native Boulenger wrote to Bateson that "I have bound myself to ignore everything published in Germany after 1914." A group of RS members, which included Ronald Ross and some leading physiologists, circulated a note (Nov. 1917) declaring that the society's Yearbook should "omit mention of German Academies and Societies."

Bateson's reservations were evident in his attitude to the "Oxford letter" of 1920. This was organized by the Poet Laureate, Robert Bridges, and signed by about 60 Oxford academics – Hardy included. It was sent to professors in the arts and sciences in German and Austrian universities. The hand of friendship and cooperation was offered. Although it published it, *The Times* (Oct. 18, 1920) immediately condemned it in a leading article as "singularly ill-advised … and reprehensible to the extreme." Most of the ensuing letters to the Editor were hostile to the Oxford letter. However, a second letter from Oxford with many names – including Huxley – that had not been on the first letter, gave "whole-hearted agreement with the desire expressed in the letter published in your columns on October 18 for the resumption of normal relations in the field of learning and research with the professors and teachers of German and Austrian universities."

Bateson's feelings were mixed. While he deplored the supercilious and self-righteous attitude of *The Times* and its correspondents, he felt the tone of the Oxford letter was too effusive and thus likely to be counterproductive. He drafted, but probably did not send (at all events it was not published) a letter in which he applauded the final sentence of *The Times*' leading article: "Let the exchange, at this time, of knowledge, pursue its quiet, impersonal course, and wait for friendship until it can be honest on both sides."

In the early 1920s, Baur proposed that the Fifth International Congress of Genetics should be held in Copenhagen in 1921 and wrote to Bateson apparently without any prior consultation with Johannsen who would most likely have had to organize it. Bateson, writing to Ostenfeld (May 5, 1920), was unhappy about this. He feared that a truly international congress (i.e., one including the Germans) held at that time might give widespread offence. Opinions were changing, however, and he thought a delay of even as little as six months might bring a more congenial climate.

In 1925, Bateson wrote to Nilsson-Ehle: "In the spring of this year I received letters from Professor Baur and from Professor Nachtsheim … suggesting … Berlin 1927 for the 5th International Congress of Genetics. I have seconded Baur's proposal." In other correspondence (April 8, 1925), Bateson wrote of Baur's change in heart: "I had a letter from him very lately and from the tone of it I got the impression that the 'black dog had come off his shoulders'. He wrote in distinctly better spirits. His purpose was to invite us all to a Genetics Conference in Berlin in 1927. He said that in 1927 'even the French might be disposed to come to Berlin,' which suggests that he anticipates a considerable 'rapprochement' from his own side." The Fifth Genetics Congress did indeed take place in 1927 in Baur's Berlin.

Chauvinism

In 1919, David Prain, the Chairman of the Council of the John Innes Institute, became RS Treasurer. Schuster, the Secretary, formally invited Bateson to act as one of the British delegates to the Inaugural General Assembly of the IRC to be held in

Brussels for 10 days beginning July 18th. After a great deal of hesitation – there were four letters from Schuster and probably also at least one meeting – Bateson agreed. Schuster's last letter (June 29) was a personal one:

> I was glad to receive your letter and should like to have a talk with you some time this week. The Biological subjects will not come up before Wednesday 23rd July. The first few days are taken up with general matters of organization, and Astronomy and Geophysics. Attendance at Brussels would I think commit you to the exclusion of Germany from all international organizations which are initiated by the Int. Res. Council. We do not of course pretend to control others. But we are working under resolutions passed by the Allied Academicians, and so far as the Royal Society has adhered to these resolutions its representatives at Brussels are bound not to go against them at the meeting. We should value your advice in all ... matters concerning [the] proposal to form organizations in Biology. We can talk this over when we meet.

Writing to Beatrice from Brussels (July 16, 1919), Bateson lamented:

> The international position is even worse than I had supposed. Every scrap of common sense is gone. The French are determined to turn the world upside down, and the rest acquiesce. If I had understood the purposes of this meeting I should not have come. It has certain real and useful objects in view, but it is being largely made an occasion to exploit science for chauvinistic purposes. I keep a quiet tongue and have only intervened when it could be done without exciting a row.

The international basis of the IRC was deliberately restricted. The former Central Powers (Germany, Austria, Bulgaria, Hungary) were excluded by statute from both the IRC and its Unions. Former neutral countries were to be admitted only by a three-quarters majority vote. The individual Unions were not granted a degree of autonomy that would permit them to override this. The IRC was thus quite openly part of the general post-war policy, spearheaded by the Treaty of Versailles, of isolating the Central Powers, of demanding from them expressions of penitence, and of ensuring that Germany never regained her old dominance in military affairs, industry, trade, or science. Ironically, Schuster became the first Secretary-General of the IRC, a post he held for nine years, and Bateson's "quiet tongue" policy may have furthered his own election to the Vice-Presidency of the IUBS. Bateson wrote to Ostenfeld: "When I accepted I thought that I was being appointed a V. P. for the *Brussels meeting*. I refused to be president of two sections, but when I understood my first mistake, I thought that to protest and withdraw then would be too 'fussy' as we say, and that I had better let things take their course."

IUBS

Although officers and a committee were elected at Brussels, and draft statutes drawn up (but not ratified), a substantive IUBS did not come into existence until 1923. In the interim, Bateson's position was quixotic: an officer of a non-existent organization with whose tenets he disagreed. When in 1924 he was offered the Presidency of the IUBS section on General Biology, he declined. Likewise, Johannsen declined the Vice-Presidency. Bateson wrote (in French) to Flahault (Jan. 21, 1924):

> To receive such a letter as that which you have sent me in the name of the Union Internationale des Sciences Biologiques is an extraordinary honour, and I greatly value this sign of the confidence which my foreign colleagues are willing to extend to me. The Presidency of the Union Internationale must be a position of distinction, and you will believe that it is not lightly or without serious consideration that I ask to be excused acceptance of that office. I have some doubt whether a general international organization in biology is yet a necessity, and whether the international organizations of the several branches of biology do not sufficiently supply what biologists require. Though doubtful on that point, I should nevertheless not regard this hesitation as ground sufficient to preclude me from accepting so flattering an invitation. I have a much graver objection.
>
> The Union Internationale des Sciences Biologiques, as at present constituted, seems to be likely rather to restrict than to promote international scientific communication. This consideration has, as no doubt you know, weighed so much with us in England that the Royal Society has declined to join. … For my own part, I deplore the introduction of the question of nationality into scientific affairs. The representatives of learning and the arts might provide a dominant influence in maintaining the sanity of the world. The force they might exert is prodigious. That they should promote further division is to me lamentable. The formation of the Union Internationale from which Germany is in effect excluded, can only increase the mischief. There is also to me something unreal and grotesque in the thought of a biological union containing no German names. After all, the developments of cellular biology have always from the beginning been very largely German. We gain nothing by refusing to recognize the facts. One of the great deprivations of the war was the suspension of intercourse with our German colleagues. I cannot be a party to any measure tending still further to alienate us.
>
> Pray do not charge me with a want of sympathy. I can understand the feelings of resentment which have led to such proposals, but I mistrust the wisdom or utility of reprisal. If in the judgment of posterity we are found at this critical time ourselves to have done nothing unworthy, it is much, and with that ambition we must be content.

In 1917, under the auspices of the RS a "Conjoint Board of Scientific Societies" (CBSS) of all the main British scientific societies had been established, and Bateson was a delegate from the Linnean Society to a special meeting (Jan. 8, 1920) to discuss "the general principles involved" in the establishment of International Unions. There was much correspondence between Prain (Kew Gardens) and Bateson before the meeting. Prain wrote (Jan. 6, 1920):

> I enclose Lotsy's letter to Scott. One thing I think is quite clear, Scott should adopt Lotsy's suggestion and send a circular letter to all those who have promised to act as guarantors telling them frankly how matters stand and giving them the opportunity of withdrawing from their obligation. I have not yet said so to Scott, but I propose to do so when I return Lotsy's letter to him. All I said to Scott today when acknowledging receipt of Lotsy's letter was that I would send that letter to you; that he should at once say to Lotsy not to hand Nyhoff the contract signed by himself and Lotsy, till Lotsy should hear from him again; and that a question bearing on the position of the Association Internationale was down for discussion by the Conjoint Board of Scientific Societies on Thursday.

Bateson replied (Jan. 7):

> When Lotsy was here I told him that there was not the smallest chance of getting the French to join. Anyone who had listened to Flahault at Brussels would know this. Indeed, to truly frustrate international collaboration was the purpose of the Brussels Meeting. Nevertheless,

> Lotsy's letter, which he thinks 'particularly nice' not unnaturally caused the explosion to somehow be more violent than it need have been. What … delicious specimen of the ways of the two types! The clumsy Teuton through sheer insensitiveness and want of humanity offering a good case so grossly and with such transparent absence of real sympathy as to make his own supporters blush, and the agile Frenchman, utterly in the wrong, all emotion and sparkling with passion, clawing his face for him! The guarantors must no doubt be told the story. Most of us, I suppose, foresaw what we were in for. I have heard of this clause in the Treaty before, but I do not know what the words are, nor by what authority they can be enforced. I suspect they refer to organizations supported by public funds. If anything can be more unwise than Lotsy's … letter to Flahault it is surely that he asks Scott to sign. Fancy choosing such a moment for threatening the French with the consequences of exclusion! Even the reference to Errera strikes as a spurious ornament. It is perhaps put in for [the] Belgians, though they will remember [the] past enough that Errera was characteristically a Jew. Better to have said in a few simple words that for us Science is a thing apart, the one unquestionable and universal good, that our aim as men of science is to promote the growth of knowledge, and that therefore we must admit to membership those likely to further that aim, without regard to their morality, nationality, and any other collateral circumstance or other Devil.

As the result of the January 8th meeting, various subject area Committees were formed, one of which was in Biology. Subsequently, the Committee on Biology reporting to the CBSS passed two resolutions. The first declared that the Committee thought "the restoration of normal conditions of scientific intercommunication with Austria and Germany is an indispensable preliminary to the formation of a Biological Union." The second, reflecting the main strand of criticism of the IRC, expressed unwillingness to commit itself to the proposed scheme for the IUBS and suggested that "the restoration of co-operative international relations in scientific work may for the present be better achieved through the medium of International Congresses and by individual initiative."

These resolutions appear close to Bateson's view. He wrote to Ostenfeld (Oct. 27, 1920): "This result was to me quite unexpected. It has given me also very great pleasure. I had no idea that I should be able to bring about this result. We hope we have heard the last of this Biological Union which nobody wants! It seems biologists have more sense than other people after all!" Some months later, with some input from Bateson, Chalmers-Mitchell published an anonymous article in *The Times* (March 8, 1921) headlined: "The progress of science – revolt against super-organisation." This attacked the IRC both on grounds of bureaucracy and of German exclusion, although the former charge was given greater prominence. Godfrey Hardy read the letter and soon guessed Bateson's involvement. He wrote to Bateson (March 30, 1921):

> I thought you must be at the bottom of it. Somehow I am glad to find it was so (or course the style of the *Times* article was visibly not yours). If there should be any further correspondence I hope you will find an opportunity of butting in. It is not that most people don't agree with us; but it isn't worth anyone's while to say anything, unless he is one of the few with a vested interest in the Council. I found that in the Lond. Math Soc. There I was really in a large majority; but I have only been able to keep them from joining (for the sake of a quiet life) by making it as Secretary a question of confidence. … I think I agree with you (in substance) about the Oxford letter – I thought too it was the wrong line. But that of course

> was Bridges. I was with him at bottom, of course: and I am sick of people who won't ever do anything because everything is done in the wrong way. So of course I signed like a lamb.

In 1922, in a further minor foray in the cause of internationalism in science, Bateson tendered his resignation (which was not accepted) from the Linnean Society, because their library had delayed restoring subscriptions to German periodicals. In 1923, the RS sectional committees of Botany, Physiology, and Zoology met conjointly to pass a recommendation very close to the CBSS resolutions of 1920, which was duly adopted by the RS Council (Nov. 30). Bateson wrote to Ostenfeld (Dec. 7):

> I am pleased to be able to say that the Biological 'Union' is – so far as England is concerned – killed at last. This time I do not think the question will be raised again. I do not know who started it, but we had it once more before the Biological Committee of the Royal Society, and it was carried *nemine contradicente* that we do not join until it is possible to admit the Germans, or words to that effect.

In May 1924, the Executive of the National Union of Scientific Workers, of which Hardy was President, passed the resolution: "That the organization of scientific unions or congresses which are described as international, but from which particular nations are excluded on political grounds, is unworthy of the spirit and injurious to the interests of science." The latter was hammered home with the point that Einstein was now excluded from "international cooperation" in astronomy.

At the second General Assembly of the IRC in July 1922, there had been little support for a Swedish proposal for the statutes to be altered so that all countries could join. At the third General Assembly in 1925 there was much more support, but the two-thirds majority required to amend the statutes was not achieved. However, in December 1925, the Executive Committee of the IRC decided to call an Extraordinary General Assembly specifically to consider the question of membership restriction.

Politics continued at all levels. Two days before the IUBS issue was to come up again at the RS, and only a few weeks before Bateson's death, Prain wrote to Bateson (Jan. 10, 1926): "Surely the Dutch and the Swiss expect too much when they ask the cooperation of English botanists in a scheme to put into the hands of the Botanical Section of the International Biological Union a mechanism intended to crush out of existence organizations created not only by their former German enemies but also by their former American allies." Needless-to-say, Britain joined the IUBS in 1929.

The following June the Extraordinary General Assembly met and resolutions to amend the statutes, and to send invitations to join to the leading academies of all the old Central Powers, were passed unanimously. Three months later, Germany was admitted to the League of Nations. It could be held that the IRC was following in the wake of public opinion, rather than (as might have been hoped) setting an example. Ironically, most German scientists, in the face of pressure from their own government, refused to countenance joining the IRC. It seemed that they were prepared to join only on conditions which could readily be construed as a German victory rather than an act of reconciliation. To persuade the French and Belgians to swallow sufficient of their pride to let Germany in was difficult enough; to expect them to swallow their pride whole, even to abase themselves by apologizing, in effect, for having excluded Germany in the first place, was too much to ask. The response in the smaller countries of the Central Powers was more favorable: Hungary joined in 1927. In 1931, the IRC was renamed the International Council of Scientific Unions (ICSU) and more autonomy was granted to the discipline-specific Unions. Bulgaria joined in 1934, but Germany and Austria stayed outside until after the Second World War [7].

Summary

With the passage of years, the snobbish nationalism of Bateson's youth gave way to cautious internationalism. He was increasingly drawn into affairs both at the national and international levels. His judicial activities concerning a priority dispute over the discovery of the malaria parasite brought home the need for senior scientists to give full credit to the contributions of their juniors. Ross noted how easy it was for false facts to pass unquestioned from textbook to textbook, so gaining a momentum that was difficult to correct; a scientific establishment with a vested interest in the mythical *Bacillus malariae* had delayed recognition of the true malaria parasite. Bateson was not overly concerned when several young bioscientists (Hogben, Crew, Huxley, Haldane) proposed a new society that might usurp the Genetical Society. However, despite the overlap, distinct constituencies were apparent, and both the new Society for Experimental Biology and the Genetical Society thrived. During the war, various scientists refused to cite the works of German scientists and pressed the RS to strike German nationals off its membership list. Bateson opposed these measures. After the war, he played a tactful role in reuniting the international scientific community and establishing supra-national scientific organizations.

Chapter 23
Degrees for Women

> *Though Bateson was deeply interested in the broader aspects of social and political questions ... some half a dozen essays are all that he has left us. But their weight is out of all proportion to their bulk, for they are the fearless pronouncement on human society as he found it by one of the most clear-sighted and critical minds of the age.*
>
> *Punnett [1]*

William Henry Bateson had been more prominent as a university politician and administrator than as a teacher or a scholar. His son William never became involved to the same extent. All the same, he did take an active part in affairs, both at college and at university level; probably more than the average Cambridge don of his day. He was Steward of his college from 1892 to 1908. At university level, he was prominently involved in the two major controversies of the period: those over degrees for women and over the retention of compulsory Greek in the Previous Examination. Each of these was not just a single campaign culminating in a vote in the Senate for or against a particular change in the statutes of the University, but an issue which, in variant forms, recurred at irregular intervals over a period which exceeded at both ends Bateson's graduate career at Cambridge (1883–1910).

Degrees for Women

The campaign for women to take degrees was merely a facet of the campaign in the country generally to improve opportunities for women in higher education and of the feminist movement as a whole [2]. It was very much a concern, not just of William, but of the whole Bateson family. His father and more especially his mother were involved from an early stage. Family links with Newnham were close: Mrs. Bateson was a member of the first College Council, serving until 1895, and William

A. G. Cock, D. R. Forsdyke, *Treasure Your Exceptions*,
https://doi.org/10.1007/978-3-030-92099-9_23

was a member from 1908 to 1911 [3]. In a wider context, Mary made an "able and trenchant speech" to the Prime Minister as one of a deputation demanding women's suffrage, while Anna included the Secretaryship of the New Forest Suffrage Society among her public activities [4]. Margaret published in 1895 *Professional Women and their Professions* – a series of 26 articles from *The Queen* magazine, each based on an interview with a woman in a particular profession. The professions (the word being liberally interpreted) ranged impressively through painting, medicine, poor law administration, stock-broking, printing, laundry work, and indexing. The book was intended partly to convince men that women could competently engage in these and other occupations, but even more to encourage other women to "go out and do likewise." It sold well enough to be reissued in a cheap edition in 1897 (Fig. 23.1).

The first stirrings at Cambridge seem to have been in 1865 when local examinations for girls under 18 were initiated. In May 1868, William Henry Bateson appears as one of 77 signatories of a "memorial" to the Vice-Chancellor urging the University to introduce examinations and certificates for female teachers. In 1870, Mrs. Bateson is named in a leaflet as Treasurer of the Executive Committee of the Committee of Management for Lectures for Women. She dispensed tickets of admission to lectures and gave parties for participants and fellow-members of the Committee in St. John's College Lodge [3]. She was also successful in soliciting funds. A report of 1878 gave a list of donations including £60 from her father James Aiken and a guinea from her brother Edward. By 1877, the Committee of Management had evolved into the "Association for Promoting the Higher Education of Women at Cambridge" with Mrs. Bateson as Honorary Secretary. This change to a more unwieldy title reflected a more militant attitude and a broadening of its concern beyond the purpose of organizing lectures. A circular announced a Conference of the Headmistresses of Public Schools and gave warning that the Association would not in future be content simply to accept whatever the University saw fit to provide.

Fig. 23.1 Margaret Bateson

The University had always recognized the existence of women if only through the rule that college fellows could not marry. In 1881, came official recognition of a more positive nature. A practice had grown up, during the preceding few years, of admitting women to the Tripos (Honors) examinations and of awarding them (unofficial) classes based on the examination results. Senate had appointed a Syndicate to look into this practice and make recommendations. Three Graces embodying the recommendations of the Syndicate gave official sanction to practices already fairly firmly established, and authorized the issue, to women successful in the examinations, of certificates showing the class gained. The Graces were passed by Senate, not without some opposition, but by a comfortable majority: 392 votes to 32. They appeared to mark an important advance in the women's cause, although they did no more than regularize an existing situation.

Opponents feared this would be merely the thin end of a wedge, but it was to be a long time before the wedge was driven in further. Up till 1881, the University had simply excluded women; henceforth, and for the next 70 years, it was to admit them, but as second-class citizens. They could attend lectures, sit examinations, and be awarded classes, but they could not, without circumlocution, call themselves graduates nor place the magic letters B.A. after their names. More importantly, from the point of view of the defense of a bastion of male privilege, since they were not graduates, they had neither the right to a vote in Senate nor any say in University government.

The 1887–1888 Campaign

During 1887–1888, the matter was raised again in a memorial signed by 120 members of Senate (and about 600 other persons) in favor of degrees for women. This soon evoked an opposing memorial, which took a form followed in subsequent campaigns: there were two parallel opposing memorials, a moderate 'A' and a more extreme 'B.' Naturally, most signatories of 'B' also signed 'A.' Memorial 'A' recorded opposition to degrees for women in the form of Cambridge degrees, but, by implication, left the way open for some kind of compromise solution. Memorial 'B,' on the other hand, was uncompromising: It simply desired that "no steps be taken by the University towards admission of women to membership and degrees in the University."

The 1887–1888 campaign was the first in which Bateson could have taken part as a graduate of the University. However, he was away in Turkestan so that, although his name appeared on the "for" memorial, he took no active part. In any case, the campaign collapsed without reaching a vote in Senate. When the memorials were presented, the Council of Senate refused to recommend any change in the *status quo* on the ground that the preponderance of opinion within the University was against change. Bateson's position, *vis-à-vis* his biological colleagues at Cambridge, was an isolated one. This was already evident in 1887–1888 and remained true in many subsequent campaigns. Bateson was the *only* resident biologist to sign the "for" memorial of 1887, although the signatories from outside Cambridge included two eminent biologists: Francis Galton and Thomas Huxley. Among the signatories of the 1888 "against" memorial were Hans Gadow, Adam Sedgwick, and Raphael Weldon, although none of these signed the more extreme 'B' memorial. Other friends of Bateson who were active in the opposite camp included the physicist Joseph Larmor and Frances Jenkinson (classicist and lepidopterist, and soon to become University Librarian).

The 1895–1896 Campaign

Battle was joined again in 1895. A committee was formed (with Bateson as Secretary) which put out a Memorial to the Vice-Chancellor urging the appointment of a Syndicate to consider "on what conditions and with what restrictions, if any, Women should be admitted to Degrees in this University." In February 1896, the Council of Senate recommended the appointment of a Syndicate, but this evoked a characteristic little piece of infighting. It was charged that the Syndicate would contain too many declared partisans. There were also too many members of Council of Senate and too few young men, but these points, one suspects, were thrown in simply for good measure. Bateson, with the Chairman of his committee, distributed a flysheet rebutting the charges. This evoked more flysheets from the other side. When the matter came before Senate in March, the Grace setting out the membership of the Syndicate was defeated. A delay, not an outright defeat, was thus achieved: a new and differently constituted Syndicate was approved in June. In the interim, there was a further spate of flysheets and memorials. The Syndicate pursued its deliberations with no great haste and issued its report in March 1897.

The solution recommended was a compromised one: that the title of the degree of BA be conferred by diploma on women who had satisfied the examiners in a final Tripos examination and had fulfilled the usual residence requirements. This was also to apply, retrospectively, to past "graduates" of Newnham and Girton, and the titles of M.A., Sc.D., and Litt.D., and of honorary degrees, could also be conferred on women, under the same terms as for men. As for the right to put letters after their names, women were to be put on an equal footing with men, but membership of the University and any share in its government were to remain an exclusively male privilege.

The vote on the Grace embodying the Syndicate's proposals was not to be taken until 21st May, but in March there was a discussion in the Senate House, which occupied three days and 66 pages in the *University Reporter*. The committee of which Bateson was Secretary recommended acceptance of the proposals: in view of the strength of opposition, particularly among non-resident members (revealed by memorials circulated in 1896), there was not the slightest hope that any more radical proposals could be passed, even if the Council of Senate could be persuaded to put forward the Graces. The opponents had, as in 1888, organized two parallel memorials A and B, but A was now the more extreme of the two. Memorial A stated that:

"We earnestly deprecate the admission of women to membership of the University or to any of the Degrees which are conferred on members," while memorial B said that: "We are prepared to support a proposal for conferring some title which does not imply membership." By June 1896, A had received almost 2,000 signatures and B over 1,300. (Over 1,100 had signed both.) One might have thought that the proposals of the Syndicate would at least have been acceptable to the signatories of memorial B, but the committee which had organized the memorial felt otherwise and recommended their followers to oppose.

Senate Speech

At the Senate House meeting, Bateson made a typically trenchant speech [5]. He began with a small factual point. An earlier speaker had alleged that a certain (unnamed) Professor had to leave out a fifth of his subject because he did not like to deal with it before women. Bateson challenged the speaker to produce the name of his authority. He pointed out that the Syndicate had issued a questionnaire to all teachers who had women in their classes: the answers given were printed in an appendix to their report, which included no answer such as that alleged. From here he broadened the attack, using a quotation from Richard Bentley: "We have to deal with a shifty adversary." One speaker said that while opposed to the title of degrees, he would be in favor of a proposal that "a certificate should be given permitting women to affix the letters 'B. A. Cambridge' to their names." If this was what the signatories of memorial B wanted, then Bateson thought that this wording would be acceptable to the great majority of those who thought with himself.

The "if," however, was the critical word: Bateson could not believe that either the substance or the tone of the speeches made by his opponents were the expression of such a difference as that between titles of degrees and certificates permitting the affixation of "B. A. Cambridge." Moreover, if one paid attention not just to flysheets, speeches, and formal documents but also to casual conversations, it was clear that many of his opponents were opposed to women being at Cambridge altogether. (Bateson's point was confirmed by cries of "Hear, Hear".) He wanted to get the views that were expressed in private expressed in public, because "we should then know how to deal with them." Bateson concluded by expressing passionately his belief in co-education, as something desirable, which would come sooner or later.

Bateson's challenge to his more extreme opponents to express their views publicly produced at least one response. James Mayo issued a flysheet in May in which he listed five grounds of objection to the proposals of the Syndicate. The last of these, quite representative of the general tenor of his flysheet, was: "Because a University course is an incident in and a part of certain professions (and of these only) which are, and by the appointment of Divine Providence must always be, exclusively virile." Indeed, Mayo specifically discounted as irrelevant two of the arguments widely canvassed by the more moderate opponents. These were that granting titles of degrees would prejudice proposals for the establishment of a separate University for women ("The Queen's University for Women" – a scheme advocated at a national level particularly by the Bishop of Stepney), and that it would serve only to encourage demands for further rights for women in the future. It was "the intrinsic viciousness of the *present* request, without speculations as to the future," to which Mayo objected.

The arguments adduced on both sides covered a wide range, and some of those put forward in favor of the proposals may have been as repugnant to many of the women clamoring for admission to the University as they would be to modern feminists. The Master of Trinity, for example, quoted Archdeacon Hare approvingly: "In giving us sisters God gave us the greatest of all moral antiseptics," and went on to suggest that the presence of women in the University was desirable as an "intellectual antiseptic." On the other side, there were echoes of the dispute about titles of degrees *versus* certificates permitting the annexation of the letters B.A.

Some objected to women having the right to use B.A. at all, since this would imply parity with men. They suggested as alternatives M. Litt. and M. Sc., neither of which were (at that time) conferred on men. Alfred Marshall had earlier (1896) suggested E. (External) B.A. or A. (Associate) B.A. The Provost of King's College was willing to support the proposals of the Syndicate, but only if they were revised to remove the compulsory residence requirements for women; he also wanted women to pay reduced fees. Much of this may strike us now as comically pedantic, but the motivation behind such objections seems clear. The proposals did not allow women to become members of the University (they did not even give them the right to attend courses in the University: professors and lecturers were to be free, as before, to admit or to refuse to admit women to their classes), but they could be interpreted as a step *towards* membership. Any difference in terminology and any extension of the differences in the regulations governing women and men would serve to diminish the plausibility of that interpretation.

As the day of voting approached, activity in the two opposing committees increased: non-residents were written to, urging them to vote and explaining the arrangements for voting (rather complicated arrangements, since there were three Graces to be voted on); special handbills giving the times of trains from and to London were sent out, and arrangements were made for the accommodation of those who needed to stay overnight. The accounts of the committee opposing change show a total expenditure over the whole 1886–1887 campaign of £369. Most of this sum must have been spent in circularizing the non-resident members (who were more numerous than the residents).

The result of the voting was an overwhelming defeat: the first and crucial Grace was defeated, *non placet* 1707, *placet* 661, and the two remaining Graces were consequently withdrawn. Obviously, many of the signatories of memorial B had either changed their minds or interpreted that memorial in a very restricted sense – as, indeed, their committee had advised. One group within the University was overjoyed at the result: the undergraduates. Over 2100 had signed a memorial opposing the Syndicate's proposals, and a poll conducted by the *Cambridge Review* had yielded 1726 against, and only 446 for. A motion debated in the Union Society condemning the Syndicate's proposals had given an even more overwhelming result: it was defeated by 1083 votes to 138. In the evening, the victory was celebrated by undergraduates with a carnival and bonfire on Market Hill. The *Cambridgeshire Weekly News* issued a special four-page edition devoted entirely to the events of voting day, with headlines such as "Triumph of Man," "The Varsity and Women's Degrees," "Carnival and Bonfire - Scenes and Incidents," and "England Expects Every M.A.n to do his Duty."

The magnitude of the defeat was a bitter blow to Bateson and his friends. They had campaigned for nearly 18 months, they had accepted a compromise which gave much less than many of them wanted, and yet they had been defeated by a majority, which left little hope that the decision could be reversed within the next decade or more. For Bateson himself, there was some slight consolation: he was no longer alone among his biological colleagues: Frances Darwin, Foster, and Harmer had joined the ranks of the supporters, but, on the other hand, there had been accretions to the opposition: MacBride, Newton, and Shipley. Bateson had married during the course of the campaign, and it must have been galling to find his brother-in-law, Herbert Durham, among his opponents.

The issue remained quiescent until 1911, when a sub-committee, charged with the task of considering the constitution of the University, issued a unanimous Report, signed by all seven members (including Punnett and Seward). Their chief – and revolutionary – proposal was that the Senate be divided into two houses, with effective power restricted to the upper Resident House, which was to consist broadly of university officers, teaching staff, and resident fellows of colleges. After two more, relatively minor, proposals they recommended – almost as an afterthought – that women be admitted as members of the University on substantially equal terms with men. The Report seems to have been lost, buried, or otherwise disposed of without occasioning any public controversy, and with it went the recommendation concerning women.

The 1920–1921 Campaign

Another prolonged campaign took place in 1920–1921. Building up to this, in April 1918, a flysheet signed by 142 resident members advocated the admission of women as full members of the University and asked that a Syndicate be set up to make detailed recommendations. When the Syndicate reported, it made two alternative sets of proposals. The first (scheme A) admitted women on virtually equal terms with men, whereas scheme B envisaged the setting up of a separate University for women at Cambridge, with Newnham and Girton as founding colleges. Most, but not all, of the supporters of scheme B were willing to have some kind of federation between the two Universities – a Senate with separate houses for men and for women (and, one presumes, some collaboration in teaching). A vote taken in December 1920 on a Grace embodying scheme A, resulted in defeat: *placet* 712, *non placet* 904 – a defeat much narrower than in preceding campaigns. Scheme B, voted on the following February, was also defeated: 50 to 146. The derisively low poll is significant; the issues were not ones that either supporters or opponents of the women's cause felt strongly about. Indeed, one suspects that most of the 50 who voted for the Grace were not supporters of degrees for women, but extremist opponents who held that if only one could get the women's colleges tied up with the idea of a separate University, this would put paid to any attempts to encroach further into *our* University.

Soon after the defeat, leading supporters and opponents tried to work out a compromise. This resulted in a flysheet, issued in April, which advocated that women be allowed to take all degrees and hold most of the salaried posts, but that their membership of the University be subject to certain restrictions. The chief of these were that they were to be subject to a separate disciplinary body; that they were not to be members of the parliamentary electoral roll, or of Senate, although they would elect two (non-voting) representatives on Senate; that there should be a limit of 500 on the number of women *in statu pupillari* (i.e., undergraduates plus those who had taken their B.A. within the past 3 years); and that a woman professor should not *ex officio* be head of her department.

This was a great deal less than "full and equal membership," but a great deal more than entitlement to degrees, and most of the former supporters seem to have accepted the proffered compromise fairly readily. In doubt was whether it would prove acceptable to the former opponents. However, a Royal Commission on the affairs of the University was imminent. If the University did not itself produce an acceptable compromise, then there was the possibility that one more radical would be imposed.

Council of Senate, with greater speed than in previous campaigns, produced a scheme (Grace 1) differing in only relatively minor details from that outlined in the flysheet of April. It also – almost inevitably – produced an alternative scheme (Grace 2), which gave titles of degrees and nothing more. Voting on the two Graces was fixed for October 20. The vote was preceded by the usual overture of flysheets and campaigning activity, but organized opposition was directed almost exclusively against Grace 1. The arguments put forward by either side were not greatly different from those of earlier campaigns: the argument that the admission of women would not be in the best interests of women themselves, who would be better served by separate institutions adapted to their own special needs, was again prominent. The old argument, that Cambridge ought not to make a move in the matter except in conjunction with Oxford, was now stood upon its head. Oxford had already admitted women, so it was now all the more important that Cambridge should retain its position as the one University still run exclusively by men for men. As in 1897 and 1920, considerable efforts were made to persuade non-residents to come up to vote.

The result was that the Grace was narrowly defeated, by 908 votes to 694 – figures remarkably close to those obtained the previous year. Grace 2, however, was passed overwhelmingly, by 1011 votes to 369. So, the women had at last gained a victory – the first since 1881. It was, like the 1881 victory, a minor one: they now had titles of degrees, but no powers or guaranteed rights within the University, either as undergraduates or as graduates. There seems to have been no celebrating comparable to the carnival and bonfire of 1897, although there was one incident. A mob of undergraduates, egged on by an M.A., damaged the bronze gates of Newnham College.

Victory

The settlement of 1921 lasted a long time. In 1924, as a result of the proposals of a Statutory Commission under Lord Ullswater, women were allowed to hold professorships and other University teaching posts. Women had of course for long taught at Cambridge, and some of them taught men as well as women, but their status was that of Fellows or College Lecturers of Newnham or Girton, not of University teaching officers. A new and delicate anomaly was hereby created: women, in increasing numbers, now became teaching officers of a University of which they were not members.

When Cambridge at long last gave way, it did so gracefully. In December 1947, a Grace was passed without opposition giving women full and equal rights in every respect with men and incorporating Newnham and Girton as colleges of the University. The existing men's colleges remained exclusively male: that was a matter that only could be changed by the colleges themselves. The ideal that Bateson envisaged in 1897 of a fully coeducational university was eventually achieved, though it probably took longer than he expected. Noting Bateson's views on supposed psychical manifestations, it would be an affront to his memory to suggest that his ghost would have been pleased. We may, however, suggest that, if he had had a ghost, the ghost would have found pleasure in the final result [6].

Compulsory Greek

In his address to the Salt Schools in 1915 (Chap. 16), Bateson described the failings of his classical education:

> I look back on my school education as a time of scarcely relieved weariness, mental starvation and despair. There came at last a moment when I was turned into a chemical laboratory and for the first time found there was such a thing as real knowledge which had a meaning and was not a mere exercise in pedantry. Our staple was of course Latin and Greek, of which I made nothing. Some emotional pleasure came towards the end of my school course from the Greek tragedies but otherwise those years were almost blank. Now what I, and thousands of other boys like me, discover in after-life is that by those very same materials, perhaps more than by any others, we might have been 'wakened to ecstasy' and to the joy of development.

Bateson fought in a cause, degrees for women, which would conventionally be regarded as "progressive." His fight, on the other hand, for the retention of Greek as a compulsory subject in the entrance examination, would ordinarily be regarded as a fight in the cause of "reaction." This combination of stances was not uncommon. From a cursory perusal of the names on flysheets and memorials one gets the impression that, while most people took a straight "progressive" or "reactionary" line on both issues, "cross-voters" were numerous. Another simplistic view would be that the dispute about the abolition of compulsory Greek was a fight between arts men in general (and classicists in particular) and scientists and engineers. While it is probably true that most arts men were for retention and most scientists for abolition, exceptions were numerous and sometimes eminent. It must have been an embarrassment to the retentionists that one of the most outspoken advocates of abolition was Henry Jackson, Greek scholar and much-loved teacher, who from 1905 was Regius Professor of Greek. Another prominent abolitionist was Mary Bateson's colleague, F. W. Maitland, the Downing Professor of Law.

The issue revolved around the syllabus for what was known as "the Previous Examination." This had been instituted in 1822 as an examination to be taken in the fifth term of residence, but in 1879, the regulations had been altered to permit it to be taken in the first term. This trend toward taking the examination earlier – toward it becoming an entrance qualification similar to the matriculation examination of other universities – continued, and in 1919 it became permissible to take the Previous Examination *before* coming into residence and matriculating. From then on, it became increasingly common not to take the Previous Examination itself, but to secure exemption from it by passing in the requisite subjects in a comparable public examination.

In 1870, when the first agitation for the abolition of compulsory Greek began, the syllabus was unchanged since its inception. There were papers on the Greek Testaments, on Paley's *Evidences of Christianity*, on a Greek and a Latin author, and on the oddly named "additional subjects" whose content was mathematical. If this seems a strange balance of subjects, we must remember that in the first half of the nineteenth century Cambridge was, as far as its teaching functions went, almost the antithesis of the modern conception of a university. The despised pass degree apart, there had been no alternative route to the B.A. save for the so-called Senate House Examination, which, during the eighteenth century, had become overwhelmingly mathematical in content. There was, it is true, a little philosophy (Locke and Clarke) and theology (Paley's *Evidences*) tacked on, but it was mathematics which formed the bulk of the examination, and the effect of performance in the other papers on the final placings was slight.

The first breach in the hegemony of mathematics took place in 1822 when the Classical Tripos was founded. Further Triposes were established in 1848 in Moral Sciences and Natural Sciences, and also Triposes in Law and Medicine soon after. In the 1970s and 1980s, there came further Triposes: Theology, History (which had previously been part, first of Moral Sciences, then of Law), Semitic Languages, Indian Languages, and Mediaeval and Modern Languages. By the 1970s, Classics had achieved approximate equality with Mathematics, at least as far as the number of students, and these two greatly exceeded the more recently founded Triposes in the numbers of students they attracted. This position of approximate equality between Mathematics and Classics was recently established, and when there was clamor for the abolition of compulsory Greek, many classicists must have felt this to be a threat to their newly gained prominence.

First discussions seem to have been sparked by a letter to the Vice-Chancellor from the Chairman of the Endowed Schools Commission. Throughout all the campaigns, the views of schoolmasters were to be extensively cited on both sides: A Syndicate, appointed in 1871, was in favor of allowing French and German as alternatives to Greek, but its report was rejected by Senate. On this occasion, little controversy seems to have been aroused, for the numbers voting in Senate were derisorily small in comparison with later campaigns: 20 to 16 for appointment of the Syndicate and 51 to 48 against acceptance of its report. Matters came to a head again in 1879–1881, following an abolitionist memorial signed by 11 Headmasters (of public schools – the Cambridge intake from other schools being at this time negligible) and by 27 eminent persons outside the university – including Thomas Carlyle, Charles Darwin, Thomas Huxley, Joseph Hooker, and John Tyndall. The poet Matthew Arnold added his powerful and informed support to the abolitionist cause.

A Syndicate was again appointed, which extended its deliberations over 18 months, but finally came up with recommendations only marginally different from those of the 1870–1871 Syndicate. Candidates need offer only one of the two classical languages, French or German being acceptable alternatives for the second. Further, this relaxation was to apply only to those who would be candidates for honors, not to pass men. The abolitionists may have hoped to pacify their opponents by allowing *either* Latin *or* Greek to be dropped, for Latin seemed at that time to rouse lesser passions, and there were fewer ready to rush to defend the educational and cultural value of Latin than of Greek. The proposals of the Syndicate were defeated by 185 votes to 145: the number voting was larger than in 1870 but was still comparatively small.

A further agitation for abolition was begun in 1891: it was brief but intensive, evoking numerous flysheets. Bateson was Secretary of the committee organizing opposition to any change. This time the matter was fought out over the preliminary issue of the appointment of a Syndicate "To consider whether it be expedient to allow alternatives ... for one of the classical languages in the Previous Examination." The Grace appointing the Syndicate was defeated by 525 votes to 185 and that was the end of the matter. As Beatrice records, Bateson and his friends were not only delighted by the result but greatly surprised: they had expected to win, but not by nearly so wide a margin. The retentionists included Newton, Harmer, C. S. Sherington, Seward, and Larmor (although Seward became an abolitionist in 1904).

The changed situation can plausibly be explained in terms of two factors. First, the smallness of the 1870 vote suggests that many people did not take the threat to Greek very seriously. The closeness of the vote in 1870 (and even the 1878 vote could hardly be regarded as a safe retentionist majority) was a shock and led many more residents of retentionist views to vote in 1891 and much greater efforts to persuade non-residents to vote. The second factor is a more general one. The confident prosperity of the mid-Victorian age was beginning to pass. Cambridge was severely affected by the agricultural depression which began in 1880, because the incomes of the colleges came predominantly from their large holdings of agricultural land. The "dividends" of the colleges (i.e., the stipend each fellow received) were severely reduced. On the other hand, the university needed more money than before, especially to build, equip, and maintain laboratories for the new and growing science and engineering departments.

Under insidious attack from outside forces which it little understood, it was a natural reaction for Cambridge to look back with nostalgia to the past, and to resist change of any kind. Thus, while the vote of 1870 was, I suggest, almost exclusively on the educational virtues of Greek narrowly conceived, in later polls, though the nominal issue remained the same, the retentionists felt themselves to be voting to protect a whole culture and way of life. Greek had become important, not so much for its own sake, but as a symbol of this. Certainly, Bateson's own speeches and writings on the matter, from 1891 to 1921, strongly support this interpretation. He devotes much less space to extolling the educational virtues of Greek than to dire predictions of the triumph of philistinism, crass materialism, and commercialism that will ensue if compulsory Greek is abolished. He seems, too, more impassioned and more sure of his ground when dwelling on the latter theme.

When the issue was resuscitated in 1903, it was in rather unusual circumstances resulting from an initiative of the Chancellor, the Duke of Devonshire. Chancellors then, as now, while not expected to act solely as figureheads, were not expected to take the initiative in the internal affairs of the University, especially on controversial matters. The Duke of Devonshire was a special case: he had become a major benefactor of the university by endowing the Cavendish Laboratory, named after his own illustrious career. A small informal meeting was called, to which a representative group of members of the university, including Bateson, was invited. The Chancellor expressed concern about the financial state of the university and the future in particular of applied science and of medicine. The University needed to be "re-endowed" to meet the conditions of the modern world, and the main hope lay in attracting private endowments from industrialists and men of commerce. Such men might be put off by the old-fashioned and "ivory tower" image which the university had acquired. The requirements for Latin and Greek were an important constituent of this and might also deter these men from sending their own sons to Cambridge.

Discussion, in the event, centred almost entirely around the Greek question. James Stuart, Professor of Mechanism (i.e., of engineering), brought forward a specific example: an endowment of £100,000 had been offered to found a School of Naval Architecture (at Cambridge, of all places!). But it was a condition of the offer that the School be "open to all comers," and the donor thought the requirements of the Previous Examination conflicted with this, so the offer was withdrawn. Bateson made the point that Stuart's was the only specific instance of a lost endowment that had been brought forward, and even this was a matter of endowing a new department, not of re-endowing existing ones, which was their main problem. Any mistrust of Cambridge among commercial men was due, Bateson thought, to "the idleness of the pace," not to classics requirements. Though for many reasons it was no longer so clear that Greek should be retained, he believed the abolition would have no sensible effect on the prospect of re-endowment.

As a direct result of the Chancellor's meeting 5 months earlier, a Syndicate was appointed in November 1903, with Bateson as one of its members. It was a full year before the Syndicate reported, and when it did, its recommendations were in essence the same as those of 1891: that only one of the two classical languages be required. During this period, and more especially in the months that followed before a vote was taken in April 1905, there was an intensive barrage of memorials and flysheets, and strenuous efforts were made to persuade non-residents to vote. The size of the total vote greatly exceeded that in any other campaigns concerning either Greek or degrees for women. Each of the four graces embodying the Syndicate's recommendations was defeated by over 1,500 votes to about 1,050.

The absolute majority of the retentionists had increased slightly since 1892, but their proportional majority had declined greatly. This perhaps contributed to a feeling that Cambridge could not continue to reject outright demands for a change in the Previous Examination. A compromise was sought. First, a memorial was circulated advocating a Syndicate to consider an alternative examination to be sat by these not taking (or gaining exemption from) the Previous Examination. This was signed by a combination of retentionists (including Punnett and Larmor) and abolitionists. Second, Senate agreed to prolong the life of the original Syndicate by a year and to appoint four additional members.

When they reported in March 1908, the majority report resurrected in a modified form the old idea suggested in 1891 of separate requirements for degrees in letters (arts) and in science. Those intending to read letters should pass in one of the classical languages and in one modern language (French or German), while those intending to read science (or mathematics) should pass in only one of these four languages. It was thus a greatly relaxed version of the 1891 proposal. The requirements for letters men were to be the same as those suggested for science men in 1891, and scientists were now to escape compulsory classical languages altogether. Nor could it be regarded as in any way a moderate compromise in comparison with the scheme rejected the previous year: the changes it proposed were more, not less, radical, both in the degree of relaxation of the language requirements, and in the distinction made between scientists and others. It is therefore not surprising that when, in May, it came to the vote, it was heavily defeated: 746 votes to 241. The greatly reduced poll may occasion some surprise, but it may simply reflect confidence of success on the retentionist side and, after the vigorous campaign of the previous year, a declining interest in the topic.

The spirit of compromise and of the need for some reform in the Previous Examination remained in the air. Punnett, G. H. Hardy (who had been two of the secretaries of the retentionist committee), and R. K. Gaye issued, only two days after the poll, a flysheet headed "Greek, Science and Reform," which advocated moderate reforms. The most important of these were that Greek, while remaining compulsory, should be treated in a literary rather than a linguistic manner; that the grammar paper should be abolished; that the paper on Paley's *Evidences* should be abolished; and that there should be a compulsory paper in elementary science. Nothing came of these suggestions, and compulsory Greek was not finally abolished until January 1919. A new Syndicate to consider this and other changes in the

Previous Examination had reported in June 1914, but matters were overtaken by the war. The reconstituted Syndicate after the war also recommended one of the other changes advocated by Punnett and Hardy – a compulsory science paper – but this was defeated. It had been a frequent prediction of retentionists, Bateson included, that if Greek went, Latin would not long survive it, but this was not to be. Latin as a compulsory element of the Previous Examination was not abolished until 1968.

The Classics and Civilization

It was not until the 1920s that Bateson gave full vent to his views on classics in education. By that time, the issue had gone far beyond the universities. In 1916, Prime Minister Asquith had appointed Committees to consider the role of the natural sciences and modern languages in the education system, and in 1919 a further Committee on Classics was formed. A synopsis of Bateson's input (June 1920) was included in Beatrice's *Memoir* (*Classical Education and Science Men*). He advised that "the thoughtful and the learned *are* a separate caste," which collectively "might exercise a prodigious authority" as "the only effectual defence against the ignorant politician and tradesman," whose disdain for classics could mean that "classical teaching becomes the prerogative of the literary caste." This would not only "break the continuity of European civilisation," but would break "any coherence among the contemporary intellectual community," for "the best hope of creating such unity is to provide a common education, literary and scientific, in which all have at one time shared." In this way, "if in a common education each had learned at least the essential rudiments of the other's business," then the thoughtful and learned, although following different pursuits, would "not dissipate their influence by again subdividing themselves."

The Committee issued a report in 1921, which Bateson reviewed very positively in *Nature* (Sept 1921). However, he was not optimistic about the outcome. "The members of the Committee are sanguine men if they expect their recommendations to be adopted. The mind of the country is set on other things." For "we are probably witnessing that rare and portentous event, a break in the continuity of civilisation. In the Press, in the arts, and, most singularly of all, in learning of various kinds, the same phenomenon appears." He warned that the Committee should not believe "that the scientific world especially concurs in its opinion." Those who had testified to it were unlikely to be representative. The world, including many scientists, "cares not a jot that 'all our modern forms of poetry, history, and philosophy' originated with the Greeks, and has only a scant curiosity as to whether Western civilisation is grounded on that of the Mediterranean or of some other coast."

In his submission, Bateson had argued that the teachers of classics were partly to blame, because they had "steadily refused to put grammar anywhere but first." But, "it is possible to know a language enough for many purposes … with very slender equipment in grammar. A normal mind learns a language first, and the grammar afterwards, if at all." The Committee responded by noting that "great stress should be laid on the subject matter and the historical background of the texts read, though not to the prejudice of exact training in the language."

Bateson's Attitudes

Bateson describes himself in the flysheet "For Greek," issued during the 1891 campaign as "one of these for whom the Classical System may be said to have failed." He goes on to develop a different theme. Noting its foreignness to ordinary life, and, above all, its uselessness, he makes the challenging claim that "it is exactly for common men in general and for Natural Science men in particular that the System should be kept. To common men a Classical Education gives the single glimpse of the side of life which is not common." The common man, Bateson claims, does not understand things which are beautiful and have no "use," but it would be dangerous for society "if he had never thus stood once in the presence of noble and beautiful things." The scientist needs the classics most of all "if only that he may know the greatness of his own calling" and know that "his problems are those which the poets have put."

A similar concern is expressed to his sister Anna in 1886 when approving her decision to be a botanist: "I don't altogether regret for your sake, that I led you in at such an infernally strait gate. … If you had ever been successful in such a subject as history, you would have ever regretted that your work did not lie a little nearer the origin of things, and was not, so to speak, '*purer*' than such work must be." (He is naturally careful to preface this with an assurance that he does not mean "to throw mud at Mary;" her success as a historian is a source of pleasure and pride, for he is sure that she will not "make mere pot-boiling stuff of it, but also real work.")

The theme of concern over the cultural status of science recurs, with emphasis on the danger of increasing the gap between the literary and. scientific traditions: the proposals of 1908 would, Bateson declared, "make the division between the Science Schools and the other Schools at Cambridge more emphatic." He thought it was already too emphatic:

> If classics cease to be generally taught and become the appanage of a few scholars, the gulf between the literary and the scientific will be made still wider. Milton will need more explanatory notes than O. Henry. Who will trouble about us poor scientific students then? We shall be marked off from the beginning, and in the world of laboratories Hector, Antigone and Pericles will soon share the fate of poor Ananias and Sapphira.

His purpose here is not simply to deride the ignorance of many scientists in classical and literary matters, but to point a parallel with the widespread ignorance of even elementary science among those with an arts background. (The context is an essay on "The Place of *Science* [not classics] in Education;" see Chap. 16).

The core, however, of Bateson's resistance is the cultural and social conservatism, to which I have already referred. Within this general theme, one may distinguish a number of different, although not fully separable strands: anti-industrialism, anti-commercialism, the cult of the "useless," intellectual snobbery, together with a substantial element of nostalgia for a golden age (whether real or merely idealized in memory). These combined to form an attitude which saw the abolition of Greek as a threat to a whole culture and civilization. Such a threat was not entirely imaginary – indeed, the First World War could be regarded as its execution – but Greek was more its symbol than its substance.

The Botanic Garden

Bateson's incursions into University politics cannot have given him much satisfaction, nor can he have felt himself to be carrying on effectively the tradition set by his father. There was, however, a brief and relatively minor episode in 1904, when it was proposed to sell to the County Council part of the Botanic Garden Estate. Bateson's intervention may have been decisive: for this reason, if for no other, all Cambridge biologists are in his debt.

The matter began in the summer of 1903, when the County Council approached the Botanic Garden Syndicate with an offer to purchase land at the northeast end of the site for the sum of £7000. The Syndicate was – and remained throughout – against any sale, although the more northerly part of the site had not been incorporated into the Garden itself but was let out as allotments. Bateson himself, until he moved to Grantchester, had been tenant of one allotment and his colleagues, most notably Saunders, still used allotments for experimental crops. Bateson's fierce opposition to sale was therefore motivated by the quite legitimate vested interest of himself and his friends, as well as by wider concerns.

The Financial Board favored sale, and when the matter came to Senate in February 1904, the Financial Board's case received widespread support, most notably from Oscar Browning. The arguments for the sale were threefold: the straitened finances of the University; the Syndicate and the Botany School put this part of their land to no good use (commensurate with the capital tied up in it); unless needed, land within the Borough was a financial burden and ought to be sold whenever possible. A Grace authorizing the sale was put down for voting on 5th March.

On 1st March, opponents of the sale distributed a "notice of non-placet" signed by H. Marshall Ward, Professor of Botany, and 13 others (including Bateson, Biffen, Francis Darwin, Doncaster, and Punnett – the signatories probably included, or were co-terminous with, the Botanic Garden Syndicate). On the same date, Bateson issued a two-page flysheet on his own behalf. While not denying – indeed, even emphasizing – the financial difficulties ("indigence" was the term he used) of the University, Bateson ironically suggested that if it were merely a matter of realizing insufficiently used assets, then the *Codex Bezae* should be for sale. The £100,000 or even £1,000,000 which this would fetch would not only make a substantial difference to the financial picture of the University – which the mere £7000 would not – it would also help indirectly, by giving publicity to the difficult circumstances of the University.

Two flysheets by way of reply to Bateson's appeared in the next two days. H. McLeod Innes advanced no new counter-arguments, but simply denied that he had advocated selling off, for £50,000, the whole of the "surplus" Botanic Garden land, bit by bit. John Willis Clark reiterated the arguments for selling any land within the borough that could be dispensed with ("I believe that the price of building-land in Cambridge has now reached high-water-mark"), and cast doubt on the Syndicate's *bona fides* ("we have had the present Botanic Garden for more than half a century, without one word being said" on the importance of the land for scientific purposes), and on the suitability of the land (with a frontage to a "dusty high road") for the cultivation of plants. Finally, he denounced Bateson's reference to the *Codex Bezae* as "wholly irrelevant," and expressed his conviction that the offer of £7,000, which Bateson had loftily characterized as a petty sum, ought to be accepted. The fourth of March brought a sober and dispassionate flysheet by Marshall Ward, outlining the history and main arguments of the controversy, together with another from Bateson – a firecracker accompanying an artillery shell. Bateson seized on a phrase which had been used by Oscar Browning, who thought the County Council's offer should be regarded as "a windfall" and eagerly accepted. "A few more such 'windfalls'," he commented, "and our orchard will be gone."

It is traditional that electioneering, in the sense of attempts to influence opinion, ceases on the eve of polling day: activities thereafter are properly directed only at "getting out the vote." Bateson did not follow tradition on this occasion: there is a small flysheet, dated March and headed "Windfalls" in which he instances the "windfall" which the University acquired in 1784 by selling a plot of land to Messrs. Mortlock for £150. "In 1897 Messrs. Mortlock had, shall I say a windraise, selling it back to us for some £2,000." After campaigning of this nature, it is anti-climactic to record that the proposal to sell was defeated that afternoon by a mere 96 votes to 53. In 1934, a large bequest was received from the estate of Reginald Cory. Using funds from this, between 1946 and 1960, the previously unplanted land was planted out and incorporated into the Botanic Garden. It would have given Bateson pleasure to know that about four acres were devoted to plots and greenhouses for research, together with a laboratory.

Summary

Bateson played an active part in the administration and politics of Cambridge University, which throws light on his personality, general attitudes, and allies. The names of those who supported him sometimes, but not always, coincided with the names of those who supported his science. He was a member at times of the University Press and Botanic Gardens Syndicates. He issued flysheets and gave speeches in Senate in three campaigns. The first two campaigns surfaced repeatedly with quiescent periods in between. He was a "progressive" on one (women's degrees), and a "reactionary" on the other (compulsory Greek): 1. *For the admission of women to degrees.* This was very much a Bateson family cause. Partly as a result of its activities, in the 1860's the Local Examinations Board (governing non-university examinations) allowed women to take exams. Following the foundation of Girton (1870) and Newnham (1872), women were able to attend lectures, albeit at the discretion of the lecturer. In 1881, women were admitted to university examinations. But it was not until 1921 that the accompanying "title," B. A., could be adopted. 2. *For the retention of compulsory Greek in the Previous Examination.* Here Bateson fought against the view held by many scientists, and lost. 3. *For preventing the sale of Botanic Gardens land.* The University had provisionally agreed to sell to the County Council the part of the land belonging to the Botanic Garden Estate. Bateson's opposition was partly motivated by the fact that Saunders used an allotment on the spare land for her genetical experiments. The campaign to stop the sale was successful.

Part V
Eclipse

Chapter 24
Bashing

The term controversial is conveniently used by those who are wrong to apply to the persons who correct them.

Bateson to Hurst (Feb. 2, 1907)

It is one thing to be buried and forgotten. It is another to be buried and have people come from far afield to stamp on your grave. Bateson's work, rather than being ignored, was actively deprecated both by those who had been his contemporaries and by those who came after. Those who, either professed to have read and understood Bateson, or to have read those who professed to have done so, came in profusion to bash our fallen hero. We are now set to determine whether, and if so to what extent, his "bad press" was justified. I conclude that, as in the cases of Butler and Romanes, he was misunderstood because the facts of evolution were misunderstood. Koestler with *The Midwife Toad* was not alone in his misrepresentations. But whereas Koestler had his amateur status as shield, the professional scientists who from their high ground confidently dismissed his worth can less readily be excused. Collectively, they made it acceptable for a kick at Bateson to become *de rigueur* in twentieth-century publications touching the history of genetics. The posthumous Bateson bashing industry went from strength to strength, attaining an apparent apogee in 1996 when the matter came to my attention.

Retirement Declined

In August 1926, Bateson would be 65. In August 1925, he wrote offering to retire to the Chairman of the John Innes Board of Trustees, David Prain, who had himself recently retired from Kew Gardens. Bateson was urged to continue (Jan. 6, 1926). He replied (Jan. 8, 1926): "I accept the invitation to go on a bit longer with pride and gratitude." But there were regrets:

A. G. Cock, D. R. Forsdyke, *Treasure Your Exceptions*,
https://doi.org/10.1007/978-3-030-92099-9_24

> When I came here first I looked forward to doing much more than has been accomplished. Nor can I blame … the bad times through which we have passed. When the centre of interest in Genetics shifted away from work of my own type, to that of the American group, I was already too old and too much fixed in my ideas to become master of so very new and intricate a development. It has taken me years even to assimilate the new things and I recognize that the Institution has a right to a younger man in my place.

He promised that in the time remaining, he would try to make up for what he admitted as a "short-coming" that was perhaps "more serious" – that of identifying and grooming a potential successor. But it was not to be.

In some ways, with the trip to Russia in 1925 Bateson's life had come full circle. Following his return to England, he was not in good health. One evening (Feb. 2), he went to The Athenaeum for dinner and a game of chess. He collapsed at a station on the way home. Hurst noticed his absence when he read a paper on polyploidy at a meeting of the Linnean Society which Bateson would normally have attended (Feb. 4). Beatrice and Gregory were nearby when he died at home in the John Innes Institute (Feb. 8). Beatrice related in her *Memoir*: "By his wish his body was cremated, and his ashes scattered at Golders Green, 12th February 1926."

Although an atheist, Bateson had left no wish that there should not be a memorial service. One was held in St. John's College chapel (Feb. 17) with the hymn "O God our help in ages past." After mentioning the family and relatives, the *Cambridge Daily News* listed those attending beginning with the Masters, Knights, and Professors, proceeding through the Doctors and Mr's, to Mrs's, and Miss's. Rona Hurst recalled:

> At the end Mrs. Bateson and her remaining son, Gregory, now up at St. John's himself, walked down the aisle alone together between the files of standing people from many facets of university life. They were all that was left now of a once flourishing family of five and all present must have felt it a most poignant ending to a long chapter of genetic endeavour.

In a will dated September 1923 Bateson left £25,000. The executors were Beatrice, his brother Edward, and Alan Barlow. He left £1000 to Caroline Pellew "in consideration of the great value of her assistance to me in my work." Barlow and his wife, Nora Darwin, arranged for the storage of the Bateson papers when Beatrice died during World War II.

Obituaries

It was to Hurst that Jackson of the Linnean Society first turned for an obituary. As Hurst related to Beatrice (Feb. 12, 1928): "Much to his annoyance, it was vetoed by the Council who substituted someone who knew little of his work at first hand and made a poor show of it. Fortunately, Laydon Jackson triumphed in the end by asking Morgan to do it and the result was excellent. I was however rather pained that I was not permitted to express my homage to a friend and colleague."

Morgan [1] recognized "the species problem" as "ever present in the background of Bateson's thought," but admitted he was not in sympathy with Bateson's concerns "relating to 'species,' and to natural selection" which "have an interest wider than the specific problems relating to Mendelian inheritance." The obituary truly revealed, in Coleman's phrase, its author's "conservative thought" [2]. Regarding "the relation of species-formation to the theory of natural selection" Morgan considered:

> While there can be no doubt that the central difficulty on which he lays so much emphasis, was, at the time, a real stumbling-block, I do not think that it now offers any substantial difficulty, but can be given a rational, even if theoretical, solution. ... Bateson's chief contention, if I understand him, is that the theory of natural selection does not explain the distinctive feature of species, because their distinctiveness rests on characters that seem often to be trivial, but are nevertheless constant. How, he asks repeatedly, could species be created by natural selection if those parts that distinguish related species from each other are concerned with parts not essential to the life of the individual. This undoubtedly raises a serious question for Darwin's theory if Bateson's views regarding species are accepted, but I venture to think that it is not so serious otherwise. In the first place it is to be remembered that while Darwin entitled his great work 'The Origin of Species,' the whole argument went to show that the attempt sharply to separate species from varieties is futile, because in most cases there is no such sharp distinction. If this is conceded, then natural selection may be an approximate solution of the situation as it exists. But Bateson takes a different view in regard to species, and believes there are distinctions, essential ones, that give species a particular hierarchy in the scale of organic life.

Morgan correctly saw the questions of "infertility between species, and the sterility of hybrids produced by species crosses" as being "intimately bound up with Bateson's argument concerning the species question." Both "were much discussed by Darwin with a fullness of information and open-mindedness never since surpassed. Personally, I believe he [Darwin] practically met the requirements of the situation. Recent work substantiates, I think, the essentials of Darwin's argument." Morgan believed he had found concessions in Bateson's Melbourne address (Chap. 14) and dismissed as "somewhat vague" that which he could not label as a concession: "Here we find the admission that natural selection may account for 'the organism as a whole' and for 'favoured races,' but 'scarcely at all' for 'the diversity of species,' which is somewhat vague; and when all is said, it is not so different from much that Darwin himself was contending for." Nevertheless, Morgan agreed with Bateson (and Romanes) that mutations affecting the physiology of gamete formation, resulting in infertility, might be independent of mutations causing changes in external characters:

> If mutants are incipient species, we should be led to expect that some of them would be infertile with the parent species but fertile with their own kind, for these are the most distinctive features of species. ... Practically all the cases of mutant changes that have been observed and studied relate to external characters that might not have had anything to do with physiological functions causing infertility. Only if they happened to have been correlated with cross-infertility would they be expected to give rise to new species with this particularity. In other words, cross-infertility in its incipient stages may be exceptional rather than the rule.

Morgan agreed that primary chromosomal changes could affect gamete formation:

> Hybrid sterility is a very variable condition. It has been shown in a number of cases to result from a failure of the conjugation of the chromosomes at maturation which leads automatically to great subsequent mortality amongst the germ-cells. If it be granted that in the course of evolution, changes in the constitution of chromosomes occur, or else in the re-arrangement of the elements (as we now have demonstrable evidence may occur), there would be no difficulty in understanding why the hybrid is in such cases sterile.

However, he drew a distinction between what he called "point mutations," which at that time would mean a change that could be mapped to a distinct chromosomal location and larger changes in chromosome structure or number:

> As far as hybrid sterility rests on imperfections in the ripening process [of gamete formation], it may be traced to differences that have involved changes in chromosome numbers, or re-arrangements. Evidence is coming in at present that promises to supply materials for the study of this problem, and we can afford to wait awhile before making the sterility of the hybrid a fatal objection to the theory of natural selection through mutational changes.

As for the "chromosome theory" which Bateson had considered to have "fallen short of the essential discovery," Morgan remained with a "point of view miles apart from that of Bateson."

Morgan's ultra-Darwinian viewpoint was echoed in 1932 in his last major treatise – *The Scientific Basis of Evolution* [3]. Here he cited only *Materials* and *RS-Report 1,* and stated that "many of the cases of discontinuity that Bateson recorded in his book on the subject [*Materials*] relate to somatic changes that would not be inherited, but are the result of changes temporarily affecting the development of the individual." Thus, Morgan believed Bateson had "opened the door to a transcendental conception of the evolutionary process." In this category, among "philosophers, metaphysicians and mystics," he included Geoffrey St. Hilaire, who, like Bateson in *Materials*, had appealed both "to the monstrous embryonic forms that so frequently appear," and to "abrupt changes, somewhat as we do today to mutations; but we now know that the malformations and embryonic monstrosities referred to by St. Hilaire are not inherited, while the chief characteristic of mutants is that they are inherited." Morgan did not make clear that mutants are classified as either lethal or non-lethal. The lethal cannot be inherited, but they are still mutants. Whether or not they affect the germline, they can provide indications of the *scope* of mutability within a species. Beyond such limits, Bateson had argued, the species cannot go without the intervention of another type of process that permits a new species to emerge [4].

Morgan's Columbia colleague, Henry Osborn, extended their "newspaper and magazine war" over the Toronto address (Chap. 17) to an entire book – *Evolution and Religion in Education*: "At the moment of the sudden and regrettable death of the Great Commoner I was pleased to recall that I had never said anything harsh of him in controversy, and that his final attack on my supposed utter ignorance as to the evolution of man, published in the *Forum* in July, 1925, was good-natured from beginning to end."

Farmer in an obituary for the RS [5] declared that Bateson "had the ... disconcerting gift of facing and concentrating on the difficulties that were too often slurred over by his contemporaries, and he set himself to probe into the fundamental assumptions that lay at the root of the theory of evolution as then understood." *Materials* was "before its time," and "the whole position was set forth with a lucidity and force that ought at once to have arrested general attention, but the rigidly scientific presentation ran counter to the facile teleology which, like a noxious weed, had overgrown the solid framework of evolutionary doctrine."

An anonymous author wrote in *The Gardeners' Chronicle* (Feb. 20, 1926):

> It was amusing to sit near him while a game, perhaps of croquet, was in progress. The players would be admonished, cheered and jeered at, and in the midst of it all he would turn to a neighbour and say, 'Yes, it is the question of inter-sterility we want to get at.' It was this almost boyish zest that made it possible for him to turn from animals to plants as soon as Mendel's work was republished, and in a very short time he got a good grasp of many branches of horticulture, and even appeared as a witness in a famous law suit when the identity of Gradus Pea was in question.

Haldane, who had known Bateson only in his dotage, wrote an obituary for *The Nation and the Athenaeum* [6]:

> The working of his mind was well illustrated by his attitude to the work of Morgan and his school in New York, who have shown that the Mendelian factors are carried in or by the chromosomes which can be seen in a dividing nucleus. For eight years Bateson attacked this theory with the utmost vigour; not because he considered it inherently improbable, but because he believed that it went beyond the evidence, and because the natural bent of his mind and his profound knowledge of the history of science led him to doubt the validity of long chains of reasoning, however convincing. When, however, the possibility of ocular demonstration arose, he went over to America, and returned a convert, though with certain reservations which I believe that the future will largely justify.

And on Bateson's greatness:

> He held that ... [Mendelism] would not ... explain evolution. It is normal for a discoverer to be obsessed by the importance of his own discoveries, and it is a thoroughly excusable weakness. There are times in the history of thought when an idea must be born, and if it is a great idea it may be expected to overwhelm and obsess the man who gave it birth. He either becomes its slave, or preserves a certain independence only by continuing to hold views incompatible with it at the expense of dividing his mind into watertight compartments. William Bateson escaped these fates because he was greater than any of his ideas.

In his own dotage, in an article entitled "The Theory of Evolution before and after Bateson," Haldane reminisced [7]:

> He could be described as an angry and obstinate old man. But his anger was largely reserved for inaccuracy and loose thinking, and for certain types of injustice. His obstinacy made it difficult to convince him of the truth of theories which had previously been asserted without adequate evidence and were now being substantiated. ... To me, at least, he showed no signs whatever of a senile failure of original thought.

Likewise, Punnett in his dotage wrote [8]:

> Dominant personality as he was, he was never domineering. Only once can I recall his having lost his temper with me, and then he had every justification because I had imported a trio

> of Silky fowls without his knowledge. I was getting a little bored with the everlasting single, pea, rose and walnut combs and I had a hunch that the queer little Silky with its unusual comb might bring in something new. That it certainly did, and out of its crosses with the Brown Leghorn came the data which enabled us to establish the doctrine of sex-linkage. So I was forgiven.

Writing to Beatrice in 1928, Punnett commented on her two books on Bateson: "I am anxious to see the 'Steppe' book. But the one I want to see most is the one that can't appear – the one with intimate judgements on contemporaries and things. We've all a bit of gossip in us." To some extent, the subsequent partial biographies by Provine and Lipset did this. However, the gossip was not so much by Bateson, as about Bateson. And it was more than mere gossip.

Fisher

Mathematicians with an interest in biological problems have tended to team up with biologists. Pearson had his Weldon. Fisher's American adversary, Sewall Wright, had his Dobzhansky (see Chap. 25). Fisher had Major Leonard Darwin. While composing his seminal work, *The Genetical Theory of Natural Selection* [9], Fisher sought Darwin's advice on Bateson's contributions to genetics. Darwin replied [10]:

> As to Bateson, if I had to write, I should write something like the following. But I am not well up in what he did do, and may well blunder .... 'In the future the great merit of Mendelism will be seen to rest on the proof that the ingredients of germ plasm on which heredity depends are located in pairs in each organism, one of each pair selected by chance disappearing at each sexual union. ... The merit for this discovery must mainly rest with Mendel, whilst among our countrymen, Bateson played the leading part in its rediscovery. Unfortunately he was unable to grasp the mathematical and statistical aspects of biology, and from this and other causes, he was not only incapable of framing an evolutionary theory himself, but entirely failed to see how Mendelism supplied the missing parts of the structure first erected by Darwin. Nothing but harm can come from following Bateson in regard to evolution theory, though his name will come to be honoured for his pioneer work in Mendelism when what he failed to do as regards theory has been accomplished.' Having written it, I daresay I should tear it up, and advise you to do ditto.

Fisher replied:

> Many thanks for the note on Bateson; it puts the point admirably, and though I have already altered the wording somewhat, it seems to me just what was wanted. The only thing to do is to commend Bateson's enthusiasm for genetics, without saying, which would rather comfort my conscience, 'while greatly retarding its progress in his own country.' ... I have just been reading Samuel Butler's *Luck or Cunning*; what a malignant knave he must have been, yet Bateson borrowed his sneers and quoted his opinions.

Disregarding Darwin's caveat, Fisher did not "tear it up." While commending Bateson for "played the leading part" in the "early advocacy of Mendelism," Fisher wrote [9]:

> Unfortunately he was unprepared to recognize the mathematical and statistical aspects of biology, and from this and other causes he was not only incapable of framing an evolutionary theory himself, but entirely failed to see how Mendelism supplied the missing parts of the structure first erected by Darwin. His interpretation of Mendelian facts was from the first too exclusively coloured by his earlier belief in the discontinuous origin of specific forms. Though his influence upon evolutionary theory was thus chiefly retrogressive, the mighty body of Mendelian researches throughout the world has evidently outgrown the fallacies with which it was first fostered. As a pioneer of genetics he has done more than enough to expiate the rash polemics of his earlier writings.

Thus, Fisher, who shortly thereafter was to cast doubt on the soundness of the data of Mendel himself [11, 12], based his attack on Bateson on the words of someone who admitted to not having thoroughly read Bateson's work. To compound the matter, Bateson's *Materials* was described by Fisher as: "A work which owed its influence to the acuteness less of its reasoning than of its sarcasm."

Defending Bateson, Alan Cock [13] concluded that: "Fisher's criticisms are ... unfair and in large part based on a misunderstanding of Bateson's views." However, even Alan subscribed to the view that "theoretical innovation was not where Bateson's strength lay" and cited the "reduplication hypothesis." In his 1974 Presidential Address to the Thirteenth International Congress of Genetics, Curt Stern (a student of Goldschmidt) noted that while the reduplication hypothesis "could not beat the [later] competition," nevertheless "it was an ingenious suggestion" to propose that "Mendelian segregation ... occurs during a somatic cell division followed by differential multiplication of the different genotypes" [14]. Indeed, the higher-than-expected incidence of certain human hereditary diseases (e.g., achondroplastic dwarfism with influence of parental age) can be explained in these terms [15].

Fisher, together with his protégé Edmund B. Ford (1901–1988), worked hard to establish a selective basis for melanism in moths ("industrial melanism"). In 1956, Haldane, who in 1924 had calculated the high selection coefficient involved (see Chap. 17), also continued to press this viewpoint [16]. As described in Chap. 6, Bateson had been skeptical. However, Ford, in collaboration with the lepidopterist Bernard Kettlewell (1907–1979), appeared to settle the issue beyond doubt, and melanism in moths became the standard textbook example of natural selection. So persuasive was the Kettlewell–Ford case that their fame and fortune were well established before a sad story of egotism and possible self-deception emerged at the end of the twentieth century. Nevertheless, while briefly mentioning the Evolution Committee study on progressive melanism, a scientific account did not mention Bateson [17], and a journalistic account mentioned him only in the context of Mendelism [18].

Muller

In 1922, Muller visited Russia and donated fruit fly samples. In 1932, he attempted suicide. Later that year, after giving a famous address at the Sixth International Congress of Genetics at Ithaca, he abandoned his Texas operations and sailed alone to Europe. While working in a Berlin laboratory, he witnessed a raid by Hitler's storm troopers. He moved on to Russia where in 1934 he spoke on "Lenin's Doctrines in Relation to Genetics" [19]. Having criticized Pearson and various others for their opposition to Mendelism, which was perceived as "the first anti-materialist wave," Muller saw Bateson's hesitance to agree to the relationship between genes and chromosomes as an expression of a deeply based anti-materialism:

> Many of those who accepted the Mendelian rules as such, adopted a second line of anti-materialist defence, in that they endeavoured to put them upon as vague as possible a basis. It was no doubt the feeling that the identification of the Mendelian units, the genes, with the chromosomes was merely a materialistic vulgarity, which, as much as anything else, held Bateson and, following him, almost the whole British school of Mendelians (excepting Lock and Doncaster) and much of continental Europe as well, aloof from taking part in this so fertile liaison.
>
> And yet it was evident very early, from the extraordinary parallelism between the methods of chromosome and gene distribution, that the former must (barring almost a miracle of scientific coincidence) constitute the visible material basis of the latter. Not to recognize this was but one way of staving off the advance of materialism longer, and it is no accident that we find the center of this reaction in England, the land where idealism probably has its firmest roots in intellectual circles ....
>
> This resistance to further genetic development even on the part of Bateson – one of those who had at first done the most in helping to establish Mendel's laws as such – probably impeded the progress of genetics more than did the opposition of the non-Mendelians, since it led away from fruitful lines more of those who could otherwise have taken a real part in genetic advance. Of course the refusal of Bateson and his school to accept the chromosomes as the basis of Mendelism was not alleged to be any opposition to materialism, but pretended to be founded in a kind of empiricism, as has so often been true in similar cases, as, for example, in the case of the Machian 'empirio-criticists' whom Lenin attacked. ...
>
> In addition to the above groups of compromisers and sabotagers of materialism, there have been some biologists who, while accepting Mendelism and genes, and in some cases even some sort of connection of the genes with the chromosomes, nevertheless, adopting a third line of defence, have stated outright that these genes must be regarded purely as 'concepts,' that is, as mental abstractions, not as real things. Chief among these were Johannsen and East. They had to guide their course, however, within such a difficult no-man's land of paradox ... that it was not possible for their following to grow very large, before the falsity of their position became manifest.

Muller, strongly supported by Huxley who had recruited him to Texas in 1913, later returned to America (Chap. 18).

Huxley

After taking over the Secretaryship of the Zoological Society from Chalmers-Mitchell in 1935, Julian Huxley became a major spokesman for science using the media, including the new television broadcasting system, to full effect. His lapidary phrases "the modern synthesis" and "the eclipse of Darwinism" struck a popular chord and constituted a serious attack on everything that Bateson had stood for [20]. "The modern synthesis" successfully marketed the notion that the new biometricians (Fisher, Wright, and Haldane) had united Darwinism and Mendelism. Thus, the Darwinian sun now shone as bright as ever after a mere "eclipse" in the early decades of the century. Bateson's ideas were of ephemeral significance and scholars of evolutionary biology could safely disregard them. To compound this, in his later autobiography [21], Huxley falsely stated that, following appropriate tutelage, Bateson had returned to the Darwinian fold: "He eventually, under R. A. Fisher's influence, came to accept a Darwinian position – natural selection by the slow selection of favourable genetic combinations, including occasional mutations, mostly of minor scale."

Darlington

In *Genetics and Man* [22], Darlington labeled Bateson as "naïve" for embracing the concept of "the unit character," and as an "immaterialist," since "our individuality rests ... not as John Locke and Bateson imagined, on an 'immaterial substance,' but ... on the genetic substances in the fertilized egg." To Darlington, this was more important than segregation:

> To us it is now evident that the great revolutionary moment was when Mendel referred to 'elements which determine.' ... According to Bateson, however, Mendel's 'essential discovery' was not the elements-which-determine but the fact-of-segregation. Bateson would scarcely admit that there were elements which determine. Thus, although Mendel had taken the plunge, Bateson still shivered on the bank – where he remained for the rest of his life.

Invoking ancient views (Aristotle, Harvey, Spallanzani), Darlington declared that Bateson's "refusal to face the assumption of material bodies in the germ cells, acting throughout life in determining the characters of the mature plant or animal, would astonish us if it had no antecedents." Furthermore, he implied that Bateson did not understand the genotype–phenotype distinction. And as for the presence and absence theory:

> This hypothesis would seem to have had two merits, merits which it still partly retains. It was a minimum assumption, strictly according to William of Occam; and it implied the existence of material particles. But for Bateson it obviously had a third merit: it sounded immaterial. The philosopher Locke had described the soul as an 'immaterial substance' and Bateson wanted something from the same bottle.

Of course, from everything in this book so far, readers will recognize that Bateson was an atheist with no place in his philosophy for the mystical or immaterial. People such as Darlington may have been misled by his references to waves and vibrations (analog not digital), but these were always considered to require some medium through which they were transmitted. Thus, in correspondence with Wheldale, Bateson pondered whether the biochemistry of chromosomes would permit their transmitting vibrations (see Chap. 17). His meristic ideas are now seen in terms of waves and gradients of morphogenic macromolecules radiating from a source within a differentiating embryo [23].

Mayr

Ernst Mayr was the most relentless of the many disparagers and misrepresenters of Bateson. He migrated from Germany in 1931 to join Osborn as curator at the American Museum of Natural History in New York. For the last 50 years of his life he was based at Harvard University where he had the opportunity to influence several generations of students, many of whom later attained authoritative positions in the biosciences. His numerous books and papers were seldom immoderate. Yet, over the years, his indictment was cumulatively the most damning.

In 2004, there was much celebration at Mayr's 100th birthday, and he died shortly thereafter. But in the Galton Laboratory, doubts were expressed [24]: "I do not wish to belittle the work of Mayr and the geneticist Theodosius Dobzhansky – but our impression that they solved the species problem is illusory. They were merely the ones who translated it from the technical literature, enunciating much more clearly than before what had become the prevailing view of species among those who had thought about the problem." Here are some samplings from an "essay review" by Mayr (1973) where Bateson and de Vries were lumped together [25]:

- It is obvious that the naturalists understood certain aspects of evolution which were consistently misunderstood or ignored by experimental scientists like Bateson and De Vries.
- The early hybridizers were completely right and the early Mendelians (Bateson, De Vries, and so on), who ignored them and believed the opposite, were quite wrong.
- Bateson was pig-headed, intemperate, and intolerant ... uncompromising ... and ... quite incapable of understanding the nature of natural populations.
- Bateson's stubborn resistance to the chromosome theory resulted in much effort by members of the Morgan school that could have been devoted to new frontiers in genetics.
- Attempts to refute the power of selection were truly a comedy of errors. Bateson tried to do so .... None of those who denied the efficiency of selection so insistently bothered to visit the animal breeders.

Mayr had a modest way of disarming potential critics. In correspondence (Nov. 9, 1973), concerning Alan Cock's impending visit to Harvard (see Prologue), Mayr mentioned his above essay review and continued:

> Needless to say, my interpretations are subjective and are likely to be wrong not infrequently, since I am only mortal. What I am trying to avoid is the uncritical copying of what others have said before. The history of biology is full of that technique. I would much rather be occasionally wrong and thereby stir up a critical analysis than behave like a parrot.

Similar dissemblances are in the essay review itself:

> If I arrive occasionally at biased evaluations, it would seem to me less a fault than to express no opinion at all and to repeat the same standard assertions that have been with us for fifty or more years. I hope that I will be refuted where my judgement is faulty, so that, in good dialectic fashion, thesis and antithesis will eventually lead to a well-balanced, objective synthesis.

Dawkins

Opposition also came from Oxford zoologist Richard Dawkins. Shaking his head in bewilderment, in *The Blind Watchmaker* he found it "extremely hard for the modern mind to respond ... with anything but mirth" to the quaint ideas of the "mutationists" Bateson and de Vries [26]. It is one thing openly to attack someone whose ideas on a problem you disagree with or even to misrepresent that person as agreeing with you (see below). However, it can be far more effective to imply that the problem is unimportant and the person who has dedicated his/her life to it has been foolishly barking up the wrong tree. As an advocate of "universal Darwinism," Dawkins adopted the latter approach [27]. For him, "the job we ask" of evolutionary theories is "explaining the evolution of organized, adaptive complexity." So, what of the problem Bateson had made his life's work?

> Some biologists ... get excited about 'the species problem,' while I have never mustered much enthusiasm for it as a 'mystery of mysteries.' For some, the main thing that any theory of evolution has to explain is the diversity of life – cladogenesis. Others may require of their theory an explanation of the observed changes in the molecular constitution of the genome. I would not presume to convert any of these people to my point of view.

By this sleight of hand, Dawkins implied that "the evolution of organized adaptive complexity" could somehow be separated from the problem which most concerned Bateson.

Dawkins' powerful advocacy of selection at the gene level led many away from group selectionist ideas, and hence from speciation construed as a form of group selection. In 1982, in *The Extended Phenotype* [28], he wrote disparagingly of "sloppily unconscious group-selectionism." Two decades later, David Sloan Wilson [29] quoted with approval Joel Peck as stating that "there is no doubt that we were too hasty in trashing group selection … the theoretical models of the 60s and 70s were very oversimplified and should be taken with a pinch of salt."

Dawkins' attack even extended to Bateson's daughter-in-law. In *Unweaving the Rainbow* [30], he described her, on the authority of anthropologist Derek Freeman, as "the gullible but immensely influential American anthropologist Margaret Mead." However, subsequent remarks of the President of the American Anthropological Association [31] suggested this may have been but one more salvo from the pen of the immensely influential English biologist Richard Dawkins:

> The 'Freeman debate' has been the subject of a number of books and scholarly articles that support her views on the importance of culture for the adolescent experience, while criticizing some details of her research. I have taught about the controversy for the last 18 years and am still impressed by the fact that a 24-year-old woman could produce a study so far ahead of its time. Dr. Freeman studied a different island 20 years after Mead's research, and his notion that biology is more determinative than culture is oversimplified. Most serious scholarship casts grave doubt on his data and theory.

So much of Dawkins' writing was profound and enlightening that it was easy to believe it was *all* profound and enlightening. So much of Dawkins' writing rightly disparaged "the vacuous rhetoric of mountebanks and charlatans" [32], that it was easy to believe that he was holier-than-thou in this respect. Yet there came signs that Dawkins was relenting. In 2004, in *The Ancestor's Tale*, Beatrice's short story fantasy of 1895 came to fruition – Bateson was posthumously knighted!

Gould

Stephen Jay Gould began his academic career by toeing the party line. Like Mayr, he tended to closet de Vries with Bateson. Their emphasis on "large mutations" had been resolved by the population geneticists in the 1930s who recognized "micromutation as the agent of evolutionary change and equated it with Darwinian variability." Alas, "not only was Bateson not a Darwinian, he even ended his career in utter confusion on how new species arise" [33]. Around 1980, Gould toyed with a fresh approach to evolution, and for a while championed the views of Goldschmidt that were deemed heretical by most contemporaries [34].

It appears that an interest in Galton was sparked by Gould's reading of Bateson. Referring to the first of Gould's many books, *Ontology and Phylogeny* (1977), Alan Cock wrote to Gould (Aug. 23. 1984): "While you allude briefly (p. 84) to precursors, genuine or alleged, of punctuated equilibrium, you don't mention one highly eligible and respectable candidate: Francis Galton's concept of 'positions of organic stability'." Gould replied (Sept. 5, 1984): "I have used Galton's metaphor of the polyhedron extensively in my own writings, but it now occurs to me that I know it only from the citations by Bateson. Since you have obviously sorted this out far more thoroughly, could you direct me to the source of Galton's own derivation of the metaphor? I have always wanted to read it in the original."

Later in his final work, *The Structure of Evolutionary Theory* (2002), Gould was generally kind to Bateson and correctly quoted him as holding that "[natural] selection is a true phenomenon, but its function is to select, not create." However, Gould considered Bateson's attack on the Panglossian preachings of the Darwinists as "bordering on meanness." Bateson was labelled as an "obstinate," "stubborn," "old fogey" who "had fallen a bit behind the times" and "had his own particular axe to grind."

The BDM Meme

Many misrepresented and dismissed Bateson, but at least they did not claim he was really on their side. In 2000, there came the "Bateson, Dobzhansky, Muller model of speciation." It is to Dobzhansky's credit that he resurrected the Romanes–Gulick view that reproductive isolation was of paramount importance for speciation (Chap. 5), but he held that, in the general case, this reproductive isolation was *initiated* (i.e., the speciation process began) through differences in gene products ("genic" isolation), a view also held by Muller. In 1996, in a paper entitled "Dobzhansky, Bateson, and the Genetics of Speciation" [35], a modern Biometrician, H. Allen Orr, cited Bateson's Centenary essay (Chap. 12) as revealing that:

> William Bateson offered the "Dobzhansky–Muller" model in 1909, just nine years after the rediscovery of Mendelism and a good quarter of a century before Dobzhansky or Muller. And when I say that Bateson offered the model, I do not mean he obliquely alluded to it. Rather, Bateson spells it out, step by step, presenting it as the likely 'secret of interracial sterility.'... Those who differ on larger issues, as Dobzhansky and Bateson surely did, can nevertheless arrive at the same conclusion. Recent work on speciation renders this coincidence all the happier: for Bateson and Dobzhansky not only arrived at the same conclusion, but at the right conclusion.

Of course, Dobzhansky and Muller, both of whom had worked with Morgan, are likely to have read enough of Bateson's writings to know, even if they did not understand him, that he most certainly did *not* support a genic model for speciation. Thus, they did not cite him in this respect. Orr suggested, however, that the reason they did not cite Bateson was that "neither Dobzhansky nor Muller knew of Bateson's model." Orr believed that "Bateson apparently never repeated his argument."

In 1998, Orr's misinterpretation was further relayed by an editor of the multiauthor text *Endless Forms. Species and Speciation* and by Jerry Coyne (with Orr) in the *Philosophical Transactions of the Royal Society of London* [36]. In 2000, the misinterpretation was dignified with an acronym "BDM," standing for the "Bateson, Dobzhansky, Muller model of speciation" [37]. Gathering momentum, in 2001, the BDM "meme" began to appear in textbooks [38–40] and a review [41]. Soon it had spread to *Science* [42] and *Nature Genetics* [43]. In 2006, it hit *Nature* itself [44].

Biohistorians

It was the biohistorian William Coleman who made the first in-depth study of the papers Beatrice and her family had conserved. His long paper "Bateson and Chromosomes: Conservative Thought in Science" [2] could readily have been turned into a book. Had such a book been written, publication would have been unlikely because the reviewers would have alerted prospective publishers to the widespread antagonism of potential readers to Bateson. Sales would likely be poor. Coleman observed that by the 1920s, "the current trend of genetic research and its public acclaim" had "made obsolete" Bateson's "objectives and virtually irrelevant his research." Coleman's influential paper led others to consider Bateson as "an archetypal conservative thinker" [45], and as one of "the most recalcitrant idealists" [46]. Many of Coleman's negative points were answered by Alan Cock in 1983 in a less influential paper [47].

Alan considered that Coleman "fails to get to grips with the various detailed objections raised by Bateson ... against chromosome theory," which were then substantial. Yet, Alan argued that, while "Bateson's view, that *between*-species differences were somehow qualitatively distinct from *within*-species differences, ... was an important source of Bateson's opposition to evolution by natural selection, it was hardly relevant to his attitude to chromosome theory" – the latter being a theory concerned with the location and disposition of genes. On the other hand, I have here argued that the species problem was highly relevant to his attitude to chromosomes. Bateson's difficulty was in seeing how chromosomes could provide the basis for a postulated qualitative distinction which was fundamental to speciation. Decades of work by the biochemical successors of Hopkins, Onslow, and Wheldale were needed before a likely solution was to emerge (see Part VI). In similar fashion, some of Einstein's objections to quantum mechanics that led to his estrangement from many in the Physics community are now being fitted into a more complete theory of space, time, and matter [48].

Some biohistorians were better able to navigate texts laden with anti-Bateson rhetoric than others. In 1971, William Provine concluded his analysis of *The Origins of Theoretical Population Genetics* with an expression of profound diffidence: "With the gap between theoretical models and available observational data so large, population genetics began and continues with a theoretical structure containing obvious internal inconsistencies." But in 1983 in his encyclopedic *The Eclipse of Darwinism*, Peter Bowler nailed his colors to the mast [49]. He praised "the modern synthesis," as an "ultimately successful theory," but conceded that Gould "may well be right when he claims that some of the more basic issues have not been completely resolved by the emergence of the modern synthesis." Indeed, Mark Adams writing on "Little Evolution, Big Evolution" (1990), stressed that the understanding of (Batesonian) macroevolution would demand "a radically new interpretation of the history of Darwinism, population genetics and the evolutionary synthesis." For "if intra- and inter-specific variation differ not in kind, but only in degree, then it is possible, by extension, to envision selection as the creator of a new species. But if varieties are fundamentally different from species – if the fundamental character of intraspecific and interspecific variation is essentially different – then the effect of selection on a population cannot explain evolution" [50].

Popular Science Writers

The media mediate. They do not usually originate. When scientists get the story wrong, then historians may be misled, and the popularizers and journalists who depend on the scientists and historians may compound the error. In a letter to Huxley with whom he and his son were collaborating on a book, H. G. Wells wrote in 1928 [21]: "I am against any further alterations of that Bateson paragraph. I know the man. My last talk with him was with Morgan in N. Y. (Chap. 17) and he has a schoolboy pleasure in making trouble and a Samuel Butler-like hatred for Darwin. Any fool can play the negative game and no doubt some of the young fools will go on with it." Here is "that Bateson paragraph" [51]:

> There remains one other temperamental type which has found expression in these discussions, and that is the brilliant sceptic as typified by the late Professor William Bateson. He accepted the facts of Evolution, if only on the palaeontological evidence, but, as the outcome of a life spent largely in the study of variation and especially of Mendelism, he developed an increasing inability to satisfy himself how any progressive variation could ever occur. He crowned his scientific career by various lectures and addresses in which he reiterated his imaginative failure. This type of agnosticism was probably the negative aspect of a passionate and unquestioning faith in the implacable unteachableness and integrity of certain Mendelian units of heredity we shall presently describe and discuss. Later work has removed much of the point of his criticisms.

A later anti-Bateson tract (2000) was Robin Henig's *The Monk in the Garden* [52]. At the outset, the reader was informed that only Bateson's "droopy eyes" saved him from appearing "self satisfied and smug." When writing about Linnaeus (the founder of taxonomy) there was some diffidence in that "he was *said to* be a personally unpleasant man" (my italics). Remarks about Bateson, however, were categorical:

- Bateson's inability to hold two competing thoughts at once, his tendency to see the world in stark blacks and whites, drove much of the debate in the years after Mendel's discovery.
- If not for Mendel, we might know – or care – very little indeed about the opinionated zoologist from Cambridge.
- Unlike Bateson, Morgan made a habit of admitting to his earlier mistakes.

Bateson's collaborator, Rebecca Saunders, was not left unscathed, being referred to as "his long-time *research assistant*" (my italics). With coauthors – particularly female coauthors – one is interested to know whether their contributions were greater than they were given credit for (e.g., the speculation about Einstein's first wife). To her rhetorical question: "Why ... would she have stuck for so long with a man who treated her coldly ...?" Henig replied suggesting a romantic attachment.

The possibility that Saunders was as excited by the scientific questions as Bateson was not considered. The biohistorian Marsha Richmond in a paper on "Women in the Early History of Genetics" [53, 54] noted that women participated both through their labors and intellectually to the Cambridge "school in genetics" which Bateson headed, and lamented that Henig "has labelled these women Bateson's 'research assistants' and presumed a romantic attachment." Kindlier, Haldane named Saunders "the 'mother' of British plant genetics."

Bateson's sometimes harsh, *but private*, comments to Beatrice concerning Saunders that appear on these pages, perhaps indicate that he wished Beatrice to have no doubts that the relationship was platonic. The few surviving letters from Saunders to Bateson give no indication that she thought otherwise. They display that she was thinking more about the here-and-now, than about the future – all letters were dated by day and month, not by year. They were factual, dealing with their common work, and generally began "Dear Mr. Bateson" and ended "Yours sincerely, E. R. Saunders." There were lists of corrections relating to spelling and interpretations, suggesting that she was not a passive partner where the writing of papers and grant applications was concerned. For example: "As it stands it seems to suggest a contradiction," and "Is not this rather à la sledgehammer … ?", and "Cannot you manage to use me [cite me] less often as (personal) communication?", and "Would it be possible to indicate the main line of descent with a thicker line?"

Relatives

As far as we know, his sisters and sisters-in-law remained much in awe of their brother William. However, brother Edward may have differed. The Cambridge ethologist, Patrick Bateson, a distant relative with a distinct physical resemblance to Gregory Bateson, related in 2002 [55]:

> When I was a boy, my parents cared for William's younger brother, Ned, when he was a widower and an old man. … Earlier in my career I had formed an unfavourable picture of William from talking to … Ned, who called him a tyrant and one of the real Victorian autocrats. In a similar vein his anthropologist son, Gregory, likened his father to the unbending Reverend Theobald Pontifex in Samuel Butler's *Way of All Flesh*. Gregory, of course, was ambivalent and also spoke with fondness and respect about his father. … This disagreeable picture of William is strongly reinforced by many historians of science (Mayr 1982).

Summary

The John Innes Council wished to renew Bateson's Directorship. He agreed regretting that many aims had not been achieved, that he had not kept up with new developments, and that he had groomed no successor. Several months before his 65th birthday, he died. In an obituary for the Linnean Society, Morgan displayed himself as an obstinate Darwinian with little patience for Bateson's "somewhat vague" insistence that there was something special about species formation. The warm sentiments of Punnett and Haldane endured when they recalled Bateson in their dotage. Haldane held that the future would justify his reservations concerning chromosome theory. But Bateson was not allowed to lie in peace. Huxley dismissed the Bateson years as a mere "eclipse" from which the Darwinian sun had emerged as "the modern synthesis." Both Muller, transiently entranced by communism, and Darlington at Oxford, saw Bateson as an antimaterialist, while others (Fisher, Mayr, Orr) questioned his competence and asserted that he had delayed progress. Dawkins considered that the species problem, to which Bateson had devoted his life, was trivial compared with that of adaptation. While recognizing speciation as a fundamental problem, mathematical geneticists praised Bateson for providing a foundation for their genic viewpoint, a viewpoint which he had in fact consistently opposed. Popular science writers leapt to relay the anti-Batesonian message.

Chapter 25
Epilogue

> *While mistakes of a factual nature in scientific results or theories are rapidly corrected, those concerning the historical development … may remain unchallenged for more than a century. Such mistakes are seldom corrected by the scientific historian, who has only a bird's-eye view of the development of the subject and often does not realize their importance. The task of correction thus falls to the scientist.*
>
> *David Keilin [1]*

In *Evolution, Old and New* (1879), the "scientist" Samuel Butler quoted extensively from the appendix of Patrick Matthew's 1831 book *On Naval Timber and Arboriculture,* which proposed evolution by natural selection [2]. Matthew's book in Bateson's library at the John Innes Centre is inscribed "W. Bateson, 1925," suggesting a relatively late interest. Bateson's papers contain a quote from an 1832 review of the book in *Gardener's Magazine*:

> An appendix of 29 pages concludes the book …. This may be truly termed, in a double sense, an extraordinary part of the book. One of the subjects discussed … is the puzzling one of the origin of species and varieties; and if the author has hereon originated no original views (and of this we are far from certain), he has exhibited his own in an original manner.

Thus, the appendix was favorably received by at least one of Matthew's contemporaries (J. C. Loudon). The papers include a letter (July 31, 1925) giving details of Matthew's Perthshire background prior to his death in 1874. However, in his popular textbook, Robert Lock dismissed the natural selection ideas of W. C. Wells (1813) and Patrick Matthew (1831) as "merely historical," since they showed "the direction in which thought was tending" [3]. Lock was in no way alarmed. This has been an attitude of busy scientists both in the past and our own time. The attribute of mere historicity implies scientists' satisfaction with the reading and understanding of the literature by those they rely on to tell them the direction thought was (or is) tending.

A. G. Cock, D. R. Forsdyke, *Treasure Your Exceptions*,
https://doi.org/10.1007/978-3-030-92099-9_25

Although this strategy often succeeds, there are many examples – of which Mendel appears the paradigm case in biology – that reveal the folly of assuming that the foundations of one's discipline are secure. This study of the life of William Bateson has described his contributions to evolutionary biology, the importance of which has become increasingly evident (see Part VI). We may conclude that there was not just one nineteenth-century "Mendel." In addition to Wells, Matthew, and Mendel himself, there were at least five others – Galton, Hering, Butler, Romanes, and de Vries. Through his work in the early decades of the twentieth century, Bateson can be seen as the ninth. Most of the "Mendels," like the original, did not (or could not) press their findings. However, against the hierarchy of complacent Darwinists, Butler and Bateson used their best weapons – their pens and sharp intelligences.

Mendelization

Three themes emerge when contemporaries ("peers") fail to read or correctly assess a person's work, hence leading to that person's possible "Mendelization." First, there is some other idea which so fascinates that thoughts cannot be reconfigured – the "exceptions" that ruffle the apparently harmonious surface are dismissed, not "treasured." Quite simply, Wells and Matthew may have been thwarted by Paley's *Evidences of Christianity* (1794). Second, having convinced themselves that one aspect of a person's work is wrong, contemporaries less readily entertain other aspects. This is a useful rule-of-thumb but can result in the loss of some profound conceptual "babies" with the academic "bath water." Finally, there is a satisfaction with the status quo reinforced by the over-selling of perceived successes – the strutting and posturing, the pinning of medals and epaulettes – that imply arrival when in fact the journey, perhaps along the wrong path, has only just begun.

Reinforcing the latter is what Mayr referred to as "the uncritical copying of what others have said before" (Chap. 24). Ross noted how easy it was for fact to "pass from text-book to text-book" without amendment and regretted that it took five years to break the momentum of the mythical *Bacillus malariae* (Chap. 22). Although Tower's Kammerer-like fraud had been exposed by 1915, his results were still being quoted in 1919 (Chap. 11). Likewise, the "BDM meme" lingers on (Chap. 24).

In 1874, Darwin wrote [4]: "False facts are highly injurious to the progress of science, for they often endure long; but false views, if supported by some evidence, do little harm, for everyone takes a salutary pleasure in proving their falseness." This view may itself be false. While false facts may not always be as "rapidly corrected" as Keilin supposed, it is false views that show the greatest propensity to "endure long." Everyone may not take pleasure in proving false a view that has abstract elements and/or is politically correct (e.g., the doctrine of natural selection). Often it is the scientist-historian, not the historian of science, who is best placed to remedy this [1].

The Ninth Mendel

When reading the copies of *Nature* that Anna had sent him in 1886 (Chap. 2), Bateson "discovered" Romanes' Linnean Society address and commented with approval. It was a discovery in the sense that someone might have read Matthew's appendix in 1831, or Mendel's paper in 1866, and cried "eureka." However, it was a discovery Bateson seems not to have consciously attributed to Romanes whose account of reproductive isolation was considered a rather obvious "straight forward, common-sense suggestion" (Chap. 5). There was no "eureka." Although referring to Romanes' "physiological selection" time and again in his lectures, Bateson seldom formally referred to Romanes or employed the phase "reproductive isolation." Romanes had to wait a century to be "rediscovered" [5].

Galton, having been consigned to oblivion by the senior Darwinians (Darwin, Wallace, Huxley, Hooker), was "rediscovered" by Romanes and Bateson in the 1880s. Bateson's rediscovery of Galton was formally acknowledged by Galton himself when reviewing *Materials* in 1894 (Chap. 4). Galton's words might have been those of Mendel had he been alive in 1900: "These … are the views that I have put forward in various publications …, but all along I seem to have spoken in empty air. I never heard nor have I read any criticism of them, and I believed they had passed unheeded and that my opinion was a minority of one." Galton had forgotten that his ideas were privately criticized both around the time of their inception by Darwin (Nov. 7, 1875) and when the 27-year-old Romanes entertained Galton and Darwin in his London home. Romanes wrote to Galton (Dec. 13, 1875):

> I was honoured this morning by a visit from Mr. Darwin. Of course we talked about your theory of heredity and … I remembered the difficulty I forgot when you were here. Mr. Darwin agreed with me that it is a serious one, so I should like to take this opportunity if stating it to you. … Why are *congenital* variations so liable to be inherited? A congenital variation is certainly an *individual* variation, and the fact that it takes place at an embryonic period of life is a fact of no significance as far as the stirp theory is concerned. … I know you will not object to my stating frankly what seems to me an objection to a theory, which … I recognize as a valuable and honest attempt at meeting one of the most burning questions of the day.

Perhaps Romanes' point was that a congenital variation, if considered to have arisen during development, could not have arisen in ancestors and so would not have been transmitted as part of the "stirp" (DNA to the modern reader) to the afflicted individual through the germline. If transferable to the offspring of the afflicted, it could be in the same category as an acquired character (e.g., increased muscular strength in a blacksmith transferred to his offspring). Their Lamarckian mode of thinking may have blinkered Romanes and Darwin from recognition that Galton had given pangenesis a novel interpretation, which the *Intracellular Pangenesis* of de Vries was to make explicit. We now understand that congenital variations can be the result of primary (or ancestrally hidden recessive) germ-line mutations in parental genes that affect development.

Weldon and Pearson were deeply impressed with the statistical approaches Galton had pioneered, but neither seemed to notice his 1870 papers. In the 1880s,

Romanes, his own attempts to verify Darwin's version of pangenesis having failed, began to see more in Galton's ideas and pressed successfully for Oxford to award him an honorary degree (Chap. 5). Championed by so many young and prominent figures, the shy and retiring Galton could hardly have failed to be appointed in 1894 to the Chairmanship of the RS "Committee for Conducting Statistical Enquiries" (Chap. 6).

Butler (and with mention of Butler, Hering is also implied) was not rediscovered by Bateson. Even Butler's most ardent scientific advocate, Marcus Hartog, admitted to being far from convinced. Like Romanes, Butler had to wait a century for rediscovery [6]. Bateson became aware of Butler's writings around 1908, but never really understood them (Chap. 19). And finally, there was Mendel himself – a rediscovery by de Vries, Correns, and Tschermak, and a near miss for Bateson (Chap. 8). So, Bateson's score stands at two: one overt, but shared, rediscovery (Galton); one covert discovery (Romanes); one near miss (Mendel); one absolute miss (Butler).

Today de Vries is primarily remembered as a rediscoverer of Mendel and for believing that his reproductively isolated evening primrose "constant hybrids" were representative of the mutational process by which species generally originate (Chap. 9). The novel aspects of his masterpiece, *Intracellular Pangenesis* (e.g., what we now know as messenger RNA), were rediscovered independently in piecemeal fashion, and their original proposer was never recognized. Not being strong in German, the novel aspects of *Intracellular Pangenesis* are likely to have passed Bateson by, and his later antipathy to de Vries probably did not make for assiduous reading of the English translation in 1910.

In 1971, the biohistorian Robert Olby asked [7]: "Why were there so many speculative theories of heredity in the latter part of the [nineteenth] century, and why did none succeed [get general recognition in the immediate or short-term]?" He concluded that de Vries' *Intracellular Pangenesis* "was not more influential" because of "the superabundance of such speculations and to the common failing of attempting to explain too much. By contrast, Mendel's ability to abstract and even to ignore is the more to be admired." While this may be true, that so many of de Vries' speculations have proved correct cannot be ignored. There is a need for better methods of discerning gold from dross. Is there some way of sifting the words of a Galton, Butler, Romanes, de Vries, or Bateson from the babble of the many? One answer has been to sift on the basis, not of the speculations themselves, but of the nature and nurture of the speculators and of their track-records relative to the opportunities they have enjoyed [8].

Two Nobodies

Only one rediscovery, Mendel, won general acceptance in Bateson's lifetime. In *Principles* (1909), Bateson pondered Nägeli's absolute miss:

> Nägeli was … especially devoted to the study of heredity, and even made it the subject of elaborate mathematical treatment. As we now know, he was in correspondence with

> Mendel, from whom he received a considerable number of letters and illustrative specimens. These must have utterly failed to arouse his interest, for when in 1884, the year of Mendel's death, he published his great treatise on heredity, no reference was made to Mendel's work. That this neglect was due to want of comprehension is evident from a passage where he describes an ... observation on cats, which as it happens, gave a simple Mendelian result. ... From the discussion which he devotes to the occurrence it is clear that Mendel's work must have wholly passed from his memory, having probably been dismissed as something too fanciful for serious consideration.

So, to what can we impute Bateson's own misses? He seems to have missed Butler for "want of comprehension" combined with the belief that this was something "too fanciful" – namely the notion that there might be something akin to a written text within cells. This contrasted strongly with his apprehension of Mendel – as soon as he became aware of Mendel's work, he embraced it (Chap. 8). For Galton, Bateson did likewise, at least at the outset. Unlike Mendel, Galton was still alive and might have done more to promote his early views had he not been distracted by eugenics and the complications of his advancing years (Chap. 15).

Romanes died in 1894 at the age of 46. Bateson's failure formally to acknowledge him seems to have stemmed neither from a "want of comprehension," nor from thinking Romanes "too fanciful." As described elsewhere [5], Romanes' ideas were not appreciated by Thomas Huxley, Thiselton-Dyer, Lankester, and Francis Darwin, and it is likely that their negative attitudes influenced Bateson. Negative attitudes can also stifle ideas by preventing their publication. Reminiscing in 1950, Punnett noted [9]:

> In spite of the success of the [1904] Cambridge meeting in getting Mendelism a hearing, the older generation of biologists endorsed Weldon's hostility and the pens of Alfred Russell Wallace, Professor Poulton and Professor J. Arthur Thomson were soon engaged in attempting its belittlement. In this they were supported by *Nature*, though by now such hostility was of less account; Mendelism had become news and the columns of secular periodicals were opened to us.

Prior to that:

> It was a difficult time for struggling geneticists when the leading journals refused to publish their contributions to knowledge, and we had to get along as best we could with the more friendly aid of the Cambridge Philosophical Society and the Reports of the Evolution Committee of the Royal Society.

The "friendly aid" was probably due more to Bateson's influential connections than to recognition of the scientific excellence of his work. As the Butler story reveals, in Victorian England, times were hard for the academically gifted who were not backed by substantial private incomes and/or had received no formal training in science (so were without mentors to smooth their paths). If the interpretations offered in this book stand, then Butler (together with Galton) will be seen as conceptually far ahead of the Darwinians.

There were also contemporaries who superceded Romanes and Bateson. Despite not being accorded the privilege of the Darwin mentorship that Romanes had enjoyed, Edmund Catchpool of Sheffield anticipated by two years the physiological selection theory of Romanes (Chap. 5). And it was a Plymouth-based genius – C. R. Crowther – who was able to spell out to Bateson in 1922, with a simple sword-scabbard analogy (Chap. 17), a mechanism for physiological selection that accords well with modern perceptions [5]. In both cases, it would seem to be the chance alignment of compliant editors and compliant reviewers that allowed these "nobodies" to express their opinions in the hallowed pages of *Nature*. The number of "nobodies" that were not so privileged is, of course, unknown (indeed, many, anticipating rejection, would not have submitted in the first place). Even "somebodies" could have a hard time. Bateson and Punnett declined an important Sturtevant paper (Chap. 13) and would not give James Wilson's criticisms "a hearing" in the *Journal of Genetics* (Chap. 14). Pearson kept Mendelism well out of *Biometrika* (Chap. 15).

The importance of private means is demonstrated by the difference between two Canadian ex-patriates, Romanes and Grant Allen [10]. The latter tried to support his scientific enquiries by writing popular fiction. In 1879, the independently wealthy Romanes organized a fund (to which Darwin contributed) to send the exhausted Allen on holiday to the south of France [11]. Without his Fellowship and Stewardship of the College kitchens, Bateson's life could have taken the same course as Allen's – either "a begger's life in Cambridge" or a move "to the metropolis" (Chap. 2).

As frequently noted here, Bateson disagreed on many issues with the Cambridge Professor of Zoology and Comparative Anatomy, Alfred Newton (1829–1907), whose personal wealth was derived from Caribbean slavery. Bateson had to tread carefully because Newton, a leading ornithologist, was of the utmost importance for his personal advancement, appointing him Deputy in 1899. This suggests a possible reason for Bateson, either consciously or unconsciously, to downplay Romanes' work (Chaps. 14 and 17). In a letter to John Gulick (September 2, 1888), Romanes related that Newton "hates me with a deadly hate." It seems that Romanes had questioned Newton's ornithological knowledge on the coloring of cuckoos' eggs (a weighty issue concerning brood parasitism) [12].

Red Herring

Bateson deprecated the slavish following of Darwin by the post-1859 biologists, but he himself slavishly followed Mendel in the post-1900 decades. In the 1920s, he regretted this. Mendelism may have cleared the ground, but it had not led to a solution to the problem that most concerned him – the origin of species. In March 1972, Alan Cock began corresponding with Gregory Bateson who was feeling burned from his interactions with Koestler. Gregory wrote (Sept 1972): "Your letter reassures me that your interest in W.B.'s work is mainly in the 'growth and form' stuff; graft hybrids, etc. He knew by 1925 that what he had been after in the 1890s was much more rewarding than Mendelism and chromosomes." Koestler in *The Midwife Toad* reported that Gregory had related that "by 1924, Bateson had come to realize, and told his son in confidence, 'that it was a mistake to have committed his life to Mendelism, that this was a blind alley which would not throw any light on the differentiation of species, nor on evolution in general'." Gregory later repeated this to Alan (Dec. 1972): "He did tell me once in the last years that 'Mendelism' had been a 'red herring.' And it was. I guess the truth is not that Darwinism distracted zoologists from genetics, but that Mendel distracted Bateson from *biology* (the study of life)."

Alan also corresponded in 1973 (Nov. 14) with Edmund Ford, who described a meeting with Bateson. Ford had met Leonard Darwin accidentally in Piccadilly early in 1922. Darwin said that he had just come from The Atheneum, where he had seen Bateson, but had not spoken to him. Ford remarked that Bateson had recently come back from America and wondered what he now thought of Morgan's work. They decided to return to The Atheneum to ask him about it. The following exchange occurred. Darwin: "So you have seen the *Drosophila* work with Morgan. Do you now believe in the chromosome basis of heredity?" Bateson: "Yes, I do – and all my life's work has gone for nothing." We should recall that Bateson was at that time profoundly depressed following Martin's suicide.

For those who, at this late hour, favor a genic basis for the usual way species initiate branching into new species, Mendelism remains highly pertinent. But those who favor a non-genic basis would agree with Bateson that Mendelism was a distractor. Nevertheless, Mendelism, like the mechanisms of gene expression that so occupied biochemists in the twentieth century, was of such importance that few would now retrospectively council the early geneticists and biochemists to do other than what they did. Had they been more attuned to Bateson and Goldschmidt, they would perhaps have kept the species problem more in view, so that when Erwin Chargaff in the 1950s discovered the "accent" of DNA (see Part VI), there could have been a faster "joining of the dots."

But one must be cautious with such "what if" (counterfactual) assertions. In 1971, biohistorian William Provine declared [13]: "If the Mendelians had worked with, instead of against, the biometricians, the synthesis of Mendelian inheritance and Darwinian selection into a mathematical model, later accomplished by population genetics, might have occurred some fifteen years earlier." Of course, even if the so-called "modern synthesis" had been earlier, it would still have been a wrong synthesis, and any early successes might have preempted and deflected the arguments of Bateson and Goldschmidt that later gained much weight (Part VI).

Lamarck

What may have been Alan Cock's final word on Bateson's abstractions is found in an unpublished article, "Pilgrimages," probably written in the 1980s: "The core of his difficulties over natural selection lay in the belief, which he had held at least since 1900, that interspecific variation was somehow of a qualitatively different nature from intraspecific variation. As he nowhere clearly explains what that difference consisted in, his position here remains rather obscure." This was as far as Alan would go. In the 1920s, Bateson's "base" or "residue" remained as a vague abstraction, so making it easier for the Lamarckist proposals of Lysenko to gain acceptance in Russia [14]. Had Vavilov been able to make a more cogent case, Lysenko's path to power might have been less smooth (Chap. 18). Indeed, many years after Lysenko's downfall, Lamarckism was still deemed attractive in some quarters. Koestler wrote insightfully in *The Midwife Toad*:

> All sorts of corrections and amendments to Darwinism have been proposed by biologists over the years with varying degrees of plausibility, and … the naïve version of Lamarckism current in Darwin's day is not the only alternative. There seems to be every reason to believe that evolution is the combined result of a whole range of causative factors, some unknown, others dimly guessed, yet others so far completely unknown. And I do not think one is justified in excluding the possibility that within that wide range of causative factors a modest niche might be found for a kind of modified 'Mini-Lamarckism' as an explanation for some limited and rare evolutionary phenomena. They must, by necessity, be rare….
>
> What the 'Weismann barrier' … really means is that a … filtering apparatus must protect the hereditary substance against the blooming, buzzing confusion of biochemical incursions, which would otherwise play havoc with the stability and continuity of the species. If every experience of the ancestors left its hereditary imprint on the progeny, the result would be chaos of forms and a bedlam of instincts. But that does not exclude the possibility that ... the 'Weismann barrier' might not be an impenetrable wall, but a very fine-meshed filter, which can only be penetrated under special circumstances.

As is related elsewhere [15], there is now evidence that such special circumstances may sometimes exist.

Papa Knows Best

Someone who can add, subtract, multiply and divide is light-years ahead of someone who cannot perform these operations. But the same may not be true for someone capable of more complex mathematical manipulations compared with someone who can just perform the lesser operations. While Bateson avowed ineptitude, mathematical thinking pervades his work. Geometry – symmetrical, radial, bilateral – is everywhere in *Materials* and in his personal notebooks. Bateson respected the power of mathematics and the skills of the adept, but he did not, like Weldon, hold it in awe. Mathematics was a method – a tool like the microtome and the microscope – not an end-in-itself. Johannsen also knew this, declaring that we must take our genetics *with* mathematics, not *as* mathematics.

Like the use of any tool, a time-cost is incurred in increasing proficiency that must be weighed against the benefits likely to accrue. If Bateson were to err, his instinct was to err on the side of less proficiency. Weldon fell into the same trap as

Romanes' ally John Gulick (Chap. 5), who in the 1890s buried himself in mathematics when he might have been promulgating the physiological selection theory [16]. When Romanes persuaded an Oxford colleague to express physiological selection in mathematical terms [5], Gulick sent from Japan a 12-page critique entitled "Remarks on Mr. Moulton's Reasonings and Calculations" [17].

Weldon's decision to collaborate with a mathematician was wise, and his efforts to increase his personal understanding of mathematics are likely to have enhanced this. More importantly, the collaboration between biologist and mathematician being overt, doubts about their individual weaknesses in the other's discipline were assuaged. When submitting a paper to Bateson for possible publication in the *Journal of Genetics*, Julian Huxley wrote (May 26, 1921): "Haldane has been over the calculations, so they should be alright … and if necessary Haldane would read over a proof, I think." The later Dobzhansky–Wright collaboration also demonstrated this synergism. Dobzhansky wrote [18]:

> Wright is very hard to read. He has a lot of abstruse, in fact almost esoteric mathematics … of a kind which I certainly do not claim to understand. I am not a mathematician at all. My way of reading Sewall Wright's papers, which I still think is perfectly defensible, is to examine the biological assumptions which the man is making and to read the conclusions which he arrives at, and hope to goodness that what comes in between is correct. 'Papa knows best' is a reasonable assumption, because if the mathematics were incorrect, some mathematician would have found out.

Judging from a later biography of Wright, "some mathematician" did not always find out, and a generation of biologists espoused Wright's "shifting balance" ideas believing them to have been underwritten by Dobzhansky [19]. Bateson expressed his high hopes for mathematical contributions to biology after evaluating a paper by Yule (Chap. 18). Bateson wrote to his co-reviewer Hardy (May 20, 1924):

> I saw your report on Yule. Your representation of my position as regards the introduction of Mathematics into Biology was, and I suppose meant to be, slightly burlesque. I am always hoping against hope that mathematics will be introduced into biology in my time, but at the right place. Such a paper as Yule's has about as good a claim to call itself science as the pictures of Van Gogh have to call themselves art.

Hardy replied (May 24, 1924): "The work [of Yule] was, on one side, quite competent, and on the other (so far as I could judge) not palpably absurd. That being so, I felt he ought to be allowed his say, and if at all, then at length, half measures being obviously futile. … How does J. Haldane's work strike you: it looks to me quite good?" Without commenting on Haldane, Bateson replied (May 28):

> I did not make clear my point about Yule. In my judgment the paper is preposterous. But I knew that others entitled to an opinion might not agree. I therefore suggested to the Committee that you should be consulted, and I am quite willing that your opinion should be adopted. I am not for burking. In what I said in my letter to you, I was not insisting on my estimate of Yule's paper, but repudiating the charge you made against me – that I resented mathematics in application to Biology. We have had some absurd attempts – mostly from biometricians – to apply mathematics to biology, but as I said my hope is still that I may live

> to see mathematics applied to biology properly. The most promising place for a beginning, I believe, is the mechanism of pattern.

Bateson wanted mathematicians to be, as they say, on tap but not on top. However, in the decades ahead, as biologists and biochemists became increasingly absorbed in the details of their disciplines, a new generation of biometricians – transformed into "biostatisticians" or "mathematical geneticists" – seized the genetical high ground. The most prominent of the born-again biometricians, Fisher, not appreciating that Mendel followed the statistics of his time, was even so bold as to throw Mendel's integrity into question [20–22]. We can here note a statement in an early statistics textbook, which Mendel may well have read [23]:

> Even the shrewdest observer with the best instruments never gets results which are totally devoid of error, and in order to come as close as possible to the truth, the only thing one can do is to repeat the operation as many times as necessary and to select from all results those which have the smallest error.

Clearly the reproducibility of Mendel's results has stood the test of time. But as the decades passed, the genetics journals of the twentieth century became increasingly mathematical while declaring the problem that most concerned Bateson – the origin of species – to be their own. Nevertheless, as Bateson had hoped, eventually mathematicians of high caliber (e.g., Alan Turing) did begin to address mechanisms of pattern formation in biological systems [24].

An Instrument to Know

A scientific discipline is like a biological species in that movement beyond a position of stability (e.g., Galton's rough stone) requires that a discontinuity be bridged in order that there be further progress. Sometimes bridging requires new ideas. But sometimes it is technology that is rate-limiting. Scientists in numerous disciplines today acknowledge their profound debt to the Biometricians who fashioned statistical methods – a new technology – that allowed them to determine more precisely whether their observations were significant and not the result of chance. There are also profound debts to those who engineered powerful microscopes and developed methods of sectioning and staining tissues. Lankester, the editor of the *Quarterly Journal of Microscopy*, greatly valued technical advances in this field (Chap. 4), and automatic microtomes facilitated Bateson's early acorn worm studies (Chap. 2). However, there is the danger that the "high priests" with their tablets of technological wisdom become so obsessed with the technology that they lose sight of the underlying goals. For example, in the 1990s, there were abundant funds for those who wanted to sequence the DNA of species of medical and/or economic importance to man, but much less for those who wished to better interpret the sequence information *already* available.

Bateson did not fall into this trap. While exploiting the new Mendelian technology (i.e., cross-breeding), the "species problem" remained ever in his thoughts. He was an open-air person who saw the seductions of technology as characteristic of the laboratory rather than of the "seed bed and the poultry yard." Writing to Anna from the Steppes (Chap. 2), he doubted "the value of oracular inscriptions on recording drums." His taste for cytology was not enhanced by the awarding of the Balfour Scholarship to Caldwell, a co-inventor of an automatic section cutter. However, from the outset, a microscope was essential to the Mendelians, if only to examine pollen grains. Bateson's insecurity with microscopes is perhaps revealed by the withdrawal of his paper on the rotation of flagellae (Chap. 16). In his 1924 address at Birkbeck College (Chap. 18), Bateson compared the "instrument makers" who developed the microscope to the astrologer Sidrophel, described by the original Samuel Butler (1612–1680) in *Hudibras*:

> I like at least to think that the [important] questions were asked before the instrument-makers came on the field. Sage Sidrophel, who 'made an Instrument to Know, If the moon shine full or no,' has been a great begetter of modern researches, but less fortuitous discoveries are the more honourable and command a warmer admiration.

The phenomenon of coupling revealed itself to the Bateson school as a fortuitous consequence of the application of the new Mendelian technology (Chaps. 8 and 9). But there were no accolades for this. These came later when the phenomenon was *explained* in terms of linkage (Chap. 13). In other words, the discovery itself counted less than its interpretation. Theory superceded fact. This can be contrasted with the modern discovery of introns ("split genes") as a fortuitous consequence of the application of the new sequencing technology in the 1970s. A Nobel Prize followed shortly thereafter, even though the phenomenon was unexplained. Discovery counted more than interpretation. Fact superceded theory.

Bulldog

In 1928, Hurst wrote to Beatrice (Jan. 20): "Few people today realize the personal persecution your husband and I had to endure in the early days of Mendelism. Curiously enough Punnett, Biffen, Doncaster, Miss Saunders, and others escaped all this in some mysterious way, and your husband and I bore the brunt of it all in England. The rest of the world welcomed our work with open arms." Indeed, most of those who survived continued to escape persecution. Saunders (1865–1945) remained active in college affairs and, in addition to many scientific papers, produced a two-volume work on *Floral Morphology* (1937, 1939). She was President of the Genetical Society (1936–1938) and was killed in a bicycle accident in her eightieth year.

Albeit much milder than in Bateson's case (Chap. 24), the persecution of Hurst continued, partly at the hands of Punnett, who was not sympathetic to cytology. Hurst elected not to work with him when beginning cytological studies on roses at

Cambridge. While nominally working with the botanists, it was the Zoology Department that eventually accorded him adequate research space (Chap. 18). In 1950, Punnett reminisced [9]: "Now Hurst was a tireless worker and full of ideas, but over-apt to find the 3:1 ratio in everything he touched. While valuing Hurst's enthusiasm for the cause, Bateson was nevertheless mistrustful of his slickness, for he knew that his critical ability had never been sharpened by passing through the scientific mill." This remark does not accord with what we have seen of Hurst here. We know that Hurst self-funded much of his research. With his knowledge of Mendelism, he could have greatly expanded his personal fortune. Instead, he pursued basic research and left himself vulnerable to the 1929 stock-market crash that severely damaged his finances. It is true there is no evidence that he understood Bateson's deeper messages, but then, neither did Punnett.

Hurst's collected papers were published by Cambridge University Press in 1925, and *The Mechanism of Creative Evolution* followed in 1932, the year of the Sixth International Congress of Genetics (Ithaca, USA), which he attended. He was less cautious than Bateson regarding eugenics and wrote on the inheritance of human intelligence [25]. His advice continued to be sought by individuals and organizations such as the Wool Breeding Council. An ambitious proposal for an Institute of Human Genetics failed to get support. At the beginning of the Second World War, he joined the Royal Observer Corps, but ill-health led to his retirement. A Royal Pension was awarded on the instigation of Daniel Hall, who in 1927 had assumed the Directorship of the John Innes Institute. Hurst died in 1947.

During the Second World War, Rona Hurst joined the staff of Christ Hospital School where she taught for 18 years. She wrote an elementary introduction to Genetics [26] in which she acknowledged help from Darlington. During the 1960s, she wrote *The Evolution of Genetics* but could not find a compliant publisher. It was updated in 1974 with material from letters that Alan Cock had found in the Coleman microfilm (see Prologue). Darlington wrote to Alan (Oct. 10, 1973): "I think Mrs. Hurst's book is one I have warned publishers not to take: the widow of a long dead charlatan is something of a hazard." Rona got her own back in a letter to Alan (Feb. 10, 1980): "Darlington is an old sinner and I think you were wise to ignore him – my husband always did, he wastes too much time on his opponents … and Julian Huxley was another. So far as I know neither have made any or very few *real* genetical experiments, and yet they have the cheek to question those who have!" She went on to relate the trouble she had had when she was assigned the role of "collaborator" with Darlington, who was in charge of the National Rose Collection project at Bayfordbury, and "didn't believe in any of our rose work."

Right-Hand Man

Punnett was Bateson's "right-hand man," an expression denoting an indispensable chief assistant. In Punnett, he chose conservatively someone with training like his own. At that time, there was need for exhaustive application of a method – cross-breeding – to a wide range of organisms. There was a need for more of the same. So he sought someone who would share his experimental approach rather than someone who would bring in a complementary approach. Their relationship was not organic, in the sense that right and left hands complement. There was no one, such as Doncaster or Lock, to provide a left-handedness sufficient to complement Bateson's theoretical insights, so giving a whole greater than the sum of its parts (Fig. 25.1).

Bateson's craving for such input was partly reflected in his sudden urges to visit continental colleagues such as Cuénot and Baur. He usually returned disappointed. On the other hand, the Mendelian "game plan" had long been evident, even before the rediscovery. What was needed was muscle to get the work done in an era when there was meager funding for "megaprojects" (c.f. the sequencing of the human genome). The muscle came by exploiting a team of underprivileged female coworkers, and from his right-hand man and woman – Punnett and Beatrice.

Punnett's first major public award was the Darwin Medal in 1922. Among his later works were *Heredity in Poultry* and the *Scientific Papers of William Bateson*. Long the Secretary of the Genetical Society, in 1930 he became its President. Throughout his life he stuck with the classical approaches he had learned with Bateson. In 1932, in a paper on linkage in the sweet pea, he showed that 18 recessive mutations fell into seven linkage groups which corresponded with the haploid chromosome number. In 1940 at age 65, he ceded his Chair to Fisher. The Editorship of the *Journal of Genetics* was passed to Haldane in 1945. These two moves symbolized the growing dominance of the biometric approach in Genetics. Punnett continued the experimental breeding of poultry at his home in Somerset until 1955, and his last poultry paper appeared in 1958. Continuing his life-long interest in sport, his last papers were in *Bridge Magazine*. Eveline Punnett died in 1965, and Punnett died two years later [27].

Fig. 25.1 Reginald Punnett

Beatrice

Bateson was conservative in his choice of staff. Spousal relationships can be viewed similarly. Some seem to thrive with a right-hand spouse (e.g., Bateson, and Hurst in the case of his second wife), some with a left-hand spouse (e.g., Punnett, and Hurst in the case of his first wife). Some seem to need no spouse at all (e.g., Butler, although there was Miss Savage and the enigmatic Madame Dumas; [28]). Had there been an early romantic attachment between Bateson and Constance Black (Chap. 2), she could only have been a left-hand in his life and he in hers. She herself needed a right-hand man and seems to have found this in Edward Garnett, the son of Richard Garnett and father of David Garnett (Chap. 9).

Among Bateson's entire writings, both published and unpublished, there is no acknowledgment of Beatrice's help. From what we know of the pair, we can assume, quite simply, that nothing needed to be formally said. On leaving the Manor House at Merton, Beatrice moved to Kensington where, as has been related here (Prologue), she devoted herself to the collection of Bateson's letters and their publication together with a selection of his writings. In 1920, Bateson had collected together ten of his "lay" articles that were "more or less lawfully begotten by Mendelism out of Common Sense, *me obstetricante*." These were classed as either "digestible", or "for the eupeptic only," or "indigestible." The two in the latter category were "Gamete and Zygote" (Chap. 16) and his 1904 Address to the B.A. Zoological Section (Chap. 9). As for a title, he related (*Memoir*) (Fig. 25.2):

> A Scotch soldier when I was lecturing in YMCA huts said 'Sir, what ye're telling us is nothing but Scientific Calvinism.' Sometimes I think that would serve, but the scientific world to whom I suppose I appeal, resents puzzling and recondite allusions, so perhaps it will have to be 'Addresses and Essays 1904–1921,' or some such words intelligible to children of 6.

But a publisher could not be found. In 1921, Macmillan, after consultation with the Editor of *Nature*, said it was looking for a more comprehensive work. In 1924, Chapman and Hall were more positive, but eventually called for a "remodeling … into a more homogeneous whole" and a title including the word "eugenics." It devolved upon Beatrice finally to publish the articles in 1928 as a supplement to her *Memoir*. As far as we know, she simply held on to Bateson's letters and papers and made no attempt to place them in a formal archive. According to her granddaughter [29], Beatrice is likely to have maintained an interest in the scientific literature, but whether any of the post-1926 Bateson bashing came to her attention is not known. She died in 1941 (Fig. 25.3).

Anna

Anna Bateson's nursery at Bashley thrived and she was active in local affairs. In 1913, she walked from Bournemouth to London on the "Suffrage Pilgrimage" organized by the National Union of Women's Suffrage Societies. During the war, she

Fig. 25.2 Beatrice and William

served upon the Military Service Tribunal, which had to decide who was indispensable for home food production. She died (May 17, 1928) while being driven in a car by Ned. When Alan Cock wrote an article on her in *Hampshire* magazine (1979), it provoked the following poem from Burnal Lane, who was probably the son of Charles Lane, her Foreman, who continued the business after her death:

> You always liked my sonnets, so here's one,
> It could well be the last that I shall write.
> I have lived seventy years beneath the sun,
> And thought I'd like to leave your name all bright.
> You left the world in nineteen twenty-eight,
> When the last pheasant eye were coming out,
> The white, strong scented one, that still blooms late,
> When dewy, yellow cowslips are about.
> Miss Anna Bateson, I owe you such a lot,
> For things you taught me a long time ago,
> About antirrhinums and forgetmenots,
> And French and poems you thought that I should know;
> I like this sonnet I have made today,
> And in it, I think, you will always stay.

Ned was Editor for the Northumberland volumes of the *Victorian County History*, and then entered the Colonial Civil Service becoming a District Judge in the Sudan. In his dotage, he lived with the descendants of one of his uncles, and, shortly before

Fig. 25.3 Anna Bateson

his death in 1955, helped their son, Patrick Bateson, get to Cambridge. Margaret left Cambridge for London in 1886 and established and edited the Women's employment section of *Queen* magazine from 1889 until 1914. Beneath a veneer of Victorian respectability, *Queen* did much to advance women's causes. Margaret returned to Cambridge in 1901 when she married Heitland. He was sympathetic to her cause, which was being opposed by the Women's National Anti-Suffrage League [30]. She died in 1938.

On Lips of Living Man

Butler waited in the wings for the Darwinians to awaken and drag him on stage. That moment never came. Did Charles Darwin, consciously or unconsciously, manipulate Romanes and Krause? Were they Darwin's catspaws? How candid was Darwin? These questions remain unanswered. In his autobiography, Darwin gave the following description of Thomas Huxley, the main opponent of "non-fecit saltum" (i.e., the main proponent of discontinuous, *non-gradual*, evolution): "He has been the mainstay in England of the principle of the *gradual* evolution of organic beings" (my italics). Should we dismiss this as the ramblings of an old man? We might also consider the words Darwin interpolated to his autobiography around 1881, when he was still smarting from the Butler episode:

> Whenever I have found out that I have blundered, or that my work has been imperfect, and when I have been contemptuously criticized, and even when I have been overpraised, so that I have felt mortified, it has been my greatest comfort to say hundreds of times to myself that 'I have worked as hard and as well as I could, and no man can do more than this'.

The dispute with Butler was not resolved at the time of Butler's death (Chap. 19). The following poem found among Butler's papers affirms that he shared with Darwin no romantic intimations concerning an afterlife [31]:

Not on sad Stygian shore, nor in clear sheen
Of far Elysian plane, shall we meet those
Among the dead whose pupils we have been,
Nor those great shades whom we had held as foes.
No meadow of asphodel our feet shall tread,
Nor shall we look each other in the face
To love or have each other being dead,
Hoping some praise, or fearing some disgrace.
We shall not argue saying "Twas this or thus,"
Our argument's whole drift we shall forget;
Who's right, who's wrong, 'twill be all one to us!
We shall not even know that we have met.
Yet we shall meet, and part, and meet again
Where dead men meet, on lips of living man.

Yet, in the mid-1880s, there had been a potential turning point. Butler's work was sympathetically reviewed in *The Atheneum* (Chap. 19). Butler wrote to Edwin Clodd (March 26, 1884): "The *Atheneum* has been a great lift to me and given me much encouragement; really I was beginning to think I had no chance, no matter what I did. Even more encouraging than the *Atheneum* itself is the fact that Romanes & Co. are taking the line which I have insisted upon, in company with others, for so long – for after all it is the theory not the person which is the thing to be thought of." With this encouragement, and with improved finances following the death of his father in 1886, Butler was poised for further work, but the Darwinians gave no quarter, and Butler's final evolution lectures in the early 1890s were mere summaries of his previous work and the ensuing disputes [32]. Butler opted to follow his other interests.

As suggested by the quotation on "greatness" from *The Way of All Flesh* (Chap. 20), Butler was quite philosophical about all this. The world could go its way, and he would go his. In his *Notebooks*, he remarked: "It is said the world knows … little of its greatest men; it might be added that its greatest men have known very little of the world. Indeed, they never can, for they and the world have nothing, or at any rate very little, in common, and cannot understand each other." Voltaire had spoken similarly in 1770 [33]: "The number of wise men will always be small. … God is always for the big battalions. It is necessary that honest people quietly stick together. There is no way their little force can attack the host of the closed-minded who occupy the high ground."

In short, since great men cannot manage the world, they must manage without the world. The counter-question – not posed by Butler or Voltaire – is whether the world can manage without its great men/women? Surely the world cannot afford to let the great languish in the shade? Surely, we need them more than they need us? Sadly, the quality of greatness can exclude various "street smart" qualities essential for survival. As Shaw pointed out (Chap. 19), too often it is merit at self-marketing, rather than true merit, that is decisive. And if capital is not inherited, it must be acquired elsewhere. Shaw privately lamented that that he would have liked to have married Constance Black but could not afford it. Had Miss Savage been a Miss Powell (Mrs. Herringham), these pages might have relayed a very different story. As it was, Butler spent much of the 1870s – his potentially most productive years – wearily attempting to salvage investments.

Philosophy

Gregory related [34] that his father would say: "If you want to put salt on a bird's tail, you will be advised not to look at the bird while you approach it. He was always trying to put salt on the tail of nature and particularly to catch that component of nature which we might as well call Mind." In his writings and addresses, Bateson was reticent concerning his deepest philosophical beliefs. In his lectures, he was more forthcoming, describing "absolute truth" as an ultimate goal of scientific enquiry (Chap. 5). From his first letter to Beatrice (Chap. 2) and from her *Memoir*, we are assured that he was a life-long atheist.

In 1924, the biomedical philosopher R. E. Lloyd sent Bateson a complimentary copy of his book *Life and Word: an Essay in Psychology*. Bateson had "not the faintest idea what it is all about. … I can only come to the conclusion, which has already been impressed on me by others, that there is no ingredient of noumenon in my composition; plenty of *phenomenon*, but no *noumenon*. James Ward used to get quite cross with me about that." If we take *noumenon* to mean an "element of thought," this suggests that Bateson was modestly implying that his work had provided observations but little insight into any deep meanings of those observations. This is true to the extent that he believed that, in probing the unknown, we do best to base ourselves on what is securely known and to frankly state the limits of that knowledge.

For example, sufficient was then known about pigments and enzymes ("ferments") that it was not fanciful to imagine how a change in a gene corresponding to an enzyme might cause a change in flower color. But regarding biological patterns – the sudden appearance of an extra digit or of a hand where each finger possessed only two phalanges (like a thumb) – Bateson was lost (Chap. 11): "I cannot at all readily conceive how any ferment or other transmissible substance can be supposed to be responsible for such a variation." To understand this, "we shall have to make an extreme demand upon the specific powers of chemical substances." If true wisdom is knowing what one does not know, then Bateson was wise. He stated where knowledge was missing. He did not dismiss the unknown with some word – like "gravity" – without pondering what that word implied apart from its semantic derivation. As to the phenomenon of extra digits, the first item on the agenda was to observe how frequently it occurred and to coin new words where necessary. Thus, in 1894, he gave us "homeosis." When twentieth-century scientists found the master genes regulating development, they named them "homeotic" genes.

Philosophers tend starkly to classify modes of thought as either materialistic or non-materialistic. The latter can easily transform into "spiritualistic." Coleman assigned Bateson to the non-materialistic category while noting that Brooks in his *Law of Heredity* had urged students to study the materialistic interpretations of Mivart (a Catholic):

> Brooks by no means welcomed all aspects of Mivart's doctrine. He did, however, agree that, according to his own conception, 'a new variation is caused in essentially the same manner which Mivart suggests as probable,' that is, the cause of variation is an 'upset of the previous rhythm of the physiological units of the living organism.' Mivart referred these events to the 'material organic world,' and Brooks to the 'gemmules' produced by the disturbed physiological units; both therefore included matter along with rhythmic forces in explanation of heredity and variation.

Coleman pointed to Butler and Ward as supporting "the memory [i.e. informational] theory of inheritance." But then, implying that memory can exist without a material basis, added [34]: "Their doing so is but further evidence of the then current widespread reaction against doctrinaire materialism."

Time

In his 64 years, Bateson achieved a great deal. In some respects, he managed his time well, not getting too involved in the politics of the day and, as Punnett tells us, reading a newspaper mainly for the information on art auctions. Radio-communications were not then developed, and people looked to those around them for entertainment. Bateson entertained and was entertained. His outspokenness created enemies, but also friends. It was probably as much his inter-personal skills as the excellence of his research that led to funding at a critical stage (Chap. 8). He traveled, often alone, through much of England and the Continent gathering information from all and sundry, which was relayed each evening by mail back to his base – his family. Yet, to many today, Bateson would be seen as amateur, not professional. His account of the long railway trip east from Toronto to Ithaca with Bridges and his like (Chap. 17) brings this out quite well. Perhaps Bateson was one of the last great amateurs in his field? Or, maybe more accurately, one of the last great generalists in a time of increasing specialization?

When Bateson was introduced to Gilman aboard R.M.S. Majestic bound for New York in 1907 (Chap. 9), would it have occurred to him that the highly professional Carnegie employee might be a plant, with the sole purpose of cajoling him to submit a proposal that would allow better assessment of the funding needs of the "local boy" – Davenport. The fact that this might waste a considerable amount of Bateson's time would not have entered into Gilman's calculations. For the writer of a "begging letter" it is a win-lose situation. For the potential donor, however, it can be a win-win situation – he gets a free basket of ideas and if the application is declined, he gets to keep his money. One potential donor, the Duke of Bedford, was able, at no cost to himself, to get the finest biological mind in the land to labor long hours over a text specifying how the operations at Malvern Abbey might be reorganized (Chap. 7). The cost to the applicant and, indeed, to the world, of this misuse of Bateson's time, would not have entered into the Duke's calculation.

Much time can be lost through ill-health. From a physical point of view, Bateson's health appeared robust until he pleaded exhaustion in 1908 (Chap. 11). His sudden urges to go traveling alone to distant parts might be partly reflective of a tendency to "black moods." Sometimes, as in the periods after the deaths of John and Martin, there were objective bases for these. A friend who would seem to have had the power to help Bateson more than he did was also subject to depression; Francis Darwin and his wife Ellen Crofts, were both depressives, a condition their daughter Frances inherited. She married Jane Harrison's star pupil, Francis M. Cornford, in 1909 [35].

Bateson's dispute with Weldon tore both apart. To simplify a highly complex situation, the Darwinians (e.g., Lankester, Thiselton-Dyer) can be seen as putting all their eggs in the basket of the independently wealthy Weldon. Against the Darwinian juggernaut, those such as Catchpool and Crowther were in no position to contend. But Bateson, with a toehold in the College kitchens, could. Quite naturally, Weldon saw the establishment's endorsement of himself as proof that he was on the right track. But by the dawn of the twentieth century, he was becoming increasingly bewildered. Overtaken by events he could not comprehend, and desperately attempting to sustain his position, he died exhausted in 1906 at the age of 46. Bateson had his first heart attack in 1913 at the age of 52.

Was all this inevitable? It is possible that some diffidence on the part of those who supported Weldon's "main-stream" thinking, and more sympathy toward Bateson's, would have, as they say, leveled the playing field, and fostered collaboration instead of confrontation. They began as friends, and they tried to maintain their friendship. They would still have had to write "begging letters," but the pressures would have been less intense. Perhaps it was "the system" that drove them apart and to their graves. In some respects, the system is unchanged today. The "Mrs. Herringhams," "Punnetts," "Hursts," "Saunders," "Beatrices," "Miss Savages," and "Festing-Joneses" still give their love and support, while the "Weldons" and "Pearsons," with noble intentions, end up making much mischief while leading eager young "Darbishires" down sterile avenues.

Where Bateson Dropped It

A recurring theme in Bateson's addresses is the arrest of breeding studies following the appearance of Darwin's *Origin of* Species. So, he concluded, "we must go back and take up the thread of the inquiry exactly where Mendel dropped it" (Chap. 8). Many of Bateson's addresses "fell dead in their hour of birth" (*Memoir*). These words might well have been those of Mendel in, say 1880, when retrospectively considering the reception of his work. The works of the other "Mendels," like Mendel's own, can now be seen as far from "merely historical." Among these were Bateson's contributions to the species problem – encapsulated in his 1909 Centenary essay (see Part VI; Chap. 29).

During Bateson's early career, Darwinism was the great distractor. Later it was the simplicity of the Mendelian paradigm itself. Today it is the genic paradigm. Bateson foresaw, it now appears correctly, a twentieth century with biochemistry dominant (Chap. 14): "Our knowledge of the nature of unorganized matter must first be increased. For a long time yet we may have to halt." So (Chap. 18): "the development of evolutionary theory has been tacitly suspended or postponed, and activity is concentrated on the exploration of genetical physiology, the theoretical evaluation of the knowledge thus gained being relegated to the future." But new developments in bioinformatics (EB; see Part VI) suggest that we can now, a century later, "go back" and "take up the thread of the enquiry," just where Bateson dropped it [15].

As one of the nine "Mendels," Bateson is an "exception" to be treasured for his contributions to the science of evolution. He also gave sage advice, at the level of both individual and society, on the art of making the best of the few precious years each human is allotted. Although in his address to the Salt Schools (1915) he mused that "we may yet find even the elixir of life," he had no illusions that this would be in his own time, and he lived deeply. There is much more to treasure in Bateson than his science. When reviewing an early draft of this book in 1974, Gregory Bateson complained that "what did not come through" was his father's "intense sense of fun" [36]. Other than this, Bateson should be treasured for being, as Beatrice remarked and as these pages have so often shown, "a rare personality."

Summary

The phenomenon of a scientist being "before his/her time" is evident not only from the German texts of Mendel, Hering, and de Vries but also from the English texts of Butler, Galton, Romanes, and Bateson. Unlike Butler, the latter three were center stage. However, their words were not read or, if read, not understood, either because of the enchantment of classical Darwinism (Galton, Romanes) or of both Darwinism and Mendelism (Bateson). While 35 years elapsed before the rediscovery of Mendel's ideas, those of de Vries' *Intracellular Pangenesis* emerged in piecemeal fashion at intervals throughout the twentieth century and were not attributed to him.

The rediscovery of Galton by Romanes and Bateson in the 1880s was not itself recognized until they themselves were rediscovered after a century of neglect. Bateson can now be seen as a colossus astride the biological sciences of the last decade of the nineteenth century and the early decades of the twentieth. A truly "modern synthesis" has been completed by the rediscovery of the information concepts advanced by Hering and Butler in the 1870s. Must the advancement of knowledge be so disorderly? With hindsight, the "rediscoveries" might have been earlier "discoveries" if their proposers had been treasured for their exceptional qualities (their natures and nurtures) by contemporaries who appreciated that scientific ideas that initially appear to glitter may not, in the course of time, prove to be gold – and vice versa. Today's fanciful may be tomorrow's certitude. Look not so much at what is being proposed, but who is proposing it, being ever mindful of the possibility of the, perhaps not so rare, "exceptions" (Catchpool, Crowther). As Wall Street knows so well, be guided by track record and hedge your bets.

Part VI
Further Rediscovery

Chapter 26
The Third Base

If I thought that by learning more and more I should ever arrive at the knowledge of absolute truth, I would leave off studying. But I believe I am pretty safe.

Samuel Butler, Notebooks

Darwin's mentor, the geologist Charles Lyell, and Darwin himself, both considered the relationship between the evolution of biological species and the evolution of languages [1]. But neither took the subject to the deep informational level of Butler and Hering. In the twentieth century, the emergence of a new science – Evolutionary Bioinformatics (EB) – was heralded by two discoveries. First, that DNA – a linear polymer of four base units – was the chromosomal component conveying hereditary information. It was our "genome." Second, that much of this information was "a phenomenon of arrangement" – determined by the sequence of these four base "letters." The first edition of this book concluded with a brief update on the new work as it appeared to relate to William Bateson's evolutionary ideas. However, imbued with true Batesonian caution ("I will believe when I must"), it was relegated to an Appendix to indicate its provisional nature. In the interim, evidence supporting a "rediscovery" of Bateson has grown to an extent that justifies replacing the Appendix with this new section (Part VI). Indeed, for those less interested in historical details, Part VI sufficiently repeats Parts I–V to permit independent reading.

There is a modern rendition of Mendelian basics in Chap. 27 that should assist the non-specialist reader who, nevertheless, can be assured of remaining "pretty safe." This is followed (Chap. 28) by further consideration of Bateson's important 1909 Centenary essay, which has often been incorrectly interpreted as demonstrating his support for genic (rather than chromosomal) hypotheses for the initiation of speciation processes. We conclude with the identification of Bateson's "residue" with chromosomal patterns of oligonucleotides (k-mers) that finely distinguish species (Chap. 29). The present chapter provides a general overview.

A. G. Cock, D. R. Forsdyke, *Treasure Your Exceptions*,
https://doi.org/10.1007/978-3-030-92099-9_26

Differences in Accent

Modern languages have similarities that indicate branching evolution from common ancestral languages [2]. We recognize early variations within a language as dialects or accents. When accents are incompatible, communication is impaired. As accents get more disparate, mutual comprehension decreases, and at some point, when comprehension is largely lost, we declare that there are two languages where there was initially one. Thus, the origin of language begins with differences in accent. If we understand how differences in accent arise, then we may come to understand something fundamental about the origin of language (and hence of a text written in that language). It was for this reason that an abstraction was introduced in Chap. 5 – the notion that hereditary information could include an "accent" that might be the key to understanding the origin of species. A distinction was made between a message itself (primary information) and the accent with which it was "spoken" (secondary information). In other words, multiple forms of information (e.g., message and accent) can share a common text, and we can consider them in hierarchical fashion (primary, secondary, etc.).

Typology Versus Synthetic Orthodoxy

In 1979, in *Mind and Nature*, Gregory Bateson referred to the hierarchical levels of hereditary information postulated by the anti-Darwinian "typologists," which contrasted with the "synthetic" views of those favoring the "modern synthesis" of Darwinian and Mendelian ideas [3]:

> It is interesting to note the current controversy between the upholders of 'synthetic' theory in evolution (the current Darwinian orthodoxy) and their enemies, the 'typologists.' Ernst Mayr, for example, makes fun of the blindness of the typologists: 'History shows that the typologist does not and cannot have any appreciation of natural selection' [4]. ... Do any of the genetic messages and static signs that determine the phenotype have the sort of syntax ... which would divide the 'typological' from 'synthetic' thinking? Can we recognize, among the very messages which create and shape the animal form, some messages more typological and some more synthetic?

Here Gregory Bateson was thinking of two physical classes of message each conveying a distinct kind of information, rather than of *one* physical class of message able to convey two kinds of information. However, he acknowledged possible conflicts between different kinds (layers) of information within one physical message: "Every evolutionary step is an addition of information to an already existing system. Because this is so, the combinations, harmonies, and discords between successive pieces and layers of information will present many problems of survival and determine many directions of change." One such direction of change could lead to establishment of a new species.

Coding Degeneracy Gives Role Flexibility

Since hereditary information is transmitted in the form of DNA, our task is twofold: First, to determine what is the "accent" of DNA. Second, to determine how changes in that accent can bring about an incompatibility between members of a species (discord), such that they do not effectively reproduce with each other. Being reproductively isolated they would be, by definition, independent species even if, in the extreme case, they showed no anatomical or physiological differences.

Such differences (changes in the conventional phenotype) are largely determined by proteins. These are linear polymers of basic units – the amino acids. The sequence of amino acids distinguishes one protein from another and determines how each folds into a unique structure. Thus, its sequence determines its properties, which may be structural (e.g., tendons) or catalytic (e.g., bringing about chemical changes so that muscles contract). What determines the sequence of amino acids determines the nature of the resulting protein, and hence determines the organism.

We can equate phenotype with protein, and genotype with DNA. The linear sequence of the bases in an organism's DNA codes for the linear sequences of amino acids in its proteins. If we know its DNA sequence, we know which proteins the organism can make. If we fully understand the properties of those proteins, we can predict the organism's anatomical and physiological characteristics. In short, DNA (the genotype) *determines* protein (the phenotype). However, protein-encoding genes only occupy part of the genome, whereas its "accent" is genome-wide. How do genes balance the evolutionary demands of protein-encoding with the demands of the genome-wide accent? How do genes play both roles?

The code which relates a particular base sequence to a particular amino acid is a triplet code. Each amino acid corresponds to a three-base sequence (codon) in DNA. For example, the amino acid glycine is encoded by GGT, and the amino acid threonine is encoded by ACT. So, a protein sequence that happened to consist just of these two amino acids – say glycine, alanine, glycine, glycine, alanine, glycine – could be encoded by the sequence:

$$\mathrm{GGTACTGGTGGTACTGGT{-}{-}{-}{-}} \tag{26.1}$$

It turns out that the code is *degenerate* so that each amino acid has four possible codons. There is coding flexibility. Glycine can be encoded by GGT, GGC, GGA, or GGG. Threonine can be encoded by ACT, ACC, ACA, or ACG. As far as an amino acid is concerned, what matters is the first *two* bases of its codon. The third base indicates the *accent* of DNA. Thus, two organisms may make the above amino acid sequence, but their corresponding DNA sequences may have different accents – for example, instead of the above sequence, one might have the sequence:

$$\mathrm{GG\underline{C}AC\underline{C}GG\underline{C}GG\underline{C}AC\underline{C}GG\underline{C}{-}{-}{-}{-}} \tag{26.2}$$

This also encodes the sequence glycine, alanine, glycine, glycine, alanine, glycine. In this case, in every codon, the third base is C (underlined). But *any* of the four bases in the third codon position will suffice as far as the "primary information" – i.e., for the amino acid sequence – is concerned. So, organisms can *vary* in their accent (secondary information) while maintaining the role of their primary information (and hence the same phenotype).

The first hint at the existence of secondary information in DNA arose in the 1940s when Erwin Chargaff observed that species differed in the relative proportions of G + C and A + T, referred to as GC% [5]. Thus, sequence 1 above could correspond to a low GC% species (third codon bases enriched in Ts). Sequence 2 could correspond to a high GC% species (third codon bases enriched in Cs).

Pairing of Chromosomes

Although somewhat simplified, the above account is uncontentious and widely agreed upon. More contentious is whether, in the *general* case, it is this difference in accent that can *initiate* reproductive isolation between two groups living as one species in the same territory. Furthermore, even if this accent difference did bring about reproductive isolation, what would be the mechanism? William Bateson, under some prompting from Crowther (Chap. 17), agreed that a most fundamental form of reproductive isolation, manifest at an early stage, would be the hybrid sterility seen when recently diverged (allied) species are crossed. Sterility was due to an incompatibility that could be seen cytologically as a failure of paternal and maternal chromosomes to pair correctly during meiosis (see Fig. 9.5 of Chap. 9). Thus, if we can understand what makes chromosomes incompatible, we can understand hybrid sterility. And if we understand hybrid sterility, we can understand the origin of species.

We seek to understand how chromosomes that are homologous (i.e., are alike) can pair with each other. Do they pair by virtue of this likeness (like-with-like) or by virtue of some key-in-lock (sword-in-scabbard) complementarity, which implies that they are not really alike? One must be the sword, the other the scabbard. This paradox seemed to be resolved when it was appreciated that chromosomal DNA is a duplex consisting of two complementary strands – a "Watson" strand and a "Crick" strand – that pair with each other because of their complementarity. So, in Crowther's terminology, potentially the sword of one chromosome could pair with the scabbard of the other (and vice versa). For this, swords must be unsheathed from their own scabbards and then each inserted into the scabbard of the other. Thus, the Watson strand of one chromosome would pair with the Crick strand of its homologous chromosome (and vice versa). This would require that the Watson strand be displaced from pairing with the Crick strand of its own chromosome. Likewise, the Crick strand of the homologous chromosome would be displaced from pairing with the Watson stand of its own chromosome. Then the cross-pairing could occur.

It was Francis Crick, a codiscoverer of the structure of DNA, who worked it out. Most pictures of DNA molecules show them as double helices with two strands of DNA containing *inward-looking* bases, which pair with each other, thus joining the two strands to form a duplex. Crick proposed that under certain circumstances, parts of the strands would separate (unpair) from each other and become *outward-looking*. In this circumstance, the outward-looking bases in the DNA of a maternal chromosome could pair with complementary outward-looking bases of a paternal chromosome [6].

What about the accent of DNA? It has been suggested that for chromosomes to pair (as when a sword "pairs" with its scabbard) their DNA's must-have similar accents [7, 8]. When accents differ (a potentially complementary scabbard has been misshapen by many small blows), pairing is disrupted. How does accent affect pairing? It has been shown that the degree of unpairing from the parental duplex would be extremely sensitive to the accent of the corresponding DNA. Specifically, the strands in a segment of DNA that is rich in the bases G and C will unpair less readily than those in a segment of DNA that is rich in the bases A and T. If (through multiple small blows, i.e., mutations) maternal and paternal DNAs have begun to differ very slightly in accent (i.e., in their degree of richness in G and C; GC%), this will disturb the synchrony of their unpairing from the conventional duplex configuration. Hence, they will not display their outward-looking bases in a manner (time and configuration) permitting cross-pairing between maternal and paternal chromosomes.

The parents are reproductively isolated from each other since a hybrid offspring is unable to continue their line, by virtue of its being sterile (whether or not it can flourish due to natural selection). This is shown in Fig. 26.1, where the "accent" is quantified as the proportion of two bases (G and C) expressed as the G + C percentage among the four bases (A, C, G, and T).

Base Composition as Elusive "Residue"

Darwin's view on hybrid sterility was regarded by Huxley as the "weak point" of his evolutionary hypothesis. Darwin regarded sterile hybrid offspring in the context of natural selection (acting on their *conventional* phenotypes) rather than as manifestations of directly interacting parental genomes (that can be regarded as having *genome* phenotypes). Galton in *Natural Inheritance* suggested that an organism has a certain stability, so that internal changes must build up over several generations (i.e., remain latent) until there is a sudden change to a new position of stability (Chap. 3). Likewise, Romanes (Chap. 5) pointed to a "collective variation" building up in "a whole race or strain." This can now be seen as a cumulation of differences in GC% building up to a threshold beyond which reproductive isolation becomes increasingly apparent. Future changes in conventional phenotype (that might be favored by natural selection) are then conserved since the continuation of effective crossing with the main type (that might result in blending of characters in fertile offspring) is impaired [5]. We will see (Chap. 29) that the decisive factor is likely not the GC% itself, but its k-mer context.

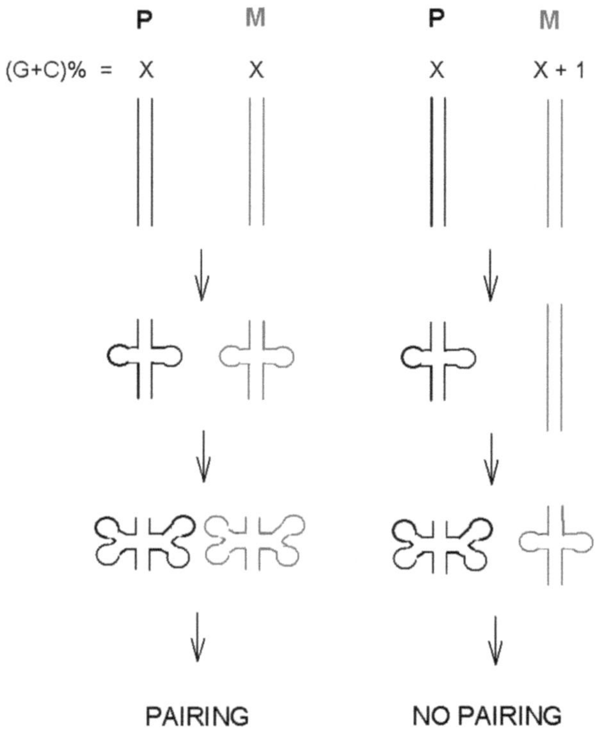

Fig. 26.1 Stem-loop extrusion from duplex DNA is exquisitely sensitive to difference in base composition (e.g., X% compared with [X + 1]%). This prevents pairing between otherwise nearly identical DNA sequences. At the left, paternal (P) and maternal (M) duplexes have the same GC% value (X). During meiosis, the strands of each duplex synchronously open and equivalent stem-loop structures are extruded. Interactions between the single-stranded loops can then progress to pairing. At the right, the duplexes differ slightly in their base compositions. The maternal duplex, of slightly higher value (X + 1), opens less readily and stem-loop structures are not equivalent. Thus, it is not the lack of identity itself, but the fact that this lack of identity is accompanied by minute differences in base composition, that leads to mispairing. In other words, differences in base composition provide a sensitive index of underlying homology differences. Small differences desynchronize the unpairing so no cross-pairing can occur, meiosis fails, and no gametes are produced

It is of interest that William Bateson, having at last found convincing the evidence that genes (primary information) were located on chromosomes (Chap. 17), looked there in vain for the location of his postulated "residue," and instead pondered a cytoplasmic location (Chap. 13). He sometimes referred to the "residue" as a "base" – but, of course, this was a different usage from that employed here. We now see that Bateson's "residue," Johannsen's "*Grundstock*" (a "great central something") and Guyer's "general substratum" are likely to be actually within DNA itself. In protein-encoding regions (exons), this "accent" role would largely be adopted by the third base of each codon, which can change without changing the encoded protein (conventional phenotype). In this context, we can see the exon of a gene not as a sequence of consecutive bases, but as a sequence of sets of two adjacent bases, each set being separated from the next by a single "residue base" that is essentially non-genic (despite its location in a gene).

Why has this solution been so elusive? Apart from the political games described in this book, the answer is that definitive evidence has been hard to obtain. A simple metaphor may help. Your newly acquired dog must be kept from running off. A barrier is needed. At first you tether it in your garden by a leash. The leash is the first barrier. Then you build a fence. So the leash is no longer necessary. If the first barrier (the leash) is damaged or lost, it may not be noticed. The second barrier (fence) should suffice. On the other hand, the leash then becomes available for other purposes. Thus, following establishment of a second barrier (between species), a first barrier (between species) may degenerate or change in a random way or may find another role. If a Sherlock Holmes then tried to discern whether there had been an earlier barrier, and what form it had taken, there might be a problem.

While a clear distinction within a codon between the two amino acid-determining bases and the third "residue" base will serve our present purposes, they can interact in subtle ways [5]. There could be opportunities for prospective mating partners to signal, through their conventional protein-determined phenotypes, the natures of their genome phenotypes (whether their genomic GC% values deviate in the same or different directions from the GC% norm that characterizes their species). This could have implications for the meiotic repair or speciation alternatives in the gonads of any hybrid offspring they might come to produce. Thus, the apparent subjectivity of beauty being "in the eye of the beholder" may have long-term evolutionary implications.

Summary

Bateson's long sought after "residue," which relates to the origin of species (Chap. 28), can be identified with Chargaff's famous GC% of chromosomal DNA that tends to differ between species. Greater species specificity is provided by differences in the oligonucleotide (k-mer) patterns associated with each GC% (i.e., context; Chap. 29). Differences in parental k-mer patterns inhibit the pairing of chromosomes at meiosis, resulting in sterility of their child (hybrid offspring). Thus, the parents, although able to produce that offspring, are reproductively isolated, in that their line cannot continue. However, each parent may find another partner within the species, from whom he/she is not reproductively isolated (i.e., compatible k-mer patterns). Thus, two independent lines may emerge and independently face the challenges of natural selection, possibly leading to two species where there had initially been only one.

Chapter 27
Mendel Basics

"Now I am going to show you my children!" With a smile Gregor Mendel (Fig. 27.1) would lead his visitors to the garden of the Augustinian monastery of St Thomas to see the peas. He was crossing varieties of peas and observing the outcome, not just qualitatively, as had long been the practice in studies of heredity, but quantitatively. He crossed a tall variety with a small variety, planted the peas from the pods, and *counted* how many tall and how many small plants then grew [1].

Analysis of these results led to the idea that contrasting characters (e.g., tallness, smallness) were often passed from generation to generation as distinct units ("formative elements"). To form a new plant, one unit (e.g., representing tallness) came from the male pollen-donating parent and one unit (e.g., representing smallness) came from the female seed-bearing parent (or vice versa). In the resulting hybrid, either one of these units was "dominant" (e.g., tallness over smallness), or the units came to a compromise (e.g., an intermediate size). Whatever the *visible* outcome, the unit representing a non-dominant ("recessive") character was *still present*. Units were constant, not variable. When new pollen and egg cells (gametes) were made, half of each type would carry units representing one of the contrasting characters, and the other half would carry units representing the other character. The distribution of characters among the next generation of plants was determined by statistical laws. The patterns of distribution became known as Mendel's laws and provided the basis for a new approach to the study of heredity.

Peasant Origins

Mendel was the only son of German-speaking peasant parents (Rosine Schwirtlich and Anton Mendel) of the village of Heinzendorf in Austrian Silesia. His academic gifts were soon evident, but although his parents and younger sister (who donated her dowry) were supportive, he was on "half rations" at Troppau High School.

A. G. Cock, D. R. Forsdyke, *Treasure Your Exceptions*,
https://doi.org/10.1007/978-3-030-92099-9_27

Fig. 27.1 Mendel, Gregor Johann. Courtesy of the National Library of Medicine, USA

The system of labor in lieu of taxes (corvée) meant that when his father was injured (1838), Mendel had to miss classes to help. In 1841 at the Olmütz Philosophical Institute, his physics professor, Friedrich Franz, recommended him to Prelate Cyrill Napp at the monastery at Altbrunn, which was part of the city of Brünn (now Brno). In a brief autobiography that accompanied his later application for entrance to an examination for high-school teachers, Mendel noted that "it was incumbent on him to enter a profession in which he would be spared perpetual anxiety about a means of livelihood." In 1843, he assumed his monastic name Gregor, which thereafter he used before his baptismal name, Johann.

Scientist, Not Parish Priest

Under Napp the monastery was a major intellectual and cultural centre. Most monks were engaged in artistic or scientific activities. From 1845 to 1848, Mendel attended Brünn Theological College. Here the Professor of Agriculture, Franz Diebl, was teaching that a non-mystical, but "unknown force" might cause new characters emerging in hybrids to remain *constant*. When hybrids of the same type were crossed, uniform offspring with the hybrid characters would be produced. They would breed true. Such hybrids would then constitute new varieties, from which some thought new species might emerge. However, in those times it often appeared, in Mendel's words, "impossible to draw a sharp line of distinction between species and varieties."

Mendel was ordained priest in 1847, and Napp recommended him as a supply teacher, writing that "he is very diligent in the study of the sciences; but he is much less fitted for work as a parish priest, the reason being that he is seized by an unconquerable timidity when he has to visit a sick-bed or to see anyone ill and in pain." However, Mendel played an active role in the political discontents that embroiled Europe in 1848, indicating less timidity on other fronts. Mendel taught mathematics and Greek at Znaim High School with much success. Despite a quip that a visiting school inspector had "more fat than understanding," Mendel was well liked and was encouraged to take examinations for a formal degree. His Vienna examiners noted "that he was devoid neither of industry nor talent. It would seem, however, that he can have had no opportunity for acquiring exhaustive knowledge and must have lacked access to the necessary means of study."

Napp then arranged for him to spend two years at Vienna University studying mathematics, physics, and biology. Teachers such as the botanist Franz Unger and the physicist Christian Doppler (discoverer of the "Doppler effect") gave Mendel what he needed, but not what the examiners needed. When he returned from an examination in physics and natural history in 1856, he "had a sore head" after a difference of opinion with the botany examiner. Although his teaching continued to receive high praise, he never reattempted the examination.

Research

To begin his studies with peas (1856–1863), Mendel first set out to characterize pure lines that would breed true from generation to generation. In his words, "they held true to type." A tall line of peas should always produce tall offspring. A small line should always produce small offspring (Fig. 27.2).

Next, he found character pairs that were *always* alternatives and would not produce a blended intermediate form (Fig. 27.3). Sometimes one of the alternatives, dubbed "recessive," would *entirely* disappear in the first generation and *all* offspring would display the other "dominant" alternative.

Then he studied crossings between members of two pure lines. This led to establishment of his famous 3:1 ratio F_2 (Fig. 27.4).

Finally came his recognition that two parentally donated units representing a character retained their individuality in the offspring and are separated during formation of gametes (Fig. 27.5). This individuality was also observed when fundamentally different character pairs (e.g., height and seed color) were studied together in offspring (independent assortment): "Simultaneously it thus is shown that the behaviour of each pair of differing characters in hybrid association is independent of other differences between the two original parental plants."

These results were reported to the Brünn Society for the Study of Natural Science in 1865 and briefly summarized in a local newspaper. They were formally published in the Society's *Transactions* (1866), which were distributed to academic centres throughout Europe. His abstract units (later known as "genes") were constant but, by their reassortment, mainly variable hybrids were generated. He came to understand why, much to the consternation of professional breeders, the plants would not

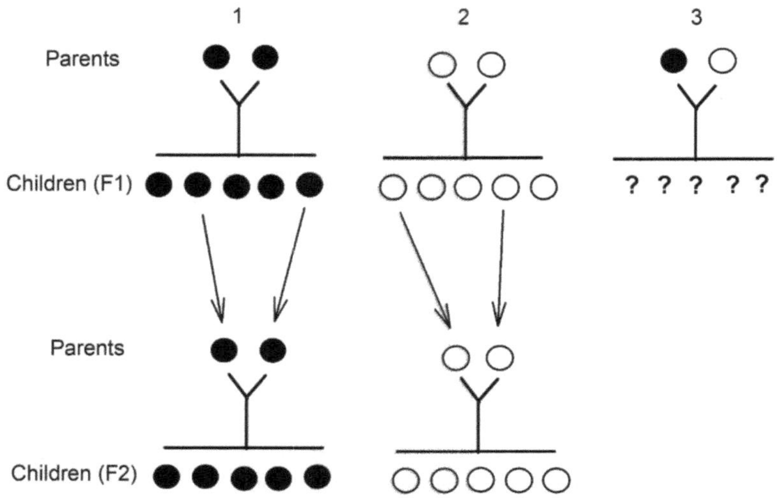

Fig. 27.2 Pure lines breed true. Mendel sought pure lines (varieties) of peas. The units (elements) that represent contrasting pea characters are here symbolized by black-or-white circles. Members of pure lines bred true when crossed among themselves (1. black pollen with black ova; 2. white pollen with white ova). On crossing, *any pairs* from the first generation (F1 hybrids) should produce a second generation (F2 hybrids) *all* members of which would *retain* the parental character. The black line never produced a white. The white line never produced a black. Mendel was then prepared to study a cross between the lines (3)

usually breed true (Figs. 27.3 and 27.4). He believed that "constant hybrids acquire the status of new species," but he had not found a path to such species. To extend his results to other plants, in 1866, he began studying the hawkweed (*Hieracium*), which was a favorite of the professor of botany at Munich, Carl Wilhelm von Nägeli. But, unknown at that time, in this case for most generations reproduction is asexual (parthenogenetic), and he was unable to clearly interpret his results.

Mendel's choice to work with plants was politically correct *theologically*, but there were also studies on mice and bees, meteorology, and sun-spots. Sadly, these records were not retained by the monastery authorities after his death. He had been able to help with the education of his sister's sons, one of whom wrote (1902): "He often told us nephews that we shall find as his heritage, papers for publication, which he could not publish during his life. But we did not receive anything from the cloister, not even a thing for remembrance." Nevertheless, the 1864 German translation of Charles Darwin's *The Origin of Species by Natural Selection*, and many other scientific books in the monastery library, contain annotations in Mendel's neat handwriting.

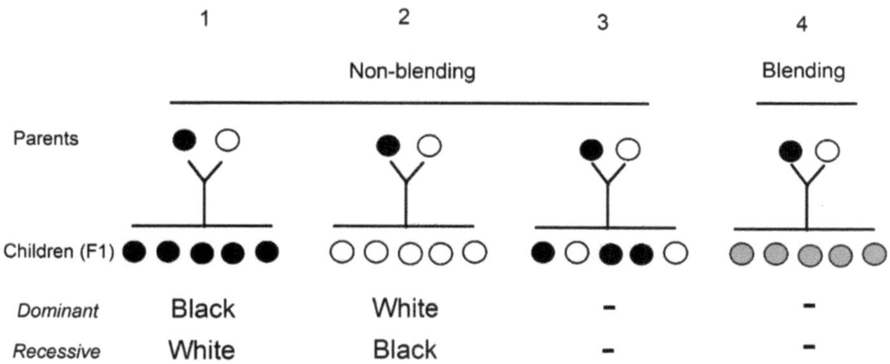

Fig. 27.3 Theoretical outcomes of crosses between different lines (symbolized as black or white). Mendel sought pure lines that when crossed displayed *non-blending* inheritance among child plants. Only one of the alternative parental characters would be displayed in a hybrid offspring (1–3). However, professional breeders were more interested in *blending* inheritance (here symbolized as grey; 4). In this case, new characters such as a better flower color in plants, or better wool in sheep, might emerge. When all hybrids displayed the black character (1), then black was deemed "dominant" and white "recessive." Conversely, when all hybrids displayed the white character (2), then white was dominant and black recessive. If a mixture resulted (3), there was neither dominance nor recessiveness

A Non-Darwinian "Other Force"

Prior to Darwin's theory, many thought that being a member of a species implied a stable "harmony of the inner nature" or "essence" that was distinct from the character differences such as tallness or smallness that were observed between species subgroups (varieties). Indeed, Mendel had observed "how risky it can be to draw inferences about the internal relatedness of hybrids from external similarities." His failure to find constant hybrids suggested that some "other force" than hybridization itself would usually be needed for a new species to emerge. Could this force be natural selection?

We do not know when (and if) they first met, but correspondence dating from 1866 to 1871 survives between Mendel and Nägeli. Opposing Darwin's theory of natural selection, Nägeli considered that "The formation of … constant varieties or races is not the consequence and the expression of outer agencies, but is brought about through inner causes." The outer agency – natural selection – worked like a gardener clipping the exuberant branching of a tree but was powerless to bring about the initial branching.

This view seems close to that of Mendel. Regarding natural selection, Mendel is quoted as saying: "This much already seems clear to me, that nature does not modify species in any such way, so some other force must be at work." Furthermore, "It remains more than probable that there is a factor in action for the variability of cultivated plants, which hitherto has received little attention." In opposition to those

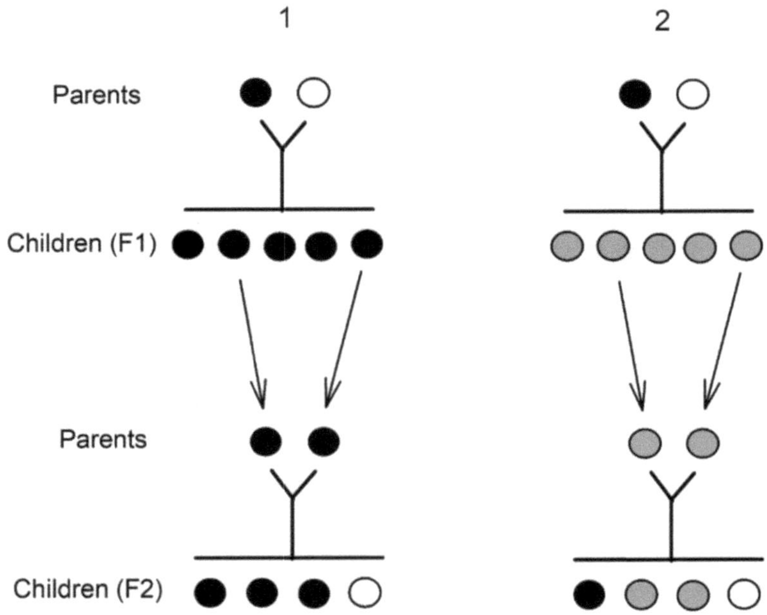

Fig. 27.4 Mendel's discovery of the 3:1 ratio when one character is dominant. In the F1 generation all hybrids appear identical (1, at left). This equals the first case in Fig. 27.3, with black symbolizing the dominant character. The crossing of any pairs selected *randomly* from the F1 generation produced an F2 generation with a minority of recessives (white) in the ratio (black:white = 3:1). Here the three black forms appear equal. With blending inheritance (2, at right), *which Mendel did not study*, there would be visible variation among the three non-white forms (explained in Fig. 27.5)

who assume species evolution to be a *continuous* process, Mendel seemed to agree with an earlier hybridist:

> Gärtner was persuaded to oppose the opinion of those naturalists who dispute the stability of plant species and assume continuous evolution of plant species. He sees in the completed transformation of one species into the other the unambiguous evidence that a species has fixed limits beyond which it cannot change.

Thus, Mendel wrote that "a confirmation deserving notice regarding … the variability of cultivated plants is found in the experiments performed by Gärtner." In other words, if one species evolves from another, then that evolution is, somehow, *discontinuous*.

Administrator, Not Advocate

In 1868, Mendel was preferred over a Czech nationalist rival (Rambousek, who had also entered the monastery in 1843) to succeed Napp as Prelate. ,He noted in a letter to Nägeli that "all at once, from my modest sphere as teacher of experimental

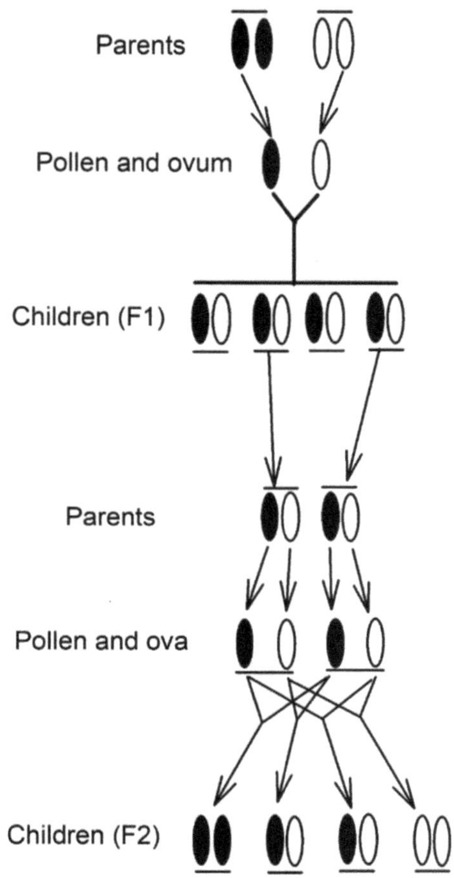

Fig. 27.5 Mendel's recognition that his chosen parental characters (here symbolized as black or white) existed in *dual forms* in each parent. These forms were separated when pollen and ova were formed (gametogenesis). In the F1 generation, all hybrids are equal (black and white are both represented), but since black is dominant, they all *appear* black. The crossing of any pairs from the F1 generation will result, statistically, in an F2 generation with, for each one white, one pure black, and two with black and white both represented but *appearing* black. Applying the same interpretation with the blending model shown in Fig. 27.4, these two would appear grey

physics, I am translated into one in which a good deal seems strange to me." He became increasingly involved in monastic and community affairs, which included membership and eventually chairmanship of the Moravian Mortgage Bank. Although his political leanings were liberal, from 1874 until his death in 1884, he was much engaged in a bitter dispute with the German Liberal Party, which had imposed severe taxes on monasteries. His advocacy on behalf of the monasteries was politically naive and counterproductive. Although he travelled widely on the continent and once to London, there is little evidence that Mendel visited other scientists or attempted to increase awareness of his pea experiments. Despite claims to the contrary, there is no evidence that Charles Darwin received a copy of Mendel's 1866 paper.

Rediscovery of Mendel's Paper

In 1881, Mendel's studies were briefly mentioned both by the German botanist Wilhelm Focke in a book, and by Charles Darwin's young research associate, George Romanes (1848–1894), in an article on hybridism in the *Encyclopaedia Britannica*. But their significance was not generally appreciated. By the date of Mendel's death (1884), Nägeli had not mentioned him in his published work. However, in the 1880s, George Romanes, who was deeply interested in hybrid sterility and had long corresponded with Focke (and quoted him in a book), began Mendelian-style crossings with rabbits and rats. He privately noted this in the year of his death (1894).

The role of chromosomes in cell division was not known in Mendel's time. Unaware of Mendel's work, from microscopic studies of the cell divisions (meiosis) during sperm formation in birds, Michael Guyer in his University of Chicago thesis (1900), was led to suggest the independent segregation of units determining characters among offspring and their association with chromosomes.

Like Romanes, certain botanists on the continent of Europe had been carrying out Mendelian-style crossings in the 1890s and later discovered, and in 1900 drew attention to, Mendel's 1866 paper. Indeed, one of these, Hugo de Vries in Amsterdam, found constant hybrids appearing among varieties of the large-flowered evening primrose, in the apparent *absence* of natural selection; many of these were later found to be due to multiple sets of chromosomes (polyploidy), which can be a cause, *but not a usual cause*, of the origin of species.

With Rebecca Saunders and Charles Hurst, the Cambridge zoologist William Bateson extended Mendel's work to various animals and plants. Familiar with Guyer's studies, he argued Mendel's case in the English-speaking world and coined many of the terms we use today (e.g., homozygote, heterozygote, F1 generation). Given intense opposition from mathematical biologists ("biometricians"), his attention was directed more to demonstrating the universality of Mendel's laws, than to pronouncing on their cellular basis. However, at a 1902 conference in New York, he observed that "we have reason to believe that the chromosomes of the father and mother plant, side by side, represent blocks of parental characters." It is likely that two of the botanists who had "rediscovered" Mendel – de Vries, and Correns in Tübingen – thought similarly.

Post-rediscovery Debates

Refusing to accept natural selection as an originator of species, and perhaps influenced by Nägeli's "inner cause" and Mendel's "other force," Bateson attributed their discontinuous origin to a "residue," which was inherited like, but was distinct from, the Mendelian units (genes). In Bateson's view, the reproductive isolation that was critical for speciation resulted from failure of the different residues in the gametes of the two parents to complement "like sword with scabbard" when their offspring were forming the gametes for the next generation (i.e., during meiosis). Although much debated, there is now growing support for this "chromosomal," rather than "genic," viewpoint.

There are also continuing debates on the extent to which Mendel appreciated the significance of his observations, why they were initially neglected, and their statistical validity. The most prominent of the born-again biometricians, Cambridge professor of genetics Ronald Fisher, even threw Mendel's integrity into question. However, an 1842 statistics text, which Mendel may well have read, states:

> Even the shrewdest observer with the best instruments never gets results which are totally devoid of error, and in order to come as close as possible to the truth, the only thing one can do is to repeat the operation as many times as necessary and to select from those results those which have the smallest error.

Thus, he followed the statistics *of his time.* Indeed, his results have withstood the test of time. Perhaps the key to understanding this nineteenth-century monk is that, like some of the innovative molecular biologists of the twentieth century, he approached problems of biology from the perspective of a physicist. He who can write of the "the unity of the evolutionary plan of organic life," of "the developmental law found for each pair of differing characters," and of "distinctive characters of two plants" that "ultimately rest only on differences in the nature and grouping of the elements that are present in their foundational cells," has some sense of what we now call a "genetic program."

Summary

Gregor Mendel (1822–1884) was a Moravian biologist from whose quantitative plant breeding studies were derived laws of inheritance that founded a new branch of biology, genetics. He crossed varieties of peas and examined the distribution of characters among offspring over successive generations. Hybrids resulting from crossing two varieties with contrasting characters did not breed true. They were not the "constant hybrids" much sought by professional breeders that might lead to independent species. In the 1880s, Mendel's work was appreciated by Wilhelm Focke and George Romanes, who initiated similar studies with animals. Studies with plants were initiated in the 1890s by European botanists who discovered Mendel's work in 1900. A relationship of hereditary units that determined characters – now known as genes – to chromosomes, was noted by cytologist Michael Guyer in 1900. The major early advocate of Mendel's work, zoologist William Bateson, agreed with Mendel that new species could emerge discontinuously.

Chapter 28
George Romanes, William Bateson, and Darwin's "Weak Point"

Many organisms repeat the cycle adult-gametes-zygote-adult through the generations (Fig. 3.2 in Chap. 3). When lineages diverge this cycle is interrupted and two separate cycles result. The interruption – a potential origin of species – can occur at any stage of the cycle, either prezygotically, when gametes meet to form a zygote, or postzygotically, when the zygote develops into an adult that can then form gametes. Since each stage requires the action of specific gene products, it follows that changes in certain genes might lead to cycle interruption and divergence into two reproductively isolated lines [1, 2]. However, "speciation genes" have been hard to identify [3, 4], and a classical example has recently been questioned [5].

Our purpose here is not to question the genic view but to remind readers of a non-genic view suggested in 1886 by George Romanes, who was Charles Darwin's research associate for eight years prior to the latter's death in 1882. This view was further developed by William Bateson, who in 1900 brought Mendel's now famous 1865 paper to the attention of the English-speaking world. In particular, Romanes and Bateson noted what Thomas Huxley had referred to as "the weak point" in Darwin's theory [6, 7]. If offspring could be obtained from an *inter-species* cross, they were usually sterile (hybrid sterility). However, no breeder had ever produced new forms from *intra-species* crosses that could be characterized in this way. Furthermore, in proposing that evolution proceeded smoothly without jumps, punctuations, or discontinuities, Darwin seemed to have imposed upon himself "an unnecessary difficulty"[8].

While the work of Mendel went unnoticed for a mere 35 years, the views of Romanes and Bateson were disparaged or went unnoticed for over a century. This was due partly to the twentieth century's obsession with genes, but also to the fact that they were struggling with novel terminology. With much of this ("allele", "homozygote") we are now familiar. But terms such as "collective variation" and "residue" – now seen as highly relevant to speciation studies [3, 4, 9] – have largely disappeared. Proceeding chronologically from the late Victorian era to modern times, we here consider the latter terms and, through textual analyses of selective quotations [10], let Romanes and Bateson retell their story [11, 12].

A. G. Cock, D. R. Forsdyke, *Treasure Your Exceptions*,
https://doi.org/10.1007/978-3-030-92099-9_28

Bateson Approved "An Additional Suggestion"

In the spring of 1886, William Bateson set off on an eighteen-month scientific expedition to the Russian Steppes, with the intention of catching up with the relevant literature on his return. However, in the autumn his sister Anna, who was then working with Francis Darwin, dispatched three consecutive issues of *Nature* containing a serialized May address to the Linnean Society. Here Romanes had made an "additional suggestion on the origin of species" that he named "Physiological Selection," or the "Evolution of Species by Independent Variation" [13]. This had raised a storm of criticism from the senior Darwinians (Alfred Wallace, Thomas Huxley, Joseph Hooker) and their junior surrogates (Francis Darwin, William Thiselton Dyer, E. Ray Lankester, Raphael Meldola). But the response of the 25-year-old Bateson (Kasalinsk, Dec. 4th, 1886) indicated tempered approval [14]:

> Thanks for those 'Natures.' I don't agree with you that Romanes' paper is poor. It seems a fair contribution and at all events does, as he says, put the whole view on a much more logical basis. The scheme thus put will at least work logically while the other, as left by Darwin, would not. Of course, as to the novelty of the suggestion I know nothing, and I don't much care. I did not suppose Romanes would ever write as good a paper. – It is a straightforward, common-sense suggestion.

Another English biologist, the 27-year-old Joseph Cunningham, was also likely to have approved (see below).

Cryptic Variability of the Reproductive System

Darwin viewed hybrid sterility as a puzzling negative trait that should have been countered by natural selection [11, 12, 15]. In his Linnean address, Romanes went beyond this, pointing out that, apart from variations between individuals that we can easily detect (e.g., differences in eye color), there are likely to be variations that we cannot easily detect. Some variations in parental gonads might ultimately be detected by their conferral of the sterility phenotype on the otherwise healthy offspring of certain pairs. This would designate the parents as reproductively isolated from each other. The *partner-specific* reproductive-isolation-phenotype of the parents would manifest as the sterility phenotype of their offspring. In Romanes' view, the establishment of this form of reproductive isolation – isolation of the *fit* from the *fit* – would in many cases have been critical for branching into two species, with or without the *subsequent* intervention of natural selection that would isolate (by elimination) the *unfit* from the *fit* (linear, non-branching, evolution).

Romanes' physiological selection was an abstraction for which there was then no cytological or biochemical basis. To give his Victorian audience an example of one form of physiological selection, Romanes pointed to a variation that might lead a plant to flower earlier than most members of its species. Early flowering forms would be reproductively isolated from late flowering forms by a temporal barrier. At an early-stage, pollen kept from an early flowering form would still be capable of fertilizing flowers of late flowering forms. Over time, incompatibilities between early pollen and late ovules would randomly arise, and this mutual incompatibility (a further barrier) would eventually define the two forms as distinct species. Today we attribute allochronic and gamete incompatibility barriers, as we attribute allopatric barriers, to differences in genes. But Romanes did not think this form of variation would have operated generally to originate species [13]:

> In many cases, no doubt, this ... variation, has been caused by the season of flowering or of pairing having been either advanced or retarded in a section of a species, or to sundry other influences of an extrinsic kind; but probably *in a still greater number of cases* it has been due to what I have called intrinsic causes, or to the 'spontaneous' variability of the reproductive system itself. [my italics]

Romanes felt no obligation to further clarify intrinsic causes: "It is enough for the explanation which is furnished by Mr. Darwin's theory of the evolution of adaptive structures by natural selection, that the variations in question take place; and similarly as to the present theory of the evolution of species by physiological selection" [13]. On the other hand, he was quite prepared to agree that this form of variation could proceed *continuously* as Darwin had proposed for the evolution of adaptations [16]:

> Starting from complete fertility within the limits of a single parent species, the infertility between derivative or divergent species, *at whatever stage in their evolution this began to occur*, must usually at first have been well-nigh imperceptible, and thenceforth have proceeded to increase stage by stage. ... Does it not therefore become ... in a high degree probable, that from the moment of its inception this isolating agency must have played the part of a segregating cause, in a degree proportional to that of its completeness as a physiological character?

Collective Variation of a Population Subset

Anticipating modern developments in population genetics, Romanes thought the slow increase of his postulated independent variation of the reproductive system might reflect its "collective" nature, in that it would affect a distinct lineage *within* a population. One of the misconceptions of his opponents was [16]:

> That I imagine physiological varieties always to arise 'sporadically,' or as merely individual 'sports' [mutations] of the reproductive system. On the contrary, I expressly stated that this is *not* the way in which I suppose the 'physiological variation' to arise, when giving origin to a new species; but that it arises, whenever it is effectual, as a 'collective variation' affecting a number of individuals simultaneously, and therefore characterizing 'a whole race or strain'.

Free intercrossing would have tended to swamp or blend the sporadically appearing variations upon which natural selection might act. But Romanes' collective "independent variation" of the reproductive system – postulated to be a "general variability ... about a mean" (e.g., as human heights form a normal distribution) – was held to militate against this [17]:

> I am guilty of no inconsistency when thus arguing for a 'collective variation' on the part of the reproductive system, after having urged the difficulty against natural selection which arises from free intercrossing – i.e. the difficulty of supposing that a sufficient number of variations of the same kind should always be forthcoming simultaneously to enable natural selection to overcome the influence of free intercrossing. For, as previously explained, this objection is only valid in the case of 'accidental,' 'sporadic,' or 'spontaneous' variations, which, *ex hypothesi*, are relatively very few in number. The objection does not apply to 'collective variations,' which, being due either to a common cause or to general variability of size, etc., about a mean, affect a number of individuals simultaneously.

Remarkably, two of Romanes' sternest critics provided valuable guidance on terminology. Thiselton Dyer [18] suggested the term "reproductive isolation." Wallace [19] noted that two individuals who had each become reproductively isolated from members of their main species (with whom they might cross to produce healthy but sterile offspring) might produce healthy and fertile offspring when mutually crossed. Hence, the parents could be considered as "physiological complements."

Another critic, Huxley, failed to recognize that his "weak point" criticism of Darwin had been met [11]. Romanes had distinguished two types of variation: *overt* sporadic variations that might lead to the evolution by natural selection of better adapted forms (which the breeder might capture and selectively pair) and *cryptic* collective variations that might lead to the evolution of species by physiological selection (which the breeder would find hard to simulate).

Sporadic and Regular Sterilities

Romanes died at age 46 in 1894, the year that Bateson published a major treatise – "Materials for the Study of Variation, Treated with Especial Regard to Discontinuity in the Origin of Species" [20]. At that time, in keeping with Huxley, Bateson saw the origin of species as a discontinuous process, similar to that manifest as the sudden arising, in one generation, of a new variety of flower with one perfect extra petal. The suddenness of the emergence implied a discontinuous underlying mechanism uninfluenced by natural selection. The book was warmly reviewed by the fish biologist Joseph Cunningham, who noted similarities to Romanes' physiological selection theory but declared Bateson's argument to be "defective" since there was "no necessity for discontinuity of variation to explain discontinuity of species" [21].

Like various continental botanists (Hugo de Vries, Carl Correns, Erich von Tschermak), in the 1890s, Bateson and E. Rebecca Saunders had begun crossing various lines of plants and animals and scoring the inheritance of characters among the progenies. Following the discovery of Mendel's work in 1900, Bateson became its leading advocate in the English-speaking world. Given the intense opposition he faced from the mathematical biologists ("biometricians"), it is not surprising that his attention was directed more to verifying Mendel's laws than to pronouncing on their cellular basis. However, in October 1902, at the Second International Conference on Plant Breeding and Hybridization in New York, he noted: "We have reason to believe that the chromosomes of the father plant and mother plant, side by side, represent blocks of parental characters" [22].

In December 1901, Bateson and Saunders had submitted a detailed 160-page report to the Evolution Committee of the Royal Society, to which they made minor amendments in March 1902 [23]. The Report noted that the classical sterility of the hybrids resulting from a cross between species that were believed to have recently diverged from a common ancestor (allied species) did not follow a Mendelian pattern of character assortment among offspring. Although the offspring might have inherited some conventional character, such as color or height, in a Mendelian fashion (e.g., 3:1 ratio), usually the same offspring would *all* inherit the sterility phenotype (Fig. 28.1).

They then had to admit [23]:

> We know of no Mendelian case in which fertility is impaired. We may, perhaps, take this as an indication that the sterility of certain crosses is merely an indication that *they cannot divide up the characters among their gametes*. If the parental characters, however dissimilar, can be split up, the gametes can be formed. ... That the sterility of hybrids is generally connected in some way with inability to form germ-cells correctly ... is fairly clear, and there is in some cases actual evidence that this deformity of pollen grains in hybrids is due to irregularity or imperfection in the processes of division from which they result.

Thus, the hybrid sterility phenotype that resulted from a cross between allied species was something intrinsic to an organism and did not relate to extrinsic environmental elements. That they were thinking of an imperfection in the pairing of homologous parental chromosomes is indicated by their citing of Guyer, who "in ignorance of Mendel's work" had independently derived Mendel's laws from the

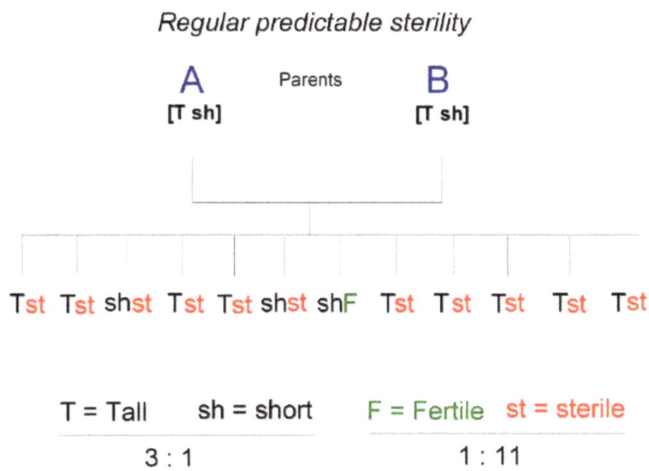

Fig. 28.1 The regular, predictable, sterility observed when members of *closely allied* species are crossed. Parents, A and B, are heterozygous for height, with an allele for tallness (**T**) and an allele for shortness (**sh**). If tallness is dominant, the ratio of tall to short offspring is 3:1, in accord with Mendelian predictions. However, the majority of offspring are sterile (**st**), not fertile (**F**)

cytology of spermatogenesis in pigeon hybrids [24]. Furthermore, early in 1902, Bateson noted [25]:

> It is impossible to be presented with the fact that in Mendelian cases the cross-bred produces on an average *equal* numbers of gametes of each kind, that is to say, a symmetrical result, without suspecting that this fact must correspond with some symmetrical figure of distribution of those gametes in the cell-divisions by which they are produced.

There was repeated reference to Guyer in Bateson's later works [26, 27]. But soon Bateson and his colleagues came across cases of infertility that followed Mendelian patterns of inheritance (Fig. 28.2). Their second report to the Evolution Committee in 1905 drew a clear distinction between this unpredictable *sporadic* sterility, seen among the offspring of crosses between *closely related* individuals, and the predictable sterility seen among the offspring of crosses between *distantly related* individuals (i.e., members of allied species) [28]:

> Contabescent [withered] anthers were seen from time to time in many families .... This sporadic sterility has not been particularly studied. It is interesting to compare this example of the definite appearance of sterility ... with the familiar occurrence of sterility in cross-breds. Such a phenomenon has often been supposed to indicate remoteness of kinship, yet here a closely comparable effect occurs in F_2 as the result of a cross between two types which must be very nearly related. Mr. Gregory in a careful examination of the pollen-genesis, found that the divisions were normal up to the reduction division, when the chromosomes form shapeless knots and entanglements, failing to divide.

Today we would interpret the latter sterility as a malfunction due to mutation in a gene determining one of the many cell components required for anther development. In this case, the component seemed to be required for the reduction division of meiosis. Whatever the ultimate basis of the sterility, Bateson and his colleagues

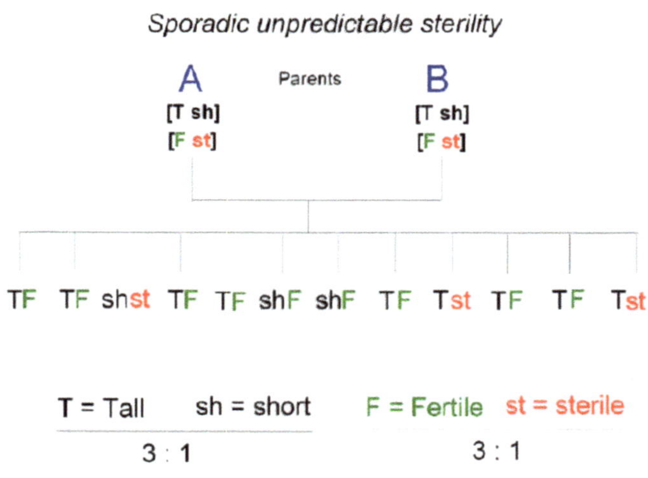

Fig. 28.2 The sporadic, unpredictable, sterility observed among the offspring of a *within-species* cross. Parents, A and B, are heterozygous for height as in Fig. 28.1. They are also (unknowingly) heterozygous for a recessive sterility gene (**st**). The ratios of dominant to recessive among offspring accord with Mendelian predictions, thus indicating that sterility can be of genic origin

pointed out that this sporadic event arising from a *within-species* cross (unpredictable, in that it had not been anticipated when the cross was first made) contrasted with the *regular* (predictable) F_1 sterility appearing when allied species in general were crossed. Only the former displayed a Mendelian (i.e., genic) pattern of inheritance. Indeed, in one case, the sterility character was found to be coupled (linked) to another character displaying Mendelian inheritance (leaf axil color; Table 28.1).

A review on chromosomes that Winge sent to Bateson in 1918 made the further distinction that polyploidy could not "cure" genic sterility but could "cure" non-genic sterility (Fig. 9.5 in Chap. 9). Bateson did not appreciate this.

In 1902, Bateson and Saunders recognized that Mendel's unit characters were "the sensible manifestations of physiological units of as yet unknown nature"[23]. Although Bateson did not begin to use the word "gene" until the second decade of the twentieth century, it is employed here when referring to the Mendelian "physiological units" that determine characters.

Chromosomes and the Residue or Irresoluble Base

Bateson was among the first to see a link between Mendelian "genes" and chromosomes and cited Guyer in this context [23]. However, as evidence for a chromosomal location became stronger, Bateson's doubts increased. It was not a question of Bateson's not seeing the wood for the trees. His idea of what constituted the wood differed from that of others. They saw a collection of genes. He saw the genes plus something else that was in some way related to the question of the origin of species. What was the something else and where was it located? Was it on the chromosomes like the Mendelian factors? Or was it elsewhere, perhaps outside the nucleus in the cytoplasm? No clear evidence on this had emerged by 1926, the year of Bateson's death. Thus, his pronouncements on chromosomes were always carefully hedged [12].

That the story was somehow incomplete was made explicit early in 1902: "Has a given organism a fixed number of unit characters? Can we rightly conceive of the whole organism as composed of such unit characters, or is there some residue – a basis – upon which the unit characters are imposed?"[23] This was expressed in a slightly different form, a few months later: "From the fact of the existence of

Table 28.1 Comparison of sterilities

Type of sterility	Genic	Non-genic
Mendelian inheritance	Yes	No
Coupling (linkage)	Yes	No
Parental relationship	Close (within species)	Distant (allied species)
Frequency	Sporadic	Regular
Incidence	Unpredictable	Predictable

interchangeable [allelic] characters we must, for purposes of treatment, and to complete the possibilities, necessarily form the conception of an *irresoluble base* [i.e. a base not resolvable by Mendelian analysis], though whether such a conception has any objective reality we have no means as yet of determining."[25] The report of Bateson and Saunders linked the concept to Darwin's great question, the nature of the origin of species – a "phenomenon" that did not "attach to" the elements (genes) that determined allelic characters [23]:

> We know, of course, that we cannot isolate this residue from the unit characters. We cannot conceive a pea, for example, that has no height, no colour, and so on; if all these were removed there would be no living organism left. But while we know that all these characters can be interchanged [through breeding experiments], we are bound to ask is there something not thus interchangeable? And if so, what is it? We are thus brought to face the further question of the bearing of Mendelian facts on the nature of Species. The conception of species, however we may formulate it, can hardly be supposed to attach to allelomorphic or analytical varieties [genes]. We may be driven to conceive 'Species' as a phenomenon belonging to that 'residue' spoken of above, but on the other hand we get a clearer conception of the nature of sterility on crossing.

Two Factors Must Complement to Produce Sterility

Bateson was considering the sterile offspring of crosses between allied species that were presumed to have derived from a common ancestral species. Since the offspring of crosses between members of that ancestral species would have been fertile (otherwise there would have been no descendants), then the hybrid sterility revealed the emergence of a reproductive barrier between the descendent species (i.e., an interruption of the unitary generational cycle). This could have been an *originating* barrier leading to the state of reproductive isolation that defined them as distinct species. With little indication that he was writing about chromosomes, the residue, or an irresoluble base, Bateson developed this theme abstractly in an essay he contributed to the Darwin centenary celebration in 1909 [29]. Here he concluded that "*Failure to divide* [to produce gametes] is … the immediate 'cause' of the sterility," so that "we are justified in supposing that there are factors which can arrest or prevent cell division." Thus, "When two species, both perfectly fertile severally, produce on crossing a sterile progeny, there is a presumption that the sterility is due to the development in the hybrid of some substance which can only be formed by the meeting of two complementary factors …" (Fig. 28.3a). Applying this line of reasoning to what would appear to be a single species, Bateson continued:

> We see that the phenomenon could only be produced among the divergent offspring of one species by the acquisition of at least *two* new factors; for if the acquisition of a single factor caused sterility the line would then end. Moreover each factor must be separately acquired by distinct individuals, for if both were present together, the possessors would by hypothesis be sterile. And in order to imitate the case of species, each of these factors must be acquired by distinct breeds. The factors need not, and probably would not, produce any other perceptible effects; they might, like the colour-factors present in white flowers, make

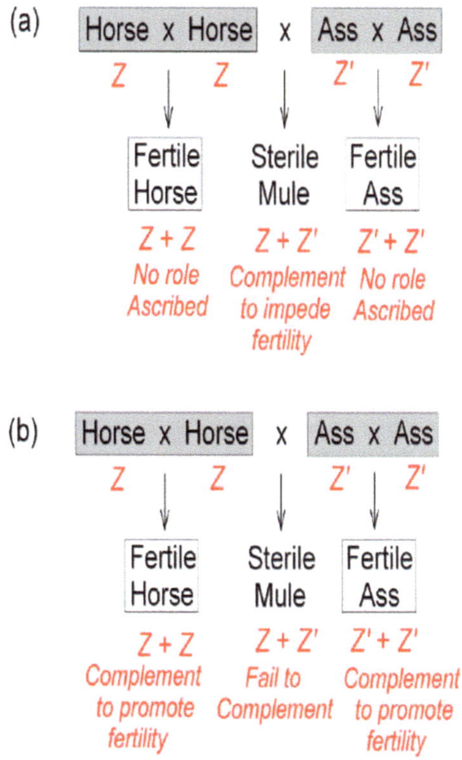

Fig. 28.3 Hybrid sterility in crosses between allied species (e.g., horse, ass) as due to hypothetical factors, **Z** and **Z'**, which either (**a**) *complement* to bring about a negative effect or (**b**) *fail to complement* to bring about a positive effect. In both cases, the result is negative (i.e., sterility). Since parents display fertility with their own kind (intraspecies cross), in 1909, Bateson deduced (**a**), that in an interspecies cross **Z** and **Z'** complement to bring about a negative effect. In 1913, Bateson entertained the converse (**b**), that **Z** and **Z'** fail to complement to bring about a positive effect

> no difference to the form of other characters. Not until the cross was actually made between the complementary individuals would either factor come into play, and the effects even then might be unobserved until an attempt was made to breed from the cross-bred.

Thus, the factors (say Z') which distinguished certain members of a species from their fellows (bearing say Z) would be latent. They would become patent, revealing a sterility-of-offspring phenotype, only if, by chance, there were a cross between certain individuals that had come to differ in the factors (Z' and Z; not Z' and Z', or Z and Z). At least one of the individuals would be rare, since the normal abundant types were generally fertile when intercrossed (i.e., no role was ascribed to Z in the presence of Z, or Z' in the presence of Z'). How often could this happen?

> If the factors responsible for sterility were acquired, they would in all probability be peculiar to certain individuals and would not readily be distributed to the whole breed. Any member of the breed into which *both* the factors were introduced would drop out of the pedigree by virtue of its sterility. Hence the evidence that various domesticated breeds, say of dogs or fowls, can when mated together produce fertile offspring, is beside the mark. The real question is, Do they ever produce sterile offspring? I think the evidence is clearly that sometimes they do, oftener perhaps than is commonly supposed.

Sterility-of-offspring being a parental phenotype, the fact that a completely sterile offspring could not be subjected to further breeding analysis would not matter. The parents of the offspring complemented each other sufficiently to produce the phenotype, but the complementary factor of one parent (say Z') might be more prone to produce the phenotype than the complementary factor of the other (say Z). Through segregation (further crossbreeding of each parent with other individuals), these complementary factors might be investigated.

In his 1909 essay, Bateson considered two sterility factors working together (complementing) to *impede* fertility (Fig. 28.3a). However, he later noted that organisms could be reproductively incompatible because the reproductive system of "each is lacking in one of two complementary elements" that *promote* fertility (Fig. 28.3b) [30]. This view, reminiscent of Romanes, was later adopted by Richard Goldschmidt, who proposed that normally parents contribute complementary factors (e.g., Z and Z) making parental chromosomes *compatible* at meiosis in their hybrids, which hence are fertile (i.e., the parental factors work together to produce a *positive* effect). When the factors are not sufficiently complementary (e.g., Z and Z'), the parental chromosomes are *incompatible* in their hybrids, which hence are sterile [31].

Complementation Like Sword and Scabbard

Bateson persisted with his Huxleyan idea that the sterility of a hybrid resulting from a cross between allied species was "a distinction in kind, of a nature other than those we perceive among our varieties" [30]. And he was still attached to the Huxleyan idea of discontinuous variation. In December 1921, he gave his famous address – "Evolutionary Faith and Modern Doubts" – at the Toronto meeting of the American Association for the Advancement of Science:

> That particular and essential bit of the theory of evolution which is concerned with the origin and nature of *species* remains utterly mysterious. We no longer feel … that the process of variation now contemporaneously occurring is the beginning of a work which needs merely the element of time for its completion; for even time cannot complete that which has not yet begun. The conclusion in which we were brought up, that species are a product of a summation of variations, ignored the chief attribute of species that the product of their crosses is frequently sterile in greater or less degree. Huxley, very early in the debate, pointed out this grave defect in the evidence, but before breeding researches had been made on a large scale no one felt the objection to be serious [32].

A subsequent interchange with C. R. Crowther is of particular interest [33]. Questioning Bateson's attitude to the mechanism of meiotic chromosome pairing (Fig. 13.1 in Chap. 13), Crowther began by noting that while parental chromosomes had to cooperate for development of the zygote from embryo to adult, a far higher degree of cooperation would be needed when the chromosomes paired (conjugated) in the gonad of that adult:

> Homologous chromosomes … have to cooperate to produce the somatic cell of the hybrid, and their co-operation [for this purpose] might be expected to require a certain resemblance; but for the production of sexual cells [gametogenesis] they must do more, they must conjugate; and for conjugation it is surely reasonable to suppose that a much more intimate

> resemblance would be needed. We might, therefore, expect, on purely theoretical grounds, that as species and genera gradually diverged, it would be increasingly difficult to breed a hybrid between them; but that, even while a hybrid could still be produced, a fertile hybrid would be difficult or impossible, since the cells of the germ-track would fail to surmount the meiotic reduction stage when the homologous chromosomes conjugate. This is exactly what happens: the cells go to pieces in the meiotic phase.

Bateson's disparagement of the idea that species might be "a product of a summation of variations" left Crowther "frankly puzzled," for "the proposition is certainly not self-evident." Surely, if the sterility of an offspring were due to a failure within that offspring of homologous chromosomes to conjugate, it mattered little whether the lack of complementarity responsible for that failure was produced by one large variation or by the summation of many smaller variations. That Crowther was thinking of primary variations occurring at the chromosomal level, rather than anatomical variations of the sterile individual, was explicit:

> If a sword and its scabbard are bent in different directions, it will happen sooner or later that the sword cannot be inserted, and the result will be the same whether the bending be effected by a single blow, or whether it be, in Dr. Bateson's words, 'a product of a summation of variations.' Is this illustration inapt? The sword and the scabbard are the homologous chromosomes. ... It seems easier to imagine sterility arising from a gradual modification, spread over a length of time, and involving many chromosomes.

Bateson conceded that discontinuity of variation was not critical [34]:

> It is ... not difficult to 'imagine' interspecific sterility produced by a gradual (or sudden) modification. That sterility might quite reasonably be supposed to be due to the inability of certain chromosomes to conjugate, and Mr. Crowther's simile of the sword and the scabbard may serve to depict the sort of thing we might expect to happen. But the difficulty is that we have never seen it happen to swords and scabbards which we know to have belonged originally to each other. On the contrary, they seem always to fit each other, whatever diversities they may have acquired.

Modern Chromosomal Views

In his 1909 essay, Bateson did not spell out the relation between his postulated complementary sterility factors and the above "residue" or "irresoluble base." In Bateson's mind, the "residue" was not only distinct from genes – the agencies determining the "transferable characters" that were the subject of Mendelian analysis – but it could also limit the extent to which organisms could exchange genes. This limitation (i.e., absence of gene flow) would be complete when the organisms belonged to independent species [35]:

> These transferable characters are attached to or implanted upon some basal organization, and the attributes of powers which collectively form that residue may perhaps be distinguished from the transferable qualities. The detection of the limits thus set upon the interchangeability of characters would be a discovery of high importance and would have a most direct bearing on the problem of the ultimate nature of Species.

Bateson agreed with Crowther that a fundamental form of reproductive isolation, manifest as the hybrid sterility seen when allied species were crossed, was due to an incompatibility that could be characterized cytologically as impaired pairing of paternal and maternal chromosomes at meiosis. It was inferred that if we can understand what makes chromosomes incompatible, then we can understand hybrid sterility. And if we can understand hybrid sterility, we can understand an origin of species. But how do chromosomes that are homologous (i.e., are *alike*) pair with each other? Do they pair by virtue of this likeness (like-with-like) or by virtue of some key-in-lock (sword-in-scabbard) complementarity, which implies that they are not really alike? One must be the sword and the other the scabbard.

This paradox was resolved when it was appreciated that hereditary information was stored and transmitted as duplex DNA with two strands – a "Watson" strand and a "Crick" strand – that paired with each other by virtue of base complementarity [36]. So, in Crowther's terminology, potentially the sword strand of one chromosome can pair with the scabbard strand of the homologous chromosome (and vice versa). For this purpose, swords must be unsheathed from their own scabbards and then each inserted into the scabbards of the other. Thus, the Watson strand of one chromosome can pair with the Crick strand of the other, and vice versa. This requires that the Watson strand be displaced from pairing with the Crick strand of its own chromosome. Likewise, the Crick strand of the homologous chromosome must be displaced from pairing with the Watson strand of its own chromosome. Then cross-pairing (homology search without strand breakage) can occur, as Crick himself proposed [37, 38].

The pairing requires complementarity of DNA base sequences. A sporadically appearing change in a certain base could, if dominant, introduce a new phenotype (e.g., an extra petal) but would not greatly affect the overall complementarity between parental chromosomes. But, over time, base changes, including some affecting the classical phenotype, could accumulate. Romanes' "collective variation" that would build up "in a whole race or strain" can now be interpreted as the variation of the base composition of DNA about some characteristic value for the species (i.e., a normal distribution). When differences between chromosomal homologues reached a critical value, meiotic pairing would be impaired. A possible mechanism for this is described elsewhere [9, 26, 39] and summarized in Fig. 26.1 in Chap. 26. Indeed, assuming experiments with laboratory strains to be relevant to natural yeast populations, such "simple sequence divergence" appears to be the predominant form of species initiation in yeast [3]. The initiation might later be reinforced by segmental chromosomal changes that, in some circumstances, could themselves be capable of initiation, as described elsewhere [40, 41].

Conclusion

In the twentieth century, the writings of Romanes and Bateson on the origin of species were either disparaged or, more usually, ignored [11, 12]. However, we can now see them as latter-day "Mendels" whose recognition of the non-genic nature of the hybrid sterility arising from crosses between allied species anticipated, by a century, modern developments in genome analysis.

While differences in genes can progressively *differentiate* species, the spark that *originates* species can be non-genic. Bateson's non-genic "residue" is not in the cytoplasm but is "attached to or implanted upon" the chromosomes where the genes reside. Here mutations corresponding to the "collective variations" of Romanes can accumulate, eventually exceeding a critical value in a population subset (the difference between Z and Z' in the terminology of Fig. 28.3). In this event, a major discontinuity – an origin of species – can result.

We can now see how this might have come about at the DNA level. Further questions – whether, in the general case, genic changes have sparked species originations more often than non-genic changes and whether the latter have involved multiple base changes more frequently than gross karyotypic changes – are not argued here. The point is that, until controverted, the Romanes–Bateson hypothesis deserves at least an equal place at the table with other hypotheses.

Summary

Thomas Huxley identified as Darwin's "weak point" the failure of breeders when crossing varieties within a species to simulate the sterility of hybrids derived from crosses between allied species. Due to the sterility, the parents of a hybrid were, in an evolutionary sense, reproductively isolated from each other, so would be members of distinct species.

In his 1886 theory of "physiological selection," Romanes postulated germline "collective variations" that accumulate in certain members of a species, who are thus "physiological complements" producing fertile offspring when mutually crossed, but sterile offspring when crossed with others. Unlike Darwin's natural selection that secured reproductive isolation of the fit by elimination of the *unfit*, physiological selection postulated variations in the reproductive system that were not targets of natural selection; these could sympatrically isolate the fit from the *fit*, leaving two species where there was initially one.

Bateson approved physiological selection in 1886. He later noted that Mendel's "unit characters" were "sensible manifestations" of what we now refer to as "genes," but postulated a "residue," distinct from genes, that might affect gene flow between organisms and so originate species. The reproductive isolation of the parents of a sterile hybrid was due to two normally complementing non-genic factors (the "residue") separately introduced into the hybrid by each parent. Modern studies, especially of yeast hybrids, support the Romanes–Bateson viewpoint.

Chapter 29
Bateson's Residue: Oligonucleotide Disharmony May Drive Speciation

We appreciate music as a harmonious relationship between a succession of notes. Within the context of a given key or chord, many can detect a discordant note (relative pitch). The possibility that music might serve as a metaphor that could assist the understanding of Darwin's great problem – the origin of species – seems to have led an early geneticist to a "pitch theory of reproductive cells" [1]. A century later, the same problem was addressed in terms of the "wavelength" of the bases of DNA [2].

The discovery by Chargaff [3] of the parity of certain bases in DNA samples (A = T; G = C) guided Watson and Crick [4] when elaborating their double helical model for duplex DNA in 1953. Lesser known than this "first parity rule" are his 3 other DNA "rules" – the second parity rule, the cluster rule, and the GC rule [5]. The latter is that the GC% of a species (reciprocating with its AT%) is characteristic of entire genomes or of large genome sectors [6]. Implicit in this is that a high GC% species has more GC-rich oligonucleotides (e.g., GCG, GAG, GCC) than AT-rich oligonucleotides (e.g., ATA, AGA, ATT). Some species have high GC%, low AT%, values. Others have low GC%, high AT%, values. However, even when they have the *same* values, the oligonucleotide (k-mer) patterns of two species lack harmony. They can greatly differ, depending on the evolutionary distance between them. The patterns between phylogenetically close taxa appear more harmonious than those between distant taxa [7]. Indeed, rather than sequence alignments, k-mer patterns are now used routinely to *correct* taxonomic information in databases [8, 9].

The possibility that speciation mechanisms might include pattern disharmony among members of a species was considered by Rogerson in 1991 [10] and by Forsdyke in 1995 [11]. The question arose as to whether a specific k-mer pattern could be a consequence of a primary evolutionary pressure acting at the GC% level, or vice versa? Which was the cart and which was the horse – mononucleotide (1-mer) or oligonucleotide (k-mer)? The same question was posed by Blake and colleagues in 1992 [12] when pondering the context-dependence of point mutations relative to immediate upstream or downstream bases, namely mutations in 2-mers from the set of 4^2 (= 16) dinucleotides:

A. G. Cock, D. R. Forsdyke, *Treasure Your Exceptions*,
https://doi.org/10.1007/978-3-030-92099-9_29

> The classic studies of Benzer (1961) [13], showing the sites of … genes of T4 mutate with different efficiencies, could be seen as reflecting different neighbor effects at the several steps in the fixation of mutations. Benzer designated sites of high efficiency as hotspots. ... An influence of the neighbor environment could also be partially responsible for the biases in neighbor frequences in sequences (Josse et al. 1961) [14]. Whether such bias is a cause or an effect of one or more of these events remains to be demonstrated ….

In other words, was fixation of a mutation in a 2-mer (from one of the 15 other 2-mers in the set) due solely to an influence at the 1-mer level or to the 2-mer relationship? A case for the primacy of higher k-mers in evolutionary processes (i.e., a role of context) was argued on theoretical grounds by Forsdyke and Bell [15]. This was supported by detailed statistical analyses of many entire genomic sequences [16–18]. I here review the latter, drawing on the example of complementary interactions between mRNA codons and tRNA anticodons as a guide to the complexities of transient "kissing" interactions between recombining DNA molecules – a process with deep speciation-related historical roots [2, 8, 19–24]. Finally, I bring to general attention the existence of archival notes on a "pitch theory" by William Bateson [1], whose century-old attempt to relate speciation metaphorically to musical composition may have anticipated later studies.

Context Puzzlement

The availability of multiple sequenced human genomes enabled a statistical analysis of contextual relationships by Aggarwala and Voight in 2016 [17]. Rather than k = 1, the k-mer optimum for mutation fixation was found to be 7 and was held to reveal "principles" that – although the underlying mechanisms remained undetermined – were deemed as "fundamental."

Along similar lines, Morozov [18] reported k-mer optimalities ranging from k = 5 to k = 7 among various species and noted: "There is no question of whether k-mer distribution … is species-specific or whether the divergence of these distributions correlates with evolutionary distances. However, there is no answer to why it does." A similar puzzlement was expressed by Liu and Samee [25] when seeking to understand their results in terms of the chemistry of DNA structural differences between k-mers:

> It is generally believed that mutations occur and retain in the population through a process of 1) a mutation-promoting event causes the mutation, 2) the DNA repair mechanism fails to repair the mutation or repairs it incorrectly, and 3) the individual with the mutation manages to reproduce and retain the mutation in the population. Although step 3 has been a recurring theme …, there is limited understanding of how the interplay of steps 1 and 2 affect mutation rates of k-mers.

Initiation of Speciation

Within-species recombination and *between*-species anti-recombination would seem to be the fundamental principles they were seeking. It has been proposed that natural selection affects recombination by acting at the behest of higher order k-mers. The quantity of 1-mers (GC%) is secondary to this. It can be supposed that primarily neutral variation generates k-mer differences. A discordant "note" in, say, a 5-mer, changes it to one of the 4^5–1 (= 1023) alternative 5-mers. This could then inhibit local within-species recombination and, if by chance aggregated genome-wide with numerous similar discords, could also collectively serve, either to initiate sympatric divergence prior to the entry of natural selection into the speciation process or to sustain the reproductive isolation that prior natural selection had brought about [2, 8, 19, 22]. Extruded DNA stem-loop structures would be involved (see below).

Although natural selection can directly influence base composition (e.g. purine-loading in keeping with the Chargaff cluster rule [26]), many differences that accumulate between species over time may be selectively neutral. These *interspecies* mutations could sometimes change genome-wide oligonucleotide (k-mer) frequencies *directionally*, for which 1-mer differences (increases or decreases in GC%) would act as crude indicators. Being neutral, there are no major changes in phenotype until natural selection comes into play. Thus, new species appear to jump, discontinuously, into existence [27].

An association of genome-wide GC% (1-mer) values with amino acid usage was reported by Sueoka in 1961 [28]. However, a greater influence of higher k-mers was elegantly shown by Fran Supek and his colleagues, who reported that, when some choice of amino acid was possible, k-mer frequencies where $k > 1$ would predict amino acid usage with greater accuracy than 1-mer frequencies [16]. They concluded that "evolutionary shifts in overall amino acid composition appear to occur almost exclusively through factors shaping the global oligonucleotide content of the genome." Similar conclusions were arrived at using the amino acid k-mers corresponding to nucleotide base k-mers rather than nucleotide base k-mers themselves [18]. Thus, oligonucleotide composition acts as a genome-wide *internal* enabling force, working in conjunction with, but distinct from, conventional, externally imposed, natural selection. Divergence into independent species can result.

RNA–RNA "Kissing" Recognition

Although early examples of k-mer control of nucleic acid interactions were at the RNA level, there is reason to suppose that DNA–DNA interactions are similar (see later). One of the most fundamental reactions in biology is the reversible "kissing" between the anticodon sequence in a loop in the stem-loop structure of a tRNA molecule and the corresponding codon in an mRNA molecule. This occurs rapidly, and it can be inferred that the intracellular environment (e.g., salt concentrations and pH) has been evolutionarily fine-tuned in its favor, so affecting interactions not only between tRNA and mRNA but also between RNA molecules in general. Although k = 3 for the specific codon–anticodon k-mers themselves, the more constant bases neighboring a codon are also involved [29]. So, it is essentially a 5-mer–5mer interaction with the constant neighbors imposing regularities in the disposition of purines and pyrimidines in the codons themselves (the "RNY rule" [27, 30]).

Conventionally, a single base (1-mer) mutation affecting a codon position is seen as the primary cause of a single amino acid change that may lead to a pathology, such as sickle cell anemia (replacement of glutamate with valine). Fixated base substitution mutations are viewed as 1-mer phenomena. However, *functionally*, we can now see these as mutations from one of 4^3 (= 64) 3-mer codons to one of the 63 alternative 3-mers. The *selective* difference created by such mutations usually relates to local base *context* (i.e., the k-mers within which a primary 1-mer mutation is embedded). It is a k-mer phenomenon. For example, at the DNA level, a central A to T codon mutation in sickle cell anemia (GAG to GTG) attracts, at the RNA level, a complementary anticodon in a loop of a valine tRNA, not of a glutamate tRNA. Thus, a hydrophobic amino acid is exchanged for a hydrophilic amino acid. On the other hand, G to T mutations in the first or third codon position should result, at the RNA level, in lethal chain termination (UAG) or a mild change from glutamate to aspartate (GAU).

Tomizawa's studies of RNA–RNA regulatory interactions [31] demonstrated the more general nature of primary loop-kissing interaction mechanisms, which could then lead to more extensive base-pairing between stem sequences. If an initial interaction were favorable – namely there was precise complementary base pairing between two loop k-mers – then neighboring duplex stem structures would unwind to explore the possibility of extending the pairing. The two RNA strands could be either in-tune or discordant.

Discordant DNA Loops

Noting Tomazawa's studies, and building on the hypothesis that the individual strands of parental meiotic DNA duplexes might "unpair" to allow homologous cross-pairing [32], in 1993 Kleckner and Weiner [33] proposed that a successful k-mer "kissing" with a complementary k-mer might be consummated by recombination. On the other hand, if the k-mers were not complementary (base-pairing discordance) recombination would fail, so acting to generate, and/or sustain, members of emerging species by preventing meiotic recombination between parental sequences [2, 8, 19, 22, 34].

Given their advantages (e.g., error-correction), it is likely that recombination mechanisms evolved by natural selection at an early stage in evolution. These mechanisms involve a substrate (DNA) and enzymes that act on that substrate. A substrate such as glucose cannot adapt to improve its metabolism by specific enzymes. However, DNA *can* adapt to assist the function of enzymes acting upon it [35].

Thus, *both* substrate and the enzymes acting on that substrate can be targeted by natural selection to improve recombination efficiency – hence genome-wide stem-loop potential (the potential to extrude single strand stem-loop structures from duplex DNA strands). Thus, recombination begins with higher order k-mers in loops seeking complementary k-mers on other loops [36, 37]. Failure to achieve complementation could initiate, and/or sustain, the speciation process.

Selectable K-mers as Arbiters

Within a nucleic acid sequence, are conserved regions and non-conserved regions. In the latter, mutations are likely to be accepted and can become fixed in a species. A region designated as "conserved" because no mutations have been scored there, could either have escaped mutation completely (i.e., real conservation), or the mutations would have been correctly repaired before scoring could be attempted (i.e., lagged conservation), or the mutations would have been lethal, so organisms with those mutations would have been eliminated from the population and would not be available for scoring (apparent conservation).

Given 4 (= 4^1) bases, designation of a region as "non-conserved" requires the scoring of just one point mutation (i.e., a change of one of the four into one of the other three). Thus, the region would have escaped from either real, lagged, or apparent conservation, the latter being a predominant assumption since evidence to the contrary (perhaps a marginal decrease in population size) is difficult to obtain. The scoring as non-conserved would have indicated departure of the ancestral base and its replacement by another base that had not resulted in lethality. The rate at which this occurs is encompassed by the term "mutation rate," which implies that the unit of mutation is the single base.

It is here implied that we should not let such focus distract us from the higher k-mer units that are often the ultimate arbiters, through their influence on selection, on the scoring of 1-mer mutations that occur within themselves. Thus, as concluded by Aggarwala and Voight [17], a "sequence context model broadly explains variability." Their value of >81% success for higher k-mers explaining substitution probabilities, admits the existence of, but leaves little space for, other factors. Furthermore, their "strong correlation between divergence and substitution probabilities" raises the possibility of a role in speciation.

Bateson's Pitch Theory

Even before noting an instrumental role of "the residue" in speciation, William Bateson was referring to "vibrations" and "waves" [38, 39]. An archived set of hand-written notes, likely dating from the early 1890s (precise date uncertain), is here brought to general attention [1] (Table 29.1). Fertility is equated with "compatibility of pitch." A sterility that can lead to divergence into new species is equated with "incompatibility of pitch." The success of his 23-point "pitch theory" in relating complex features of musical harmony or disharmony to mechanisms for species conservation or origination, as inferred here, is left for the reader to decide. Musical notes being quantifiable in terms of frequencies and wavelengths, a metaphor of radiowave frequencies [2] was similar, but perhaps less successful. The best metaphor would seem to be Bateson's, where the notion of one off-key note in a diatonic scale or chord is easily transferrable to the notion of an "off-key" base in a k-mer oligonucleotide. Thus, Nature senses disharmony by assessing *relative* pitch (comparing with neighboring notes), not *absolute* pitch (no reliance on neighbors).

Table 29.1 Pitch theory of reproductive cells [Transcript by DRF of William Bateson's notes (possibly early 1890s); Originals in Queen's University Archives G5e: 11a, 11b]

1	Species	Melodies
2	Reproductive cells (units from which species start)	Keynotes. Units of vibration from which melodies start. Definite units of vibration, bearing definite relation to all other notes in the melody
3	Relation of reproductive cells of one species to reproductive cells of another species of same genus	Relation of Tone to its Overtones
4	Relation of reproductive cells of one species to those of another with which it can be cooped and yield fertile offspring	Relation of Tone to its nearest overtones
5	Hybrids can be obtained from the reproductive cells of two species which yield fertile offspring	Chords in harmony can be formed from fundamental notes struck with their nearly altered overtones, and further two melodies may be worked together if their fundamental tones are in simple relations to each other to form a counterpoint
6	Hybrids are for the most part obviously compounded of their parents. Often (usually?) definite pieces of each parent appear (though sometimes intermediate between the two in part at all events)	Two voice Counterpoints are obviously composed of two melodies. Pieces have sometimes to be interpolated or left out to make the parts fit together (?) yet the two melodies are still apparent
7	[empty, but WB maintained the spacing and numbering, so that the two sets of sheets could be aligned for comparison. DRF]	The figures made by sand on glass plates when made to sound to certain tones. Their outward resemblance to appearance of segmenting eggs, transverse sections of stems, veins in leaves, etc.
8	Often an hybrid can be produced, between two outwardly very similar species	No counterpoint can be made of two voices singing the same melody in unrelated or distant keys
9	Nature abhors a hybrid	Musical students abhor counterpoint
10	A species may be varied to any extent, yet as a rule, varieties breed with their parent species no matter how varied they have become. Yet some species, though more nearly allied, to all appearances as a rule will not or do not breed together	A melody may be varied to any extent yet remain the same melody and stand in the same relation to other melodies unless its ground tone is altered in pitch. Then though it is still the same melody it does not bear the same relation [to] its original melody nor to other melodies

(continued)

Table 29.1 (continued)

1	Species	Melodies
11	Reciprocal hybrids (such as mules and hinnies)	Twisted intervals or inverted counterpoints (?are there such things)
12	The power which a male cell has of making a female cell of same species segment	The power which vibrations from one tuning fork have of making another of the same pitch originally at rest sing out its note
13	Entire prepotency of one parent over another. Direct effect of male over female (as in oranges)	The power which a swinging pendulum possesses of coercing one swinging beside it, into its own period of oscillation
14	The power which certain species of the same genus have of breeding together	The power which tones not of the same pitch as the resonant body have of throwing resonant bodies into vibration, if they contain the right tone amongst their upper partials (overtones)
15	The fact that certain species are more fertile with pollen of other species than their own (i.e., dimorphic plants and self-sterile plants)	The power which certain resonant bodies (circular membranes) possess of responding only to sounds more acute than their own note
16	Species of the same genus	Notes on a diatonic scale
17	Possible varieties filling in between species	Chromatic or not "well-tempered" intervals in sound
18	A variety cooped with its parent reproduces within itself or its parent (although are said by Darwin to be not different from hybrids in kind. Is that so? They do seem different really)	A melody played with its variation forms a unison of melody and variation not a counterpoint
19	Any number of varieties may cross with each other yet the original species remains the same, with perfect fertility	Any number of instruments of different Timbre may play the same melody together, yet there is only unison, no counterpoint
20	Fertility	Compatibility of pitch
21	Sterility	Incompatibility of pitch
22	Hybrids are either totally sterile or reproduce one or other of the parents, but not the hybrid form; do not seem capable of producing intermediate reproductive cells. The means of reproduction are present as a rule the cells only are wanting	Two melodies started in different keys and played together in counterpoint have still two different keynotes, not a keynote intermediate between the two
23	Species are thought to be derivable one from the other. Should there be anything in this pitch theory of reproductive cells, it is conceivable that a change in what corresponds to timbre in living things may occasionally take place suddenly right through an organism, whereby their reproductive cells should change in timbre also and "synchronize"[?] with the cells of their equally changed brethren? (Rather wild!)	Diatonic intervals are derived one from the other There are definite overtones belonging to every fundamental tone. Certain instruments make overtones more apparent to our senses than others (this is a matter of timbre in the instrument itself). Also if the lowest tone of a resonance box does not correspond with the prime tone but with some of the upper partials, the corresponding upper partial predominates

Summary

A biological explanation for the dependence of genome-wide mutation-rate variation on local base context is now becoming clearer. The proportion of G + C relative to A + T – expressed as GC% – is a species-specific DNA character. The frequencies of these single bases (1-mers) correlate with frequencies of corresponding oligonucleotides (k-mers) that are *more-sensitive* indicators of species specificity. Thus, when k = 3, there are 64 possible trinucleotide sequences and a GC%-rich species has a high frequency of GC-rich 3-mers (e.g., GCG, CGG, CGC, etc.). Closely related species have similar k-mer patterns. For distantly related species, even if happening to have the *same* GC% values, k-mer patterns (where k > 1) are widely discordant. Conventionally, a base substitution mutation is viewed as a chemical 1-mer phenomenon. However, the selective difference by which the corresponding mutation is scored usually relates to context (i.e., to the k-mers within which a 1-mer mutation is embedded). Thus, an A to U codon mutation in sickle cell anemia (GAG to GUG), rather than attracting an anticodon in a structural loop of glutamate tRNA, pairs with the more complementary anticodon loop of valine tRNA. Likewise, a recent statistical analysis of genome-wide mutation-rate variation supports the view that a failure of discordant DNA loops to pair at meiosis can initiate and/or sustain the speciation process. Such disharmony among base patterns may have unknowingly been hinted at by geneticist William Bateson who invoked a music metaphor when considering the role of his mysterious "residue" in speciation mechanisms.

Publications of William Bateson

Bateson's papers are in Punnett (1928). *Scientific Papers of William Bateson*, vols. 1 and 2, Cambridge University Press. Some are available in Beatrice Bateson's *Memoir* (1928) and in Forsdyke's web-pages:Archive-It – Faculty Research. Another resource is the collected papers of Charles Hurst (1925) *Experiments in Genetics*, Cambridge University Press.

1. Bateson, W. (1884a) The early stages in the development of *Balanoglossus*. *Quarterly Journal of Microscopical Science* 24, 208–236.
2. Bateson, W. (1884b) On the development of Balanoglossus. *Annals and Magazine of Natural History* 13, 65.
3. Bateson, W. (1885) The later stages in the development of *Balanoglossus kowalevskii*, with a suggestion as to the affinities of the Enteropneusta. *Quarterly Journal of Microscopical Science* 25(supplement), 81–122.
4. Bateson, W. (1886a) Continued account of the later stages in the development of *Balanoglossus kowalevskii*, and of the morphology of the Enteropneusta. *Quarterly Journal of Microscopical Science* 26, 511–533.
5. Bateson, W. (1886b) The ancestry of the chordata. *Quarterly Journal of Microscopical Science* 26, 535–571.
6. Bateson, W. (1888) Suggestion that certain fossils known as *Bilobites* may be regarded as casts of *Balanoglossus*. *Proceedings of the Cambridge Philosophical Society* 6, 298.
7. Bateson, W. (1889a) On some variations of *Cardium edule*, apparently correlated to the conditions of life. *Philosophical Transactions of the Royal Society, B.* 180, 297–330.
8. Bateson, W. (1889b) On some variations of *Cardium edule*, apparently correlated to the conditions of life. *Proceedings of the Royal Society, B.* 46, 204–211.
9. Bateson, W., & Bourne, G. C. (1889) Notes and memoranda: Notes on the senses and habits of some Crustacea. *Journal of the Marine Biological Association* 1, 211–217.
10. Bateson, W. (1890a) The sense organs and perceptions of fishes: With remarks on the supply of bait. *Journal of the Marine Biological Association* 1, 225–256.
11. Bateson, W. (1890b) On some cases of abnormal repetition of parts in animals. *Proceedings of the Zoological Society* 579.
12. Bateson, W. (1890c) On some skulls of Egyptian mummified cats. *Proceedings of the Cambridge Philosophical Society* 7, 68. [Abstract].
13. Bateson, W. (1890d) On the nature of supernumerary appendages in insects. *Proceedings of the Cambridge Philosophical Society* 7, 159. [Abstract].
14. Bateson, W. (1890e) Application for Deputy Linacre Professorship, [See C. B. Bateson *Memoire* (1928).]
15. Bateson, W. (1890f) *For Greek*. Cambridge University Press. [Flysheet].
16. Bateson, W., & Bateson, A. (1891) On variations in floral symmetry of certain plants having irregular corollas. *Journal of the Linnean Society, Botany* 28, 386–421.
17. Bateson, W. (1892a) On variation in the colour of cocoons of *Eriogaster lanestris* and *Saturnia carpini* (pp. 45–52). *Transactions of the Entomological Society*.

A. G. Cock, D. R. Forsdyke, *Treasure Your Exceptions*,
https://doi.org/10.1007/978-3-030-92099-9

18. Bateson, W. (1892b) On numerical variation in teeth, with a discussion of the conception of homology. *Proceedings of the Zoological Society* 102–115.
19. Bateson, W. (1892c) On variation in the colour of cocoons, pupae and larvae. Further Experiments (pp. 205–214). *Transactions of the Entomological Society*.
20. Bateson, W. (1892d) The alleged "aggressive mimicry" of *Volucellae*. *Nature* 46, 585.
21. Bateson, W. (1892e) The alleged "aggressive mimicry" of *Volucellae*. *Nature* 47, 77.
22. Bateson, W. (1892f) Exhibition of, and remarks upon, some crab's limbs bearing supernumary claws. *Proceedings of the Zoological Society* 76. [A five-line note.].
23. Bateson, W., & Brindley, H. H. (1892) On some cases of variation in secondary sexual characters statistically examined. *Proceedings of the Zoological Society* 585–594.
24. Bateson, W. (1893) Exhibition of and remarks upon an abnormal foot of a calf. *Proceedings of the Zoological Society* 530–531.
25. Bateson, W. (1894a) *Materials for the Study of Variation, Treated with Especial Regard to Discontinuity in the Origin of Species*. Macmillan.
26. Bateson, W. (1894b) Exhibition of specimens of the common pilchard (*Clupea pilchardus*) showing variation in the number and size of the scales. *Proceedings of the Zoological Society* 164.
27. Bateson, W. (1894c) Exhibition of specimens and drawings of a phytophagus beetle, in illustration of discontinuous variation in colour. *Proceedings of the Zoological Society* 391.
28. Bateson, W. (1894d) On two cases of colour-variation in flat-fishes, illustrating principles of symmetry. *Proceedings of the Zoological Society* 246–249.
29. Bateson, W. (1895a) Note in correction of a paper on colour-variation in flat-fishes. *Proceedings of the Zoological Society* 890–891.
30. Bateson, W. (1895b) The origin of the cultivated Cineraria. *Nature* 51, 605–607.
31. Bateson, W. (1895c) The origin of the cultivated Cineraria. *Nature* 52, 29.
32. Bateson, W. (1895d) The origin of the cultivated Cineraria. *Nature* 52, 103.
33. Bateson, W. (1895e) Notes on hybrid Cinerarias produced by Mr. Lynch and Miss Pertz. *Proceedings of the Cambridge Philosophical Society* 10, 308.
34. Bateson, W. (1895f) On the colour variations of a beetle of the family *Chrysomelidae* statistically examined. *Proceedings of the Zoological Society* 850–860.
35. Bateson, W. (1896) Exhibition of, and remarks upon, three pigeons showing webbing between the toes. *Proceedings of the Zoological Society* 989–990.
36. Bateson, W. (1897a) Habits of *Zygaena exulens*. *Entomological Record* 9, 328. [Short note on date of emergence and relation to snow.].
37. Bateson, W. (1897b) On progress in the study of variation. *Science Progress* 6, 554–568.
38. Bateson, W. (1898a) On progress in the study of variation. *Science Progress* 7, 53–68.
39. Bateson, W. (1898b) Experiments in the crossing of local races of Lepidoptera. *Entomological Record* 10, 241. [Report of an exhibition of *napi, egeria,* etc., at the Fourth International Zoological Congress, Cambridge. This, and a passage in *Problems in Genetics*, form the only published account of this work.].
40. Bateson, W. (1898c) Protective colouration of Lepidopterous pupae. *Entomological Record* 10, 285.
41. Bateson, W. (1900a) Hybridization and cross-breeding as a method of scientific investigation. *Journal of the Royal Horticultural Society* 24, 59–66. [From the First International Conference on Plant Breeding and Hybridization, London, 1899.].
42. Bateson, W. (1900b) On a case of homoeosis in a crustacean of the genus *Asellus*. Antennule replaced by a mandible. *Proceedings of the Zoological Society* 268–271.
43. Bateson, W. (1900c) British lepidoptera. *Entomological Record* 12, 231. [Review of volume 2 of Tutt's *British Lepidoptera*.].
44. Bateson, W. (1900d) Collective inquiry as to progressive melanism in moths. *Entomological Record 12*, 140. [Memorandum from the Evolution Committee of the Royal Society, London; also in *Entomology Monthly Magazine*, p. 139.].
45. Bateson, W., & Pertz, D. (1900) Notes on the inheritance of variation in the corolla of *Veronica buxhaumii*. *Proceedings of the Cambridge Philosophical Society* 10, 78–92.

46. Bateson, W. (1901a) Problems of heredity as a subject for horticultural investigation. *Journal of the Royal Horticultural Society* 25, 54–61. [Read May 8, 1900].
47. Bateson, W. (1901b) Heredity, differentiation, and other conceptions of biology: a consideration of Professor Karl Pearson's paper 'On the principle of Homotyposis.' *Proceedings of the Royal Society* 69, 193–205.
48. Bateson, W. (1902a) Introductory note to the translation of 'Experiments in plant hybridization' by Gregor Mendel. *Journal of the Royal Horticultural Society* 26, 1–3.
49. Bateson, W. (1902b) *Mendel's Principles of Heredity. A defence*. Cambridge University Press. [Reprinted (1903) as: *The Problems of Heredity and Their Solution*. Smithsonian Museum, Washington.].
50. Bateson, W. (1902c) Note on the resolution of compound characters by cross-breeding. *Proceedings of the Cambridge Philosophical Society* 12, 50–54.
51. Bateson, W. (1902d) British lepidoptera. *Entomological Record* 14, 320. [Review of volume 3 of Tutt's *British Lepidoptera*.].
52. Bateson, W., & Saunders, E. R. (1902) Experimental studies in the physiology of heredity. *Reports to the Evolution Committee of the Royal Society* I, 1–160.
53. Bateson, W. (1903a) *Variation and Differentiation in Parts and Brethren*. Cambridge University Press.
54. Bateson, W. (1903b) On Mendelian heredity of three characters allelomorphic to each other. *Proceedings of the Cambridge Philosophical Society* 12, 153–154.
55. Bateson, W. (1903c) The present state of knowledge of colour-heredity in mice and rats. *Proceedings of the Zoological Society, London* 2, 71–99.
56. Bateson, W. (1903d) Mendel's principles of heredity in mice. *Nature* 67, 462. and 585; *68*, 33.
57. Bateson, W. (1904a) Practical aspects of the new discoveries in heredity. *Memoirs of the Horticultural Society of NewYork* 1, 1–9. [Opening paper at the Second International Conference on Plant Breeding and Hybridization, New York, September, 1902.].
58. Bateson, W. (1904b) Presidential Address to Section D (Zoology) of the British Association, Cambridge. [Reprinted in part as: Heredity and evolution. *Popular Science Monthly*, New York. pp. 522–531.]
59. Bateson, W. (1904c) Albinism in Sicily. A correction. *Biometrika* 3, 471–472. [Letter correcting Weldon.].
60. Bateson, W. (1904d) A natural history of British lepidoptera. *Entomological Record* 16, 234. [Review of volume 4 of Tutt's *British Lepidoptera*.].
61. Bateson, W. (1904e, Feb 18) Exhibition of a series of *Primula sinensis*. *Linnean Society. Report of General Meeting* pp. 2–3.
62. Bateson, W. (1905a) *Experimental studies in the physiology of heredity. Report to Committee*. British Association 1904, pp. 346-348. [74th meeting, Cambridge]
63. Bateson, W. (1905b, June 24) Evolution for amateurs. *The Speaker*. [A review of Weismann's *The Evolution Theory* (1904), Edward Arnold, London.]
64. Bateson, W. (1905c) Compulsory Greek at Cambridge. *Nature* 71, 390.
65. Bateson, W. (1905d, Oct. 14) Heredity in the physiology of nations. *The Speaker*. [A review of Archdall Reid's *The principles of heredity*.]
66. Bateson, W. (1905e) Practical aspects of the new discoveries in heredity. *Journal of the Royal Horticultural Society* 29, 417–419. [Abstract of paper from Second Conference on Plant-Breeding and Hybridization, New York, 1902.].
67. Bateson, W. (1905f) The exhibition of, and remarks upon specimens of fowls, illustrating peculiarities in the heredity of white plumage. *Proceedings of the Zoological Society, London* 2, 3.
68. Bateson, W. (1905g) Albinism in Sicily. A further correction. *Biometrika* 4, 231.
69. Bateson, W., & Punnett, R. C. (1905) A suggestion as to the nature of the 'walnut' comb in fowls. *Proceedings of the Cambridge Philosophical Society* 13, 165–168.
70. Gregory, R. P., & Bateson, W. (1905) On the inheritance of heterostylism in Primula. *Proceedings of the Royal Society, B.* 76, 581–586.

71. Bateson, W., Saunders, E. R., & Punnett, R. C. (1905) Experimental studies in the physiology of heredity. *Reports to the Evolution Committee of the Royal Society* II, 1–131. [In his listing in *Scientific Papers*, Punnett adds H. Killby to the author list, and omits C. C. Hurst whose "Experiments with Poultry," although separate, formed part of the report (pp. 131–154).].
72. Bateson, W. (1906a, April 14) Science of sorts. *The Speaker.* [Review of Burke's *The Origin of Life*.]
73. Bateson, W. (1906b, Aug. 4) The progress of genetic research. *Gardeners' Chronicle*. [An Inaugural address to the Third Conference on Hybridism and Plant Breeding. Also published (1907) as the *Report of the Third International Conference 1906 on Genetics*. Edited by W. Wilks. Royal Horticultural Society, London. pp. 90–97].
74. Bateson, W. (1906c) Mendelian heredity and its application to man. *Brain* 29, 157–179. [February 1st address to Neurological Society, London.].
75. Bateson, W. (1906d) The progress of genetics since the rediscovery of Mendel's papers. *Progressus Rei Botanicae* 1, 368-418. [Jena]
76. Bateson, W. (1906e) A textbook of genetics. *Nature* 74, 146. [Review of Lotsy's *Vorlesungen über Descendenz-theorien*. I.].
77. Bateson, W. (1906f) Coloured tendrils of sweet peas. *Gardeners' Chronicle (Series 3)* 39, 333.
78. Bateson, W., Saunders, E. R., & Punnett, R. C. (1906a) Further experiments on inheritance in sweet peas and stocks: preliminary account. *Proceedings of the Royal Society, B.* 77, 236–238.
79. Bateson, W., Saunders, E. R., & Punnett, R. C. (1906b) Experimental studies in the physiology of heredity. *Reports to the Evolution Committee of the Royal Society* III, 1–53.
80. Bateson, W. (1907a) Facts limiting the theory of heredity. *Science* 26, 649–660. [Also (1912) *Proceedings of the International Congress of Zoology* (Boston, 1907) 7, 306–319.].
81. Bateson, W. (1907b) Trotting and pacing: dominant and recessive. *Science* 26, 908.
82. Bateson, W. (1908a) *The Methods and Scope of Genetics*. Cambridge University Press. [Inaugural lecture, Oct. 23].
83. Bateson, W. (1908b) Lectures on evolution. *Nature* 78, 386. [Review of Lotsy's *Vorlesungen über Descendenz-theorien,* II.].
84. Bateson, W. (1908c) British Association discussion on sex-determination. Correction. *Nature* 78, 665.
85. Bateson, W., & Punnett, R. C. (1908) The heredity of sex. *Science* 27, 785–787.
86. Bateson, W., Saunders, E. R., & Punnett, R. C. (1908) Experimental studies in the physiology of heredity. *Reports To the Evolution Committee of the Royal Society* IV, 1–59. [H. B. Killby is a coauthor in Saunder's section. F. M. Durham has sections on coat color in mice (pp. 41–53), and sex in canaries (pp. 57–59). L. Doncaster has a section on sex-inheritance in moths (pp. 53–57).].
87. Bateson, W. (1909a) *Summary of "Mendel's Principles."* Harmsworth's World's Great Books.
88. Bateson, W. (1909b) Heredity and variation in modern lights. In A. C. Seward (Ed.), *Darwin and Modern Science* (pp. 85–101). Cambridge University Press.
89. Bateson, W. (1909c) *Mendel's Principles of Heredity*. Cambridge University Press.
90. Conklin, E. G., et al. (1909) William Keith Brooks. A sketch of his life by some of his former pupils and associates. *Journal of Experimental Zoology* 9, 1–51.
91. Wheldale, M., Marryat, D. C. E. & Sollas, I. B. J. (1909) Experimental studies in the physiology of heredity. *Reports to the Evolution Committee of the Royal Society* V, pp. 1-78. [Bateson will have read and approved this report.]
92. Bateson, W. (1910) Boyle Lecture. Delivered at Oxford 1909. [Manuscript not found.]
93. Bateson, W. (1911a) An appreciation of J. W. Tutt. *Entomological Record* 23, 123–124.
94. Bateson, W. (1911b) The John Innes Horticultural Institute Report for 1910. *Gardeners' Chronicle (Series 3)* 49, 179. [Subsequent annual reports were privately printed for the Council of the Institution.].
95. Bateson, W. (1911c) Recent advances in the genetics of plants. *Nature* 88, 36–37. [Review of Baur's *Einführung in die experimentelle Vererbungslehre*.].

96. Bateson, W. (1911d) Presidential address to the Agricultural Subsection, British Association, Portsmouth. *British Association Report* 587–596.
97. Bateson, W., & de Vilmorin, P. L. (1911) A case of gametic coupling in *Pisum*. *Proceedings of the Royal Society, B.* 84, 9–11.
98. Bateson, W., & Punnett, R. C. (1911a) On the interrelations of genetic factors. *Proceedings of the Royal Society, B.* 84 3–8.
99. Bateson, W. & Punnett, R. C. (1911b) Reduplication of terms in series of gametes. In *Proceedings of the 4th Conference Internationale de Génétique*. [Conference Proceedings published in 1913, pp. 99–100.]
100. Bateson, W., & Punnett, R. C. (1911c) The inheritance of the peculiar pigmentation of the silky fowl. *Journal of Genetics* 1, 185–203.
101. Bateson, W., & Punnett, R. C. (1911d) On gametic series involving reduplication of certain terms. *Journal of Genetics, 1* 293–302. [Also in *Verh. Naturforsch. Verein. Brünn* 49, 324–334.].
102. Bateson, W. (1912a) *Biological Fact and the Structure of Society*. Clarendon Press. [Herbert Spencer Lecture, delivered February 28.].
103. Bateson, W. (1912b) Lectures to Royal Institution (Fullerian Professorship). *Gardeners' Chronicle* 51, 57, 74, 89, 104, 120, 123, 139.
104. Bateson, W. (1913a) Problems of the cotton plant. *Nature* 90, 667–668. [Review of Balls' *The Cotton Plant in Egypt*.].
105. Bateson, W. (1913b) *Problems in Genetics*. Yale University Press. [Silliman Memorial Lectures.].
106. Bateson, W. (1913c) Heredity. *British Medical Journal* 2, 359–362. *Lancet* 2, 451–454. [Address to Seventeenth International Congress of Medicine.].
107. Bateson, W. (1913d) *Oenothera* crosses. *Gardeners' Chronicle, Series* 3, 54, 406.
108. Bateson, W. (1913e) Discussion of Lotsy's theory of evolution by hybridization. *Proceedings of the Linnean Society* 89. [A paragraph.].
109. Bateson, W. (1914a) Presidential addresses, British Association Meetings at Melbourne and Sydney. *Nature* 93, 635–642. and 674–681. [Also (1916) *Smithsonian Report*, pp. 359–394.].
110. Bateson, W. (1914b) Royal Institution and Fullerian Lectures. *Gardeners' Chronicle* 55, 74, 92, 112, 131, 149, 171, 332–33.
111. Bateson, W. (1915) Address to salt schools. [In C. B. Bateson, *Memoire*, 1928.]
112. Bateson, W., & Pellew, C. (1915a) On the genetics of "rogues" among culinary peas, *Pisum sativum*. *Journal of Genetics* 5, 13–36.
113. Bateson, W., & Pellew, C. (1915b) Note on an orderly dissimilarity in inheritance from different parts of a plant. *Proceedings of the Royal Society, B.* 89, 174–175.
114. Bateson, W. (1916a) Review of *The mechanism of Mendelian heredity* by Morgan et al. *Science* 44, 536–543.
115. Bateson, W. (1916b) Notes on experiments with flax at the John Innes Horticultural Institution. *Journal of Genetics* 5, 199–201.
116. Bateson, W. (1916c) Root-cuttings, chimaeras and "sports.". *Journal of Genetics* 6, 75–80.
117. Bateson, W. (1917a) Gamete and zygote. A lay discourse. [Henry Sidgwick Lecture, Newnham College. In C. B. Bateson, *Memoire*, (1928).]
118. Bateson, W. (1917b) Philippe Leveque de Vilmorin. *Proceedings of the Linnean Society of London, Session, 130* 44–46.
119. Bateson, W. (1917c) *The Place of Science in Education. Cambridge Essays on Education*. Cambridge University Press.
120. Bateson, W. (1917d) The ear of Dionysius. *The Times Literary Supplement*. May 3 and May 17.
121. Bateson, W. (1917e) Is variation a reality? *Nature* 99, 43. [Review of Lotsy's *Evolution by Means of Hybridization*.].
122. Bateson, W., & Thomas, R. H. (1917) Note on a pheasant showing abnormal sex-characters. *Journal of Genetics* 6, 163–164.

123. Bateson, W. (1918, February 15) Gamete and zygote. *Proceedings of the Royal Institute of Great Britain* 1–4. [Abstract. Also in C. B. Bateson, *Memoire*, 1928.].
124. Bateson, W. (1919a) Science and nationality. *Edinburgh Review* 229, 123-138. [Inaugural address to the Yorkshire Natural Science Association.]
125. Bateson, W. (1919b) Progress in Mendelism. *Nature* 104, 214.
126. Bateson, W. (1919c) Linkage in the silk worm. A correction. *Nature* 104, 315.
127. Bateson, W. (1919d) Studies in variegation. I. *Journal of Genetics* 8, 93–99.
128. Bateson, W. (1919e) Dr. Kammerer's testimony to the inheritance of acquired characters. *Nature* 103, 344.
129. Bateson, W., & Sutton, I. (1919) Double flowers and sex-linkage in *Begonia*. *Journal of Genetics* 8, 199–207.
130. Bateson, W. (1920a, June) Classical education and science men. Précis of evidence offered to the Prime Minister's Committee on Classics. [In C. B. Bateson, *Memoire*, (1928).]
131. Bateson, W. (1920b) Genetic segregation. *Proceedings of the Royal Society, B.*, 91, 358–368. [Croonian Lecture. Also in (1920) *Nature* 105, 531 [abstract] with a correction (1921) *Nature* 107, 233; and in (1921) *American Naturalist* 55, 1–19.].
132. Bateson, W. (1920c) Organization of scientific work. *Nature* 105, 6. [Letter concerning State organization of science in India.].
133. Bateson, W. (1920d) Prof. L. Doncaster, FRS. *Nature* 105, 461–462.
134. Bateson, W., & Pellew, C. (1920) The genetics of "rogues" among culinary peas, *Pisum sativum*. *Proceedings of the Royal Society, B.*, 91, 186–195.
135. Bateson, W. (1921a) Common sense in racial problems. *Eugenics Rev* 13, 325–338. [Galton Lecture.].
136. Bateson, W. (1921b) Leonard Doncaster, 1877–1920. *Proceedings of the Royal Society, B.* 92, 41–46.
137. Bateson, W. (1921c) Root cuttings and chimaeras. II. *Journal of Genetics* 11, 91–97.
138. Bateson, W. (1921d) The determination of sex. *Nature* 106, 719. [Review of Goldschmidt's *Mechanismus und Physiologie der Geschlechtsbestimmung*.].
139. Bateson, W. (1921e) Variegation in a fern. *Nature* 107, 233. [Correction to Croonian Lecture.].
140. Bateson, W. (1921f) Classical and modern education. *Nature* 108, 308. [Review of *Classics in Education*, the report of the Prime Minister's Committee to inquire into the position of classics in the educational system of the United Kingdom.].
141. Bateson, W., & Gairdner, A. E. (1921) Male sterility in flax, subject to two types of segregation. *Journal of Genetics* 11, 269–275.
142. Bateson, W. (1922a) Articles on "Genetics," "Mendelism," and "Sex." In the twelfth edition of the *Encyclopedia Britannica*.
143. Bateson, W. (1922b) Evolution and education. In *Ideals, Aims and Methods in Education*. The New Educator's Library. Isaac Pitman Ltd. [Written 1915.]
144. Bateson, W. (1922c) Evolutionary faith and modern doubts. *Science* 55, 55–61. [Address to the American Association for the Advancement of Science, Toronto. Also in *Nature* 109, 553–556.].
145. Bateson, W. (1922d) Genetical analysis and the theory of natural selection. *Science* 55, 373. [A correction of misquotation from *Historia Plantarum*, and a reply to Osborn.].
146. Bateson, W. (1922e) Interspecific sterility. *Nature* 110, 76. [Controversy following the Toronto address. Response to Crowther.].
147. Bateson, W. (1922f, April 13) Darwin and evolution, limits and variation. *The Times*. [Letter to the Editor concerning the Toronto address.]
148. Bateson, W. (1923a) Area of distribution as a measure of evolutionary age. *Nature* 111, 39. [Review of Willis's *Age and Area*. Cambridge University Press, 1922.].
149. Bateson, W. (1923b) Somatic segregation in plants. In *Report of International Horticultural Congress* (pp. 155–156). [Amsterdam]
150. Bateson, W. (1923c) The revolt against the teaching of evolution in the United States. *Nature* 112, 313–314.

151. Bateson, W. (1923d) Note on the nature of plant chimaeras. *Studia Mendeliana, Brünn* 9–12.
152. Bateson, W. (1923e) Dr. Kammerer's Alytes. *Nature* 111, 738, 878.
153. Gregory, R. P., de Winton, D., & Bateson, W. (1923) Genetics of *Primula sinensis*. *Journal of Genetics* 13, 219–253.
154. Bateson, W. (1924) Progress in biology. *Nature* 113, 644–646 and 681–682. [Address given (March 12) at the Centenary of Birkbeck College.].
155. Bateson, W. (1925a) Huxley and evolution. *Nature* 115, 715–717.
156. Bateson, W. (1925b) Mendeliana. *Nature* 115, 827–830. [Review of Corren's collected papers and Iltis's *Life of Mendel*.].
157. Bateson, W. (1925c) Evolution and intellectual freedom. *Nature* 116, 78. [Supplement]
158. Bateson, W. (1925d) Science in Russia. *Nature* 116, 681–683. [Account of visit.]
159. Bateson, W., & Bateson, G. (1925) On certain aberrations of the red-legged partridges *Alectoris rufa* and *saxatilis*. *Journal of Genetics* 16, 101–123.
160. Bateson, W. (1926) Segregation: being the Joseph Leidy Memorial Lecture of the University of Pennsylvania, 1922. *Journal of Genetics* 16, 201–235.
161. Bateson, C. B. (1928a) *Letters from The Steppe Written in the Years 1886-1887 by William Bateson*. Cambridge University Press.
162. Bateson, C. B. (1928b) *William Bateson, F. R. S. Naturalist. His Essays and Addresses Together with a Short Account of his Life*. Cambridge University Press. [Include's Beatrice's *Memoir*.].
163. Hall, A. D. (1928) Bateson's experiments on bolting in sugar beet and mangolds. *Journal of Genetics* 20, 219–232.

References and Notes

In works of this kind there is usually a detailed list of references. But much of the early literature is now available online, in word- or phrase-searchable form, through such operations as Robert Robbin's Electronic Scholarly Publishing (ESP) project (history of genetics), the JSTOR scholarly journals archive, the Project Gutenberg, and Google Scholar. Furthermore, primary sources such as the Darwin Correspondence Online Database, and archives, such as that at the John Innes Centre, can be accessed for searching. In this circumstance other considerations, such as the need for conciseness, come into play. Accordingly, the minimal information for retrieving most references is given in the text (usually the date of a letter, or of one of Bateson's publications). Other pertinent references are listed below on a chapter-by-chapter basis. Full texts of some references and much supplementary material can be found in Forsdyke's webpages.

Prologue

1. Darwin, C. (1859) *The Origin of Species by Natural Selection, or The Preservation of Favoured Races in the Struggle for Life*. John Murray, London.
2. Mendel, G. (1866) *Verhandlungen des naturforschended Vereines in Brunn* 4, 3–47.
3. Forsdyke, D. R. (2001) *The Origin of Species, Revisited*. McGill-Queen's University Press, Montreal.
4. Forsdyke, D. R. (2016) *Evolutionary Bioinformatics*, 3rd edition. Springer, New York.
5. Gould, S. J. (2002) *The Structure of Evolutionary Theory*. Harvard University Press.
6. Lesch, J. E. (1975) The role of isolation in evolution. George J. Romanes and John. T. Gulick. *Isis* 66, 483–503.
7. Provine, W. B. (1986) *Sewall Wright and Evolutionary Biology*. The University of Chicago Press, pp. 207.
8. Bateson, C. B. (1928) *William Bateson, F. R. S. Naturalist. His Essays and Addresses Together with a Short Account of his Life*. Cambridge University Press. [Beatrice's *Memoir* constitutes the first 160 pages.]
9. Bateson, C. B. (1928) *Letters from The Steppe Written in the Years 1886–1887 by William Bateson*. Cambridge University Press.
10. Coleman, W. (1968) The Bateson Papers. *The Mendel Newsletter* 2, 1–3.
11. Provine, W. (1971) *The Origins of Theoretical Population Genetics*. The University of Chicago Press.
12. Coleman, W. (1970) Bateson and chromosomes: Conservative thought in science, *Centaurus* 15, 228-314.
13. Lipset, D. (1980) *Gregory Bateson. The Legacy of a Scientist*. Prentice-Hall, Englewood Cliffs.
14. Cock, A. G. (1973) William Bateson, Mendelism and Biometry. *Journal of the History of Biology* 6, 1–36.
15. Cock, A. G. (1977) The William Bateson Papers. *Mendel Newsletter* 14, 1–4.
16. Pais, A. (1982) *The Science and Life of Albert Einstein*. Oxford University Press.
17. Forsdyke, D. R. (2000) *Tomorrow's Cures Today?* Harwood Academic, Newark.
18. Harrison, E. (1987) Whigs, prigs and historians of science. *Nature* 329, 213–214.
19. Bateson, C. B. (1927) Letter to C. C. Hurst, 13 June. In: Hurst, R. (1974) *The Evolution of Genetics*. Unpublished manuscript in Cambridge University Library.

A. G. Cock, D. R. Forsdyke, *Treasure Your Exceptions*,
https://doi.org/10.1007/978-3-030-92099-9

20. Wyhe, J. van (2007) Mind the gap: did Darwin avoid publishing his theory for many years? *Notes and Records of the Royal Society* 61, 177–205.
21. Adams. M. B. (2021) Little Evolution, BIG Evolution: Rethinking the History of Darwinism, Population Genetics, and the "Synthesis." In: *Natural Selection. Revisiting its Explanatory Role in Evolutionary Biology*. Edited by R. G. Delisle. Springer Nature, Switzerland, pp. 195–230.
22. Otis, L. (1994) *Organic Memory. History and the Body in the Late Nineteenth and Early Twentieth Centuries*. University of Nebraska Press, Lincoln. [The story is traced back to Michel de Montaigne, circa 1580.]
23. Adams, M. B. (1990) La génétique des populations était-elle une génétique évolutive? In: *Histoire de la Génétique*. Edited by Jean-Louis Fischer and William H. Schneider. ARPEM, Paris, pp. 153–171.
24. Provine, W. (2001). *The Origins of Theoretical Population Genetics with a New Afterword.* The University of Chicago Press.
25. Forsdyke, D. R. (2010) George Romanes, William Bateson, and Darwin's "Weak Point." *Notes and Records of the Royal Society* 64, 139–154.
26. Forsdyke, D. R. (2021) Complementary oligonucleotides rendered discordant by single base mutations may drive speciation. *Biological Theory* 16, 237–241.
27. Forsdyke, D. R. (2018) Mendel, Gregor Johann. *Encyclopedia of Life Sciences*. Wiley, Chichester.
28. Punnett, R. C. (1926) William Bateson. *Edinburgh Review* 244, 71–86.
29. Edwards, A. W. F. (2020) De-naming Francis Galton. *Oxford Magazine*. Trinity term, pp. 14–15. [A University College "Commission of Enquiry into the History of Eugenics at UCL" recommended "UCL to rename the following: the Pearson Building, the Galton Chair [i.e. the Galton Professorship of Human Genetics] and the Galton Lecture Theatre – these should lose their current names as soon as possible."]
30. Bodmer, W., Bailey, R. A., Charlesworth, B., Eyre-Walker, A., Farewell, V., Mead, A. & Senn, S. (2021) The outstanding scientist, R.A. Fisher: his views on eugenics and race. *Heredity* 126:565–576.
31. Harper, P. (2009) Alan G. Cock, Donald R. Forsdyke: Treasure your exceptions. The science and life of William Bateson. *Human Genetics* 125, 233–234. ["This book puts the record straight and one is left with a feeling of admiration for Bateson as both a scientist and a man, and with no doubt that he, more than anyone else, was largely responsible for the rapid progress in modern genetics during the first decade of the twentieth century."]
32. Wade, M. J. (2009) William Bateson: variation, heredity and speciation. *Evolution* 63, 2768–2770. ["I strongly recommend this book for its information on an important central figure and for its bringing to life the several controversies at the origins of Genetics. It greatly illuminates the conceptual foundations of evolutionary genetics."]
33. Carlson, E. A. (2009) "Treasure your exceptions." The science and life of William Bateson. *Quarterly Review of Biology* 84, 398–399. ["This volume will be of enormous benefit to historians of science who like to follow how ideas are born or die and why participants of different sides of each controversy held such rigid views of their own work and saw little merit in their competitor's research."]
34. Schwartz, J. S. (2009) Choice 46 (7) March. ["This work includes key events in Bateson's career and is strengthened by discussion of the rediscovery of Mendelian principles by early 20th-century geneticists. Its basic premise is that Bateson was not merely one of the founders of genetics but played a pivotal role in the development of evolutionary biology by serving as the link between Victorian naturalists and early 20th-century biologists."]
35. Joshi, A. (2010) *Journal of Genetics* 89, 555–557. ["There is much in this book that I was not aware of, despite a formal training in genetics and a strong amateur interest in the history of biology.... It is certainly a book that any serious student of either genetics or evolution would greatly benefit from reading."]

1 A Cambridge Childhood

1. Jones, H. F. (1919) *Samuel Butler Author of Erewhon (1835–1902). A Memoir*, Vols. 1 and 2. Macmillan, London.
2. Weismann, A (1909) Charles Darwin, *Contemporary Review* 96, 1–22.
3. Huxley, J. (1961) The evolutionary vision. In: *Evolution After Darwin*, Vol. 3. Edited by S. Tax & C. Callender. University of Chicago Press, Chicago, pp. 246–361.
4. Smocovitis, V. B. (1999) The 1959 Darwin Centennial celebration in America. *Osiris* 14, 274–323.
5. Bateson, M (1895) *Professional Women and their Professions*, Horace Cox, London.
6. Heitland, W. E. (1926) *After Many Years*. Cambridge University Press.
7. Forsdyke, D. R. (2001) *The Origin of Species, Revisited*. McGill-Queen's University Press, Montreal.
8. Geison, G. L. (1978) *Michael Foster and the Cambridge School of Physiology*. Princeton University Press.
9. Scharpey-Schafer, E. (1927) *History of the Physiological Society during its First Fifty Years*. The Physiological Society, London.
10. Harris, H. (1999) *The Birth of the Cell*. Yale University Press.
11. Haeckel, E. (1909) Charles Darwin as an anthropologist. In: *Darwin and Modern Science. Essays in Commemoration of the Centenary of the Birth of Charles Darwin and of the Fiftieth Anniversary of the Publication of the Origin of Species*. Edited by A. C. Seward. Cambridge University Press, pp. 137–151.
12. Lankester, E. R. (1890) The history and scope of zoology. In: *The Advancement of Science*. Macmillan, London. [Reproduced from the ninth edition of the *Encyclopaedia Britannica*.]
13. Raverat, G. (1952) *Period Piece. A Cambridge Childhood*, Faber and Faber, London.
14. Heitland, W. E. (1926) William Bateson 1861–1926. *The Eagle* 44, 227–230. [St. John's College magazine]
15. Garnett, R. (1991) *Constance Garnett: A Heroic Life*. Sinclair-Stevenson, London.

2 From Virginia to the Aral Sea

1. Brooks, W. K. (1883) *The Law of Heredity. A Study of the Cause of Variation and the Origin of Living Organisms*. Murphy, Baltimore.
2. Hall, B. K. (2005) Betrayed by *Balanoglossus*: William Bateson's rejection of evolutionary embryology as the basis for understanding evolution. *Journal of Experimental Zoology* 304B, 1–17.
3. Punnett, R. C. (1926) William Bateson. *Edinburgh Review* 244, 71–86.
4. Conklin, E. G. et al. (1910) William Keith Brookes: a sketch of his life by some of his former pupils and associates. *Journal of Experimental Zoology* 9, 1–51.
5. Eden J. (2013) 'The first that ever burst, etc.': William Bateson in the Steppe, 1886–1887. *Studies in Travel Writing* 17, 1-21.
6. Bidder, G. P. (1941) Mr. W. H. Caldwell. *Nature* 148, 557–558.
7. Lankester, E. R. (1890) The history and scope of zoology. In: *The Advancement of Science*. Macmillan, London. [Reproduced from the ninth edition of the *Encyclopaedia Britannica*.]
8. Robins, L. (2005) The platypus frontier: eggs, Aborigines and empire in 19th century Queensland. In: *Dislocating the Frontier: Essaying the Mystique of the Outback*. Edited by D. B. Rose & R. Davis. E Press, Australian National University, Canberra, pp. 99–120. [In 1891 Richard Semon followed the trail Caldwell had opened in Australia.]
9. Garnett D. (1954) *The Golden Echo*. Vol 1. *Intimations of Mortality*. Chatto and Windus, London.

10. Fowler, H. (1996) *Cambridge Women. Twelve Portraits*. Edited by E. Shils & C. Blacker, Cambridge University Press.
11. Lankester, E. R. (1890) The scientific results of the international fisheries expedition. In: *The Advancement of Science*. Macmillan, London.

3 Galton

1. Wallace, A. R. (1889) *Darwinism: an Exposition of the Theory of Natural Selection with Some of its Applications*. Macmillan, London.
2. Galton. F. (1889) *Natural Inheritance*, John Murray, London.
3. Weismann, A. (1889) *Essays upon Heredity and Kindred Biological Problems*. Translated by E. B. Poulton, S. Schönland and A. E. Shipley. Clarendon Press, Oxford. [Bateson's copy is inscribed "From A. E. Shipley, June 4, 1889."]
4. Perhaps Bateson is here referring to the fact that in ancient Greek the "n" of pan changes to "m" before a word beginning with "m", so it should not be "panmixia," but "pammixia." There is no Greek noun corresponding to this. There is a Greek adjective from the same roots – pammiges – meaning "mixed of all sorts" as in an army made up of men from different social statuses and ethnic origins.
5. Galton, F. (1872) On blood-relationship. *Proceedings of the Royal Society* 20, 394–402.
6. Galton, F. (1865) Hereditary talent and character. *Macmillan's Magazine* 12, 157–166, 318–327.
7. Galton, F. (1875) A theory of heredity. *Contemporary Review* 27, 80–95.
8. Galton, F. (1875) A theory of heredity. *Journal of the Anthropological Institute* 5, 329–348.
9. Romanes, G. J. (1893) *An Examination of Weismannism*. Open Court Publishing, Chicago.
10. Wilson, E. B. (1896) *The Cell in Development and Inheritance*. Macmillan, New York.
11. Darwin, C. (1868) *The Variation of Animals and Plants under Domestication*. John Murray, London.
12. Sanders, A. (1870) On Mr. Darwin's hypothesis of pangenesis as applied to the faculty of memory. *Journal of Anthropology* 1, 144–149.
13. Galton, F. (1871) Experiments in pangenesis, by breeding from rabbits of a pure variety, into whose circulation blood taken from other varieties had previously been largely transfused. *Proceedings of the Royal Society* 19, 393–410.
14. Darwin, C. (1871) Pangenesis. *Nature* 3, 502–503.
15. Romanes, E. (1896) *The Life and Letters of George John Romanes*. Longmans, Green and Co., London.
16. Betteridge, K. 2003. A history of farm animal embryo transfer and some associated techniques. *Animal Reproduction Science* 79, 203–244.
17. Galton, F. (1897) The average contribution of each of several ancestors to the total heritage of offspring. *Proceedings of the Royal Society* 61, 401–413.
18. Jenkin, F. (1867) The origin of species. *The North British Review* 46, 277–318.
19. Galton, F. (1892) *Finger Prints*. Macmillan, London.
20. Yule, G. U. (1902) Mendel's laws and their probable relations to intra-racial heredity. *The New Phytologist* 1, 193–207, 222–238.
21. De Vries, H. (1889) *Intracellulare Pangenesis*. Fischer, Jena. [The English translation by C. S. Gager was published (1910) by Open Court Publishing, Chicago.]
22. Roll-Hansen, N. (1978) The genotype theory of Wilhelm Johannsen and its relation to plant breeding and the study of evolution. *Centaurus* 22, 201–235.
23. Johansson, W. (1903) *Über Erblichkeit in Populationen on reinen Linien. Ein Beitrag zur Beleuchtung schwebender Selektionsfragen*. [On heredity in pure lines and populations. A contribution to pending questions of selection] Fischer, Jena.

24. Johansson, W. (1903) About Darwinism, seen from the point of view of the science of heredity. *Tilskueren*, pp. 525–541. Translation by N. Roll-Hansen, Johannsen-Om Darwinismen (bshs.org.uk)
25. Weldon, W. F. R. & Pearson, K. (1903) Inheritance in *Phaseolus vulgaris*, *Biometrika* 2, 499–503.
26. Yule, G. U. (1904) Professor Johannsen on heredity, *Nature* 69, 223–224.
27. Shull, G. H. (1905) Galtonian regression in the 'pure line' *Torreya* 5, 21–25.

4 Variation

1. Barnes, E. (1998) *The Early Career of George John Romanes 1867–1878*. Unpublished undergraduate thesis, Cambridge University.
2. Romanes, E. (1896) *The Life and Letters of George John Romanes*. Longmans, Green and Co., London.
3. Romanes, G. J. (1873) Permanent variation of colour in fish. *Nature* 8, 101.
4. Romanes, G. J. (1896) *A Selection from the Poems of George John Romanes*. Edited by T. H. Warren. Longmans, Green and Co. London.
5. Bateson, A. & Darwin, F. (1888) The effect of stimulation on turgescent vegetable tissues. *Linnean Journal (Botany)* 24, 1–27. [Read at the Linnean Society on January 20th, 1887; *Proceedings of the Linnean Society* 99, 8.]
6. Bateson, A. & Darwin, F. (1888) On a method of studying geotropism. *Annals of Botany* 2, 65–68.
7. Bateson, A. (1888) The effect of cross-fertilization of inconspicuous flowers. *Annals of Botany* 1, 255–261.
8. Bateson, A. (1889) On the change of shape exhibited by turgescent pith in water. *Annals of Botany* 4, 117-–125.
9. Bateson, W. & Bateson, A. (1891) On the variation in the floral symmetry of certain plants having irregular corollas. *Linnean Society Journal* 28, 386–421.
10. Raverat, G. (1952) *Period Piece. A Cambridge Childhood*, Faber and Faber, London.
11. Romano, T. M. (2002) *Making Medicine Scientific. John Burdon Sanderson and the Culture of Victorian Science*. Johns Hopkins University Press, Baltimore.
12. Lankester, E. R. (1890) The history and scope of zoology. In: *The Advancement of Science*. Macmillan, London. [Reproduced from the ninth edition of the *Encyclopaedia Britannica*.]
13. Huxley, T. H. (1859) Letter to Lyell, 25 June. In: *Life and Letters of Thomas Henry Huxley*, Appleton and Company, New York, 1901.
14. Darwin, C. (1875) *The Variation of Animals and Plants under Domestication*. John Murray, London. [Here Darwin cites Owen's contrast between "metamorphosis" and "metagenesis." In the latter case "the new parts are not moulded upon the inner surfaces of the old ones. The plastic force has changed its mode of operation. *The outer case, and all that gave form and character to the precedent individual, perish, and are cast off; they are not changed* into the corresponding parts of the new individual. These are due to a new and distinct developmental process."]
15. Goldschmidt, R. (1940) *The Material Basis of Evolution*. Yale University Press. [Silliman Lectures]
16. MacAlister, A. (1894) Materials for the study of variation. *The Expositor* 53, 375–380.
17. Galton, F. (1894) Discontuity in evolution. *Mind* (new series) 3, 362–372.
18. Pearson, K. 1906. Walter Frank Raphael Weldon 1860–1906. *Biometrika* 5, 1–52.
19. Jones, H. F. (1919) *Samuel Butler Author of Erewhon (1835–1902). A Memoir*. Macmillan, London.
20. Weldon, W. F. R. (1890) The variations occurring in certain decapod crustacean: 1. *Crangon vulgaris*. *Proceedings of the Royal Society* 47, 445–453.

21. Weldon, W. F. R. (1892) On certain correlated variations in *Crangon vulgaris*. *Proceedings of the Royal Society* 51, 2–21.
22. Delboeuf, J. (1877) A law of mathematics applicable to the theory of transformation. *Kosmos* 2, 105–127. [An English translation of this paper is in Forsdyke's web-pages.]
23. Provine, W. B. (1986) *Sewall Wright and Evolutionary Biology*. University of Chicago Press.
24. Weldon, W. F. R. (1894) The study of animal variation. *Nature* 50, 25–26.

5 Romanes

1. Forsdyke, D. R. (2001) *The Origin of Species, Revisited*. McGill-Queen's University Press, Montreal.
2. Romanes, G. J. (1886) Physiological selection: an additional suggestion on the origin of species. *Nature* 34, 314–316, 336–340, 362–365.
3. Romanes, G. J. (1897) *Darwin, And After Darwin. Vol. 3. Isolation and Physiological Selection*. Longmans, Green and Co., London.
4. Romanes, G. J. (1874) Rudimentary organs. *Nature* 9, 441–442; Disuse as a reducing cause in species. *Nature* 10, 164.
5. Wells, H. G. 1934. *Experiment in Autobiography*. Macmillan, New York
6. Catchpool, E. (1884) An unnoticed factor in evolution. *Nature* 31, 4.
7. Romanes, E. 1896. *The Life and Letters of George Romanes*. Longmans, London. [Apart from this biography by his wife, other personal aspects of Romanes may be found in Turner, G. (1931) *Unorthodox Reminiscences*. John Murray, London, pp. 142–167.]
8. Crichton, M. (1990) *Jurassic Park*. Knopf, New York.
9. Dyer, W. T. (1888) Opening Address to the British Association, Section D. *Nature* 38, 473-480.
10. Darwin, F. (1886) Physiological selection and the origin of species. *Nature* 34, 407.
11. Belt, T. (1874) *A Naturalist in Nicaragua*. Dent, London.
12. Milner, R. (1999) Huxley's bulldog: the battles of E. Ray Lankester (1846–1929). *The Anatomical Record* 257, 90–95.
13. Meadows, A. J. (1972) *Science and Controversy. A Biography of Sir Norman Lockyer*. MIT Press, Cambridge. [The Lankester letter (Sept. 25) was probably written in 1888.]
14. Lankester, E. R. (1889) Darwinism. *Nature* 40, 566–570.
15. Lankester, E. R. (1905) Nature and Man. *Nature* 72, 184. [Abstract of his Romanes Lecture.]
16. Huxley, T. H. (1888) Obituary Notices of Fellows Deceased. *Proceedings of the Royal Society*, vol. 44. [Obituary of Charles Darwin; also in *Darwiniana, Collected Essays*. Macmillan, London, 1893.]
17. Huxley, T. H. (1894) Letter to Bateson, Feb. 20. [Nephelococcygia derives from *The Birds*, a play by Aristophanes (414 B. C.), where birds imagine shapes in clouds.]
18. Ralston, H. J. (1944) Neuromuscular Physiology. G. J. Romanes on the Excitability of Muscle. *Science* 100, 123-124.
19. Eimer, G. H. T. (1890) *Organic Evolution as the Result of the Inheritance of Acquired Characters According to the Laws of Organic Growth*. Translated by J. T. Cunningham. Macmillan, London.
20. Cunningham, J. T. (1895) The Origin of Species Among Flat-Fishes. *Natural Science* 6, 169–177, 234–239.
21. Lankester, E. R. (1896) The utility of specific characters. *Nature* 54, 365–366.
22. Buch, C. L. de (1825) *Physikalische Beschreibung der Canarische Inseln*. Berlin.
23. Gulick, J. T. (1905) *Evolution, Racial and Habitudinal*. Carnegie Institute, Washington.
24. Romanes, G. J. 1895. *Darwin, and After Darwin. Vol. 2. Post-Darwinian Questions. Heredity and Utility*. Longmans, Green and Co., London.

6 Reorientation and Controversy

1. Poulton, E. B. (1890) The Colours of Animals: Their Meaning and Use, Especially Considered in the Case of Insects. Kegan Paul, Trench, Trubner and Co., London.
2. Pasteur, G. (1982) A classificatory review of mimicry systems. *Annual Reviews of Ecology and Systematics*, 13, 169–199.
3. Weldon, W. F. R. (1895) Variation in animals and plants. *Nature* 51, 449–450.
4. Dyer, W. T. (1895) Variation and specific stability. *Nature* 51, 459–461.
5. Darwin, C. (1876) *The Effects of Cross and Self Fertilization in the Vegetable Kingdom*. John Murray, London. [Footnote: pp. 335]
6. Weldon, W F. R (1895) Attempt to measure the death rate due to the selective destruction of *Carcinus moenas* with respect to a particular dimension. *Proceedings of the Royal Society* 57, 360–379.
7. Pearson, K. (1906) Walter Frank Raphael Weldon 1860–1906. *Biometrika* 5, 1–52.
8. Pearson, K. (1930) *The Life, Letters and Labours of Francis Galton*. Cambridge University Press.
9. Edelston, R. S. (1864) *Amphidasis betularia. Entymologist* 2, 150.
10. Tutt, J. W. (1896) *British Moths*. Routledge, London.
11. Doncaster, L. (1906) Collective inquiry as to progressive melanism in lepidoptera. *The Entymologist's Record* 18, 1–16.
12. Majerus, M. E. N. (1998) *Melanism. Evolution in Action*. Oxford University Press.
13. Pearson, K. (1901) On the principle of homotyposis and its relation to heredity, to the variability of the individual, and to that of the race. Part 1. Homotyposis in the vegetable kingdom. *Philosophical Transactions of the Royal Society, A*. 197, 285–379.
14. Davenport, C. B. (1899) *Statistical Methods with Special Reference to Biological Variation*. Wiley, New York.
15. Pearson, K. (1902) On the fundamental conceptions in biology. *Biometrika* 1, 320–344.

7 What Life May Be

1. Lago, M. (1996) *Christiana Herringham and the Edwardian Art Scene*. Lund Humprhies, London.
2. Rothenstein, W. (1931, 1939). *Men and Memories*. Faber, London.
3. Blackman, F. F. (1926) *The Eagle* 44, 225–233. [St. John's College magazine.]
4. Romanes, G. J. (1887) Mental differences between men and women. *Nineteenth Century* 21, 654–672.
5. Richmond, M. L. (1997) A lab of one's own. The Balfour biological laboratory for women at Cambridge University, 1884-1914. *Isis* 88, 422–455.
6. Durham, C. B. (1895) At a conversazione. *The English Illustrated Magazine*. September, p. 551.
7. Bedford, H. A. R. & Pickering, P. S. U. (1919) *Science and Fruit Growing*. Macmillan, London.
8. Darby, R. (2005) *The Surgical Temptation. The Demonization of the Foreskin and the Rise of Circumcision in Britain*. University of Chicago Press.
9. Cock, A. G. (1977) Bernard's symposium. The species concept in 1900. *Biological Journal of the Linnean Society* 9, 1–30.

8 Rediscovery

1. Olby, R. C. (1985) Origins of Mendelism. Second edition. University of Chicago Press.
2. Olby, R C. (1987) William Bateson's introduction of Mendelism to England: a reassessment. *British Journal of the History of Science* 20, 399–420.
3. Olby, R. C. (2000) Horticulture: the font for the baptism of genetics. *Nature Reviews Genetics* 1, 65–70.
4. Vorzimmer, P. J. (1968) Darwin and Mendel. The historical connection. *Isis* 59, 77–82.
5. Naudin C. 1856. Observations constatant le retour simultané de la descendance d'une plante hybride aux types paternels et maternels. *Compte Rendu Academie Sciences, Paris* 42: 628–636.
6. Romanes, G. J. (1881) Hybridism. *Encyclopaedia Britannica*. Ninth edition.
7. Romanes, E. (1896) *The Life and Letters of George Romanes*. Longmans, London.
8. Romanes, G. J. (1894) Letter to Schafer. May 18. Welcome Museum of the History of Medicine, London.
9. Cock, A. G. (1973) William Bateson, Mendelism and biometry. *Journal of the History of Biology* 6, 1–36.
10. Weldon, W. F. R. (1902) Mendel's laws of alternative inheritance in peas. *Biometrika* 1, 228–254.
11. Guyer, M. F. (1900) Spermatogenesis in hybrid pigeons. *Science* 21, 248–249; 312.
12. Hurst, R. (1974) *The Evolution of Genetics*. Unpublished manuscript in Cambridge University Library.
13. Lipset, D. (1980) *Gregory Bateson. The Legacy of a Scientist*. Prentice-Hall, Englewood Cliffs.
14. Keynes, G. (1981) *The Gates of Memory*. Clarendon Press, Oxford.
15. Wilmott, A. J. (1926) Obituary. William Bateson. *The Journal of Botany* 64, 78–80.
16. Lago, M. (1996) Christiana Herringham and the Edwardian Art Scene. Lund Humprhies, London.
17. Punnett, R. C. (1950) Early days of genetics. *Heredity* 4, 1–10.

9 Mendel's Bulldog

1. Mitchison, N. (1968) Beginnings. In: *Haldane and Modern Biology*. Edited by K. R. Dronamraju. Johns Hopkins University Press, Baltimore.
2. Yule, G. U. (1906) On the theory of inheritance of quantitative compound characters on the basis of Mendel's laws – a preliminary note. *Report on the Third International Conference 1906 on Genetics*. Edited by W. Wilks. Royal Horticultural Society, London, pp. 140-142.
3. Fisher, R. A. (1918) The correlation between relatives on the supposition of Mendelian inheritance. *Transactions of the Royal Society of Edinburgh* 52, 399–433.
4. Castle, W. E. (1903) The laws of heredity of Galton and Mendel, and some laws governing race improvement by selection. *Proceedings of the American Academy of Arts and Sciences* 39, 223–242.
5. Weldon, W. F. R. (1902) On the ambiguity of Mendel's categories. *Biometrika* 2, 44–55.
6. Yule, G. U. (1902) Mendel's laws and their probable relations to intra-racial heredity. *The New Phytologist* 1, 193–207, 222–238.
7. Ankeny, R. A. (2000) Marvelling at the marvel: the supposed conversion of A. D. Darbishire to Mendelism. *Journal of the History of Biology* 33, 315–347.
8. Weldon, W. F. R. (1903) Mr. Bateson's revisions of Mendel's theory of heredity. *Biometrika* 2, 286–298.
9. Garrod, A. E. (1902) Incidence of alkaptanuria. A study of chemical individuality. *Lancet* 2, 1616.
10. Bearn, A. G. (1994) Archibald Edward Garrod. The Reluctant Geneticist. *Genetics* 137, 1–4.

11. Gregory, R. P. (1903) The seed characters of *Pisum sativum*. *The New Phytologist* 2, 226–228.
12. Linnean Society (1904) Report of meeting (Feb. 18).
13. Punnett, R. C. (1950) Early days of genetics. *Heredity* 4, 1–10.
14. Pearson, K. (1906) Walter Frank Raphael Weldon 1860–1906. *Biometrika* 5, 1–52.
15. Rushton, A. R. (2000) Nettleship, Pearson and Bateson: the biometric-Mendelian debate in a medical context. *Journal of the History of Medicine and Allied Sciences* 55, 134–157.
16. Lankester, E. R. (1906) Inaugural address by Professor E. Ray Lankester, President of the Association. *Nature* 74, 321–335. [Address to the BA at York.]
17. Schuster, E. H. J. (1906) Results of crossing grey (house) mice with albinos. *Biometrika* 4, 1–12.
18. Darbishire, A. D. (1905) The mutation theory of the origin of species. *Nature* 72, 314–316. [Review of *Species and Varieties* by de Vries (1905).]
19. Darbishire, A. D. (1906) On the differences between physiological and statistical laws of heredity. *Memoirs and Proceedings of the Manchester Literary and Philosophical Society* 50, 1–44.
20. Schuster, E. H. J. (1913) *Eugenics*. Collins, London.
21. Lister, J. J. (1906) The life history of the Foraminifera. *Nature* 74, 400–404. [Address to the BA at York.]
22. Pearson, K. (1906) The latest critic of biometry. *Nature* 74, 465–466.
23. Lister, J. J. (1906) Biometry and biology: a reply to Professor Pearson. *Nature* 74, 584–585.
24. Pearl, R. (1911) Inheritance of fecundity in the domestic fowl. *American Naturalist* 45, 321–345.
25. Galton, F. (1901) On the probability that the son of a very highly gifted father will be no less gifted. *Nature* 65, 79.
26. Weldon, W. F. R. (1902) Professor de Vries on the origin of species. *Biometrika* 1, 365–374.
27. Mayr, E. (1982) *The Growth of Biological Thought*. Harvard University Press, Cambridge, MA.
28. Dawkins, R. (1986) *The Blind Watchmaker*. Longmans Scientific, Harlow.
29. Spillman, 1909. Application of some principles of heredity to plant breeding. *Bulletin* no. 165. US Department of Agriculture.
30. Anonymous (1906) Review by "N. R. C" of J. B. Burke's *The Origin of Life* (May 10) *The Cambridge Review*, pp. 375–376.

10 Bateson's Bulldog

1. Haldane, J. B. S. (1926) *The Nation and the Atheneum*. Feb. 20, p. 713.
2. Spillman, W. (1901) *Bulletin 115*. Office of Experimental Stations, Washington, DC, pp. 88–98.
3. Hurst, C. C. (1903) Mendel's principles applied to wheat hybrids. *Journal of the Royal Horticultural Society* 27, 876–893.
4. Davenport, C. B. (1901) Mendel's law of dichotomy in hybrids. *Biological Bulletin* 2, 307–310.
5. Spillman, W. J. (1902) Exceptions to Mendel's law. *Science* 16, 794–496.
6. Carlson, L. (2005) *William J. Spillman: Scientific Agriculture in the Progressive Era*. University of Missouri Press.
7. Hurst, C. C. (1897) Curious crosses. *The Orchid Review* 5, 179–180. [This and other "curiousities of orchid breeding," are in Hurst's collected papers *Experiments in Genetics* (1925) Cambridge University Press, pp. 1–53.]
8. Hurst, C. C. (1900) Notes on some experiments in hybridization and cross-breeding. *Journal of the Royal Horticultural Society* 24, 90–126.
9. Hurst, C. C. (1904) Notes on Mendel's methods of cross-breeding. *Memoirs of the Horticultural Society of New York* 1, 11–15. [Read at the Second International Conference on Plant Breeding and Hybridization, New York, September 1902.]

10. Roberts, H. F. (1929) *Plant Hybridization before Mendel*. Princeton University Press.
11. Hurst, C. C. (1903) Mendel's principles applied to orchid hybrids. *Journal of the Royal Horticultural Society* 27, 614–624.
12. Hurst, C. C. (1904) Mendel's discoveries in heredity. *Transactions of the Leicester Literary and Philosophical Society* 8, 121–134.
13. Dawkins, R. (1976) *The Selfish Gene*. Oxford University Press.
14. Castle, W. E. 1903. The heredity of Angora coat in mammals. *Science* 18, 760–761.
15. Hurst, C. C. (1905) Experimental studies on heredity in rabbits. *Journal of the Linnean Society, Zoology* 29, 283–324.
16. Hurst, C. C. (1905) Experimental studies on heredity in rabbits. *Transactions of the Leicester Literary and Philosophical Society* 9, 110–117.
17. Pearson, K. & Lee, A. (1900) Mathematical contributions to the theory of evolution, VIII. On the inheritance of characters not capable of exact quantitative measurement. Part 1. Introductory. Part 2. On the inheritance of coat colour in horses. Part 3. On the inheritance of eye colour in man. *Philosophical Transactions of the Royal Society, A* 195, 79–150.
18. Pearson, K., Yule, G. U., Blanchard, N. & Lee, A. (1903) Law of ancestral heredity. *Biometrika* 2, 214–215.
19. Hurst, R. (1974) *The Evolution of Genetics*. Unpublished manuscript in the Cambridge University Library. [See also: Hurst, R. (1975) The Hurst collection of genetic letters now in the Cambridge University Library. *Heredity* 34, 279–280; (1977) Further note on the Hurst collection of genetical material in the Cambridge University Library. *Heredity* 38, 259–160.]
20. Hurst (1906) On the inheritance of coat colour in horses. *Proceedings of the Royal Society, B*. 77, 388–394.
21. Punnett, R. C. (1950) Early days of genetics. *Heredity* 4, 1–10.
22. Weldon, W. F. R. (1906) Note on the offspring of thoroughbred chestnut mares. *Proceedings of the Royal Society, B* 77, 394–398.
23. Hurst, C. C. (1907) Mendelian characters in plants and animals. *Report of the Third International Conference 1906 on Genetics*. Edited by W. Wilks. Royal Horticultural Society, London. pp. 114–128.
24. Shull, G. H. (1909) The 'presence and absence' hypothesis. *American Naturalist* 43, 410–419.
25. Hurst, C. C. (1910) Mendel's law of heredity and its application to horticulture. *Journal of the Royal Horticultural Society* 36, 22–52.
26. Shull, G. H. (1910) Germinal analysis through hybridization. *Proceedings of the American Philosophical Society* 49, 281–290.
27. Shull, G. H. (1915) Genetic definitions in the new standard dictionary. *American Naturalist* 49, 52–59.
28. Shull G. H. (1911) Reversible sex mutants in *Lychnis dioica*. *Botanical Gazette* 52, 329–368.
29. Johannsen, W. (1911) The genotype conception of heredity. *American Naturalist* 45, 129–159.
30. Shull, G. H. (1912) 'Genotypes,' 'Biotypes,' 'Pure Lines,' and 'Clones.' *Science* 35, 27–29.
31. Wilson, E. B. (1914) Croonian Lecture: The bearing of cytological research on heredity. *Proceedings of the Royal Society of London, B* 88, 333–352.
32. Punnett, R. C. (1905) *Mendelism*. Macmillan and Bowes, Cambridge.
33. Darbishire, A. D. (1907) *Nature* 76, 73–74. [Anonymous review of second edition of reference 32]
34. Hardy, G. H. (1908) Mendelian proportions in a mixed population. *Science* 28, 49–50.
35. Weinberg, W. (1908) Ueber den Nachweis der Vererbung beim Menchen. *Jahreshefte des Vereins fürVaterländische Naturkunde in Württemberg* 64, 368–382.
36. Lock, R. D. (1906) *Recent Progress in the Study of Variation, Heredity and Evolution*. John Murray, London.
37. Carlson, E. A. (1966) *The Gene: A Critical History*. W. B. Saunders, Philadelphia.
38. Edwards, A. W. F. (2013) Robert Heath Lock and his textbook of genetics, 1906. *Genetics* 194, 529–537.
39. Glendinning V. (2006) *Leonard Woolf. A Biography*. McClelland & Stewart, Toronto.

11 On Course

1. Weinstein, A. (1980) A note on W. L. Tower's *Leptinotarsa* work. In: *The Evolutionary Synthesis*. Edited by E. Mayr and W. Provine. Harvard University Press.
2. Hurst, C. C. (1908) On the inheritance of eye-colour in man. *Proceedings of the Royal Society, B* 80, 85–96.
3. Davenport, C. B. (1907) Heredity of eye colour in man. *Science* 26, 589–592.
4. Report concerning a Chair in Genetics. (1908) *Cambridge University Reporter*, March 3, p. 632.
5. Olby, R. (1989) Scientists and bureaucrats in the establishment of the John Innes Horticultural Institution under William Bateson. *Annals of Science* 46, 497–510.
6. Pearson, K. (1930) *The Life, Labour, and Letters of Francis Galton*. Volume 3, pp. 340.
7. Wallace, A. R. (1908) The Present Position of Darwinism. *Contemporary Review* 94, 129–141.
8. Poulton, E. B. (1908) *Essays on Evolution, 1889–1907*. Clarendon Press, Oxford.
9. Punnett, R. C. (1908) Old Bottles. *The New Quarterly*, October.
10. Rothenstein, W. (1939) *Men and Memories*. Volume 2. Faber, London.
11. Morgan, T. H. (1909) What are "factors" in Mendelian explanations? *American Breeders Association Reports* 5, 365–368.
12. Anonymous (1908) The Influence of Heredity on Disease, with special reference to Tuberculosis, Cancer and Diseases of the Nervous System. A Discussion. *Royal Society of Medicine General Report*, pp. 9–142.

12 Darwin Centenary

1. Haldane, J. B. S. (1957) The theory of evolution, before and after Bateson. *Journal of Genetics* 56, 11–27.
2. Nogler, G. A. (2006) The lesser known Mendel: his experiments with Hieracium. *Genetics* 172, 1–6.
3. Bateson. W. (1909) Boyle lecture. [There is no indication that this was formally published. The manuscript has not been found.]
4. Richmond, M. (2006) The 1909 Darwin celebration. Reexamining evolution in the light of Mendel, mutation and meiosis. *Isis* 97, 447–484.
5. Rothenstein, W. (1931) *Men and Memories*. Volume 1. Faber, London.
6. Meldola, R. (1909) Evolution: old and new. *Nature* 80, 481–485.
7. Gulick, J. T. (1872) On the diversity of evolution under one set of external conditions. *Journal of the Linnean Society (Zoology)* 11, 496–505.
8. Gould, S. J. & Lewontin, R. C. (1979) The spandrels of San Marco and the Panglossian paradigm: a critique of the adaptionist program. *Proceedings of the Royal Society of London. B.* 205, 581–598.
9. Forsdyke, D. R. (2001) *The Origin of Species, Revisited*. McGill-Queen's University Press, Montreal.
10. Orr, H. A. (1996) Dobzhansky, Bateson and the genetics of speciation. *Genetics* 144, 1331–1335.
11. Shipley, A. E. (1910) *Report of the British Association for the Advancement of Science 1909*, Murray, London, pp. 484–502.
12. Haldane, J. B. S., Sprunt. A. D. & Haldane. N. M. (1915) Reduplication in Mice. Preliminary Communication. *Journal of Genetics* 5, 133–135.
13. Haldane, J. B. S. (1926) *The Nation and the Atheneum*. Feb. 20, p. 713.

13 Chromosomes

1. Boveri, T. H. (1888) Zellen-Studien II. Die Befruchtung und Teilung des Eies von *Ascaris megalocephala. Zeitschrift fur Med. und Nat.* 22, 685–882.
2. Boveri, T. H. (1889) Translated by T. H. Morgan (1893) An organism produced sexually without characteristics of the mother. *American Naturalist* 27, 222–232.
3. Roux, W. (1883) *Uber die Bedeutung der Kerntheilungsfiguren*. Leipzig.
4. Stern, C. (1950) Boveri and the early days of genetics. *Nature* 166, 446.
5. Cremer, T. & Cremer, C. (1988) Centennial of Wilhelm Waldeyer's introduction of the term "chromosome" in 1888. *Cytogenetics and Cell Genetics*. 48, 65–67.
6. Farmer, J. B. & Moore, J. E. S. (1905) On the meiotic phase (reduction division) in animals and plants. *Quarterly Journal of Microscopical Science* 48, 489–557.
7. Weismann, A. (1887) On the significance of the polar globules. *Nature* 36, 607–609.
8. Mitchell, P. C. (1892) The new volume of Weismann. *Nature* 46, 558–559.
9. Romanes, G. J. (1893) *An Examination of Weismannism*. Open Court Publishing, Chicago.
10. Wilson E. B. (1900) *The Cell in Development and Inheritance*. Second edition. Macmillan, New York.
11. Cannon, W. A. (1904) Some cytological aspects of hybrids. *Memoires of the Horticultural Society of NewYork* 1, 89–92. [Read at the International Conference on Plant Breeding and Hybridization, New York, October 1st, 1902; see p. 123 for Bateson's remark on chromosomes.]
12. Sutton, W. S. (1900) The spermatogonial divisions in *Brachystola magna. Kansas University Quarterly* 9, 135–160.
13. Sutton, W. S. (1902) On the Morphology of the Chromosome Group in *Brachystola magna. Biological Bulletin* 4, 24–39.
14. Cannon, W. A. (1902) A cytological basis for Mendelian Laws. *Bulletin of the Torrey Botanical Club* 29, 657–661.
15. Sutton, W. S. (1903) The Chromosomes in Heredity. *Biological Bulletin* 4, 231–251.
16. McCusick, V. A. (1960) Walter Sutton and the physical basis of Mendelism. *Bulletin of the History of Medicine* 34, 487–497.
17. de Vries, H. (1910) Fertilization and Hybidization. In the English translation of *Intracellular Pangenesis* (1910) Open Court Publishing, Chicago, pp. 217–263.
18. Boveri, T. H. (1904) *Ergebnisse über die Konstitution der chromatischen Substanz des Zellkerns*. G. Fischer, Jena.
19. Montgomery, T. H. (1901) A study of the germ cells of the metazoa. *Transactions of the American Philosophical Society* 20, 154–236.
20. Paulmier, F. C. (1899) *The Spermatogenesis of Anasa tristis*. Ginn & Co., Boston.
21. McClung, C. E. (1899) A peculiar nuclear element in the male reproductive cells of insects. *Zoological Bulletin*. 2, 187.
22. Stevens, N. M. (1905) Studies in spermatogenesis with especial reference to the "accessory chromosome." *Carnegie Institute of Washington Publication* 36, 1–32.
23. Boveri, T. H. (1901) *Das Problem der Befruchtung*. Fischer, Jena.
24. Boveri, T. H. (1902) Über mehrpolige Mitosen als Mittel zur Analyse des Zellkerns. *Verh. Phys.-med. Ges.* 35, 67–90.
25. Herringham, W. P. (1888) Muscular atrophy of the peroneal type affecting many members of a family. *Brain* 11, 230–236.
26. Doncaster, L. & Raynor, G. H. (1906) Breeding experiments with lepidoptera. *Proceedings of the Zoological Society London* 1, 125–133.
27. McKusick, V. A. (1962) On the X chromosome of man. *Quarterly Review of Biology* 37, 69–175.
28. Wilson, E. B. (1909) Recent researches on the determination and heredity of sex. *Science* 29, 53–70.
29. Smith, G. (1906) *Fauna und Flora, Golf von Neapel. Rhizocepbala*. 29, 89.

30. Hurst, C. C. (1909) Mendelism and sex. In: *Experiments in Genetics*. Cambridge University Press (1925), pp. 331–350.
31. Gregory, R. P. (1904) The reduction division in ferns. *Proceedings of the Royal Society* 73, 86–92.
32. Guyer, M. F. (1900) Spermatogenesis in hybrid pigeons. *Science* 11, 248–249; 312.
33. Guyer, M. F. (1899) Ovarian structure in an abnormal pigeon. *Zoological Bulletin* 2, 211–224.
34. Guyer, M. F. (1902) Hybridism and the Germ Cell. *Bulletin Number 21 of the University of Cincinnati*. Series 2. Volume 2. [Based on Guyer's thesis (1900) *Spermatogenesis in Normal and Hybrid Pigeons*. University of Chicago.]
35. Bungener, P. & Buscaglia, M. (2003) Early connection between cytology and Mendelism: Michael F. Guyer's contribution. *History and Philosophy in the Life Sciences* 25, 27–50.
36. Guyer, M. F. (1903) The germ cell and the results of Mendel. *The Cincinnati Lancet Clinic* 53, 490–491.
37. Guyer, M. F. (1907) Do offspring inherit equally from each parent? *Science* 25, 1006–1010.
38. Guyer, M. F. (1911) Nucleus and cytoplasm in heredity. *American Naturalist* 45, 284–305.
39. Guyer, M. F. (1909) Deficiencies of the chromosome theory of heredity. *University Studies*, Series 2, Vol. 5, no. 3. University Press, Burnet Woods, Cincinnati, pp. 1–19. [Report of paper presented in Boston in 1907.]
40. Godlewski, E. (1906) Die hybridisation der Echiniden und Crinoiden familie. *Archiv für EntwicklungsMechanik der Organismen*. 20, 579.
41. Spillman, W. J. (1910) Heredity. *American Naturalist* 44, 504–512.
42. Whitman, C. O. (1909) cited in Riddle, O. (1919) *Orthogenic Evolution in Pigeons. Posthumous works of C. O. Whitman*. The Carnegie Institute of Washington.
43. Hagemann, R. (2000) Erwin Baur or Carl Correns: who really created the theory of plastid inheritance? *Journal of Heredity* 91, 435–440.
44. Fangerau, H. & Müller, I. (2007) Scientific exchange: Jaques Loeb and Emil Godlewski as representatives of a transatlantic developmental biology. *Studies in the History and Philosophy of Biology and Biomedical Sciences*. 38, 608–617.
45. Johannsen, W. (1923) Some remarks on units in heredity. *Hereditas* 4, 133-141.
46. Forsdyke, D. R. (2016) *Evolutionary Bioinformatics*, 3rd edition. Springer, New York.
47. Shull, G. H. (1909) The 'presence and absence' hypothesis. *American Naturalist* 43, 410–419.
48. Moore, A. R. (1910) A biochemical conception of dominance. *University of California Publications in Physiology* 4, 9–15.
49. Forsdyke, D. R. (1994) The heat-shock response and the molecular basis of genetic dominance. *Journal of Theoretical Biology* 167, 1–6.
50. Galton, F. (1897) The average contribution of several ancestors to the total heritage of the offspring. *Proceedings of the Royal Society* 61, 401–413.
51. Pearson, K. (1906) Walter Frank Raphael Weldon 1860–1906. *Biometrika* 5, s1–52.
52. Castle, W. (1906) Effects of inbreeding, cross-breeding and selection on the fertility and variability of Drosophila. *Proceedings of the American Academy of Arts and Sciences* 41, 731.
53. Davenport, C. B. (1941) The early history of research with Drosophila. *Science* 93, 305–306.
54. Morgan, T. H. (1910) Chromosomes and heredity. *American Naturalist* 44, 449–496.
55. Morgan, T. H. (1909) For Darwin. *Popular Science Monthly* 70, 367–380.
56. Morgan, T. H. (1911) Random segregation versus coupling in Mendelian inheritance. *Science* 34, 384.
57. Haldane, J. B. S. (1919) The combination of linkage values and the calculation of distance between linked factors. *Journal of Genetics* 8, 299–309.
58. Bridges, C. B. (1914) The chromosome hypothesis of linkage applied to cases of sweet peas and primula. *American Naturalist* 48, 524–534.
59. Sturtevant, A. H. (1914) The reduplication hypothesis as applied to Drosophila. *American Naturalist* 48, 535–549.
60. Muller, H J. (1914) A gene for the fourth chromosome of Drosophila. *J. Experimental Zoology* 17, 325–336.

61. Sturtevant, A. H. (1915) The behaviour of the chromosomes as studied through linkage. *Zeitschrift für induktiv Abstammungs- und Vererbungslehre* 13, 234–287.
62. Punnett R. C. (1923) Linkage in the sweet pea. *Journal of Genetics* 13:101–123.

14 Passages

1. Butler, S. (1912) *The Notebooks of Samuel Butler*. Edited by H. F. Jones. Fifield, London.
2. Olby, R. (1989) Scientists and bureaucrats in the establishment of the John Innes Horticultural Institution under William Bateson. *Annals of Science* 46, 497–510.
3. Rothenstein, W. (1939) *Men and Memories*. Volume 2. Faber, London.
4. Lago, M. (1996) *Christiana Herringham and the Edwardian Art Scene*. Lund Humprhies, London.
5. Anonymous (1910) Our Portrait. *The Gownsman*, Cambridge, May 5.
6. Crew, F. A. E. (1967) Reginald Punnett. *Biographical Memoirs of Fellows of the Royal Society* 13, 309–326.
7. Lipset, D. (1980) *Gregory Bateson. The Legacy of a Scientist*. Prentice-Hall, Englewood Cliffs, p. 55.
8. Balls, W. L. (1912) *The Cotton Plant in Egypt*. Macmillan, London.
9. Olby, R. (1991) Social imperialism and state support for agricultural research in Edwardian Britain. *Annals of Science*, 48, 509–526.
10. Hurst, C. C. (1911) The government grant for horse-breeding: the question of eugenics. *The Times*, March 4.
11. Button, C. (2018) James Cossar Ewart and the origins of the animal research department in Edinburgh, 1895-1920. *Journal of the History of Biology* 51, 445-477.
12. Hurst, C. C. (1911) The application of genetics to horse-breeding. In: *Experiments in Genetics*. Cambridge University Press, 1925, pp. 374-376.
13. Hurst, C. C. (1912) Mendelian experiments with thoroughbred horses. *Bloodstock Breeders' Review* 1, 86-90.
14. Wilson, J. (1914) Unsound Mendelism developments, especially as regards the presence and absence theory. *Scientific Proceedings of the Royal Dublin Society* 8, 399-421.
15. Cunningham, J. T. (1914) Biological Agnosticism. [This is a newspaper cutting the source of which has not been identified.]
16. Harwood, J. (1993) *Styles of Scientific Thought. The German Genetics Community 1900–1933*. University of Chicago Press.
17. Gates, R. R. (1915) *The Mutation Factor in Evolution*. Macmillan, London.
18. Goldschmidt, R. (1955) *Theoretical Genetics*. University of California Press, p. 478.
19. Winge, Ö. (1917) The chromosomes. Their numbers and general importance. *Comptes Rendus des Travaux du Laboratoire Carlsberg* 13, 131–275.
20. Smolin, L. (2007) The other Einstein. *The New York Review of Books*. 54 (10), 76–83.

15 Eugenics

1. Galton, F. (1865) Hereditary talent and character. *Macmillan's Magazine* 12, 157–166, 318–327.
2. Galton, F. (1873) On the causes which operate to create scientific men. *Fortnightly Review* 13, 345–351.

3. Fancher, J. (1979) A note on the origin of the term "nature and nurture" *Journal of the History of Behavioural Sciences* 15, 321–322.
4. Pearson, K. (1903) Inheritance of psychical and physical characters of man. *Nature* 68, 607–608. [Abstract of his Huxley lecture; Oct. 16.]
5. Gilham, N. W. (2001) *A Life of Sir Francis Galton*. Oxford University Press.
6. Harper, P. S. (2005) Julia Bell and the *Treasury of Human Inheritance*. *Human Genetics* 116, 422–432.
7. Lipset, D. (1980) *Gregory Bateson. The Legacy of a Scientist.* Prentice-Hall, Englewood Cliffs.
8. Reid G. A. (1907) The interpretation of Mendelian phenomena. *Nature* 76, 566.
9. Lawrence, W. (1823) *Lectures on Physiology, Zoology and the Natural History of Man.* London. Third edition.
10. Hurst, C. C. (1908) Mendel's law of heredity and its application to man. *Transactions of the Leicester Literary and Philosophical Society* 12, 35–48.
11. Hurst, C. C. (1912) Mendelian heredity in man. *The Eugenics Review* 4, 1–25.
12. Fisher R. A. (1911) Address on "Heredity" recorded in the minute book of the Cambridge Eugenics Society meeting held at Trinity College on 10th November 1911. In: Norton, B. & Pearson, E. S. (1976) A note on the background to, and refereeing of, R. A. Fisher's 1918 paper 'On the correlation between relatives on the supposition of Mendelian inheritance'. *Notes and Records of the Royal Society of London* 31, 151–162.
13. Darwin, L. (1912) The first international eugenics congress. *Nature* 89, 558–561.
14. Punnett, R. C. (1917) Eliminating feeblemindedness. *Journal of Heredity* 8, 464–465.
15. Pilkington, D. (1996) *Follow the Rabbit-Proof Fence*, University of Queensland Press.
16. Baur, E, Fischer, E. & Lenz F. (1921) *Grundriss der menschlichen Erblichkeitslehre und Rassenhygiene*. J. F. Lehmanns Verlag, Munich.
17. Forsdyke, S. (2021) *Slaves and Slavery in Ancient Greece*. Cambridge University Press.
18. Govier, M. (1999) The Royal Society, slavery, and the island of Jamaica: 1660–1700. *Notes and Records of the Royal Society, London* 53, 203–217.
19. Davis, D. B. (2014) *The Problem of Slavery in the Age of Emancipation*. Knopf, New York.
20. Jablonski, N. G. (2012) *Living Color. The Biological and Social Meaning of Skin Color.* University of California Press, Berkeley.
21. Lurie, E. (1954) Louis Agassiz and the races of man. *Isis* 45, 227–242.
22. Morton, S. G. (1847) Hybridity in animals, considered in reference to the question of the unity of the human species. *American Journal of Science and Arts* 3, 39–50, 203–212.
23. Darwin, C. (1847) Letter to Charles Lyell. In: *The Correspondence of Charles Darwin. Vol. 4, 1847–1850.* Edited by F. Burkhardt and S. Smith. Cambridge University Press (1989), pp. 45–47.
24. Forsdyke, D. R. (2016) William Bateson, black slavery, eugenics and speciation: The relative roles of politics and science. *Social Sciences Research Network*; https://doi.org/10.2139/ssrn.2783480

16 War

1. Butler, S. (1912) *The Notebooks of Samuel Butler*. Edited by H. F. Jones. Fifield, London.
2. Cock, A. G. (1983) Chauvinism and internationalism in science: the international research council, 1919–1926. *Notes and Records of the Royal Society of London*. 37, 249–288.
3. Haldane, J. B. S. (1925) *Callinicus: a Defence of Chemical Warfare*. Kegan Paul, London.
4. Smith, N. (1916) The place of natural science in secondary education. *The Times Educational Supplement* (May 6).
5. Morgan, T. H., Sturtevant, A. H., Muller, H. J. & Bridges, C. B. (1915) *The Mechanism of Mendelian Heredity*. Henry Holt & Co., New York.

6. Bridges, C. B. (1916) Non-disjunction as proof of the chromosome theory of heredity. *Genetics* vol. 1 pp. 1–52 and 107–163.
7. Lotsy J. P. (1916) *Evolution by Hybridisation*. M. Nijhoff, The Hague.
8. Brink, R. A. (1958) Paramutation at the *R* locus in maize, *Cold Spring Harbor Symposium on Quantitative Biology* 23, 379–391.
9. Stam, M. & Mittelsten Scheid, O. (2005) Paramutation: an encounter leaving a lasting impression. *Trends in Plant Science* 10, 283–290.
10. Reese, V. R. & Forsdyke, D. R. (2016) Meiotic pairing inadequacies at the levels of X chromosome, gene, or base: epigenetic tagging for transgenerational error-correction guided by a future homologous duplex. *Biological Theory* 11, 150–157.
11. Gregory R. P. (1914) On the genetics of tetraploid plants in *Primula sinensis*. *Proceedings of the Royal Society, B*. 87, 484–492.
12. Wilson, E. B. (1914) The bearing of cytological research on heredity. *Proceedings of the Royal Society, B*. 88, 333–352.
13. Winge, Ö. (1917) The chromosomes. Their numbers and general importance. *Comptes Rendus des Travaux du Laboratoire Carlsberg* 13, 131–275.
14. Rosenberg, O. (1907) Cytological investigations of plant hybrids. *Report of the Third International Conference 1906 on Genetics*. Edited by W. Wilks. Royal Horticultural Society, London. pp. 289–291.
15. Herringham, W. P. (1919) *A Physician in France*. Edward Arnold, London.

17 My Respectful Homage

1. Haldane, J. B. S. (1919) The combination of linkage values and the calculation of distances between the loci of linked factors. *Journal of Genetics* 8, 299–309.
2. Forsdyke, D. R. (2000) Haldane's rule: hybrid sterility affects the heterogametic sex first because sexual differentiation is on the path to species differentiation. *Journal of Theoretical Biology* 204, 443–452.
3. Provine, W. (1971) *The Origins of Theoretical Population Genetics*. The University of Chicago Press.
4. Haldane, J. B. S. (1924) A mathematical theory of natural and artificial selection. *Transactions of the Cambridge Philosophical Society* 23, 19–41.
5. Wilson, E. B. & Morgan, T. H. (1920) Chiasmatype and crossing over. *American Naturalist* 54, 193–219.
6. Johannsen, W. (1923) Some remarks about units of heredity. *Hereditas* 4, 133–141.
7. Muller, H. J. (1934) Lenin's doctrines in relation to genetics. In: Graham, L. R. (1972) *Science and Philosophy in the Soviet Union*, A. Knopf, New York, pp. 451–469.
8. Darlington C. D. (1979) Morgan's crisis. *Nature* 278, 786–787.
9. Provine, W. B. (1986) *Sewall Wright and Evolutionary Biology*. The University of Chicago Press, p. 104.
10. Schwartz, J. (2008) *In Pursuit of the Gene. From Darwin to DNA*. Harvard University Press, p. 326.
11. Cock, A. G. (1989) Bateson's two Toronto addresses, 1921. 1. Chromosomal skepticism. *Journal of Heredity* 80, 91–95.
12. Cock, A.G. (1989) Bateson's two Toronto addresses. 1921, 2. Evolutionary faith. *Journal of Heredity* 80, 96–99.
13. Osborn, H. F. (1922) William Bateson on Darwinism. *Science* 55, 194–197.
14. Mitchell, P. C. (1922) Unfashionable Darwinism: No Alternative Solution. *The Times*, April 11.
15. Mitchell, P. C. (1922) Darwin and evolution. *The Times*. April 15.
16. Cunningham, J. T. (1922) Species and adaptations. *Nature* 109, 775-777.
17. Crowther, C. R. (1922) Evolutionary faith and modern doubts. *Nature* 109, 777.

18. Cock, A. G. (1983) William Bateson's rejection and eventual acceptance of chromosome theory. *Annals of Science* 40, 19–59.
19. Harman, O. S. (2004) *The Man Who Invented the Chromosome. A Life of Cyril Darlington.* Harvard University Press, Cambridge.
20. Forsdyke, D. R. (2001) *The Origin of Species, Revisited.* McGill-Queen's University Press, Montreal.
21. Dobell, C. (1925) The chromosome cycle of the sporozoa considered in relation to the chromosome theory of heredity. *La Cellule* 25, 167–192.
22. Harwood, J. (1993) *Styles of Scientific Thought. The German Genetics Community 1900–1933.* University of Chicago Press.
23. Johannsen, W. (1922) Hundert Jahre Vererbungsforschung. *Verhandlungen der Gesellschaft deutscher Naturforscher und Arzte* 87, 70–104.
24. Russell, E. S. (1930) *The Interpretation of Development and Heredity.* Clarendon Press, Oxford.
25. Punnett, R. C. (1922) *Nature* 110, 85–86. [Obituary of Onslow.]

18 Limits Undetermined

1. Haldane, J. B. S. (1926) *The Nation and Atheneum.* Feb. 20. p. 713.
2. Hurst C. C. (1921) The genetics of egg-production in white leghorns and white wyandottes. *The National Poultry Journal.* Sept. 2 and Dec. 9.
3. Hurst C. C. (1923) The genetics of fecundity in the domestic hen. In *Eugenics, Genetics and the Family.* Williams and Wilkins, Baltimore, pp. 212–217.
4. Hurst, R. (1974) *The Evolution of Genetics.* Unpublished manuscript in Cambridge University Library.
5. Hurst, C. C. (1925) Chromosomes and characters in Rosa and their significance in the origin of species. In: *Experiments in Genetics*, Cambridge University Press, pp. 534–550.
6. Roll-Hansen, N. (2014) Johannsen's criticism of the chromosome theory. *Perspectives in Biology and Medicine* 57, 40–56.
7. Williams, G. C. (1966) *Adaptation and Natural Selection.* Princeton University Press.
8. Forsdyke, D. R. (2016) *Evolutionary Bioinformatics*, 3rd edition. Springer, New York.
9. Bateson, W. (1925) Science in Russia. *Nature* 116, 681–683.
10. Roll-Hansen, N. (2005) The Lysenko effect: undermining the autonomy of science. *Endeavour* 29, 144–147.

19 Butler

1. Butler S. (1887) *Luck or Cunning as the Main Means of Organic Modification?* Trübner, London.
2. Forsdyke, D. R. (2006) Heredity as transmission of information: Butlerian "intelligent design." *Centaurus* 48, 133–148.
3. Forsdyke, D. R. (2009) Samuel Butler and long-term memory. Is the cupboard bare? *Journal of Theoretical Biology* 258, 156–164.
4. Forsdyke, D. R. (2015) "A vehicle of symbols and nothing more." George Romanes, theory of mind, information, and Samuel Butler. *History of Psychiatry* 26, 270–287.
5. Forsdyke, D. R. (2016) *Evolutionary Bioinformatics.* 3rd ed. Springer, New York.
6. Darwin C. (1867) Letter to Canon Farrar, March 5. In: *More Letters of Charles Darwin*, Vol. 2. Edited by F Darwin. Appleton, New York, 1903, pp. 441–442. [Comments on a lecture the Canon had given at the Royal Institution "On Some Defects in Public School Education," a transcript of which he had sent to Darwin.]

7. Butler, S. (1896) *The Life and Letters of Dr. Samuel Butler*. Volume 1. John Murray, London, p. 353, pp. 376 –391.
8. Butler, S. (1903) *The Way of All Flesh.* Grant Richards, London.
9. Heitland, W. E. (1926) *After Many Years*. Cambridge University Press.
10. Jones, H. F. (1919) *Samuel Butler Author of Erewhon (1835–1902). A Memoir*, Vols. 1 and 2. London: Macmillan.
11. Bowlby, J. (1990) *Charles Darwin. A New Life.* Norton, New York.
12. Iltis, H. (1932) *Life of Mendel.* Allen and Unwin, London.
13. Orel, V. (1984) *Mendel*. Translated by S. Finn. Oxford University Press.
14. Butler, S. (1914) *A First Year in Canterbury Settlement with Other Earlier Essays*. Edited by R. A. Streatfeild. Fifield, London, pp. 149–185.
15. Butler, S. (1872) *Erewhon or Over the Range*. Trübner, London.
16. Butler, S. (1880) *Unconscious Memory*. David Bogue, London.
17. Romanes, G. J. (1881) Unconscious memory. *Nature* 23, 285–287.
18. Butler, S. (1914) *The Humour of Homer and Other Essays*. Kennerley, New York.
19. Forsdyke, D. R. (2001) *The Origin of Species, Revisited.* McGill-Queen's University Press.
20. Butler, S. (1878) *Life and Habit*. Trübner, London. ["We are such stuff as dreams are made on," are the words of Prospero in Shakespeare's *The Tempest*.]
21. Hering E. (1870) *Über das Gedachtniss als eine allgemeine Function der organisirten Materie*. Karl Gerold's Sohn, Vienna. ["On memory as a universal function of organized matter." A lecture delivered at the anniversary meeting of the Imperial Academy of Sciences at Vienna. Butler learned German in order to understand Hering, and there is a translation in Butler's *Unconscious Memory*.]
22. Butler, S. (1912) *The Notebooks of Samuel Butler*. Edited by H. F. Jones. Fifield, London.
23. Brooks, W. K. (1883) *The Law of Heredity. A Study of the Cause of Variation and the Origin of Living Organisms*. Murphy, Baltimore, pp. 33–41.
24. Mivart, St. G. J. (1871) *On the Genesis of Species*. Macmillan, London.
25. Newman, S. A. (2007) William Bateson's physicalist ideas. In: *From Embryology to Evo-Devo: a History of Evolutionary Development*. Edited by M. Laubichler and J. Maienschein, MIT Press, Cambridge, MA, pp. 83–107.
26. Galton, F. (1875) A theory of heredity. *Contemporary Review* 27, 80–95.
27. Forsdyke, D. R. & Ford, P. M. (1983) Segregation into separate rouleaux of erythrocytes from different species: evidence against the agglomerin hypothesis of rouleaux formation. *Biochemical Journal* 214, 257–260.
28. Haeckel, E. (1876) *Die Perigenesis der Plastidule oder die Wellenzeugung der Lebenstheilchen*. Reimer, Berlin.
29. Darwin, C. (1876) Letter to Romanes, May 29. In: *More Letters of Charles Darwin*, Volume 1. Edited by F. Darwin. Appleton, New York, 1903, pp. 363–365.
30. Lankester, E. R. (1876) Perigenesis versus pangenesis – Haeckel's new theory of heredity. *Nature* 14, 235–238.
31. Clodd, E. (1916) *Memories*. Chapman & Hall, London.
32. Brücke, E. (1861) Die Elementarorganismen. *Sitzungsberichte der Akademie der Wissenschaften Wein, Mathematische-wissenschaftliche Classe*, 44, 381–406.
33. Butler, S. (1879) *Evolution, Old and New*. Hardwicke and Bogue, London.
34. Lamarck, J. B. (1809) *Philosophie Zoologique*. Translated as *Zoological Philosophy.* University of Chicago Press, 1984.
35. Butler, S. (1884) *Selections from Previous Works, with Remarks on Mr. G. J. Romanes' "Mental Evolution in Animals," and a Psalm of Montreal*. Trübner, London.
36. Forsdyke, D. R. (2020) Aging, DNA information and authorship: Medawar, Schrödinger and Samuel Butler. *Biological Theory* 15, 50–55.
37. Bernstein, C, & Bernstein, H. (1991) *Aging, Sex and DNA Repair.* Academic Press, San Diego.
38. Romanes, E. (1896) *The Life and Letters of George Romanes*. Longmans, London, p. 132, 153.

39. Haeckel, E. (1909) Charles Darwin as an Anthropologist. *Darwin and Modern Science*. Cambridge University Press, pp. 137–151.
40. Waterhouse, P. M., Wang, M.-B., & Lough, T. (2001) Gene silencing as an adaptive defence against viruses. *Nature* 411, 834–841.
41. Dawkins, R. (1976) *The Selfish Gene*. Oxford University Press.
42. Wallace, A. R. (1879) Organization and intelligence. *Nature* 19, 477–480.
43. Ward, J. (1879) Review of *Evolutiosn, Old and New*. *The Atheneum*, July 26, pp. 115–117. [Anonymous author inferred by Pauly P. J. (1982) *Victorian Studies* 25, 161–180.]
44. Romanes, G. J. (1884) *Mental Evolution in Animals, with a Posthumous Essay on Instinct by Charles Darwin*. Appleton, New York.
45. Ward, J. (1884) Review of *Mental Evolution in Animals*. *The Atheneum*, March 1, pp. 282–283.
46. Ward, J. (1884) Review of *Selections from Previous Works*. *The Atheneum*, March 23, pp. 378–379.
47. Romanes, G. J. (1884) The Darwinian theory of instinct. *Nineteenth Century* 16, 434–450.
48. Salisbury, Lord (1994) Inaugural address of the Most Hon. the Marquis of Salisbury, K. G., D. C. L., F. R. S., Chancellor of the University of Oxford, President. *Nature* 50, 339–343. [Salisbury (Robert Cecil) ceded his Prime Ministership to his nephew, Arthur Balfour, in 1902; in other words, "Bob's your uncle."]
49. Darwin, C. (1958) *The Autobiography of Charles Darwin 1809-1882. The First Complete Version*. Edited by Nora Barlow. Collins, London.
50. Keynes, G. & Hill, B. (1935) *Letters between Samuel Butler and Miss E. M. A. Savage*. Jonathan Cape, London. [These authors also facilitated the production of Bartholomew's *Further Extracts from the Note-Books of Samuel Butler*. (1934)]
51. Darwin, F. (1878) The analogies of plant and animal life. *Nature* 17, 388-391, 411–414.
52. Darwin, F. & Pertz, D. F. M. (1892) On the artificial production of rhythm in plants. *Annals of Botany* 6, 245--264.
53. Hartog, M. (1892) Problems of reproduction: conjugation, fertilization and rejuvenescence. *Contemporary Review* 62, 92–104.
54. Hartog, M. (1897) Fundamental principles of heredity. *Natural Science* 11, 233–239, 305–316.
55. Hartog M. (1914) Samuel Butler and recent mnemic biological theories. *Scientia* 15, 38–52.
56. Russell, E. S. (1916) *Form and Function*. John Murray, London, pp. 335–344.
57. Dawkins, R. (2003) *The Ancestor's Tale*. Weidenfeld & Nicolson, London. [Dawkins was unaware of Semon's word "mneme" when he suggested the word "meme" as a self-replicating element of culture, passed on by imitation. Sigmund Freud declared he had "learned from Semon's book [1904] only that the Greek word for memory is *mneme*. ... Only he who had new things to say is entitled to coin new terms" (*Minutes of the Vienna Psychoanalytic Society* 2: 48–51).]
58. Jones, H. F. (1911) *Charles Darwin and Samuel Butler. A Step Towards Reconciliation*. Fifield, London.
59. Galton, F. (1910) *Kantsaywhere* (unpublished novel). Galton Archives, University College, London.
60. Bateson, M. C. (1984) *With a Daughter's Eye. A Memoir of Margaret Mead and Gregory Bateson*. W. Morrow, New York.
61. Bateson, G. (1962) The role of somatic change in evolution. *Evolution* 17, 529–539.
62. Bateson, G. (1971) A re-examination of Bateson's rule. *Journal of Genetics* 60, 230–240.
63. Bateson, G. (1979) *Mind and Nature. A Necessary Unity*. Dutton, New York.
64. Bateson, G. & Bateson, M. C. (1987) *Angels Fear. Towards an Epistemology of the Sacred*. Macmillan, New York.
65. Cock, A. C. (1975) Taped interview with Gregory Bateson. Queen's University Archives, Kingston.
66. Bateson, G. (1970) The Alfred Korzybski memorial lecture 1970. *General Sematics Bulletin* 37. [A lecture given at the Harvard Club of New York, 9th January.]

20 Pilgrimages

1. Radick, G. (2016) Presidential address. Experimenting with the scientific past. *British Journal of the History of Science* 49, 153–172.
2. Cock, A. G. (1980) William Bateson's pilgrimages to Brno. *Folia Mendeliana* 15, 243–250. [Volume 65 in the series *Acta Musei Moraviae*.]
3. Cock, A. G. (1982) Bateson's impressions at the unveiling of the Mendel monument at Brno in 1910. *Folia Mendeliana* 17, 217–223. [Volume 67 in the series *Acta Musei Moraviae*.]
4. Hurst, C. C. (1911) Mendelian characters in plants, animals, and man. In: "Festschrift zum Andenken an Gregor Mendel," *Verhandlungen des Naturforshenden Vereines in Brünn* 49, 192–213.
5. Brozek, A., Haskovec, L., & Ruzieka, V. eds (1925) *Memorial Volume in Honor of the 100th Birthday of J. G. Mendel*. Fr. Borovy, Prague.
6. Correns, C. (1924) *Carl Correns*. Julius Springer, Berlin. [Collected papers.]
7. Iltis, H. (1924) *Gregor Johann Mendel, Leben, Werk und Wirkung*. Julius Springer, Berlin.
8. Nivet, C. (2004) Une maladie énigmatique dans la vie de Gregor Mendel. *Médecine Sciences* 20, 1050–1053.
9. Nivet, C. (2006) 1848: Gregor Mendel, le moine qui voulait être citoyen. *Médecine Sciences* 22, 430–433.
10. Nivet, C. (2020) Gregor Mendel fut-il soumis à la corvée avant de devenir moine en 1843? *Médecine Sciences* 36, 63–68
11. Adams, M. B. (2001) Networks in action: the Krushchev era, the cold war, and the transformation of Soviet science. Edited by G. E. Allen and R.M. MacLeod. *Science, History and Social Activism: A Tribute to Everett Mendelsohn*, Kluwer Academic, Boston, pp. 255–276.
12. Orel, V. (1965) Editorial. *Folia Mendeliana* volume 1.

21 Kammerer

1. Lamarck, J. B. (1809) *Philosophie Zoologique*. Translated as *Zoological Philosophy*. University of Chicago Press, 1984.
2. Noble, G. K. (1926) Kammerer's *Alytes*. *Nature* 118, 209–210.
3. Koestler, A. (1971) *The Case of the Midwife Toad*. Hutchinson, London.
4. Gould, S. J. (1972) Zealous advocates. *Science* 176, 623–625.
5. Bateman, K. G. (1959) The genetic assimilation of four venation phenocopies. *Journal of Genetics* 56, 443–474.
6. Cannon, H. G. (1959) *Lamarck and Modern Genetics*, Charles C Thomas, Springfield, Illinois.
7. Aronson, L. R. (1975) The case of *The Case of the Midwife Toad*. *Behavior Genetics* 5, 115–125.
8. Gliboff, S. (2005) 'Protoplasm … is soft wax in our hands': Paul Kammerer and the art of biological transformation. *Endeavour* 29, 162–167.

22 Science and Chauvinism

1. Nuttall, G. H. F. (1901) On the question of priority with regard to certain discoveries upon the aetiology of malarial diseases. *Quarterly Journal of Microscopical Science* 44, 429–441.
2. Sleigh, M. A. & Sutcliff, J. F. (1966) The Origins and History of the Society for Experimental Biology. [A 32-page booklet printed for the Society for Experimental Biology by Tollfree Ltd., London.]

3. Cock, A. G. (1979) The Genetical Society in 1924 – a near demise. *Heredity* 42, 113–117.
4. Cock, A. G. (1983) Chauvinism and Internationalism in Science. *Notes and Records of the Royal Society of London* 37, 249–288.
5. Larmor, J. (1917) Papers in Royal Society Collection. Letter from H. E. Armstrong to Larmor (July 8).
6. Hampson, G. (1918) A classification of the Pyralidae. *Proceedings of the Zoological Society*, pp. 55–131.
7. Jones, H. S. (1960) The early history of the ICSU. *ICSU Review* 2, 169–187.

23 Degrees for Women

1. Punnett, R. C. (1926) William Bateson. *Edinburgh Review* 244, 71–86.
2. Sidgwick, E. (1897) *University Education for Women.* Macmillan and Bowes, Cambridge.
3. Gardner, A. (1921) *A Short History of Newnham College*. Bowes and Bowes, Cambridge. [See also *Newnham College Register*.]
4. Cock, A.G. (1979) Anna Bateson of Bashley – Britain's first professional woman gardener. *Hampshire* 19, 59–62.
5. Bateson, W. (1897) *Cambridge University Reporter*. March 26, pp. 797–799.

6. This chapter was one of the first completed by Alan Cock on Bateson since "it was relatively easy to write up in isolation." In letter to Darlington (Mar. 8, 1977) Alan said that it had been sent, as a sample chapter, to a potential publisher (Cambridge University Press). Having seen an early draft, Gregory Bateson commented (Sept. 3, 1974): You seem unaware that WB almost totally reversed his position on feminism at the end of his life. He became aware that women's colleges were a mix of matrimonial market with blue-stocking, and he did not approve of either as *components in a university*. For better or worse, he enjoyed the flavor of male monasticism in St. Johns. He knew very well that the flavor of female monasticism would be very different. He was after all surrounded by female celibacy – his sisters, Beatrice's sisters, Saunders, Pellew, etc., etc. His own marriage had been prevented for some years by the crazy "Dick" (Edith Durham) – and so on. I recall (but cannot date) his giving me a little lecture on women in the university – that of course they could learn whatever they might want to – but *not* to be a part of the institution which might decide the values and style of the institution. If and when the women's colleges install cellars of good wine like other colleges, then they might be accepted as policy makers in the university. 'Why should the men invite into their society people who will certainly change the character of that society'.

24 Bashing

1. Morgan, T. H. (1926) William Bateson. *Proceedings of the Linnean Society*, Session 138, pp. 66–74.
2. Coleman, W. (1970) Bateson and Chromosomes: Conservative Thought in Science. *Centaurus* 15, 228–314.
3. Morgan, T. H. (1932) *The Scientific Basis of Evolution*. Faber and Faber, London.
4. Forsdyke, D. R. (2001) *The Origin of Species, Revisited.* McGill-Queen's University Press, Montreal.
5. Farmer, J. B. (1927) Obituary notice of fellows deceased. 1927. William Bateson, 1861–1926. *Proceedings of the Royal Society, B.* 101, i–v.
6. Haldane, J. B. S. (1926) *The Nation and the Atheneum*. Feb. 20, p. 713.

7. Haldane, J. B. S. (1957) The theory of evolution before and after Bateson. *Journal of Genetics* 56, 11–27.
8. Punnett, R. C. (1950) Early days of genetics. *Heredity* 4, 1–10.
9. Fisher, R. L. (1930) *The Genetical Theory of Natural Selection.* Oxford University Press.
10. Darwin, L. (1929) Letter to Fisher, Jan. 19. In: *Natural Selection, Heredity and Eugenics.* Edited by J. H. Bennett. Oxford University Press (1983), pp. 95–96.
11. Fisher, R. A. (1936) Has Mendel's work been rediscovered? *Annals of Science* 2, 225–237.
12. Cock, A. G. (1985) Mendel's honesty. *The Biologist* 35, 5.
13. Cock, A. G. (1973) William Bateson, Mendelism and Biometry. *Journal of the History of Biology* 6, 1–36.
14. Stern, C. (1974) Domain of genetics. Presidential address. *Genetics* 78, 21–33
15. Qin, J., Calabrese, P., Tiemann-Boege, I., Shinde, D. N., Yoon, S.-R., Gelfand, D., Bauer, K., & Arnheim, N. (2007) The molecular anatomy of spontaneous germline mutations in human testis. *PLOS Biology* 5, e224 [The technology was promoted by Bateson and Punnett (1911) who suggested that mutant cells would clonally localize to particular gonadal segments.]
16. Haldane, J. B. S. (1956) The theory of selection for melanism in Lepidoptera. *Proceedings of the Royal Society, B.* 144, 217–220.
17. Majerus, M. E. N. (1998) *Melanism. Evolution in Action.* Oxford University Press.
18. Hooper, J. (2002) *Of Moths and Men. An Evolutionary Tale.* Norton, New York.
19. Muller, H. J. (1934) Lenin's doctrines in relation to genetics. In: Graham, L. R. (1972) *Science and Philosophy in the Soviet Union*, A. Knopf, New York, pp. 451–469.
20. Huxley, J. (1942) *Evolution the Modern Synthesis.* Allen & Unwin, London.
21. Huxley, J. (1970) *Memories.* Allen & Unwin, London.
22. Darlington, C. D. (1964) *Genetics and Man.* Allen and Unwin, London.
23. Weiss, K. (2002) Good vibrations: the silent symphony of life. *Evolutionary Anthropology* 11, 176–182.
24. Mallet, J. (2004) Species problem solved 100 years ago. *Nature* 430, 503.
25. Mayr. E. (1973) The recent historiography of genetics. *Journal of the History of Biology* 6, 125–154.
26. Dawkins, R. (1986) *The Blind Watchmaker.* Longmans, Harlow.
27. Dawkins, R. (1983) Universal Darwinism. In: *Evolution from Molecules to Man.* Edited by D. S. Bendell. Cambridge University Press.
28. Dawkins, R. (1982) *The Extended Phenotype.* W. H. Freeman, San Francisco.
29. Wilson, D. S. (2001) Evolutionary biology. Struggling to escape exclusively individual selection. *Quarterly Review of Biology* 76, 199–205.
30. Dawkins, R. (1998) *Unweaving the Rainbow.* Houghton Mifflin, New York.
31. Lamphere, L. (2001) Letter from the President of the American Anthropological Association. *New York Times* (Aug. 12).
32. Dawkins, R. (1998) Postmodernism disrobed. A review of *Intellectual Impostures. Nature* 394, 141–143.
33. Gould, S. J. (1972) Zealous advocates. *Science* 176, 623–625.
34. Gould, S. J. (1980) Is a new and general theory of evolution emerging? *Paleobiology* 6, 119–130.
35. Orr, H. A. (1996) Dobzhansky, Bateson and the genetics of speciation. *Genetics* 144, 1331–1335.
36. Coyne, J. A. & Orr, H. A. (1998) The evolutionary genetics of speciation. *Philosophical Transactions of the Royal Society* 353, 287-305.
37. Lynch, M. & Force, A. G. (2000) The origin of interspecies genomic incompatibility via gene duplication. *American Naturalist* 156, 590-605.
38. Wilkins, A. (2001) *The Evolution of Developmental Pathways.* Sinauer, Sunderland.
39. Gavrilets, S. (2004) *Fitness Landscapes and The Origin of Species*, Princeton University Press.
40. Coyne, J. A. & Orr, H. A. (2004) *Speciation.* Sinauer, Sunderland.

41. Johnson, N. A. (2002) Sixty years after "Isolating mechanisms, evolution and temperature": Muller's legacy. *Genetics* 161, 939-944.
42. Navarro, A. & Barton, N. (2003) Chromosomal speciation and molecular divergence – accelerated evolution in rearranged chromosomes. *Science* 300, 321–324.
43. Crow, J. F. (2005) Herman Joseph Muller, evolutionist. *Nature Genetics* 6, 941–945.
44. Scannell, D. R., Byrne, K. P., Gordon, J. L., Wong, S. & Wolfe, K. H. (2006) Multiple rounds of speciation associated with reciprocal gene loss in polyploid yeasts. *Nature* 440, 341–344.
45. Mackenzie, D. (1978) Statistical theory and social interests" *Social Studies in Science* 8, 35–83.
46. Allen, G. E. (1978) Opposition to Mendelian-chromosome theory: the physiological and developmental genetics of Richard Goldschmidt. *Journal of the History of Biology* 7, 49–92.
47. Cock, A. G. (1983) William Bateson's rejection and eventual acceptance of chromosome theory. *Annals of Science* 40, 19–59.
48. Smolin, L. (2007) The other Einstein. *The New York Review of Books*. 54 (10), 76–83.
49. Bowler, P. (1983) *The Eclipse of Darwinism. Anti-Darwinian Evolution Theories in the Decades around 1900.* John Hopkins University Press, Baltimore.
50. Adams, M. B. (1990) La génétique des populations était-elle une génétique évolutive? In: *Histoire de la Génétique*, pp. 153–171. Edited by Jean-Louis Fischer and William H. Schneider (Paris: ARPEM). [There is also: "Little evolution, big evolution. Rethinking the history of population genetics," a 34-page unpublished manuscript kindly sent to DRF by MBA in 2003. This was eventually published: Adams. M. B. (2021) Little evolution, BIG evolution: Rethinking the history of Darwinism, population genetics, and the "synthesis." In: *Natural Selection. Revisiting its Explanatory Role in Evolutionary Biology*. Delisle R. G (ed). Springer Nature, Switzerland, pp. 195–230. The emphasis "BIG" is by MBA.]
51. Wells, H. G., Huxley, J. & Wells, G. P. (1931) *The Science of Life*. Cassell, London.
52. Henig, R. M. (2000) *The Monk in the Garden. The Lost and Found Genius of Gregor Mendel, the Father of Genetics*. Houghton Mifflin, New York.
53. Richmond, M. (2001) Women in the early history of genetics. William Bateson and the Newnham College Mendelians, 1900–1910. *Isis* 92, 55–90.
54. Richmond. M. (2006) The "domestication of heredity:" the family organization of geneticists at Cambridge University, 1895–1910. *Journal of the History of Biology* 39, 565–605.
55. Bateson, P. (2002) William Bateson. A biologist ahead of his time. *Journal of Genetics* 81, 49–58.

25 Epilogue

1. Keilin, D. (1966) *The History of Cell Respiration and Cytochrome*. Cambridge University Press.
2. Matthew, P. (1831) *Naval Timber and Arboriculture*. Longmans and Co., London.
3. Lock, R. D. (1906) *Recent Progress in the Study of Variation, Heredity and Evolution*. John Murray, London.
4. Darwin, C. (1874) *The Descent of Man*. John Murray, London.
5. Forsdyke, D. R. (2001) *The Origin of Species, Revisited.* McGill-Queen's University Press, Montreal.
6. Forsdyke, D. R. (2006) Heredity as transmission of information: Butlerian "intelligent design." *Centaurus* 48, 133–148.
7. Olby, R. (1971) The influence of physiology on heredity theories in the nineteenth century. *Folia Mendeliana* 6, 99–103.
8. Forsdyke, D. R. (2000) *Tomorrow's Cures Today?* Harwood Academic, Amsterdam.
9. Punnett, R. C. (1950) Early days of genetics. *Heredity* 4, 1–10.

10. Forsdyke, D. R. (2004) Grant Allen, George Romanes, Stephen Jay Gould and the evolution establishments of their times. *Historical Kingston* 52, 95–103.
11. Morton, P. (2005) *The Busiest Man in England: Grant Allen and the Writing Trade, 1875–1900*. Palgrave-Macmillan, New York.
12. Forsdyke, D. R. (2016) William Bateson, black slavery, eugenics and speciation: The relative roles of politics and science. *Social Sciences Research Network*; https://doi.org/10.2139/ssrn.2783480
13. Provine, W. (1971) *The Origins of Theoretical Population Genetics*. The University of Chicago Press.
14. Crowther, J. G. (1952) *British Scientists of the Twentieth Century*. Routledge & Kegan Paul, London.
15. Forsdyke, D. R. (2016) *Evolutionary Bioinformatics*, 3rd edition. Springer, New York.
16. Gulick, A. (1932) *John Thomas Gulick. Evolutionist and Missionary*. University of Chicago Press.
17. Gulick, J. T. (1891) Remarks on Mr. Moulton's reasonings and calculations. Unpublished paper (142/2B). Galton Archives, University College, London.
18. Dobzhansky, T. (1962–1963) Oral memoir quoted in Provine, W. B. (1994) The origin of Dobzhansky's *Genetics and the Origin of Species*. In: *The Evolution of Theodosius Dobzhansky*. Edited by M. B. Adams, Princeton University Press.
19. Provine, W. B. (1986) *Sewall Wright and Evolutionary Biology*. The University of Chicago Press.
20. Cock, A. G. (1980) Faking and the intent to deceive. *British Medical Journal* 281, 1214–1215.
21. Meijer, O. G. (1982) The essence of Mendel's discovery. In: *Gregor Mendel and the Foundation of Genetics, the Past, and Present of Genetics*. Edited by V. Orel. The Mendelianum of the Moravian Museum, Brno, p. 173–200.
22. Cock, A. G. (1988) Mendel's honesty. *The Biologist* 35, issue 1.
23. Baumgartner, A, & Ettinghausen, A. von (1842) *Die Naturlehre nach ihrem gegenwärtigen Zustand mit Rucksicht auf mathematische Begrundung*. Carl Gerold, Wein.
24. Turing, A. M. (1952) The chemical basis of morphogenesis. *Philosophical Transactions of the Royal Society of London. B.* 237, 37–72.
25. Hurst, C. C. (1932) A genetical formula for the inheritance of intelligence in man. Proceedings of the Royal Society, B. 112, 80–97.
26. Hurst, R. (1951) *What's All This About Genetics*. Watts & Co, London.
27. Crew, F. A. E. (1967) Reginald Crundall Punnett. *Biographical Memoirs of Fellows of the Royal Society* 13, 309–326.
28. Muggeridge, M. (1936) *The Earnest Athiest*. Eyre & Spottiswoode, London.
29. Bateson, M. C. (2006) Personal communication to DRF.
30. Heitland, W. E. (1908) *A Letter to a Lady, or a Word with the Female Anti-Suffragists*. E. Johnson, Trinity Street, Cambridge.
31. Jones, H. F. (1919) *Samuel Butler Author of Erewhon (1835–1902). A Memoir*, Vols. 1 and 2. Macmillan, London.
32. Butler, S. (1914) *The Humour of Homer and Other Essays*. Kennerley, New York.
33. Voltaire (1770) Letter to F. L. H. Leriche. In: *The Complete Works of Voltaire*. Vol. 120. The Voltaire Foundation, Banbury, 1975, p. 18.
34. Coleman, W. (1970) Bateson and chromosomes: Conservative thought in science, *Centaurus* 15, 228–314.
35. Fowler, H. (1996) In: *Cambridge Women.Twelve Portraits*. Edited by E. Shils & C. Blacker. Cambridge University Press.
36. Cock, A. G. (1980) Obituaries of Rona Hurst and Gregory Bateson. *Mendel Newsletter* 19, 5–7.

26 The Third Base

1. Lyell, C. (1863) *The Geological Evidences of the Antiquity of Man with Remarks on Theories of the Origin of Species by Variation*. G. W. Childs, Philadelphia.
2. Körner, K. (1983) *Linguistics and Evolution Theory (Three Essays by August Schleicher, Ernst Haeckel and Wilhelm Bleek)*. J. Benjamin, Amsterdam.
3. Bateson, G. (1979) *Mind and Nature*. E. P. Dutton, New York.
4. Mayr, E. (1970) *Populations, Species and Evolution*. Harvard University Press, Cambridge, MA.
5. Forsdyke, D. R. (2016) *Evolutionary Bioinformatics*. 3rd edition. Springer, New York.
6. Crick, F. (1971) General model for chromosomes of higher organisms. *Nature* 234, 25–27
7. Forsdyke, D. R. (2007) Calculation of folding energies of single-stranded nucleic acids: conceptual issues. *Journal of Theoretical Biology* 248, 745–753.
8. Forsdyke, D. R. (2007) Molecular sex. The importance of base composition rather than homology when nucleic acids hybridize. *Journal of Theoretical Biology* 249, 325–330.

27 Mendel Basics[1]

1. Forsdyke (2018) Mendel, Gregor Johann. *Encyclopedia of Life Sciences (eLS)*. John Wiley, Chichester.

28 George Romanes, William Bateson, and Darwin's Weak Points[2]

1. Dobzhansky, T. (1937) *Genetics and the Origin of Species*, Columbia University Press, New York, pp. 228–258.
2. Muller, H. J. (1942) Isolating mechanisms, evolution and temperature. *Biology Symposium* 6, 71–125.
3. Louis, E. J. (2009) Origins of reproductive isolation. *Nature* 457, 549–560.
4. Greig, D. (2009) Reproductive isolation in *Saccharomyces*. *Heredity* 102, 39–44.
5. Schartl, M. (2008) Evolution of *Xmrk*: an oncogene, but also a speciation gene? *BioEssays* 30, 822–832.
6. Huxley, T. H. (1863) Letter to Kingsley, April 30th. *Life and Letters of Thomas Henry Huxley*. Edited by L. Huxley, vol. 1. Appleton, New York, p. 257. Also (1860) Darwin on the origin of species. *Westminster Review* 17 (n.s.), 715–717.
7. Bateson, W. (1925) Huxley and evolution, *Nature* 115, 715–717.
8. Huxley, T. H. (1859) Letter to Darwin, November 23rd. *Life and Letters of Thomas Henry Huxley*, vol 1. Edited by L. Huxley. Appleton, New York, pp. 188–189.
9. Forsdyke, D. R. (2016) *Evolutionary Bioinformatics*, 3rd edition. Springer, New York, pp. 123–180.
10. Today's readers cannot easily turn to nineteenth century texts without guidance. For example, the words "genetics" and "virus" were used in various contexts several decades before attaining their modern meanings. Yet failure to read early texts (as did many biologists in the 35

[1] Adapted from Forsdyke (2018) Mendel, Gregor Johann. *Encyclopedia of Life Sciences (eLS)*. John Wiley, Chichester. [See article for references.]

[2] Adapted from Forsdyke (2010) *Notes and Records of the Royal Society* 64, 139–154.

years prior to the 'rediscovery' of Mendel's work), can lead to the reinvention of many wheels (e.g. de Vries, Correns and Tschermak in the 1890s).
11. Forsdyke, D. R. (2001) *The Origin of Species, Revisited. A Victorian who Anticipated Modern Developments in Darwin's Theory*. McGill-Queen's University Press, Montreal.
12. Cock, A. G. & Forsdyke, D. R. (2008) *"Treasure your Exceptions." The Science and Life of William Bateson*. 1st edition. Springer, New York.
13. Romanes, G. J. (1886) Physiological selection. An additional suggestion on the origin of species, *Journal of the Linnaean Society (Zoology)* 19, 337–411 and *Nature* 34, 314–316, 336–340, 362–365.
14. Bateson, W. (1886) Letter to Anna Bateson, 4th December. *The William Bateson Papers*, section G1b (item 26 in folder 101–150), Archives of Queen's University, Kingston.
15. Darwin, C. (1875) *The Variation of Animals and Plants under Domestication*, vol. 2, pp. 170–171 (John Murray, London). [Darwin 's error was to emphasize hybrid sterility as a character of a sterile individual, rather than of the parents of that individual.]
16. Romanes, G. J. (1897) *Darwin, and After Darwin*, vol. 3. Longmans, Green & Co., London, pp. 43-44.
17. Romanes, G. J. (1887) Physiological selection. *Nineteenth Century* 21, 59-80.
18. Thiselton Dyer, W. T. (1888) Mr. Romanes' paradox. *Nature* 39, 7–9.
19. Wallace, A. R. (1886) Physiological selection and the origin of species, *Nature* 34, 467–468.
20. Bateson, W (1894) *Materials for the Study of Variation, Treated with Especial Regard to Discontinuity in the Origin of Species*. Macmillan, London.
21. Cunningham, J. T. (1895) The origin of species among flat-fishes, *Natural Sci*ence 6, 169–177, 234-239.
22. Bateson, W. 1904. Practical aspects of the new discoveries in heredity. *Memoires of the Horticultural Society of New York* 1, 1–9. [The conference ran from September 30th to October 2nd 1902. Bateson spoke on the first day. His comment on chromosomal localization was made in the discussion following a paper presented for W. A. Cannon on 1st October (p. 123).]
23. Bateson W. & Saunders, E. R. (1902) Experimental studies on the physiology of heredity. *Reports to the Evolution Committee of the Royal Society* 1, 1-–160.
24. Guyer, M. F. (1900) Spermatogenesis in hybrid pigeons, *Science* 21, 248–249, 312.
25. Bateson, W. (1902) *Mendel's Principles of Heredity. A Defence*. Cambridge University Press, Cambridge.
26. Forsdyke, D. R. (1999) Two levels of information in DNA. Relationship of Romanes' "intrinsic" variability of the reproductive system and Bateson's "residue" to the species-dependent component of the base composition, (C+G)%. *Journal of Theoretical Biology* 201, 47–61. [The term "residue" is likely derived from Bateson's Granchester neighbor, Alfred North Whitehead, who perhaps pointed out William James's 1890 article "The hidden self" (*Scribner's Magazine* 7, 361–374).]
27. Bungener P. & Buscaglia, M. (2003) Early connection between cytology and Mendelism: Michael F. Guyer's contribution. *History and Philosophy of the Life Sciences* 25, 27–50.
28. Bateson, W., Saunders, E. R. & Punnett R. C. (1905) Experimental studies on the physiology of heredity. *Reports to the Evolution Committee of the Royal Society* 2, 1–131.
29. Bateson, W. (1909) Heredity and variation in modern lights. *Darwin and Modern Science*. Edited by A. C. Seward. Cambridge University Press, Cambridge, pp. 85–101.
30. Bateson, W. (1913) *Problems of Genetics*. Yale University Press, New Haven, p. 241.
31. Forsdyke, D. R. (2003) William Bateson, Richard Goldschmidt, and non-genic modes of speciation. *Journal of Biological Systems* 11, 341–350.
32. Bateson, W. (1922) Evolutionary faith and modern doubts, *Nature* 109, 553–556.
33. Crowther, C. R. (1922) Evolutionary faith and modern doubts, *Nature* 109, 777. [Perhaps C. R. Crowther M.D., B.Ch., a surgeon at the Devon and Cornwall Ear and Throat Hospital, who published a theory "On the Somatic Origin of Malignant Disease" in the *British Medical Journal* in 1926, with a follow-up in the *Lancet* in 1927.]
34. Bateson, W. (1922) Interspecific sterility, *Nature* 110, 76.
35. Bateson, W. (1909) *Mendel's Principles of Heredity*. Cambridge University Press, Cambridge, p. 73.

36. Watson, J. D. & Crick, F. H. C. (1953) Genetical implications of the structure of deoxyribonucleic acid. *Nature* 171, 964–967.
37. Crick, F. (1971) General model for the chromosomes of higher organisms. *Nature* 234, 25–27.
38. Wilson, J. H. (1979) Nick-free formation of reciprocal heteroduplexes: a simple solution to the topological problem. *Proceedings of the National Academy of Sciences USA* 76, 3641-3645.
39. Forsdyke, D. R. (2007) Molecular sex. The importance of base composition rather than homology when nucleic acids hybridize. *Journal of Theoretical Biology* 249, 325–330.
40. White, M. J. D. (1978) *Modes of Speciation.* Freeman, San Francisco, 1978.
41. King, M. (1993) *Species Evolution. The Role of Chromosome Change.* Cambridge University Press, Cambridge.

29 Complementary Oligonucleotides Rendered Discordant by Single Base Mutations May Drive Speciation[3]

1. Bateson, W. (1894) Pitch theory of reproductive cells. Bateson's Notes on a Pitch Theory of Reproductive Cells (archive-it.org)
2. Forsdyke, D. R. (1996) Different biological species "broadcast" their DNAs at different (G+C)% "wavelengths". *Journal of Theoretical Biology* 178, 405–417.
3. Chargaff, E. (1951) Structure and function of nucleic acids as cell constituents. *Federation Proceedings* 10, 654–659.
4. Watson, J. D. & Crick, F. H. C. (1953) A structure for deoxyribose nucleic acid. *Nature* 171, 738–740.
5. Forsdyke, D. R. & Mortimer, J. R. (2000) Chargaff's legacy. *Gene* 261, 127–137.
6. Bernardi, G. (1993) The vertebrate genome: isochores and evolution. *Molecular Biology & Evolution* 10, 186–204.
7. Röhling, S., Linne, A., Schellhorn, J., Hosseini, M., Dencker, T. & Morgenstern, B. (2020) The number of *k*-mer matches between two DNA sequences as a function of *k* and applications to estimate phylogenetic distances. *PLoS ONE* 15, e0228070.
8. Forsdyke, D. R. (2019) Success of alignment-free oligonucleotide (k-mer) analysis confirms relative importance of genomes not genes in speciation. *Biological Journal of the Linnean Society* 128, 239–250.
9. Bize, A., Midoux, C., Mariadassou, M., Schbath, S., Forterre, P. & Da Cunha, V. (2021) Exploring short k-mer profiles in cells and mobile elements from Archaea highlights the major influence of both the ecological niche and evolutionary history. *BMC Genomics* 22, 186.
10. Rogerson, A. C. (1991) There appear to be conserved constraints on the distribution of nucleotide sequences in cellular genomes. *Journal of Molecular Evolution* 32, 24–30.
11. Forsdyke, D. R. (1995) Relative roles of primary sequence and (G+C)% in determining the hierarchy of frequencies of complementary trinucleotide pairs in DNAs of different species. *Journal of Molecular Evolution* 41, 573–581.
12. Blake, R. D., Hess, S. T. & Nicholson-Tuell, J. (1992) The influence of nearest neighbors on the rate and pattern of spontaneous point mutations. *Journal of Molecular Evolution* 34, 189–200.
13. Benzer, S. (1961) On the topography of the genetic fine structure. *Proceedings of the National Academy of Sciences USA* 47, 403–415.
14. Josse, J., Kaiser, A.D. & Kornberg, A. (1961) Enzymic synthesis of deoxyribonucleic acid. VIII. Frequencies of nearest neighbor base sequences in deoxyribonucleic acid. *Journal of Biological Chemistry* 236, 864–871.

[3]Adapted from Forsdyke (2021) *Biological Theory* 16, 237–241.

15. Forsdyke, D. R. & Bell, S. J. (2004) Purine-loading, stem-loops, and Chargaff's second parity rule: a discussion of the application of elementary principles to early chemical observations. *Applied Bioinformatics* 3, 3–8.
16. Brbić, M., Warnecke, T., Kriško, A. & Supek, F. (2015) Global shifts in genome and proteome composition are very tightly coupled. *Genome Biology and Evolution* 7, 1519–1532.
17. Aggarwala, V. & Voight, B. F. (2016) An expanded sequence context model broadly explains variability in polymorphism levels across the human genome. *Nature Genetics* 48, 349–355.
18. Morozov, A. A. (2017) *k*-mer distributions of aminoacid sequences are optimised across the proteome. *bioRxiv*: https://doi.org/10.1101/190280 (18 September)
19. Forsdyke, D. R. (2014) Implications of HIV RNA structure for recombination, speciation, and the neutralism-selectionism controversy. *Microbes and Infection* 16, 96–103.
20. Forsdyke, D. R. (2017) Speciation: Goldschmidt's chromosomal heresy, once supported by Gould and Dawkins, is again reinstated. *Biological Theory* 12, 4–12.
21. Forsdyke, D. R. (2017) Base composition, speciation, and why the mitochondrial barcode precisely classifies. *Biological Theory* 12, 157–168.
22. Forsdyke, D. R. (2019) Hybrid sterility can only be primary when acting as a reproductive barrier for sympatric speciation. *Biological Journal of the Linnean Society* 128, 779–788.
23. Forsdyke, D. R. (2020) Classics in Biological Theory. Revisiting George Romanes' "Physiological Selection" 1886. *Biological Theory* 15, 143–147.
24. Forsdyke, D. R. (2021) Neutralism versus selectionism: Chargaff's second parity rule, revisited. *Genetica* 149, 81–88.
25. Liu, Z. & Samee, A. H. (2021) Mutation rate variations in the human genome are encoded in DNA shape. *bioRxiv*: https://doi.org/10.1101/2021.01.15.426837 (15 January).
26. Lao, P. J. & Forsdyke, D. R. (2000) Thermophilic bacteria strictly obey Szybalski's transcription direction rule and politely purine-load RNAs with both adenine and guanine. *Genome Research* 10, 228–236.
27. Forsdyke, D. R. (2016) *Evolutionary bioinformatics*, 3rd edition. Springer, New York, pp. 125–127, 189.
28. Sueoka, N. (1961) Compositional correlation between deoxyribonucleic acid and protein. *Cold Spring Harbor Symposium in Quantitative Biology* 26, 35–43.
29. Bossi, L. & Roth, J. R. (1980) The influence of codon context on genetic code translation. *Nature* 286, 123–127.
30. Shepherd, J. C. W. (1981) Method to determine the reading frame of a protein from the purine/pyrimidine genome sequence and its possible evolutionary justification. *Proceedings of the National Academy of Sciences USA* 78, 1596–1600.
31. Eguchi, Y., Itoh, T. & Tomizawa, J. (1991) Antisense RNA. *Annual Reviews of Biochemistry* 60, 631–652.
32. Crick, F. (1971) General model for the chromosomes of higher organisms. *Nature* 234, 25–27.
33. Kleckner, N. & Weiner, B. M. (1993) Potential advantages of unstable interactions for pairing of chromosomes in meiotic, somatic and premeiotic cells. *Cold Spring Harbor Symposium in Quantitative Biology* 58, 553–565.
34. Bozdag, G. O., Ono, J., Denton, J. A., Karakoc, E., Hunter, N., Leu, J.-Y. & Greig, D. (2021) Breaking a species barrier by enabling hybrid recombination. *Current Biology* 31, R161–R185.
35. Paz, A., Kirzhner, V., Nevo, E. & Korol, A. (2006) Coevolution of DNA interacting proteins and genome 'dialect.' *Molecular Biology and Evolution* 23, 56–64.
36. Kleckner, N. (1997) Interactions between and along chromosomes during meiosis. Harvey Lec*tures* 91, 21–45.
37. Zickler, D. & Kleckner, N. (2015) Recombination, pairing, and synapsis of homologs during meiosis. *Cold Spring Harbor Perspectives in Biology* 7, a016626.
38. Newman, S. A. (2007) William Bateson's physicalist ideas. *From Embryology to Evo-Devo: A History of Evolutionary Development*. Edited by M. Laubichler & J. Maienschein. MIT Press, Cambridge, MA, pp 83–107.
39. Cock, A. G. & Forsdyke, D. R. (2008) *"Treasure Your Exceptions". The Science and Life of William Bateson*. First edition. Springer, New York

Index

A. G. Cock, D. R. Forsdyke, *Treasure Your Exceptions*,
https://doi.org/10.1007/978-3-030-92099-9

S

MIX
Papier aus verantwortungsvollen Quellen
Paper from responsible sources
FSC® C105338

If you have any concerns about our products,
you can contact us on
ProductSafety@springernature.com

In case Publisher is established outside the EU,
the EU authorized representative is:
Springer Nature Customer Service Center GmbH
Europaplatz 3, 69115 Heidelberg, Germany

Printed by Libri Plureos GmbH
in Hamburg, Germany